Cable and Wireless Networks

Theory and Practice

Cable and Wireless Networks

Theory and Practice

Mário Marques da Silva

CRC Press
Taylor & Francis Group
Boca Raton   London   New York

CRC Press is an imprint of the
Taylor & Francis Group, an **informa** business

CRC Press
Taylor & Francis Group
6000 Broken Sound Parkway NW, Suite 300
Boca Raton, FL 33487-2742

© 2016 by Taylor & Francis Group, LLC
CRC Press is an imprint of Taylor & Francis Group, an Informa business

No claim to original U.S. Government works

Printed on acid-free paper
Version Date: 20151201

International Standard Book Number-13: 978-1-4987-4681-6 (Hardback)

Visit the Taylor & Francis Web site at
http://www.taylorandfrancis.com

and the CRC Press Web site at
http://www.crcpress.com

Contents

Summary xv
Laboratorial Introductory Notes xvii
Author xix

Chapter 1   Introduction to Data Communications and Networking 1

    1.1   Fundamentals of Communications 1
        1.1.1   Analog and Digital Signals 2
        1.1.2   Modulator and Demodulator 3
        1.1.3   Transmission Mediums 4
        1.1.4   Synchronous and Asynchronous Communication Systems 4
        1.1.5   Simplex and Duplex Communications 5
        1.1.6   Communications and Networks 6
        1.1.7   Switching Modes 6
                1.1.7.1   Circuit Switching 6
                1.1.7.2   Packet Switching 7
        1.1.8   Connection Modes 8
                1.1.8.1   Connection-Oriented Service 8
                1.1.8.2   Connectionless 9
        1.1.9   Network Coverage Areas 10
        1.1.10 Network Topologies 10
        1.1.11 Classification of Media and Traffic 13
    1.2   Present and the Future of Telecommunications 14
        1.2.1   Convergence 15
        1.2.2   Collaborative Age of the Network Applications 16
        1.2.3   Transition toward the Collaborative Age 17
    Chapter Summary 19
    Review Questions 20
    Lab Exercises 20

Chapter 2   Network Protocol Architectures 21

    2.1   Introduction to the Network Architecture Concept 21
    2.2   Open System Interconnection—Reference Model 24
        2.2.1   Seven-Layer OSI-RM 25
                2.2.1.1   Physical Layer 25
                2.2.1.2   Data Link Layer 26
                2.2.1.3   Network Layer 27
                2.2.1.4   Transport Layer 28
                2.2.1.5   Session Layer 29
                2.2.1.6   Presentation Layer 29
                2.2.1.7   Application Layer 29
        2.2.2   Service Access Point 29
    2.3   Overview of the TCP/IP Architecture 33
        2.3.1   Application Layer 35
        2.3.2   Transport Layer 35
        2.3.3   Internet Layer 37

        2.3.4   Data Link Layer 40
        2.3.5   Physical Layer 42
    Chapter Summary..43
    Review Questions 44
    Lab Exercises 44

Chapter 3   Channel Impairments...45
    3.1   Shannon Capacity..45
    3.2   Nyquist Sampling Theorem..47
    3.3   Attenuation 48
    3.4   Noise Sources 49
        3.4.1   Atmospheric Noise 50
        3.4.2   Man-Made Noise 52
        3.4.3   Extraterrestrial Noise 52
        3.4.4   Thermal Noise 53
        3.4.5   Electronic Noise 54
    3.5   Influence of the Transmission Channel 55
        3.5.1   Delay and Phase Shift 56
        3.5.2   Distortion...57
        3.5.3   Equalization 58
    3.6   Interference Sources 61
        3.6.1   Intersymbol Interference 61
            3.6.1.1   Nyquist ISI Criterion.......................................63
        3.6.2   Multiple Access Interference...67
        3.6.3   Co-Channel Interference 68
        3.6.4   Adjacent Channel Interference 69
    Chapter Summary..71
    Review Questions 71
    Lab Exercises 72

Chapter 4   Cable Transmission Mediums 73
    4.1   Twisted Pairs 73
        4.1.1   Characteristics..74
        4.1.2   Types of Protection...75
        4.1.3   Categories..76
        4.1.4   Connectors and Cables..78
    4.2   Coaxial Cables..80
        4.2.1   Characteristics..80
    4.3   Optical Fibers 81
        4.3.1   Characteristics..82
        4.3.2   Categories..82
        4.3.3   Connectors and Cables..87
    4.4   Interference Parameters in Metallic Conductors.........................89
        4.4.1   Near End Crosstalk 90
        4.4.2   Far End Crosstalk 90
        4.4.3   Attenuation to Crosstalk Ratio 91
        4.4.4   Equal Level Far End Crosstalk 91
        4.4.5   Other Performance Indexes of Crosstalk 92

Chapter Summary 92
Review Questions 93
Lab Exercises 94

Chapter 5   Wireless Transmission Mediums 95

5.1   Wireless Propagation 95
    5.1.1   Direct Wave Propagation 95
        5.1.1.1   Free Space Path Loss 96
        5.1.1.2   Link Budget Calculation 97
        5.1.1.3   Carrier-to-Noise Ratio Calculation 100
        5.1.1.4   Bit Error Probability Calculation 101
    5.1.2   Wireless Propagation Effects 102
        5.1.2.1   Reflection 102
        5.1.2.2   Diffraction 107
        5.1.2.3   Scattering 110
    5.1.3   Fading 110
        5.1.3.1   Shadowing Fading 112
        5.1.3.2   Multipath Fading 113
    5.1.4   Groundwave Propagation 115
    5.1.5   Skywave Propagation 116
5.2   Satellite Communication Systems 121
    5.2.1   Physical Analysis of Satellite Orbits 121
    5.2.2   Characteristics of Different Orbits 123
        5.2.2.1   Geostationary Earth Orbit 124
        5.2.2.2   Medium and Low Earth Orbit 124
        5.2.2.3   Highly Elliptical Orbit 126
    5.2.3   Satellite's C/N Ratio Analysis 126
5.3   Terrestrial Microwave Systems 131
Chapter Summary 132
Review Questions 133
Lab Exercises 133

Chapter 6   Source Coding and Transmission Techniques 135

6.1   Source Coding 136
    6.1.1   Voice 137
        6.1.1.1   Analog Audio 137
        6.1.1.2   Digital Audio 137
    6.1.2   Video 142
        6.1.2.1   Analog Video 142
        6.1.2.2   Digital Video 143
6.2   Differential and Nondifferential Transmission 143
6.3   Line Coding Schemes 144
    6.3.1   Return to Zero 146
    6.3.2   Nonreturn to Zero 146
    6.3.3   Nonreturn to Zero Inverted 146
    6.3.4   Bipolar Alternate Mark Inversion 146
    6.3.5   Pseudoternary 147
    6.3.6   Manchester 147

	6.3.7	Differential Manchester	147
	6.3.8	Two Binary One Quaternary	148
6.4	Modulation Schemes		148
	6.4.1	Amplitude Shift Keying	149
	6.4.2	Frequency Shift Keying	149
	6.4.3	Phase Shift Keying	150
	6.4.4	M-QAM Constellations	151
6.5	Coding Efficiency of a Symbol		154
6.6	Scrambling of Signals		154
6.7	Multiplexing		155
	6.7.1	Frequency Division Multiplexing	156
	6.7.2	Time Division Multiplexing	160
Chapter Summary			161
Review Questions			162
Lab Exercises			163

Chapter 7 Advanced Transmission Techniques to Support Current and Emergent Multimedia Services 165

7.1	Advances in Wireless Systems and Their Technical Demands		165
7.2	Spread-Spectrum Communications		166
7.3	Code Division Multiple Access		167
	7.3.1	General Model	169
	7.3.2	Narrowband CDMA	171
	7.3.3	Wideband CDMA	174
7.4	Orthogonal Frequency Division Multiplexing		177
7.5	Single-Carrier–FDE		182
	7.5.1	IB-DFE Receivers	183
7.6	Diversity Combining Algorithms		185
	7.6.1	Selection Combining	186
	7.6.2	Maximal Ratio Combining	186
	7.6.3	Equal Gain Combining	187
	7.6.4	MSE-Based Combining	187
7.7	RAKE Receiver		188
7.8	Multiple Input Multiple Output		190
	7.8.1	Space-Time Coding	193
		7.8.1.1 STBC for Two Antennas	194
		7.8.1.2 STBC for Four Antennas	195
	7.8.2	Selective Transmit Diversity	196
	7.8.3	Multilayer Transmission	197
	7.8.4	Space Division Multiple Access	199
	7.8.5	Beamforming	200
	7.8.6	Multiuser MIMO	202
7.9	Advanced MIMO Applications		202
	7.9.1	Base Station Cooperation	203
		7.9.1.1 CoMP Transmission	204
		7.9.1.2 Macrodiversity	205
	7.9.2	Multihop Relay	207
		7.9.2.1 Adaptive Relaying	207
		7.9.2.2 Configurable Virtual Cell Sizes	208
		7.9.2.3 Multihop Relay in 3GPP	209

7.9.3   Multiresolution Transmission Schemes 211
7.9.4   Green Radio Communications 213
Chapter Summary 216
Review Questions 217
Lab Exercises 217

Chapter 8   Services and Applications 219

8.1   Web Browsing 221
        8.1.1   Hypertext Transfer Protocol 221
8.2   E-Mail 223
        8.2.1   Simple Mail Transfer Protocol 223
8.3   File Transfer 224
        8.3.1   File Transfer Protocol 224
8.4   IP Telephony and IP Videoteleconference 224
        8.4.1   ITU-T H.323 225
        8.4.2   Session Initiation Protocol 228
8.5   Network Management 230
        8.5.1   Simple Network Management Protocol 231
8.6   Names' Resolution 232
        8.6.1   Domain Name System 232
Chapter Summary 234
Review Questions 234
Lab Exercises 235

Chapter 9   Transport Layer 237

9.1   Transmission Control Protocol 239
        9.1.1   TCP Properties 239
        9.1.2   TCP Segment Header Format 240
        9.1.3   TCP Handshaking 242
9.2   User Datagram Protocol 243
        9.2.1   UDP Properties 243
        9.2.2   UDP Datagram Header Format 243
9.3   Integrated and Differentiated Services 244
        9.3.1   Integrated Services 244
        9.3.2   Differentiated Services 245
Chapter Summary 245
Review Questions 246
Lab Exercises 247

Chapter 10   Internet Layer: Addressing and Configuration 249

10.1   IP Version 4 250
        10.1.1   IPv4 Classful Addressing 250
        10.1.2   IPv4 Classless Addressing 253
                    10.1.2.1   Variable Length Subnet Mask 262
        10.1.3   IPv4 Datagram 263
10.2   IP Version 6 265
        10.2.1   IPv6 Addressing 266
        10.2.2   IPv6 Packet 270

10.3   Cisco Internetwork Operating System 273
        10.3.1   Introduction to Cisco IOS 274
        10.3.2   Basic Configuration of Cisco Routers and Switches 279
                 10.3.2.1   Configuration Mode 279
                 10.3.2.2   Line Configuration Submode 280
                 10.3.2.3   Interface Configuration Submode (IPv4 and IPv6) 280
        10.3.3   Dynamic Host Configuration Protocol 283
                 10.3.3.1   DHCP Configuration 285
        10.3.4   Network and Port Address Translation..286
                 10.3.4.1   Dynamic NAT and PAT Configuration 287
                 10.3.4.2   Static NAT Configuration..291
Chapter Summary 292
Review Questions 292
Lab Exercises 294

Chapter 11 Internet Layer: Routing and Configuration 297

11.1   Administrative and Metric Distances 299
11.2   Route Summarization..301
11.3   Static Routing and Flooding...302
        11.3.1   Static Route Configuration 303
        11.3.2   Floating Route Configuration 305
        11.3.3   Default Route Configuration..306
11.4   Adaptive Routing Algorithms and Protocols 307
        11.4.1   Classification of Adaptive Routing Protocols.............................309
        11.4.2   Distance Vector Protocols and Their Configuration 309
                 11.4.2.1   Routing Information Protocol 310
                 11.4.2.2   Enhanced Interior Gateway Routing Protocol............316
        11.4.3   Link-State Protocols and Their Configuration 325
                 11.4.3.1   Open Shortest Path First...327
11.5   Internet Control Message Protocol...335
11.6   Fragmentation and Reassembling 335
11.7   IPv6 Transition and Configuration 337
        11.7.1   Transition from IPv4 into IPv6 337
        11.7.2   IPv6—RIPng Configuration Using Cisco IOS.............................338
11.8   Cisco Discovery Protocol...339
Chapter Summary 340
Review Questions 340
Lab Exercises 341

Chapter 12 Data Link Layer 345

12.1   LAN Devices...346
        12.1.1   Hub...346
        12.1.2   Bridge...347
        12.1.3   Switch 348
        12.1.4   Spanning Tree Protocol 350

12.2	LLC Sublayer	354
	12.2.1 Error Control Techniques	356
	12.2.1.1 Error Detection Codes	357
	12.2.1.2 Error Correction Codes	363
	12.2.2 Automatic Repeat Request	367
	12.2.2.1 Stop-and-Wait ARQ	368
	12.2.2.2 Go-Back-N ARQ	369
	12.2.2.3 Selective Reject ARQ	370
	12.2.3 Flow Control Techniques	371
	12.2.3.1 Stop and Wait	372
	12.2.3.2 Sliding Window	372
12.3	LLC Protocols	373
	12.3.1 IEEE 802.2 Protocol	373
12.4	MAC Sublayer	374
12.5	MAC Protocols	375
	12.5.1 IEEE 802.3 Protocol	376
	12.5.1.1 Maximum Collision Domain Diameter	379
	12.5.1.2 Physical Layer Employed in IEEE 802.3 Networks	381
	12.5.2 IEEE 802.5 Protocol	383
	12.5.3 Fiber Distribution Data Interface Protocol	387
	12.5.4 Digital Video Broadcast Standard	388
12.6	Virtual Local Area Networks	388
	12.6.1 Configuration of Virtual Local Area Networks	391
	12.6.1.1 Configuration of the Management VLAN	395
	12.6.1.2 Configuration of the VLAN Default Gateway	396
	12.6.2 Inter-VLAN Routing	396
	12.6.2.1 Configuration of the Native VLAN	398
	12.6.2.2 Inter-VLAN Connectivity with a Router	399
	12.6.2.3 Inter-VLAN Connectivity with a Multilayer Switch	401
	12.6.3 VLAN Trunking Protocol	403
	Chapter Summary	406
	Review Questions	406
	Lab Exercises	407
Chapter 13	**Structured Cabling System**	**411**
13.1	Hierarchical Network Design	412
13.2	Structured Cabling Elements and Subsystems	413
	13.2.1 Centralized Optical Architecture	415
13.3	Structured Cabling Standards and Specifications	416
	13.3.1 North American Standards	416
	13.3.2 International Standards	418
	13.3.3 European Standards	424
13.4	Hardware and Accessories	424
	13.4.1 Rack	424
	13.4.2 Patch Panel	425
	13.4.3 Electrical Supply	427
	13.4.4 Labeling Distributors and Cables	428

Chapter Summary 428
Review Questions 429
Lab Exercises 430

Chapter 14 Transport Networks and Protocols 431

14.1   Circuit Switching Transport Networks 432
         14.1.1   Frequency Division Multiplexing Hierarchy 433
         14.1.2   Plesiochronous Digital Hierarchy 433
         14.1.3   Synchronous Digital Hierarchies 436
                  14.1.3.1   SDH/SONET Network 437
                  14.1.3.2   SDH/SONET Frame Format 441
         14.1.4   Digital Subscriber Line 443
         14.1.5   Data Over Cable Service Interface Specification 448
14.2   Packet Switching Transport Networks and Protocols 449
         14.2.1   Asynchronous Transfer Mode 449
                  14.2.1.1   B-ISDN Reference Model 450
                  14.2.1.2   ATM Network 452
                  14.2.1.3   ATM Cell Format 453
         14.2.2   Multiprotocol Label Switching 453
                  14.2.2.1   MPLS Network 454
                  14.2.2.2   MPLS Packet Format 456
         14.2.3   HDLC Protocol 458
                  14.2.3.1   HDLC Configuration Using Cisco IOS 460
         14.2.4   Point-to-Point Protocol 461
                  14.2.4.1   PPP Configuration Using Cisco IOS 464
         14.2.5   Frame Relay 465
                  14.2.5.1   Frame-Relay Configuration Using Cisco IOS 468
Chapter Summary 471
Review Questions 472
Lab Exercises 473

Chapter 15 Cellular Communications and Wireless Standards 477

15.1   Cellular Concept 477
         15.1.1   Macrocell 481
         15.1.2   Microcell 481
         15.1.3   Picocell 481
         15.1.4   Femtocell 482
         15.1.5   Power Control 482
15.2   Evolution of Cellular Systems and the New Paradigm of 4G 483
         15.2.1   Evolution from 3G Systems into Long-Term Evolution 485
         15.2.2   IEEE 802.16 Protocol (WiMAX) 486
         15.2.3   LTE-A and IMT-Advanced 488
15.3   IEEE 802.11 Protocol (Wi-Fi) 488
Chapter Summary 491
Review Questions 492
Lab Exercises 493

Chapter 16 Network Security...495

16.1   Overview of Network Security 495
16.2   Information Gathering..496
16.3   Security Services and Attack Types 497
         16.3.1   Confidentiality...498
                     16.3.1.1   Eavesdropping 498
                     16.3.1.2   Snooping..500
                     16.3.1.3   Interception..500
                     16.3.1.4   Trust Exploitation 501
         16.3.2   Integrity...502
                     16.3.2.1   Man-in-the-Middle 502
         16.3.3   Availability...503
                     16.3.3.1   Denial of Service 503
         16.3.4   Authenticity..505
                     16.3.4.1   Replay Attack 507
         16.3.5   Accountability..508
                     16.3.5.1   Identification...508
                     16.3.5.2   Authentication 508
                     16.3.5.3   Authorization...509
                     16.3.5.4   Access Control 509
                     16.3.5.5   Monitoring...509
                     16.3.5.6   Registration 510
                     16.3.5.7   Auditing...510
16.4   Malware 510
16.5   Physical and Environmental Security...510
16.6   Risk Management 511
16.7   Security Plan..512
16.8   Protective Measures...514
         16.8.1   Symmetric Cryptography...515
                     16.8.1.1   Symmetric Cryptographic Systems..........517
         16.8.2   Asymmetric Cryptography...520
                     16.8.2.1   Asymmetric Cryptographic Systems 520
         16.8.3   Digital Signature 521
         16.8.4   Digital Certificates 523
         16.8.5   Public Key Infrastructure..525
         16.8.6   Combined Cryptography..527
                     16.8.6.1   SSL and TLS...528
                     16.8.6.2   Security Architecture for an Internet Protocol 532
16.9   Security in IEEE 802.11 Wireless Networks...535
         16.9.1   Wired Equivalent Privacy 536
         16.9.2   Wi-Fi Protected Access...537
         16.9.3   Wi-Fi Protected Access 2 and the IEEE 802.1i.......................537
         16.9.4   EAP and IEEE 802.1x..538
16.10   Cisco Switch Security..539
         16.10.1   Password Configuration 539
         16.10.2   SSH Configuration 539
         16.10.3   Port Security Configuration 540

16.11 Network Architectures 542
    16.11.1 Single Network Architecture 542
    16.11.2 Double Network Architecture 543
    16.11.3 Network Architecture with DMZ 544
16.12 Intrusion Detection System 546
16.13 Virtual Private Networks 547
16.14 Firewalls 549
    16.14.1 Packet Filtering Configuration 553
        16.14.1.1 ACLs—Stateless Inspection 553
        16.14.1.2 Reflexive ACLs 557
        16.14.1.3 Context-Based Access Control 560
    16.14.2 Firewall Cisco ASA 561
Chapter Summary 561
Review Questions 563
Lab Exercises 564

Appendix I 569

Appendix II 573

Appendix III 575

Appendix IV 583

Appendix V 587

Appendix VI 663

References 671

Index 681

Summary

This book presents a comprehensive approach to networking, cable and wireless communications, and networking security. It describes the most important state-of-the-art fundamentals and system details in the field, as well as many key aspects concerning the development and understanding of current and emergent services.

Three of the author's earlier books, *Transmission Techniques for Emergent Multicast and Broadcast Systems*, *Transmission Techniques for 4G systems*, and *MIMO Processing for 4G and Beyond: Fundamentals and Evolution*, focused on the transition from 3G into 4G and 5G cellular systems, including the fundamentals of multi-input and multi-output (MIMO) systems, and therefore, they spanned a wide range of topics. Another book by the author, *Multimedia Communications and Networking*, focused on networking.

In this book, the author gathers in a single volume his point of view on current and emergent cable and wireless network services and technologies. Different bibliographic sources cover each one of these topics independently, without establishing the natural relationships between the topics. The advantage of the present work is twofold: on the one hand, it allows the reader to learn quickly, thereby helping the reader to master the topics covered, providing a deeper understanding of their interconnection; on the other hand, it collects in a single source the latest developments in the area, which are generally only within reach of an active researcher, such as the author, with a committed research career of several years and regular participation in conferences and international projects.

Each chapter illustrates the theory of cable and wireless communications with relevant examples, contains hands-on exercises suitable for readers with a BSc degree or an MSc degree in computer science or electrical engineering, and ends with review questions. This approach makes the book well suited for higher education students in courses such as networking, telecommunications, mobile communications, and network security. Finally, the book serves as a good reference book for academic, institutional, or industrial professionals with technical responsibilities in planning, design and development of networks, telecommunications and security systems, and mobile communications, as well as for Cisco CCNA and CCNP exam preparation.

Laboratorial Introductory Notes

The lab exercises included in this book focus on three tools: the Emona Telecoms Trainer 101 (ETT-101), for which a dual-channel 20 MHz oscilloscope is required; the free network analyzer, Wireshark; and the Cisco Packet Tracer network simulator.

Emona ETT-101 consists of a telecommunications modeling system that brings block diagrams to life, with real hardware modules and real electrical signals, which are employed in this book to demonstrate the theory about telecommunications. As alternatives to ETT-101, two other pieces of laboratory equipment can be used: the Emona TIMS 301-C Telecommunications Teaching System and the Emona net*TIMS Telecommunications Teaching System. Emona TIMS 301-C corresponds to ETT-101 with extended capabilities. Emona net*TIMS allows implementation of the same experiments that TIMS 301-C does, but these can be built and controlled remotely by students across a LAN or the Internet (multiple students can do their lab work at any time and from any location in the world). Appendix VI lists experiments that can be implemented with Emona TIMS (both 301-C and net*TIMS), indicating the chapters that discuss each experiment. The free network analyzer, Wireshark, is used to demonstrate the theory on networking, namely signaling, message formats, and network procedures. The Cisco Packet Tracer simulator is used to build networks, to configure them, and to simulate their responses. Some chapters focus on telecommunications, and therefore ETT-101 is used extensively. Other chapters focus on networking, and discuss the utilization of network Wireshark and Packet Tracer. Since the ETT-101 laboratory manual already describes many experiments, the lab exercises presented in chapters on telecommunications simply refer to the different ETT-101 experiments. In this case, the student should refer to the descriptions existing in the ETT-101 laboratory manual, namely Volume 1—*Experiments in Modern Analog and Digital Telecommunications*; Volume 2—*Further Experiments in Modern Analog & Digital Telecommunications*; and Volume 3—*Advanced Experiments in Modern Analog & Digital Telecommunications*.

Author

Mário Marques da Silva (marques.silva@ieee.org) is an associate professor and the director of the Department of Sciences and Technologies at Universidade Autónoma de Lisboa, Lisbon, Portugal. He is also a researcher at Instituto de Telecomunicações in Lisbon, Portugal. He received his BSc degree in electrical engineering in 1992, and MSc and PhD degrees in electrical and computer engineering (telecommunications) in 1999 and 2005, respectively, both from Instituto Superior Técnico, University of Lisbon, Portugal.

From 2005 to 2008, he was with the NATO Air Command Control and Management Agency in Brussels, Belgium, where he managed the deployable communications of the new Air Command and Control System Program. He has been involved in multiple networking and telecommunications projects. His research interests include networking and mobile communications, namely Internet protocol (IP) technologies and network security, block transmission techniques, interference cancellation, MIMO systems, and software-defined radio. He is also a Cisco Certified Network Associate (CCNA) instructor.

He is the author of four books published by CRC Press, *Multimedia Communications and Networking, Transmission Techniques for Emergent Multicast and Broadcast Systems, Transmission Techniques for 4G Systems*, and *MIMO Processing for 4G and Beyond: Fundamentals and Evolution*. He has authored dozens of journal and conference papers, is a member of IEEE and AFCEA, and has been a reviewer for a number of international scientific IEEE journals and conferences. He has also chaired many conference sessions and has been serving in the organizing committee of relevant EURASIP and IEEE conferences.

1 Introduction to Data Communications and Networking

LEARNING OBJECTIVES

- Describe the fundamentals of communications.
- Identify the key components of networks and communication systems.
- Describe different types of networks and communication systems.
- Identify the differences between a local area network (LAN), a metropolitan area network (MAN), and a wide area network (WAN).
- Identify the different types of media and traffic.
- Define the convergence and the collaborative age of the network applications.

1.1 FUNDAMENTALS OF COMMUNICATIONS

Communication systems are used to enable the exchange of data between two or more entities (humans or machines). As can be seen from Figure 1.1, data consists of a representation of information source, whose transformation is performed by a source encoder. An example of a source encoder is a thermometer, which converts temperatures (information source) into voltages (data). A telephone can also be viewed as a source encoder, which converts the analog voice (information source) into a voltage (data), before being transmitted along the telephone network (transmission medium). In case the information source is analog and the transmission medium is digital, a CODEC (COder and DECoder) is employed to perform digitization. A VOCODER (VOice CODER) is a codec specific for voice, whose functionality consists of converting analog voice into digital at the transmitter side, and the reciprocal at the receiver side.

The emitter of data consists of an entity responsible for the insertion of data into the communication system and for the conversion of data into signals. Note that signals are transmitted, rather than data. Signals consist of an adaptation* of data, such that their transmission is facilitated in accordance with the used transmission medium. Similarly, the receiver is responsible for converting the received signals into data.

The received signals correspond to the transmitted signals subject to attenuation and distortion, and added with noise and interferences. These channel impairments originate that the received signal differs from that transmitted. In the case of analog signals, the resulting signal levels do not exactly translate the original information source. In the case of digital signals, the channel impairments originate corrupted bits. In both cases, the referred channel impairments originate a degradation of the signal-to-noise plus interference ratio (SNIR).† A common performance indicator

---

* Signals can be, for example, a set of predefined voltages that represent bits used in transmission.
† In linear units, the SNIR is mathematically given by $\text{SNIR} = S/(N + I)$, where S stands for the power of signal, N expresses the power of noise, and I denotes the power of interferences. For the sake of simplicity, the SNIR is normally only referred to as SNR (signal-to-noise ratio), but where the interference is also taken into account (in this case N stands for the power of noise and interferences). Furthermore, both SNIR (or SNR) are normally expressed in logarithmic units as $\text{SNIR}_{dB} = 10 \log_{10}(S/[N + I])$.

FIGURE 1.1   Generic block diagram of a communication system.

of digital communication systems is the bit error rate (BER). This corresponds to the number of corrupted bits divided by the total number of transmitted bits over a certain time period.

A common definition associated with information is knowledge. It consists of a person's ability to have access to the right information, at the right time. The conversion between information and knowledge can be automatically performed using information systems, whereas information can be captured by sensors and distributed using communication systems.

1.1.1   ANALOG AND DIGITAL SIGNALS

Analog signals present a continuous amplitude variation over time. An example of an analog signal is voice. Contrarily, digital signals present amplitude discontinuities (e.g., voltages or light pulses). An example of digital data includes the bits[*] generated in a workstation. The text is another example of digital data. Figure 1.2 depicts examples of analog and digital signals.

Digital signals present several advantages (relating to analog) such as the following:

* Error control is possible in digital signals: corrupted bits can be detected and/or corrected.
* Because they present only two discrete values, the consequences of channel impairments can be more easily detected and avoided (as compared to analog signals).
* Digital signals can be regenerated, almost eliminating the effects of channel impairments. Contrarily, the amplification process of analog signals results in the amplification of signals, noise, and interferences, keeping the SNR relationship unchanged.[†]
* The digital components are normally less expensive than the analog ones.
* Digital signals facilitate cryptography and multiplexing.
* Digital signals can be used to transport different sources of information (voice, data, multimedia, etc.) in a transparent manner.

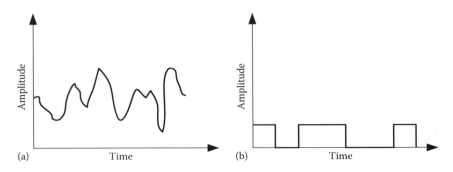

FIGURE 1.2   Example of (a) analog and (b) digital signals.

---

[*] With logic states 0 or 1.

[†] In fact, the amplification process results even in a degradation of the SNR, as it adds the amplifier's internal noise to the signal at its input. This subject is detailed in Chapter 3.

However, digital signals present an important disadvantage:
- For the same information source, the bandwidth required to accommodate a digital signal is typically higher than the analog counterpart.[*] This results in a higher level of attenuation and distortion.

1.1.2 MODULATOR AND DEMODULATOR

As can be seen from Figure 1.3, when the source (e.g., a computer) generates a digital stream of data and the transmission medium is analog, a MODEM (MOdulator and DEModulator) is employed to perform the required conversion. The modulator converts digital data into analog signals, whereas the demodulator (at the receiver) converts analog signals into digital data. An example of an analog transmission medium is radio transmission, whose signals consist of electromagnetic waves (present a continuous variation in time).

A modem (e.g., asynchronous digital subscriber line [ADSL] or cable modem) is responsible for modulating a carrier wave with bits, using a certain modulation scheme.[†] The reverse of this operation is performed at the receiver side. Moreover, a modem allows sending a signal modulated around a certain carrier frequency, which can be another reason for using such a device.

In case the data is digital and the transmission medium is also digital, a modem is normally not employed, as the conversion between digital and analog does not need to be performed. In this case, a line encoder/decoder (sometimes also referred to as a *digital modem*, nevertheless not accurately) is employed. This device adapts the original digital data to the digital transmission medium,[‡] adapting parameters such as levels and pulse duration. Note that, using such a digital encoder, the signals are transmitted in the baseband.[§]

The output of a line encoder consists of a digital signal, as it comprises discrete voltages that encode the source logic states. Consequently, it can be stated that the line encoder is employed when the transmission medium is digital. On the other hand, the output of a modulator consists of an analog signal, as it modulates a carrier that is an analog signal.

In the case of high data rate, the required bandwidth necessary to accommodate such a signal is also high.[¶] In this scenario, the medium may originate a high level of attenuation or distortion at limit frequency components of the signal. In such a case, it can be a good choice to use a modem that allows the modulation of the signal around a certain carrier frequency. The carrier frequency can be carefully selected such that the channel impairments in the frequencies around it (corresponding to the signal bandwidth) do not seriously degrade the SNR.

The reader should refer to Chapter 6 for a detailed description of the modulation schemes used in modems, as well as for the description of digital encoding techniques.

FIGURE 1.3 Generic communication system incorporating a modem.

---

[*] As an example, analog voice is transmitted in a 3.4 kHz bandwidth, whereas the digital pulse code modulation (PCM) requires a bandwidth of 32 kHz (64 kbps).

[†] Using amplitude, frequency, or phase shift keying. Advanced modems make use of a combination of these elementary modulation schemes.

[‡] Using line codes such as return to zero, nonreturn to zero, and Manchester (as detailed in Chapter 6).

[§] Instead of carrier modulated (bandpass), as performed by a modem.

[¶] According to the Nyquist theorem, as detailed in Chapter 3.

1.1.3 TRANSMISSION MEDIUMS

Transmission mediums can be classified as cable or wireless. The examples of cable transmission mediums include twisted pair cables, coaxial cables, multimode or single mode optical fiber cables, and so on.

In the past, LANs were made of coaxial cables. These cables were also used as a transmission medium for medium- and long-range analog communications. Although coaxial cables were replaced by twisted pair cables in LANs, the massification of cable television enabled their reuse.

As a result of telephone cables, twisted pairs are still the dominant transmission medium in houses and offices. These cables are often reused for data. With the improvement in isolators and copper quality, as well as with the development of shielding, the twisted pair has become widely used for providing high-speed data communications, in addition to the initial use for analog telephony.

Currently, multimode optical fibers have been increasingly installed at homes, allowing reaching throughputs of the order of several gigabits per second (Gbps). Moreover, single mode optical fibers are the most used transmission medium in transport networks. A transport network consists of the backbone (core) network, used for transferring large amount of data among different main nodes. These main nodes are then connected to secondary nodes and finally connected to customer nodes.

A radio or wireless communication system is composed of a transmitter and a receiver, using antennas to convert electric signals into electromagnetic waves and vice versa. These electromagnetic waves are propagated over air. Note that wireless transmission mediums can be either guided or unguided. In the former case, directional antennas are used at both the transmitter and the receiver sides, such that electromagnetic waves propagate directly from the transmit into the receive antenna.

The reader should refer to Chapter 4 for a detailed description of cable transmission mediums, while Chapter 5 introduces the wireless transmission mediums.

1.1.4 SYNCHRONOUS AND ASYNCHRONOUS COMMUNICATION SYSTEMS

Synchronous and asynchronous communications refer to the ability or inability to have information about the start and end of bit instants.*

Using asynchronous communications, the receiver does not achieve perfect time synchronization with the transmitter, and the communication accepts some level of fluctuation. Consequently, start and stop bits are normally included in a frame† in order to periodically achieve bit synchronization of the receiver with the transmitter. Note that between the start and the stop bit, the receiver of an asynchronous communication suffers from a certain amount of time shift. The referred periodic synchronization using start and stop bits is normally included as part of the functionalities implemented by a modem, when establishing a communication in asynchronous mode of operation. Normally, asynchronous communications do not accommodate high-speed data rates. They are typically used for random (not continuous) exchange of data (at low rate).

On the other hand, synchronous communications consider a receiver that is bit synchronized with the transmitter. This bit synchronization can be achieved using one of the following methods:

- By sending a clock signal multiplexed with the data or using a parallel dedicated circuit
- When the transmitted signal presents a high zero crossing rate, such that the receiver can extract the start and end of bit instants from the received signal

Synchronous communications are normally employed in high-speed lines, and for the transmission of large blocks of data. An example of a synchronous communication system is the synchronous digital hierarchy (SDH) networks, used for the transport of large amounts of data in a backbone.

---

* Nevertheless, frame synchronization is required in either case.
† A group of exchanged bits.

1.1.5 SIMPLEX AND DUPLEX COMMUNICATIONS

A simplex communication consists of a communication between two or more entities where the signals flow only in a single direction. In this case, one entity only acts as a transmitter and the other(s) as a receiver. This can be seen from Figure 1.4. Note that the transmitter may be transmitting signals to more than one receiver.

When the signals flow in a single direction, but with alternation in time, it is stated that the communication is half-duplex. Therefore, although both entities act simultaneously as a transmitter and as a receiver (at different time instants), instantaneously, each host acts as either a transmitter or a receiver [Stallings 2010]. The half-duplex communication is depicted in Figure 1.5.

Finally, when the communication is simultaneously in both directions, it is in full-duplex mode. In this case, two or more entities act simultaneously as both a transmitter and a receiver. The full-duplex communication is depicted in Figure 1.6. Full-duplex communications normally require two parallel transmission mediums (e.g., two pairs of wires): one for transmission and another for reception.

FIGURE 1.4   Simplex communication.

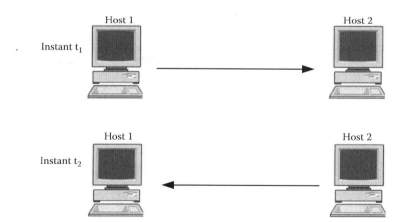

FIGURE 1.5   Half-duplex communication.

FIGURE 1.6   Full-duplex communication.

1.1.6 COMMUNICATIONS AND NETWORKS

A point-to-point communication establishes a direct connection (link) between two adjacent end stations, between two adjacent network nodes (e.g., routers), or between an end station and an adjacent node.

A network can be viewed as a concatenation of point-to-point communications, composed of several nodes and end stations, where each node is responsible for switching the data, such that an end-to-end connection is established between two end stations. The examples of point-to-point communications and of a network are depicted in Figure 1.7. An end-to-end network connection consists of a concatenation of several point-to-point links, where each of these links can be implemented using a different transmission medium (e.g., satellite and optical fiber).

A node of a network can be a router or a private automatic branch exchange (PABX). The former device switches packets (packet switching), while the latter is responsible for physically establishing permanent connections, such that a phone call between two end entities is possible (circuit switching). This subject is detailed in Section 1.2.

Depending on the number of destination stations of data involved in a communication, this can be classified as unicast, multicast, or broadcast. Unicast stands for a communication whose destination is a single station. In case the destination of data is all the network stations, the communication is referred to as *broadcast*. Very often broadcast communications are established in a single direction (i.e., there is no feedback from the receiver into the transmitter). Finally, when the destination of the data is more than a single station, but less than all network stations, the communication is referred to as *multicast*.

1.1.7 SWITCHING MODES

1.1.7.1 Circuit Switching

Circuit switching establishes a permanent physical path between the source and the destination. This switching mode is used in classic telephone networks. Only after startup, is allowed a synchronous

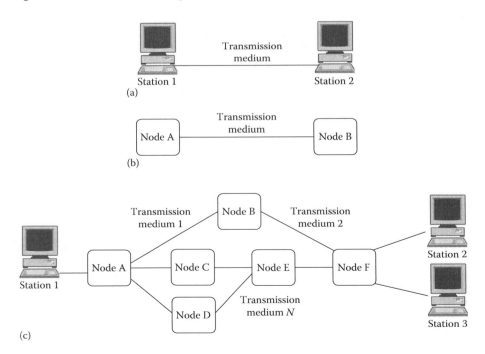

FIGURE 1.7 Examples of (a, b) a point-to-point communication and (c) a network.

exchange of data. This end-to-end path (circuit) is permanently dedicated until the connection ends. The time to establish the connection is long, but a delay is assured only because of the propagation speed of signals. This kind of switching is ideal for delay-sensitive communications, such as voice. If the connection cannot be established because of lack of resources, it is said that the call was blocked, but once established, congestion does not occur. All the bandwidth available is assigned to a certain connection that, for long time periods, may not be used and, in other periods, may not be enough (e.g., if that connection is sending variable data rates). For this reason, it is of high cost. In telephone networks, switching is physically performed by operators using PABX. This consists of a switch whose functionality is typically achieved using space and/or time switching. Space switching consists of establishing a physical shunt between one input and one output. Because digital networks normally incorporate multiplexed data into different time slots[*] (each telephone connection is transported in a different time slot), there is a need to switch a certain time slot from one physical input into another time slot of another physical output. This is performed by the time and space switching functionality of a digital PABX. An example of a circuit switching (telephony) network is depicted in Figure 1.8.

1.1.7.2 Packet Switching

With the introduction of data services, the notion of packet switching has arrived. Packet switching considers the segmentation of a message into parts, where each part is referred to as a *packet* (with fixed[†] or variable[‡] length). As can be seen from Figure 1.9, a digital message is composed of many bits, while a packet consists of a small number of these bits.

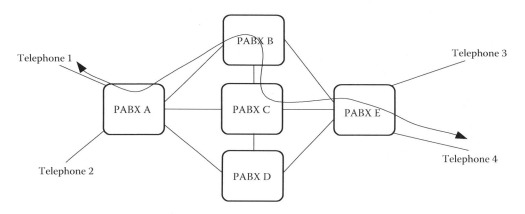

FIGURE 1.8 Example of a circuit switching (telephone) network.

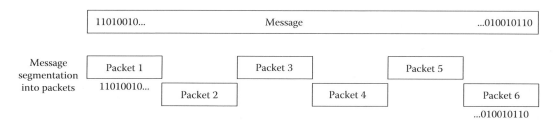

FIGURE 1.9 The segmentation of a message into packets.

---

[*] This is normally referred to as *time division multiplexing*.
[†] For example, asynchronous transfer mode (ATM).
[‡] For example, multiprotocol label switching or Internet protocol (IP).

Packets are forwarded and switched independently through the nodes of a network, between the source and the destination. Each packet transports enough information to allow its routing (end destination address included in a header).

While the nodes of a circuit switching network establish a permanent shunt between one input and one output, because packet switching considers a number of bits grouped into a packet, the nodes of a packet switching network only switch data for the duration of a packet transmission. The following packet that uses the same input or output of a node may belong to a different end-to-end connection. This is depicted in Figure 1.10. Consequently, packet switching networks make much better usage of the network resources (nodes) than circuit switching. Note that a node of a packet switching network is typically a router.

Each node of the network is able to store packets, in case it is not possible to send it because of temporary congestion. In this case, the time for message transmission is not guaranteed, but this value is kept within reasonable limits, especially if quality of service (QoS) is offered. Packet switching is of lower costs than circuit switching, and is ideal for data transmission, because it allows a better management of the resources available (a statistical multiplexing is performed). Moreover, with packet switching, we need not assign all of the available resources (i.e., bandwidth) to a certain user who, for long periods, does not make use of them, the network resources being shared among several users, as a function of the resources available and of the users' need. The network resources are made available as a function of each user's need and as a function of the instantaneous network traffic.

There are different packet switching protocols, such as ATM, IP, frame relay, and X.25. The IP version 4 (IPv4) does not introduce the concept of QoS, because it does not include priority rules to avoid delays or jitter (e.g., for voice). Moreover, it does not avoid loss of data for certain types of services (e.g., for pure data communication), and it does not allow the assignment of higher bandwidth to certain services, relating to other (e.g., multimedia vs. voice). On the other hand, ATM and IP version 6 (IPv6) have mechanisms to improve the QoS.

1.1.8 CONNECTION MODES

Depending on the end-to-end service provided, the connection modes through networks can be of two types: connectionless and connection oriented. These modes are used in any of the layers of a network architecture, such as in the Open System Interconnection reference model, or in the transmission control protocol/IP (TCP/IP) stack.

1.1.8.1 Connection-Oriented Service

In order to provide a connection-oriented service, there is a need to previously establish a connection before data is exchanged, and to terminate it after data exchange. The connection is established

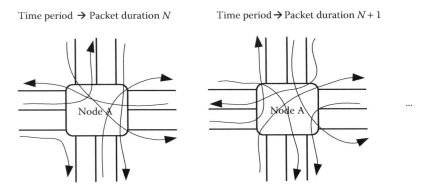

FIGURE 1.10 Switching of packets in different instants.

between entities, incorporating the negotiation of the QoS and cost parameters of the service being provided. The communication is bidirectional, and the data is delivered with reliability. Moreover, in order to prevent a faster transmitter to overload a slower receiver, flow control is employed (to prevent overflow situations). An example of a connection-oriented service is the telephone network, where a connection is previously established before voice exchange. In the telephone network, taking, as a reference, two words transmitted one after the other, we do not experience an inversion of the correct sequence of these words (e.g., receiving the second word before the first one). The TCP of the TCP/IP stack is an example of a connection-oriented protocol.

A connection-oriented service is always confirmed,[*] as the transmitter has information about whether or not the data reached the receiver free of errors, correcting the situation in case of errors. This can be performed using positive confirmation, such as the positive acknowledgment with retransmission (PAR) procedure, or using negative confirmation, such as the negative acknowledgment (NAK).

In the PAR case, when the transmitter sends a block of data, it initiates a chronometer and expects for the correct reception of an acknowledgment (ACK) message from the receiver within a certain time frame. In case the ACK message is not received in time, the transmitter assumes that the message was received corrupted and performs the retransmission of the block of data. In case the ACK message is received, the transmitter proceeds with the remaining transmission of data. The advantage of this procedure is that the ACK message sent by the receiver to the transmitter allows two confirmations: (1) the data was properly received (error control) and (2) the receiver is ready to receive more data (flow control).

In the case of the NAK, the receiver only sends a message in case the data is received with errors; otherwise, the receiver does not send any feedback to the transmitter. The advantage is the lower amount of data exchanged. The disadvantage is that in the PAR case, flow control is performed together with error control, whereas in the NAK situation, only error control is performed.

The reader should refer to Chapter 2 for a detailed description of the service primitives used in connection-oriented services.

1.1.8.2 Connectionless

The connectionless mode does not perform the previous establishment of the connection, before data is exchanged. Therefore, data is directly sent, without prior connection establishment.

As the connection-oriented mode requires a handshaking between the transmitter and the receiver,[†] this introduces delays in signals. Consequently, for services that are delay sensitive, the connectionless mode is normally employed. The connectionless mode is also utilized in scenarios where the experienced error probability is reduced (such as in the transmission of bits in an optical fiber).

Depending on whether the service is confirmed or not, data reliability may or may not be assured. Even though if data reliability is not assured, such functionality can be provided by an upper layer of a multilayer network architecture. In such a scenario, there is no need to execute the same functionalities twice.

The connectionless mode can provide two different types of services:

- Confirmed service
- Nonconfirmed service

In the case of the nonconfirmed service, the transmitter does not have any feedback about whether or not the data reached the receiver free of errors. Contrarily, in the case of the confirmed service, although a connection establishment is not required before the data is exchanged (as in the case of

---

[*] On the other hand, the connectionless service can be confirmed or nonconfirmed.
[†] For example, implementing data retransmission, in order to assure data reliability.

the connection-oriented service), the transmitter has feedback from the receiver about whether or not the data reached the receiver free of errors. The reader should refer to the description of the confirmation methods used in confirmed services presented for the connection-oriented service, namely the PAR and NAK.

As an example, Internet telephony (IP telephony) is normally supported by the nonconfirmed service, specifically, by the user datagram protocol (UDP), which is connectionless. However, in IP telephony, the reordering of packets is performed by the application layer.[*] Another example of a nonconfirmed connectionless mode is the IPv4 protocol, which does not provide reliability to the delivered datagrams and which does not require the previous establishment of the connection before data is sent. In case such reliability is required, the TCP is utilized as an upper layer (instead of the UDP). The serial line IP is an example of a data link layer protocol that is nonconfirmed and connectionless.

The reader should refer to Chapter 2 for a detailed description of the service primitives used in confirmed and nonconfirmed connectionless services.

1.1.9 NETWORK COVERAGE AREAS

Packet switching networks may also be classified as a function of the coverage area. Three important areas of coverage exist: LANs, MANs, and WANs.

A LAN consists of a network that covers a reduced area such as a home, office, or small group of buildings (e.g., an airport), using high-speed data rates. A MAN consists of a backbone (transport network) used to interconnect different LANs within a coverage area of a city, a campus, or similar. This backbone is typically implemented using high-speed data rates. Finally, a WAN consists of a transport network (backbone) used to interconnect different LANs and MANs, whose area of coverage typically goes beyond 100 km. While the transmission medium used in a LAN is normally the twisted pair, optical fiber, or wireless, the optical fiber is among the most used transmission medium in a MAN and WAN.

1.1.10 NETWORK TOPOLOGIES

A network topology is the arrangement of the devices within a network. The topology concept is applicable to a LAN, a MAN, or a WAN. In the case of a LAN, such a topology refers to the way hosts and servers are linked together, while in the MAN and WAN cases, this refers to the way nodes (routers) are linked together. For the sake of simplicity, this description refers to hosts and servers (in the case of LAN) and nodes (in the case of MAN and WAN) just as hosts.[†]

A bus topology is the topology where all hosts are connected to a common and shared transmission medium. This topology is depicted in Figure 1.11. In this case, the signals are transmitted to all hosts

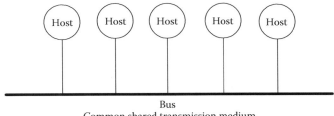

Bus
Common shared transmission medium

FIGURE 1.11   Bus topology.

---

[*] These functions are carried out by the real-time protocol (RTP).
[†] In fact, both workstations and routers are hosts.

and, because the host's network interface cards (NIC) are permanently listening to the transmitted data, they detect whether or not they are the destination of the data. In case the response is positive, the NIC passes the data to the host; otherwise, the data is discarded [Monica 1998]. This topology presents the advantage that, even though if a host fails, the rest of the network keeps running without problems. The main disadvantage of this topology relies on the high overload of the whole network (including all network hosts) that results from the fact that all data is sent to all network hosts.

In a ring topology, the cabling is common to all the hosts, but the hosts are connected in serial. This topology is depicted in Figure 1.12. Each host acts as a repeater: each host retransmits in a termination, and the data received in the other termination. The main disadvantage of this topology is that if a host fails, the rest of the network is placed out of order. This topology is normally utilized in SDH networks (MAN and WAN), where double rings are normally utilized to improve redundancy. The token ring technology used in LAN is also based on the ring topology.

A star topology includes a central node connected to all other hosts. The central node repeats or switches the data from one host into one or more of the other hosts. Because all data flows through this node, this represents a single point of failure. This topology is depicted in Figure 1.13.

A tree topology is a variation of the star topology. In fact, the tree topology consists of a star topology with several hierarchies. This topology is depicted in Figure 1.14. The central node is responsible for repeating or switching the data to the hosts within each hierarchy. In case the destination of the data received by a certain central node refers to another hierarchy, such central node forwards the data to the corresponding hierarchy central node, which is then responsible for forwarding the data to the destination host.

Finally, in a mesh topology each host is connected to all[*] or part[†] of the other hosts in the network. This topology is depicted in Figure 1.15. The advantage of such configuration relies on the existence of many alternative pathways for the data transmission. Even though if one or more paths are interrupted or overloaded, the remaining redundancies represent alternative paths for the data transmission. The drawback of such a topology is that the large amount of cabling is necessary to implement it.

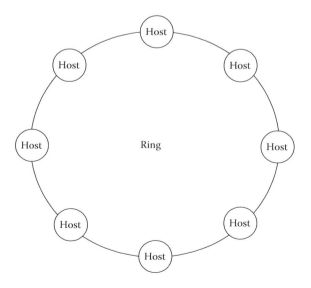

FIGURE 1.12 Ring topology.

---

[*] Complete mesh topology.
[†] Incomplete mesh topology.

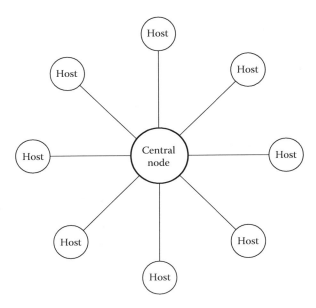

FIGURE 1.13   Star topology.

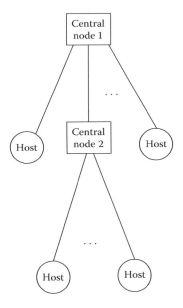

FIGURE 1.14   Tree topology.

It is worth noting that there are two different types of topologies: physical topology and logical topology. The physical topology refers to the real cabling distribution along the network, while the logical topology refers to the way the data is exchanged in the network. A physical star topology with a repeater (a hub)* as a central node presents a medium common and shared by all network hosts. In such a case, the logical topology is the bus topology (common and shared medium).

---

* A hub/repeater repeats in all other outputs the bits received in one input. In addition, it acts as a regenerator.

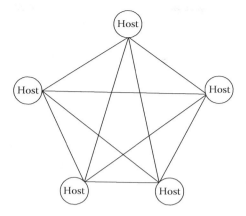

FIGURE 1.15 Mesh topology.

On the other hand, a physical star topology with a switch* as a central node corresponds, as well, to a logical star topology. Moreover, a logical ring topology corresponds to a physical star topology with a central node that rigidly switches the data to the adjacent host (left or right).

1.1.11 CLASSIFICATION OF MEDIA AND TRAFFIC

Different media can be split into three groups [Khanvilkar et al. 2005]:

- *Text*: Plaintext, hypertext, ciphered text, and so on.
- *Visuals*: Images, cartography, videos, videoteleconference (VTC), graphs, and so on.
- *Sounds*: Music, speech, other sounds, and so on.

While the text is inherently digital data (mostly represented using a string of 7-bit ASCII characters), the visuals and sounds are typically analog signals, which need to be digitized first, in order to allow its transmission through a digital network, such as an IP-based network (e.g., the Internet or an intranet). As can be seen from Figure 1.16, the multimedia is simply the mixture of different types of media, such as speech, music, images, text, graphs, and videos.

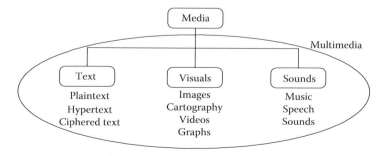

FIGURE 1.16 Basic types of media.

---

* A switch only switches data to the output where the destination host is located. This is performed based on the address of the destination.

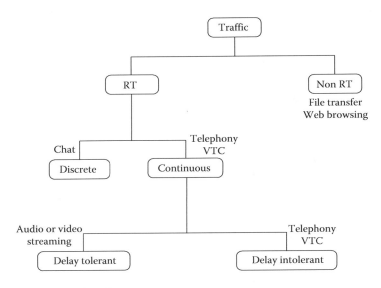

FIGURE 1.17 Classification of traffic.

When media sources are being exchanged through a network, it is generically referred to as *traffic*. As depicted in Figure 1.17, the traffic can be considered as real time (RT) or non-real time (NRT). While RT traffic is delay sensitive, NRT media is not. An example of RT traffic is telephony or VTC, whereas a file transfer or the web browsing can be viewed as NRT traffic.

RT traffic can also be classified as continuous or discrete. Continuous RT traffic consists of a stream of elementary messages with interdependency. An example of continuous RT traffic is telephony, whereas the chat is an example of discrete RT traffic.

Finally, RT continuous traffic can still be classified as delay tolerant or delay intolerant. RT continuous delay-tolerant traffic can accommodate a certain level of delay in signals, without sudden performance degradation. Such tolerance to delays results from the use of a buffer that stores in memory the difference between the received data and the played data. In case the transfer of data is suddenly delayed, the buffer accommodates such delay, and the media presented to the user does not translate such delay introduced by the network. Video streaming is an example of a delay-tolerant media. Contrarily, the performance of delay-intolerant traffic degrades heavily when the data transfer is subject to delays (or variation of delays). An example of RT continuous and delay-intolerant media is telephony or VTC. IP telephony or VTC allows a typical maximum delay of 200 ms, in order to achieve an acceptable performance.

1.2 PRESENT AND THE FUTURE OF TELECOMMUNICATIONS

Current and emergent communication systems tend be IP based and are meant to provide acceptable QoS in terms of speed, BER, end-to-end packet loss, jitter, and delays for different types of traffic.

Many technological achievements have been made in the last few years in the area of communications and others are planned for the future to allow the new and emergent services. However, whereas in the past new technologies pushed new services, nowadays the reality is the opposite: end users want services to be employed on a day-by-day basis, whatever the technology that supports it. Users want to browse over the Internet, get e-mail access or use the chat, establish a VTC, regardless of the technology used (e.g., fixed or mobile communications). Thus, services must be delivered following the concept of *anywhere* and at *anytime*. Figure 1.18 presents the bandwidth requirements for different services.

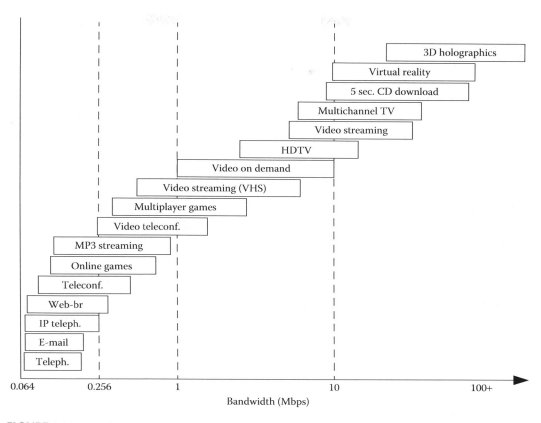

FIGURE 1.18 Bandwidth requirements of the different services.

1.2.1 CONVERGENCE

The main objective of the telecommunications industry is to create conditions to make the convergence a reality [Raj et al. 2010]. The convergence of telecommunications can be viewed in different ways. It can be viewed as the convergence of services, that is, the creation of a network able to support different types of service, such as voice, data (e-mail, web browsing, database access, files transfer, etc.), and multimedia, in an almost transparent way to the user [Raj et al. 2010].

The convergence can also be viewed as the complement between telecommunications, information systems, and multimedia, in a way to achieve a unique objective: make the information available to the user with reliability, speed, efficiency, and at a low price. According to the Gilder law, the speed of telecommunications will increase three times every year in the next 20 years, and according to the Moore law, the speed of microprocessors will duplicate every 18 months.

The convergence can be viewed as the integration of different networks in a single one, in a transparent way to the user. It can also be viewed as the convergence between fixed and mobile concepts [Raj et al. 2010], as the mobile is covering indoor environments (e.g., femtocells* of long-term evolution [LTE]) allowing data and television/multimedia services, traditionally provided by fixed services, whereas fixed telecommunications are giving mobility with the cordless systems, whose example is the Digital European Cordless Telephone standard. There are terminals that are able to

---

* A femtocell is a cellular base station for use at home or in offices that creates an indoor cell, in locations where cellular coverage is deficient, inexistent, or to provide high-speed data services. It typically interconnects with the service provider via broadband xDSL or cable modem.

operate as cellular phones or as fixed network terminals. New televisions not only receive the TV broadcast but also allow browsing over the Internet.

The convergence is viewed by many people as the convergence of all the convergences, which will lead to a deeply different society, whose results can already be observed nowadays with the use of the following services:

- Telework
- Telemedicine
- Web-TV
- E-Banking
- E-Business
- Remote control over houses, cars, offices, machines, and so on
- VTC

Human lives, organizations, and companies will tend to increase their efficiency, with the new communication means, and with the increase of the available information, as well as with multicontact.

With technological evolutions—increase of user data rates, improved spectral efficiency, better performances (lower BER), increase of network capacity, and decrease of latency (RT communications)—and with the massification of telecommunications as a result of lower prices (as a result of technological evolution and the increase of competition), it is expected that virtual reality and 3D holographic will be a reality in the near future.

1.2.2 Collaborative Age of the Network Applications

While the convergence approach was based on the ability to allow information sharing using a common network infrastructure, the new approach consists of the use of the network as an enabler to allow sharing of knowledge. It consists of the ability to provide the right information to the right person at the right time. For this to be possible, a high level of interactivity made available to each Internet user is required. In parallel, business intelligence is an important platform that allows decision makers to receive the filtered information* required for the decision to be made in a correct moment. The concept of Internet of Things enables the knowledge by making available a large amount of data captured by multiple machines and sensors, and by enabling machine-to-machine communications. Moreover, to enable the sharing of knowledge, there is a need to complement the Internet of Things with the processes and applications. This is required to process the data captured by sensors and machines.†

We observe, nowadays, an explosion of ad hoc applications that allow any Internet user to inject nonstructured information (e.g., Wikipedia) into the Internet world, in parallel with an increase of mobile-cloud and peer-to-peer applications such as Torrent, eMule, and IP telephony. Social networks are currently being used by millions of people that allow the exchange of unmanaged multimedia by groups of people just to share information or by groups interested in the same subject. Note that this multimedia exchange can be text, audio, video, multiplayer games, and so on. This can only be possible with the ability of the IP to support all types of services in parallel with the provision of QoS by the network, that is, with the convergence as a support platform. This is the new paradigm of the modern society: the collaborative age. The collaborative age of the Internet can also be viewed as the transformation of man-to-man communication into man-to-machine and machine-to-machine communication, using several media, and where the source or destination party can be a group instead of a single entity (person or equipment).

Figure 1.19 shows the evolution of the network usage. Initially, this was viewed merely for the data applications. Afterward, as referred to in Section 1.2.1, convergence was an important issue to allow

---

* For example, key performance indicators.
† This also presents a relationship with big data analysis.

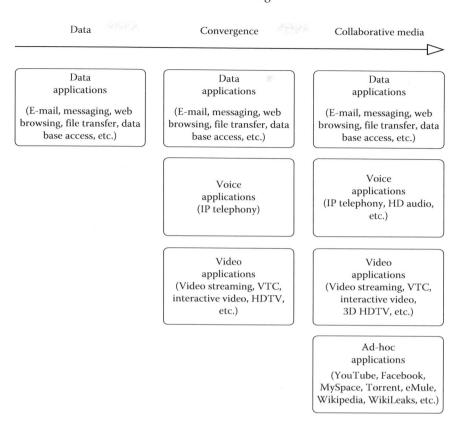

FIGURE 1.19 Evolution of network applications: from data to collaborative tool.

a better usage of the network. An increase in the level of Internet users interactivity made the Internet world a space for deep collaboration between entities, but with a higher level of danger as well.

1.2.3 TRANSITION TOWARD THE COLLABORATIVE AGE

To reach demands of the modern society, in terms of both convergence and collaborative services, several problems need to be solved from the scientific and industrial community. From Section 1.2.2, we may conclude that the convergence can be viewed as an important requirement to support the collaborative services.

Although we observe an enormous demand for convergence, we see that there are still problems that need to be solved. An example is the universal mobile telecommunication system (UMTS), which still treats voice and data in different ways, as data is IP based whereas voice is still circuit switching based. The LTE is the cellular standard that deals with this issue and makes the all-over-IP a reality.

From the point of view of services, the total digitalization of several information sources and the use of efficient encoding and compressing data algorithms are very important. The information sources can be voice, fax, images, music, videoconference, e-mail, web browsing, positioning systems, high-definition television, and pure data transmission (database access, file transfer, etc.). Different services need different transmission rates, different margin of latencies and jitter, different performances, or even fixed or variable transmission rates. Several MPEG protocols for voice or video, those already existent and those that are still in the research and development phase, intend to perform an adaptation of several information sources to the transmission media, allowing a reduction of the number of encoded bits to be transmitted.

Different services present different QoS requirements, namely:

- Voice communications are delay sensitive, but are low sensitive to loss of data, and require low data rate but approximately constant.
- Iterative multimedia communications (e.g., web browsing) are sensitive to loss of data, requiring considerable data rate, with a variable transmission rate, and are moderately delay sensitive.
- Pure data communications (e.g., database access and file transfer) are highly sensitive to loss of data, requiring relatively variable data rate, without sensitivity to delay.

Jitter is defined as the delay variation through the network. Depending on the application, jitter can be a problem, or jitter issues can be disregarded. For instance, data applications that only deliver their information to the user if the data is completely received (reassembling of data) pay no attention to the jitter issues (e.g., file transfer). This is totally different if voice and video applications are considered; those applications degrade immediately if jitter occurs.

The transmission of data services (e.g., pure data communications and web browsing) through most of the reliable mediums (e.g., optical fiber and twisted pair) usually considers error detection algorithms jointly with automatic repeat request[*] (ARQ), instead of error correction (e.g., block coding or forward error correction). This happens because these services present very rigid requirements in terms of BER, whereas not very demanding in terms of delay sensitivity (in this case, stopping the transmission and requesting for repetitions are not crucial). Note that the utilization of error correction requires more redundant bits per frame than error detection (amount of additional data beyond the pure information data). This can be seen from Figure 1.20.

Nevertheless, the transmission of data services through a nonreliable medium (e.g., wireless) is normally carried out using error correction, as the number of repetitions would be tremendous, creating much more overhead (and corresponding reduction of performance due to successive repetitions) than the overhead necessary to encode the information data with error correction techniques. A similar principle is applied to services that are delay sensitive (voice), where, to reduce latency, error correction is normally a better choice, instead of error detection.

These are the notions that introduce the QoS concept, implying that each service will impose certain requirements. For the convergence to become a reality, the network should be able to take all these requirements into account.

Taking into account all the previously described factors, one that presents a great contribution to support the new collaborative services is the maximum transmission rate, as it is associated with the user data rate. The factors that limit the use of higher transmission rates are several sources of interference and noise. The effects of noise can be minimized through the use of regenerators, as

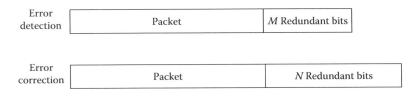

FIGURE 1.20 Types of error control: the error detection and error correction. Their differences in terms of the amount of redundant bits ($N > M$).

---

[*] ARQ works associated with error detection. The transmitter sends groups of bits (known as frames), which are subject to an encoding in the transmitter. The decoding process performed in the receiver allows this station to gain knowledge about whether or not there was an error in the propagation of the frame. In the case of error, the receiver requests a repetition of the frame from the transmitter.

well as advanced detection algorithms (e.g., matched filters). Interferences tend to increase with the increase in the used bandwidth (which corresponds to an increase of transmission rates), this being the main limitation of the use of higher data rates.

The challenge facing the today's telecommunications industry is how to continually improve the end-user experience, to offer appealing services through a delivery mechanism that offers improved speed, service attractiveness, and service interaction. In order to deliver the required services to the users with the minimum cost, the technology should allow better and better performances, higher throughputs, improved capacities, and higher spectral efficiencies.

What can be done in order to increase the throughput of a wireless communication system? One can choose a shorter symbol duration T_S. This, however, implies that a larger fraction of the frequency spectrum will be occupied, because the bandwidth required by a system is determined by the baud rate $1/T_S$. Wireless channels are normally characterized by multipath propagation caused by reflections, scattering, and diffraction in the environment. The shorter symbol duration might therefore cause an increased degree of intersymbol interference (ISI) and thus performance loss. As an alternative to the shorter symbol duration, one may choose using a multicarrier approach, multiplexing data into multiple narrow subbands, as adopted by orthogonal frequency division multiplexing (OFDM) [Marques da Silva et al. 2010]. The OFDM technique has been selected for LTE, as opposed to wideband code division multiple access that is the air interface technique that has been selected by European Telecommunications Standard Institute for UMTS. Thus, the problem of ISI can be mitigated. But still, the requirement for increased bandwidth remains, which is crucial with regard to the fact that the frequency spectrum has become a valuable resource. This imposes the need to find schemes able to reach improved spectral efficiencies, such as higher order modulation schemes, the use of multiple antennas at transmitter and at receiver such as multiple input multiple output systems, more efficient error control, and so on [Marques da Silva et al. 2010].

CHAPTER SUMMARY

This chapter provided an introduction to multimedia communications and networking, including the study of most important fundamentals of communications and future trends.

It was described that digital signals allow regeneration, multiplexing, and error control, functionalities not possible when analog signals are employed. Nevertheless, it was viewed that digital signals tend to require a higher bandwidth than the analog counterpart.

It was also viewed that the modem is employed when the transmission medium is analog, whereas the digital encoder (also referred to as the *line encoder*) is employed with digital transmission mediums. Moreover, the modem sends carrier modulated signals, that is, signals modulated around a certain carrier, whereas the digital encoder sends baseband signals, that is, signals modulated around a null frequency.

It was shown that transmission mediums can be cable or wireless. In the latter case, the difference between guided and unguided wireless transmission mediums was described. Among the cable transmission mediums, the optical fiber is the most resistant to interferences, and supports the higher bandwidth. Moreover, single mode optical fibers support higher bandwidths than multimode optical fibers.

We have viewed that synchronous communications allow higher data rates than asynchronous communication systems. Synchronous communications extract the synchronism reference from the received signal, or using an additional transmission pair or channel. Contrarily, asynchronous communication systems need to periodically use start and stop bits for allowing the receiver to determine the bit transition instants.

It was shown that simplex communications send signals only in a single direction, whereas duplex communications allow bidirectional communications. In the case of full duplex, two channels are required to allow simultaneous bidirectional communications.

It was described that a network is composed of a concatenation of point-to-point links, which can be of different types. In this case, intermediate nodes are responsible for linking the required sequence point-to-point links.

We have also viewed that circuit switching uses all assigned resources during the connection, whereas packet switching allows a more efficient use of the network resources. In case of packet switching, this can be of two modes: connection oriented or connectionless. The connection-oriented mode provides flow control and error control, which allows the service being confirmed. It was described that the connectionless mode can provide a confirmed service, or a nonconfirmed service.

The difference between a LAN, a MAN, and a WAN was described. The LAN is used within an office, or a house. A MAN is used to cover typically a city, or a university campus, being used to interconnect different LANs. Finally, a WAN corresponds to a network that typically covers a wide territory, such as a country, being also employed to interconnect different LANs.

The logical topology corresponds to the way data is interchanged, whereas the physical topology corresponds to the way network devices are physically interconnected. Bus, star, ring, tree, and mesh are examples of topologies that can be employed.

It was described that the traffic consists of the exchange of media sources through a network, or through a communication system. Moreover, media can be text, visual, or sounds. In addition, traffic can be RT, or NRT, discrete, or continuous. In the case of RT, and continuous traffic, this can be delay tolerant, or delay intolerant.

REVIEW QUESTIONS

1. What are the advantages of using digital communications relating to analog communications? What are the disadvantages?
2. What are the reasons that may imply the use of a modem?
3. What is the difference between simplex, half-duplex, and full-duplex communication?
4. What is the physical topology used to implement a logical bus? In such a case, what is the central node?
5. What is the difference between unicast, multicast, and broadcast communication?
6. What is the difference between an analog and a digital signal?
7. What is the difference between a LAN, MAN, and WAN?
8. What is the difference between a connectionless and connection-oriented service?
9. What is the difference between a point-to-point communication and a network?
10. What is the difference between a circuit switching and a packet switching network? Give examples of networks based on these two switching types.
11. What are the most important QoS requirements?
12. What is the convergence of telecommunications?
13. What is the difference between physical topology and logical topology?
14. Which types of media do you know?
15. How can the different types of traffic be grouped?
16. What is the collaborative age of the telecommunications?

LAB EXERCISES

1. Using the Emona Telecoms Trainer 101 laboratory equipment, and volume 1 of its laboratory manual, perform experiment 1—Setting up an oscilloscope.
2. Using the Emona Telecoms Trainer 101 laboratory equipment, and volume 1 of its laboratory manual, perform experiment 2—An introduction to Telecoms Trainer 101.

2 Network Protocol Architectures

LEARNING OBJECTIVES

- Describe the network protocol architecture concept.
- Describe the Open System Interconnection—Reference Model (OSI-RM).
- Describe the transmission control protocol/Internet protocol (TCP/IP).
- Describe the functions of each layer of the OSI-RM and TCP/IP.

2.1 INTRODUCTION TO THE NETWORK ARCHITECTURE CONCEPT

The problem of interconnecting terminals in a network is a complex task. The approach of trying to solve all the problems without segmentation of functions in groups becomes an equation with a very difficult solution. Therefore, the traditional solution is to group functionalities into different layers and allocate each group to a different layer. This is called the *network protocol architecture*, also commonly known as the *network architecture*. This approach only defines *what* is to be done by each layer, but not *how* such functionalities are to be implemented by the layer, whose responsibility belongs to the protocol of the layer individually. This approach leaves room for a layer to improve (due to, e.g., technological evolutions), without implications in the remaining layers, as long as the interface between a certain layer and its adjacent layers is kept as specified by the network protocol architecture. In this sense, the network architecture defines the number of layers, what is to be done by each layer, and the interface between different layers. Note that a network architecture not based on layers would not allow changing the *how to do* without changing the architecture itself and without changing the remaining functions of the network architecture.

There are different network architectures. The International Organization for Standardization (ISO) created the widely known OSI-RM, as depicted in Figure 2.1. Layers can also be identified by numbers, starting from the lower (physical layer—layer 1) up to the upper layer (application layer—layer 7).

This seven-layer architecture model defines and describes a group of concepts applicable to communication between real systems composed of hardware, physical processes, application processes, and human users [Stallings 2010].

This architecture can be split into two groups: the four lower layers (from the physical up to the transport layer) being responsible for assuring a reliable communication of data between terminal equipment; the three upper layers (from the session up to the application layer), with a higher level of logical abstraction, interfacing with the user application. Note that the OSI-RM is only a reference model, and the systems implemented use more or less parts of this model. Therefore, we may view the TCP/IP stack, in use in the Internet world, as the most used real implementation closer to the OSI-RM.

To better understand this network architecture concept, let us consider Peter in Boston who writes a letter to Christine in Bristol. The letter is written using a specific protocol, starting by *Dear Christine* and ending by *Warm regards* and signature. The letter is inserted into an envelope, to

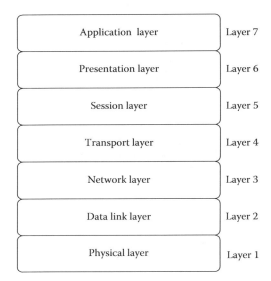

FIGURE 2.1   Layers of the OSI-RM.

which the destination address (Christine's address) is written in a specific location. The envelope is transported and delivered to the post office, in Boston, where a stamp is added in a specific location. This post office will send the letter through the post office network, which may use several means of transportations (van, bus, airplane, train, etc.) to allow the delivery of the letter at the post office in Bristol. In Bristol, the letter will be distributed between zones based on the destination address, and delivered to the postman, who will post the letter at Christine's postbox.

We may view the transportation of the letter as a process composed of several layers. An upper layer (application layer) that corresponds to the communication between Peter and Christine using a specific protocol (the letter starts by *Dear Christine* and ends by *Warm regards* and signature). This protocol specifies what and where in the letter it is to be written and only refers to the agents of this layer (Peter and Christine), not any of the intermediate agents (e.g., post office, plane, and postman). This is control data, that is, overhead. Moreover, this communication follows a protocol that consists of a set of procedures that are to be followed between the two entities.

Although the communication between them is supported by the lower layers (envelopes, post, airplane, etc.), one may say that there is a virtual circuit between Peter and Christine. In Figure 2.2, the application layer of the source and destination is linked by a dashed line, representing a virtual circuit. As in the case of the circuit between Peter and Christine, there is no direct connection between them. The lower layer (presentation) is used as a service (that is also supported by another lower layer) to allow the data arriving at the destination.

In the second stage, the letter was inserted into an envelope, and the destination address was written on the proper location. The second stage also has its own protocol, which includes the added overhead (destination address) to be read by the post office network (lower layer), essential to allow the letter being forwarded to the destination. Once the letter reaches the post office in Boston, a stamp is added to the envelope on a specific location. This is a procedure that is recognized by the worldwide post office protocol, essential to allow the letter being forwarded from the United States/Boston to the destination address in United Kingdom/Bristol. An employee at the post office in Boston will take the letter (together with many others) into the airport using a van as a mean of transportation. In Bristol, another employee will collect the letter from the airport, and transport it to the post office by a van.

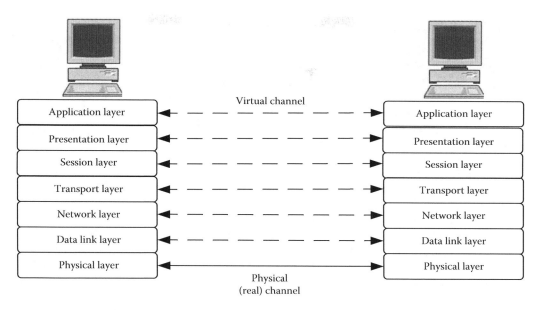

FIGURE 2.2   Communication of two workstations using the OSI-RM.

The address is composed of two parts: the country/city and the street name and number. The country/city pair is the information necessary to decide about the route to use in order to forward the letter from the source to the destination, through the worldwide post office network. This can be viewed as the network layer control data (overhead). A decision has to be made about whether to follow a direct flight from Boston to Bristol (in case there is one) or to send it through London. In the latter case, London acts as an intermediate node (router). Such a node receives the letter, reads the destination address, and decides about how to forward it to Bristol. This is the function of a router, which belongs to the network layer (layer 3). It receives the data from the physical layer (bits), and it goes up the several layers up to the network layer, removing the layer 1 and layer 2 control data. It reads the destination address (country/city) and decides about the best output interface to forward the data into the destination. Again, the router adds new layer 2 and layer 1 control data and sends it through the physical layer.

Returning to our example, the street name/number pair is the information necessary to decide about how to forward the letter within the destination city (Bristol). Note that this additional control data is to be processed by a lower layer. In this case, one may say that this control data is to be processed by the data link layer (DLL; switch), which is responsible for forwarding the letter within the city (point to point connection between Bristol post office and Christine's house). A switch belongs to the DLL and is responsible for forwarding data within a LAN, whereas a router is responsible for forwarding data between different LANs. The switch receives the data from the physical layer (bits), removes the layer 1 control data, and reads the layer 2 control data. Using the example, it reads the destination address (street name/number) and decides about the best output interface to forward the data to the destination. We may conclude that the country/city pair can be viewed as a LAN (layer 3 address), and the interconnection between different LANs (cities) is performed at layer 3 routing. Similarly, the street name/number can be viewed as the physical address of the terminal (layer 2 address), and the interconnection within the city (i.e., between streets and numbers) is performed at layer 2 switching.

From the example we conclude that each layer of a network architecture performs different functionalities, each presenting its own protocol, and a different overhead.

2.2 OPEN SYSTEM INTERCONNECTION—REFERENCE MODEL

Section 2.1 described that the network protocol architecture deals with functionalities performed by each layer, as well as the type of interfaces between different layers. It was seen that a specific layer provides services to the upper layer and makes use of the services provided by the lower layer. The definition of how these functionalities are carried out by each layer is not specified by the network architecture. This is specified by the protocol adopted by each layer.

The OSI-RM consists of an abstract network architecture model, being implemented in part by different network protocol architectures. As will be described, the TCP/IP includes many of the concepts specified by the OSI-RM.

As can be seen from Figure 2.3, the message format of each layer is referred to as the protocol data unit (PDU) preceded by a letter corresponding to the layer. The message format of the application layer is the application PDU (APDU). The message format of the presentation layer is the presentation PDU (PPDU). The message format of the session layer is the session PDU (SPDU). The message format of the transport layer is the transport PDU (TPDU).* The message format of the network layer is the network PDU (NPDU), also known as *packet*.† The message format of the DLL is the LPDU, also known as *frame*.‡ Finally, the message format of the physical layer is the bit. As can be seen from Figure 2.3, the nth layer service data unit (n-SDU) corresponds $(n + 1)$-PDU received from the upper layer, being encapsulated into the n-PDU. Moreover, the nth layer protocol control information (n-PCI) corresponds to the overhead generated in the nth layer, which may include fields such as source and destination addresses, redundant bits for error detection or correction, and acknowledgment numbers for error control and flow control.

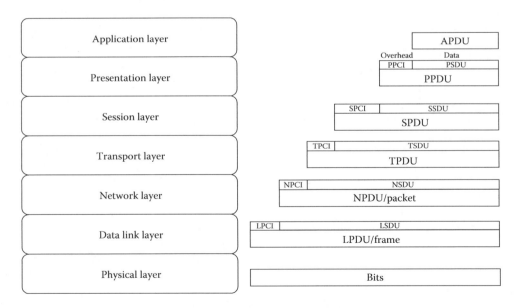

FIGURE 2.3 Protocol data units of different layers.

---

* In the TCP/IP, TPDU is called *segment*.
† Its main purpose is to allow it being forwarded throughout the entire network (i.e., between LANs).
‡ A frame is composed of a group of information bits to which control bits are added in order to allow performing error control and flow control.

2.2.1 SEVEN-LAYER OSI-RM

Figure 2.4 depicts several layers of the OSI-RM, describing generically the functions provided by each layer.

A brief description of each layer is provided in the following.

2.2.1.1 Physical Layer

The physical layer is layer one of the seven-layer OSI model of computer networking, also known as the OSI-RM. This layer is responsible for the transmission of the data received from the upper layer (DLL), in the form of bits, between adjacent nodes (point-to-point[*]). As shown in Figure 2.5, the link

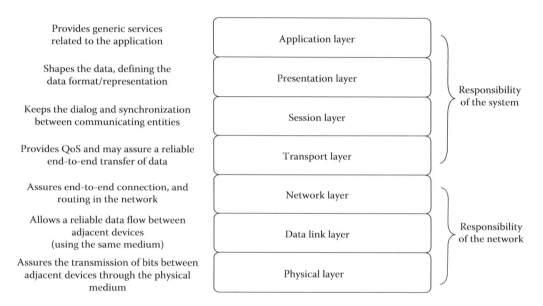

FIGURE 2.4 Generic description of the OSI-RM layers.

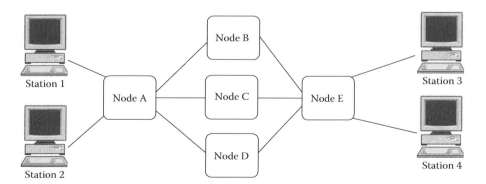

FIGURE 2.5 Example of a packet switching network.

---

[*] In the sense of a network, point-to-point refers to the interconnection between two adjacent routers (nodes), between a host and an adjacent router, or between two adjacent hosts.

between adjacent nodes can be the link between station 1 and node A, or between node A and node B, and so on. It is responsible for the representation of bits to be transmitted through a transmission medium.[*] Such representation includes the type of digital encoding or modulation scheme to use (voltages, pulse duration, amplitude, frequency or phase modulation, etc.) [Marques da Silva et al. 2010]. This layer also specifies the type of interface between the equipment and the transmission medium, including the mechanical interfaces (e.g., RJ45 connector).

Synchronization issues are also dealt with by this layer. This includes the ability of a receiver to synchronize with a transmitter (start and end of bit instants), before bits are transferred.

This layer aims to optimize the channel capacity as defined by Shannon[†] [Shannon 1948], making use of encoding techniques (or modulation schemes), multiple transmit and receive antennas, regenerators, equalizers, and so on. Although the physical layer may use error control, the provision of reliability to the exchanged data is normally a functionality to be provided by the DLL.

2.2.1.2 Data Link Layer

This layer is responsible for providing reliability to the data exchanged by the physical layer. This reliability is provided by the use of error control and flow control. Note that the DLL (as well as the physical layer) focuses on the point-to-point exchange of data.[‡]

The exchange of bits performed by the physical layer is subject to noise, interferences, distortion, and so on. All of these channel impairments may originate corrupted bits, which degrades the performance. The DLL makes use of error control techniques to keep the errors at an acceptable level. Depending on the medium that is being used to exchange data, error control can be performed using either error detection or error correction techniques. In the case of error detection, codes such as cyclic redundancy check (CRC) or parity bits are used to allow errors being detected on the receiver side, and the receiver may request the retransmission of the frame. However, if the medium is highly subject to noise and interferences (e.g., wireless medium), the choice is normally the use of error correction. In the latter case, the level of overhead per frame is higher, but it avoids successive retransmissions, which also translates in a decrease of overhead. Note that, in both cases, the DLL handles blocks of bits to which the corresponding overhead is added (redundant bits to allow error detection or error correction, as well as this layer address). These blocks of bits, with a specific format depending on the protocol of the DLL, are the previously mentioned LPDU, commonly known as *frame*.

Figure 2.6 shows the decomposition of an LPDU, as composed of an NPDU (packet received from the upper layer on the transmitting side) plus this layer overhead. The LPDU overhead is the startup flag, the source and destination address, the redundant bits for error control, and the end flag. Flags are used to allow synchronization, that is, for the receiver to identify the beginning and end of a frame. It is composed of a sequence of bits with a low probability of occurrence in the information part of a frame.

In the example of Section 2.1, the letter follows different types of transportation corresponding to each point-to-point connection (DLL). Peter walked from home to the post office, then a van took the letter into the airport, a flight was taken from Boston airport until Bristol airport, another van to Bristol post office, and so on. Each different type of transportation has its own protocol. Similarly, the end-to-end path is composed of a concatenation of links (DLL). Each link may have a different

Start flag	Address	Packet	Redundant bits	End flag

FIGURE 2.6 Frame decomposition.

---

[*] This transmission medium can be a twisted pair, coaxial cable, fiber optic cable, satellite link, wireless, and so on.
[†] Shannon capacity is defined in Chapter 3.
[‡] For example, between adjacent routers, or between a host and a router.

transmission medium[*] and different DLL protocol[†] running over it. Note that the DLL also includes the interchange of data between hosts within the same LAN. This interconnection of hosts within a LAN is achieved using a hub (repeater of bits), a switch, or a bridge. The functions of these devices rely on allowing the distribution of data within a LAN. These devices are defined in Chapter 12.

It is worth referring that another important functionality of the DLL is the flow control. Normally, the transmitter can transmit faster than the receiver is able to receive. To avoid loss of bits, the receiver needs to send feedback (control data) to the transmitter about whether or not it is ready to receive more frames. This is achieved through flow control. The flow control protocol can be associated with the error control protocol as follows: when a receiver checks the existence of errors (using, e.g., CRC or parity bits) and sends a feedback message to the transmitter informing that it is ready to receive the following frame (meaning that the previously received frame was free of errors), the two protocols (error control and flow control) work together.

2.2.1.3 Network Layer

This layer relies mainly on routing of packets along the network as well as addressing issues. The network layer is the first one (from the bottom) that takes care of end-to-end issues.

As shown in Figure 2.5, end-to-end connection is the connection between station 1 and station 3, or between station 2 and station 4, and so on.

Let us focus again on the example of Section 2.1. The letter had to be sent from Boston to Bristol. The post office in Boston had to decide about the best way to make the letter arrive in Bristol. It could be using a direct flight (in case there is one), or through London, and so on. This is the decision that has to be made by the network layer. Allow the NPDU to reach the destination through the best path. In case there is no direct flight, an intermediate *router* in London would have to read the destination address (country/city pair) and decide about the next hop to reach Bristol. Therefore, the network layer is responsible for the end-to-end routing of the NPDU in the network. There are two different basic modes of routing:

- Datagram[‡]
- Virtual circuit[§]

In the datagram mode, each NPDU carries the destination address and each node (router) has to decide about the best way to forward the NPDU in order to reach the destination. On the other hand, in the virtual circuit mode, each NPDU has information about only the virtual circuit to which it belongs. Several channels flowing in the network would belong to the same virtual circuit, and the node (router) only has to know the output interface corresponding to a certain virtual circuit.

The virtual circuits are established in advance, before the data is exchanged. In this case, all NPDUs of a certain connection follow the same predefined path. Contrarily, in the datagram mode, each node decides the following path, and different NPDUs of the same connection may follow different paths.

In the virtual circuit mode, the routing tends to be faster as the amount of decision that has to be taken by routers is lower. Different packets with different destination addresses may belong to the same virtual circuit in a specific part of the path. Looking into the network depicted in Figure 2.5, let us consider that station 1 needs to send data to station 3. A possibility could be sending packets through node A, node B, and node E. The NPDU has an identifier that identifies the virtual circuit, and the node only needs to read this identifier to find the output interface to use, not having to know the final destination address of the packet. Note that the router does not make any decision. Routers

---

[*] For example, wireless, twisted pair, satellite, and optical fibers.
[†] For example, IEEE 802.11, IEEE 802.3, and point-to-point protocol (PPP).
[‡] For example, IP.
[§] For example, X.25 or MPLS.

only have to read the virtual channel identifier, which is shorter (in number of bits) than the destination address, and thus, the level of overhead is reduced.

In case the datagram mode is in use, node A receives packets from station 1, reads the destination address (a field within the packet whose length is longer than the virtual channel identifier), and decides about the following node to use in order to make the packet arrive the final destination. Note that, in the datagram case, as the network changes dynamically in time, different packets may follow different paths, and the packets may reach the destination out of order. In this case, another layer would have to make the reordering of packets.[*]

In both cases, routers make use of routing tables. In the datagram mode, a routing table stores information about the output interface to which packets should be sent to in order to reach a certain destination address. In the virtual circuit mode, a routing table stores information about the output interface that corresponds to a certain virtual circuit. Note that the virtual circuit mode allows data to be forwarded faster, but the construction of the routing table is more complex (and requires a higher level of overhead) than in the case of a datagram. The IP is based on a datagram mode.

2.2.1.4 Transport Layer

This layer is responsible for making sure that end-to-end data delivered by the network layer has the required quality of service (QoS) (reliability, delay, jitter, bandwidth, etc.). In other words, it is responsible for providing the desired service to the upper layers. Depending on the classification of the service provided, there are two different types of connections that influence the provision of this layer QoS:

- Connection oriented
- Connectionless

A connection-oriented service is a service provided by a layer that comprises three different phases: (1) connection setup, (2) data exchange, and (3) connection termination. A connection-oriented service assures that packets that reach the receiver follow the transmission order. In addition, it makes use of error control techniques to provide reliability to the delivered data.[†] Although connection-oriented services bring benefits in terms of data reliability, it demands more processing from both transmitter and receiver, which translates in additional resources and time (e.g., time to establish before sending data, delay due to request for repetition of packets, and reordering of data at the receiver). On the other hand, a connectionless service is minimalist in terms of processing, but the delay is also minimized. In this mode, there is no need to make a connection setup before transmission. Reordering of packets is not performed at the receiver.[‡] Let us consider the IP telephony service. As previously described in Section 1.2, an important characteristic of the voice communication service is that it is delay sensitive, whereas not very sensitive to loss (or corrupted) of data. Therefore, the use of a connection-oriented transport layer does not seem to be a good idea, as it may introduce delays.[§] Therefore, the IP telephony service is normally supported in connectionless mode as it minimizes the delay, whereas the errors that may occur are normally not critical for the message to be understood.[¶]

---

[*] Normally, for services that require the reordering of packets, this is performed by the transport layer. Nevertheless, in some cases, this can be done by another upper layer (e.g., IP telephony service).

[†] In the TCP/IP stack, the connection-oriented service of the transport layer is implemented using the TCP.

[‡] In the TCP/IP stack, the connectionless service of the transport layer is implemented using the user datagram protocol (UDP). Moreover, the IP (layer 3) is based on the connectionless mode.

[§] In fact, this delay is subject to fluctuations, which is called jitter. The level and variation of delay would depend on the amount of requests for repetition.

[¶] Typically, a voice service has an acceptable quality with a bit error rate of the order of 10^{-3}, whereas most of other data services require much more reliable data.

Let us now consider a file transfer between two terminals through the network. This service is highly sensitive to loss of data (otherwise the file would be corrupted), whereas not very sensitive to delay. It is important to make sure that packets arrive the destination in the correct order, and free of errors. Therefore, it is clear that the service provided by the transport layer should be connection oriented.

In addition to the above-mentioned transport functions, this layer may also offer other functionalities: let us consider the case where the network layer has a 512 kbps connection established, and where the session layer is requesting a 1024 kbps connection. In this situation, the transport layer may establish two 512 kbps network connections and, in a transparent manner, offer a 1024 kbps connection to the session layer. A similar function may be offered when the maximum NPDU length is lower than the NPDU length being requested by the session layer (SPDU). In this situation, the transport layer performs the segmentation of the SPDU into two (or more) NPDUs.

2.2.1.5   Session Layer

This layer allows the mechanisms for setting up, managing, and closing down sessions between end-user application processes. While supported in the transport layer, it consists of requests and responses between applications. Logical sessions need to be properly managed in terms of which station can send data and when. As an example, although the lower layers can be full-duplex, some applications are specified to communicate in half-duplex. Another function of the session layer consists of the ability to reestablish the connection in case of failure. In this context, synchronism points are inserted such that, in case of failure, the session is reestablished from the last synchronism point correctly processed. Furthermore, the session layer may ask its counterpart (destination session layer) about whether or not the data received before a certain synchronism point has been properly processed.

2.2.1.6   Presentation Layer

This layer is responsible for formatting and delivering of data to the application layer for processing. Different workstations use internally different representations of data. However, the data exchanged along a network needs to follow a common representation; otherwise, the communication could not succeed. This definition is performed by the presentation layer. The interconnection of different networks that use different presentation layers requires a gateway. A gateway can be seen as a device that is able to understand two (or more) languages and is able to translate one language into another. It is still worth noting that, although it can be performed by other layers, because encryption can be viewed as a different way of representing data, it is typically performed by the presentation layer.

2.2.1.7   Application Layer

This layer of the network architecture is responsible for interfacing with the application program of the user. Note the difference between an application layer of a network architecture and an application program. The former has some attributes in the exchange of data in the network, whereas the latter is only a specific application resident in hosts. These attributes include the definition of fields and rules of interpretation for these fields. It provides services specific to each kind of application (file transfer, web browsing, e-mail, etc.). As an example, Microsoft Outlook is an e-mail application, whereas CCITT X.400 is an e-mail application layer that defines the way functionalities are carried out.

2.2.2   Service Access Point

Each layer has its own addressing format. The purpose of an address is to allow the corresponding layer to identify whether or not a certain host is the destination of a PDU. In addition, the source address allows the identification of who was the sender of such message that is circulating in the network.

The address of each layer, referred to as the *service access point* (SAP), is preceded by a letter corresponding to the layer, and is part of the layer control data. As previously mentioned, a specific layer (N) communicates with the upper layer ($N + 1$) to offer services, and communicates with the lower layer ($N - 1$) to use its services. This concept can be seen in Figure 2.7. A SAP can be viewed as the interface between adjacent layers. Note that a layer may communicate with adjacent layers using more than one SAP.

The SAP of layer N is the address of the interface between layer N and layer $N + 1$. The address between the application layer and the presentation layer is the presentation SAP (PSAP). The address between the presentation layer and the session layer is the session SAP (SSAP). The address between the session layer and the transport layer is the transport SAP (TSAP).[*] The address between the transport layer and the network layer is the network SAP (NSAP). The same principle applies to the other layer SAP. Note that the physical address is the DLL address used in the interface between the DLL and the network layer.

Returning to the example of Section 2.1, the two post office employees, in Boston and Bristol, respectively, have a virtual circuit between them (dashed lines in Figure 2.2), but this virtual circuit is supported by a lower layer, which is the airplane. Therefore, we may view the airplane as the physical layer (lower layer), being the unique that represents a real circuit [Forouzan 2007].

As can be seen from Figure 2.8, the communication between two terminals is performed by different layers. Each different layer will develop a specific function, and have its own protocol.

Note that each upper layer uses the services made available by the lower layer to establish a virtual circuit with its counterpart layer at the destination address. Furthermore, on the transmitter side,[†] each layer adds specific control data (overhead), essential to allow the message (or part of it) being forwarded to its counterpart layer at the destination address. On the transmitter side, the lower layer receives the user data and control data from the upper layer and considers all of this data as user data (this layer control data is also to be added). The control data of a specific layer is only of interest to the corresponding layer at the receiver, and follows a specific protocol depending on the

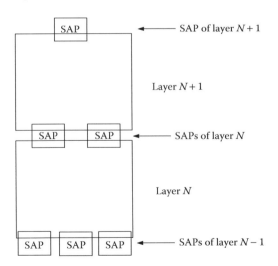

FIGURE 2.7   Service access point as the interface between layers.

---

[*] In the TCP/IP stack, TSAP is known as the port number, the NSAP is known as the IP address, and LSAP is known as the medium access control (MAC) address, hardware address, or physical address.

[†] Note that each station normally acts simultaneously as a transmitter and as a receiver. Nevertheless, for the sake of simplicity, in this description, we assume that a station acts as a transmitter and another one as a receiver. In reality, their functions alternate in time (half-duplex) or both stations may even act simultaneously as a transmitter and a receiver (full-duplex).

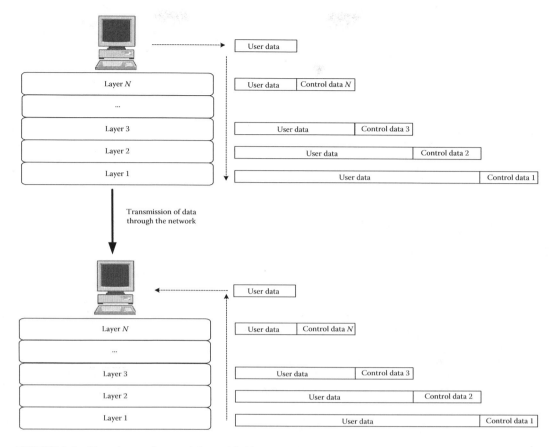

FIGURE 2.8 User data and control data added by several layers.

specifications of the layer. The receiver side removes the control data previously added by the corresponding layer at the transmitter and delivers the user data to the upper layer.

The service provided by a layer to its upper layer is defined as a group of elementary services. Each of these elementary services is implemented using service primitives. As can be seen from Figure 2.9, the four basic service primitives considered by the OSI-RM are as follows:

- Request
- Indication
- Response
- Confirm

The request is a service primitive sent by a source upper layer ([$N + 1$]-layer) to its adjacent lower layer (N-layer), being transmitted along the network until the destination counterpart layer as an indication primitive (from the destination N-layer to the destination [$N + 1$]-layer). In case the service is confirmed,[*] the service primitive response is sent from the destination upper layer ([$N + 1$]-layer) to its adjacent lower layer (N-layer), meaning that such indication message was properly received by the destination layer, whereas on the source side this response is delivered from the lower layer (N-layer) to its adjacent upper layer ([$N + 1$]-layer) in the form of a confirm primitive.

---

[*] An example of a confirmed service is the TCP, whereas the UDP is not confirmed. Both of these protocols are layer 4 protocols of the TCP/IP stack.

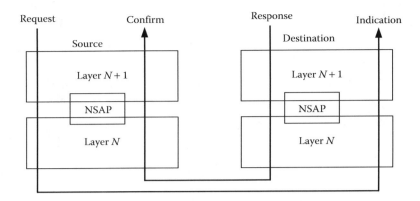

FIGURE 2.9 Service primitives.

As can be seen from Figure 2.10, a connection-oriented service comprises the exchange of the four service primitives for each of the following elementary operations:

• Connection setup
• Exchange of each group of bits (frame or segment)
• Connection termination

As described in Chapter 1, a connectionless service can be confirmed or nonconfirmed. As can be seen from Figure 2.11, a confirmed connectionless service includes only the service primitives associated with the transmission of data, including response and confirm primitives, necessary for

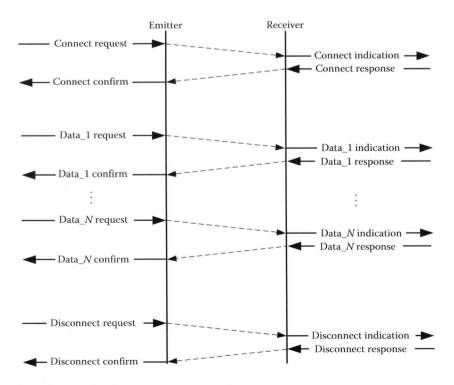

FIGURE 2.10 Service primitives of connection-oriented services.

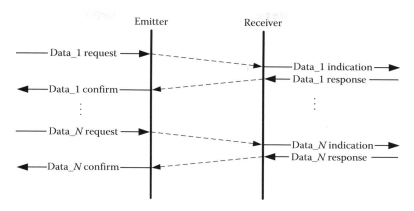

FIGURE 2.11   Service primitives of confirmed connectionless services.

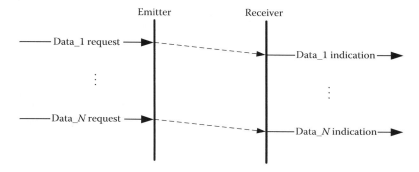

FIGURE 2.12   Service primitives of nonconfirmed connectionless services.

the confirmation of the service. Contrarily, as can be seen from Figure 2.12, a nonconfirmed connectionless service makes use of only the first two primitives (request and indication).

The reader should refer to Chapter 1 for the description of a nonconfirmed service.

2.3   OVERVIEW OF THE TCP/IP ARCHITECTURE

The TCP/IP architecture adopted by the Internet* is the most used real implementation of the OSI-RM. Nevertheless, while the basis is the same, there are some differences between these two architectures. While the OSI-RM is a seven-layer architecture, the TCP/IP model is composed of only five layers. This can be seen from Figure 2.13.

Similar to the OSI-RM, the layer N of the TCP/IP model uses the services made available by the layer $N - 1$, and provides services to the layer $N + 1$. In addition, the layer N on the transmitting side has a virtual circuit with the layer N on the receiving side. Naturally, this virtual circuit is established making use of the services provided by the lower layers (the only real circuit is the one established by the physical layer). In addition, Figure 2.14 shows the control data added by each different layer of the TCP/IP, on the transmitter side. Note that this control data is inversely removed on the receiver side.

The TCP/IP application layer includes functionalities assigned to the application, presentation, and session layers of the OSI-RM (see Figure 2.13). In addition, some bibliography refers to the two

---

* The TCP/IP architecture is, sometimes, also known as the *Internet model.*

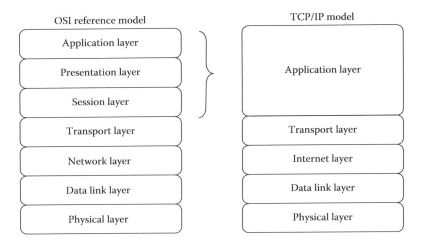

FIGURE 2.13   Comparison between the OSI-RM and the TCP/IP model. (Data from Marques da Silva, M., *Multimedia Communications and Networking*, 1st edition, CRC Press, Boca Raton, FL, March 2012.)

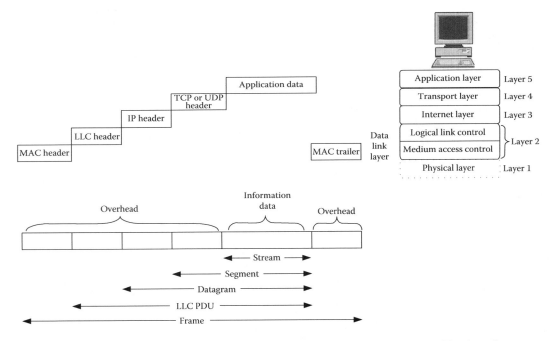

FIGURE 2.14   Description of overhead added by different TCP/IP layers and identification of message formats.

lower layers of the TCP/IP (physical layer and DLL) as the *network access layer*. In fact, the functionality of these two layers is to provide access to the network.

The following subsections provide a generic description of each of the layers of the TCP/IP architecture. This description will start by the layer with a higher level of abstraction (application layer), as it is closer to the hosts that make use of the network protocol architecture to interchange data with a remote host.*

---

* The reader should refer to Chapters 8 through 13 for a detailed description of each of the layers and their protocols.

The OSI-RM is defined such that each layer can only make use of services made available by the adjacent lower layer. In the TCP/IP architecture, the level of flexibility is much higher as any layer may invoke a service of any of the other layers, not only the lower layers but also the upper layers.[*]

2.3.1 APPLICATION LAYER

The TCP/IP application layer incorporates most of the functions defined for the three upper layers of the OSI-RM (session, presentation, and application layers). In this case, the application layer deals with all of the issues related to communication of user processes, whereas the OSI-RM splits these functionalities into three layers.

The reader should refer to the session, presentation, and application layers of the OSI-RM. In addition, the reader should refer to Chapter 8 for a detailed description of the TCP/IP application layer.

2.3.2 TRANSPORT LAYER

As described for the OSI-RM, this layer is responsible for the provision of QoS. Such functionalities include the implementation and control of service requirements such as reliability of data, delay, jitter, low or high bit rate, and constant or variable bit rate. From Figure 2.14, it is seen that the application layer generates data that is segmented and delivered to the transport layer. The message format of the transport layer of the TCP/IP is called *segment*, and can be of two different types:

- UDP
- TCP

While the TCP is connection oriented, it requires the setup of the connection before the data is exchanged. Making use of error detection and CRC, it provides reliability to the packets delivered. The provision of reliability is performed through the use of error detection (CRC codes) associated with the positive acknowledgment with retransmission (PAR) and the sliding window protocol.[†] In this sense, a receiving station acknowledges the good reception of packets. In case an error is detected, the transmitter *timer* reaches the timeout[‡] without receiving the acknowledgment, and the packet is retransmitted. Note that this successive repetition of packets introduces variable delay (jitter) in signals.

In fact, the TCP performs other functions, namely it assures that:

- Data is delivered with reliability.[§]
- Packets are received in the correct sequence.
- Packet losses are detected and corrected.
- The duplication of packets is avoided.

Therefore, this mode is ideal for services that require reliable data, without presenting sensitivity to delay or jitter. Figure 2.15 depicts several protocols and technologies that can be used by different TCP/IP layers. From this figure, it can be seen that some application layers supported by the TCP are file transfer protocol, telnet, simple mail transfer protocol (SMTP), or hypertext transfer protocol.

---

[*] As an example, the open shortest path first (OSPF) of the TCP/IP architecture is a layer 3 protocol that is used to create routing tables. This protocol invokes the UDP, which is a layer 4 protocol.

[†] Refer to Chapter 12 for a detailed description of the sliding window protocol.

[‡] The transmitter starts the *chronometer* (i.e., a timer) whenever a packet is transmitted.

[§] In fact, there are limits to reliability of data. The TCP does not guarantee 100% of error-free packets, but keeps it at an acceptable level for the more demanding services in terms of error sensitivity.

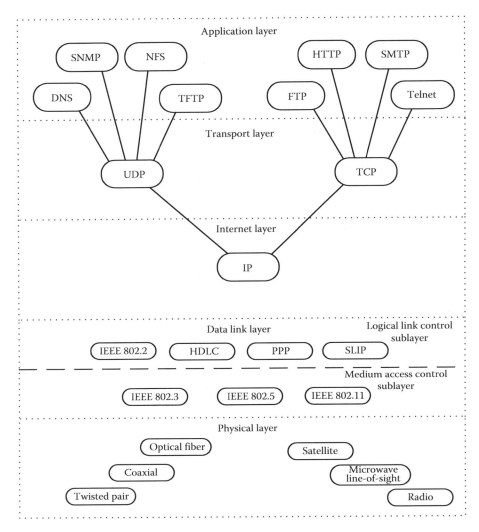

FIGURE 2.15 Example of some protocols and technologies used by different TCP/IP layers. (Data from Marques da Silva, M., *Multimedia Communications and Networking*, 1st edition, CRC Press, Boca Raton, FL, March 2012.)

The UDP transport protocol is connectionless and provides a nonconfirmed service, and therefore the data is exchanged without requiring the preliminary setup of the connection. In this case, the delivery of data to the end station is based on the best effort. While not providing data reliability, it presents the advantage of not introducing delay in signals. Therefore, this is the ideal mode for services that can resist to some level of errors. Such examples are the simple network management protocol (SNMP), the domain name system, or the network file system. Note that although the UDP does not provide reliability, such reliability can be provided by the upper layers, or by other means than error detection and retransmission. The SNMP uses UDP because transmission of data is very redundant in time (repeated from time to time). In the case of the IP telephony, reordering of packets is performed by the application layer.

The TSAP of the TCP/IP stack is the concatenation of the IP address with the port number. The port number is composed of a 16-bit address (between 0 and 65 535), referenced using the decimal notation.

The reader should refer to Chapter 9 for a detailed description of the TCP/IP transport layer.

2.3.3 INTERNET LAYER

Similar to the network layer of the OSI-RM, the Internet layer of the TCP/IP model is the most important, and deals with routing issues between different networks. Nevertheless, actions are to be taken by each node (router) of a network.

The IP has been developed and standardized in different versions. This protocol started with version 1. Versions 2 and 3 were defined but were replaced by version 4 (IPv4), which has been the most used version of the IP. Version 5 was developed, being a specific protocol optimized for data streams (voice, video, etc.). Finally, the new IP, initially entitled *Internet protocol of the next generation* (IPng) during the development phase, was standardized by the RFC 2460 (request for comments) as the IP version 6 (IPv6). It is worth noting that some authors refer to the Internet layer as the network layer. In this book, these terms are used interchangeably.

In the example of Section 2.1, where Peter sent a letter to Christine, the interconnection of different cities was viewed as the interconnection of LANs, as we can view a city as a LAN. Note that, between two cities, a letter may follow different paths (e.g., direct flight and through a third city). Furthermore, the forwarding of packets within a city (i.e., between houses) was viewed as the layer 2 switching.

We have seen in the definition of the network layer of the OSI-RM that it can be based on a datagram or a virtual circuit mode, depending on whether all packets between the source and destination follow the same path (virtual circuit) or, eventually, different paths (datagram). The IP is based on the datagram method, being connectionless. Therefore, because the network changes dynamically in time, different packets may follow different paths, and the packets may reach the destination out of order. In case reordering of packets is necessary for the service to be supported, another layer (normally this is performed by the TCP in layer 4) would have to perform it.

In the IP, each node (router) has to decide about the best way to forward packets, and these packets transport information about the destination address. From Figure 2.14, we see that a datagram[*] is composed of the layer 4 data (segment) plus the IP header. In the case of the OSI-RM, different control data is added at the source and removed at the destination. The exchange of data along the network depicted in Figure 2.16 is as follows:

- In the source host (leftmost host), the application layer of the TCP/IP stack receives the application data, formats it for transmission, establishes a session with the remote host (rightmost host), segments it, and transfers it to the transport layer.
- The transport layer receives the data from the application layer, adds the necessary overhead, including the source and destination port address (in the case of the TCP, this includes the addition of CRC redundant bits, as well as the sequence number), and transfers it to the Internet layer.

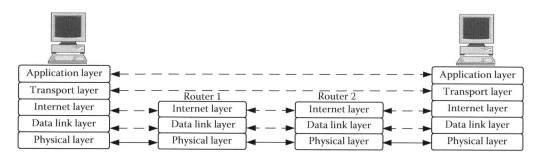

FIGURE 2.16 Example of a network using the TCP/IP.

---

[*] Note that IPv4 designates NPDU (packet) as a datagram. IPng, known as IPv6, returns the packet designation.

- The Internet layer of the source host adds the layer 3 overhead, namely the source and destination IP address, and passes it to the DLL for point-to-point transfer.
- The DLL receives the packet from layer 3 and adds the layer 2 overhead (source and destination MAC address, flags, error control redundant bits,* and other control data). Afterward, the frame (layer 2 PDU) is transferred to layer 1.
- The physical layer of the source host receives the frame from the upper layer, formats the bits for transmission† (type of modulation scheme or digital encoding, bit synchronization, etc.), and starts the transmission of symbols to the leftmost router (being directly connected to the host).
- The physical layer of the router receives the symbols, performs equalization (if applicable), converts the symbols into bits, and transfers them to layer 2 of the router.
- The DLL of the router groups the bits into the frames, and checks for errors. If an error is detected in a frame, there are two possibilities:
  - If error correction is used by layer 2, the frame is corrected by layer 2 of the router.
  - If error detection and automatic repeat request is used by layer 2, a frame is sent from the router to the source host, requesting for the retransmission of the corrupted frame. This frame has to go to layer 1 of the router for transmission. Afterward, the corresponding bits are received by the host, and passed to layer 2. The host layer 2 understands that this frame corresponds to a request for repetition, and starts the retransmission of the frame, being then transferred into layer 1 for transmission. In the router, its layer 1 receives the bits and passes them again to layer 2 for another error check.
- Once the frame is assumed to be correctly received by layer 2 of the router, this layer overhead is removed and the packet is transferred to layer 3.
- The router's layer 3 reads the destination address and consults its routing table. From this table, it extracts the output interface that should be used to send the packet. In the example of Figure 2.16, this is the next router. Therefore, the packet is transferred again to layer 2, where the new control data is added, and passed to layer 1 for transmission. This process is repeated until the packet reaches the destination host.
- At the destination host (rightmost host of Figure 2.16), the data goes up the several layers until the transport layer. This layer may have two different procedures:
  - In case the TCP is in use, the protocol checks for segment errors. If an error is detected, as the PAR is used by the TCP, it does not acknowledge the reception of the frame and layer 4 of the source host reaches the timeout and starts the retransmission of such packets. Furthermore, once the packet is correctly received by layer 4 of the destination host, it checks the correct sequence of packets (to avoid the wrong sequence of packets, duplication of packets, or the absence of packets), corrects it (if necessary), removes the layer 4 overhead, and transfers it to the application layer.
  - In case the UDP is in use, it removes only the layer 4 overhead and transfers it to the application layer.
- The application layer reassembles the several packets received from the source host (transferred from the transport layer), makes the necessary conversion of data, and transfers it to the application process.

The NSAP of the TCP/IP stack is the IP address. The Internet Assigned Numbers Authority (IANA) is responsible for the global coordination of the IP addressing, including the assignment of IP

---

* The TCP/IP does not define the type of error control technique to be used by the DLL. It depends on the data link protocol adopted in each point-to-point connection. It can be based on error detection (e.g., parity bits, or CRC) or based on error correction (e.g., forward error correction).
† On the transmitter side, each bit, or group of bits, is encoded into one or more symbols. Conversely, on the receiver side, symbols are converted into bits. Symbols are transmitted, not bits. Symbols are generated using a modulator or digital encoder technique.

address groups. As can be seen from Figure 2.17, an IPv4 address is composed of 32 bits, grouped into four octets,* that is, four groups of eight binary numbers. Nevertheless, for the sake of simplicity, it is normally displayed in four groups of decimal numbers, using the dotted-decimal notation.

The IPv4 address space is divided into classes, from A to E (see Table 2.1). There are different possible ways to identify the class of an IPv4 address. Performing the conversion of the leftmost octet from decimal into binary, and observing the position of the leftmost zero, from Table 2.1, the address class can be identified. Class A has the leftmost zero in the most significant bit (MSB). Class B has the leftmost zero in the second position, and the MSB is 1. Class C has the leftmost zero in the third position, and the two leftmost bits are 1.

A router is a layer 3 device that is used to interconnect two or more networks. In addition, its two or more interfaces are normally of different types, that is, the DLL protocols in each of the interfaces are different. As an example, a router is normally used to interconnect a domestic Ethernet LAN with a WAN, which allows reaching the Internet service provider (ISP).† The router is connected to adjacent devices using different links‡ (at the DLL level). In the example of Figure 2.18, the router has two network interface cards (NIC). Each NIC is connected to each of the networks to which the router is connected to. Moreover, each NIC presents a different IP address.

Routing algorithms as well as IPv4 and IPv6 protocols are described in Chapters 10 and 11.

FIGURE 2.17 An example of an IPv4 address in both binary and dotted-decimal notation.

TABLE 2.1

Mapping between the Address Class and the Leftmost Octet Value

Class	Binary Range of the Leftmost Octet
Class A	0XXXXXXX
Class B	10XXXXXX
Class C	110XXXXX
Class D	1110XXXX
Class E	11110XXX

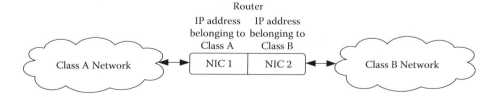

FIGURE 2.18 Example of a router with two NIC.

---

* The theoretical limit of the IPv4 address space is 2^{32}, corresponding to 4 294 967 296. Because of the rapid growth of the Internet, the available address space is depleted. IPv6 solves this problem, as its address is composed of 128 bits, which makes a much wider address space available for the Internet world.

† Using, for example, an ADSL or cable modem.

‡ For example, it connects to a LAN using the IEEE 802.3 protocol and it connects to the ISP through a WAN using, for example, the PPP.

2.3.4 Data Link Layer

As already mentioned for the OSI-RM, this layer refers to point-to-point communication between devices. The DLL is composed of two sublayers (see Figure 2.19):

- Logical link control (LLC) that deals with flow control and error control. This layer aims to mitigate the negative effects of the channel impairments experienced in the physical layer, such as attenuation, noise, interferences, and distortion.
- MAC that determines when a station is allowed to transmit within a LAN. Note that this sublayer only exists in the case of a LAN (and some types of MAN). When stations share the transmission medium in a LAN, it is said that the access method is with collisions (e.g., Ethernet). In this case, the MAC sublayer is responsible for defining when a station is allowed to transmit in such a way that collisions among transmissions from different stations is avoided (which originate errors). On the other hand, when stations of a LAN do not share the transmission medium, it is said that the method is without collisions (e.g., token ring).

In the example of Section 2.1, the layer 3 switching (routing) is responsible for finding the best path in order to forward the letter between cities. This could be a direct flight, through London, and so on. On the other hand, each elementary link of the full path between the source and the destination can be viewed as a DLL. Each of the links may use a different mode of transportation: car, van, bus, plane, ship, and so on. Moreover, we have seen that a city can be viewed as a LAN, and the distribution of letters within the city can be viewed as a layer 2 switching. Note that end-to-end (layer 3) forwarding is based on the country/city part of the address, whereas the distribution within a city is based on the street/number part of the address. Similarly, the DLL is responsible for the connection between two successive routers, or between a router and a host (even though if it is performed through a switch or a hub). Each connection may use a different layer 2 protocol (e.g., HDLC, PPP, and IEEE 802.3[*]), and a different communication medium (e.g., satellite, optical fiber, and twisted pair).[†]

In order to allow a better understanding of the differences between layer 2 (DLL) and layer 3 (Internet layer), let us analyze Figure 2.20.

The connection between router 1 and router 2 refers to layer 2 (point-to-point connection). The same applies to the connection between router 1 and the hosts in its LAN (197.139.18.0). If the host with the IP address 193.139.18.2, in the LAN connected to router 1, needs to exchange data with a host connected to router 2 (e.g., 197.139.18.2), the layer 3 protocol is used to forward the packets between the source and the destination. It deals with routing of packets in intermediate nodes

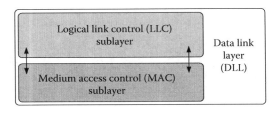

FIGURE 2.19 Data link layer and its sublayers.

---

[*] IEEE 802.3 corresponds to a standardization of the widely used Ethernet technology. This technology was developed by the consortium Digital, Intel, and Xerox. For this reason, Ethernet was also known as DIX. The standard IEEE 802.3 presents some variations to the Ethernet, being, for this reason, also known as Ethernet II or DIX II. This standard defines the physical (type of cable) and the MAC sublayer. This standard is detailed in Chapter 12.

[†] In this case, the communication mediums refer to the physical layer.

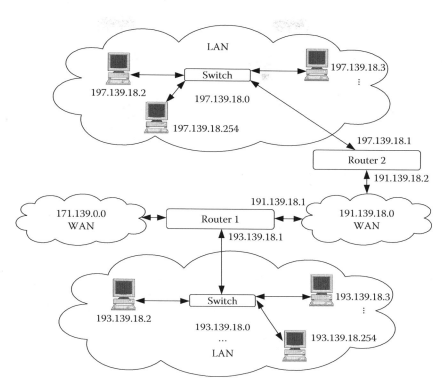

FIGURE 2.20 Layer 2 and layer 3 switching.

(router 1 and router 2), based on the destination IP address (197.139.18.2). Packets use different point-to-point connections, with different layer 2 protocols. The LAN connected to router 1 may use the IEEE 802.3/IEEE 802.2 protocols,* the connection between the two routers may use the PPP, and the connection between router 2 and hosts may be supported on the IEEE 802.5 (token ring)/ IEEE 802.2 protocols.

A switch is responsible for the distribution of frames (instead of packets) within a LAN, based on the destination MAC address† (instead of the IP address). Note that the MAC address is composed of 48 bits and represented by six groups of two hexadecimal digits (e.g., 00-1f-33-ac-c5-bb). In fact, a switch learns the MAC address of each device connected to each interface. Every time a frame is received in a certain interface, if this interface is not associated with any MAC address, the mapping is added to the table. Such a MAC address table maps interfaces to MAC addresses.

In addition, before a host or a router sends frames to a certain destination, that host needs to find the MAC address that corresponds to the destination IP address included in the packet. This is required because a packet is encapsulated into a frame for local transmission, and one of the frame's overhead is the destination MAC address. Such mapping is listed in the address resolution protocol (ARP) table. If the IP address is not listed in the ARP table, then the ARP procedure needs to be implemented as follows [RFC 826]: when a host has a packet to send or to relay, it tries to find the destination IP address in the ARP table, in order to extract the corresponding MAC address. In case there is no entry table corresponding to such an address, it broadcasts (in the LAN) an ARP packet that contains information about a desired IP address. The station with such an IP address answers with a hello packet and the station extracts its MAC address, inserting a new line into the ARP table

---

* IEEE 802.2 is an LLC protocol, whereas IEEE 802.3 consists of a MAC protocol.
† This is also called the *physical address* or the *hardware address*.

with the mapping. The entry to this table is kept for a certain period of time. After a certain period without traffic to be passed to (or received from) this station, the entry to this table is removed, and the procedure is restarted, when required. Figure 2.21 shows an example of an ARP table.

Although most of the LANs being implemented nowadays use a switch as the central node, the IEEE 802.3 protocol was standardized for use with a hub as a central node.* While the switch performs the layer 2 switching based on the destination MAC address, the hub simply acts as a repeater: it repeats in all other outputs the bits received in one of its inputs. Therefore, the medium becomes common, and when a station sends data, all other stations in the same LAN receive such data.

The carrier sense multiple access with collision detection† is used to define when a host is allowed to transmit within a LAN.

In terms of performance, because the hub broadcasts data through the LAN, the amount of collisions tends to be high, for medium to high traffic load (typically above 20%). On the other hand, the switch mitigates this problem as it allows having half of the stations transmitting to the other half of the stations in the LAN (considering a half-duplex network).

The DLL is composed of two sublayers: the LLC sublayer that deals with error control and flow control, and the MAC sublayer that determines when a station is allowed to transmit and using which format. Note that the MAC sublayer exists only in LANs. In the example of Figure 2.20, the LAN connected to router 1 could be an IEEE 802.3 LAN. In the interconnection between the two routers, because there is only one pair of stations (router 1 and router 2), there is no need to coordinate authorizations to transmit, and therefore, the MAC sublayer does not exist.

The reader should refer to Chapter 12 for a detailed description of several DLL and sublayer protocols.

2.3.5 Physical Layer

This is the lowest layer of the TCP/IP stack and, as the DLL, also refers to point-to-point interchange of data. As defined for the OSI-RM, this is the only layer where data is physically moved across the nodes. On the other hand, the other layers only create messages that are passed to lower (on the transmitter side) or to upper (on the receiver side) layers. The type of data interchanged by the physical layer‡ is bits. This layer is subject to all impairments of the transmission medium such as interference, noise, distortion, and attenuation. In addition, this layer deals with transmission parameters (that may mitigate the above-mentioned channel impairments) such as modulation schemes or digital encoding techniques, bandwidth, transmission rate, transmission power, equalization, and advanced receivers to mitigate channel impairments.

The physical layer is only responsible for the transmission of bits, whereas error control is typically provided by layer 2. Nevertheless, in some cases, the physical layer may also adopt some error control techniques. Such an example is the transmission of bits through a wireless medium. In this

Internet Address	Physical Address	Type
192.168.0.1	00-1f-33-ac-c5-bb	Dynamic
192.168.0.255	ff-ff-ff-ff-ff-ff	Static
224.0.0.23	01-00-5e-27-43-18	Static
239.255.255.251	01-00-5e-14-08-fd	Static
239.255.255.249	01-00-5e-7f-fc-fa	Static
255.255.255.255	ff-ff-ff-ff-ff-ff	Static

FIGURE 2.21 Example of a MAC address table.

---

* This corresponds to the worst-case scenario.
† This protocol is defined in detail in Chapter 11.
‡ Physical layer is also referred to as PHY.

case, because of the high probability of error, this layer may adopt the FEC,* which is a type of error correction technique.

The important functionalities of the physical layer include the following:

- *Encoding of signals*: The way bits are sent over the network. Such functionalities include the decision about the type of modulation scheme or digital encoding technique to use, and the voltages and powers.
- *Transmission and reception of bits*: The choice of the bandwidth to use, the transmission rate, whether or not an equalizer is adopted on the receiver side, whether or not regenerators are necessary in the transmission path, decision about the use of multiple transmit and receive antennas, and so on.
- *Mechanical specifications*: The definition of the type of connectors (such as RJ45, RJ11, and BNC) and cables to use (e.g., UTP twisted pair, STP twisted pair, coaxial cable, and optical fiber).
- *Physical topology of the network*: The definition of a physical topology to use within a network such as star, ring, tree, or mesh. It also includes the definition of whether cabling should be half- (e.g., one cable pair) or full-duplex (e.g., two cable pairs).

For a detailed description of the physical layer, the reader should refer to Chapters 3 through 7.

CHAPTER SUMMARY

This chapter provided a view about the network protocol architectures.

We provided an introduction to the network protocol architecture concept, with a view about its aim, and how encapsulation and deencapsulation of user data are performed. It was described that the network protocol architecture approach only defines *what* is to be done by each layer, but not *how* such functionalities are to be implemented by the layer, whose responsibility lies with the protocol of the layer individually. Moreover, it was observed that the network protocol architecture defines the number of layers, what is to be done by each layer, and the interface between different layers. This approach leaves room for a layer to improve, due to technological evolutions or other reasons, without implications in the remaining layers.

An introduction to the OSI-RM was also given. It was observed that this reference model was created by the ISO. The seven layers of the OSI-RM were described, including the functionalities performed by each layer. It was observed that this architecture can be split into two groups: the four lower layers, from the physical up to the transport layer, being responsible for the reliable communication of data between terminal equipment, and the three upper layers, from the session up to the application layer, with a higher level of logical abstraction, interfacing with the user application. Moreover, it was observed how service access points are employed to interconnect different layers.

An introduction to the TCP/IP architecture was given. The TCP/IP architecture, adopted by the Internet, is the most used real implementation of the OSI-RM. Nevertheless, while the basis is the same, there are some differences between these two architectures. While the OSI-RM is a seven-layer architecture, the TCP/IP model is only composed of five layers: the application layer, the transport layer, the Internet layer, the DLL, and the physical layer. It was described that the application layer of the TCP/IP model includes functionalities assigned to the application, presentation, and session layers of the OSI-RM. Moreover, the application layer performs the segmentation of a message into multiple streams.

The application layer transfers a stream into the transport layer, adding its own overhead, namely the TCP or UDP header. The stream received from the application data, together with the transport layer header, is encapsulated into a segment. The segment is the message format of the transport layer. A similar encapsulation procedure is performed by the lower layers on the transmitter side.

---

* The reader should refer to Chapter 12 for a description of the FEC.

The segment is transferred into the Internet layer, added with the IP header, and encapsulated into a datagram. The IP header, among other fields, is composed of the source and destination IP addresses, being composed of 32 bits for IPv4. The datagram is the message format of the IP.

The DLL is composed of two sublayers: the upper sublayer, entitled LLC, and the lower sublayer, entitled MAC. The message format of the LLC sublayer is the LLC PDU, whereas the message format of the MAC sublayer is the frame. The frame header includes, among other fields, the source and destination MAC address, that is, the physical or hardware address of the sender and receiver. The MAC address is a 48-bit address field. Note that the DLL deals with point-to-point communications.

Finally, the physical layer also refers to point-to-point interchange of data. Similar to the DLL, this layer refers to a point-to-point link among adjacent network nodes. This is the only layer where data is physically moved across the nodes. The bits are the type of data interchanged by the physical layer. The physical layer deals with all impairments of the transmission medium, such as interference, noise, distortion, or attenuation. Moreover, this layer also deals with transmission parameters, such as modulation schemes, or digital encoding techniques, bandwidth, transmission rate, transmission power, equalization, or advanced receivers to mitigate the channel impairments.

REVIEW QUESTIONS

1. To which layer of the TCP/IP model do TCP and UDP belong?
2. Which of the OSI-RM layer is responsible for forwarding packets along the several nodes (routers) of the network?
3. Let us consider that a router needs to send packets to a host in a LAN to which it is connected to, and that the corresponding MAC address is unknown. Which protocol is used and what is the sequence of packets expected to be exchanged?
4. Which protocol of the TCP/IP architecture ensures that a connection is previously established before data is exchanged, and ensures that the correct sequence of packets is maintained on the receiver side?
5. How can a switch make a better usage of the LAN bandwidth?
6. What are routing tables used for?
7. What is the difference between a router and a switch?
8. For which purpose is the ARP used for?
9. What is the difference between the UDP and TCP? Enumerate services that use either protocol.
10. Which of the OSI-RM layer is responsible for the end-to-end forwarding of data?
11. What are the advantages of using a network architecture model based on several layers? What is the most implemented network architecture model based on layers?

LAB EXERCISES

1. Download and install the free network analyzer* Wireshark. Open the application in a PC and select the interface connected to the Internet. You will see the IP datagrams that are being exchanged by the NIC of the PC, with the source IP address, destination IP address, and protocol type, along with other information. Select one of these packets. In a window that appears at the bottom, visualize the content of the frame, the MAC sublayer, the IP datagram, and the layer 4 message (e.g., TCP segment). Verify that a segment is encapsulated into a datagram, and that a datagram is encapsulated into a frame. Verify the addresses of the different layers, namely the MAC address, the IP address, and the port.

---

* A network analyzer is also commonly referred to as packet sniffer.

3 Channel Impairments

LEARNING OBJECTIVES

- Describe the different channel impairments experienced in telecommunication systems.
- Define the Shannon capacity.
- Describe the concept of attenuation.
- Describe the different types of noise.
- Describe the effects of distortion and the use of equalization to mitigate it.
- Identify the different sources of interference.

In a communication system, signals are subject to a myriad of impairments that accumulate over the channel path between the transmitter and the receiver (see Figure 3.1). These signals are used to allow the exchange of messages between these two parties. For the transmitted signal to be properly extracted at the receiving side, the received signal must have a signal-to-noise plus interference ratio (SNIR) higher than a certain threshold. Otherwise, the message cannot be properly understood by the receiving party.

Several impairments degrade the SNIR. The SNIR degradation occurs in two different ways: (1) by decreasing the signal level S and (2) by increasing the noise (N) and interference (I) levels.

Attenuation is the factor that originates a decrease in the signal level, whereas an increase in the noise and interference levels is caused by different factors, namely:

- Different noise sources
- Distortion
- Other interferences

3.1 SHANNON CAPACITY

While in analog communications, the degradation of a signal is approximately linear with the decrease in the SNIR level, in the case of digital signals, the bits degrade heavily below a certain threshold, originating an abrupt increase in the bit error rate (BER). This can be seen from Figure 3.2, whose curve is valid for the binary phase shift keying (BPSK) and the quadrature phase shift keying (QPSK) modulation scheme.[*]

The acceptable BER threshold depends on the service under consideration (the threshold level for voice is different from that for file transfer). For voice, the tolerated bit error probability is approximately 10^{-3}, meaning that the BPSK or QPSK modulation requires a minimum of 7 dB of bit signal-to-noise ratio (SNR) (E_b/N_0) (see Figure 3.2).

Depending on the source of impairments and whether the signal is analog or digital, there are different measures that can be used to mitigate it. Because the currently most used type of communications is digital, the description of this chapter focuses on this type of transmission.

---

[*] As detailed in Chapter 6, the BER for BPSK is the same as that for QPSK. However, this is an exception, as for M-QAM modulation schemes, increasing the modulation order M leads to a degradation of the BER. This occurs because the Euclidian distance between constellation points decreases and, consequently, the modulation becomes more sensitive to noise and interferences.

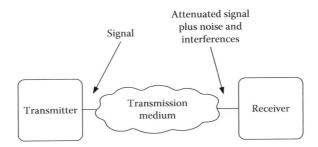

FIGURE 3.1   Generic chain of a communication system.

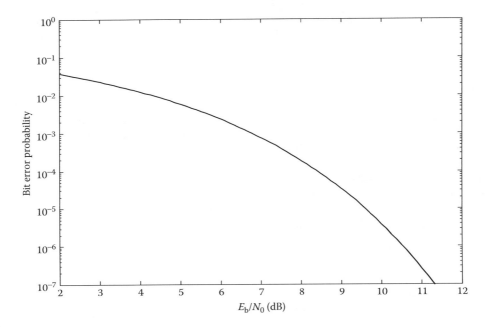

FIGURE 3.2   Bit error probability for BPSK and QPSK as a function of E_b/N_0.

The capacity limit of any telecommunications system is taken to be the resulting throughput obtained through the full usage of the allowed spectrum. For an Additive White Gaussian Noise channel, Claude Shannon derived, in 1948, the following capacity formula [Shannon 1948]:

$$C = W \log_2 \left(1 + \frac{S}{N} \right) (\text{bps})$$ (3.1)

This equation provides information about the maximum theoretical rate at which the transfer of information bits[*] can be achieved, with an acceptable quality, in a certain transmission medium that has a channel bandwidth W (in Hz), power of noise N, and a transmit signal power S (both in watts). Dividing both sides of the above equation by the channel bandwidth W, we obtain the spectral efficiency [Marques da Silva et al. 2012]:

---

[*] Note that the Shannon capacity refers to the maximum rate of information bits, that is, overhead and control bits are not included.

$$\frac{C}{W} = \log_2\left(1 + \frac{S}{N}\right)\left(\text{bit/s/Hz}\right) \qquad (3.2)$$

This is expressed in bits per second per hertz and gives us an indication of how many bits per second can be transported in each hertz of the channel bandwidth.

Examining the voice-grade twisted pair, which has a typical channel bandwidth W of 3.1 kHz, and assuming an SNR[*] of 3.7 dB,[†] from Equation 3.1, we conclude that the maximum speed of information bits is 38 103 bps. Therefore, the solution to accommodate higher transmission rates must correspond either to an increase of the available medium channel bandwidth or to an increase of the signal power, or even to a decrease of the power of noise and interferences.

If multiple transmit and receive antennas are employed,[‡] the capacity may be raised. If there are a sufficient number of receive antennas, it is possible to resolve all messages, as long as the channel correlation between the antennas is not too high. The pioneer work of Foschini and Gans [1998] established the mathematical roots from the information theory field that, with multipath propagation, multiple antennas at both the transmitter and the receiver can establish essentially multiple parallel channels that operate simultaneously on the same frequency band and at the same time. In a more practical case of a time variable and randomly fading wireless channel, the capacity is written as follows:

$$C = W \log_2\left[1 + \frac{S}{N} \cdot \left|\mathbf{H}^2\right|\right] \qquad (3.3)$$

where $\mathbf{H}^2$ is the normalized channel-power transfer function. $\mathbf{H}$ is an $M \times N$ power complex Gaussian amplitude of the $M \times N$ channel, where M stands for the number of transmit and N for the number of receive antennas. Multiple antenna systems are described in Chapter 7. It is worth noting that, in Equations 3.1 and 3.3, the letter N refers generically to all sources of noise and interference (not only the noise). Therefore, it is important to define all of these impairments that influence this parameter. Depending on the system and transmission medium, there are different factors, which are defined in the following sections.

3.2 NYQUIST SAMPLING THEOREM

The Nyquist sampling theorem states that the minimum sampling rate used in the digitization process that assures a digital distortionless signal is given by

$$f_{min} = 2 \times B_{max} \qquad (3.4)$$

where B_{max} corresponds to the maximum frequency component present in the signal.

Using a sampling rate equal to or higher than f_{min} in the digitization process, it is assured that the analog signal reconstructed from its digital samples is not subject to a specific type of distortion, entitled aliasing. Let us suppose that there is a voice signal, with frequency components within 300 Hz up to 3.4 kHz. According to the Nyquist sampling theorem, these signals must be digitized with a sampling rate equal to or higher than $f_{min} = 2 \times 3.4$ kHz $= 6.8$ kHz.

---

[*] Equation 3.1, while generic, establishes a relationship between the capacity and the SNR (as it corresponds to S/N). However, the generic noise designation corresponds, in the sense of this equation, to noise plus interference (including distortions, interferences, etc.). In other words, the S/N of Equation 3.1 should be understood as $S/(N + I)$.

[†] Because we are dealing with logarithmic units, this is calculated as $\text{SNR}_{dB} = 10 \log_{10}\left(S/N\right)$.

[‡] Multiple antennas are adopted by the air interface of IEEE 802.11n and IEEE 802.11ac standards, WiMAX standard, long-term evolution (LTE) cellular systems, and so on.

3.3 ATTENUATION

The propagation of a signal in any transmission medium is subject to attenuation. This effect corresponds to a decrease in the signal strength. The level of attenuation depends on the medium, but, as a rule of thumb, it increases with the distance at a variable scale. Figure 3.3 shows an example of a transmitted signal and the corresponding attenuated signal, as a result of the propagation losses through a medium.

As an example, let us consider the propagation of electromagnetic waves in a free space. In this case, the attenuation is due to the path loss, being known as free space path loss (FSPL). Therefore, the propagation can be viewed as a sphere, whose surface area increases with the increase of the distance from the transmit antenna. While the power can be considered as constant, because it spreads out over the surface area of the sphere, and because its radius corresponds to the propagation distance, its power spatial density decreases with the distance.

From Chapter 5, it is seen that the FSPL is quantified by $FSPL = \left(4\pi df/c\right)^2$ [Proakis 1995], where d stands for the distance from the transmitter, f represents the frequency, and c denotes the speed of light. Therefore, in the special case of electromagnetic waves, and assuming free space propagation, the attenuation increases with the square of the distance. In the case of a coaxial cable, the dependence of the distance is different, as the attenuation tends to increase with the increase of the logarithm of the distance.

Moreover, as a rule of thumb, the attenuation increases with the frequency of the signal being transmitted. This is generically applicable to most of the transmission media.

Note that the attenuation depends on the type of propagation between the transmitter and the receiver. Electromagnetic propagation is normally composed of several paths, namely a direct line of sight, and several reflected, diffracted, and scattered waves (e.g., in buildings and trees) [Burrows 1949]. As a result, in real propagation scenarios (other than free space), the attenuation depends on distance with a higher rate than its square. Normally, an exponent between 3 and 5 is experienced in scenarios subject to shadowing and multipath.[*] Consequently, we conclude that the attenuation experienced in real scenarios is higher than the FSPL. This results from the three basic propagation effects [Theodore 1996]: (1) reflection, (2) refraction, and (3) scattering. These propagation effects are characterized in Chapter 5.

Having a signal subject to attenuation, there is a need to increase its level such that detection is possible at the receiver. For detection to be possible, the received signal needs to be above the receiver's sensitivity threshold. The signal level is increased by an amplifier at the receiver side. Note that, similarly, the signal at the receiver's input must be above the amplifier's sensitivity threshold, or otherwise the signal cannot be amplified. Therefore, for long distances, there is a need to use amplifiers in the transmission path, before the distance attenuation is too high, and before the signal is below the amplifier's and receiver's sensitivity threshold.

Because this device amplifies its input signal, being composed of signal plus noise and interferences, amplifying this signal does not add any value in terms of SNR gain.[†] Moreover, an amplifier

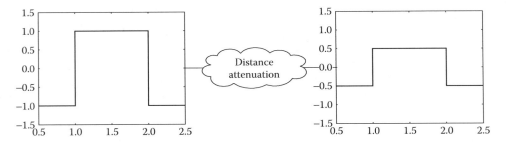

FIGURE 3.3 Transmitted signal and attenuated version caused by the propagation channel.

---

[*] The shadowing effect is detailed in Chapter 5. The multipath effect is detailed later in Section 3.6.1.
[†] It amplifies the signal, the noise, and the interferences present at the input, leaving the SNR unchanged.

also introduces additional noise, leading to a noise factor[*] higher than 1. This means that the SNR suffers a degradation after the amplification process.

An important advantage of digital signals relies on the ability to implement regeneration, which is a more effective process than amplification, as it allows a gain in terms of SNR. A regenerator[†] includes a detector followed by an amplifier. Therefore, inserting regenerators along the transmission chain allows recovering the original signal (performed by its detector), before the signal is amplified. This enables partially removing the negative effects of noise and interferences, and thus, this tends to lead to an improvement in the SNR value. In addition, as in the case of the amplifier, a regenerator has to be placed in locations along the transmission chain such that the distance attenuation does not degrade the signal level below the regenerator's sensitivity threshold. As an example, note that synchronous digital hierarchy (SDH) networks makes use of regenerators, typically every 60 km of optical fibers.

3.4   NOISE SOURCES

One of the most important impairments of a telecommunication medium is noise. Noise is always present with higher or lower intensity. A receiver detects the desired (attenuated) signal superimposed with noise and interferences. Therefore, noise can be defined as unwanted impairments superimposed on a desired signal, which tends to obscure its information content.

As previously described, a signal needs to be received with an SNR higher than a certain threshold to allow a good service quality.

Figure 3.4 shows an attenuated signal (on the left), due to the distance attenuation, equal to the one considered in Figure 3.3. Because this signal does not present any kind of noise or interferences, its SNR is infinite. This signal is then received together with the noise present at the receiver's antenna location (signal on the right). The resulting SNR is now degraded. Although the noise has been added to the signal, because its power is not too high, we observe that the shape of the envelope is similar to the original signal without noise.

A receiver uses a certain instant within the pulse duration to perform sampling. Based on the sampled signal at the sampling instant, a decision is made about whether the received signal is assumed as a symbol +1 or −1.[‡] In this case, the hard decision should be as follows: if the sampled signal is above 0, it is assumed as +1; otherwise (with the sampled signal below 0), it is assumed that the received symbol is a −1 (as the estimated transmitted symbol).

Figure 3.5 presents the same signals, but the plot on the right includes a signal subject to a stronger noise power. Note that, although the transmitted signal between instants 1 and 2 was +1, depending on the exact sampling instant, the sampled received signal can be decided as a −1 (because the

FIGURE 3.4   Addition of low power noise to an attenuated signal (possible received signal).

---

[*] The noise factor is defined by Equation 3.10.
[†] A regenerator is also known as a repeater.
[‡] In this example, we assume the use of amplitude shift keying (ASK) as defined in Chapter 6. A +1 level may correspond to a logic state 1, whereas a −1 level may correspond to a logic state 0. In fact, the +1 and −1 levels can be any value, depending on the transmitting power.

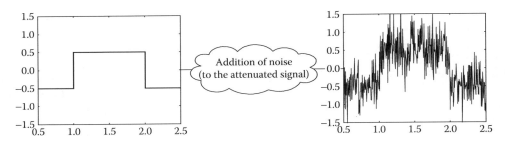

FIGURE 3.5   Addition of high power noise to an attenuated signal (possible received signal).

sampled value can be below zero, which is assumed as a decision threshold). In such a situation, a symbol error occurs. This is because, for the same signal level as the one in Figure 3.4, the noise power is much more intense, resulting in a lower SNR value. In the case of digital signals, the resulting bit error probability becomes higher, resulting in a degraded signal.

Depending on its sources, noise can be of different types. The most important types of noise can be grouped as follows:

- External noise:
  - Atmospheric
  - Man-made
- Extraterrestrial noise
- Internal noise:
  - Thermal
  - Electronic

The total noise power, resulting from all different noise sources, is seen at the receiver's detector. The different types of noise are defined in the following.

3.4.1 ATMOSPHERIC NOISE

Atmospheric noise consists of an electromagnetic disturbance, being caused by a natural atmospheric phenomenon, such as lightning discharges in thunderstorms. It consists of cloud-to-ground and cloud-to-cloud flashes. While more intense in tropical regions, it consists of a high-power and low-duration current discharge, resulting in a high-power electromagnetic impairment. These flashes occur approximately 100 times a second, on a worldwide scale, and the sum of all these flashes results in the random atmospheric noise.

In an area surrounding thunderstorms, the noise presents an impulsive profile (i.e., very low duration but high intensity). Because the pulse is very narrow in the time domain, its bandwidth is very wide. This means that the noise is experienced by many nearby communication systems that make use of different parts of the electromagnetic spectrum.

The combination of all distant thunderstorms (low-duration pulses) results in white noise[*] that is felt, at distant locations, with continuity over time but with a lower power level. Its power varies with the season and proximity of thunderstorm centers. In addition, because this phenomenon is more frequent in tropical regions, the atmospheric noise tends to decrease with the increase of the latitude.

As described by the FSPL equation (Section 3.3), electromagnetic attenuation increases with frequency. Consequently, the higher frequency components of the noise are subject to higher attenuation levels. This is the reason why atmospheric noise is felt at a long distance with higher power at lower frequencies, and with lower power at higher frequencies. Consequently, the atmospheric noise dominates at the VLF and LF bands (frequency bands are defined in Table 3.1). This can be seen from Figure 3.7.

---

[*] White noise presents a constant power spectral density (PSD).

TABLE 3.1

Definition of Frequency Bands and Their Limits

	Frequency Bands								
	VLF	LF	MF	HF	VHF	UHF	SHF	EHF
Designation	Very low frequency	Low frequency	Medium frequency	High frequency	Very high frequency	Ultrahigh frequency	Superhigh frequency	Extremely high frequency
Lower limit	3 kHz	30 kHz	300 kHz	3 MHz	30 MHz	300 MHz	3 GHz	30 GHz
Upper limit	30 kHz	300 kHz	3 MHz	30 MHz	300 MHz	3 GHz	30 GHz	300 GHz
Applications	Navigation, maritime, and submarines communications	Navigation, maritime, and submarines communications	Radio broadcast and maritime	Radio broadcast and maritime	Radio and TV broadcast and maritime	TV broadcast, maritime, and cellular communications	Satellite communications	Satellite and microwave communications

Moreover, as described in Chapter 5 for the groundwave propagation, at low frequencies, electromagnetic waves with horizontal polarization experience higher attenuation levels than vertically polarized waves. Consequently, vertically polarized atmospheric noise tends to be more intense than horizontally polarized noise.

3.4.2 MAN-MADE NOISE

Man-made noise is electromagnetic, being caused by human activity, namely by the use of electrical equipment, such as car ignitions, domestic equipment, or vehicles. The intensity of this kind of noise varies substantially with the region. Urban man-made noise tends to be more intense than rural noise. This noise is characterized by the emission of low-duration and high-power pulses, when the corresponding source is activated (e.g., when the car ignition is activated). Figure 3.6 shows the typical man-made noise along the frequency spectrum, for different environments (business, residential, rural, and quiet rural). While very intense in the HF band, its noise figure (F_{am}) decreases at higher frequencies.

3.4.3 EXTRATERRESTRIAL NOISE

Extraterrestrial noise is a type of electromagnetic noise that comes from certain limited zones of the cosmos and galaxies.

Extraterrestrial noise is also known as galactic noise and solar noise. An antenna directed toward certain regions of the sky, such as the sun or other celestial objects, may experience powerful wideband noise. Note that this type of noise depends on the relative orientation of the antenna's radiation pattern. This orientation varies along the day because of the earth's rotation, and therefore, attention needs to be paid to a sudden increase of noise experienced by some types of stations, such as by a satellite earth station. The pattern of extraterrestrial (galactic) noise can be seen from Figure 3.7.

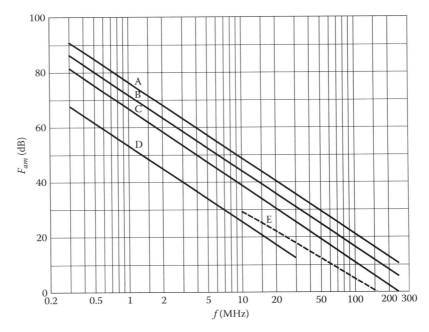

FIGURE 3.6 Noise factor for A—business, B—residential, C—rural, D—quiet rural, E—galactic. (Data from Lawrence, D.C., *CCIR Report 322 Noise Variation Parameters*, Technical Document 2813, June 1995.)

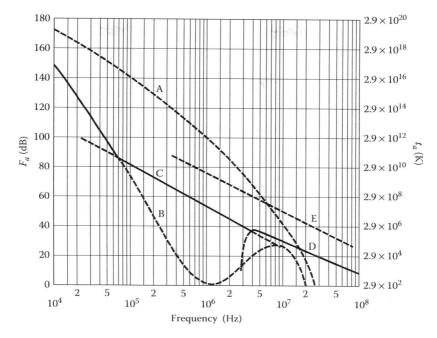

FIGURE 3.7 Comparison of noise figure (F_a [dB]) and temperatures (t_a [K]), for different types of electromagnetic noise: A—percentile 0.5 of atmospheric noise, B—percentile 0.5 of atmospheric noise, C—median man-made noise for business, D—median man-made noise for galactic, and E—median man-made noise for rural. (Data from Lawrence, D.C., *CCIR Report 322 Noise Variation Parameters*, Technical Document 2813, June 1995.)

Figure 3.7 shows different electromagnetic noise contributions in different environments. From this figure, it is noticeable that atmospheric (galactic) noise dominates at the VLF and LF bands, whereas man-made noise is more intense in the HF band.

Because noise is a random process, its measure is normally performed using statistical tools. Therefore, noise is normally expressed in percentile. As an example, percentile 10 is a value characterized by having 90% of the samples with a value above this percentile 10 value, and having 10% of the samples with a value below the percentile 10 value. Noise may also be expressed in median. Median corresponds to the percentile 50 (50% of the samples are above the median value, and the other 50% of the samples below).

3.4.4 Thermal Noise

Thermal noise is experienced inside electrical conductors (wires, electrolytes, resistors, etc.), being caused by thermal agitation of charges at the amplifier's input resistance. In the case of radio communications, thermal noise presents a wide variation of amplitude, depending on the temperature viewed by the receive antenna.

The frequency profile of thermal noise presents a PSD approximately constant along the frequency spectrum, that is, thermal noise is approximately white.

The noise power P_n captured by its amplifier's input resistance is given by [Carlson 1986]

$$P_n = k_B T_n B \tag{3.5}$$

where:

k_B is the Boltzmann constant, with $k_B = 1.3806503 \times 10^{-23}$ J K^{-1} (expressed in Joules per Kelvin)
T_n is the resistor's absolute temperature (expressed in Kelvin)
B is the receiver's bandwidth (expressed in Hertz)

Note that, in statistical terms, the noise power P_n corresponds to the noise variance σ^2.

The PSD N_0 corresponds to the noise power divided by the receiver's bandwidth B, being given by

$$N_0 = P_n/B$$
$$= k_B T_n \tag{3.6}$$

The root mean square (RMS) voltage of thermal noise, generated in a given amplifier's input resistance R (expressed in ohm), is given by

$$v_n = \sqrt{4 k_B T_n R B} \tag{3.7}$$

In the case of wireless communications, the value of T_n captured by the amplifier's input resistance depends on the orientation of the antenna's radiation pattern. The thermal noise of a satellite earth station pointing toward the sky is typically very low,[*] as the temperature of the sky is also low (200 K > T > 80 K). On the other hand, the thermal noise of a satellite transponder is typically high, as it is pointing toward the earth, whose temperature is also high (T > 300 K).

Another way to express the noise level is using the noise factor f_a coefficient, which is defined by

$$f_a = \frac{P_n}{k T_o B} \tag{3.8}$$

The noise factor is defined as the ratio between the received noise power and the noise power delivered by a charge with the reference noise temperature of 300 K (T_o). Expressing this value in logarithmic units leads us to the noise figure,[†] defined by

$$F_a = 10 \log_{10} f_a \tag{3.9}$$

3.4.5 Electronic Noise

Electronic noise is a type of internal noise generated in active elements (e.g., transistors) in the interior of active equipment, namely in amplifiers or in active filters. As a consequence, the SNR at the output of active equipment is lower than that at its input. Similar to thermal noise, the electronic noise level is typically quantified by the noise factor as follows:

$$f_a = \frac{\text{SNR}_{\text{IN}}}{\text{SNR}_{\text{OUT}}} \tag{3.10}$$

Alternatively, this can also be expressed in logarithmic units using Equation 3.9. Using this equivalence, from f_a, and knowing the received SNR at the input of a satellite transponder (SNR_{IN}), the computation of the SNR at its output (SNR_{OUT}) is straightforward.

In the case of a cascade of N electronic devices (e.g., amplifiers and filters), the resulting noise factor f_{OUT} is

$$f_{\text{OUT}} = f_1 + \frac{f_2 - 1}{g_1} + \frac{f_3 - 1}{g_1 g_2} + \frac{f_4 - 1}{g_1 g_2 g_3} + \cdots + \frac{f_N - 1}{\prod_{i=1}^{N-1} g_i} \tag{3.11}$$

---

[*] Except if it is pointing toward an extraterrestrial object, such as the sun.
[†] Note that Figures 3.6 and 3.7 express noise using the noise figure notation.

where:

g_i stands for the device gain

f_i $(i = 1, \ldots, N)$ stands for the noise factor of the ith electronic device

Still in the case of a cascade of N electronic devices, the resulting SNR_{OUT} is

$$
\begin{aligned}
\text{SNR}_{\text{OUT}} &= \frac{S_{\text{OUT}}}{N_{\text{OUT}}} \\
&= \frac{g_{\text{OUT}} \cdot S_{\text{IN}}}{g_{\text{OUT}} \cdot f_{\text{OUT}} \cdot N_{\text{IN}}} \\
&= \frac{S_{\text{IN}}}{f_{\text{OUT}} \cdot N_{\text{IN}}}
\end{aligned} \tag{3.12}
$$

where:

g_{OUT} is $g_{\text{OUT}} = \prod_{i=1}^{N} g_i$

f_{OUT} is as defined by Equation 3.11

Using Equation 3.12, we reach an equivalence similar to Equation 3.10, but for a cascade of electronic devices,

$$
f_{\text{OUT}} = \frac{\text{SNR}_{\text{IN}}}{\text{SNR}_{\text{OUT}}} \tag{3.13}
$$

This allows us to compute the resulting SNR_{OUT} from SNR_{IN} and from f_{OUT}, for a cascade of electronic devices.

Note that other types of noise may also be viewed as the noise generated in electronic devices, but where the device gain should correspond to, for example, the path loss attenuation and the noise figure of the device is, for example, the noise figure of the thermal noise. Therefore, with such an approach one could use Equation 3.11 to compute f_{OUT} and then use Equation 3.12 to compute SNR_{OUT} from the initial SNR_{IN}, that is, without computing all intermediate SNR.

3.5 INFLUENCE OF THE TRANSMISSION CHANNEL

An ideal transmission channel would be the one in which the received signal is equal to the transmitted one. Nevertheless, it is known that real channels introduce an attenuation and a phase shift to signals. When the attenuation is constant over the signal's frequency components (i.e., over the entire signal's bandwidth), it is said that the channel does not introduce amplitude distortion to signals. Similarly, when the phase shift is linear over the signal's frequency component, it is said that the channel does not introduce phase distortion to signals. In these cases, only the amplification process can be used before detection.

Nevertheless, very often the channel introduces different attenuations and nonlinear phase shifts at different frequency components of signals. This is more visible when the signal bandwidth is relatively high.[*] In this case, the equalization process may also be required at the receiver side before amplification and detection.

The channel's frequency response $H(f)$, also known as the channel's transfer function, translates mathematically the way the channel processes different frequency components of the signal that crosses it, both in terms of the signal's attenuation and phase shift. Note that $H(f)$ corresponds to the Fourier transform of the channel's impulse response $h(t)$.

---

[*] Namely higher than the coherence bandwidth of the channel.

The signal at the output of the channel is given by

$$V_R(f) = V_E(f) \cdot H(f) \tag{3.14}$$

where:

$V_E(f)$ stands for the Fourier transform of the transmitted time domain signal $v_E(t)$
$V_R(f)$ stands for the Fourier transform of the received time domain signal $v_R(t)$ (see Figure 3.8)

Appendix I presents a short description of the Fourier transform theory.

3.5.1 DELAY AND PHASE SHIFT

Let us consider a carrier-modulated signal in the carrier frequency f_c, being propagated through the transmission medium between a transmitter and a receiver. The phase θ of the signal varies by 2π radians for every propagation distance corresponding to a wavelength λ (see Figure 3.9). Moreover, the propagation delay τ increases by $\tau = 1/f_c$ for every propagation distance corresponding to a wavelength λ (f_c is the carrier frequency and τ is the carrier period). Similarly, the phase shift at the instant $t = \tau$ is given by $\theta = 2\pi f_c \tau$.

As the propagation distance increases, the phase shift and the delay vary accordingly. In fact, the received signal is composed of a superposition of components, including direct, reflected, diffracted, and scattered components.

FIGURE 3.8 Generic communication system with the signals depicted in the time domain.

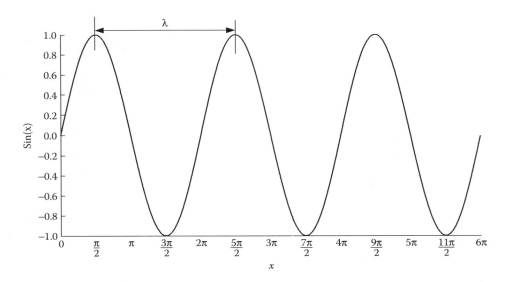

FIGURE 3.9 Sinusoidal wave and its characteristics.

Because each of these elementary waves experiences a different distance,[*] each component arrives at the receive antenna with a different phase shift and delay before they are superimposed.

As can be seen from Figure 3.10, the ideal[†] phase shift response of a channel corresponds to a slope, with a constant gradient across different frequencies.[‡] Note that the time delay is related to the phase shift by $\tau(f) = -(1/2)(d\theta(f)/df)$, where both $\tau(f)$ and $\theta(f)$ are a function of the frequency f [Marques da Silva et al. 2010]. As a result, the superposition of waves from different paths results in either constructive or destructive interference, amplifying[§] or attenuating the signal power seen at the receiver (relating to the line-of-sight wave propagation). The variation of the signal level received from one or more multipaths originates a variation of the resulting superimposed signal level known as fading.

Assuming a variation of distance between a transmit and a receive antenna, it is observed that the envelope of the received signal level presents a cyclic period for every $\lambda/2$ distance variation. In fact, this fluctuation of the signal level across distance can be viewed in the frequency spectrum as distortion, which is explained in the following sections.

3.5.2 DISTORTION

Transmitted signals are not composed of a single frequency. On the contrary, signals are composed of a myriad of frequency components, presenting a certain bandwidth. As an example, an audible spectrum spans from around 20 Hz up to around 20 kHz.

As previously described, transmission media tend to introduce different attenuations at different frequencies. As a rule of thumb, the attenuation level tends to increase with an increase in the frequency. Moreover, channels tend to introduce different delays[¶] and nonlinear phase shifts at different frequency components. Therefore, the signal after propagation through a medium is subject to

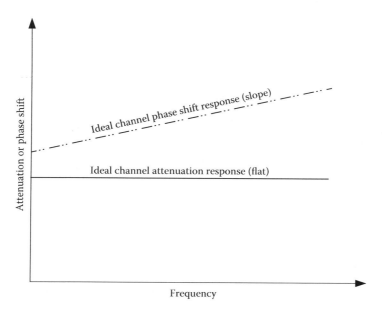

FIGURE 3.10 Ideal channel's frequency response (distortionless).

---

[*] Note that a reflected wave experiences a path longer than the directed wave.
[†] Which does not distort signals.
[‡] A phase shift profile defined by a curve introduces distortion.
[§] Relating to the line-of-sight version of the signal.
[¶] As a result of different delays at different frequency components.

different attenuations and nonlinear phase shifts at different frequency components, and hence, the received signal is different from the transmitted one. This effect is known as distortion[*] (Figure 3.11).

3.5.3 EQUALIZATION

It was described in the previous subsection that channels tend to introduce higher attenuation at higher frequency components of the signal, whose effect is known as distortion. In this scenario, the equalization process aims to mitigate the negative effects of distortion, by introducing higher amplification gains at higher frequency components of the signal, and lower amplification gains at lower frequency components of the signal (see Figure 3.12). Ideally, this results in a signal equally attenuated after the combined system composed of the propagation channel plus the equalization process.

The same principle is applicable to the phase shift introduced by the channel, that is, the equalizer aims to transform the nonlinear phase shift introduced by the channel into a linear phase shift introduced by the combined system composed of the propagation channel and the equalizer.

The equalizer is normally a part of the receiver, being especially important for signals with higher bandwidths. Note that, from Equation 3.22, higher digital transmission rates correspond to higher bandwidths. A frequency response[†] of a channel is approximately constant for a very narrowband signal (low transmission rate). In the case of a very narrowband signal, its frequency span tends to zero. In this case, the curve of the channel's frequency response tends to its tangent at the reference point. On the other hand, a wideband signal (high transmission rate) suffers heavily from distortion, as the signal spans over a wide bandwidth. Therefore, increasing the transmission rate demands higher processing from the receiver to perform the required equalization, and this processing is never optimum, resulting in some residual level of distortion.

Figure 3.13 shows an example of the attenuation introduced by a channel, as a function of the frequency. Ideally, the equalizer's frequency response should be such that the attenuation of the

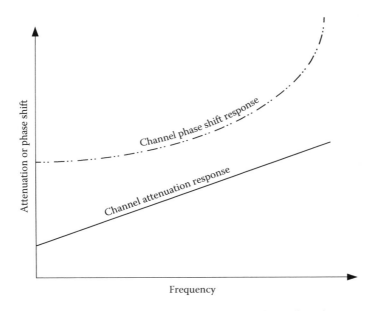

FIGURE 3.11   Example of a channel's frequency response that introduces distortion.

---

[*] Note that distortion may also be introduced by devices, such as amplifiers and filters.
[†] A system's frequency response is a graphic that shows the attenuations and phase shifts introduced by the system at different frequency components.

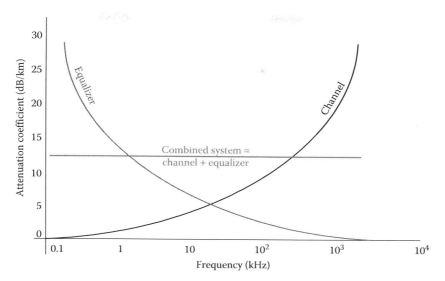

FIGURE 3.12 Example of a channel's frequency response, with the corresponding ideal equalizer.

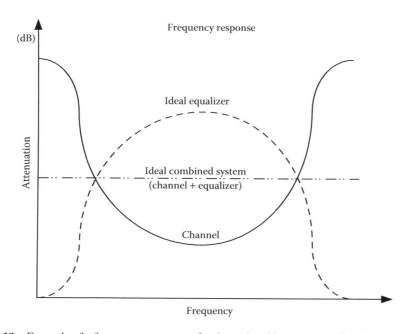

FIGURE 3.13 Example of a frequency response of a channel and its corresponding ideal equalizer.

combined system that results from cascading the channel and the equalizer is a straight line, that is, a continuous attenuation over the bandwidth of interest (absence of distortion). Note that this figure only depicts the attenuation as a function of the frequency, but the same principle is applicable to phase shift, where the ideal phase shift response is a straight line, instead of a curve (see Figure 3.10).

From the description of a signal from Section 3.5, it is known that the transfer function of a zero force (ZF) equalizer is given by

$$\text{Eq}(f)_{\text{ZF}} = \frac{1}{H(f)} \tag{3.15}$$

where $H(f)$ stands for the channel transfer function. Similarly, the transfer function of a minimum mean square error equalizer is given by

$$\text{Eq}(f)_{\text{MMSE}} = \frac{\left[H(f)\right]^*}{1/\text{SNR} + \left[H(f)\right]^2} \tag{3.16}$$

where $\text{Eq}(f)$ stands for the equalizer's transfer function, that is, the Fourier transform of the equalizer's impulse response.

Figure 3.14 depicts the concatenation of the channel and the equalization (part of the receiver). Note that $X(f)$ represents the Fourier transform of the transmitted signal $x(t)$.

An equalizer requires being tuned to the channel. This is normally performed using pilots or training sequences. A pilot or training sequence consists of using a predefined signal (sequence of known symbols), which is periodically transmitted. While the receiver has knowledge about the transmitted signal, it computes the difference between the transmitted signal and the received one. This difference is a function of the distortion introduced by the channel. From this received pilot sequence, the receiver may extract the channel coefficients (in terms of attenuations and phase shifts at different frequencies), which are then utilized to implement the equalization process.

As previously described, a wireless channel is typically subject to fluctuations (fading). This signal variation is the result of variations of the distance between transmit and receive antennas, variation of the environment surrounding the receiver, variations of the refraction index, and so on.

Note that the channel may present flat fading or frequency selective fading. In the case of flat fading, different frequency components of the signal experience constant attenuation and linear phase shifts introduced by the propagation channel. In the case of frequency selective fading, different frequency components of the signal experience different attenuations and nonlinear phase shifts introduced by the propagation channel (distortion).* Moreover, in either types of fading, the attenuation and phase shifts may suffer variations over time. In this case, the equalizer should be able to follow the channel parameters, requiring the periodic transmission of pilots in order to allow the implementation of the equalization process, performed adaptively. This process is called adaptive equalization. Note that some advanced receivers may also perform channel estimation directly from the received modulated signal [Proakis 1995].

It is finally worth noting that the equalization process is never optimum. This results in some residual level of distortion, which degrades the performance. In case the channel suffers rapid variations over time, the equalizer tends to have more difficulties to find the channel coefficients, and therefore, the equalization process tends to be poorer.

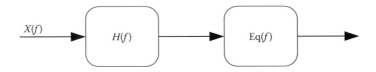

$X(f) \rightarrow \boxed{H(f)} \rightarrow \boxed{Eq(f)} \rightarrow$

FIGURE 3.14   Generic communication including the channel and the equalizer at the receiver.

---

* Note that frequency selective fading corresponds to a type of distortion.

3.6 INTERFERENCE SOURCES

The received SNIR needs to be higher than a certain threshold to achieve an acceptable BER performance. In Section 3.4, different types of noise were described. This section focuses on the description of different types of interferences. The main sources of interferences can be grouped into four main categories:

- Intersymbol interference (ISI)
- Multiple access interference (MAI)
- Co-channel interference (CCI)
- Adjacent channel interference (ACI)

3.6.1 INTERSYMBOL INTERFERENCE

Intersymbol interference occurs in digital transmissions of symbols when the channel is characterized by the existence of several paths (see Figure 3.15), where the delay of relevant signal replicas[*] that arrive at the receiver's antenna corresponds to a delay higher than the symbol period. Note that the RMS delay spread corresponds to the delay of the RMS signal replicas that arrive at the receiver's antenna.

In other words, ISI exists when the signal is propagated through a channel whose RMS delay spread of the channel is higher than the symbol period. In this case, this effect can be viewed in the frequency domain as having two sinusoids with frequency separation greater than the channel coherence bandwidth, being affected differently by the channel (in terms of attenuation and delay/phase shift) [Benedetto et al. 1987]. This corresponds to distortion, but applied to digital signals.

The channel coherence bandwidth is the bandwidth above which the signal presents frequency selective fading, that is, different attenuations, different delays, and nonlinear phase shifts at different frequencies (the signal is severely distorted by the channel). In the case of frequency selective fading, and observing this effect in the time domain, it can be concluded that different digital symbols suffer from interference from each other, whose effect is usually known as intersymbol interference. This can be viewed as a type of distortion applicable to digital transmissions. This effect can be seen from Figure 3.16. Note that ISI tends to increase with the increase of the signal's bandwidth (increase of data rates, according to the Nyquist theorem).

On the other hand, if the signal's bandwidth is within the channel coherence bandwidth, the channel is said to be frequency nonselective and the type of fading is characterized as flat fading.

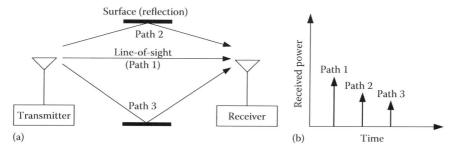

FIGURE 3.15 Propagation of a signal in a multipath environment: (a) diagram with multipaths; (b) average received power by each path.

---

[*] Except those replicas whose average power is 30 dB, below the normalized average power.

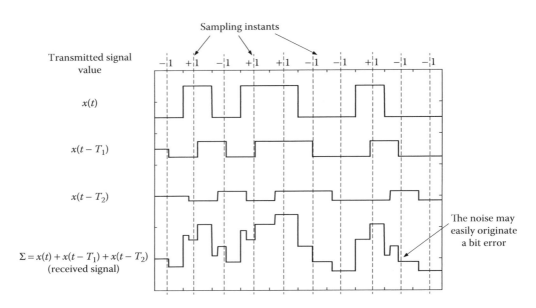

FIGURE 3.16   Plot of received signals through different paths, and the resulting received signal.

In this case, all frequencies fade in unison (i.e., different signal frequency components present the same attenuations and linear phase shifts). In this case, the channel does not originate ISI.

Because, in real radio propagation scenarios, there are always some levels of multipaths, the decision to evaluate the type of channel depends on the average power of the received multipaths. When the channel profile has a normalized average power below 30 dB for nondirect path replicas, although these multipaths may present a delay higher than the RMS delay spread, it is normally assumed that we are in the presence of a single-path channel (frequency nonselective fading or flat fading). On the other hand, when the channel profile has an average power of nondirect multipaths higher than this threshold, it is normally assumed that the channel presents selectivity in frequency. In this case, ISI is experienced in the digital transmission of symbols when the symbol rate is sufficiently high. Note that no serious ISI is likely to be experienced if the symbol duration is longer than several times the delay spread. On the contrary, higher symbol rates correspond to higher levels of ISI.

Figure 3.16 depicts the received signal subject to interference (represented by $\Sigma = x(t) + x(t - \tau_1) + x(t - \tau_2)$) being composed of the cumulative sum of the signal received through direct path (represented by $x(t)$) and the signals received through two multipaths (represented by $x(t - \tau_1)$ and $x(t - \tau_2)$, respectively). Note that noise is not shown in this figure. Each multipath consists of a reflected ray in a certain surface, and the corresponding signal level depends on the reflection index $\Gamma (0 < \Gamma < 1)$. In the example of Figure 3.16, it was assumed that $\Gamma = 0.5$ for path 2 and the reflection coefficient assumed for path 3 was $\Gamma = 0.3$. Moreover, τ_1 corresponds to the delay between the first multipath and the direct path, whereas τ_2 corresponds to the delay between the second multipath and the direct path.

A decision is to be taken by the receiver at sampling instants. As can be seen from Figure 3.16, in some cases, the resulting signal at sampling instants has a level above the one received through the direct path, which results in a constructive interference of the multipaths. Nevertheless, in other cases, the resulting signal has a level below the one received through the direct path, corresponding to a destructive interference caused by ISI. In these cases, a low noise power may be enough to make this sample resulting in an erroneous symbol estimation.

There are measures that can be implemented to mitigate the effects of ISI, namely the use of equalization, channel coding with interleaving, antennas diversity, and frequency diversity [Proakis 1995].

In code division multiple access (CDMA) networks, the presence of multipaths is used by the receiver (RAKE receiver[*]) in order to exploit multipath diversity.

Finally, it is worth noting that the variation of distance between a transmitter and a receiver also originates the Doppler effect that results in a variation of the received carrier frequency relating to the transmitted one. The Doppler frequency is given by $f_D = d\theta/dt$ (variation of the wave's phase).

3.6.1.1 Nyquist ISI Criterion

As previously described, ISI can be originated by the frequency selective channel. Moreover, ISI can also be caused by the nonoptimum sampling instant of the detector.

Typically, the transfer function of the channel and the transmitted pulse shape are specified, and the problem is to determine the transfer functions of transmit and receive filters so as to reconstruct the original data symbol. The receiver extracts and then decodes the corresponding sequence of channel coefficients from the output $v(t)$ (see Figure 3.17). The extraction process involves sampling the output $v(t)$ at time $t = iT_S$, where T_S stands for the symbol period. The decoding requires that the weighted pulse contribution $h(t) \cdot p(iT_S - kT_S)$ for $k = i$ be free from ISI because of the overlapping tails of all other weighted pulse contributions represented by $k \neq i$. This, in turn, requires that it controls the overall time domain pulse shape $p(t)$ as follows:

$$p(iT_S - kT_S) = \begin{cases} 1, & i = k \\ 0, & i \neq k \end{cases} \tag{3.17}$$

where $p(0) = 1$, by normalization. If $p(t)$ satisfies the condition of Equation 3.17, the receiver output $v(t_i)$ (with $t_i = T_S$) implies zero ISI, that is, $v(t_i) = \mu(h)_i$ (for all i).

Because it is assumed that the pulse $p(t)$ is normalized such that $p(0) = 1$, the condition for zero ISI is satisfied if [Proakis 1995]

$$\sum_{n=-\infty}^{+\infty} P(f - nR_S) = T_S \tag{3.18}$$

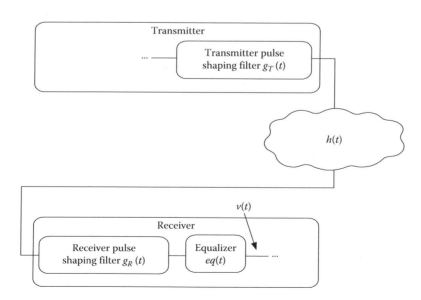

FIGURE 3.17 Location of the transmitter and receiver pulse shaping filters in the communication chain.

---

[*] The RAKE receiver is described in Chapter 7.

Therefore, the Nyquist criterion for distortionless baseband transmission in the absence of noise can be stated as follows: the frequency function $P(f)$ eliminates ISI, namely that caused by the nonoptimum sampling instant, for samples taken at intervals T_S provided that it satisfies Equation (3.18). Note that $P(f)$ refers to the overall system, incorporating the transmit filter, the channel, the receive filter, and so on, such that

$$\mu P(f) = G_T(f)H(f)G_R(f) \qquad (3.19)$$

Here, $G_T(f) = \mathcal{F}[g_T(t)]$ and $G_R(f) = \mathcal{F}[g_R(t)]$, where $g_T(t - nT)$ stands for the transmitting pulse shaping filter and $g_R(t - nT)$ for the receiving pulse shaping filter ($\mathcal{F}[x]$ is the Fourier transform of x). $H(f)$ stands for the channel transfer function. The above system and signal description refers to Figure 3.17, where the carrier modulator and demodulator, as well as the symbol modulator and demodulator, were not depicted for the sake of simplicity.

Besides the nonoptimum sampling instant, the ISI generated because of the frequency selective fading (i.e., intense multipath channel) is usually not completely removed by pulse shaping but by different techniques such as equalization.

To ensure that ISI is not present at the receiver because of the nonoptimum sampling instant, the Fourier transform of the signal at the equalizer's output $V(f)$ must be described by a function that satisfies the Nyquist ISI criterion. In other words, if a communication channel satisfies the Nyquist ISI criterion, the received signal is free of ISI originated by the nonoptimum sampling instant. A possibility of following the Nyquist ISI criterion consists of assuring that the signal $V(f)$ presents a pulse shape that follows the raised cosine function, defined as follows [Proakis 1995]:

$$P(f) = \begin{cases} T_S & |f| \leq f_N(1-\alpha) \\ \dfrac{T_S}{2}\left[1 - \sin\left(\dfrac{2\pi f_N}{\alpha}|f| - \dfrac{\pi}{2\alpha}\right)\right] & f_N(1-\alpha) \leq |f| \leq f_N(1+\alpha) \\ 0 & |f| \geq f_N(1-\alpha) \end{cases} \qquad (3.20)$$

where α stands for the rolloff factor, taking values between 0 and 1; it indicates the excess bandwidth over the ideal solution, $f_N = 1/(2T_S)$, and is usually expressed as a percentage of the Nyquist frequency f_N. Specifically, the bandwidth of a baseband transmission of absolute values is defined by $B'_T = [(1+\alpha)/(2T_S)]$. However, in the case of bandpass transmissions (carrier modulated), the previously negative baseband part of the spectrum becomes positive, being also transmitted. Therefore, the transmitted bandwidth is $f \in [-W\ +W]$ (i.e., it is doubled), which is defined by $B_T = 2 \cdot B'_T$, and becomes

$$B_T = \left[\frac{1+\alpha}{T_S}\right] \qquad (3.21)$$

For the special case of single sideband (SSB) transmissions, only the positive (or negative) part of the baseband spectrum is transmitted, which results in $B_T = B'_T = [(1+\alpha)/(2T_S)]$.

We can use the Nyquist theorem to deduct the relationship between the minimum bandwidth B_{min} of a transmission medium and the symbol rate $R_S = 1/T_S$. The minimum bandwidth B_{min} of the transmission medium is obtained by taking $\alpha = 0$, being related to the symbol rate R_S,[*] which is transmitted in a baseband signal, through the following equivalence [Proakis 1995]:

$$B_{min} = \frac{R_S}{2} \qquad (3.22)$$

---

[*] The symbol rate is also referred to as the *transmission rate*, being expressed in symbols per second (symb/s) or in baud.

Note that this equivalence corresponds to the maximum symbol rate that can be accommodated in the above-mentioned bandwidth. In the case of a bandpass signal (carrier modulated), this bandwidth is doubled and becomes

$$B_{\min} = R_S \tag{3.23}$$

As an example, let us consider a satellite link with a 2-MHz bandwidth. From Equation 3.23, we conclude that the maximum symbol rate (bandpass signal) that can be transmitted within this bandwidth is 2 Msymbols/s. In case one needs to transmit 2 Mbps, the symbol constellation should be such that each symbol transports 1 bit.[*] Alternatively, if one intends to transmit 4 Mbps in the referred satellite link bandwidth, a symbol constellation that accommodates 2 bits per symbol is the solution (e.g., QPSK).

As can be seen from Figure 3.18, for the rolloff factor $\alpha = 0$ and bandpass SSB transmission, we obtain the minimum bandwidth capable of transmitting signals with zero ISI defined by

$$
\begin{aligned}
B_{\min} &= f_N \\
&= \frac{1}{2T_S} \\
&= \frac{R_S}{2} \\
&= \frac{R_b}{2} \log_2 M
\end{aligned}
\tag{3.24}
$$

where R_S stands for the symbol rate and R_b stands for the bit rate. Moreover, M stands for the symbols constellation order and $\log_2 M$ for the number of bits transported in each symbol.

Furthermore, the spectrum of the pulse shaping filter depicted in Figure 3.18 corresponds to the Fourier transform of its impulsive response depicted in Figure 3.19. As can be seen from Table A.1, the Fourier transform of the sin c function corresponds to the rectangular pulse, that is, $\mathcal{F}\left[\sin c(2Wt)\right] = (1/2W)\Pi(f/2W)$ (valid for $\alpha = 0$). Because, for $\alpha \neq 0$, the pulse in the time domain is a variation of the *sinc* function, its Fourier transform may be viewed as a variation of the rectangular pulse.

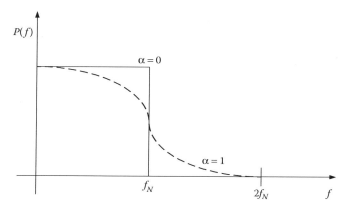

FIGURE 3.18 Raised cosine pulse in the frequency domain for $\alpha = 0$ and 1.

---

[*] Note that modulation schemes are dealt with in Chapter 6.

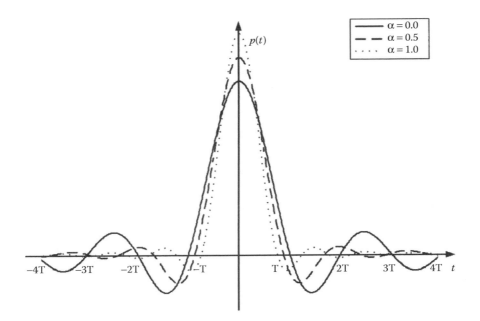

FIGURE 3.19 Raised cosine pulse in the time domain, for $\alpha = 0$, 0.5, and 1.

It is also worth defining the spectral efficiency. Assuming a baseband signal of an M-ary constellation, the spectral efficiency becomes

$$\varepsilon = \frac{R_b}{B_T}$$

$$= \frac{2\log_2(M)}{1+\alpha}$$

(3.25)

Naturally, for a bandpass signal (carrier modulated), the spectral efficiency becomes $\varepsilon = \left(\log_2(M)\right)/(1+\alpha)$.

In the case of a baseband signal and binary transmission (each symbol transports a single bit), the minimum channel bandwidth is $R_b/2$. Naturally, the transfer function that leads to the minimum bandwidth with $\alpha = 0$ is not physically realizable. Consequently, the transmission bandwidth is always higher than the minimum bandwidth B_{min}.

The function $p(t)$ consists of the product of two factors: the factor $\mathrm{sinc}(\pi t/T_S)$ characterizing the ideal Nyquist channel and a second factor that decreases as $1/|t|^2$ for large $|t|$. The first factor ensures zero crossing of $p(t)$ at the desired sampling instants of time $t = iT_S$, with i an integer (positive and negative). The second factor reduces the tails of the pulse considerably below that obtained from the ideal Nyquist channel, so that the transmission of binary waves using such pulses is relatively insensitive to sampling time errors. In fact, for $\alpha = 1$, this leads to the most gradual rolloff in that the amplitudes of the oscillatory tails of $p(t)$ are smallest. Thus, the amount of ISI resulting from timing error decreases as the rolloff factor α increases from zero to unity. The special case with $\alpha = 1$ is known as the full-cosine rolloff characteristic. This response exhibits two interesting properties:

- At $t = \pm T_S/2 = \pm 1/4W$, we have $p(t) = 0.5$; that is, the pulse width measured at half amplitude is exactly equal to the bit duration T_S.
- There are zero crossings at $t = \pm 3T_S/2, \pm 5T_S/2, \ldots$ in addition to the usual zero crossings at the sampling times $t = \pm T_S, \pm 2T_S, \ldots$

These two properties are extremely useful in extracting a timing signal from the received signal for the purpose of synchronization. However, the price paid for this desirable property is the use of a channel bandwidth double that required for the ideal Nyquist channel corresponding to $\alpha = 0$.

Because of the smooth characteristics of the raised cosine spectrum, it is possible to design practical filters for the transmitter and the receiver that approximate the overall desired frequency response. In the special case of an ideal channel, that is, with $H(f) = 1$, $|f| \le W$,

$$P(f) = G_T(f) G_R(f) \tag{3.26}$$

where $G_T(f)$ and $G_R(f)$ are the frequency responses (transfer function) of the transmit and receive filters, respectively, and $P(f)$ is the frequency response of the raised cosine pulse. Assuming that the receive filter is matched to the transmit filter leads to $P(f) = G_T(f) G_R(f) = |G_T(f)|^2$. Ideally,

$$G_T(f) = \sqrt{|P(f)|} \, e^{-j2\pi f t_0} \tag{3.27}$$

and $G_R(f) = G_T^*(f)$, where t_0 is some nominal delay that is required to ensure physical implementation of the filter. Thus, the overall raised cosine spectral characteristic is split evenly between the transmit and the receive filters. Note also that an additional delay is necessary to ensure the physical realization of the receive filter. Moreover, note that the PSD is proportional to $|G_T(f)|^2 = P(f)$, that is, PSD $\propto |G_T(f)|^2 = P(f)$.

The pulse shaping filter in the transmitter has the main function to allow the symbols formatting (in order to avoid ISI, as previously described) and to limit the spectrum inside the desired band, whereas the receive filter intends not only to contribute to format the symbols jointly with the transmitting pulse shaping filter but also to eliminate the noise outside the signal's bandwidth, allowing only the reception of the noise inside the signal's bandwidth.

3.6.2 MULTIPLE ACCESS INTERFERENCE

Multiple access interference occurs in networks that make use of multiple access techniques. This type of interference is experienced when there is no perfect orthogonality between signals from different users, viewed at the receiver's antenna of a certain user. In time division multiple access (TDMA) networks, this orthogonality is normally assured through guard periods, which avoids the overlapping of signals transmitted in different time slots (from different users). In frequency division multiple access (FDMA) networks, this orthogonality is assured through the use of guard bands (see Figure 3.20), and through the use of filters that reject undesired in-band interferences.

In CDMA networks, this kind of interference is normally present in real scenarios, and represents the main limitation of CDMA networks. MAI exists in CDMA networks because of the following:

- The use of spreading sequences that are not orthogonal (nonzero cross-correlation between different spreading sequences).
- Even using orthogonal spreading sequences, the orthogonality between spreading sequences is not assured when the network is not synchronized.[*] For this reason, it is often preferable to use quasi-orthogonal spreading sequences,[†] especially when in the presence of an asynchronous network.

---

[*] The uplink of a cellular network is normally asynchronous, that is, the transmission of symbols from different mobiles does not start at the same instant.

[†] Quasi-orthogonal spreading sequences present some level of cross-correlation. However, they present better autocorrelation properties in asynchronous networks than orthogonal spreading sequences (e.g., gold sequences).

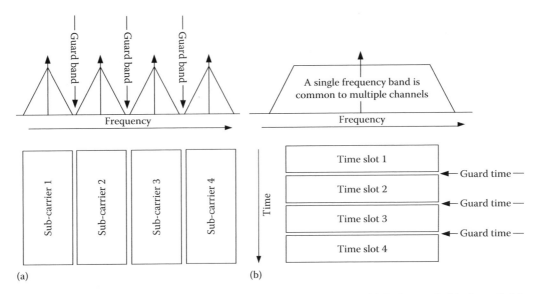

FIGURE 3.20 Separation of channels in (a) frequency division multiplexing and (b) time division multiplexing.

- Even in the downlink of a cellular network where synchronism normally exists between different transmissions, and even with the use of orthogonal spreading sequences, the multipath channel profile originates a relative level of asynchronism in the network. This originates nonzero cross-correlation values between superimposed signals received from different multipaths, especially when the channel presents frequency selectivity. Consequently, MAI is also present in this scenario.

In CDMA networks, MAI is directly related to the received power from different users. A certain user, with an excessive power, originates a level of MAI higher than others. Therefore, in CDMA networks, it is essential to use an effective power control to mitigate the fading effect, as well as to mitigate the near-far problem.* Also, in these networks, MAI can be reduced by using multiuser detection (MUD), power control, as well as sectored/adaptive antennas [Marques da Silva et al. 2010].

3.6.3 Co-Channel Interference

Co-channel interference occurs when two different communications using the same channel interfere with each other. In a cellular environment, this occurs when a communication is interfered by another communication being transmitted in the same carrier frequency but typically coming from an adjacent cell. In cellular networks using TDMA/FDMA, this type of interference can be mitigated by avoiding the use of the same frequency bands in adjacent cells, introducing the concept of a frequency reuse factor higher than 1.† In CDMA networks, this kind of interference is always present because the whole spectrum is typically reused in all cells, making the reuse factor as 1. The reuse factor refers to the reutilization of the same frequency bands in adjacent cellular networks. In Figure 3.21a, different letters in different cells mean that different sets of frequency bands are utilized in different cells. In this case, because each group of seven cells use different frequency

---

* Near-far problem: because of path loss, a received signal originated from a transmission in the neighborhood is much more powerful than a received signal originated from a transmission made at a long distance.
† See Chapter 15.

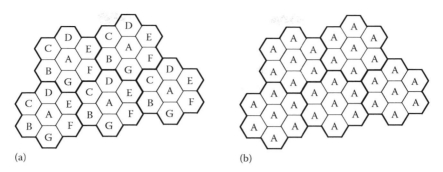

(a)   (b)

FIGURE 3.21 Two cellular environments with different reuse factors. (a) Reuse factor 7 and (b) reuse factor 1.

bands, and the overlapping of frequency bands is only reutilized in another adjacent group of seven cells, the reuse factor is 7.

The reuse factor 1 is normally adopted in CDMA networks. The reuse factor 1 means that all frequency bands are utilized in all cells (see Figure 3.21b). This factor is adopted because, although CCI occurs, leading to a decrease of performance, the gain in capacity is higher than the decrease of performance. Moreover, CCI can be mitigated through the use of MUD and adequate power control. In CDMA networks, this kind of interference is also known as MAI, being, however, generated by users located in adjacent cells.

3.6.4 Adjacent Channel Interference

Adjacent channel interference consists of an inadequate bandwidth overlapping of adjacent signals. This is due to inadequate frequency control, transmission with spurious, broadband noise, intermodulation distortion (IMD), transmission with a bandwidth greater than the one to which the operator is authorized, and so on. Guard bands are measures that are normally used to minimize the inadequate frequency control (see Figure 3.22).

Because any transmitter's oscillator presents a certain level of broadband noise, direct interference may be generated from this source. The power generated by an oscillator presents typically a Gaussian shape around the desired transmitting frequency. Therefore, the energy out of the desired signal's spectrum is considered as broadband noise. As can be seen from Figure 3.23, broadband noise consists of an unwanted signal transmitted in a frequency adjacent to the desired transmitted signal. This noise has a power very much below the transmitted signal (typically −155 dBc/Hz @ 3 MHz offset from the carrier). Nevertheless, receiving a signal at a short distance and very close in frequency to a transmitted signal may result in direct interference. This type of direct interference

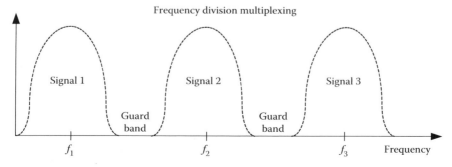

FIGURE 3.22 Frequency control using guard bands between different channels.

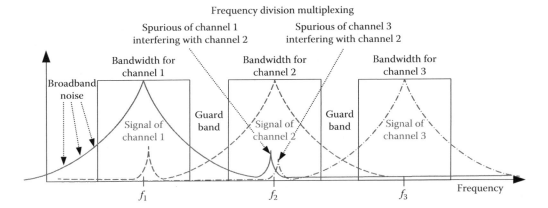

FIGURE 3.23 Broadband noise and spurious.

can be mitigated by using pre- and postselector filters, and by increasing the distance between adjacent transmit and receive antennas.

Spurious is another type of direct interference normally generated in transmitters. As can be seen from Figure 3.23, spurious consists of an undesired transmission in a frequency band different from the one reserved for sending the signal. A receiver located at a short distance from such a transmission with spurious may result in a high-power interfering signal that may block one or more channels. The measure that can be used to mitigate this direct interference consists of keeping transmit and receive antennas sufficiently spaced apart, to assure the required isolation. Moreover, spurious transmission filtering is normally mandatory from frequency management regulators. Pre- and postselector filters may also mitigate the negative effects of spurious.

IMD is another type of interference consisting of undesired signal generated within nonlinear elements (such as in a transmit amplifier, receiver, low noise amplifier, and multicoupler). Two or more signals present at such a nonlinear element are processed, and additional signals are generated, at the sum and difference of multiple frequencies, being known as *intermodulation products* (IMPs). As can be seen from Figure 3.24, IMPs can be of third order, fifth order, seventh order

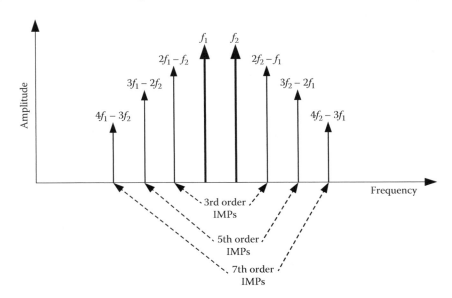

FIGURE 3.24 Intermodulation products.

and so on. Assuming that two isolated carriers f_1 and f_2 are present at a nonlinear element, the generated third-order IMPs become $(2f_1 - f_2)$ and $(2f_2 - f_1)$. Finally, it is worth noting that IMD is considered as indirect interference, and its effect can be worse than direct interference. This can be mitigated with the use of postselector filters (at the transmitter), as well as preselector filters (at the receiver).

CHAPTER SUMMARY

This chapter provided a description of channel impairments, experienced in both cable and wireless transmission media. It was viewed that channel impairments accumulate over the channel path, between the transmitter and the receiver, degrading the SNR. Note that for a transmitted signal to be properly extracted at the receiving side, this should present an SNR higher than a certain threshold. In the case of transmission of digital signals, it was described that the degradation of the SNR translates in a degradation of the BER.

This chapter defined the Shannon capacity, which corresponds to the maximum capacity that a channel can support. Then, the attenuation channel impairment was defined. It was observed that the attenuation results in a degradation of the SNR, by reducing the power of the signal.

The different electromagnetic noise sources were exposed in this chapter, degrading the SNR by increasing the power of noise. This includes atmospheric noise, being generated by the atmosphere at a long distance, caused by thunderstorms in tropical regions. The atmospheric noise is more intense in the VLF and LF bands, and stronger in the vertically polarized electromagnetic waves, than in horizontally polarized waves. Then, man-made noise was also defined, being generated by human activity, namely by the use of electrical equipment, such as car ignitions, domestic equipment, or vehicles. The intensity of man-made noise varies substantially with the region. It was described that this noise tends to be more intense in urban than in rural environments. Extraterrestrial noise was defined, such as galactic noise. Extraterrestrial noise can be intense when a receive antenna is pointed toward a planet or a star, such as the sun. Moreover, thermal noise was also defined, being more intense when an antenna is pointed toward a location with higher temperature. For example, a receive antenna of a satellite transponder pointed toward the hearth experiences higher thermal noise than that of a satellite ground station. Finally, electronic noise was also exposed, being generated in active elements, such as transistors.

This chapter described the influence of the transmission channel on the signal quality. It was described that the transmission channel may originate multiple effects, including delay and phase shift, as well as distortion of signals. Then, equalization was described, comprising a process that aims to mitigate the negative effects of the channel distortion.

The different sources of interference were also described, which include the following: (1) the intersymbol interference, whose effects increase with the increase of the symbol rate; (2) the multiple access interference, which consists of an interference experienced when the channel is shared between different users; (3) the co-channel interference, which results from the reutilization of the same frequency bands, but typically from transmissions coming from adjacent locations; and finally, (4) the adjacent channel interference, which is an interference type experienced when the frequency bands of adjacent channels partially overlap.

REVIEW QUESTIONS

1. Which kinds of channel impairments do you know?
2. What are the effects of channel impairments in either analog or digital signals?
3. Which kinds of noise do you know?
4. What does intersymbol interference stand for?
5. What is the difference between adjacent channel interference and co-channel interference?
6. What is multiple access interference?

7. Assuming baseband transmission, what is the minimum bandwidth necessary to accommodate a digital signal with a symbol rate of R_S?
8. Assuming bandpass transmission (carrier modulated), what is the minimum bandwidth necessary to accommodate a digital signal with a symbol rate of R_S?
9. What does distortion stand for?
10. Which types of distortion do you know?
11. What does spurious stand for? How can we mitigate the negative effects of spurious?
12. What does thermal noise stand for?
13. How can we quantify the FSPL?
14. What is the meaning of the Shannon capacity?
15. Assuming a twisted pair with a 1 MHz bandwidth, and an SNR of 5 dB, according to the Shannon capacity limit, what is the maximum speed of information bits that can be transmitted?
16. How are intermodulation products generated? What can be their negative effects? How can we mitigate them?
17. What is the difference between atmospheric noise and man-made noise? Characterize these two types of noise.
18. What does noise factor stand for?
19. How can we compute the noise factor of a system composed of a cascade of N electronic devices?
20. What is the ideal frequency response of a channel, in terms of phase shift and attenuation?
21. How can we mitigate the effects of the nonideal frequency response of a system, in terms of phase shift and attenuation?
22. According to the Nyquist theorem, what is the minimum sampling rate that can be employed to digitize a signal with a spectrum in the range 8–60 kHz?

LAB EXERCISES

1. Using the Emona Telecoms Trainer 101 laboratory equipment, and volume 1 of its laboratory manual, perform experiment 4—amplitude modulation (AM).
2. Using the Emona Telecoms Trainer 101 laboratory equipment, and volume 2 of its laboratory manual, perform experiment 1—AM (method 2) and product detection.
3. Using the Emona Telecoms Trainer 101 laboratory equipment, and volume 2 of its laboratory manual, perform experiment 2—noise in AM communications.
4. Using the Emona Telecoms Trainer 101 laboratory equipment, and volume 2 of its laboratory manual, perform experiment 9—signal-to-noise ratio and eye diagrams.

4 Cable Transmission Mediums

LEARNING OBJECTIVES

- Identify and describe the different cable transmission mediums.
- Describe the different types of twisted pairs.
- Describe the different types of coaxial cables.
- Describe the different types of optical fibers.
- Identify and describe the interference parameters in metallic conductors.

The transmission medium that still dominates houses and offices is the twisted pair. In the past, local area networks (LANs) were made of coaxial cables, which were also used as a transmission medium for medium- and long-range analog communications. Although their use in a LAN was replaced by the twisted pair, the development of the cable television made the coaxial cable reused. With the improvement of isolators and copper quality, as well as with the development of shielding, the twisted pair became widely employed for providing high-speed data communications, in addition to the initial use for analog telephony. Currently, most of the companies use IP telephony with the same physical infrastructure as the one used for data, which represents a convergence between voice and data. We observe an increase in the demand for optical fibers, in LAN, MAN, and WAN segments, because of their immunity to electromagnetic interferences and extremely high bandwidth.

As a rule of thumb, the attenuation of cable transmission mediums increases with the increase of the distance, but at a different rate for different transmission mediums (i.e., it is different for twisted pair, coaxial cable, and optical fiber). Moreover, the attenuation and phase shift also tend to increase with the increase of the frequency, whose effect is more visible at longer distances. This results in distortion, and in the case of digital communications, it is viewed as intersymbol interference.[*] Consequently, it can be stated that the available bandwidth decreases with the increase of the link distance. Decreasing the link distance, the attenuation and phase shift at limit frequencies also reduce, resulting in a higher throughput supported by the cable. Table 4.1 shows the typical bandwidths for different cable transmission mediums.

As described in Section 3.2, when the transmitter and the receiver are sufficiently far apart, amplifiers (used for analog signals) or regenerators (used for digital signals) need to be employed at regular intervals, in order to improve the signal-to-noise ratio, as well as to allow keeping the signal strength above the receiver's sensitivity threshold. The maximum distance where regenerators need to be placed depends on the characteristics of the cable transmission medium and on the bandwidth under consideration. Higher bandwidths require regenerators at shorter distances.

The following sections describe each of these important transmission mediums, in terms of use, bandwidth, attenuation, distortion, resistance to interference, and so on.

4.1 TWISTED PAIRS

Low-bandwidth twisted pairs, normally referred to as voice-grade twisted pairs, have been widely used for decades, at home and in offices, for analog telephony. Twisted pairs were also widely used to link houses and offices with local telephone exchanges. In order to reduce distortion, inductors (load coils) can be added to voice-grade twisted pairs, at certain distance intervals. This results in a

---

[*] As previously described, a way to mitigate this effect is by employing an equalizer at the receiver.

TABLE 4.1
Bandwidths of Different Cable Transmission Mediums

	Voice-Grade Twisted Pair (Grade 1)	Twisted Pair Category 6	Coaxial	Optical Fiber
Bandwidth	3.4 kHz	250 MHz	500 MHz	150 THz

flatter frequency response of the twisted pair over the analog voiceband (300 Hz to 3.4 kHz), translating in a lower attenuation level. Note that a twisted pair with loading cannot be used to flatten the frequency response of twisted pair cables used for data, as the bandwidth of data communications (several MHz or even GHz) is typically much higher than that of analog voice.

Because of the pre-existence of voice-grade twisted pairs in houses and offices, their use for data communications became a viable and inexpensive solution. Nevertheless, as they are very susceptible to noise, interferences, and distortion, these cables could not allow the data rates in use by most of LANs. The improvement of twisted pair technology (such as shielding, twisting length, and cable materials) increased the resistance to these impairments, leading to an increased bandwidth. Consequently, these improved twisted pair characteristics, added to its reduced cost, resulted in a massive use of this physical infrastructure, instead of the previously used coaxial cable.

Nevertheless, comparing the twisted pair with coaxial and optical fiber, the distances and bandwidths reached with the initial twisted pair were less than those obtained with coaxial and optical fiber.

4.1.1 CHARACTERISTICS

As can be seen from Figure 4.1, a twisted pair is considered as a transmission line, being composed of two isolated and twisted conductors in a spiral pattern. The proper selection of the twisting length of these conductors leads to a reduction of low-frequency interferences and crosstalk.

Crosstalk consists of an electromagnetic coupling of one conductor into another (wire pairs or metal pins in a connector). The electromagnetic field received by an adjacent conductor generates an interfering current, being superimposed on the signal's current. This originates a degradation of the signal-to-noise plus interference ratio.

The material employed in conductors is normally copper, whereas polyethylene is normally used in isolators. In order to improve the crosstalk properties, twisted pairs are normally twisted and bundled in two pairs (four wires): one pair for transmission and another pair for reception (full duplex). In order to further improve the crosstalk properties, and to optimize the cables, two groups of two pairs (i.e., eight wires) are twisted and wrapped together, using a protective sheath. This results in cables composed of four pairs.

The quality of the twisted pair depends on several factors, such as the material and width of the isolator, the copper wire purity and width (typically between 0.4 mm and 0.9 mm), the twist length, the type of shielding (when used), and the number of pairs twisted together. All of these parameters define the impedance of the twisted pair, which results in a certain attenuation coefficient (expressed in dB/km) and phase shift coefficient, both as a function of the signal's frequency. The maximum bandwidth and distance supported by a certain type of cable depend on these parameters.

FIGURE 4.1   Twisted pair as composed of two isolated copper wires properly twisted.

Naturally, because of increased resistance to interference, multipair cabling presents a bandwidth higher than single pair.

4.1.2 TYPES OF PROTECTION

Twisted pairs are normally grouped as unshielded twisted pairs (UTPs), foiled twisted pairs (FTPs), shielded twisted pairs (STPs), or as screened STPs (S/STPs) [ANSI/TIA/EIA-568].

As the name refers, UTP cabling is not surrounded by any shielding, whereas STP presents a shielding with a metallic braid or sheathing, applied to each individual pair of wires, that protects wires from noise and interferences.

The attenuation coefficient of a 0.5 mm copper wire UTP cable can be seen from (as a function of the frequency).

The resulting attenuation, expressed in decibel, is given by

$$A_{db} = l \cdot \alpha(f) \tag{4.1}$$

where:

l stands for the cable length (in km)

$\alpha(f)$ stands for the attenuation coefficient (in dB/km)

UTP cabling is employed in Ethernet and telephone networks, being normally installed during building construction. Because it does not present any shielding, UTP cabling is very subject to noise and external interferences, presenting typically an impedance of 100 Ω. UTP cabling is typically less expensive than STP, and also less expensive than coaxial and fiber optic cables. Furthermore, STP and FTP are more difficult to handle than UTP. Consequently, a cost–benefit analysis needs to be done before a decision is made about the type of cabling to employ.

STP cabling includes a metal grounded shielding surrounding each pair of wires, presenting a typical impedance of 150 Ω. STP supports up to 10 Gbps, being considered in 10GBASET technology employed to implement the IEEE802.3 LAN.

When the shielding is applied to multiple pairs, instead of a single pair of wires, it is referred to as screening. This is the case of FTP, being also referred to as screened UTP (S/UTP). It consists of UTP cabling whose shielding surrounds the cable (screened), not presenting shielding in each pair of copper wires. Consequently, while this cabling presents good resistance to interferences originated from outside of the cable, the crosstalk properties (interference between different pairs of cabling) are typically poorer than STP.

Finally, screened FTP (S/FTP) cabling, also referred to as S/STP cabling, presents a shielding surrounding both individual pairs and the entire group of copper pairs, and therefore, it is both externally and internally protected from adjacent pairs (crosstalk). The above description is summarized in Table 4.2.

TABLE 4.2

Protection Type for Different Twisted Pair Cablings

	Shielding	Screening
UTP	No	No
FTP	No	Yes
STP	Yes	No
S/STP (S/FTP)	Yes	Yes

4.1.3 CATEGORIES

Another way to characterize twisted pair cablings is to group them into different categories, from 1 to 7. As can be seen from Table 4.3, the increased cabling category results in a higher bandwidth and data rate. The better performance is achieved at the cost of better and thicker copper wires, isolation, improved shielding, or improved twisting. Consequently, higher bandwidths tend to correspond to higher costs. With the exception of the voice-grade twisted pair, cables of the other categories comprise four pairs of conductors.

Although the standards [ANSI/TIA/EIA-568-A; ISO/IEC 11801] recognize only categories 3 up to 6, categories (or grade) 1, 2, and 7 are also listed in Table 4.3, as these designations are normally assigned to cabling configurations. This table lists the cabling and connectors, as well as the corresponding bandwidths and maximum data rates. Note that the copper connectors listed in Table 4.3 are defined in Tables 4.4 and 4.5, respectively, for T568A* and T568B† terminations of 8P8C modular connectors (commonly referred to as *RJ45*).

As previously described, the maximum bandwidth supported by a cable transmission medium depends on the link distance. Shorter distances allow accommodating higher bandwidths, and vice

TABLE 4.3

Characteristics of Different Twisted Pair Categories (@100 m)

	Category 1	Category 2	Category 3	Category 4	Category 5	Category 6	Category 7
Cable type	Voice-grade TP	UTP (multipair voice graded)	UTP (multipair voice graded)	UTP	UTP/FTP	UTP/STP/FTP	S/STP/S/FTP
Copper connector/ termination	Not specified	8P8C/T568A	8P8C/T568B	Not specified	8P8C/ T568B	8P8C/T568A or 8P8C/ T568B	8P8C/ T568B
Bandwidth	3.4 kHz	6 MHz	16 MHz	20 MHz	100 MHz	250 MHz	600 MHz
Maximum data rate	Analog voice	4 Mbps	12 Mbps	16 Mbps	100 Mbps	1 Gbps	40 Gbps @ 40 m

TABLE 4.4

T568A Termination of 8P8C Modulator Connectors (*RJ45*)

Pin	Pair	Color
1	3	White/green
2	3	Green
3	2	White/orange
4	1	Blue
5	1	White/blue
6	2	Orange
7	4	White/brown
8	4	Brown

Source: ANSI/TIA/EIA-568-A, *Commercial Building Telecommunications Standard*, 1995.

---

* T568A is the designation of the 8P8C termination standardized by ANSI/TIA/EIA-568-A [1995].
† T568B is the designation of the 8P8C termination standardized by ANSI/TIA/EIA-568-B [2001].

TABLE 4.5

T568B Termination of 8P8C

Modulator Connectors (*RJ45*)

Pin	Pair	Color
1	2	White/orange
2	2	Orange
3	3	White/green
4	1	Blue
5	1	White/blue
6	3	Green
7	4	White/brown
8	4	Brown

Source: ANSI/TIA/EIA-568-B, *Commercial Building Telecommunications Standard*, 2001.

versa. This results from the fact that different frequencies present different attenuation coefficients and different delays, whose effect is more visible at longer distances. In fact, the attenuation of a twisted pair increases approximately exponentially with the increase of the frequency. The effect that results from this impairment is known as distortion (attenuation and/or phase distortion) and, in the case of digital communications, results in intersymbol interference. Improved twisted pair quality results in longer distances for the same bandwidth, or higher bandwidth for the same distances, as compared to lower quality twisted pair. As a rule of thumb, for digital signals, there is a need to use regenerators at a distance interval of 2–3 km of twisted pair cable.

The bandwidths listed in Table 4.3, for different categories, are those specified for 100 m of distance (90 m of cable plus 10 m of patch cord). These values may be exceeded for shorter distances.

Category 1 UTP consists of low-quality twisted pairs specified for analog voice only (without external isolation). The twisting of multiple pairs, previously individually twisted, into the same cable originates the category 2 UTP. Some authors also consider categories 2 and 3 as a voice-grade twisted pair. Nevertheless, the exceeding bandwidth, besides that of the analog voice, is used for data communications. Categories 3 and 4 UTP cabling is similar to category 2, but with improved copper and isolation, as well as using a twisting length that improves the resistance against noise and interferences. This results in the ability to support a bandwidth of 16 MHz for category 3 and 20 MHz for category 4 twisted pair cabling. Category 3 cables are considered in the LAN standard IEEE 802.3 (at 10 Mbps). Moreover, categories 3 and 4 are considered in the LAN standard IEEE 802.3u at 100 Mbps, using several parallel pairs.

While consisting of either UTP or FTP, category 5 is currently installed during construction in most offices. It supports a bandwidth of 100 MHz, being considered by the LAN standard IEEE 802.3u (at 100 Mbps) and by IEEE 802.3ab (at 1 Gbps), in the latter case using four negotiated parallel pairs for transmitting or receiving. Category 5e (enhanced) refers to category 5 cabling with an improved shielding performance in terms of near end crosstalk (NEXT), attenuation to crosstalk ratio (ACR), equal level far end crosstalk (ELFEXT), and so on (see Section 4.4). These improved characteristics make the full-duplex operation possible in each pair, requirement that is important for the implementation of IEEE 802.3 at 1 Gbps (i.e., IEEE 802.3z).

Category 6 supports 250 MHz of bandwidth and a data rate of 1 Gbps, being based on UTP (adopted by the LAN standard IEEE 802.3ab), STP, or FTP cabling (IEEE 802.3z). There is a variation, referred to as category 6a, which allows twice the bandwidth of category 6, that is, 500 MHz.

Finally, category 7 is defined to support 600 MHz of bandwidth and data rates as high as 40 Gbps (contrary to the other categories, the listed value refers to 40 m). This is achieved using S/STP, which makes use of double shielding, resulting in a high level of immunity to noise and interferences. There is a variation, referred to as category 7a, defined to support frequencies up to 1 GHz.

Let us focus on Figure 4.2. For avoiding distortion, the bandwidth of the signal should be carefully selected such that the frequency response is approximately flat, or such that the receiver's equalizer is able to counteract the frequency selectivity of the channel. Note that this figure plots in the ordinates the attenuation coefficient. This means that the attenuation value is a function of the link distance. Therefore, one can conclude that a certain twisted pair cable can support higher bandwidths at shorter distances, and lower bandwidths at longer distances. The maximum bandwidths supported by certain link distances depend on these attenuation coefficient curves. Table 13.3 shows the maximum link distances that can be supported by different twisted pair categories, for different signal bandwidths. In fact, depending on the consumed bandwidth, that is, depending on the transmission rate, the traffic generated by hosts can be split into class types.

It is worth noting that the link distance may, under specific circumstances, be limited by other than the attenuation factor. In fact, factors such as the ACR or the ELFEXT may also be the link bottleneck. As described at the end of this chapter, these factors should be positive for the link to be viable. A specific link, with a certain length and signal bandwidth, may not be constrained by the attenuation factor, but by the corresponding NEXT (or ELFEXT), whose value should not be negative. In this case, the link length may have to be decreased in order to make this factor positive.

4.1.4 CONNECTORS AND CABLES

Twisted pair cables may use different types of connectors. The IEEE 802.3z at 1 Gbps (1000BASE-CX) uses the DB9 or HSSDC connectors. Most of the other physical layers of the IEEE 802.3 network use the 8P8C modulator connector, also referred to as the ISO 8877 connector (sometimes also generically called *RJ45* connectors).

Tables 4.4 and 4.5 list the two most common termination layouts utilized in 8P8C modulator connectors, namely T568A and T568B.

Using one or another connector layout is not relevant, as long as the whole installation is coherent. Nevertheless, it is worth noting that the standard ANSI/TIA/EIA-T568-A is the most common. In fact, the ISO 8877 connector adopts ANSI/TIA/EIA-T568-A, leaving the standard ANSI/TIA/EIA-T568-B as an option.

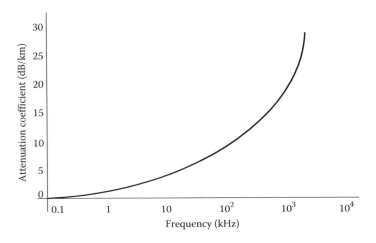

FIGURE 4.2 Attenuation coefficient of a twisted pair with 0.5 mm copper wires as a function of frequency.

In terms of cabling, although more than four pairs could be employed in horizontal cabling, the latter configuration is rare. The four pairs can be employed to serve two terminals in full-duplex operation. In contrast, in vertical cabling, depending on the number of terminals to serve, a cable with more than four pairs can be employed.

In the most common four-pair configuration, the color codes employed in four pair cables are listed in Table 4.6.

IEEE 802.3 cabling, also referred to as Ethernet cabling, may have two different basic configurations:

- Straight-through cable is typically used to interconnect:
  - A switch/hub to a computer (PC or server) or network printer
  - A router to a modem
  - A router to a switch/hub
- Crossover cable is typically used to interconnect:
  - A router to a computer
  - Two computers
  - Two switches/hubs
  - Two routers

The visual identification of straight-through and crossover cables is simple. The wire arrangement of both cable terminals of straight-through cables (and patch cables used to interconnect, e.g., a patch panel to a switch) is the same, whereas the wire arrangement of the two sides of crossover cables is different. Pins 1 and 2[*] are used for transmit, whereas pins 3 and 6[†] are used for receive. The transmit pins in one terminal of a crossover cable becomes the receive pins in the other terminal, and vice versa. Consequently, pins 1 and 2 in terminal A of a crossover cable become pins 3 and 6 of the terminal B, and pins 3 and 6 of terminal A become pins 1 and 2 of terminal B. While a straight-through cable (either T568A or T568B) has the same wire arrangements at both terminals, a crossover cable can be viewed as a cable with T568A wire arrangement on one terminal and with T568B wire arrangement on the other terminal. The straight-through termination of T568B can be memorized with the mnemonic OGBB (orange, green, blue, brown), where the first wire of each pair has a full color (example: wire 1 is full orange), and the second wire of each pair has white with lists (example: wire 2 is orange with lists). In the case of T568A, the mnemonic becomes GOBB, with the same remaining rules.

When configuring Cisco equipment from a workstation using a console cable for the interconnection between these two pieces of equipment, the wire arrangement is different from the two described above. In this case, the wire arrangement is reversed, that is, wire 1 on one termination becomes wire 8 on the other, wire 2 on one termination becomes wire 7 on the other, wire 3 on one termination becomes wire 6 on the other, and so on. Very often, a console cable connects to the

TABLE 4.6
Color Code of Four-Pair Twisted Pair Cables

Pair	Color	Conductor 1	Conductor 2
1	Blue	Blue	White with blue lists
2	Orange	Orange	White with orange lists
3	Green	Green	White with green lists
4	Brown	Brown	White with brown lists

---

[*] Pair 3 of T568A termination or pair 2 of T568B termination.
[†] Pair 2 of T568A termination or pair 3 of T568B termination.

Cisco equipment using an ISO 8877 connector (commonly referred to as *RJ45*), while the workstation connects to the Cisco equipment using a DB-9 connector (RS-232). In this case, an adapter from ISO 8877 into DB-9 is required.

4.2 COAXIAL CABLES

Coaxial cables were the main transmission medium of long-range analog transmission, namely between local telephones exchange. They supported high capacity communications as defined by ITU-T G.333 recommendation (10 800 telephone channels, with a maximum frequency of 60 MHz). The utilization of coaxial cables in telephone networks has been replaced by optical fibers. The initial use of coaxial cables as LAN infrastructure was also replaced, in this case by twisted pair cables. Currently, coaxial cables are used for either analog or digital signals. Its use in cable television networks (analog and digital) is still of high importance. Furthermore, this physical infrastructure is currently used for video distribution between devices, namely between an antenna and a television receiver or a DVD recorder. Its use in a LAN is limited to special applications. Nevertheless, the current trend is to use more and more optical fibers.

4.2.1 CHARACTERISTICS

A coaxial cable consists of two concentric conductors: an inner conductor and an outer conductor. As can be seen from Figure 4.3, the inner conductor is isolated and centered with the help of a dielectric material or an isolator. The outer conductor is cylindrical in shape, being covered with a shield or jacket. Coaxial pairs are grouped in low number, being wrapped into a cable, protected from traction forces.

The design of a coaxial cable leads to good electromagnetic isolation, which translates into good resistance properties against noise and interferences. Consequently, the bandwidth of a coaxial cable is typically higher than that of a regular twisted pair. The characteristic impedance of coaxial cables is 75 Ω.

The structure of coaxial cables almost eliminates the possibility of coupling between parallel coaxial pairs of the same cable. Albeit galvanic coupling may exist between external conductors, the crosstalk properties of this transmission medium are excellent for operating frequencies above 60 kHz. Below this frequency, coaxial cables cannot be used as it captures high level of external interferences (high crosstalk) and suffers heavily from distortion.

It can be shown that the attenuation is minimized for a relationship between the outer conductor diameter d_2 and the inner conductor diameter d_1 equal to 3.6. As can be seen from Table 4.7, coaxial cables normalized by ITU-T present a relationship close to that value.

The attenuation coefficient of a 2.6/9.5 mm coaxial cable, standardized by ITU-T G.623, is approximated by

$$\alpha \cong 0.01 + 2.3\sqrt{f} + 0.003f \qquad (4.2)$$

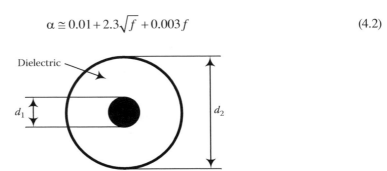

FIGURE 4.3   Coaxial cable as composed of two isolated and concentric conductors.

TABLE 4.7
Dimensions of Normalized Coaxial Cables

ITU-T Recommendation	ITU-T G.623	ITU-T G.622	ITU-T G.621
d_1 (mm)	2.6	1.2	0.7
d_2 (mm)	9.5	4.4	2.9
d_2/d_1	3.65	3.67	4.14

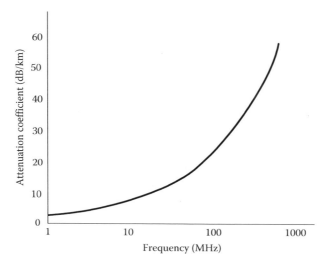

FIGURE 4.4   Attenuation coefficient of a 2.6/9.5 mm coaxial cable [ITU-T G.623] as a function of frequency.

where:
   The attenuation coefficient α is expressed in dB/km
   The frequency f is expressed in MHz

Note that Equation 4.2 is an approximation valid for frequencies above 1 MHz. The resulting function is plotted in Figure 4.4.

Having as a reference an attenuation coefficient variation from very low values up to 30 dB/km, we observe from Figure 4.4 that this corresponds to an approximate bandwidth of 400 MHz, whereas in the case of a twisted pair cable, this corresponds to about 2 MHz. From these results, we conclude that the level of distortion typically introduced by a coaxial cable is much lower than that of a twisted pair. Consequently, the bandwidth available in a coaxial cable is higher than that of a twisted pair. The typical bandwidth provided by a coaxial cable is 500 MHz. Note that because the attenuation coefficient of a coaxial cable carrying a high bandwidth is high (30 dB/km or higher), the distance between adjacent repeaters or regenerators is reduced. Conversely, carrying a lower bandwidth in a coaxial cable leads to lower attenuation levels, translating in a longer distance between adjacent repeaters.

4.3   OPTICAL FIBERS

Since the 1970s optical fibers have become one of the most important transmission mediums used in medium- and long-range communications. Their use at a short range has also taken place, instead of the previously used coaxial cables and twisted pairs. Typical use of optical fibers includes WAN and MAN networks, namely as the transmission medium of SDH, synchronous optical network (SONET), or ATM. Their applications also include the interconnection between different local telephone exchanges, as well as the interconnection of local telephone exchanges and cable television with a street cabinet that is typically located within few hundred meters from homes (fiber to the curb [FTTC]), with the a patch cord box just outside of a home (fiber to the home [FTTH]) or even with a home itself or a small business (fiber to the premises [FTTP]).

Optical fibers have been of high importance to implement the collaborative era of telecommunications principle defined in Chapter 1. In the LAN environment, we have observed the gradual replacement of twisted pairs by optical fibers. IEEE 802.3z (namely, 1000Base-SX and 1000Base-LX) is an example of the LAN standard that includes optical fibers as a transmission medium. In fact, optical fibers have already been considered in 100 Gbps standards, and its use in 400 Gbps networks is under consideration [Winzer 2010]. Finally, optical fibers are started being used to interconnect different modules of electronic equipment (e.g., of a radio transmitter or receiver). This results in high-performance equipment.

4.3.1 CHARACTERISTICS

The emergency of optical fibers, as compared to other cable transmission mediums, was due to the following reasons:

- *Extremely high bandwidth*: This translates in high throughputs (typically 150 THz).
- *Attenuation coefficient*: They have low attenuation coefficient, as compared to twisted pairs, or even to coaxial cables. Moreover, as opposed to other cable transmission mediums, the attenuation coefficient of an optical fiber varies very smoothly with the frequency, which translates in a low level of distortion.
- *Longer repeating distances*: These result from the low level of attenuation. As an example, SDH networks include typically regenerators only every 60 km distance, whereas the regeneration distance of twisted pairs and coaxial cables is about 1–2 km
- *Immunity to electromagnetic interferences*: This results from the fact that optical fibers are typically made of fiber of silica glass (S_iO_2). Because this substance is not a metallic conductor, optical fibers do not suffer from crosstalk.
- *Small dimensions and low weight*: An optical fiber presents typically a diameter 10 times smaller than that of a coaxial cable and its weight is typically 30 times lower than that of a coaxial cable.
- *Greater capacity*: This results from the fact that a duct previously used for coaxial cables can accommodate 10 times more optical fibers, and because the capacity of an optical fiber is much higher than that of a coaxial cable (typically 300 000 times higher).
- *Reduced cost*: Silica is one of the most abundant materials in the earth, which makes it much less expensive than copper.

An optical fiber is made of a dielectric guide capable of guiding an optical ray, composed of a core made of glass or plastic with refraction index n_1 and a cladding with refraction index n_2, where $n_1 > n_2$ [Keiser 1991] (see Figure 4.5). For protection, this dielectric guide is surrounded by a jacket, normally made of plastic. A light pulse is propagated along the core of a fiber, between a light emitter and a light receiver.

4.3.2 CATEGORIES

There are two main types of optical fibers: single mode and multimode. The former type of fibers supports a single propagation mode, whereas the latter supports several modes. As can be seen from Figure 4.6, in the single mode, a single path exists over the fiber length, between the emitter of the light pulse and the corresponding receiver. In this case, there are no rays propagated in any direction

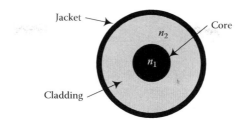

FIGURE 4.5 Optical fiber as composed of a core and a cladding, surrounded by a jacket.

different from the normal to the fiber section. Conversely, in the multimode fiber, several paths exist between the light emitter and the receiver. As can be seen from Figure 4.6, multimode fibers have several rays propagated with different angles, from the normal to the fiber section until a certain critical angle above which reflection does not occur. Different rays are reflected at the border between the core and the cladding, following the reflection principles described in Chapter 5, with a reflected angle equal to the incident angle. Consequently, an emitted pulse reaches the receiver through several different paths, with different delays. This results in a pulse dispersion, which can be viewed as intersymbol interference. This effect is referred to as modal dispersion. Because the level of dispersion increases with the link distance, multimode fibers are normally limited to short-range applications. This reduced range is also a result of the higher attenuation of multimode fibers, as compared to single mode (see Figure 4.7). Multimode optical fibers are typically characterized by the modal bandwidth, as defined at the end of this subsection. The modal bandwidth characterizes the effect of modal dispersion. This factor relates the available bandwidth with the optical fiber length, that is, with the link distance.

There are two different types of multimode fibers: step index and graded index. The core of multimode step index fiber presents a single refraction index, whereas the graded index fiber presents an index that decreases from the center to the extremities of the core. The refraction index profile is depicted in Figure 4.6. In the graded index fiber, different paths still exist, but they converge to the same point. Therefore, as in the case of a step index fiber, the receiver of a graded index fiber detects a signal that is composed of the superimposition of several replicas (i.e., the same signal from different paths), but these replicas tend to be aligned in time, that is, they tend to be synchronized.

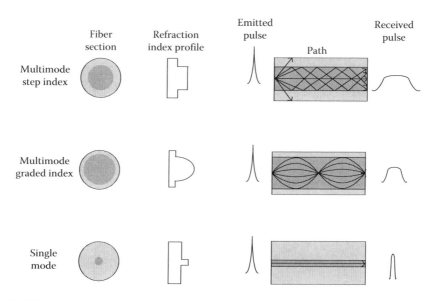

FIGURE 4.6 Characteristics of different types of optical fibers.

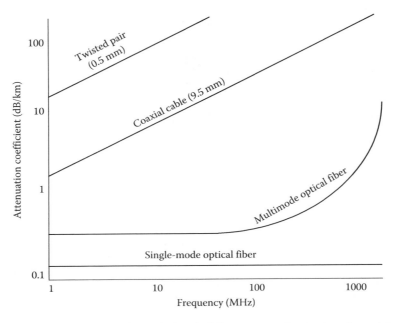

FIGURE 4.7   Attenuation coefficient of a typical optical fiber versus twisted pair and coaxial cable.

The level of alignment is never perfect, resulting in a certain level of pulse dispersion. However, the level of pulse dispersion is smaller than in the case of step index fiber, resulting in an increased bandwidth. Note that, as a result of the gradual refraction index profile, the rays suffer refraction or bending effects, as they propagate from the center to the extremities of the fiber's core. The refraction phenomenon is modeled by Snell's law, defined by

$$n_1 \sin \theta_1 = n_2 \sin \theta_2 \tag{4.3}$$

where:
   n_1 and n_2 stand for the refraction indexes of the core at two different distances from the center of a graded index optical fiber, respectively
   θ_1 and θ_2 stand for the wave's direction, measured from the normal to the border between the two material points (see Figure 4.8)

A wave that crosses a border between the medium n_1 into the medium n_2 suffers refraction, that is, deviation or bending in the wave's direction. After refraction, the new angle of the wave's direction becomes

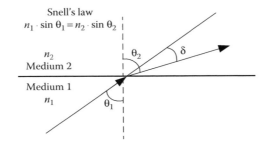

FIGURE 4.8   Refraction effect.

$$\theta_2 = \arcsin\left(\frac{n_1 \sin \theta_1}{n_2}\right) \tag{4.4}$$

Assuming that $\theta_2 = \theta_1 - \delta$, with δ the deviation angle, we have

$$\delta = \theta_1 - \arcsin\left(\frac{n_1 \sin \theta_1}{n_2}\right) \tag{4.5}$$

Single-mode fibers present a core diameter smaller than that of multimode fibers, making it more difficult to handle than multimode fibers. Moreover, as can be seen from Figure 4.7, the attenuation coefficient in any of the optical fiber's transmission windows presents a very low variation, as compared to the attenuation coefficient variation of twisted pairs or coaxial cables.[*]

This results in a low level of distortion, which translates in a high bandwidth. Furthermore, because a single mode is propagated over a single path, the level of dispersion is very low, confirming the enormous bandwidth made available by this type of optical fibers. This makes single-mode fibers well fitted for long-range links. Note that the bandwidth of the single-mode fibers is much higher than that of multimode. Nevertheless, single-mode fibers are more difficult to handle as the core diameter is smaller.

The typical diameter of the cladding is 125 μm, whereas the core diameter depends on the type of optical fiber: the multimode optical fiber diameter may span from 50 up to 65 μm; the diameter of single-mode optical fiber may span from 3 up to 10 μm.

As previously described, the received pulse is typically wider than the emitter pulse. This is due to dispersion, being classified as intermodal and intramodal dispersion. Intermodal dispersion results from the fact that different propagation modes present different propagation times, whereas the intramodal is due to the fact that different wavelengths present different propagation times. Single-mode optical fibers only present intramodal dispersion, whereas multimode fibers experience both types of dispersion.

The devices used to emit a beam of light in optical fibers can be either the light emitting diode (LED) or the injection laser diode (ILD). In high data rate links, the ILD is normally adopted, but it is more expensive. The modulation of these devices is achieved by varying the polarization current at their terminals. The variation of the light's intensity is directly proportional to the current variation. The data exchange is achieved in optical fibers with a variation of the light intensity. The most used modulation scheme is amplitude shift keying, where the different amplitudes correspond to different light intensities. Note that the emitted light is not monochromatic, presenting a certain spectral width σ_λ, whose value is lower for ILD than LED.

A photodiode is normally employed as a detector, to convert light energy into electrical energy. There are two main types of photodiodes: P intrinsic N (PIN) photodiode and avalanche photodiode (APD). PIN photodiodes are less sensitive and less expensive than APD.

The intramodal dispersion is a consequence of the spectral width σ_λ. The resulting optical pulse dispersion is given by [Keiser 1991]

$$\sigma_t = |D_\lambda| \cdot l \cdot \sigma_\lambda \tag{4.6}$$

where:
D_λ is a dispersion parameter characteristic of the fiber under consideration
l is the optical fiber length

The optical bandwidth of a fiber at −3 dB is [Keiser 1991]

---

[*] Note that the scale of the ordinate of Figure 4.7 is logarithmic.

$$B_0 = \frac{0.187}{\sigma_t} \qquad (4.7)$$

Because, from the Nyquist ISI criterion, the minimum electrical baseband bandwidth B_e that can accommodate the symbol rate R_S is given by $B_e = R_S/2$, and once the electrical bandwidth is related to the optical bandwidth by $B_0 = \sqrt{2} \cdot B_e$, the relationship between the symbol rate and the optical pulse dispersion becomes

$$R_S = \frac{0.264}{\sigma_t} \qquad (4.8)$$

Figure 4.9 shows the attenuation coefficient for different wavelengths of a typical fiber. Depending on the application and fiber type, there are different wavelengths used in data transmission. Note that these transmission windows were selected based on two important requirements: flatness of the attenuation profile over different wavelengths and low level of attenuation that translates in high ranges without the need to use regenerators.

The first transmission window uses a wavelength surrounding the 850 nm, being typically considered by multimode fibers in LAN applications. The second window, also referred to as the S band, uses a wavelength around 1300 nm, being typically considered by single-mode fibers in various applications. The third and fourth windows (C and L bands) use a wavelength around 1550 and 1600 nm being normally considered by single-mode fibers in long-range links [Stallings 2010]. Note that 1300 and 1550 nm windows correspond to infrared light.

It is worth noting that a carrier wave transmitted in optical fibers is identified by its wavelength, instead of its frequency. In electromagnetic propagation, we typically assume that the propagation speed is $v = c$, that is, we assume that the propagation speed of the wave in the medium under consideration is the same as the propagation speed of the wave in the vacuum. Consequently, the approximation that relates wavelength and frequency becomes $c = \lambda \cdot f$, with $c = 3 \times 10^8$ m/s. In fact, the true relationship between the wavelength λ and the frequency f is $v = \lambda \cdot f$, where $n = c/v$. Because the refraction index n of an optical fiber is sufficiently different from 1 (the refraction index in the vacuum), the propagation speed of the light in an optical fiber, v, is sufficiently different from the propagation speed of the light in the vacuum, c. In fact, different fibers have different refraction

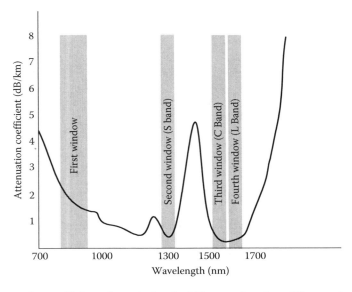

FIGURE 4.9   Attenuation coefficient of a typical optical fiber as a function of the wavelength.

indexes. Consequently, the equivalence that should be used is $v = \lambda \cdot f$. Therefore, referring to the frequency unit in optical fibers does not have the same meaning as in the case of electric or electromagnetic waves. Consequently, the wavelength unit has been adopted for optical fibers. Note that a certain frequency value has different meanings in two optical fibers with different refraction indexes. Contrarily, a certain wavelength value is always applicable, regardless of the refraction index.

The aforementioned reasons also justify the fact that the multiplexing performed in optical fibers is known as wavelength division multiplexing (WDM), instead of frequency division multiplexing (FDM).[*] Note that WDM is the multiplexing technique normally employed in MANs/WANs.

Similar to twisted pair cables, the maximum distance where an optical fiber can be employed depends on the used bandwidth. In fact, for multimode optical fibers, the maximum distance is normally calculated from the modal bandwidth value (commonly expressed in MHz km). Because of the low level of modal dispersion, this effect is not taken into account in single-mode optical fibers. In case of single-mode optical fibers, an important factor is the minimum cut wavelength. This corresponds to the wavelength below which[†] the fiber stops behaving as a single mode, and starts behaving as a multimode optical fiber.

Returning to multimode optical fibers, modal bandwidth is one of the most important specifications of an optical fiber. The modal bandwidth is a scalar value that corresponds to the multiplication of the bandwidth supported by a cable by the link distance. From the modal bandwidth value, one can calculate the maximum bandwidth that can be supported by a fiber, from a certain link distance. Alternatively, one can calculate the maximum link distance from a certain bandwidth of an optical fiber.

Let us consider a certain optical fiber (multimode) with a modal bandwidth of 600 MHz km. In this case, we may write $BW \times L = 600$ MHz km, where BW stands for the bandwidth and L for the length. Considering a link length of 2 km, the maximum bandwidth that this fiber can support is $BW = 600/L$ MHz, that is, 300 MHz.

The reader should refer to Chapters 12 and 14 for the description of the physical layer employed in different LAN and MAN technologies.

4.3.3 CONNECTORS AND CABLES

Optical fiber cables may use different types of connectors. The most common type of connectors employed in optical fibers is the subscriber connector (SC), the straight tip (ST) connector, and the Lucent connector (LC). While multimode optical fibers may use any of these types, single-mode optical fibers use either SC or LC connectors (SC is currently more frequent). In addition to these two types of connectors, there are others, namely the FC connectors and MU connectors. Regardless of the physical differences between these connector types (dimensions, shape, etc.), the link between two fiber sections is performed by the plug using a ferrule. Made of stainless steel, ceramic, or plastic, the ferrule holds the end of the fiber and precisely aligns it to the socket. If the alignment between the fiber end and the ferrule is not good, light reflections occur, degrading the performance of the optical fiber. The connection between the fiber and the ferrule can be performed in different ways, namely crimped, or cemented with an epoxy or adhesive.

Contrary to twisted pair cablings where the termination can be performed in male or female connectors, optical fibers always terminate in male connectors. In this case, the interconnection between an interconnection cable and a female connector requires a female-to-female adapter.

With regard to the type of optical fiber cables, these can be grouped into two important categories:

- Loose-tube
- Tight-buffered

---

[*] The reader should refer to Chapter 6 for a description of FDM.
[†] This corresponds to a frequency above which the fiber stops behaving as a single mode, and starts behaving as a multimode optical fiber.

The loose-tube cable accommodates multiple parallel loose tubes, where each loose tube is employed to accommodate multiple optical fibers. As can be seen from Figure 4.10, the loose-tube optical fiber cable comprises a central tensor to keep the cable straight. Moreover, a mechanical and anti-humidity protection wraps all loose tubes and provides mechanical protection, as well as protection from humidity. Finally, an external jacket surrounds the mechanical and antihumidity protection, and provides another level of protection.

When the most important requirement is mechanical protection instead of capacity, then the preferable optical cable is the tight buffer. As can be seen from Figure 4.11, the level of mechanical protection is higher than that of the loose-tube. This is achieved by sacrificing the number of optical fibers available in a cable. The individual buffer and the individual mechanical and antihumidity protection give extra resistance, as compared to the loose-tube. Note that the buffer is commonly made of silicon.

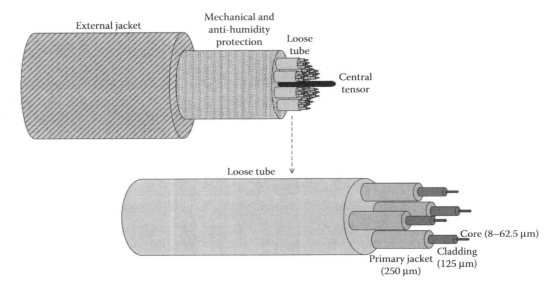

FIGURE 4.10   Loose-tube optical fiber cabling.

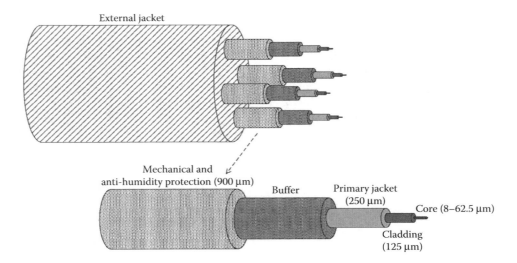

FIGURE 4.11   Tight buffer optical fiber cabling.

4.4 INTERFERENCE PARAMETERS IN METALLIC CONDUCTORS

A generic description of different interference types was given in Chapter 3. Nevertheless, this chapter provides a detailed description of those interferences, when applicable to metallic cable transmission mediums (twisted pairs, coaxial cables, etc.).

It is known from the electromagnetism fundamentals that, because of the Lorentz force, a current (I_1) that crosses an isolated conductor (conductor 1) generates an electromagnetic force around it. Moreover, if another conductor (conductor 2) is placed within this electromagnetic field, another current (I_2, which corresponds to the interfering current) will be generated, as long as there is a variation in time of the magnetic flow density. This variation in time of the magnetic field is experienced when

- Conductor 1 is moving.
- Conductor 2 is moving.
- The current I_1 suffers a variation in time (e.g., alternate current).

This effect is known from the electromagnetism fundamentals as mutual inductance, being plotted in Figure 4.12. In the telecommunications field, this interfering current generated in conductor 2 is referred to as *crosstalk interference.*

Twisted pair cables are normally composed of four pairs. These four pairs are employed to transmit four independent signals, which can be in any direction (transmit or receive). As previously described, the twisted pair may have two different types of protections: shielding or screening, as described in Table 4.2. The shielding refers to a protection around each twisted pair. The shielding provides protection mainly against crosstalk interferences, that is, interferences coming from other twisted pairs of the same cable. On the other hand, screening provides protection mainly from sources external to the twisted cable.

The crosstalk interference increases with the increase of the frequency. Consequently, increasing the throughput, the crosstalk effect also increases. Moreover, the crosstalk interference is also a function of the physical characteristics of the cable, including the type of protection (e.g., quality of shielding or screening), twisting length, and isolation. In addition, installation procedures may also affect the crosstalk properties of a cable, including the radius of a cable curvature, the traction force applied to a cable, mechanical aggression of a cable, and the type of connectors.

A common performance index of crosstalk is the crosstalk attenuation (Att_{dB}), expressed in decibel (dB), and defined as

$$Att_{dB} = 10 \log_{10} \left\{ \frac{P_{received}}{P_{reference}} \right\} \tag{4.9}$$

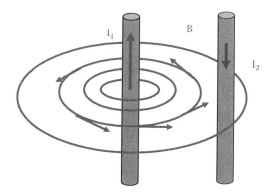

FIGURE 4.12 Mutual inductance effect.

where $P_{reference}$ stands for the transmitted power (the power of the signal that crosses conductor 1 of Figure 4.12).

Having a crosstalk of −3 dB means that the received power is half of the transmitted one. Naturally, as the crosstalk interference increases, the crosstalk attenuation decreases. As previously described, the crosstalk interference degrades the signal-to-noise ratio. In the case of digital signals, this originates a degradation of the bit error rate.

4.4.1 NEAR END CROSSTALK

Let us consider a host connected to a switch in full-duplex mode. Such host needs to use two pairs of conductors: one for transmission and another for reception. However, when the host starts a transmission, depending on the isolation/coupling properties (in conductors, connectors, etc.), a certain amount of the transmitted signal is inducted in the adjacent pair, which is employed to receive another signal in the opposite direction. Naturally, this interfering signal (transmitted signal) is superimposed on the received one, degrading its signal-to-noise ratio. This effect is referred to as near end cross talk (NEXT). Note that the receiver of reference (receiver 2) is located close to the transmitter of reference (transmitter 1). This effect is plotted in Figure 4.13. The NEXT is quantified in decibel, as it refers to the isolation between the transmitted signal and the received one. Metallic conductors of better quality typically correspond to a higher NEXT value.

4.4.2 FAR END CROSSTALK

Let us now consider that the transmission rate is split into two parallel lines. Assuming a 100 Mbps transmission, we could have 50 Mbps being transmitted in each parallel transmission chain. This method is sometimes employed because each pair is limited in terms of bandwidth. Therefore, splitting the transmission rate into two pairs, we use half of the bandwidth. As in the NEXT case, we have part of a transmitted signal in one of the conductors coupled into the other one. Nevertheless, because the other signal is being received far from the transmitter of reference, this effect is referred to as far end crosstalk (FEXT). As shown in Figure 4.14, the transmitter of reference is transmitter 1, whereas the receiver of reference is receiver 2, which is far from the transmitter.

Similar to the NEXT, the FEXT is also quantified in decibel, as it refers to the isolation between the transmitted signal and the received one. Metallic conductors of better quality correspond to a higher FEXT value.

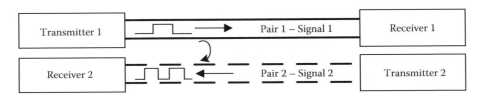

FIGURE 4.13   Near end crosstalk.

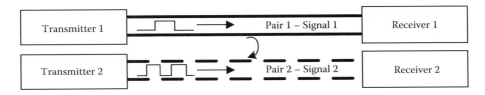

FIGURE 4.14   Far end crosstalk.

It is worth noting that signals that flow into the two parallel conductor pairs can be a certain signal that was split to achieve a lower bandwidth (as in the example above), but these signals can also be two independent signals. In the latter situation, the FEXT is equally applicable.

4.4.3 ATTENUATION TO CROSSTALK RATIO

As shown in Figure 4.13, the receiver of reference is receiver 2, which is where the NEXT effect is experienced. Therefore, the interfering signal is added to a desired signal, degrading the quality of signal 2, that is, degrading its signal-to-noise ratio. Naturally, the signal-to-noise ratio increases with the increase of the NEXT, and with a decrease of the attenuation of the desired signal (signal 2) along the transmission chain. The attenuation to cross talk ratio (ACR) refers to the difference, in decibel, between the NEXT and the attenuation, that is,

$$ACR_{dB} = NEXT_{dB} - Att_{dB} \tag{4.10}$$

As the NEXT decreases and the attenuation increases, the resulting ACR becomes smaller, which corresponds to a signal at the receiver of reference (receiver 2) with a lower signal-to-noise ratio. As can be seen from Figure 4.15, the NEXT decreases with the increase of the frequency, whereas the attenuation increases with the frequency. A negative ACR is not acceptable as it would correspond to a desired signal with a power lower than the interfering signal.

4.4.4 EQUAL LEVEL FAR END CROSSTALK

Let us now focus again on Figure 4.14, where two parallel lines are employed to split a high rate transmission into two smaller rate transmissions. The receiver of reference is receiver 2, which is far from the transmitter of reference (transmitter 1). Along the transmitting chain, the transmitted signal (desired signal) reaches the reference receiver with attenuation. Moreover, it is superimposed with the interfering signal (crosstalk), as quantified by the FEXT. Therefore, the performance index similar to ACR but measured far from the transmitter of reference is referred to as equal level far end crosstalk (ELFEXT). The ELFEXT refers to the difference, in decibel, between the FEXT and the attenuation, that is,

$$ELFEXT_{dB} = FEXT_{dB} - Att_{dB} \tag{4.11}$$

As the FEXT decreases and the attenuation increases, the resulting ELFEXT becomes smaller, which corresponds to a signal at the receiver of reference (receiver 2) with a lower signal-to-noise ratio. Similar to the ACR, a negative ELFEXT is not acceptable as it would correspond to a desired signal with a power lower than the interfering signal.

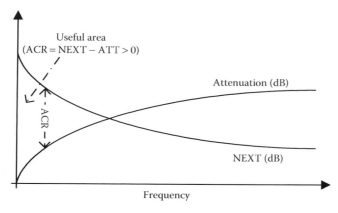

FIGURE 4.15   Attenuation to crosstalk ratio.

4.4.5 OTHER PERFORMANCE INDEXES OF CROSSTALK

Sections 4.4.1 through 4.4.4 describe the most important elementary crosstalk parameters such as NEXT, FEXT, ACR, and ELFEXT. However, a number of other crosstalk parameters exist, being applicable to more complex environments.

Let us consider the 100BASE-T4 physical implementation of an IEEE 802.3 LAN. As described in Chapter 12, this physical implementation comprises four twisted pairs, from which one pair is utilized for transmission, another for reception, while the remaining two can be employed for transmission or for reception, depending on the transmission rates in each direction (upstream and downstream). Let us suppose that we are transmitting with three pairs, whereas the reception is performed with a single pair. In this scenario, the three transmitting pairs generate crosstalk interference in the receiving pair. In the exposed scenario, the cumulative crosstalk interference, measured near the transmitter of reference, is referred to as power sum NEXT (PSNEXT). Note that the PSNEXT does not correspond to the sum of the elementary NEXTs (that would translate in an attenuation level higher than the elementary ones), but its value depends on the cumulative sum of the elementary crosstalk interference powers. Naturally, the resulting PSNEXT is quantified by an attenuation value lower than that of elementary NEXTs. Moreover, the power sum ACR (PSACR) corresponds to the ACR but where the NEXT is replaced by the PSNEXT, that is,

$$PSACR_{dB} = PSNEXT_{dB} - Att_{dB} \tag{4.12}$$

Let us again focus on the 100BASE-T4 physical implementation, and assume the same scenario with three pairs of conductors for transmission and a pair for reception. Let us consider that the reference receiver is one of the three receivers placed far from the transmitter. In this case, the cumulative effect of the crosstalk interference generated by two of the transmitters into the receiver one is referred to as power sum FEXT (PSFEXT). Moreover, the power sum ELFEXT (PSELFEXT) corresponds to the ELFEXT, but where the FEXT is replaced by the PSFEXT, that is,

$$PSELFEXT_{dB} = PSFEXT_{dB} - Att_{dB} \tag{4.13}$$

CHAPTER SUMMARY

This chapter provided a view about the cable transmission mediums, including the study of the most common types of interferences experienced in metallic cable transmission mediums.

The twisted pair was described, including the study of its characteristics, namely types of materials employed in twisted pairs and types of channel impairments experienced, such as attenuation, interferences, noise, and distortion.

This chapter described two types of twisted pair protections: shielding and screening. It was viewed that shielding refers to protection surrounding each pair of conductors, allowing a reduction in the crosstalk interference. It was also described that screening refers to protection surrounding the whole cable, allowing protection from the external noise and interferences.

It was viewed that depending on their characteristics, twisted pair cablings can be grouped into categories, defined from 1 to 7. This is a function of the materials employed in conductors and isolators, the types of protections, the twisting length, among other parameters. These parameters are determinant in the bandwidth made available by cabling.

This chapter introduced the UTP, the STP, and the FTP. It was viewed that the UTP refers to a twisted pair cabling without any type of protection. The STP employs shielding, that is, a protection surrounding each pair of conductors. The FTP employs screening, that is, protection surrounding the whole cable. Finally, it was viewed that the S/FTP, also referred to as S/STP, employs both shielding and screening. Moreover, the layout terminations employed in 8P8C connectors was also studied.

The coaxial cable was also studied in this chapter. The characteristics of coaxial cables were exposed, including types, dimensions, and parameters employed in the two isolated and concentric conductors. It was viewed that, depending on the protection, different bandwidths are made available by the coaxial cabling.

The characteristics of optical fibers were exposed, including the advantages of optical fibers as compared to metallic transmission mediums. It was viewed that optical fibers allow a bandwidth much higher than twisted pairs and coaxial cables. Moreover, the immunity of optical fibers to electromagnetic interference was also described. This results from the fact that optical fibers are typically made of silica glass, which is not a metallic material, and therefore, it does not induct interferences. There are two main types of optical fibers: multimode and single mode. It was described that multimode step index presents higher modal dispersion than multimode graded index. It was also viewed that the single-mode optical fiber presents a lower amount of modal dispersion, which translates in the ability to support a higher bandwidth.

Finally, the most important type of interference parameters relating to metallic cable transmission mediums was studied. It was viewed that the NEXT refers to the attenuation between a transmitted signal and the interference generated in a receiver that is co-located with the transmitter. Moreover, it was viewed that the FEXT refers to the attenuation between a transmitted signal and the interference generated in a receiver that is on the opposite side of the transmitter. It was also viewed that the ACR refers to the difference between the NEXT and the attenuation introduced into the desired signal along the transmission line. The ACR must be positive for the received signal to have a power strength higher than that of interferences.

Finally, it was viewed that the ELFEXT corresponds to the ACR, but where the NEXT is replaced by the FEXT. Therefore, the ELFEXT corresponds to the difference between the FEXT and the attenuation introduced into the desired signal along the transmission line.

REVIEW QUESTIONS

1. What are the advantages of optical fibers relating to twisted pairs?
2. What are the most important cable transmission mediums that you studied?
3. In terms of isolation, which types of twisted pair cabling do you know?
4. Which categories of twisted pair cabling exist?
5. What does NEXT stand for?
6. What is the difference between the FTP and the STP cabling?
7. What is a configuration where the PSELFEXT parameter needs to be taken into account?
8. What is the difference between the STP and the S/STP cabling?
9. What are the key parameters that allow different twisted pair cablings presenting different bandwidths and attenuations?
10. What is the difference between the ACR and the PSACR?
11. What is the relationship between the link distance and the bandwidth of cabling transmission mediums?
12. Which type of optical fiber cable presents a higher mechanical resistance, while sacrificing the capacity (number of optical fibers)?
13. What does ELFEXT stand for?
14. Assuming an optical fiber with a modal bandwidth of 300 MHz km, calculate the maximum bandwidth supported for a link distance of 3 km.
15. Which types of optical fibers do you know?
16. How does a single-mode optical fiber achieve a higher bandwidth than multimode?
17. Which types of multimode optical fibers do you know? Which one presents higher bandwidth and how is it possible?
18. What are the typical dimensions of the components of a multimode optical fiber?
19. What is a configuration where the FEXT parameter needs to be taken into account?
20. What are the typical dimensions of the components of a single-mode optical fiber?

LAB EXERCISES

1. Using the Emona Telecoms Trainer 101 laboratory equipment, and volume 1 of its laboratory manual, perform experiment 11—sampling and reconstruction.
2. Using the Emona Telecoms Trainer 101 laboratory equipment, and volume 1 of its laboratory manual, perform experiment 14—bandwidth limiting and restoring digital signals.
3. Using the Emona Telecoms Trainer 101 laboratory equipment, and volume 4 of its laboratory manual, perform experiment 1—an introduction to fiber optic transmission and reception.
4. Using the Emona Telecoms Trainer 101 laboratory equipment, and volume 4 of its laboratory manual, perform experiment 10—fiber optic bidirectional communications.
5. Using the Emona Telecoms Trainer 101 laboratory equipment, and volume 4 of its laboratory manual, perform experiment 11—WDM.

5 Wireless Transmission Mediums

LEARNING OBJECTIVES

- Identify and describe the different wireless transmission mediums.
- Describe the different wireless propagation modes.
- Define the physical analysis of satellite orbits.
- Describe the different types of satellite orbits.
- Quantify the satellite link budget.
- Describe the terrestrial microwave communication systems.

A typical radio communication system is composed of a transmitter and a receiver that use antennas to convert electric signals into electromagnetic waves and vice versa. These electromagnetic waves are propagated over the air.

In case the communication system is digital, a modem needs to be added at each end of the link. The modem is responsible for modulating and demodulating the bits using a certain modulation scheme (e.g., amplitude shift keying, frequency shift keying, phase shift keying, and quadrature amplitude modulation). In addition, the modem is normally responsible for the implementation of error correction techniques, as well as bit and frame synchronization. The output of the modem in the transmission chain consists of a certain bandpass (carrier modulated) signal (analog) that serves as an input signal to the radio transmitter. The radio transmitter is responsible for centering this signal on a certain carrier frequency and amplifying it before delivery to the transmit antenna. The reverse of these operations is performed in the receiving chain.

Figure 5.1 shows a conventional client–server communication system established over a radio channel (i.e., over the air).

The propagation channel can be of any type, namely using direct wave propagation, groundwave propagation, or ionospheric propagation. Moreover, the communication system can be of any type, namely a radio broadcast, a terrestrial microwave system, a satellite communication system, and so on. This chapter deals with different types of electromagnetic propagation, as well as different types of radio communication systems.

5.1 WIRELESS PROPAGATION

5.1.1 DIRECT WAVE PROPAGATION

Direct wave propagation, also referred to as *space wave propagation*, refers to the propagation between a transmit and a receive antenna that are within the horizon path. In direct wave propagation, the different rays that propagate from a transmit into a receive antenna may travel through line of sight from the transmit into the receive antenna, or may also include reflected, diffracted, and scattered rays. Figure 5.2 depicts the direct wave propagation composed of a line-of-sight component (ray) and a reflected component. In this case, the received signal is composed of the sum of these two independent components.

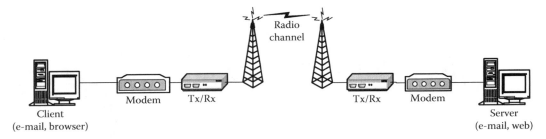

FIGURE 5.1   Conventional client/server communication system established over a radio channel.

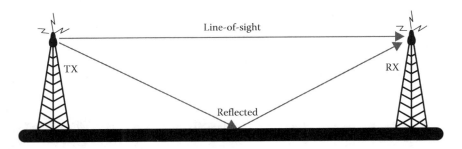

FIGURE 5.2   Direct wave propagation composed of a line-of-sight component and a reflected component.

5.1.1.1   Free Space Path Loss

Free space path loss (FSPL) is defined as the loss in the electromagnetic signal strength due to free space propagation (line of sight) between a transmit and a receive antenna. Before FSPL is deducted, it is worth defining the power spatial density S as a function of the distance d and the transmit power P_E.[*] The power spatial density corresponds to the power per square meter of a surface, measured at a certain distance d. Assuming the spherical propagation of electromagnetic waves, the power spatial density[†] becomes [Carlson 1986]

$$S = P_E \frac{1}{4\pi d^2} \tag{5.1}$$

In the scope of S, it is assumed that the radiated power P_E is isotropic, as a specific antenna gain refers to a specific direction. In this sense, the transmit antenna gain is not taken into account. Note that $4\pi d^2$ is the surface area of the sphere. It is viewed from Equation 5.1 that the transmit power is spread out over the space. Moreover, this space (surface area of the sphere) increases with the distance d.

A common definition normally used in link budget calculations is the equivalent isotropic radiated power (EIRP). It is defined as the amount of power captured by an isotropic receive antenna[‡] at a distance d, assuming a transmit power P_E and a transmit antenna gain g_E (in the direction of the maximum antenna gain):

---

[*] P_E is expressed in watt.
[†] S is expressed in watt/m^2.
[‡] With antenna aperture $A_R = (\lambda^2/4\pi)$ in all directions.

$$\text{EIRP} = g_E S \frac{\lambda^2}{4\pi}$$

$$= P_E g_E \left(\frac{\lambda}{4\pi d} \right)^2 \tag{5.2}$$

where λ^* stands for the wavelength.

The FSPL refers to the losses between the radiated power $P_E g_E$ and the EIRP at a certain distance d, which becomes

$$\text{FSPL} = \left(\frac{4\pi d}{\lambda} \right)^2 \tag{5.3}$$

Note that $A_R(r)$ (with $A_R(r) = g_R(r)(\lambda^2/4\pi) \Leftrightarrow g_R(r) = 4\pi(A_r(r)/\lambda^2)$) corresponds to the receiving antenna aperture in the direction of the emitter (r), relating to an isotropic antenna, whereas $g_R(r)$ corresponds to the receiving antenna gain in the direction of the emitter (r), relating to an isotropic antenna.

Using the approximation that electromagnetic waves propagate at the speed of light c in vacuum ($c \simeq 3 \times 10^8$ m/s), and using the equivalence $c = \lambda f$, the FSPL becomes [Carlson 1986]

$$\text{FSPL} = \left(\frac{4\pi d f}{c} \right)^2 \tag{5.4}$$

From Equation 5.4, we conclude that the path loss is proportional to the square of the distance d and the square of the frequency f.

Because link budget calculations are normally performed in logarithmic units (decibel), instead of linear units, it is worth expressing FSPL in decibel as follows:

$$\text{FSPL}_{\text{dB}} = 10 \log_{10} \text{FSPL}$$

$$= 10 \log_{10} \left[\left(\frac{4\pi d f}{c} \right)^2 \right]$$

$$= 20 \log_{10} \left(\frac{4\pi d f}{c} \right) \tag{5.5}$$

$$= 20 \log_{10} \left(\frac{4\pi}{c} \right) + 20 \log_{10} (d) + 20 \log_{10} (f)$$

$$= -147.55 + 20 \log_{10} (d) + 20 \log_{10} (f)$$

Expressing d in km and f in MHz, Equation 5.5 becomes

$$\text{FSPL}_{\text{dB}} = 32.45 + 20 \log_{10} (d)_{\text{km}} + 20 \log_{10} (f)_{\text{MHz}} \tag{5.6}$$

5.1.1.2 Link Budget Calculation

Assuming free space propagation, and knowing that the received power corresponds to the EIRP multiplied by the receive antenna gain g_R, according to the Friis formula, the received power becomes [Carlson 1986]

---

* λ is expressed in m.

$$P_R = \text{EIRP} \cdot g_R$$
$$= g_E S \frac{\lambda^2}{4\pi} g_R \tag{5.7}$$
$$= P_E g_E \left(\frac{\lambda}{4\pi d} \right)^2 g_R$$

Alternatively, we could express the received power as a function of the frequency as follows:

$$P_R = P_E g_E \left(\frac{c}{4\pi f d} \right)^2 g_R \tag{5.8}$$

Another way to compute the received signal power in logarithmic units is as follows:

$$\left(P_R \right)_{\text{dBW}} = 10 \log_{10} \left[P_E g_E \left(\frac{c}{4\pi f d} \right)^2 g_R \right] \tag{5.9}$$
$$= \left(P_E \right)_{\text{dBW}} + G_E + G_R + 147,56 - 20 \log_{10} f - 20 \log_{10} d$$

where $\left(P_E \right)_{\text{dBW}}$ and $\left(P_R \right)_{\text{dBW}}$ correspond to P_E and P_R expressed in dBW, respectively.[*]

Moreover, in Equation 5.9, G_E and G_R are the transmitting and receiving antenna gains expressed in decibel, that is, $G_E = 10 \log_{10} \left(g_E \right)$ and $G_R = 10 \log_{10} \left(g_R \right)$, respectively (Figure 5.3).

Expressing d in km and f in MHz, and using Equation 5.6, Equation 5.9 becomes

$$\left(P_R \right)_{\text{dBW}} = \left(P_E \right)_{\text{dBW}} + G_E + G_R - \left(\text{FSPL} \right)_{\text{dB}} \tag{5.10}$$
$$= \left(P_E \right)_{\text{dBW}} + G_E + G_R - 32.45 - 20 \log_{10} \left(d \right)_{\text{km}} - 20 \log_{10} \left(f \right)_{\text{MHz}}$$

From Equation 5.10, it is clear that the increase in the carrier frequency leads to a reduced received power that results from an increased path loss (see Equation 5.6). This limitation can be overcome by using transmit and receive antennas with higher gains.

The link budget calculation of a real system should also include additional losses (such as cable losses and connector losses). Taking these parameters into account, Equation 5.10 becomes

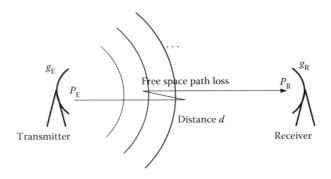

FIGURE 5.3 Generic diagram of a communication system with a line-of-sight propagation.

---

[*] A dBW is the decibel relating to the watt, that is, $\left(P_E \right)_{\text{dBW}} = 10 \log_{10} P_E$, where P_E is expressed in watt. Similarly, a dBm is the decibel relating to the milliwatt, that is, $\left(P_E \right)_{\text{dBm}} = 10 \log_{10} P_E$, with P_E expressed in milliwatt.

$$\left(P_R\right)_{dBW} = \left(P_E\right)_{dBW} + G_E + G_R + Att_{dB}$$

$$= \left(P_E\right)_{dBW} + G_E + G_R + \left[\underbrace{-32.45 - 20\log_{10}\left(d\right)_{km} - 20\log_{10}\left(f\right)_{MHz} + Att_{add\_dB}}_{Att_{dB}}\right] \quad (5.11)$$

where Att_{dB} is a negative value, defined as the sum of all elementary attenuations in decibel, namely the path loss and the additional attenuations. Moreover, Att_{add_dB} is also a negative value that stands for the additional attenuations (cable losses plus connector losses, rain attenuation, etc.), expressed in decibel. Note that the FSPL does not take into account any effect such as shadowing or multipath (i.e., diffraction, reflection, and scattering effects are not considered), as the current description refers to free space propagation. The FSPL has a distance decay rate of 2,[*] whereas in real scenarios, this value varies between 3 and 5. Rural scenarios present typically a decay rate of the order of 3, whereas the decay rate varies between 4 and 5 for urban scenarios, depending on the shadowing effects, multipath propagation scenario, and so on. Therefore, depending on the propagation environment, the received power calculation in Equation 5.11 should also take this modified path loss parameter into account.

The previous description was made taking into account free space propagation, that is, including only the line-of-sight component, and reflected, diffracted, and scattered wave components were not considered. Moreover, antenna heights were also not taken into account in the calculations. Therefore, assuming the flat earth (i.e., neglecting the earth's curvature), but that the received signal is composed of a line-of-sight component, to which one reflected path in the ground is summed (as depicted in Figure 5.2), and taking into account the antenna heights, Equation 5.11 becomes [Parsons 2000]

$$\left(P_R\right)_{dBW} = \left(P_E\right)_{dBW} + G_E + G_R + 20\log_{10}\left(h_E\right)_m + 20\log_{10}\left(h_R\right)_m$$
$$- 120 - 40\log_{10}\left(d\right)_{km} + Att_{add\_dB} \quad (5.12)$$

where $\left(h_E\right)_m$ and $\left(h_R\right)_m$ stand for the transmitting and receiving antenna heights (in meters), respectively. As in Equation 5.11, Att_{add_dB} is a negative value that stands for the additional attenuations (cable losses, connector losses, rain attenuation, etc.) in decibel.

Alternatively, we can express Equation 5.12 in linear units as follows:

$$P_R = \frac{P_E g_E g_R Att_{add}\left(h_E h_R\right)^2}{d^4} \quad (5.13)$$

Note that the flat earth model used for Equations 5.12 and 5.13 can be considered for short distances d such that $(10 \cdot h_E \cdot h_R)/\lambda < d < d_h$, where the radio horizon d_h is approximated by $d_h \simeq \sqrt{2r_E h}$ (r_E stands for the earth's radius, $r_E = 6373$ km). For distances $d > d_h$, the earth's curvature also needs to be taken into account in the calculations and, therefore, expression 5.12 loses validity.

It is worth noting that, taking into account one reflected ray, the received power strength presents a decay rate of 4 with the distance (see Equation 5.13), whereas in the free space model, the received power presents a decay rate of 2 (see Equation 5.8). Consequently, one can conclude that the presence of reflected waves represents a negative effect in the received signal strength. Establishing a link over a ground that presents bad reflection properties makes the negative effect of reflected waves less visible, reducing the decay rate of the received power strength with the distance.

---

[*] That is, the FSPL increases linearly with the increase of the square of the distance.

For a receiver to be able to decode a signal, two important conditions must be achieved:

- The received power strength (defined by Equation 5.8 or 5.11) must be higher than the receiver's sensitivity threshold.
- The E_b/N_0 level of the received signal must be higher than the minimum required for the service being transported. As an example, for voice, the bit error rate should be lower than 10^{-3}. From the graphic depicted in Figure 3.2, we extract that the E_b/N_0 level should be higher than 7 dB.[*]

5.1.1.3 Carrier-to-Noise Ratio Calculation

Assuming free space propagation, from Equation 5.8, and taking the power of thermal noise into account (as defined in Chapter 3), the carrier-to-noise ratio (C/N) becomes [Carlson 1986]

$$\frac{C}{N} = \frac{\text{EIRP} \cdot g_R \cdot A_{tt}}{k_B T_n B} \tag{5.14}$$

where:
k_B is the Boltzmann constant
T_n is the resistor's absolute temperature (expressed in kelvin)
B is the receiver's bandwidth.

In Equation 5.14, the carrier-to-noise ratio is $(C/N) = (P_R/P_N)$, with the power of the carrier $C = P_R = \text{EIRP} \cdot g_R \cdot A_{tt}$, and where $N = P_N = k_B T_n B$ stands for the power of noise at the receiver (see Chapter 3). Note that the C/N differs from the signal-to-noise ratio (SNR) because the former refers to the power of a modulated carrier, whereas the latter refers to the signal power after carrier demodulation. The conversion between C/N and SNR depends on the modulation scheme under consideration [Carlson 1986]. In fact, the SNR is normally the performance indicator adopted in analog communications, whereas in digital communications the performance indicator considered is E_b/N_0.

Still assuming free space propagation, we may express Equation 5.14, in logarithmic units, as follows:

$$\left(C/N\right)_{dB} = 10\log_{10}\left(\frac{P_R}{P_N}\right)$$

$$= \left(\text{EIRP}\right)_{dBW} + \left(G_R/T_n\right)_{dB} - 10\log_{10}\left(k_B B\right) + A_{tt\,dB} \tag{5.15}$$

where $\left(G_R/T_n\right)_{dB}$ is expressed in decibel, which stands for the receiver's merit factor, defined as follows [Carlson 1986; Ha 1990]:

$$\left(G_R/T_n\right)_{dB} = G_R - 10\log_{10} T_n \tag{5.16}$$

Moreover, in Equation 5.15, the attenuation $A_{tt\,dB}$ is a negative value. Because A_{tt} is a coefficient lower than 1, its logarithmic value $A_{tt\,dB}$ becomes negative.

Entering with the Boltzman constant $k_B = 1.3806503 \times 10^{-23}$ J/K, as defined in Chapter 3, Equation 5.15 can be rewritten as follows:

$$\left(C/N\right)_{dB} = \left(\text{EIRP}\right)_{dBW} + \left(G_R/T_n\right)_{dB} - 10\log_{10}\left(B\right) + A_{tt\,dB} + 228.6 \tag{5.17}$$

The antenna gain is defined as follows [Burrows 1949]:

$$g = \eta D \tag{5.18}$$

---

[*] dB stands for decibel.

where:

  η stands for the antenna performance
  D for the antenna directivity

In the case of a parabolic antenna, D becomes [Burrows 1949]

$$D = \left(\frac{\pi d_a}{\lambda} \right)^2 \tag{5.19}$$

with d_a the parabolic antenna (dish) diameter. In this case, the antenna gain becomes [Burrows 1949]

$$g = \eta \left(\frac{\pi d_a}{\lambda} \right)^2 \tag{5.20}$$

5.1.1.4 Bit Error Probability Calculation

Taking the C/N values expressed by Equation 5.14, we can calculate the bit error probability (P_e)* for M-ary phase shift keying (M-PSK) modulation,[†] valid for $M > 2$, as follows [Proakis 1995]:

$$P_e = \frac{1}{\log_2 M} \mathrm{erfc} \left(\sin \left(\frac{\pi}{M} \right) \sqrt{\frac{C}{N}} \right) \tag{5.21}$$

where erfc is the complementary error function.

  We can also express Equation 5.21 for M-ary PSK (valid $M = 2$ or 4, i.e., for BPSK or QPSK) as a function of E_b/N_0 [Proakis 1995],

$$P_e = Q \left(\sqrt{\frac{2E_b}{N_0}} \right) \tag{5.22}$$

where:

  E_b stands for the bit energy
  N_0 for the power spectral density of noise[‡]

Similarly, the bit error probability for M-ary PSK, valid for $M > 2$, is approximated by [Proakis 1995]

$$P_e \approx \frac{2}{\log_2(M)} Q \left(\sqrt{2\log_2(M)\sin^2\left(\frac{\pi}{M} \right) \frac{E_b}{N_0}} \right) \tag{5.23}$$

and for M-QAM (quadrature amplitude modulation) or M-PAM (pulse amplitude modulation) by [Proakis 1995]

$$P_e \approx \frac{2}{\log_2(M)} \left(1 - \frac{1}{M} \right) Q \left(\sqrt{\frac{6\log_2(M)}{M^2-1} \cdot \frac{E_b}{N_0}} \right) \tag{5.24}$$

---

[*] The bit error probability P_e is also known as bit error rate.

[†] The modulation schemes are detailed in Chapter 6.

[‡] In fact, N_0 in Equation 5.22 refers to $N_0 + I_0$, that is, the sum of the power spectral density of noise and the power spectral density of interferences, as long as the power spectral density of interference has a Gaussian behavior. For the sake of simplicity, in the system description of this chapter it is only assumed N_0.

In Equations 5.22 through 5.24, the bit energy E_b becomes [Carlson 1986]

$$E_b = P_R \cdot T_B$$

$$= \frac{P_R}{R_B} \tag{5.25}$$

where:
 R_B is the transmitted bit rate
 T_B is the transmitted bit period

In the case of M-ary modulation, this corresponds to $R_B = R_S \cdot \log_2 M$, where R_S stands for the transmitted symbol rate and $\log_2 M$ for the number of bits transported in each symbol (M stands for the number of constellation points in the modulation scheme). The power spectral density of noise is [Carlson 1986]

$$N_0 = \frac{P_N}{B} \tag{5.26}$$

$$= k_B T_n$$

From Equations 5.25 and 5.26, and knowing that $C = P_R$ and $N = P_N$, the relationship between E_b/N_0 and C/N is deduced as follows:

$$\frac{E_b}{N_0} = \frac{C \cdot T_b}{N/B}$$

$$= \frac{C \cdot T_b \cdot B}{N} \tag{5.27}$$

$$= \frac{C}{N} \cdot \frac{B}{R_b}$$

where B/R_b stands for the inverse of the minimum spectral efficiency.[*]

5.1.2 WIRELESS PROPAGATION EFFECTS

As depicted in Figure 5.4, a received electromagnetic wave can be viewed as a result of several propagation effects, namely line of sight, reflection, diffraction, and scattering. When multiple components are present, the received signal is composed of the sum of all these components. In the following, each of these propagation effects is characterized.

5.1.2.1 Reflection

Reflection consists of a change in the direction of propagation of a wave as a result of a collision into a surface. Electromagnetic waves are typically reflected in buildings, vehicles, or streets.

As can be seen from Figure 5.5, the reflected wave presents the same angle as the incident wave (relating to the normal of the surface).

An electromagnetic wave progresses in three perpendicular axes. The electric field progresses in one of the axes, and the magnetic field progresses in another, perpendicular to the first. Finally, the third axis refers to the direction of propagation of a wave. When the electric field has vertical

---

[*] From the Nyquist ISI criterion, the spectral efficiency is $\varepsilon = (R_b/B) = ((2\log_2 M)/(1+\alpha))$. The minimum spectral efficiency is achieved for $\alpha = 0$, leading to $\varepsilon_{min} = 2\log_2 M = (1/((1/2T_s)T_b)) = (2T_s/T_b)$.

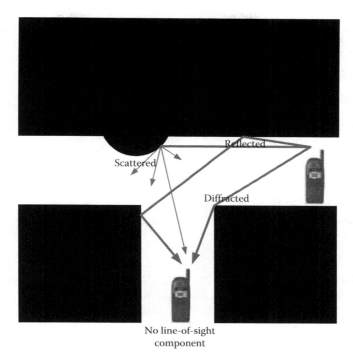

FIGURE 5.4 Example of the propagation environment with diffracted, reflected, and scattered waves.

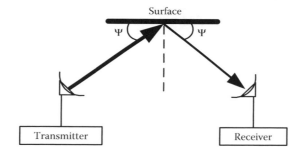

FIGURE 5.5 Reflection effect.

polarization (and consequently, the magnetic field is horizontal), it is said that the wave presents vertical polarization. On the other hand, when the electric field has horizontal polarization (and consequently, the magnetic field is vertical), it is said that the wave presents horizontal polarization. This can be seen from Figure 5.6. Finally, a wave can also present an oblique polarization, having both electric and magnetic fields with vertical and horizontal components.

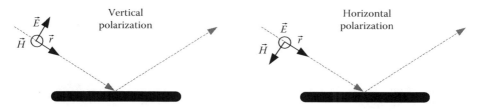

FIGURE 5.6 Vertical and horizontal polarizations.

The intensity of the reflected wave ($\vec{E}_{\text{REFL}}$ and $\vec{H}_{\text{REFL}}$) depends on the intensity of the incident wave ($\vec{E}_{\text{INCID}}$ and $\vec{H}_{\text{INCID}}$) and of the Fresnel coefficient (Γ_{H} and Γ_{V}) as follows:

$$\vec{E}_{\text{REFL}} = \vec{E}_{\text{INCID}} \Gamma_{\text{H}} \tag{5.28}$$

for the horizontally polarized electromagnetic waves with $\vec{E}_{\text{INCID}} = E_{\text{TX}}(e^{-jkd}/d)^*$ and as

$$\vec{H}_{\text{REFL}} = \vec{H}_{\text{INCID}} \Gamma_{\text{V}} \tag{5.29}$$

for the vertically polarized electromagnetic waves with $\vec{H}_{\text{INCID}} = H_{\text{TX}}(e^{-jkd}/d)$.

The power spatial density of an electromagnetic wave depends on both the electric and magnetic fields as follows:

$$S = \frac{1}{2}|E||H|$$
$$= P_E g_E \frac{1}{4\pi d^2} \tag{5.30}$$

As the incident components of both electric and magnetic fields depend inversely on the distance d, the power spatial density depends inversely on the square of the distance d^2. It is worth noting that the amplitudes of electric and magnetic fields are related through

$$E = Z \cdot H \tag{5.31}$$

where Z stands for the wave's impedance.

The wave's impedance in the vacuum is quantified as $Z_0 = 120\pi\ \Omega$.

Let us define the Fresnel coefficients as a function of the reflection index n of the surface (ground). The Fresnel coefficient for the horizontal polarization becomes [NBS 1967]

$$\Gamma_{\text{H}} = \frac{\vec{E}_{\text{REFL}}}{\vec{E}_{\text{INCID}}}$$
$$= \frac{\sin\Psi - \sqrt{n^2 - \cos^2\Psi}}{\sin\Psi + \sqrt{n^2 - \cos^2\Psi}} \tag{5.32}$$

and the Fresnel coefficient for the vertical polarization is given by [NBS 1967]

$$\Gamma_{\text{V}} = \frac{\vec{H}_{\text{REFL}}}{\vec{H}_{\text{INCID}}}$$
$$= \frac{\vec{E}_{\text{NORMAL\_REFL}}}{\vec{E}_{\text{NORMAL\_INCID}}} \tag{5.33}$$
$$= \frac{n^2\sin\Psi - \sqrt{n^2 - \cos^2\Psi}}{n^2\sin\Psi + \sqrt{n^2 - \cos^2\Psi}}$$

where it was assumed that the reflected direction is in the same plane as the incident direction.

The reflection index n of the surface (ground) is given by [NBS 1967]

$$n = \sqrt{\frac{\varepsilon'_G}{\varepsilon_0}} \tag{5.34}$$

---

* Spherical wave is considered in this formulation.

where:

ε'_G stands for the dielectric constant of the ground defined as $\varepsilon'_G = \varepsilon_G - j(\sigma_G/\omega)$

ε_0 stands for the dielectric constant of the air

Furthermore, ε_G is the ground permittivity (real part of the dielectric constant), σ_G stands for the ground conductivity (imaginary part of the dielectric constant), and ω is the angular speed defined as $\omega = 2\pi f = (2\pi/T)$ (f is the frequency of the electromagnetic wave and T its period). Note that the real part of the reflection index n is responsible for the reflection, whereas its imaginary part is responsible for the absorption of electromagnetic waves by the ground surface. Therefore, the part of the energy subject to absorption is the amount not subject to reflection, and vice versa.

We may also express Equation 5.34 as follows [Burrows 1949]:

$$
\begin{aligned}
n &= \sqrt{\frac{\varepsilon_G - j(\sigma_G/\omega)}{\varepsilon_0}} \\
&= \sqrt{\frac{\varepsilon_G \left(1 - j(\sigma_G/\omega\varepsilon_G)\right)}{\varepsilon_0}} \\
&= \sqrt{\frac{\varepsilon_G}{\varepsilon_0}\left(1 - j(\sigma_G/\omega\varepsilon_G)\right)} \\
&= \sqrt{\varepsilon_R \left(1 - j\cdot\mathrm{tg}(\delta)\right)}
\end{aligned}
\tag{5.35}
$$

where:

ε_R is the relative permittivity of the ground (relating to the air)

δ is the component responsible for the phase shift

From Equations 5.32, 5.33, and 5.35, we conclude that the Fresnel coefficients are a function of the incident angle Ψ, of the surface's characteristics (i.e., dielectric constant), and of the frequency f. Figure 5.7 depicts the reflection coefficient as a function of the incidence angle Ψ and for different frequencies f, for the seawater. Note that, in Figure 5.7, $R_V = |\Gamma|$ and $\arg(\Gamma) = \pi - C_V \Leftrightarrow C_V = \pi - \arg(\Gamma)$. Different curves for different reflection materials can be found in NBS [1967].

Focusing on Figure 5.7, where the flat earth model was considered, the received electric field E_{REC} is composed of the sum of the line-of-sight component and a reflected component as follows:

$$
\begin{aligned}
\vec{E}_{REC} &= \vec{E}_{DIR} + \vec{E}_{REFL} \\
&= \vec{E}_{DIR} + \vec{E}_{DIR} \cdot \Gamma
\end{aligned}
\tag{5.36}
$$

Assuming that the distance between the two antennas is sufficiently high, Equation 5.36 becomes

$$
\begin{aligned}
\vec{E}_{REC} &\simeq E_{TX}\frac{e^{-jkr_{DIR}}}{r_{DIR}} + E_{TX}\frac{e^{-jkr_{REFL}}}{r_{REFL}}\Gamma \\
&\simeq E_{DIR}\left(1 + \Gamma \cdot e^{-jk\Delta r}\right)
\end{aligned}
\tag{5.37}
$$

where the propagation constant is $k = 2\pi/\lambda$, and Δr is the difference between the reflected path distance r_{REFL} and the direct path distance r_{DIR}, that is, $\Delta r = r_{DIR} - r_{REFL}$. Moreover, in Equation 5.37, the direct wave electric field is $E_{DIR} = E_{TX}(e^{-jkr_{DIR}}/r_{DIR})$ and E_{TX} stands for the electric field measured at 1 m from the transmit antenna, defined as $E_{TX} = \sqrt{30 P_E G_E}$. Note that we have assumed in Equations 5.36 and 5.37 that the transmitting antenna gain G_E is the same in the direction of the direct path and of the reflected path.

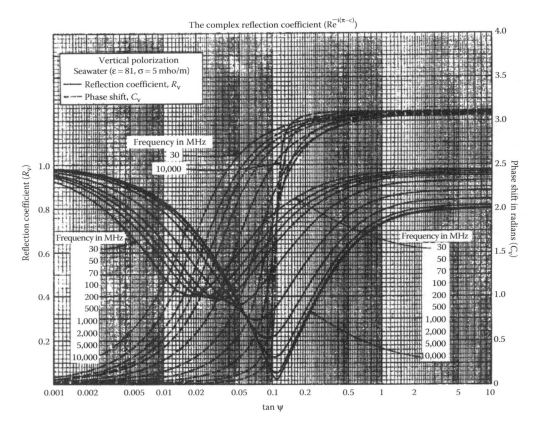

FIGURE 5.7 Reflection coefficients as a function of the different angles and frequencies for the seawater. (Data from Rice, P. L. et al., *Transmission Loss Predictions for Tropospheric Communication Circuits*, NBS Technical Note 101, Vol. II, U.S. Department of Commerce, National Bureau Standards, Washington, DC, 1967.)

For distances sufficiently high, and assuming that the refraction index is approximated by $\Gamma = -1$, Equation 5.37 can be approximated by [Parsons 2000]

$$\vec{E}_{REC} \simeq 2E_{DIR} \left| \sin \left(k \cdot h_E \cdot h_R / d \right) \right| \qquad (5.38)$$

Note that Equation 5.38 corresponds to Equation 5.13 with the difference that the Equation 5.38 refers to the electric field strength whereas Equation 5.13 corresponds to the received power strength. Using Equation 5.30, one can easily convert one into another.

Figure 5.8 depicts the received electric field strength using Equation 5.38, as being composed of the sum of the direct component (line of sight) and one reflected component, for distances between the transmit and the receive antenna between 1 m and 10 m, considering the following parameters: 60 MHz ($\lambda = 5$ m); $\Gamma = -1$ (this parameter is already included in the deduction of Equation 5.38); $h_E = h_R = 10$ m. As can be seen, the resulting received electric field strength fluctuates with the distance as a function of the interference between the direct and reflected components. Note that the type of interference between direct and reflected waves alternates between constructive and destructive. This type of interference is known as multipath interference. It is seen that the difference between two consecutive maximums (constructive interference) and consecutive minimums (destructive interference) corresponds to $\Delta r = r_{REFL} - r_{DIR} = n\lambda$. In fact, consecutive maximums and minimums occur at path variation $\Delta r = n\lambda/2$, where Δr depends on the distance d and on the antenna heights (h_E and h_R).

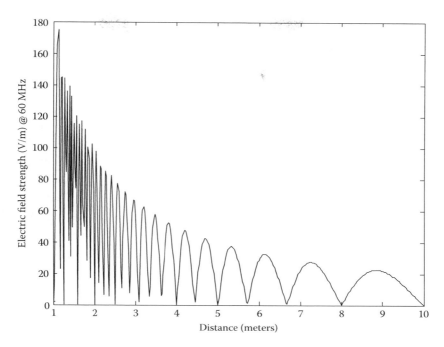

FIGURE 5.8 Plot of the received field strength for distances between 1 m and 10 m ($\Gamma = -1$ and $h_E = h_R = 10$ m).

For even values of n, destructive interference occurs between direct and reflected waves and, for odd values of n, constructive interference occurs between those component waves. Moreover, as can be seen from Equation 5.38, increasing the antenna heights, or decreasing the distance, the amplitude of the maximums and minimums increases, that is, the link becomes more subject to multipath interference.

Considering the direct component and one reflected wave, the decay rate of the electric field strength envelope with the distance is 2. This corresponds to a power decay rate 4 of the received envelope with the distance, as viewed from Equation 5.13.

Figure 5.9 depicts the received electric field strength using Equation 5.38, in the same conditions as for Figure 5.8, but for distances between 10 m and 500 m. As before, the field strength decreases with the distance, but the signal fluctuations stop for distances beyond a certain value (in this scenario, beyond around 50 m).

The level of interference generated by the reflected signal depends on several factors such as the antenna directivity. The use of an antenna with low antenna gain in the direction of the reflected wave reduces the level of interferences and, consequently, the signal fluctuations caused by fading, as well as the decay rate of the received signal strength with the distance. Alternatively, the selection of a path that blocks the reflected wave also leads to a reduction in the level of interference, reducing the decay rate, as well. Moreover, for longer distances between a transmit and a receive antenna, the signal fluctuations tend to decrease, but the level of attenuation tends to be higher than that in free space propagation (i.e., only direct path). Transmitting electromagnetic waves over a soil that presents low refraction index also leads to a low level of interference between the direct and reflected path, and a lower decay rate. Finally, using diversity such as MIMO systems avoids the fading effects, improving very much the performance of communications.

5.1.2.2 Diffraction

Diffraction occurs when a wave faces an obstacle that does not allow it reaching the receive antenna in line of sight. In this case, even in the absence of line of sight, a bending effect of waves is experienced, allowing the waves reaching the receive antenna, but properly attenuated (Figure 5.10).

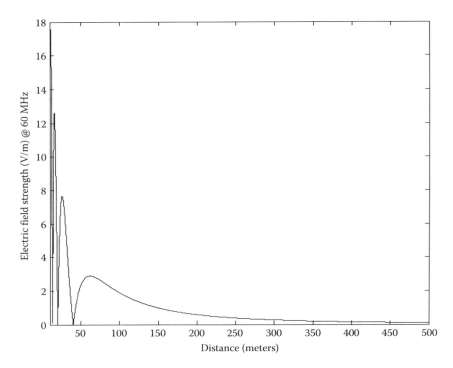

FIGURE 5.9   Plot of the received field strength for distances between 10 m and 500 m ($\Gamma = -1$ and $h_{\rm E} = h_{\rm R} = 10$ m).

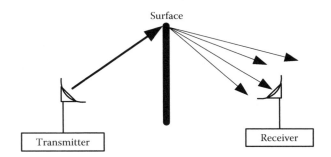

FIGURE 5.10   Diffraction effect.

This phenomenon is normally quantified using the knife-edge model. The knife-edge model considers a semi-infinite plan, located between a transmit and a receive antenna, in the object's position (Figure 5.11).

To calculate the level of attenuation introduced by a semi-infinite plan, let us focus on the geometry depicted in Figure 5.12.

First, the value $\bar{x}$ is defined as follows [Parsons 2000]:

$$\bar{x} = \frac{x_{\rm E} d_{\rm R} + x_r d_{\rm E}}{d_{\rm R} + d_{\rm E}} \tag{5.39}$$

Note that, in the example depicted in Figure 5.12, $\bar{x}$ is placed below the semi-infinite plan. This means that $\bar{x}$ is negative.

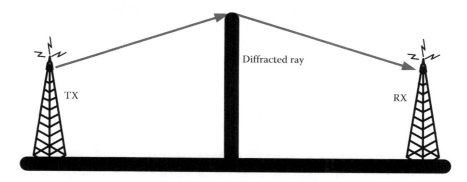

FIGURE 5.11 Propagation path between a transmit and a receive antennas achieved through diffraction.

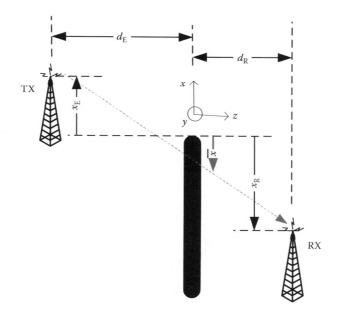

FIGURE 5.12 Geometry of the knife-edge model.

The equivalent height h_E is defined as follows [Parsons 2000]:

$$h_E = \sqrt{k \frac{(d_E + d_R)}{\pi d_E d_R} \bar{x}}$$

$$= \sqrt{\frac{2\pi}{\lambda} \frac{(d_E + d_R)}{\pi d_E d_R}} \bar{x} \qquad (5.40)$$

$$= \sqrt{\frac{2\pi f}{c} \frac{(d_E + d_R)}{\pi d_E d_R}} \bar{x}$$

Taking the value for h_E, we are now in a position to calculate the level of attenuation using the Euler formula as follows [Parsons 2000]:

$$A(h_E) = \frac{1}{2}\left[\frac{1}{2} + C(h_E)\right]^2 + \frac{1}{2}\left[\frac{1}{2} + S(h_E)\right]^2 \qquad (5.41)$$

where $S(x)$ and $C(x)$ stand for the Fresnel sine integral and Fresnel cosine integral functions, respectively.

We may conclude that the received signal level increases with the decrease in the carrier frequency, the increase of the horizontal distance between the receive antenna and the semi-infinite plan (which represents the obstacle), and with the decrease in the depth of the receive antenna.

This model is very useful in many different scenarios. One common application of this model is when one wants to quantify the attenuation introduced by an obstacle or by the earth's curvature in a microwave line-of-sight link.

5.1.2.3 Scattering

Scattering occurs when a wave is reflected by an obstacle that is not flat. Because the incident wave covers a certain area (a group of points in the surface), and because each point of this area has a different normal to the obstacle, the scattering effect corresponds to an amount of successive reflections, each at each point of the surface's obstacle covered by the incident wave (Figure 5.13).

Because of the high complexity of this phenomenon, its characterization here is not dealt with. Nevertheless, a detailed description of this phenomenon can be found in Parsons [2000].

5.1.3 FADING

In mobile communications, the channel is one of the most limiting factors for achieving a reliable transmission. Fading is characterized by random variations of the received signal level. This is caused by several factors, such as atmospheric turbulences, movement of the receiver or the transmitter, movement of the environment that surrounds the receive antenna, and variations of the atmospheric refraction index.

Depending on the depth of the received signal fluctuations, and the reasons of these fluctuations, there are two types of fading (Figure 5.14):

- Slow fading (shadowing)
- Fast fading (multipath fading)

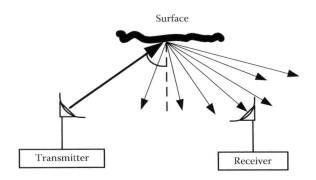

FIGURE 5.13 Scattering effect.

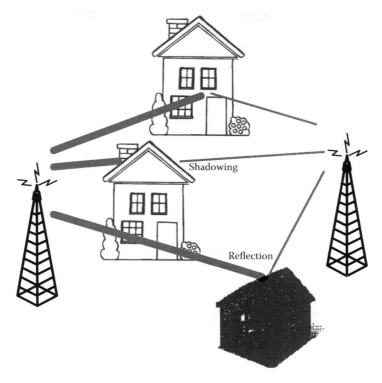

FIGURE 5.14   Shadowing and multipath effects.

Slow fading is characterized by slow variations of the received signal level, which is mainly caused by the terrain contour between the transmitter, and the receiver, which is directly related to the presence of obstacles in the signal path that avoid line-of-sight propagation. This effect can be compensated for with power control schemes.

Fast fading is characterized by fast variations of the received signal level, being caused by the reflection of the signal in various objects (buildings, trees, vehicles, etc.) that originate multiple replicas of the signal reaching the receive antenna through different paths. These replicas arrive with different delays and attenuations, superimposed in such a way that they will interfere with each other, either constructively or destructively. Because of the mobility of the transmitter or receiver, and of the surrounding objects, the replicas are subject to variations on their paths, and hence in their delays and attenuations, leading to great oscillations on the envelope of the received signal.

Because multiple replicas of the signal arrive with different delays, there will be a temporal dispersion of the received signal. This means that if a Dirac impulse is transmitted, the received signal will have a noninfinitesimal duration, that is, the received signal shape will not be of impulsive type. This temporal dispersion can be represented using a power delay profile (PDP), $P(\tau)$, which represents the average received power as a function of the delay τ. Figure 5.15 shows an example of a PDP. Summarizing, there are two main types of fading:

* Shadowing fading
* Multipath fading

Depending on the depth of received power fluctuations and fluctuation rate, there are two statistical models and types of fading, defined in the following [Fernandes 1996].

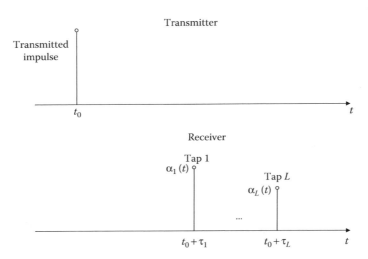

FIGURE 5.15   Discrete impulsive response of a multipath channel.

5.1.3.1   Shadowing Fading

Shadowing fading (slow fading) is characterized by slow variations of the received signal level. This is caused by an obstruction of the line of sight caused by one or multiple objects.

The factors that influence the depth of these slow signal variations are as follows:

- The movement of the receiver (although in a lower scale than the variation caused by the multipath).
- The nature of the terrain.
- The nature, density, and orientation of buildings (or other obstacles), as well as width and orientation of the streets.

This effect is experienced when there is no direct line of sight between the transmitter and the receiver, and therefore the propagation is characterized by diffraction.[*] Figure 5.14 shows the shadowing effect caused by a building between the transmitter and the receiver. The average value of the received signal level follows a log-normal distribution[†] (the logarithm of the amplitude of the field follows a normal distribution).

Higher attenuations have been experienced in urban zones with higher building densities. The standard deviation σ increases with

- Increase in the considered area
- Increase in the building proportions
- Increase in the carrier frequency

Typical values for σ, in cellular environments, are between 6 dB and 18 dB.

It is worth noting that the variation of the refraction index, ducts, rain, and fog may originate similar effects as those described as shadowing fading.

---

[*] The diffraction effect is normally quantified using the knife-edge model previously introduced.
[†] A log-normal distribution is defined by Proakis [1995] as $f(x) = (1/\sigma\sqrt{2\pi})(1/x)\exp\left[-(1/2)\left((\ln(x)-\mu)/\sigma\right)^2\right]$.

5.1.3.2 Multipath Fading

Multipath fading (fast fading) is characterized by fast changes of the received signal level. This is caused by variations such as turbulence of the local atmosphere, variation of the distance between the transmitter and the receiver, and variations of the environment surrounding the receiver. This type of fading depends on the carrier frequency, the environment surrounding the antennas, and so on.

The above-mentioned causes of fast fading originate variations of the interferences, namely constructive or destructive interferences between different propagation paths (line-of-sight, reflected, diffracted, and scattered).*

When the receiver moves a distance corresponding to about one wavelength, or when the environment surrounding the receiver moves, the intensity of the received signal experiences a deep fading of the order of 30–40 dB.

Figure 5.15 shows an example of an impulsive response of a multipath channel. In this figure, t_0 corresponds to the instant of transmission (of a pulse), and $t_0 + \tau_1$ and $t_0 + \tau_L$ correspond to instants at which the transmitted pulse was received, after being propagated and reflected in the environment.

In the case of digital transmission, if the delay spread of the channel defined by Equation 5.45, caused by the multipath environment, is higher than the symbol period, this means that the signal bandwidth is higher than the channel coherence bandwidth (defined by Equation 5.44). In this case, the channel presents frequency selectivity and the receiver experiences intersymbol interference.

Assuming a discrete multipath propagation channel with L paths, the complex equivalent low pass of the channel impulse response becomes[†]

$$h(\tau,t) = \sum_{i=0}^{L-1} a_i(t)e^{j\theta_i(t)}\delta(t - \tau_i) \tag{5.42}$$

where:

$a_i(t), \theta_i(t)$, and τ_i stand for the attenuation, phase shift, and delay of the ith multipath, respectively

$\delta(t)$ is the Dirac function

The frequency response of the channel becomes

$$H(f,t) = \sum_{i=0}^{L-1} a_i(f)e^{j\theta_i(f)}e^{-j2\pi f \tau_i} \tag{5.43}$$

Note that the time delay is related to the phase shift by [Marques da Silva et al. 2010] $\tau(f) = -(1/2\pi)(d\theta(f)/df)$, both parameters being a function of the frequency f.

Depending on the depth of the fast fading, there are two statistical models characterizing these effects [Fernandes 1996]:

- *Rayleigh model—fast and deep variations*: it is typically employed when there is no line of sight between the transmitter and the receiver, that is, there is only interference between several reflected, diffracted, and scattered multipaths. This is normally experienced in urban environments. Considering that, for each delay τ_i, a large number of scattered waves arrive from random directions, then, in accordance with the central limit theorem, $a_i(t)$ can be modeled as a complex Gaussian process with zero mean. This means that the phase $\theta_i(t)$ will follow a uniform distribution in the interval $\left[0,\ 2\pi\right]$ and the fading amplitude $|h(\tau,t)|$ will follow a Rayleigh distribution.[‡]

---

* This is caused by a variation in amplitude or delay (or both) of one or more of the received multipaths.

† This equation is generic, which is valid for both Rayleigh and Ricean models.

‡ A Rayleigh distribution is defined by Proakis [1995] as $f(x) = (x/\sigma^2)\exp\left[-\left(x^2/2\sigma^2\right)\right]$, whose probability density function (PDF) is expressed by Proakis [1995] as $p_R(|\alpha|) = (2|\alpha|/E[|\alpha|^2])\exp(-|\alpha|^2/E[|\alpha|^2])$.

- *Rice model—fast but weak variations*: it is typically employed in the presence of a line of sight between the transmitter and the receiver, to which several multipaths are added at the receiver. In this case, there is interference between the line-of-sight and several reflected paths. This effect is defined by a Rice distribution. It consists of a sum of a constant component (line-of-sight) and several reflected paths (defined by a Rayleigh distribution). This effect is typical of rural or indoor environments. Assuming the presence of a line-of-sight component with amplitude A arriving at the receiver, then $a_i(t)$ will be a complex Gaussian process with nonzero mean, and thus the fading amplitude $|h(\tau,t)|$ will follow a Rice distribution.[*]

The rapid movement between the transmitter and the receiver originates, with the variation of the propagation path distance, a change in the corresponding interferences. In addition, this movement may originate a change in the environment surrounding the receive antenna, which also originates fast fading. Finally, the turbulence of the local atmosphere is also a cause of the multipath fading, as it consists of the fast and random variation in the refraction index, originating a similar variation in the interference between the multipaths (Figure 5.16).

A cellular architecture includes three different types of cells: macrocells (rural and urban), microcells (of lower dimensions), and picocells (coverage of an office, a room, etc.).

Urban macrocell is the environment where the shadowing effect is experienced with higher intensity. This is caused by two main reasons:

- There is no line-of-sight component between the transmitter and the receiver and, therefore, the link is established through reflected, diffracted, and/or scattered rays.
- The high rate of constructions.

The absence of line of sight between the transmitter and the receiver also originates that the multipath fading of urban macrocells is characterized by a Rayleigh distribution.

On the other hand, because of the presence of line of sight, rural macrocells and picocells tend to be characterized by a Ricean distribution.

The coherence bandwidth is defined as the bandwidth above which the signal starts presenting a frequency selective fading. In other words, a signal with a bandwidth higher than the coherence

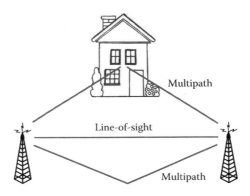

FIGURE 5.16 Propagation environment with line-of-sight and multipath components.

---

[*] A Rice distribution is defined by Proakis [1995] as $f(x) = (x/x_{\mathrm{ef}}^2)\exp\left[-((x^2+A^2)/2x_{\mathrm{ef}}^2)\right]\cdot I_0\left((x\cdot A)/x_{\mathrm{ef}}^2\right)$, whose PDF is expressed by Proakis [1995] as $p_R(|\alpha|) = \left(2|\alpha|/E\left[|\alpha|^2\right]\right)I_0\left(\left(2|\alpha|A\right)/E\left[|\alpha|^2\right]\right)\exp(-(|\alpha|^2+A^2)/E[|\alpha|^2])$, where I_0 is the modified Bessel function of zero order and where x_{ef} stands for the root mean square value of x.

bandwidth presents different attenuations and nonlinear phase shifts[*] at different frequencies. This effect is known as distortion. As exposed in Chapter 3, in the case of digital transmission, the distortion is viewed in the time domain as originating intersymbol interference.

The coherence bandwidth is obtained by [Marques da Silva et al. 2010]

$$\left(\Delta f\right)_C = \frac{1}{2\pi S} \tag{5.44}$$

where S is the RMS delay spread defined as follows [Marques da Silva et al. 2010]:

$$S = \sqrt{\frac{\int_0^\infty \left(\tau - D\right)^2 P\left(\tau\right) d\tau}{\int_0^\infty P\left(\tau\right) d\tau}} \tag{5.45}$$

where:

$P\left(\tau\right)$ is the PDP

D is the average delay defined as [Marques da Silva et al. 2010] follows:

$$D = \frac{\int_0^\infty \tau P\left(\tau\right) d\tau}{\int_0^\infty P\left(\tau\right) d\tau} \tag{5.46}$$

There are different measures that can be adopted to combat the fading effects, such as the use of multiple spaced antennas (spatial diversity), sectored antennas, matched filter equalizer, channel coding with interleaving,[†] MIMO systems, block transmission techniques (such as OFDM), or the use of frequency diversity.[‡]

5.1.4 GROUNDWAVE PROPAGATION

There are three basic propagation modes: direct wave, skywave, and groundwave.

The coverage through direct wave propagation was already dealt with in Section 5.3. Groundwave propagation can be used to cover areas that go beyond the direct line-of-sight coverage, but whose coverage is typically lower than that obtained with skywave propagation.

Figure 5.19 depicts the coverage by groundwave and skywave. Groundwave coverage may extend up to about 400 km from the transmit antenna. The coverage depends on factors, such as the carrier frequency, the characteristics of the terrain, the polarization, or the absence of obstacles between transmit and receive antennas.

The groundwave propagation is the sum of several elementary waves: (a) the surface component (wave), whose electromagnetic waves are guided over the earth's surface; (b) the direct wave; and (c) the reflected wave in the ground.

The surface component can be viewed as a result of diffraction of low-frequency electromagnetic waves by the earth's surface. As known from the knife-edge model, the diffraction effect is experienced with higher intensity at lower frequencies (as lower frequencies are less subject to attenuation by objects).

---

[*] That is, the phase shift response as a function of the frequency is not linear (it is a curve).
[†] To avoid bursts of errors and allow the channel coding to correct a certain number of corrupted bits per frame.
[‡] Transmit the same signal in different frequency bands, with a separation higher than the channel coherence bandwidth, in order to behave as uncorrelated.

The surface wave propagates mainly using vertical polarization, as horizontal polarization experiences high attenuation levels [Burrows 1949].

With regard to direct and reflected waves, as previously described, they are present in line-of-sight, reflected, or scattered paths between transmit and receive antennas. Because the reflection in the ground at a short distance tends to originate a phase inversion, the combination of these two components in line-of-sight coverage is normally destructive at low frequencies.

At a long range, the groundwave is normally only composed of the surface wave, as the other two components are not present. The groundwave attenuation corresponds approximately to the FSPL, added by 20 dB attenuation per decade. This decay becomes exponential with the increase of the distance d after a critical distance d_c, expressed in kilometers as

$$d_c = \frac{80}{\sqrt{f_{MHz}}} \tag{5.47}$$

The terrain permittivity and conductivity are determinant for the surface's wave propagation. Lower losses are achieved above surfaces with higher conductivity. Note that the sea water is highly favorable for the surface's wave propagation, because of the high rate of salinity, which improves the conductivity.

Figure 5.17 shows the field strength curves as a function of the distance for several different frequencies, for seawater with $\sigma = 5$ S/m and $\varepsilon = 70$. Similar curves for different terrains can be obtained from ITU-R P.368-7. As can be seen from Figure 5.17, the groundwave propagation is normally achieved with frequencies that span from few kilohertz up to around 3 MHz. Frequencies higher than this upper limit are subject to high attenuations, and therefore, their range becomes limited.

Using the graphic of Figure 5.17, we can calculate the received field strength expressed in dB µV/m (abscissa) at a certain distance from a 1 kW transmitter, for different frequencies. Alternatively, we can extract the range obtained with certain received field strength.

When a different transmitting power is used, a correction factor needs to be taken into account in the calculations. In order to better understand this calculation, let us consider an example of a 5 kW transmitter, using the 5 MHz frequency, and a distance of 500 km. The transmitting power is 7 dB above the 1 kW reference transmitter. Entering in the graphic of Figure 5.17, from the 5 MHz curve, and for a 500 km distance, we obtain a received field strength of 30 dB µV/m. Because this value corresponds to a reference transmitter of 1 kW, and because our transmitter has a power 7 dB above, the received field strength corresponds to $30 + 7 = 37$ dB µV/m.

An alternative way to calculate the range or signal strength, entering with the same input parameters, can be performed using the GRWAVE simulator.

In case the path between a transmitter and a receiver is composed of different sections with different terrains, the calculation can be performed as a combination of different paths with different terrains. This method is known as the Millington method [Burrows 1949].

5.1.5 SKYWAVE PROPAGATION

Long-range radio communications can be achieved by different means. While modern types of long-range communications are normally achieved using satellite communications, this can also be implemented using the so-called short wave communications, whose waves propagate at a long range using the ionosphere. In fact, skywave is the propagation type that supports long-range communications using frequencies from few hundreds of kilohertz up to few dozens of megahertz.

Skywave propagation consists of successive refraction in the ionosphere layers and successive reflection in the earth's surface. This can be seen from Figure 5.18. In fact, according to Snell's law,

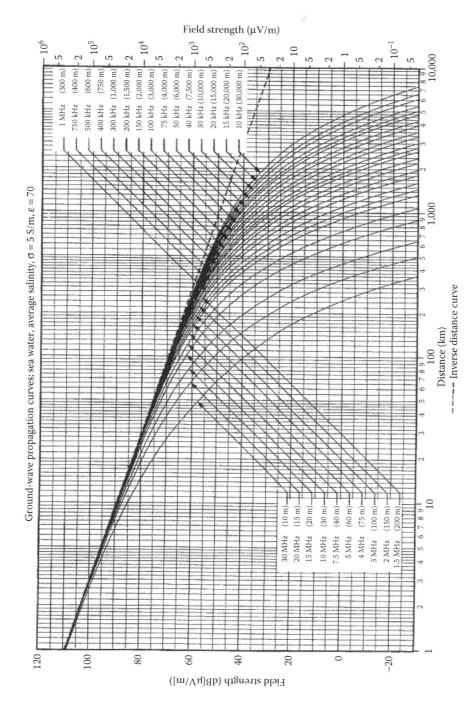

FIGURE 5.17 Field-strength curves as a function of distance with frequency for groundwave propagation (for seawater with σ = 5 S/m and ε = 70). (Data from ITU-R P.368-7, ITU-R Recommendation P.368-7, *Ground-Wave Propagation Curves for Frequencies Between 10 kHz and 30 MHz*, 1992.)

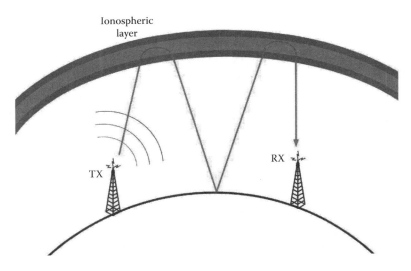

FIGURE 5.18   Refraction of electromagnetic waves in the ionospheric layers.

the gradual and successive refraction in the ionosphere layers can also be viewed as a reflection phenomenon.

Using the skywave propagation, choosing the correct carrier frequency, time of the day, and angle of incidence, a communication link can be established between any two points in the earth. The price to pay is the reduced bandwidth[*] that typically characterizes the skywave (as well as the groundwave).

The ionosphere is normally viewed as plasma with low level of ionization, composed of free electrons in a medium where they can collide with heavier particles. This plasma is normally characterized by two physical parameters: the number of electrons per volume unity and the number of collisions that electrons suffer per time unit. Moreover, the number of electrons per volume unity (N_e/m^3) shows a high variability over a day, seasons, and solar cycle. Note that the solar cycle corresponds to 11 years.

A complete reflection of an electromagnetic wave is experienced in the ionosphere if its frequency is equal to the critical frequency, as long as the propagation direction is perpendicular to the ionosphere. This characteristic is normally used by an ionosonde. Note that the critical frequency is a function of the level of ionization experienced in such part of the ionosphere.

Carrier frequencies higher than the critical frequency will cross the ionosphere. Furthermore, because of Martyn's law, the refraction occurs when the incident angle is not perpendicular to the ionosphere, even with a carrier frequency higher than the critical frequency.

Martyn's law is defined by

$$\omega = \omega_c \sec \theta_0 \tag{5.48}$$

where:

θ_0 is the incident angle (measured from the perpendicular to the ionosphere layer)
ω is the angular carrier frequency ($\omega = 2\pi f$, and f is the carrier frequency)
ω_c is the critical angular frequency

Note that Equation 5.48 shows that the ionosphere reflects more higher frequencies with an oblique propagation than with a vertical propagation. This is the ideal condition to achieve long distances.

---

[*] Which translates in a reduced data rate.

As described in Chapter 4, the refraction phenomenon is modeled by Snell's law as $n_1 \sin \theta_1 = n_2 \sin \theta_2$, where n stands for the refraction index of the medium and θ stands for the angle from the vertical. A wave that crosses a border between a medium with refraction index n_1 into a medium with refraction index n_2 suffers refraction, that is, deviation or bending in the wave's direction corresponding to $\theta_2 - \theta_1$.

The ionosphere is structured in layers, with different characteristics, namely D, E, and F layers (by ascending order of altitude). Some layers may present sublayers and some may be absent in certain periods of the day. Note that stronger ionization normally occurs at altitudes between 200 and 400 km. Figures 5.20 and 5.21 depict the typical ionospheric layers, respectively, for the day and night.

During the day, the typical layers are layer D (between 50 and 90 km), layer E (between 90 and 140 km), layer F1 (between 140 and 200 km), and layer F2 (above 230 km). Note that layer F2 is the most important, as it is present even in the absence of the sun (during the night). Furthermore, because it is placed at higher altitudes, it also allows establishing communication links at longer distances. Moreover, it refracts higher frequencies.

Figure 5.21 depicts the common layers present at night. It is seen that during the night, layers F1 and F2 merge to create a single layer F. Moreover, while layer E tends to remain present during the night, layer D is normally absent.

In addition to the above-described layers present during the day and night, E sporadic may also be present under certain conditions of the ionosphere (during the day or night). This layer may refract the same frequencies as layer F.

As can be seen from Figure 5.19, for a certain carrier frequency, very low or very high angles of incidence result in an absence of reflection[*] by the ionospheric layer. A high angle (from the normal

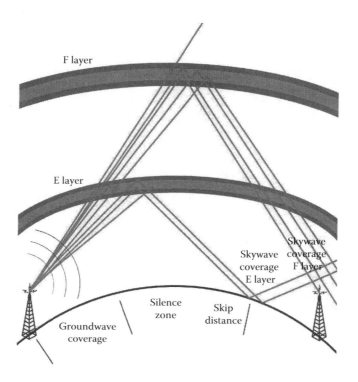

FIGURE 5.19 Propagation of electromagnetic waves using the ionospheric layers.

---

[*] In fact, the ionospheric layer does not reflect electromagnetic waves. Successive refraction is experienced in accordance with Snell's law. Macroscopically, this can be viewed as a reflection.

to the layer) originates wave absorption by the ionosphere, whereas a low angle originates that the wave crosses the layer without having been reflected.

Carrier frequencies that are extremely high or low result in rays that are not reflected. In the example of Figure 5.19, a higher frequency signal is used to achieve layer F. This frequency signal is not reflected by layer E, except at very high angles of incidence.

For a wave to reach a certain destination, the frequency and angle need to be properly selected. The optimum angles and frequencies present several levels of variation:

- *It varies from year to year*: Depending on the sunspot number as a function of the sun intensity, it varies yearly. A higher sun activity results in a higher electronic density, requiring higher frequencies (to reach a certain destination, and using a certain angle of incidence).
- *It varies with season*: Because of higher sun intensity, the ionospheric layers are more intense in the summer. Consequently, the summer frequencies are typically higher than those in the winter.
- *It varies with the time of the day*: As seen from Figures 5.20 and 5.21, because of the absence of the sun, the level of ionization is lower during the night. This results in a lower frequency for the same distance and angle of incidence.
- *It varies with the latitude*: Because the solar incidence angle is lower at high latitudes, the electronic density of layers decreases at high latitudes.

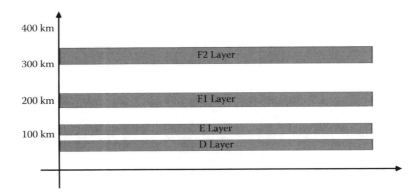

FIGURE 5.20   Representation of the ionospheric layers (day).

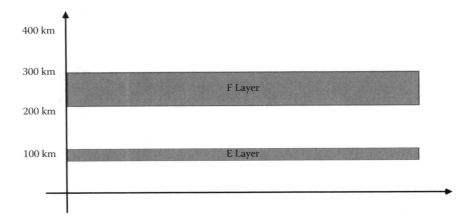

FIGURE 5.21   Representation of the ionospheric layers (night).

5.2   SATELLITE COMMUNICATION SYSTEMS

One of the most important advantages of a satellite relies on its wide coverage that translates in service availability in remote areas. Satellites can be used for many different purposes. They can be used for broadcast of radio or television channels, for point-to-point or point-to-multipoint communications, for the capture of images, or for meteorological purposes.

The moon was, in 1958, the first satellite used for communications. An electromagnetic beam was sent toward a specific position in the moon, which reflected it backward to the earth.

Afterward, in 1962, Echo I and Telstar incorporated an onboard active repeater. First, satellite applications consisted of intercontinental transmissions of television and communications with ships at sea. Later on, satellites started being used for intercontinental exchange of voice, and finally for data, positioning systems, and imagery.

The basic principles of satellite communications were not deeply modified over time. A satellite has one or several transponders, each one operating in a different frequency band. A transponder consists of a receiver, followed by a frequency translator, an amplifier, and a transmitter. The frequency translator is necessary because the uplink and downlink frequencies are different. The main functionality consists of receiving a signal, amplifying it, sending it back to the earth, and therefore, acting as a repeater.

Depending on the orbit altitude and attitude, there are different types of orbits: geostationary earth orbit (GEO), medium earth orbit (MEO), low earth orbit (LEO), and highly elliptical orbit (HEO). The LEO altitude varies between 300 and 2,000 km, whereas the MEO orbit corresponds to an altitude between 5,000 and 15,000 km. The GEO altitude is typically 35,782 km. Finally, the HEO presents an elliptical orbit with perigee at very low altitudes (typically 1,000 km) and with the apogee at high altitudes (between 39,000 and 53,600 km). The different orbits are plotted in Figure 5.22.

5.2.1   PHYSICAL ANALYSIS OF SATELLITE ORBITS

In order to understand how satellites stand in the space, it is worth introducing some physical concepts. This analysis allows us deducting the altitude, speed, and period for each different orbit.

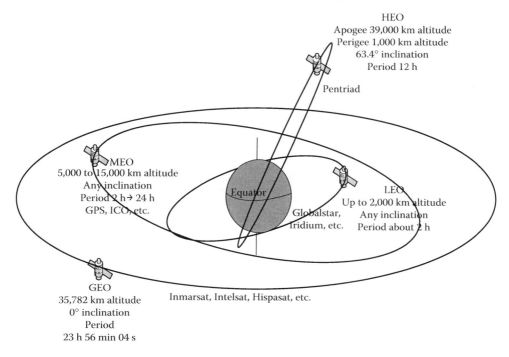

FIGURE 5.22   Different satellite orbits.

The widely known Newton's law of universal gravitation establishes the attraction force between two objects due to gravity. Particularly, considering that the two objects are the earth and a satellite, the law of universal gravitation becomes

$$f_a = G \frac{m_E \cdot m_{SAT}}{d^2} \tag{5.49}$$

where:

m_E as the mass of the earth

m_{SAT} as the mass of the satellite

G stands for the gravitational constant (6.674×10^{-11} N m^2/kg^2)

d stands for the distance between the mass center of the earth and the mass center of the satellite

On the other hand, the magnitude of the centripetal force of an object of mass m_{SAT} moving at a speed v_{SAT} along a path with a curvature radius r_{SAT} is given by

$$f_C = m_{SAT} \cdot \frac{v_{SAT}^2}{r_{SAT}} \tag{5.50}$$

Using the equality $v_{SAT} = \omega_{SAT} \cdot r_{SAT}$, Equation 5.50 becomes

$$f_C = m_{SAT} \cdot \omega_{SAT}^2 \cdot r_{SAT} \tag{5.51}$$

where ω_{SAT} stands for the angular speed of the satellite.

For a satellite to stand in the sky, the attraction force defined by Equation 5.49 needs to equal the centripetal force defined by Equation 5.51. Equaling the two forces, and making $d = r_{SAT}$, we obtain

$$G \frac{m_E \cdot m_{SAT}}{r_{SAT}^2} = m_{SAT} \cdot \omega_{SAT}^2 \cdot r_{SAT} \tag{5.52}$$

which leads to

$$\omega_{SAT} = \sqrt{\frac{G \cdot m_E}{r_{SAT}^3}} \tag{5.53}$$

Because $\omega_{SAT} = 2\pi/T_{SAT}$, we arrive at the third Kepler's law:

$$T_{SAT}^2 = \frac{4\pi^2}{G \cdot m_E} r_{SAT}^3 \tag{5.54}$$

where T_{SAT} is the orbit's period of the satellite.

Entering with the earth's mass $m_E = 5.9737 \times 10^{24}$ kg and with the gravitational constant G into Equation 5.54, we can finally enunciate the relationship between the radius of the satellite's orbit and its period as follows:

$$T_{SAT}^2 = 99,022 \times 10^{12} \cdot r_{SAT}^3 \tag{5.55}$$

Expressing the distance in kilometers and the period in hours, and isolating the satellite radius, Equation 5.55 becomes [Kadish and East 2000]

$$r_{SAT(km)} = 5076 \cdot T_{SAT\_(hour)}^{2/3} \tag{5.56}$$

Noting that r_{SAT} stands for the radius of the satellite's orbit, we know that

$$r_{SAT} = r_E + h_{SAT} \tag{5.57}$$

where:

r_E corresponds to the earth's radius

h_{SAT} stands for the orbit's altitude*

Entering with the earth's radius $r_E = 6373$ km, Equation 5.56 can now be expressed as a function of the orbit's altitude h_{SAT}, in kilometers, as follows:

$$h_{SAT(km)} = r_{SAT(km)} - r_E$$
$$= 5076 \cdot T_{SAT\_(hour)}^{2/3} - 6373 \tag{5.58}$$

Table 5.1 shows different orbit periods, expressed in hours, for several orbit's altitudes and radius. The LEO altitude is placed between 300 and 2000 km, whose period is around 2 h. The orbit altitude of the MEO is between 5,000 and 15,000 km, with an orbit period of about 4 h. Finally, the GEO orbit is at the altitude of 35,782 km, whose orbit period equals the day period (23.9327 h). At this altitude, GEO satellites go around the earth in a west to east direction at the same angular speed as the earth's rotation.

5.2.2 CHARACTERISTICS OF DIFFERENT ORBITS

Satellite communications can be viewed as a type of cellular communications, whose coverage is much higher (due to higher altitude of the satellite). This can be seen from Figure 5.23.

TABLE 5.1
Orbit Period as a Function of Orbit Altitudes and Radius

Altitude (km)	Orbit Radius (km)	Orbit Period (h)
0 (earth's surface)	6,373	1.4068
300	6,673	1.5073
2,000	8,373	2.1186
5,000	11,373	3.35
35,782	42,155	23.9327

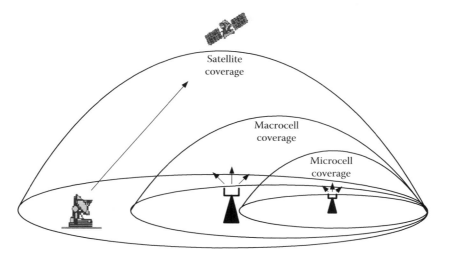

FIGURE 5.23 Indicative difference between satellite and cellular coverage areas.

* In the case of GEO, the orbit's altitude is typically 35,782 km.

Depending on the satellite altitude and attitude, there are different types of satellite orbits. Nevertheless, there are two layers around the earth where locating satellites should be avoided, because of high electromagnetic radiation, which may deteriorate the satellite equipment. These two layers are called *Van Hallen layers*, whose altitudes are 2,000–5,000 and 15,000–20,000 km, respectively.

5.2.2.1 Geostationary Earth Orbit

The mostly known type of orbits is the geostationary earth orbit. This orbit is called geostationary because the position relative to any point in the earth is kept stationary. The first satellites being used for mobile communications were launched in 1970 and were of GEO type. The GEO altitude is typically 35,782 km above the equator, and, because it is geostationary, its period equals the earth rotation period. Because they have a geostationary orbit, they are relatively easy to control. Moreover, because of the high altitude, the coverage is maximized, corresponding to approximately one-third of the earth's surface, making the communication service available to a wide number of potential users.

GEO-type satellites were unable to provide services to small mobile terminals, such as the existing cellular telephones, because of the following main reasons:

- High path loss, which results from the enormous distance from the earth.
- Low antenna gain of the satellite transponder, to allow covering a wide area of the earth's surface (typically one-third of the earth's surface). An indicative throughput available by a transponder is typically limited to 72 Mbps, covering a wider area, which means that the throughput per user[*] is very much reduced.
- Low power spectral density as a result of the low power available on board the transponder (typical 10 dBW) and the enormous distance from the earth.

Therefore, high power and high antenna gain was a basic requirement of the earth stations to allow the connection establishment with a GEO satellite, which translates in low mobility and high dimensions.

In addition to the limitations of GEO satellites, because GEO orbits are located at an altitude of around 36,000 km, the round trip distance is approximately 72,000 km. This distance corresponds, at a speed of light, to a delay of 240 ms, which is much higher than in the case of an MEO or an LEO. This represents a high latency introduced in signals. In case the two terminals are not served by the same satellite, a double hop may be necessary. In this case, this latency increases to approximately one-half of a second, which is a value that may bring problems to voice or for data communications.[†]

With the enormous growth of the telecommunications industry and the development of the new services, such as the Internet and multimedia applications, the satellite operators viewed the MEO and LEO with great potential business.

It is worth noting that the latest developments already allow some GEO-type satellites working with higher power spectral densities, providing, however, low data rate services for terminals with omnidirectional antennas (e.g., the Fleetphone used by Inmarsat constellation). This is mainly achieved with the implementation of advanced antenna systems that enables satellite antenna gains higher than 40 dBi.

5.2.2.2 Medium and Low Earth Orbit

With altitudes below that of the GEO, the MEO and LEO overcame many of the limitations experienced by GEO satellite communications.

Although the LEO and MEO constellations present lower footprints than the GEO, requiring more satellites to allow adequate area coverage, other technological achievements facilitated the

---

[*] The throughput per user is the total throughput divided by the number of potential users within the coverage area.
[†] With such delay, the stop-and-wait data link layer protocol results in a very inefficient use of the channel.

implementation of these satellite constellations, such as direct connection between different satellite's transponders, onboard switching and routing, or advanced antenna systems.

As can easily be concluded from the orbit's name, the MEO satellites are placed at medium altitude (between 5,000 and 15,000 km) and the LEO satellites are located at low altitude (between 300 and 2000 km). These orbits can be seen in Figure 5.22. In contrast to the GEO, these orbits are not geostationary and, as expressed by Equation 5.58, their period decreases with the decrease in the orbit's altitude. Consequently, as these satellites are permanently moving around the earth, the coverage of a certain region needs to make use of several different satellites. Note that the LEO and MEO can be of any type around the earth, namely above the equator, above a meridian, or with any inclination. In any case, the center of the orbit is always the center of the earth. Figure 5.24 depicts indicative footprints for GEO and MEO satellites, as well as the coverage made available by a cellular base station in a part of the east coast of the United States. The arrow connected to the MEO footprint circle represents the direction of the satellite's motion, which corresponds to the direction of the footprint's motion.

Below an altitude of 200 km, it is not possible to keep satellites in the sky, because of the enormous heating and deterioration that satellites are subject to, as well as because of the great tendency to leave the desired orbit. This would translate in an enormous use of engines and fuel to correct the orbits. Because the amount of fuel onboard satellite is limited, this is not a viable solution.

From the telecommunications point of view, a lower altitude of a satellite translates in lower path loss. While the establishment of a link with a GEO satellite requires typically a parabolic antenna (high gain), a link can normally be established with an MEO or an LEO satellite, making use of an omnidirectional antenna. As the satellites and their footprints are permanently moving (they are not stationary), a connection can be initiated with a satellite and, after a certain period of time, the connection can be handed over to another satellite. Therefore, the level of complexity necessary to manage such handover is increased, related to the GEO. Furthermore, the moving orbits require a higher level of adjustments from the control station.[*]

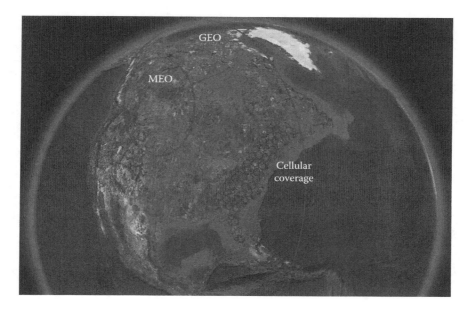

FIGURE 5.24 Example of GEO and MEO footprints against cellular coverage.

---

[*] A control station is a station in the earth that communicates with the satellite to send orders to adjust the orbit, the speed, the coarse, altitude, and so on. These orders also relate to adjustments in transmitting power, frequencies, antenna direction, and so on.

5.2.2.3 Highly Elliptical Orbit

Another type of orbit is the highly elliptical orbit. It presents an elliptical orbit with perigee at very low altitudes (typically 1000 km) and with the apogee at high altitudes (between 39,000 and 53,600 km). These orbits can be utilized for military observation or for meteorological purposes, with the perigee above the region to observe. Moreover, because GEO satellites do not cover the poles (they are located above the equator), HEO satellites are useful for providing communication services to regions with high latitudes. Nevertheless, because they are not stationary, the service is only available when the satellites pass over the region of interest, which occurs close to the perigee. Table 5.2 presents a comparison among different satellite constellations.

5.2.3 SATELLITE's C/N RATIO ANALYSIS

As can be seen from Figures 5.25 and 5.26, satellites can be used for telecommunications in two basic modes: point-to-point or point-to-multipoint modes. In the point-to-point mode, a satellite acts

---

TABLE 5.2
Advantages and Disadvantages of Different Orbits

	Advantages	Disadvantages
LEO	1. Can operate with low power levels and reduced antenna gains 2. Reduced delays	1. Complex control of satellites 2. Frequent handovers 3. High Doppler effect 4. High number of satellites
MEO	1. Acceptable propagation delay and link budget, but worse than in the LEO case	
GEO	1. Reduced number of satellites and, consequently, simplest solution 2. No need for handover	1. Requires high antenna gains and powers to overcome increased path loss 2. Difficult to operate with handheld terminals 3. High delays (240 ms) 4. Reduced minimum elevation angles for high latitudes that translate in high fading effects
HEO	1. High minimum elevation angles even for high latitudes 2. Enables coverage of very specific regions	1. Requires high antenna gains and powers to overcome increased path loss 2. Extremely high delays, except in the perigee

---

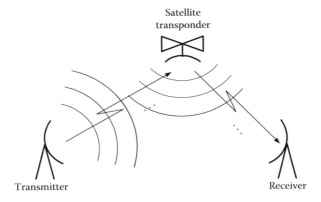

FIGURE 5.25 Generic diagram of a point-to-point satellite communication system.

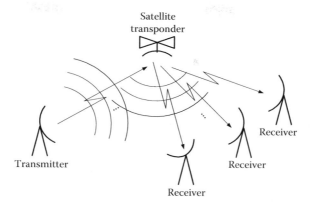

FIGURE 5.26 Generic diagram of a point-to-multipoint (broadcast) satellite communication system.

as a repeater between two terminals. In this case, the exchange of data is normally performed in both directions (bidirectional). In the point-to-multipoint mode, a satellite acts as a repeater between a transmitting station and many receiving stations. This is normally used for broadcast, such as television or radio broadcast (unidirectional).

The selection of frequencies for use in satellite communications is based on several factors. The frequency must be sufficiently high such that the desired directivity and bandwidth are achieved. Remember that a higher frequency presents a greater coherence bandwidth, supporting, in principle, signals with higher bandwidth. As can be seen from Equation 5.19, the directivity increases with the decrease in the wavelength, that is, it increases with the frequency. On the other hand, increasing the frequency also increases the path loss (see Equation 5.4), which is normally a negative effect. Figure 5.27 depicts the attenuation as a function of the frequency for a GEO satellite (distance corresponding to 36,000 km).

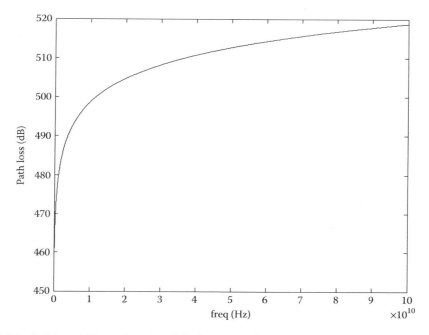

FIGURE 5.27 Path loss (dB) as a function of the frequency (Hz) for a GEO (36,000 km altitude).

As expected, higher frequencies correspond to higher attenuation levels, which may bring link budget limitations. In any case, the main objective relies on maximizing the received signal power, as defined by Equation 5.8, such that it is above the receiver's sensitivity threshold, and such that the C/N defined by Equation 5.14 is maximized. As per the Shannon capacity equation (see Chapter 3), a higher SNR allows transmitting at higher data rates. Note that the relationship between the SNR and the C/N depends on the modulation schemes [Carlson 1986].

The frequency bands normally assigned to satellite communications are the L band (1–2 MHz), the C band (4–6 MHz), the X band (7–8 MHz), the Ku band (12–14 MHz), and the Ka band (18–22 MHz). The mostly used band is the C band, using the 6 GHz band in the uplink,* whereas the 4 GHz band is normally adopted for the downlink.† Because this pair of frequency bands is currently saturated, the second mostly used spectrum for satellite communications is currently the Ku band: 14 GHz in the uplink and 12 GHz in the downlink. Moreover, because higher carrier frequencies present typically higher bandwidths, the Ku band tends to be the preferable choice. However, it is more subject to rain attenuation.

Note that, in order to avoid interference, uplink and downlink frequencies should be different. Moreover, the uplink frequency band is typically higher than the downlink frequency band. Because the earth station has typically a higher transmitting power available than the satellite transponder (typically 10 dBW), higher attenuations originated by the higher frequency in the uplink are overcome by this additional transmitting power. Furthermore, the use of higher uplink frequencies leads to higher satellite receiver's antenna gain, which allows maximizing the SNR. On the other hand, because the satellite's beam is directed toward the earth (high noise temperatures), the satellite's receiver experiences higher noise power than earth stations.

The signal received from the satellite's earth station has a power defined by Equation 5.11 (see Figures 5.28 and 5.29). Figure 5.28 depicts the propagation path between the satellite and the two end stations that are subject to the propagation impairments, as described in Section 5.1.1 for the direct wave propagation. From the earth station side, a satellite link is typically modeled by a Rice

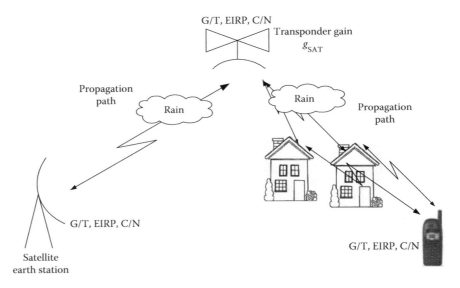

FIGURE 5.28 Typical satellite link with a satellite earth station and a mobile station.

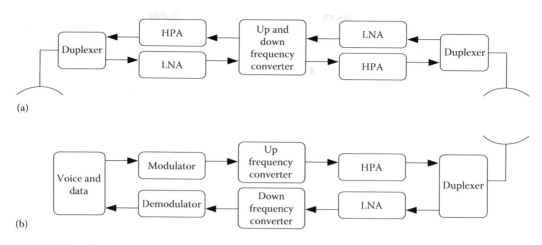

FIGURE 5.29 Scheme of (a) satellite's transponder and (b) satellite's earth station.

distribution. As previously described, it consists of a line-of-sight component to which a Rayleigh distribution is added. The Rayleigh distribution models several reflected, diffracted, and scattered rays in buildings, street, trees, earth's surface, and so on. At high latitudes, as the satellite's inclination decreases, the strength of reflected waves becomes more predominant (and the line-of-sight components become weaker), and the resulting signal is more subject to fading. Moreover, shadowing can also be important, especially in urban scenarios.

Note that additional attenuations such as antenna misalignments, rainfall, fog, diffraction caused by buildings, scattering originated by trees, or reflections caused by buildings and streets may also be quantified, and taken into account in the computation of the receiving signal power, according to the above description. The received C/N can be computed using Equation 5.17, which already takes into account the receiver's merit factor G_R/T_n specified for the satellite transponder, namely the thermal noise captured by the receive antenna and the noise introduced by the low noise amplifier and introduced by the high power amplifier.

Because the uplink frequency is higher than the downlink frequency, the up-down frequency converter depicted in Figure 5.29 is responsible for the down frequency conversion.

Because a satellite consists basically of a repeater placed at high altitude, the satellite's downlink transmitting power P_{E_D} equals the uplink (satellite) received signal power P_{R_U} multiplied by the transponder's amplification gain g_{SAT}:

$$P_{E\_D} = P_{R\_U} \cdot g_{SAT} \tag{5.59}$$

Usually, the most critical link is the downlink, because of the limited satellite transmitting power P_{E_D} and the low satellite antenna gain (limited by its size). It is worth noting that, in Equation 5.59 and in the following expressions, the transponder's amplification gain g_{SAT} has embedded the transmit antenna gains of the transponder. The most critical receiver is the earth station. The downlink received signal power P_{R_D} becomes

$$P_{R\_D} = P_{R\_U} \cdot g_{SAT} \cdot A_{tt\_D}$$
$$= P_{E\_D} \cdot A_{tt\_D} \tag{5.60}$$

where A_{tt_D} stands for the downlink attenuation (including the contribution of the receive antenna gain of the earth station).

As described in Section 3.4.4, the total received noise power N_{TOTAL} in the downlink becomes

$$N_{\text{TOTAL}} = N_U \cdot f_{\text{SAT}} \cdot g_{\text{SAT}} \cdot A_{\text{tt\_D}} + N_D \tag{5.61}$$

where:
N_U stands for the uplink noise power
N_D stands for the downlink noise power
f_{SAT} stands for the satellite's noise factor

Note that N_{TOTAL} includes the contribution of the noise in the uplink and downlink paths.
Consequently, the received C/N becomes

$$\left(\frac{C}{N}\right)_{\text{TOTAL}} = \frac{P_{\text{R\_D}}}{N_{\text{TOTAL}}} \tag{5.62}$$

Alternatively, we can also compute $(C/N)_{\text{TOTAL}}^{-1} = 1/(C/N)_{\text{TOTAL}}$ as follows:

$$\left(\frac{C}{N}\right)_{\text{TOTAL}}^{-1} = \frac{N_{\text{TOTAL}}}{P_{\text{R\_D}}}$$

$$= \frac{N_U \cdot f_{\text{SAT}} \cdot g_{\text{SAT}} \cdot A_{\text{tt\_D}} + N_D}{P_{\text{R\_U}} \cdot g_{\text{SAT}} \cdot A_{\text{tt\_D}}}$$

$$= \frac{N_U \cdot f_{\text{SAT}}}{P_{\text{R\_U}}} + \frac{N_D}{P_{\text{R\_D}}} \tag{5.63}$$

$$= \frac{f_{\text{SAT}}}{(C/N)_U} + \frac{1}{(C/N)_D}$$

$$= \frac{f_{\text{SAT}}}{(C/N)_U} + \frac{f_{\text{SAT}}}{f_{\text{SAT}} \cdot (C/N)_D}$$

and therefore, the received C/N can also be computed as follows:

$$\left(\frac{C}{N}\right)_{\text{TOTAL}} = \frac{(C/N)_U + f_{\text{SAT}} \cdot (C/N)_D}{f_{\text{SAT}}} \tag{5.64}$$

In the case of double hop satellite link, the computation of the resulting C/N can be obtained from re-using Equation 5.64 iteratively, as follows:

- Compute the C/N at the input of the earth station, as previously described
- Compute the C/N at the input of the second satellite's transponder (using the earth station's noise factor f_{EARTH})
- Compute the C/N at the input of the second earth station (using the second satellite's noise factor $f_{\text{SAT\_2}}$)

Alternatively, as defined in Section 3.4.4, one could compute $(C/N)_{\text{OUT}} = ((C/N)_{\text{IN}}/f_{\text{OUT}})$, where the overall noise factor is $f_{\text{OUT}} = f_1 + ((f_2 - 1)/g_1) + ((f_3 - 1)/g_1 g_2) + ((f_4 - 1)/g_1 g_2 g_3) + \cdots + ((f_N - 1)/\prod_{i=1}^{N-1} g_i)$. Note that we may view the path loss attenuation as a device gain and the thermal noise as the noise generated in electronic equipment. Consequently, we may process jointly different propagation paths and electronic components (e.g., satellite's transponder and satellite earth station), using the same principle, and computing the resulting noise factor f_{OUT}, which is then used to compute the resulting $(C/N)_{OUT}$.

5.3 TERRESTRIAL MICROWAVE SYSTEMS

A terrestrial microwave system consists of a bidirectional radio link between two sites that use directional antennas (typically parabolic shape, as shown in Figure 5.30). Because it consists of a radio link, all the link budget and C/N calculations defined in Section 5.1.1 are also applicable.

Moreover, in case the path between two interconnecting sites is not in line of sight, a repeater can be incorporated. This can be required because of the earth's curvature or the existence of obstacles. In this case, as a repeater may also be viewed as a satellite, the same principles as those deducted for the satellite link are also applicable.

Typical parameters used in microwave systems are as follows:

* Carrier frequency band: 10 GHz or below
* Transmitting power: 1 W
* Antenna gains: 35 dBi
* Link distance: up to around 30 km (can be extended using repeaters)

Microwave systems have been widely employed to interconnect different sites, such as cellular base stations or local area networks, for the exchange of television or radio channels between broadcast stations. Therefore, it is used for any type of media, such as voice, data, or television. Microwave systems can be viewed as an alternative to optical fibers or coaxial cables, because of their implementation simplicity.

Taking into account the earth's curvature, the radio horizon of a microwave link is limited to

$$d_h = \left[\left(r_E + h \right)^2 - r_E^2 \right]^{1/2} \approx \sqrt{2 r_E h} \tag{5.65}$$

where $r_E = 6373 \text{ km}$, which stands for the earth's radius, and h stands for the antenna heights (it is assumed that both antennas are placed at the same heights).

To identify whether or not a terrestrial microwave link is clear of obstacles, one needs to analyze the Fresnel ellipsoids (see Figure 5.31). Assuming that $r_1, r_2 \gg D_n$ we have

$$D_n = \sqrt{\frac{4 n \lambda r_1 r_2}{r_1 + r_2}} \tag{5.66}$$

where D_n stands for the nth-order ($n = 1, 2, \ldots, N$) diameter of the Fresnel ellipsoid.

Note that Equation 5.66 was deducted making the difference of path distance between the direct wave and the reflected wave as $\Delta d = d_D - d_R = n\lambda/2$ (see Figure 5.32). Even values of n correspond

FIGURE 5.30 Terrestrial microwave system.

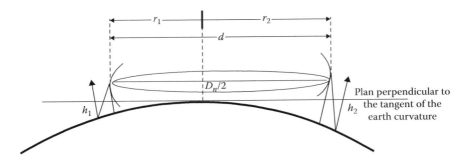

FIGURE 5.31 Terrestrial microwave link with important dimensions.

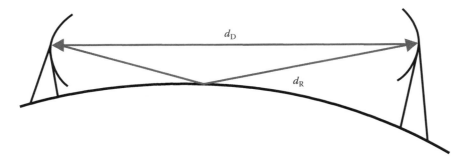

FIGURE 5.32 Direct and reflected waves of a microwave link.

to destructive interference between direct and reflected waves, whereas odd values of *n* correspond to constructive interference.

In order to assure that the received signal level is not more than 1 dB below the signal that would be received in free space, the first Fresnel ellipsoid should be clear of obstacles. Note that the distance where $D_n/2$ is taken into account refers to the position between r_1 and r_2 (see Figure 5.31). This refers to any position in the path, where an obstacle may exist, and where one intends to identify whether or not it interferes with the terrestrial microwave link. Naturally, in case the link is clear of obstacles, the only limitation is the earth's curvature. In this case, assuming that both antenna heights are the same, the bottleneck occurs typically at the midway of the link, that is, for $r_1 = r_2$.

CHAPTER SUMMARY

This chapter provided a description of wireless transmission mediums. A description of wireless propagation was also given. This included a view about direct wave propagation, such as the FSPL, the link budget calculation, and the C/N calculation. Moreover, a description of wireless propagation effects was provided, such as reflection, diffraction, and scattering.

Then, fading effects were described, including shadowing fading and multipath fading. It was viewed that the multipath fading corresponds to fast fading, whereas the shadowing corresponds to slow fading. Moreover, it was described that the multipath fading can be fast and deep fading, described by a Rayleigh distribution, or fast but not deep, as described by a Rice distribution.

Groundwave propagation was studied, as well as skywave propagation. It was viewed that the groundwave propagation corresponds to the surface wave, direct wave, and reflected wave. Moreover, it was described that a signal that is aimed to be sent using the surface wave uses vertical polarization, as the horizontal polarization is highly attenuated. Furthermore, it was described that the skywave comprises an alternative to satellite communications, to cover long ranges. Higher ranges are normally achieved with higher frequencies, as these signals are reflected by ionospheric layers located at higher altitudes.

In a second stage, the satellite communication systems were described. This included the study of the physical analysis of satellite orbits, as well as the characteristics of different orbits. The GEO, was studied, as well as the MEO, LEO, and the HEO. Then, the satellite C/N analysis was performed. It was described that the Ku band, used in satellite communications, presents an advantage over the C band, as it supports a higher bandwidth. Nevertheless, the Ku band is more subject to rain attenuation than the C band.

Finally, a view about terrestrial microwave systems was given. It was described that to assure that the received signal level is not more than 1 dB, below the signal that would be received in free space, the first Fresnel ellipsoid should be clear of obstacles.

REVIEW QUESTIONS

1. Which types of fading do you know? Characterize each one.
2. Which types of satellite orbits do you know?
3. From the known orbits, which one has the lower orbit?
4. What are the advantages and disadvantages of the LEO relating to the GEO orbit?
5. What is the difference between the C/N and the SNR?
6. What is the typical performance measure used in digital communications?
7. What are the differences between reflection, diffraction, and scattering?
8. Which type of communication can be viewed as an alternative to satellite communication, in order to achieve a long range?
9. What is the difference between the groundwave and the surface wave?
10. What is the difference between the surface wave and the skywave?
11. What are the common ionospheric layers present during day? And during night?
12. What is the relationship between the altitude of the ionospheric layer and range?
13. What is the relationship between the altitude of the ionospheric layer and frequency?
14. What is the relationship between E_b/N_0 and C/N?
15. What is the effect of a reflected wave in the propagation, as compared to free space propagation? Is it constructive or destructive?
16. Describe the model used to quantify the diffraction effect.
17. Which parameters improve the received signal strength of a receiver subject to diffraction?
18. Which measures can be used to mitigate the negative effects of a reflected wave?
19. For both free space propagation and a propagation model with a reflected wave, what is the relationship between the received power strength and the distance?
20. What is the free space path loss equation?
21. What is the relationship between bit energy, bit period, and received power?
22. What are the statistical distributions that characterize the fast fading? Characterize them.
23. What is the statistical distribution that characterizes the slow fading?
24. What does silence zone stand for in the context of skywave propagation?
25. According to the Friis formula, what is the received power, for a 1 kW transmit power, a 10 km distance and using both isotropic antennas?
26. For the surface propagation, what is the range obtained with a voice signal, assuming a transmit power of 10 W, a frequency of 2 MHz, a fading margin of 3 dB, and a noise level in the environment of the receiver of 32 dB μV/m?

LAB EXERCISES

1. Using the Emona Telecoms Trainer 101 laboratory equipment, and volume 1 of its laboratory manual, perform experiment 3—using the Telecoms-Trainer 101 to model equations.

6 Source Coding and Transmission Techniques*

LEARNING OBJECTIVES

- Identify and describe the source coding and transmission techniques used in communication systems.
- Identify and describe the different line coding schemes.
- Identify and describe the different modulation schemes.
- Describe the different multiplexing techniques.

As described in Chapter 3, one of the most important advantages of digital transmission relies on the ability to perform regeneration, although at a cost of a higher signal bandwidth, as compared to analog signals. Regeneration of signals allows partially removing the effects of channel impairments, such as noise, interference, attenuation, or distortion.

Depending on whether signals are transmitted in analog or in digital form, and assuming that the source data is digital, signals require passing through a process of modulation[†] or line coding,[‡] respectively (see Figure 6.1). This is performed using a specific modulation scheme or digital coding that aims to optimize and adapt the source signal to the transmission medium.

The transmission of signals using line coding techniques is performed in baseband, whereas the transmission associated to modulated signals makes use of a carrier to modulate a signal around a certain carrier frequency (bandpass). This may represent an advantage, as it is possible to select a transmitting frequency band whose channel frequency response presents less channel impairments.

The output of a line encoder is a digital signal, as it consists of discrete voltages that encode the source logic states. Consequently, it can be stated that the line coding is employed when the transmission medium is digital. On the other hand, the output of a modulator is an analog signal, as it modulates a carrier, which is an analog signal. Therefore, a modulator is a device that converts digital signals into analog, whereas a demodulator converts analog signals into digital.

Regardless of whether the generated bits are transmitted in digital or analog, these bits are subject to conversion into symbols (by either a line encoder or a modem), before the transmission process takes place. At the transmitter side, each bit, or group of bits, is encoded into symbols. Conversely, at the receiver side, symbols are converted into bits. A symbol with M different signal levels, allows $\log_2 M$ bits being encoded (by each symbol). It is known, from Chapter 5, that the symbol rate R_S, expressed in symbols per second (symb/s) or baud, the bit rate R_B, expressed in bits per second (bps), and the number of constellation points M are related by

$$R_S = \frac{R_B}{\log_2 M} \tag{6.1}$$

---

* The author acknowledges Dr. Francisco Monteiro for his assistance in providing three figures of this chapter.
[†] Modulation and demodulation is performed using a modem, which implements the digital to analog conversion (modulation) at the transmitter side, and the opposite at the receiver side (demodulation).
[‡] Line coding is also referred to as *digital coding*.

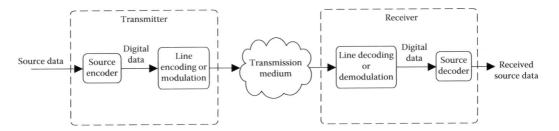

FIGURE 6.1   Generic block diagram of a digital communication chain.

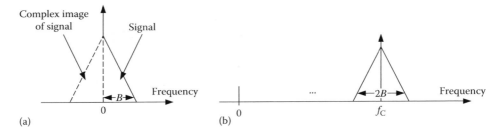

FIGURE 6.2   Spectrum of (a) baseband and (b) bandpass signals.

The bit rate R_B is a function of the source encoder. A more efficient source encoder allows transmitting more information source using a lower bit rate (for the same quality).

Knowing, from Chapter 3, that $B_{min} = R_S/2$ for a baseband signal (signal sent by a line encoder), and from Equation 6.1, we can express the minimum bandwidth by (double sideband signal)

$$B_{min} = \frac{R_S}{2} = \frac{R_B}{2 \times \log_2 M} \qquad (6.2)$$

As can be seen from Figure 6.2, in the case of a bandpass (carrier modulated) signal, the negative baseband part of the spectrum is also transmitted. Therefore, knowing, from Chapter 3, that $B_{min} = R_S$ for a bandpass signal (signal sent by a MODEM), and from Equation 6.1, we can express the minimum bandwidth by (double sideband signal):

$$B_{min} = R_S = \frac{R_B}{\log_2 M} \qquad (6.3)$$

Assuming a baseband signal, we know that a signal with a symbol rate of 1 Msymbol/s requires a minimum baseband bandwidth of $B_{min} = 10^6/2 = 500$ kHz. Naturally, assuming a raised cosine pulse,* the required bandwidth is maximized with a roll-of factor $\alpha = 1$, which corresponds to twice the minimum bandwidth, that is, $B_{max} = R_S$. This means that the baseband bandwidth is bounded by $R_S/2 < B < R_S$.

6.1   SOURCE CODING

The source coding is the process of transforming information media (source) into digital data. The information source can be voice, audio, text, video, images, and so on. Different source coding techniques exist, with different efficiencies and performances. Some more efficient source

---

* The used bandwidth can be higher for other pulses shapes (e.g., rectangular).

coding techniques allow for encoding the same information media at the cost of a lower bit rate R_B. According to Equation 6.3, for a certain modulation, a lower bit rate results in a reduced bandwidth.

6.1.1 VOICE

The voice is an analog information source, consisting of the most important information source used by humans to communicate. Its frequency domain spreads from around 80–12,000 Hz,* with the most important frequency components in the band 150–8,000 Hz. However, it has been proven that its quality is acceptable and understandable if we restrict the signal to the band of 300–3400 Hz. This consists of a tradeoff between quality/intelligibility and transmission bandwidth.

Moreover, due to its irregular structure, the voice presents many pauses. This originates that the voice signal is only present approximately 40% of the time. In channels presenting scarce bandwidth, these pauses can be used to interpolate other voice signals, using a mechanism known as *time assignment speech interpolation*.

6.1.1.1 Analog Audio

Analog data take continuous values of intensity within certain intervals. Legacy communication systems such as analog telephony or analog radio broadcast use analog voice. In this case, the analog signal is not subject to a digitization process. Nevertheless, in order to reduce its bandwidth and decrease the level of channel impairments (e.g., attenuation and distortion), a bandpass filtering† is normally applied to the signal.

Voice or audio signals can be directly transmitted as a continuous analog signal. Alternatively, a process widely employed in time division multiplexing (TDM) consists of simply transmitting a sample of an analog signal, at fixed intervals corresponding to the sampling period. This source encoding technique is known as *pulse amplitude modulation* (PAM). Each PAM sample has, at sampled instants, the amplitude of the sampled analog signal. The sampling period T_a is obtained from the sampling rate f_a as $f_a = 1/T_a$. Since a sample signal only occupies a part of the time of the channel, the remaining time can be used to send samples of other analog signals. The generation of a PAM sample from an analog signal is shown in Figure 6.3.

Knowing that a bandpass filtered voice signal has a spectrum in the range of 300 Hz to 3.4 kHz, according to the Nyquist sampling theorem, the minimum sampling rate is $f_a \geq 2 \times 3.4 = 6.8$ ksamples/s. This results in a minimum sampling period of $T_a \leq \left[1/(6.8 \times 10^3) \right] \approx 147.06$ μs. In voice signals, the sampling period of $T_a = 125$ μs is widely adopted, translating in a sampling period of $f_a = 8$ ksamples/s, that is, $f_a = 8$ kHz.

Other analog source encoding techniques exist, such as the pulse position modulation or the pulse duration modulation. While in the PAM, each sample presents the amplitude of the sampled analog signal, the pulse position modulation samples present constant amplitude, but the amplitude of the analog signal is encoded by varying the position of the pulse position modulation sample. In the case of the pulse duration modulation, the amplitude of the sample is also kept fixed, while the width of the pulse is used as a variable to encode the amplitude of the analog signal.

6.1.1.2 Digital Audio

Contrary to analog signals, digital signals take discrete values within certain intervals. Examples of digital signals are numbers, text, or binary signals. A binary signal is a special case of a digital signal, where the discrete values are only two (logic state 0 or 1, encoded with certain voltages). As described in Section 1.1.1, an advantage of binary signals relies on their improved resistance against the channel impairments and regeneration.

---

* The ear response is typically 20 Hz to 20 kHz for a young person, degrading with the age.
† For analog telephony, the bandpass filtering is applied within the range of 300 Hz to 3.4 kHz.

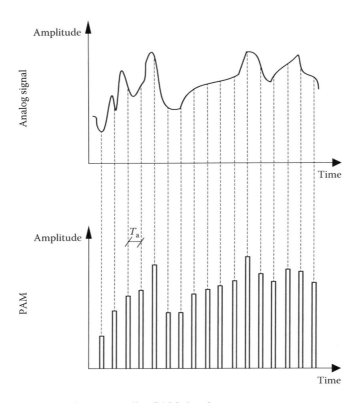

FIGURE 6.3 Analog signal and corresponding PAM signal.

6.1.1.2.1 Pulse Code Modulation

Pulse code modulation (PCM) is a source coding technique that comprises the digitization of analog voice, as well as the encoding of each sample of voice signal using a total of 8 *bi*nary digi*t*s (bits). The source encoding and decoding is performed with a device named *codec*, being responsible for performing the following individual operations (after bandpass filtering the speech within 300 Hz to 3.4 kHz):

- Sampling
- Quantization
- Coding

The sampling of an analog signal must follow the steps previously described by the sampling Nyquist theorem for PAM. Using a sampling rate equal or higher than the double of the maximum frequency component present in the analog signal assures that the recovered analog signal from its samples is not subject to distortion. The minimum sampling rate of a bandpass filtered voice signal is $f_a = 2 \times 3.4 = 6.8$ ksamples/s. Nevertheless, the PCM uses a sampling rate of 8 ksamples/s.[*] The signal resulting from the sampling operation corresponds to the PAM signal that is depicted in Figure 6.3.

The quantization process considers the approximation of the sampled analog signal[†] to a certain finite value.[‡] The quantization process results in a certain level of quantization noise, which results

---

[*] Some authors express sampling rate in Hz, instead of samples/s (samples per second).
[†] This may take an infinite range of values.
[‡] The closer finite value to the analog value.

from the difference between the input signal and its quantization levels. The level of quantization noise can be minimized by increasing the number of finite values L (quantization levels). The number of N_b bits is given by $N_b = \log_2 L$, where L corresponds to the number of quantization levels. However, since the quantization levels are encoded by a certain number of bits, increasing the number of quantization levels results in the increase of the number of bits, which translates to an increase in data rate. The PCM technique considers a total of 256 quantization levels.

Finally, coding is the process of transforming a quantization level (finite value) into a certain number of bits by following a certain coding law. The PCM technique considers 8 bits to encode a total of $2^8 = 256$ levels,[*] with the A- or μ-law. Knowing that the bit rate is given by $R_b = f_a \times N_b$, and that the sample rate of PCM is 8 ksamples/s, the resulting bit rate of the PCM technique becomes $R_b = 8000 \times 8 = 64$ kbps. Figure 6.4 depicts the principles employed in each individual PCM operation.

Music is another signal of interest. The sample rate typically employed in music signals is 44 ksamples/s, whereas its uniform coding comprises 16 bits per sample. This results in a bit rate of $R_b = 44,000 \times 16 = 704$ kbps.

The above description of the PCM corresponds to the uniform PCM. By uniform PCM, we mean that the quantization levels are uniformly distributed in amplitude. Nevertheless, it is known that lower amplitudes present a higher probability of occurrence than higher amplitudes. In order to minimize the quantization noise without increasing the number of quantization levels, a solution is to consider narrower quantization intervals at lower amplitudes and wider quantization intervals at higher amplitudes. This adapts the quantization levels to the statistic of the voice signal, and therefore, it allows reducing the quantization noise without having to increase the number of quantization levels. Alternatively, this allows a reduction in the source data rate for the same level of quantization noise. This technique is known as *nonuniform PCM*. Such nonuniform quantization is obtained using a logarithmic compression characteristic, as applied in the real PCM systems normalized by the ITU-T Recommendation G.711. Two main techniques exist [ITU-T G.711]: the μ-law used in the United States and Japan, and the A-law, adopted by Europe.

The μ-law is defined by

$$y = \text{sgn}(x) \frac{\ln(1 + \mu|x|)}{\ln(1 + \mu)} \tag{6.4}$$

where:
x stands for the signal amplitude of the input signal $(-1 \le x \le 1)$
$\text{sgn}(x)$ is the sign function of x
μ is a parameter used to define the compression degree

A typical value for μ is 255. Finally, y is the quantized level (output).

The A-law is defined by

$$y = \begin{cases} \text{sgn}(x) \dfrac{A|x|}{1 + \ln(A)} & 0 \le |x| \le \dfrac{1}{A} \\[3mm] \text{sgn}(x) \dfrac{1 + \ln|Ax|}{1 + \ln(A)} & \dfrac{1}{A} \le |x| \le 1 \end{cases} \tag{6.5}$$

where A has a typical value of $A = 87.6$.

---

[*] Since $N_b = \log_2 L \Leftrightarrow L = 2^{N_b}$.

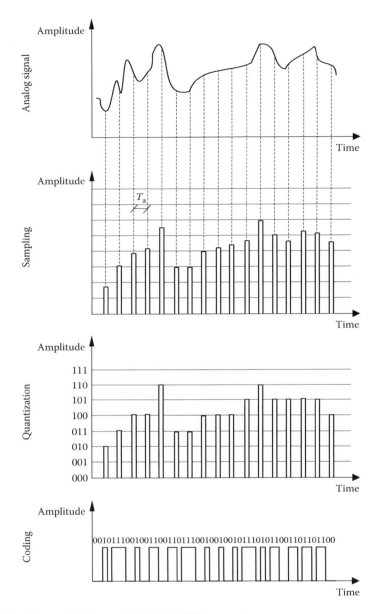

FIGURE 6.4   Principles applied in each individual PCM operation.

From Equation 6.5 and from Figure 6.5, we can conclude that there are two regions in the A-law:

- Low amplitudes region $(x \leq 1/A)$, with a linear variation
- High amplitudes region $(1/A \leq x \leq 1)$, with a logarithmic variation (see Figure 6.5)

It is worth noting that the μ-law leads to a slightly less amount of the quantization noise[*] than the A-law, for lower signal levels, whereas the A-law tends to achieve better performance at higher amplitudes.

---

[*] This translates into a higher SNR for lower signal levels.

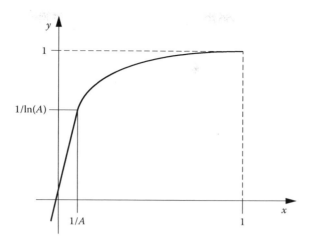

FIGURE 6.5 A-law compression characteristic.

The A-law uses 13 segments, with seven positives and seven negatives. Those two segments that cross the origin are linear, and therefore, they are accounted for as a single segment. Each 8-bit PCM word is encoded as follows[*]:

- The first bit represents the polarity of the input analog sample.
- The following three bits represent the segment.[†]
- The following (and last) four bits encode the linear quantization within each segment.

Comparing logarithmic A-law or μ-law against the linear algorithm, a reduction in the dynamic range from 12 into 8 bits is achieved. This translates in a throughput reduction from 96 to 64 kbps.

6.1.1.2.2 Differential Source Coding Techniques

An alternative solution to encode a source analog signal relies on encoding the difference between the actual signal sample and the previous signal sample (or, alternatively, the predicted signal sample[‡]). This is especially useful for signals such as voice or video, as those signals tend to suffer a small variation from sample to sample. Since the variation range of the difference signal is lower than that of the samples, a lower number of bits are necessary to encode the signal, in order to achieve the same performance.

A widely known differential voice coding technique is the delta modulation. This transmits a single bit to encode a sample. Its logic state is 1 if the signal increases its level (comparing to the previous sample) or 0 otherwise. In order to achieve an acceptable performance, the sampling frequency needs to be higher than that of the Nyquist sampling theorem. An important advantage of the delta modulation relies on the simplicity of the hardware employed to implement it.

Contrary to delta modulation, which considers the transmission of the difference signal between the actual and the previous signal sample, the differential PCM transmits the difference signal between the actual and the predicted next signal sample. The prediction is obtained by extrapolation from previous samples. A common implementation of the extrapolation is by making use of a transversal filter. The length of the transversal filter corresponds to the number of previous samples that are accounted in the prediction of the actual sample. Employing a transversal filter whose coefficients adapts to the variation of the signal normally improves the differential PCM technique.

---

[*] The μ-law uses 15 segments, instead of 13.
[†] The group of bits 000 and 001 are used by the central segment.
[‡] The predicted signal sample is obtained statistically making use of the previous samples.

The resulting technique is known as *adaptive differential PCM* [ITU-T G.726]; it allows a signal quality equivalent to that obtained with the PCM, but using data rates from 16 up to 40 kbps.[*]

6.1.1.2.3   Linear Predictive Coding

The previously described codec involves the transmission of the signal itself or, alternatively, of the difference signal. There is another group of codec that are specific to voice signals (vocoder), which simply involve the transmission of a certain number of parameters characteristic of the voice signal. Such parameters allow obtaining a replica of the original signal, by synthesis, at the receiver side. The analysis and generation of the key parameters at the transmitter side is normally employed by making use of a transversal filter. A vocoder includes two individual steps: voice analysis at the transmitter side and synthesis of the voice from the received parameters at the receiver side.

The linear predictive coding (LPC) includes a transversal filter that models the transfer function of the voice, as well as the type of excitation associated with voice signals. LPC is employed in the global system for mobile communications (GSM) for the transmission of voice at 8.0 kbps. A specific type of LPC, entitled LPC-10, allows the transmission of voice, in digital form, at a rate of 2.4 kbps over HF radio channels.

Another vocoder technique employed in radio channels is the mixed-excitation linear predictive, using rates of 2.4 kbps [MIL-STD-3005], 1.2 kbps [Wang et al. 2000], and 600 bps [Chamberlain 2001]. Its advantage relies on the high quality of the voice obtained at a reduced data rate,[†] as well as the improved performance obtained even under noisy conditions.

The moving picture experts group (MPEG) layer III (commonly referred to as MP3) is widely used to compress music, at a low data rate of 112–128 kbps, while enabling CD-quality music wideband audio (10–22 kHz). This is a very economic way of storing and transferring music. The layer III of the MPEG protocol (see Section 6.1.2) defines the audio compression technique that is sent or stored in a synchronized manner with the MPEG video.

6.1.2   VIDEO

A video is widely used as an information source to communicate. Examples of systems that make use of videos are analog and digital television, videoteleconference, interactive television, and high-definition television, among others.

The video information source consists of successive images. In order to give the idea of continuity, a minimum of 20 images per second is required. In television, 25–30 images per second are transmitted. Moreover, in order to give an idea of continuity, each image must be composed of a sufficient number of lines. A low-resolution image has 300–400 lines, whereas a high-resolution image has more than 1000 lines.

Finally, a line is composed of a number of image elements, known as *pixels*.[‡] Each pixel is characterized by its relative position, luminance, and color.

6.1.2.1   Analog Video

The European video system is entitled *phase alternation line* (PAL). It considers a total of 625 transmitted lines, of which only 575 lines are visible. Moreover, the PAL system considers 25 images per second. The number of transmitted image elements per second are $M = ABC$, where A corresponds to the number of lines (625), B corresponds to the number of pixels per line (572), and C stands for the number of images per second (25). This leads to $M = 625 \times 572 \times 25 = 8,937,500$ image elements per second. Since each image element can be viewed as a signal sample, according to the

---

[*] In this case, the sample rate corresponds to the Nyquist sampling rate.
[†] The voice quality obtained with the mixed-excitation linear predictive is much better than that obtained with LPC-10.
[‡] In fact, an image is composed of a number of pixels, which correspond to the number of pixels in a line multiplied by the number of lines per image.

Nyquist sampling theorem we can conclude that the PAL system requires a minimum bandwidth of $B = M/2 = 4.47$ MHz. Taking into account the return of the beam, the required bandwidth becomes 6 MHz* for luminance. Adding to this value, the bandwidth required for audio and color, the required bandwidth of a video system using the PAL standard, becomes approximately 8 MHz.

6.1.2.2 Digital Video

The transmission of data in the digital format allows a higher level of data integrity, which translates in a better reproduction of the original information at the receiver side. In addition, the Internet world demands for the total digitization of the different information sources.

A video signal is analog by nature, and a digital video is obtained from an analog video by digitization. We have seen that an analog video signal (PAL system) requires a bandwidth of approximately 8 MHz. According to the Nyquist sampling theorem this corresponds to a total of $M = 16$ Msamples/s.

Assuming that each image element is quantized and encoded using 10 bits per sample, we obtain the data rate of 160 Mbps. Once again, using the Nyquist sampling theorem, we conclude that the minimum bandwidth required to accommodate a digital video signal corresponds is 80 MHz. Such value is 10 times higher than the required bandwidth necessary to accommodate an analog video signal. This is the price to pay for the additional video quality, which is inherent to digital signals. Nevertheless, there are algorithms that allow the digitization of video at a lower data rate, as described in the following paragraphs.

6.1.2.2.1 Moving Picture Experts Group

Similar to the ITU-T G.711 codec used for voice, there are several codec used for video. MPEG is the commonly known video algorithm, presenting different versions. MPEG [Watkinson 2001] comprises a suite of standards that define how video is digitized and compressed. These standards encode the video using a differential algorithm.† Different MPEG standards present different bandwidth requirements, compression procedures, and resolutions. On the other hand, the digital video broadcast (DVB) defines the physical layer and data link layer employed in the distribution of digital video. Since this section refers to source coding, DVB is not included here, which is described in Chapter 12.

MPEG-2 is a video compression algorithm using data rates from 3 up to 100 Mbps, comprising different resolutions from 352×240 up to 1920×1080. DVB and high-definition television use this standard.

MPEG-4 is another ISO standard for video and audio encoding and compression. It also includes other types of media such as images, text, and graphics. The compression supports the transmission of MPEG-4 video and audio in 64 kbps channels (such as in an ISDN B channel).

Since the MPEG encoder is much more complex than the decoder, the MPEG is referred to as an *asymmetric coding mechanism*. This is an important characteristic as the MPEG is aimed to be widely used for video broadcast, and where the receiver is intended to be kept with low complexity and cost. Moreover, only the decoder is standardized, whereas the encoder is kept as a proprietary algorithm.

6.2 DIFFERENTIAL AND NONDIFFERENTIAL TRANSMISSION

The transmission of signals through metallic cables can be performed in two different ways:

- Differential
- Nondifferential

---

* This value is much higher than the 3.4 kHz bandwidth required for telephony.
† Differential algorithms translate in lower data rates. Since the variation range of the difference signal is lower than that of the samples, a lower number of bits are necessary to encode the signal.

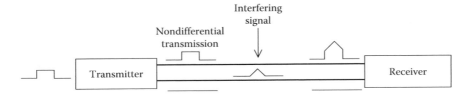

FIGURE 6.6 Nondifferential transmission.

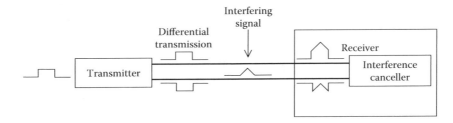

FIGURE 6.7 Differential transmission.

Nondifferential transmission considers transmission of a signal (positive or negative) in one conductor, while the other conductor is connected to the ground (Figure 6.6).

Contrarily, differential transmission considers the transmission of a signal (positive or negative) in one conductor, while the other conductor carries the same signal but with inverted polarity (negative or positive). It is worth noting that transmitted signals can be either analog or digital. In the latter case, it can comprise any line coding technique from those defined in the line coding section (e.g., nonreturn to zero, pseudoternary, and Manchester). The advantage of differential transmission relies on its inherent ability to identify interfering signals, and partially cancel them.

Let us consider the transmission chain as plotted in Figure 6.7, which refers to a differential transmission. As can be seen, the transmitted signal is initially clear of noise or interferences. Somewhere between the transmitter and the receiver, an interfering signal is added to the transmitted one, thereby degrading the signal-to-noise ratio (SNR). The resulting signal is plotted at the input of the receiver. The interference canceller performs its function by detecting that the received signals in the two conductors are not opposite. The interference canceller employs a method that relies on performing the sum of these two signals. If the result is zero, then the signal is not subject to any noise or interference. Contrarily, if the result of the sum is any other value other than zero, then such result corresponds to the double of the interfering or noise signal, which can then be subtracted from the received signals in the two conductors. Naturally, the interfering or noise signals are not pure signals, but a sum of many elementary signals. Moreover, these signals are always subject to variations over time, and the interference canceller may not be able to follow these timely variations. Consequently, even with a differential transmission, it is only partially possible to mitigate the negative effects of these impairing signals.

6.3 LINE CODING SCHEMES

Line coding is employed for the transmission of signals in baseband. Line coding is also referred to as *digital baseband modulation* or simply as *digital coding*. A block diagram composed of a line encoder/decoder is plotted in Figure 6.8. As can be seen, the line encoder/decoder normally has the error control capability embedded in itself. This is implemented using an error correction code or an error detection code associated with the retransmission of data. Both these techniques are covered in Chapter 12.

FIGURE 6.8 A communication chain including a line encoder/decoder.

The selection of the digital encoding technique depends on the following different criteria:

- *Required bandwidth*: According to the Nyquist theorem, for a certain fixed throughput, increasing the number of bits encoded by a symbol (i.e., for higher values of *M*) results in a reduced signal bandwidth.
- *Level of immunity to the channel impairments*: If the level of the transmitted symbol that encodes logic state 1 is sufficiently different from the level that encodes logic state 0, then the level of noise or interference necessary to originate a corrupted bit also needs to be high. Nevertheless, the required transmit power level increases with the increase in the difference between the levels of the two logic states.
- *Self-synchronization of the receiver from the received signal*: Using a digital encoding technique that presents zero crossing in all symbols allows the receiver to extract a clock signal for synchronism[*] purposes. This avoids the spending of bandwidth for the transmission of a synchronism signal using a different channel or multiplexed with the data. Nevertheless, these types of zero-crossing symbols are typically associated with a higher bandwidth.
- *Error detection capability*: Even without the use of an error detection code, some sequence of levels is not allowed by certain digital encoding techniques. As an example, a Manchester symbol considers always a transition at the mid instant of the symbol period.[†] The absence of such transition signalizes the receiver that the symbol is subject to an error.
- *Zero value average level*: The use of symbols with zero-value average level facilitates the amplification process of transmitters.

The symbols used by line codes may be grouped into different classifications:

- *Unipolar*: A logic state is represented by a positive or a negative voltage, whereas the other logic state corresponds to an absence of voltage. An advantage of this technique relies on its simplicity and the energy saving, which results from the absence of signal corresponding to one of the logic states.
- *Polar*: A logic state is represented by a positive voltage, whereas the other logic state corresponds to a negative voltage. According to the central limit theorem, the probability of occurrences of the two logic states is equal. This results in a signal with zero mean value.
- *Bipolar*: A certain logic state is represented by alternating positive and negative voltages, whereas the other logic state is represented by an absence of voltage. Note that the pole of the voltage *V* varies sequentially. This represents an advantage as an absence of the expected pole variation allows the receiver to detect an error.
- *Biphase*: Each logic state is represented by a transition from a positive to a negative voltage, or the reverse. This assures zero crossing in all symbols, which is an important characteristic for synchronism purposes.

---

[*] This corresponds to physical layer synchronism, that is, the ability of a receiver to know the exact instance of the start and the end of a symbol.
[†] From +V into −V (logic state 0) or from −V into +V (logic state 1).

6.3.1 RETURN TO ZERO

The return-to-zero line coding belongs to the group of unipolar techniques. It is characterized by representing logic state 1 with a certain voltage (+*V*) for half of the bit duration and an absence of voltage during the remaining half of the bit duration. Moreover, logic state 0 is represented by an absence of voltage; see Figure 6.9. Alternatively, the representation of the two logic states can be reversed, or the voltage *V* can be a negative, instead of a positive value.

6.3.2 NONRETURN TO ZERO

The nonreturn-to-zero line coding belongs to the group of unipolar techniques. It is characterized by representing logic state 1 with a certain voltage (+*V*) and logic state 0 with an absence of voltage. Alternatively, logic state 0 can be represented by a certain voltage (+*V*), while logic state 1 is represented by an absence of voltage. Note that the voltage remains constant, that is, it does not return to zero during the bit duration. Figure 6.10a and b shows both the cases.

6.3.3 NONRETURN TO ZERO INVERTED

The nonreturn-to-zero inverted technique belongs to the group of unipolar line coding techniques. It is characterized by representing logic state 1 with a transition (from 0 to *V* or from *V* to 0) and logic state 0 by an absence of transition. The transition, or its absence, occurs at the start of the bit duration. Since the encoded signal is a function of the difference between the previous and the following logic state, this technique is also referred to as *differential*. Alternatively, the representation of the two logic states can be reversed. Note that the voltage *V* can be either a positive or a negative value (Figure 6.11).

6.3.4 BIPOLAR ALTERNATE MARK INVERSION

The bipolar alternate mark inversion (bipolar AMI) line coding belongs to the group of bipolar line coding technique. As shown in Figure 6.12, this is characterized by representing logic state 0 with

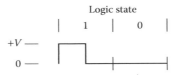

FIGURE 6.9   Unipolar return to zero.

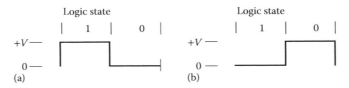

FIGURE 6.10   Unipolar nonreturn to zero (both options [a] and [b] are possible).

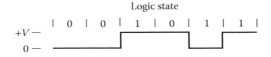

FIGURE 6.11   Nonreturn to zero inverted.

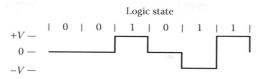

FIGURE 6.12 Bipolar AMI.

an absence of voltage and logic state 1 by alternating positive and negative voltages (+V or −V). Two important characteristics of this line coding technique rely on its inherent ability to detect errors (two successive +V or −V are impossible conditions) and on its zero mean value.

6.3.5 PSEUDOTERNARY

The pseudoternary line coding technique corresponds to the biphase AMI with the difference that logic state 1 is represented by an absence of voltage, whereas logic state 0 is alternately represented by a positive and a negative voltage (+V or −V).

The pseudoternary corresponds to the bipolar AMI; however, the bit that alternates is logic state 0, instead of logic state 1. Similar to the bipolar AMI, if two successive logic state 0 bits present the same voltage, an error is detected. This is depicted in Figure 6.13.

6.3.6 MANCHESTER

The Manchester line coding scheme belongs to the group of biphase line coding technique. It is characterized by representing logic state 1 with a positive transition at half of the bit duration (from 0 to +V) and logic state 0 with a negative transition at half of the bit duration (from +V to 0). Alternatively, the transitions can be reversed for the two logic states, or the considered voltage can be negative, instead of being positive; see Figure 6.14. An important advantage of the Manchester technique relies on its inherent synchronism capability that results from the fact that at least one transition exists per bit duration. Moreover, an important disadvantage of the Manchester scheme relies on its excessive required bandwidth that results from the excessive number of transitions.

6.3.7 DIFFERENTIAL MANCHESTER

Similar to the Manchester technique, differential Manchester belongs to the group of biphase line coding technique. It is characterized by the existence of a transition at half bit duration. However,

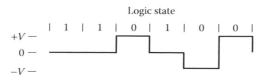

FIGURE 6.13 Pseudoternary.

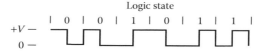

FIGURE 6.14 Manchester.

this transition is not used to encode bits; it is only used for synchronism purposes. The transition at the beginning of the bit duration is the one that encodes the bits as follows: the existence of a transition corresponds to logic state 0, and the absence of a transition represents logic state 1; see Figure 6.15.

6.3.8 TWO BINARY ONE QUATERNARY

Two binary one quaternary (2B1Q) consists of a line coding technique that encodes groups of two bits into a single voltage. The increased code efficiency of the 2B1Q results in an important advantage, as compared to previous line coding techniques. To allow this, a total of four voltages is required (two positive and two negative), as described in Table 6.1.

The 2B1Q line coding technique is plotted as shown in Figure 6.16.

6.4 MODULATION SCHEMES

A modem implements modulation at the transmitter side, whereas the demodulation process is carried out at the receiver side. The modulation involves the process of encoding one or more source bits into a modulated carrier wave. An important advantage of using a modem, instead of line coding, relies on the ability to select a frequency band where the channel impairments are less destructive to the transported signal. Contrarily, line encoding techniques always transmit the signals in the baseband, even though if the level of distortion or attenuation is high. Let us consider a voice-graded twisted pair as an example, whose bandwidth is limited to the frequency range of 300 Hz to 3.4 kHz. Above this upper frequency, the attenuation level becomes too high, translating into a high level of distortion. If one intends to transmit a symbol rate whose bandwidth is higher

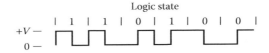

FIGURE 6.15   Differential Manchester.

TABLE 6.1
2B1Q Line Coding

Logic State of 2-Bit Groups	Signal Level
00	$-V_2$
10	$+V_2$
11	$+V_1$
01	$-V_1$

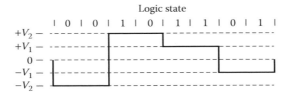

FIGURE 6.16   2B1Q line coding.

than that required for voice, the solution is to select a carrier frequency to place the signal (band-pass signal) where the channel impairments are less intense than those experienced in baseband. A dial-up or a digital subscriber line modem typically implements this operation. A block diagram of a communication chain, including a modem, is plotted as shown in Figure 6.17. As can be seen, a modem typically has the error control capability embedded in itself, consisting of error correction or error detection (normally associated with retransmission). Error control techniques are covered in Chapter 12.

The three elementary modulation schemes include amplitude, frequency, or phase of a carrier, respectively, as parameters to encode source bits [Proakis 1995], as defined in Sections 6.4.1 through 6.4.3.

6.4.1 Amplitude Shift Keying

The amplitude shift keying (ASK) is an elementary modulation scheme that uses amplitude of a carrier as the parameter to encode bits. As shown in Figure 6.18, logic state 1 is represented by the transmission of a carrier wave with certain amplitude, and logic state 0 is represented by the transmission of the same carrier, but with a different amplitude (alternatively, one of the logic states can be represented by an absence of carrier transmission). Alternatively, the encoded logic states can be reversed.

A predefined parameter is the frequency of the carrier wave. Such parameter is known by both the transmitter and the receiver. Moreover, the amplitude is another predefined parameter. A high amplitude level is more resistant to a corruption of bit,[*] but requires a higher power from the transmitter.

The ASK can easily be extended to $M = 4$ by adopting four amplitude levels (one of this level may correspond to the absence of carrier). This results in 4-ASK modulation, where each symbol encodes $\log_2 4 = 2$ bits.

6.4.2 Frequency Shift Keying

The frequency shift keying (FSK) is an elementary modulation scheme that uses frequency of a carrier as the parameter to encode bits.

As shown in Figure 6.19, logic state 1 is represented by the transmission of a carrier wave with a frequency f_1, while logic state 0 is represented by the transmission of a carrier wave with a

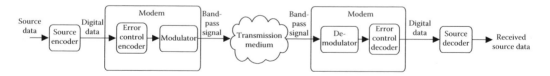

FIGURE 6.17 A communication chain including a modem.

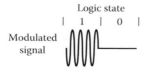

FIGURE 6.18 Amplitude shift keying.

---

[*] That is, receive logic state 1 when the transmitted logic state was 0, or the reverse.

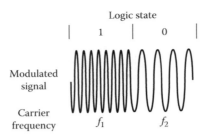

FIGURE 6.19 Frequency shift keying.

frequency f_2. In order to minimize the occurrence of corrupted bits, the frequencies f_1 and f_2 should be sufficiently far apart in order to avoid being affected by frequency oscillations or Doppler effects.

6.4.3 PHASE SHIFT KEYING

The phase shift keying (PSK) is an elementary modulation scheme that uses the phase of a carrier as the parameter to encode bits. As shown in Figure 6.20, logic state 1 is represented by the transmission of a carrier wave with the phase 0 and logic state 0 is represented by the transmission of a carrier wave with the phase π. Alternatively, the encoded logic states can be reversed.

The described modulation scheme corresponds to binary phase shift keying (BPSK*), as there are only two phases (each symbol represents a single bit). However, we may use a symbol to encode more than 1 bit. The constellation and the signal mapping of the quadrature phase shift keying (QPSK) is plotted as shown in Figure 6.21. A QPSK symbol presents four discrete values ($M = 4$), allowing the encoding of 2 bits ($\log_2 4 = 2$). While the BPSK only uses the real representation (phases 0 and π), the QPSK uses both real and imaginary parts. This results in a higher spectral

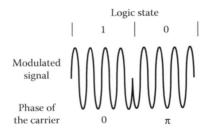

FIGURE 6.20 Binary phase shift keying.

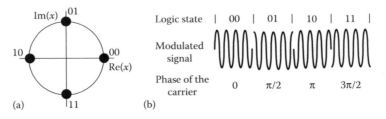

FIGURE 6.21 Quadrature phase shift keying: (a) constellation and (b) signal mapping.

---

* The BPSK modulation corresponds to 2-PSK ($M = 2$).

efficiency (bit/s/Hz), that is, a higher bit rate transported in the same signal bandwidth, as defined by (6.1) and (6.3). The price paid by this additional bit rate is the worse bit error probability that results from the decreased minimum Euclidian distance, as compared to BPSK modulation. The Euclidian distance is the distance between two constellation points. For a lower minimum Euclidian distance, a lower level of noise is enough to originate a corrupted bit. This is the reason why higher order modulations require higher levels of SNR. While in the case of BPSK, the minimum Euclidian distance is 2; in the case of the QPSK such value is reduced to $\sqrt{2}$.

The bit error probability for M-PSK modulations is presented in Chapter 5, as a function of both E_b/N_0 and C/N.

6.4.4 M-QAM Constellations

Multilevel quadrature amplitude modulation (M-QAM) is considered as an attractive technique to achieve high throughput within a limited spectrum, due to its high spectral efficiency. Therefore, it has been proposed for wireless systems by several authors [Webb and Hanzo 1994; Webb and Steele 1995; Goldsmith and Chua 1997]. In fact, 16-QAM modulation has already been standardized for the high speed downlink packet access mode of the universal mobile telecommunications system (UMTS) by 3GPP [Marques da Silva et al. 2009].

M-QAM constellations can be viewed as a mix of both ASK and M-PSK, as both the amplitude and the phase suffer a variation for different M-QAM constellation points. While the M-PSK considers all different constellation points with the same amplitude but with different phases, different constellation points of M-QAM modulation present different amplitudes, as well as different phases.

Figure 6.22 shows the constellation points of the 16-QAM modulation. The $M = 16$ levels allows each symbol to encode a total of $\log_2 16 = 4$ bits. As can be seen, moving from one constellation point to an adjacent point only changes 1 bit. The mapping of bits onto symbols that follow this rule is known as *Gray mapping*. Most frequent channel impairments (noise and interference) only originate during the movement from one constellation point to its adjacent, resulting in a single corrupted bit. As an example, the constellation point 1101 is characterized by having the phase $\pi/4$ and the amplitude $\sqrt{2}a/2$ (where a is the Euclidian distance, that is, the minimum distance between two

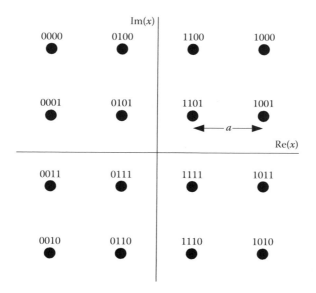

FIGURE 6.22   Constellation of 16-QAM.

adjacent constellation points). On the other hand, the constellation point 0000 is characterized by having the phase $3\pi/4$ and the amplitude $3\sqrt{2}a/2$.

With regard to M-QAM modulation, each constellation point is defined by a certain phase, as well as by a certain amplitude. Nevertheless, due to noise or interferences, a certain modulation symbol is commonly received in a position that is different from the one transmitted. The received signals correspond to the transmitted ones, plus the amplitude of noise, interference, distortion, or other channel impairments. A common strategy used in detectors relies on making a symbol decision based on the constellation point that is closer to the received one.

As the power of noise and interference increases, that is, as the SNR decreases, the distance between the received signal and the constellation point increases. In the case of high power noise or interferences, if the received signal becomes closer to a constellation point that is different from the one transmitted, the detector makes the wrong estimate for the received symbol. This translates in a corrupted symbol, and, consequently, in one or more corrupted bits (Figure 6.23).

Figure 6.24 shows a 64-QAM modulation with a SNR of 30 dB. In grey, it is shown a number of possible received signals, whereas the crosses refer to constellation points, that is, to possible transmitted symbols.

Figure 6.25 shows the same, but for a SNR of 20 dB. Note that decreasing the SNR corresponds to increasing the power of noise or interferences. This results in an increased distance between the possible received signals (dark points), relating to the transmitted signal (crosses). In other words, as the SNR decreases, the power of noise or interferences increases, resulting in a higher scattering of the received signals.

It is worth noting that the modulation scheme may adapt dynamically to the channel conditions. A noisy channel,* or a channel that is more subject to any type of interference, should use a lower order modulation scheme (e.g., QPSK), in order to achieve the same performance. Contrarily, a channel whose level of impairments is lower allows using a higher order modulation scheme (e.g., 16-QAM). As described in Chapter 12, the adaptive modulation and coding (AMC) technique changes the modulation scheme/order and/or code rate dynamically as a function of the SNR. The bit error probability for the M-QAM modulation is presented in Chapter 5.

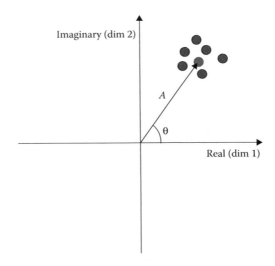

FIGURE 6.23   The transmitted constellation point (central point) and possible received signals due to noise or interference.

---

* This corresponds to a lower SNR level.

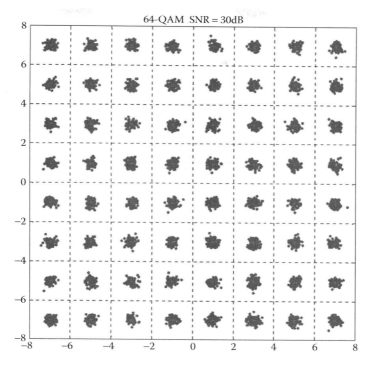

FIGURE 6.24 Scattering of the received signal around the transmitted constellation points (central crosses), for an SNR of 30 dB of 64-QAM.

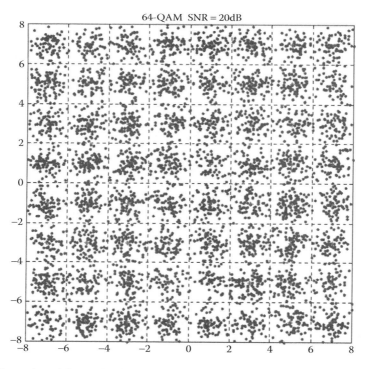

FIGURE 6.25 Scattering of the received signal around the transmitted constellation points (central crosses), for an SNR of 20 dB of 64-QAM.

6.5   CODING EFFICIENCY OF A SYMBOL

The coding efficiency of a symbol employed in a line coding or a modem consists of a quotient between the source bit rate R_B and the modulation frequency f_M used to encode the source bit rate. The modulation frequency corresponds to the number of different discrete levels per second (amplitude, frequency, phase, or a combination thereof) that is used to encode the source bit rate. We may express the coding efficiency as

$$\eta = R_B/f_M \qquad (6.6)$$

Using a code whose symbol encodes more than 1 bit (e.g., QPSK[*]) results in a coding efficiency higher than one. Contrarily, a code that needs more than one level to represent 1 bit (e.g., Manchester[†]) presents a coding efficiency lower than one.

6.6   SCRAMBLING OF SIGNALS

The scrambling operation aims to improve the signal quality by changing a sequence and logic state of bits. This is achieved by splitting a long sequence of bits into groups, and applying the scrambling operation to each group of bits individually. This operation presents the following advantages:

- *Synchronism*: Since it breaks a long sequence of bits with the same logic state, it increases the number of logic state transitions. This results in an improved capability of the receiver to extract the clock signal from the received signal.
- *Error control*: After scrambling of signals, some sequence of bits become impossible. Detecting an impossible sequence of bits gives the receiver the ability to detect an error.
- *Security*: A third party who intercepts a message cannot decode the data without having knowledge about the generator polynomial. Therefore, the scrambling operation can be viewed as a type of encryption.

Figure 6.26 shows an example of a scrambling encoder (scrambler). A scrambling encoder is implemented using feedback shift registers. The output of the scrambler depicted in Figure 6.26 is given by

$$b_n = a_n \oplus a_{n-2} \oplus a_{n-5} \qquad (6.7)$$

Note that the symbol $\oplus$ in the equation stands for modulo 2 adder (XOR).

The corresponding generator polynomial is

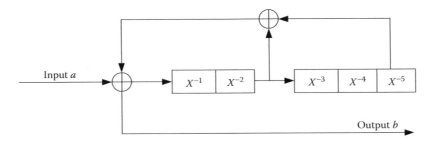

FIGURE 6.26   An example of a scrambling encoder.

---

[*] The coding efficiency of the QPSK modulation scheme is 2.
[†] The coding efficiency of the Manchester code is 0.5.

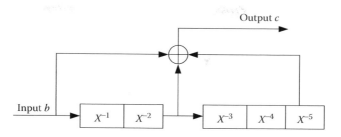

FIGURE 6.27 An example of a scrambling decoder.

$$P(x) = 1 + x^{-2} + x^{-5} \tag{6.8}$$

The encoder is normally initialized by filling all registers of the shift register with zero value bits. Then, for each data bit that is fed, the shift register shifts once to the right. An important parameter of a scrambler is the constraint length. It corresponds to the number of previous input bits that a certain output bit depends on; this corresponds to the number of memory registers.

Figure 6.27 shows an example of a scrambling decoder (descrambler). It is implemented using feed-forward shift registers. The output of the descrambler depicted in Figure 6.27 is given by

$$c_n = b_n \oplus b_{n-2} \oplus b_{n-5} \tag{6.9}$$

This corresponds to the decoder of the scrambling encoder depicted in Figure 6.26.

6.7 MULTIPLEXING

Multiplexing is a mechanism that allows the sharing of communication resources among different channels. It includes a multiplexer (MUX) at the transmitter and a demultiplexer (DEMUX) at the receiver side. Multiplexing is the operation of encapsulating different channels into a single structured signal for transmission into a common transmission medium. Figure 6.28 shows a generic block diagram of a link using multiplexing.

Depending on whether different channels are transported in different frequency subcarriers or different time slots, the multiplexing technique is referred to as *frequency division multiplexing* (FDM) or as TDM. Figure 6.29 depicts the frequency and time characteristics of both FDM and TDM signals. Independent channels, transmitted in different frequency bands or different time slots, are uncorrelated. The uncorrelation is assured by making use of guard bands or guard times between adjacent subcarriers or time slots (see Figure 6.29).

If the resources are directly shared by different users,[*] instead of channels, the multiplexing terminology is equivalent, but the designation is followed by the word *access*, that is, frequency

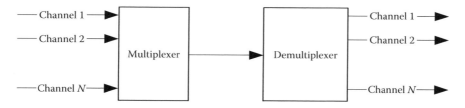

FIGURE 6.28 Generic block diagram of a link using multiplexing.

---

[*] Where each user has a different signal.

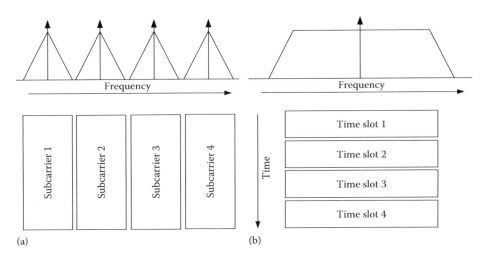

FIGURE 6.29 Characteristics of (a) FDM and (b) TDM signals.

division multiple access or time division multiple access. When different users transmit simultaneously, using the same bandwidth, but using the spread spectrum transmission technique with different spreading codes, the multiple access technique is referred to as *code division multiple access*. This topic is covered in Chapter 7. Another multiple access technique widely employed in local area networks and metropolitan area networks is the carrier sense multiple access with collision detection or the carrier sense multiple access with collision avoidance. These multiple access techniques are employed to perform a statistical multiplexing of the resources, instead of rigidly allocating resources to users (that, instantaneously, may not require). Both carrier sense multiple access with collision detection and carrier sense multiple access with collision avoidance are described in Chapter 12.

6.7.1 FREQUENCY DIVISION MULTIPLEXING

FDM considers the simultaneous transmission of different channels in different frequency bands. The uncorrelation of signals (channels) among different sub-bands is assured using guard bands. A guard band consists of a frequency band that does not include the transmission of any signal. This is especially important because the limits of frequency bands used for the transmission of signals are not abrupt. This depends on the cut-off response of the bandpass filters and amplifiers.

FDM signals are currently used by cable television operators to distribute analog or digital television channels. In the past, FDM was widely implemented in coaxial cables, which was the transmission medium employed to interconnect different telephonic switching nodes.

Figure 6.30 shows the processing of an FDM transmitter, assuming double sideband (DSB) signals. As can be seen, it consists of modulating different channels with different subcarriers ($f_1, f_2, ..., f_N$), followed by an adder module (i.e., summer at signal level). Afterward, the resulting signal is carrier modulated around a certain carrier frequency f_C, and bandpass filtered in order to remove the (negative) frequencies below the carrier frequency f_C.

If the channel consists of an analog telephony channel, the bandwidth B corresponds to 3.4 kHz. In the case of digital signals, the minimum bandpass signal bandwidth B_{min} can be deduced from (6.1) as[*]

---

[*] Note that the maximum bandpass signal bandwidth is calculated for rollof factor $\alpha = 1$, resulting in $B_{max} = (2R_B / \log_2 M)$.

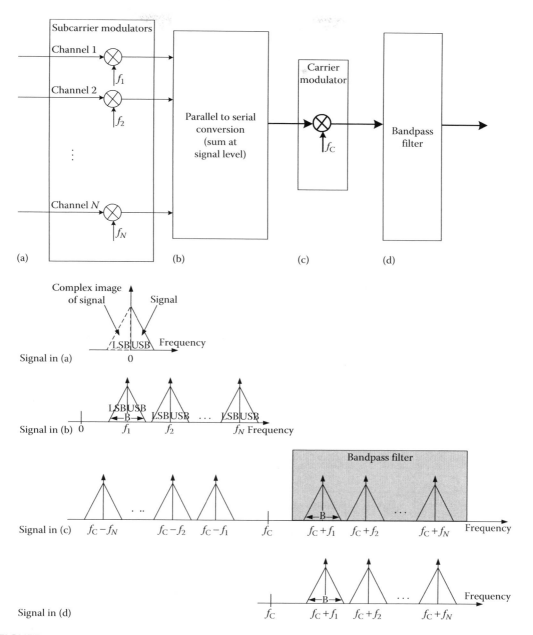

FIGURE 6.30 Processing of a FDM transmitter with the corresponding signals (DSB signals are assumed) at (a) subcarrier modulator; (b) parallel to serial conversion; (c) carrier modulator; and (d) bandpass filter.

$$B_{min} = \frac{R_B}{\log_2 M} \tag{6.10}$$

The total transmitting signal has a bandwidth corresponding to the sum of the elementary signal bandwidths plus the sum of the guard bands existing between adjacent sub-bands.

As can be seen from the plotted figure, the carrier modulation operation translates the baseband signal from the frequency 0 into the frequency f_C, generating frequencies above and below the carrier frequency. This results from the fact that the carrier modulation operation corresponds to a multiplication of a signal in the time domain by a sinusoid $\cos(jw)$. Since $\cos(jw) = (1/2)e^{jw} + (1/2)e^{-jw}$, and since $\mathcal{F}\left[v(t)e^{jw}\right] = V(f - f_c)$, we conclude that the Fourier transform (spectrum) of a time domain signal $v(t)$ multiplied by a complex carrier results in the spectrum of the signal $V(f)$ translated in the spectrum as $V(f - f_c)$. Taking into account the frequency translation with both e^{jw} and e^{-jw}, we obtain the plotted signal as shown in Figure 6.30c.

To avoid duplication of signals, the bandpass filtering operation assures that only frequency components above the carrier frequency are transmitted. This translates in power and spectrum saving. The transmitted signal corresponds to the plotted signal as shown in Figure 6.30d.

Note that the signals of elementary channels can be either DSB or single sideband (SSB). In the latter case, after the subcarrier modulation process, one of the sidebands needs to be filtered (using a sideband filter*). SSB signals only consider the transmission of the lower sideband or upper sideband, resulting in a power and a spectrum saving [Carlson 1986]. The generation of SSB signals is depicted in Figure 6.31, especially the upper sideband.

Figure 6.32 shows the generic block diagram of an FDM receiver alongside with the corresponding signals. The first operation consists of bandpass filtering the received signal with a bandwidth corresponding to approximately the bandwidth occupied by all transported channels. This filter is also referred to as the *receiving filter*, and its main objective consists of removing all the noise present outside the band of the signal of interest. After the filtering operation, the carrier demodulation operation is performed. This consists of performing a translation of the bandpass signal from the carrier frequency f_C into the baseband (i.e., frequency zero). This is followed by an operation of filtering centered in the subcarrier frequency. This isolates each of the signals, after

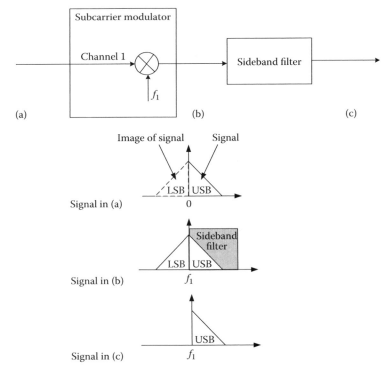

FIGURE 6.31 Generation of a SSB signal (upper sideband) at (a) input of subcarrier modulator; (b) sideband filter; and (c) output of subcarrier modulator.

---

* A sideband filter can be viewed as a bandpass filter.

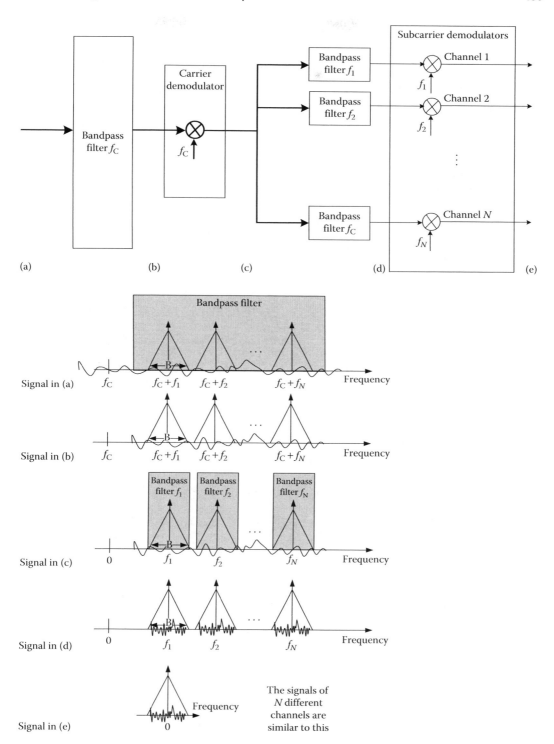

FIGURE 6.32 Processing of a FDM receiver with the corresponding signals at (a) input to bandpass filter; (b) input to carrier demodulator; (c) input to bandpass filters; (d) input to subcarrier demodulators; and (e) output to subcarrier demodulators.

which independent subcarrier demodulation is performed, in order to recover the replicas of the transmitted signals (corresponding to different channels).

6.7.2 TIME DIVISION MULTIPLEXING

TDM considers the transmission of different channels in different uncorrelated time slots, but using a single-carrier frequency. Each time slot carries a sample of a channel. Moreover, the channels or time slots 1 to N comprise a frame, and the frame is repeated at regular intervals. Therefore, the receiver n is able to receive a sample of the channel n every at regular intervals. From these signal samples, the receiver can reconstruct the original signal.

Since a synchronism among different channels in the transmitter and the receiver is required, this multiplexing technique is also referred to as *synchronous TDM*.

Figure 6.33 shows the block diagram of a TDM transmitter and a receiver.

The channels 1 to N can be either analog or digital. In the case of analog, each channel is typically encoded using the PAM. In this case, one sample of a channel (analog signal) is transmitted in each time slot.

In the case of digital signals, there are two different interposition methods: bit interposition and word interposition. The former method includes the transmission of a single bit in each time slot. On the other hand, the word interposition comprises the transmission of a group of bits[*] in each time slot.

Since the scan of the receiver (scan B) is synchronized with the scan of the transmitter (scan A), one sample of channel n is transmitted in time slot n, being extracted at the receiver side. In the case of digital signal, one or more bits are transmitted in each time slot. A common source coding technique is the PCM. For both analog and digital signals, the sampling period $T_a = 1/f_a$ needs to follow the Nyquist sampling theorem, that corresponds to $T_a \leq 1/2B$. Assuming a sampling frequency of 8 ksamples/s, the sampling period comes $T_a = 125$ μs. Consequently, the period of the frame is also 125 μs, because the receiver n needs to receive a signal sample every 125 μs.

The total throughput of the shared transmission medium is equal or higher than the sum of elementary channel's throughputs. As shown in Figure 6.29, the uncorrelation between adjacent time slots is assured by using guard times. This corresponds to time periods between adjacent time slots without data being transmitted. As shown in Figure 6.33, the group of all time slots (1, 2, ..., N) is referred to as a *frame*, and its transmission is repeated for every frame interval. This frame consists of a transport frame, not a data link layer frame.[†] This frame does not include any kind of

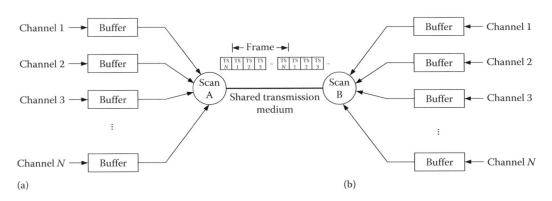

FIGURE 6.33 Processing of a TDM (a) transmitter and (b) receiver.

---

[*] Using the PCM coding, a word consists of a group composed of 8 bits (one signal sample).
[†] The data link layer is described in Chapter 12.

error control (detection or correction). Error control capability is performed for the data link layer frame.* Nevertheless, the transport frame needs to include some additional bits for signalization at the beginning and at the end of a frame (using, e.g., a flag). This allows scan A and scan B devices to maintain synchronization.

CHAPTER SUMMARY

This chapter provided a view about the source coding and transmission techniques employed in both wired and wireless communications, including the study of the difference between of a baseband and a bandpass signal.

The source coding was introduced, as well as the voice coding techniques. It was shown that the PAM consisted of an analog source coding technique, whereas PCM comprised the digitization of voice signals at a rate of 64 kbps. Moreover, some advance voice coding techniques were described. The MP3 is widely employed to encode music at a low rate of around 120 kbps, while enabling CD-quality music wideband audio.

Video coding techniques were also dealt with in this chapter. It was shown that the video coding system used in Europe is PAL. The PAL system comprises 25 images per second, with 625 lines per image and 572 pixels per line. It was described that the MPEG2 is a video compression algorithm used in DVB and in high-definition TV, using data rates from 3 to 100 Mbps, depending on the resolutions. In addition, it was viewed that the MPEG4 comprises a compression of video, audio, images, and text, in 64 kbps channels.

This chapter introduced differential and nondifferential transmissions employed in wired transmission media. It is shown that differential transmission allows the mitigation of negative effects of the noise and interferences.

Line coding techniques were introduced, being employed in wired transmission media using baseband transmission. The return-to-zero coding technique is characterized by representing the logic state 1 with a certain voltage for half of the bit duration and an absence of voltage during the remaining half of the bit duration (the same for the logic state 0, but with a different voltage). On the other hand, nonreturn to zero is characterized by representing the logic state 1 with a certain voltage and the logic state 0 with an absence of voltage (or alternatively, the voltage of the logic state 0 can be different). Naturally, the voltage employed can be either positive or negative.

The nonreturn to zero inverted defines logic state 1 by a transition from 0 V to a certain voltage or from a certain voltage to 0 V. Moreover, logic state 0 is characterized by an absence of transition. The AMI considers logic state 1 characterized by alternating positive and negative voltages, whereas logic state 0 corresponds to an absence of voltage. The pseudoternary corresponds to the AMI; however, the two logic states are reversed here. The Manchester technique considers logic state 1 characterized by a positive transition at half of the bit duration, while logic state 0 presents a negative transition at half of the bit duration. The differential Manchester technique encodes the logic state 1 with a transition that can be either positive or negative at the start of the bit duration. Logic state 0 is encoded with an absence of transition at the start of the bit duration. Finally, the 2B1Q line coding technique comprises four voltages to encode groups of two bits. Consequently, 2B1Q is well fitted to support higher throughputs.

Modulation schemes were also studied in this chapter, being employed in both wireless and wired transmission media, using bandpass signals. Three elementary modulation schemes were studied: ASK, FSK, and PSK. Moreover, some advanced modulation schemes were also described, comprising a mixture of these elementary modulation schemes. It was shown that increasing the modulation order allows a higher throughput and higher spectral efficiency. Nevertheless, since the Euclidian distance decreases, this translates in a degradation of performance, in terms of bit error rate.

---

* For example, the PPP and the HDLC data link layer protocols use CRC as error detection codes.

The coding efficiency of symbols was also addressed in this chapter. It was shown that different modulation symbols and different line coding symbols present different coding efficiency. This determines the required bandwidth necessary to accommodate a certain bit rate, which can also be regarded as the spectral efficiency.

The scrambling technique was also studied, being widely employed to improve the synchronism characteristics of the signal, for error control purposes, or for security reasons.

Finally, two basic multiplexing techniques were also studied: FDM, widely employed in analog communications, and TDM, typically used in digital communications.

REVIEW QUESTIONS

1. What is the difference between a line coder and a modem?
2. What are the kinds of line encoding techniques you studied?
3. What is the difference between FDM and TDM?
4. List the kinds of multiplexing techniques you know.
5. What is a scrambler used for?
6. What does coding efficiency stands for?
7. What is the minimum bandwidth required to accommodate a baseband signal that comprises a symbol rate of 2 Msymbol/s?
8. What is the minimum bandwidth required to accommodate a signal at the output of a modem that comprises a symbol rate of 2 Msymbol/s?
9. Describe the block diagram of an FDM transmitter.
10. Describe the block diagram of an FDM receiver.
11. How can we generate an SSB signal?
12. What are the advantages of transmitting signals using a modem, instead of a line encoder?
13. What is the bandpass filter used for in an FDM receiver?
14. What does Euclidian distance stands for?
15. Why should the minimum Euclidian distance be maximized?
16. Define the Manchester line encoding technique.
17. Define the bipolar AMI line encoding technique.
18. Define the nonreturn-to-zero inverted line encoding technique.
19. What is the difference between ASK, FSK and PSK?
20. Give an example of a modulation scheme that comprises both ASK and PSK
21. What does a differential codec stands for? What is its advantage relating to a nondifferential codec?
22. What are the differences between the ITU-T G.711 standard versions used in the United States and in Europe?
23. What is the advantage of using a logarithmic voice codec, relating to a uniform voice codec?
24. How does the ITU-T G.711 standard (A-law) use 8 bits to encode a sample signal of analog voice?
25. What is the difference between an MPEG and a DVB?
26. Describe the PAL video system. What is the bandwidth of a PAL video signal?
27. What is the sampling rate employed in PCM?
28. Consider a QPSK modulation used to transport a bit rate of 20 Mbps. What is the resulting symbol rate? What is the minimum bandwidth required to accommodate it?
29. Consider a 64-QAM modulation used to transport a bit rate of 10 Mbps. What is the resulting symbol rate? What is the minimum bandwidth required to accommodate it?
30. Consider a unipolar nonreturn-to-zero line coding technique is used to transport a bit rate of 20 Mbps. What is the resulting symbol rate? What is the minimum bandwidth required to accommodate it?

LAB EXERCISES

1. Using the Emona Telecoms Trainer 101 laboratory equipment, and volume 1 of its laboratory manual, perform experiment 5—DSB modulation.
2. Using the Emona Telecoms Trainer 101 laboratory equipment and volume 1 of its laboratory manual perform experiment 6—amplitude demodulation.
3. Using the Emona Telecoms Trainer 101 laboratory equipment and volume 1 of its laboratory manual perform experiment 7—DSB demodulation.
4. Using the Emona Telecoms Trainer 101 laboratory equipment and volume 1 of its laboratory manual perform experiment 8—SSB modulation and demodulation.
5. Using the Emona Telecoms Trainer 101 laboratory equipment and volume 2 of its laboratory manual perform experiment 18—observations of amplitude and DSBSC signals in the frequency domain.
6. Using the Emona Telecoms Trainer 101 laboratory equipment and volume 1 of its laboratory manual perform experiment 12—PCM encoding.
7. Using the Emona Telecoms Trainer 101 laboratory equipment and volume 1 of its laboratory manual perform experiment 13—PCM decoding.
8. Using the Emona Telecoms Trainer 101 laboratory equipment and volume 1 of its laboratory manual perform experiment 15—ASK.
9. Using the Emona Telecoms Trainer 101 laboratory equipment and volume 1 of its laboratory manual perform experiment 16—FSK.
10. Using the Emona Telecoms Trainer 101 laboratory equipment and volume 1 of its laboratory manual perform experiment 17—BPSK.
11. Using the Emona Telecoms Trainer 101 laboratory equipment and volume 1 of its laboratory manual perform experiment 18—QPSK.
12. Using the Emona Telecoms Trainer 101 laboratory equipment and volume 2 of its laboratory manual perform experiment 3—PCM and TDM.
13. Using the Emona Telecoms Trainer 101 laboratory equipment and volume 2 of its laboratory manual perform experiment 16—delta modulation and demodulation.
14. Using the Emona Telecoms Trainer 101 laboratory equipment and volume 2 of its laboratory manual perform experiment 22—PAM and TDM.
15. Using the Emona Telecoms Trainer 101 laboratory equipment and volume 2 of its laboratory manual perform experiment 15—line coding and bit clock regeneration.
16. Using the Emona Telecoms Trainer 101 laboratory equipment and volume 3 of its laboratory manual perform experiment 2—line coding and decision making.

7 Advanced Transmission Techniques to Support Current and Emergent Multimedia Services

LEARNING OBJECTIVES

- Define the emergent multimedia services.
- Describe the advanced transmission techniques that support emergent multimedia services.
- Describe the wideband code division multiple access.
- Describe different block transmission techniques.
- Describe the different types of multiple input multiple output.
- Define the concept of advanced transmit and receive diversity.
- Describe the energy efficiency in wireless communications.

7.1 ADVANCES IN WIRELESS SYSTEMS AND THEIR TECHNICAL DEMANDS

The challenge facing the mobile telecommunications industry today is how to continually improve the end-user experience, to offer appealing services through a delivery mechanism that offers improved speed, service attractiveness, and service interaction. Furthermore, to deliver the required services to the users with the minimum cost, the technology should continually allow better performances, higher throughputs, improved capacities, and higher spectral efficiencies. The following sections describe several measures to reach such desiderates.

The bandwidth requirements for future wireless systems present a considerable challenge since multipath propagation leads to severe time-dispersion effects, and this effect tends to increase with the increase of the symbol rate. In this case, conventional time-domain equalization schemes are not practical. Block transmission techniques, with appropriate cyclic prefixes (CPs) and employing frequency-domain equalization (FDE) techniques, have been shown to be suitable for high data rate transmission over severely time-dispersive channels [Falconer et al. 2002], and therefore presenting advantages for use with emergent wireless systems. Orthogonal frequency division multiplexing (OFDM) technique is the most popular modulation based on this technique. OFDM technique has been selected for LTE, as opposed to wideband code division multiple access (WCDMA), which is the air-interface technique that has been selected by European Telecommunications Standard Institute (ETSI) for universal mobile telecommunications system (UMTS). Single-carrier modulation using FDE is an alternative approach based on this principle [Sari et al. 1994; Falconer et al. 2002]. Owing to the lower envelope fluctuations of the transmitted signals (and, implicitly a lower PMEPR [peak-to-mean envelope power ratio]), single-carrier-FDE (SC-FDE) schemes (also entitled as single-carrier-FDMA [SC-FDMA]) are especially interesting for the uplink transmission [Falconer et al. 2002]. Moreover, multiple-input and multiple-output (MIMO) systems enhanced with state-of-the-art receivers is also normally associated to multimedia broadcast and multicast service (MBMS) in order to improve the overall system performance in terms of capacity, spectral efficiency, and coverage.

7.2 SPREAD-SPECTRUM COMMUNICATIONS

Digital communications employing spread-spectrum signals are characterized by using a bandwidth, B_{wd}, much greater than the information bit rate (see Figure 7.1). This means that for a spread-spectrum signal, we have $\text{SF} = B_{wd}/R_b \gg 1$, where R_b is the information bit rate and SF is the bandwidth expansion factor (called *spreading factor* or *processing gain*).

This spread-spectrum bandwidth expansion can be accomplished through different spread-spectrum techniques [Glisic and Vucetic 1997; Proakis 2001]:

- *Direct sequence* (*DS*): Figure 7.2 shows the general scheme for a DS spread-spectrum system. The information symbols are encoded and then modulated in combination with a pseudorandom sequence. In the receiver, the demodulator removes the pseudorandom impression from the received signal. DS techniques can be performed in two distinct ways. In the most common technique, the transmitted bits are encoded using a code with a given coding rate (say 1/2 or 1/3) and then the multiplication by the pseudorandom sequence increases the transmitted signal bit rate. This can be seen as applying a repetition code to the encoded bits and then adding the pseudorandom signature to the signal [Proakis 2001]. An alternative way of implementing DS techniques corresponds to performing all the spreading of the signal using a low rate code and then adding the pseudorandom signature to the signal. This approach is usually referred to as *code spread* (CS).
- *Frequency hopping* (*FH*): In this case, the available bandwidth is divided into several contiguous sub-bands frequency slots. A pseudorandom sequence is used for selecting the frequency slot for transmission in each signaling interval.
- *Time hopping* (*TH*): In this method, a time interval is divided into several time slots and the coded information symbols are transmitted in a time slot selected according to a pseudorandom sequence. The coded symbols are transmitted in the selected time slot as blocks of one or more code words.
- *Hybrid techniques*: DS, FH, and TH can be combined to obtain other types of spread-spectrum signals. For example, a system can use a combination of DS and FH, where the transmitted signal is associated with two code sequences. One of the sequences is multiplied by the signal to be transmitted, whereas the second is used for selecting the frequency slot for transmission in each signaling interval.

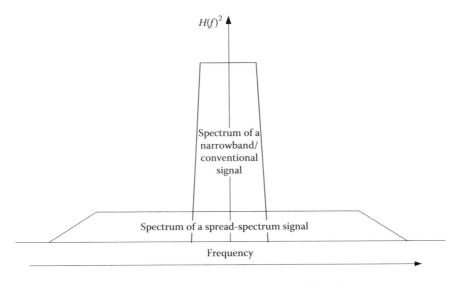

FIGURE 7.1   Spectrum of spread-spectrum signal versus narrowband signal.

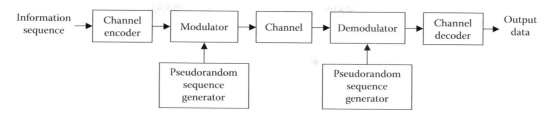

FIGURE 7.2   General scheme of a direct sequence spread-spectrum system.

In Section 7.3 we will only consider DS spreading techniques. One common application of DS spread-spectrum signals is CDMA communications, where several users share the same channel bandwidth for transmitting information simultaneously. If the spreading of each user is the joint effect of the channel encoder and the spreading code, then the transmission is designated as DS-CDMA. If the bandwidth expansion is performed by the channel encoder alone, then the transmission is referred to as CS-CDMA. In Viterbi [1990], it is shown that this technique can achieve maximum theoretical performance. In fact, DS-CDMA can be regarded as a special case in CS-CDMA. All the users can transmit in the same frequency band, and at the same time, and may be distinguished from each other by using a different pseudorandom sequence. If all the signals are transmitted with the symbols synchronized between them, and we have a flat fading channel, then for each symbol of a user, there will be only one interfering symbol from each of the other users. This is called synchronous CDMA transmissions and is usually employed in the downlink connection, between a BS and the users in a cell. If each transmitted symbol interferes with two symbols of any other user, then the transmission is asynchronous. This is the common method in the uplink connection.

7.3   CODE DIVISION MULTIPLE ACCESS

Code division multiple access system consists of different spread-spectrum transmissions [Ojanpera and Prasad 1998; Holma and Toskala 2000], each one associated with a different user's transmission using a different spreading sequence (ideally orthogonal).

While in frequency division multiple access (FDMA), different users' signals are transmitted in different carrier frequencies, and in time division multiple access (TDMA), different users' signals are sent in different orthogonal time slots, using CDMA, different users' signals are transmitted simultaneously, in the same carrier frequency, but properly subject to a spreading, using a different spreading sequence. Ideally, these spreading sequences of different users are orthogonal. This orthogonality is normally assured in a synchronous environment. Nevertheless, even using orthogonal spreading sequences, the signals that reach the receiver lose orthogonality because of the multipath propagation environment, or because the uplink transmissions are asynchronous (signals sent by different mobile stations are not synchronized). Therefore, very often it is preferable to make use of quasi-orthogonal spreading sequences.

Narrowband code division multiple access system was adopted in the 1990s by IS-95 standard, in the United States. Afterward, UMTS proceeds with the utilization of CDMA, in this particular case wideband CDMA.

Figure 7.3 depicts a simplified block diagram of a spread-spectrum transmitter, where the first block consists of a symbol modulator (mapper), responsible for the conversion of source bits into symbols. The resulting signal is then sampled for spreading purposes. The spreader consists of a block that performs the multiplication of the sampled versions of the symbols with the samples of the spreading sequence. Finally, the signal is passed through a band-limited pulse shaping filter and carrier modulator. Note that, in the case of a CDMA transmission, the pulse shaping filter is applied to the chips, instead of symbols, using a shaping such as the root raised cosine. At the receiver, the signal is convolved again with exactly the same spreading sequence.

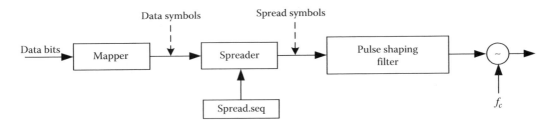

FIGURE 7.3 Generic block diagram of a spread-spectrum transmitter.

The operations performed by a CDMA transmitter can be, as seen in Figure 7.4, applied to a binary phase shift keying (BPSK) modulated signal. In this figure T_S and T_C stand for the symbol and chip period, and where SF stands for the spreading factor.

The main characteristics of spread-spectrum systems are the additional resistance to interference and the possibility to take advantage of the multipath channel in order to exploit multipath diversity. This leads to an improved performance and spectral efficiency, as compared to narrowband signals. The relationship between the power spectral density of a spread-spectrum signal and that of the original signal corresponds to the spreading factor. Figure 7.1 depicts the exposed concept.

Assuming that different signals use orthogonal spreading sequences, the corresponding signals are also orthogonal and the same spectrum can be shared between different signals.

A narrowband CDMA receiver comprises a decorrelator, which performs the correlation operation between the received signals with the user's spreading sequence. This allows extracting the desired signal, from the total received signal, which comprises the sum of the different users' signals.

Wideband CDMA consists of a CDMA system whose spread bandwidth is typically higher than the coherence bandwidth of the channel. This allows a better exploitation of multipath diversity, but requires higher spectrum availability. In this case, the receiver is normally not composed of a simple decorrelator. The typical WCDMA receiver is a RAKE receiver [Glisic and Vucetic 1997], which has several fingers to detect different multipaths of the channel. In multipath environment, since each finger of the RAKE receiver discriminates a different multipath, the combination of

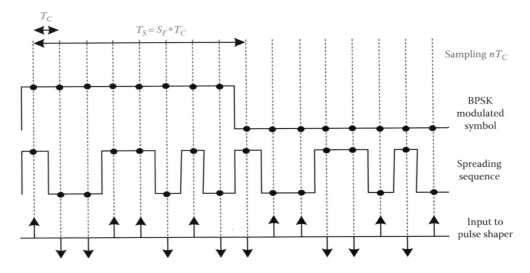

FIGURE 7.4 Generation of a BPSK spread-spectrum signal (spreading factor = 8).

the signals from different fingers with a maximum ratio combiner tends to achieve a performance improvement, as compared to a single decorrelator. For this reason, it is normally stated that a WCDMA system jointly with RAKE receiver is able to exploit multipath diversity. As the spreading factor increases, the resolution of the RAKE receiver also increases, allowing better discrimination of the several propagation paths, increasing the diversity order and, potentially, improving the performance.

Assuming correct synchronization, the received signal at the RAKE receiver output is the original signal plus higher frequency components, which is not part of the original signal, and is to be filtered. If there is any undesired interfering signal at the receiver, the spread-spectrum signal will affect it just as it did to the original signal at the transmitter, spreading it to the bandwidth of the spread-spectrum signal. Thus, the neglecting effect of the interfering signal is less powerful than in conventional narrowband signals. Furthermore, if the spreading sequences of the different signals that share the same spectrum are not perfectly orthogonal at the receiver side (owing to lack of orthogonality itself, of synchronization, to the multipath effect, etc.), then the resulting signal will be composed of the desired signal plus the noise and a component called *multiple access interference*,[*] which can be mitigated using a multiuser detector for WCDMA signals [Marques da Silva and Correia 2003a; Marques da Silva et al. 2005a].

7.3.1 General Model

To model a DS-CDMA system, we will first consider single user transmission. Figure 7.5 represents the basic communication model.

According to this model a sequence of symbols, s_i, representing the information to be transmitted, enters the modulator which outputs a sequence of ideal pulses $\delta(t - jT_c)$ modulated by the spreaded symbols $\sqrt{E_c} \cdot s_{\lfloor j/\text{SF} \rfloor} \cdot c_j$ (T_{chip} is the chip duration, c_j is the jth chip of the spreading sequence, and E_c is the average chip energy). This sequence passes through a shaping filter with impulse response $h_T(t)$ and frequency response $H_T(f)$, resulting in a transmitted signal that can be written in the following equivalent low-pass form

$$x(t) = \left[\sqrt{E_c} \sum_{j=0}^{\infty} s_{\lfloor j/\text{SF} \rfloor} \cdot c_j \cdot \delta(t - jT_{\text{chip}}) \right] * h_T(t) = \left[\sqrt{E_c} \sum_{j=0}^{\infty} s_{\lfloor j/\text{SF} \rfloor} \cdot c_j \cdot h_T(t - jT_{\text{chip}}) \right] \quad (7.1)$$

The signal is then transmitted through the channel, being modeled by a time invariant linear system with channel impulse response (CIR) $h_C(t)$ (channel frequency response $H_C[f]$), using, for example, the model defined in Appendix I. At the receiver, the signal passes through a filter with impulse response $h_R(t)$ (frequency response $H_R[f]$). The resulting signal is then obtained as the convolution of the transmitted signal with $h_C(t)$ and $h_R(t)$, that is,

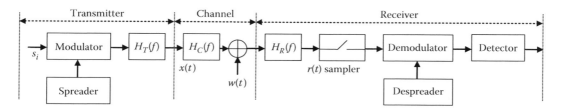

FIGURE 7.5   Basic DS-CDMA communication link.

---

[*] See Chapter 3.

$$r(t) = x(t) * h_C(t) * h_R(t) + n(t) \tag{7.2}$$

where $n(t)$ is the noise component at the output of the receiver filter, that is, it is given by

$$n(t) = w(t) * h_R(t) \tag{7.3}$$

where $w(t)$ denotes the white noise at the receiver input.

To design the shaping filter, $H_T(f)$, and the reception filter, $H_R(f)$, we will consider an ideal channel, that is, $h_C(t) = \delta(t)$ [$H_C(f) = 1$]. Therefore, the received signal can be written as follows:

$$r(t) = \sqrt{E_c} \sum_{j=0}^{\infty} s_{\lfloor j/\text{SF} \rfloor} \cdot c_j \cdot p(t - jT_{\text{chip}}) + n(t) \tag{7.4}$$

where $p(t)$ corresponds to the time response of the cascade of the transmitter and receiver filters

$$p(t) = h_T(t) * h_R(t) \tag{7.5}$$

After the filter the signal is sampled at rate f_{chip} ($f_{\text{chip}} = 1/T_{\text{chip}}$). The sequence of samples r_k can be represented as (considering no delay in the transmission)

$$r_k \equiv r(t = kT_{\text{chip}}) = \sqrt{E_c} \sum_{j=0}^{\infty} s_{\lfloor j/\text{SF} \rfloor} \cdot c_j \cdot p(kT_{\text{chip}} - jT_{\text{chip}}) + n(kT_{\text{chip}}) \tag{7.6}$$

or

$$
\begin{aligned}
r_k &= \sqrt{E_c} \sum_{j=0}^{\infty} s_{\lfloor j/\text{SF} \rfloor} \cdot c_j \cdot p_{k-j} + n_k \\
&= \sqrt{E_c} s_{\lfloor k/\text{SF} \rfloor} \cdot c_k \cdot p_0 + \sum_{\substack{j=0 \\ j \neq k}}^{\infty} s_{\lfloor j/\text{SF} \rfloor} \cdot c_j \cdot p_{k-j} + n_k
\end{aligned}
\tag{7.7}
$$

In the last passage, we admitted that $p_0 = 1$. The second term of the previous equation represents the intersymbol interference (ISI) that is not desired. The condition for no ISI is

$$p(kT_{\text{chip}}) = p_k = \begin{cases} 1, & k = 0 \\ 0, & k \neq 0 \end{cases} \tag{7.8}$$

According to *Nyquist pulse shaping criterion* the necessary and sufficient condition for $p(t)$ to obey the above condition is that its Fourier transform $P(f)$ satisfies [Proakis 2001]

$$P_{eq}(f) = \sum_{l=-\infty}^{\infty} P\left(f + \frac{l}{T_{\text{chip}}}\right) = T_{\text{chip}} \tag{7.9}$$

where

$$P(f) = H_T(f)H_R(f) \tag{7.10}$$

There are several functions satisfying the Nyquist criterion. One of the most common functions used for $P(f)$ is the family of raised cosine functions, which can be expressed as

$$P(f) = \begin{cases} T_{\text{chip}}, & f \leq \dfrac{1-\beta}{2T_{\text{chip}}} \\[2mm] \dfrac{T_{\text{chip}}}{2}\left\{ 1 + \cos\left[\dfrac{\pi T_{\text{chip}}}{\beta}\left(|f| - \dfrac{1-\beta}{2T_{\text{chip}}} \right) \right] \right\}, & \dfrac{1-\beta}{2T_{\text{chip}}} < f \leq \dfrac{1+\beta}{2T_{\text{chip}}} \\[2mm] 0, & |f| > \dfrac{1+\beta}{2T_{\text{chip}}} \end{cases} \tag{7.11}$$

where β is called the rolloff factor with $0 \leq \beta \leq 1$. This factor represents the fraction of excess bandwidth beyond the Nyquist frequency $1/(2T_{\text{chip}})$, and so $P(f)$ has a bandwidth of

$$B_{\text{wd}} = \frac{\beta+1}{2T_{\text{chip}}} \tag{7.12}$$

The corresponding impulse response in the time domain of the raised cosine function $P(f)$ is

$$p(t) = \text{sinc}\left(\frac{\pi t}{T_{\text{chip}}} \right) \frac{\cos\left(\pi\beta t/T_{\text{chip}} \right)}{1 - \left(2\pi t/T_{\text{chip}} \right)^2} \tag{7.13}$$

The receiver filter can be matched to the transmitter filter, that is, $h_R(t) = h_T^*(-t)$ and thus $H_T(f)$ and $H_R(f)$ will both have root raised cosine responses, that is,

$$|H_T(f)| = |H_R(f)| = \sqrt{|P(f)|} \tag{7.14}$$

As an example, in UMTS the shaping filter $h_T(t)$ is a root raised cosine with a rolloff factor $\beta = 0.22$ [3GPP 2004].

Therefore, if $p(t)$ satisfies the *Nyquist pulse shaping criterion* then Equation 7.7 reduces to

$$r_k = \sqrt{E_c}\, s_{\lfloor k/\text{SF} \rfloor} \cdot c_k + n_k \tag{7.15}$$

The sequence of samples r_k is then despreaded by multiplying it by the conjugate of the spreading sequence and averaging over sets of SF chips. The resulting decision variables z_i for each information symbol, i, can be expressed as

$$z_i = \frac{1}{\text{SF}} \sum_{k=i \cdot \text{SF}}^{(i+1) \cdot \text{SF}-1} c_k^* \cdot r_{0,k} \tag{7.16}$$

Although the channel effect was considered ideal for obtaining Equation 7.14, in a real system the receiver has to compensate its effect. This can be accomplished either by including $h_C(t)$ in the function $p(t)$ (resulting $p[t] = h_T[t] * h_R[t] * h_C[t]$), and then computing the receiver filter as $H_R(f) = P(f)/\left[H_T(f)H_C(f) \right]$ (this corresponds to channel equalization) where $P(f)$ satisfies the Nyquist criterion (using Equation 7.11, for example) or using some other channel compensation processing technique, as will be shown further in Section 7.3.2.

7.3.2 NARROWBAND CDMA

CDMA systems can be considered narrowband or wideband depending on the mobile propagation conditions according to the definitions presented above. An example of a cellular system that employs narrowband CDMA technique is the IS-95 standard. If the transmitted signal bandwidth is lower than the coherence bandwidth of the channel for the environments for which the system was designed, then there will be only one distinguishable received replica of the signal. In this case, the system is narrowband CDMA.

Let us consider now a multiuser environment where $N_u + 1$ users are transmitting simultaneously as shown in Figure 7.6.

In this case, the signal transmitted by each user u can be expressed in a form similar to Equation 7.1

$$x_u(t) = \sqrt{E_c} \sum_{j=0}^{\infty} s_{u,\lfloor j/\text{SF} \rfloor} \cdot c_{u,j} \cdot h_T(t - jT_{\text{chip}}) \qquad (7.17)$$

where:

$s_{u,\lfloor j/\text{SF} \rfloor}$ are the information symbols
$c_{u,j}$ are the symbols of the spreading sequence for user u

First, we will admit that the CDMA system is narrowband and thus the CIR of each user can be modeled as follows:

$$h_{C,u}(t) = \alpha_u \delta(t - \tau_u) \qquad (7.18)$$

where α is a complex attenuation (α_u is a random process dependent of the time but since we are admitting that $f_D T_{\text{chip}} \ll 1$, with f_D denoting the Doppler frequency [$f_D = f_c v / c$, with f_c denoting the carrier frequency, v the terminal speed, and c the speed of light] we can assume it is approximately constant during a symbol period)

In Figure 7.7 a basic receiver for user 0 is shown. First, the received signal passes through the matched filter $h_R(t)$. From Equation 7.2, this signal can be written as

$$r(t) = \sqrt{E_c} \sum_{u=0}^{N_u} \sum_{j=0}^{\infty} s_{u,\lfloor j/\text{SF} \rfloor} \cdot c_{u,j} \cdot \alpha_u \cdot p(t - jT_{\text{chip}} - \tau_u) + n(t) \qquad (7.19)$$

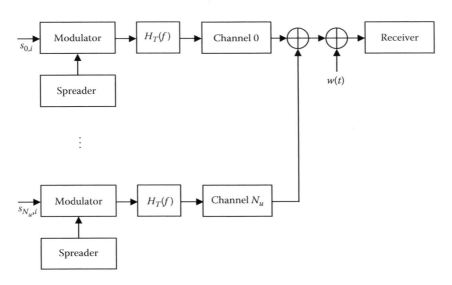

FIGURE 7.6   CDMA scheme in a multiuser environment.

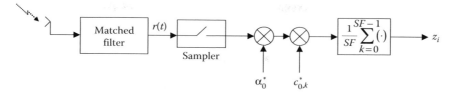

FIGURE 7.7   DS-CDMA receiver scheme for user 0.

The filter output is then sampled at times $t = kT_{\text{chip}} + \tau_0$ for extracting user 0 according to

$$
r_{0,k} \equiv r(t = kT_{\text{chip}} + \tau_0) = \sqrt{E_c} \sum_{u=0}^{N_u} \sum_{j=0}^{\infty} s_{u,\lfloor j/\text{SF} \rfloor} \cdot c_{u,j} \cdot \alpha_u \cdot p(kT_{\text{chip}} + \tau_0 - jT_{\text{chip}} - \tau_u) + n(kT_{\text{chip}} + \tau_0)
$$

$$(7.20)$$

$$
= \sqrt{E_c} \cdot s_{0,\lfloor k/\text{SF} \rfloor} \cdot c_{0,k} \cdot \alpha_0 + \sqrt{E_c} \sum_{u=1}^{N_u} \sum_{j=0}^{\infty} s_{u,\lfloor j/\text{SF} \rfloor} \cdot c_{u,j} \cdot \alpha_u \cdot p(kT_{\text{chip}} + \tau_0 - jT_{\text{chip}} - \tau_u) + n_{0,k}
$$

Note that it was taken into account that $p(t)$ satisfies the *Nyquist pulse shaping criterion* and thus Equation 7.8 is valid. In Equation 7.20, the second term represents the interference component, from the other users (MAI) which, for an asynchronous transmission, has contributions from all the transmitted chips of those users since generally $\tau_0 - \tau_u \neq aT_{\text{chip}} \ (\forall a \in \mathbb{Z})$.

The channel is compensated by multiplying the sampled sequence by the complex conjugate of α_u (considering perfect channel knowledge). The resulting sequence is then despreaded by multiplying it by the complex conjugate of the respective spreading sequence, $c_{0,j}$, and each set of *SF* samples belonging to the same information symbol are summed. The decision variable $z_{0,i}$ for the *i*th information symbol, can be expressed as

$$
z_{0,i} = \frac{1}{\text{SF}} \sum_{k=i\cdot\text{SF}}^{(i+1)\cdot\text{SF}-1} \alpha_0^* \cdot c_{0,k}^* \cdot r_{0,k}
$$

$$
= \frac{\sqrt{E_c} \cdot |\alpha_0|^2}{\text{SF}} \sum_{k=i\cdot\text{SF}}^{(i+1)\cdot\text{SF}-1} s_{0,\lfloor k/\text{SF} \rfloor} \cdot |c_{0,k}|^2
$$

$$
+ \frac{\sqrt{E_c}}{\text{SF}} \sum_{k=i\cdot\text{SF}}^{(i+1)\cdot\text{SF}-1} \sum_{u=1}^{N_u} \sum_{j=0}^{\infty} \alpha_0^* \cdot c_{0,k}^* \cdot s_{u,\lfloor j/\text{SF} \rfloor} \cdot c_{u,j} \cdot \alpha_u \cdot p(kT_{\text{chip}} + \tau_0 - jT_{\text{chip}} - \tau_u)
$$

$$
+ \frac{1}{\text{SF}} \sum_{k=i\cdot\text{SF}}^{(i+1)\cdot\text{SF}-1} \alpha_0^* \cdot c_{0,k}^* \cdot n_{0,k} \tag{7.21}
$$

$$
= \sqrt{E_c} \cdot |\alpha_0|^2 \cdot s_{0,i} + \frac{\sqrt{E_c}}{\text{SF}} \sum_{u=1}^{N_u} \sum_{j=0}^{\infty} \alpha_0^* \cdot s_{u,\lfloor j/\text{SF} \rfloor} \cdot \alpha_u
$$

$$
\times \sum_{k=i\cdot\text{SF}}^{(i+1)\cdot\text{SF}-1} c_{0,k}^* \cdot c_{u,j} \cdot p(kT_{\text{chip}} + \tau_0 - jT_{\text{chip}} - \tau_u)
$$

$$
+ \frac{1}{\text{SF}} \sum_{k=i\cdot\text{SF}}^{(i+1)\cdot\text{SF}-1} \alpha_0^* \cdot c_{0,k}^* \cdot n_{0,k}
$$

In this equation, it was assumed that $|c_{0,k}|^2 = 1$. In Equation 7.21, the second term represents the interference from the other users (MAI) that depends on the cross-correlation between the desired spreading code and the spreading code of interferer u weighted by the pulse shaping function $p(t)$ that depends on the relative delays. In a downlink synchronous transmission, the fading coefficients and the delays are all the same, $\alpha_0 = \alpha_u$ and $\tau_0 = \tau_u$ ($u = 1 \ldots N_u - 1$), and Equation 7.21 simplifies to

$$z_{0,i} = \sqrt{E_c} \cdot |\alpha|^2 \cdot s_{0,i} + \frac{\sqrt{E_c}}{SF} \cdot |\alpha|^2 \cdot \sum_{u=1}^{N_u} s_{u,i} \cdot \sum_{k=i \cdot SF}^{(i+1) \cdot SF-1} c_{0,k}^* \cdot c_{u,k} + \frac{\alpha^*}{SF} \sum_{k=i \cdot SF}^{(i+1) \cdot SF-1} c_{0,k}^* \cdot n_{0,k} \qquad (7.22)$$

In this case, to minimize the multiuser interference, it is only necessary to employ spreading sequences with low cross-correlation values. For some values of SF (e.g., if SF is a power of 2), it is possible to design sequences where

$$\sum_{k=0}^{SF-1} c_{0,k}^* \cdot c_{u,k} = 0 \qquad (7.23)$$

that is, orthogonal sequences, in which case there will be no multiuser interference (in a synchronous and for the single path propagation channel). This is what is done in the downlink connection of UMTS [3GPP 2004].

7.3.3 WIDEBAND CDMA

If the transmitted signal bandwidth is greater than the coherence bandwidth of the channel then it will be possible to resolve several multipath components and it will correspond to a wideband CDMA system. The UMTS system, which employs DS-CDMA transmission techniques, is a wideband system. The CIR for WCDMA systems can be described using a tapped delay line model [Silva et al. 2003] where the L multipaths are considered discrete. Considering that the channel is time invariant, the CIR for each user can be written as

$$h_{c,u}(t) = \sum_{l=1}^{L} \alpha_{u,l} \cdot \delta(t - \tau_{u,l}) \qquad (7.24)$$

where:
  $\alpha_{u,l}$ is the fading coefficient affecting the transmitted signal of user u for the lth propagation path
  $\tau_{u,l}$ is the respective time delay with $\tau_{u,l} = \tau_{u,1} + m_l T_{chip}$, $m_l \in \mathbb{Z}$ (the relative path delays for the same user are integer multiples of T_{chip})

Since in wideband CDMA systems the multipath replicas arriving with relative delays higher than the chip duration carry information about the transmitted signal, there is the possibility that when a replica is severely attenuated due to fading, the others may be received in more favorable conditions. Owing to the good autocorrelation properties of the spreading codes usually used in CDMA systems, it is possible to distinguish and extract the strongest replicas (with relative delays higher than the chip duration) present in the received signal and thus obtain diversity. These replicas can be extracted and combined using a RAKE receiver [Price and Green 1958] to help recovering the transmitted signal, as shown in Figure 7.8. To maximize the resulting signal-to-noise ratio (SNR) at the receiver, the extracted replicas are weighted using the complex conjugate of the respective fading coefficient and then these replicas are properly summed. This technique is designated as maximal ratio combining [Proakis 2001].

According to the scheme shown in Figure 7.8, first, the received signal goes through a matched filter similar to what was done in the narrowband CDMA receiver. From Equations 7.2, 7.17, and 7.24 the filtered signal $r(t)$ can be expressed as

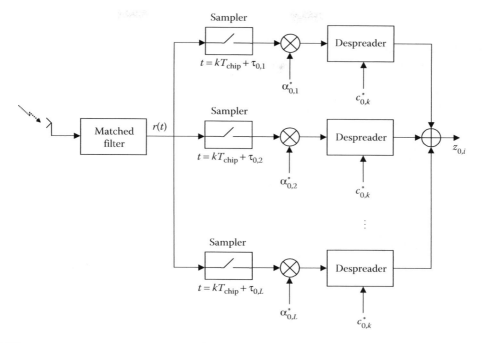

FIGURE 7.8 RAKE receiver scheme for user 0.

$$r(t) = \left[\sum_{u=0}^{N_u-1} x_u(t) * h_{C,u}\left(t\right) \right] * h_R(t) + n\left(t\right)$$

$$= \sqrt{E_c} \sum_{u=0}^{N_u} \sum_{l=1}^{L} \sum_{j=0}^{\infty} \alpha_{u,l} \cdot s_{u,\lfloor j/SF \rfloor} \cdot c_{u,j} \cdot p(t - jT_{chip} - \tau_{u,l}) + n(t)$$

(7.25)

The rest of the RAKE is composed of up to L parallel branches denoted as fingers. Each finger extracts one of the received replicas for the target user. The processing steps inside each finger are similar to the ones performed in the receiver for narrowband CDMA system. So, first the filter output is sampled at times $t = kT_{chip} + \tau_{0,f}$ for extracting user 0 in finger f. This can be expressed as

$$r_{0,k,f} \equiv r(t = kT_{chip} + \tau_{0,f}) = \sqrt{E_c} \sum_{u=0}^{N_u} \sum_{l=1}^{L} \sum_{j=0}^{\infty} \alpha_{u,l} \cdot s_{u,\lfloor j/SF \rfloor} \cdot c_{u,j} \cdot p(kT_{chip} + \tau_{0,f} - jT_{chip} - \tau_{u,l})$$

$$+ n(kT_{chip} + \tau_{0,f})$$

$$= \sqrt{E_c} \cdot s_{0,\lfloor k/SF \rfloor} \cdot c_{0,k} \cdot \alpha_{0,f} + \sqrt{E_c} \sum_{\substack{l=1 \\ l \neq f}}^{L} \sum_{j=0}^{\infty} \alpha_{0,l} \cdot s_{0,\lfloor j/SF \rfloor} \cdot c_{0,j} \cdot p(kT_{chip} + \tau_{0,f} - jT_{chip} - \tau_{0,l})$$

$$+ \sqrt{E_c} \sum_{u=1}^{N_u} \sum_{l=1}^{L} \sum_{j=0}^{\infty} \alpha_{u,l} \cdot s_{u,\lfloor j/SF \rfloor} \cdot c_{u,j} \cdot p(kT_{chip} + \tau_{0,f} - jT_{chip} - \tau_{u,l}) + n_{0,k,f}$$

(7.26)

$$= \sqrt{E_c} \cdot s_{0,\lfloor k/SF \rfloor} \cdot c_{0,k} \cdot \alpha_{0,f} + \sqrt{E_c} \sum_{\substack{l=1 \\ l \neq f}}^{L} \alpha_{0,l} \cdot s_{0,\lfloor (k+(\tau_{0,f}-\tau_{0,l})/T_{chip})/SF \rfloor} \cdot c_{0,k+(\tau_{0,f}-\tau_{0,l})/T_{chip}}$$

$$+ \sqrt{E_c} \sum_{u=1}^{N_u} \sum_{l=1}^{L} \sum_{j=0}^{\infty} \alpha_{u,l} \cdot s_{u,\lfloor j/SF \rfloor} \cdot c_{u,j} \cdot p(kT_{chip} + \tau_{0,f} - jT_{chip} - \tau_{u,l}) + n_{0,k,f}$$

To obtain this equation it was assumed that $p(t)$ satisfies the *Nyquist pulse shaping criterion* and thus Equation 7.8 is valid.

The channel is compensated by multiplying the sampled sequence by the complex conjugate of $\alpha_{u,f}$ (assuming perfect channel knowledge). The resulting sequence is then despreaded and the result is summed with the outputs of the other fingers. The decision variable $z_{0,i}$ for the ith information symbol can be expressed as

$$
\begin{aligned}
z_{0,i} &= \frac{1}{\mathrm{SF}} \sum_{f=1}^{L} \sum_{k=i\cdot\mathrm{SF}}^{(i+1)\cdot\mathrm{SF}-1} \alpha_{0,f}^{*} \cdot c_{0,k}^{*} \cdot r_{0,k,f} \\
&= \sqrt{E_c} \cdot s_{0,i} \cdot \sum_{f=1}^{L} \left| \alpha_{0,f} \right|^2 \\
&+ \frac{\sqrt{E_c}}{\mathrm{SF}} \sum_{f=1}^{L} \sum_{k=i\cdot\mathrm{SF}}^{(i+1)\cdot\mathrm{SF}-1} \sum_{\substack{l=1 \\ l\neq f}}^{L} \alpha_{0,f}^{*} \cdot c_{0,k}^{*} \cdot \alpha_{0,l} \cdot s_{0,\left\lfloor (k+(\tau_{0,f}-\tau_{0,l})/T_{\mathrm{chip}})/\mathrm{SF} \right\rfloor} \cdot c_{0,k+(\tau_{0,f}-\tau_{0,l})/T_{\mathrm{chip}}} \\
&+ \frac{\sqrt{E_c}}{\mathrm{SF}} \sum_{u=1}^{N_u} \sum_{f=1}^{L} \sum_{k=i\cdot\mathrm{SF}}^{(i+1)\cdot\mathrm{SF}-1} \sum_{l=1}^{L} \sum_{j=0}^{\infty} \alpha_{0,f}^{*} \cdot c_{0,k}^{*} \cdot \alpha_{u,l} \cdot s_{u,\left\lfloor j/\mathrm{SF} \right\rfloor} \cdot c_{u,j} \cdot p(kT_{\mathrm{chip}} + \tau_{0,f} - jT_{\mathrm{chip}} - \tau_{u,i}) \\
&+ \frac{1}{\mathrm{SF}} \sum_{f=1}^{L} \sum_{k=i\cdot\mathrm{SF}}^{(i+1)\cdot\mathrm{SF}-1} \alpha_{0,f}^{*} \cdot c_{0,k}^{*} \cdot n_{0,k,f}
\end{aligned}
\tag{7.27}
$$

In this equation, the second term represents interference caused by the user's own signal due to the delayed multipath replicas and the third term represents the interference component from all the multipath replicas of the other users. These interference components can be reduced by using multiuser detector schemes.

Much research has been undertaken in the area of the multiuser detectors [Marques da Silva and Correia 2000, 2003a] for WCDMA technology. Optimal MUDs, usually known as maximum likelihood sequence detectors, are too complex for practical application. Its complexity increases exponentially with the increase of the number of users. On the other hand, suboptimal MUD has a complexity that increases linearly with the increase of the number of users. The latter can take two different forms: linear MUD (e.g., Decorrelating [Marques da Silva et al. 2005a], minimum mean square error [MMSE]) [Glisic and Vucetic 1997] and subtractive MUD (e.g., successive interference cancellation [SIC] and parallel interference cancellation [PIC] [Glisic and Vucetic 1997; Marques da Silva and Correia 2003a]).

Linear suboptimal detectors apply a linear transformation to the bank output of the conventional detectors, in order to decrease the level of MAI seen by each user.

There are two main types of implementations for subtractive MUDs: the PIC [Varanasi and Aazhang 1990] and the SIC [Patel and Holtzman 1994; Johansson and Svensson 1999]. The main difference between both schemes relies on how the interference subtraction is performed on the received signal. While the SIC detector relies on removing the interfering signals from the received signal, one at a time as they are detected, the PIC removes all the interfering signals simultaneously after they are detected.

Nevertheless, a MUD is normally employed at the BS (uplink) where there is enough power processing capability and where it is easier to know/estimate the uplink CIR and the spreading and scrambling sequences of interfering users. Such power processing capability is normally not available at the mobile station (MS) side, because of this reason it is very important to employ alternative schemes to improve the performance.

If the transmission is synchronous and comes from a BS (downlink) the channel coefficients and delays do not depend on the user, $\alpha_{0,l} = \alpha_{u,l}$ and $\tau_{0,l} = \tau_{u,l}$ $(u = 1 \ldots N_u - 1, l = 1 \ldots L)$, and this equation simplifies to

$$
z_{0,i} = \sqrt{E_c} \cdot s_{0,i} \cdot \sum_{f=1}^{L} |\alpha_f|^2 + \frac{\sqrt{E_c}}{\text{SF}} \sum_{u=1}^{N_u} s_{u,i} \cdot \sum_{f=1}^{L} |\alpha_f|^2 \cdot \sum_{k=i\cdot\text{SF}}^{(i+1)\cdot\text{SF}-1} c_{0,k}^* \cdot c_{u,k}
$$

$$
+ \frac{\sqrt{E_c}}{\text{SF}} \sum_{u=0}^{N_u} \sum_{f=1}^{L} \sum_{k=i\cdot\text{SF}}^{(i+1)\cdot\text{SF}-1} \sum_{\substack{l=1 \\ l \neq f}}^{L} \alpha_f^* \cdot \alpha_l \cdot c_{0,k}^* \cdot c_{u,k+(\tau_f-\tau_l)/T_{\text{chip}}} \cdot s_{u,\lfloor (k+(\tau_f-\tau_l)/T_{\text{chip}})/\text{SF} \rfloor} \tag{7.28}
$$

$$
+ \frac{1}{\text{SF}} \sum_{f=1}^{L} \sum_{k=i\cdot\text{SF}}^{(i+1)\cdot\text{SF}-1} \alpha_f^* \cdot c_{0,k}^* \cdot n_{0,k,f}
$$

In Equation 7.28, the interference is grouped in a different form. The second term represents interference from the signals of the other users time aligned with the desired signal, whereas the third term represents the interference caused by the multipath replicas of all signals. These two terms are considered as multiple access interference (see Chapter 3). The second term can be cancelled if orthogonal spreading sequences are employed. In this case, Equation 7.23 is valid and Equation 7.28 simplifies to

$$
z_{0,i} = \sqrt{E_c} \cdot s_{0,i} \cdot \sum_{f=1}^{L} |\alpha_f|^2
$$

$$
+ \frac{\sqrt{E_c}}{\text{SF}} \sum_{u=0}^{N_u} \sum_{f=1}^{L} \sum_{k=i\cdot\text{SF}}^{(i+1)\cdot\text{SF}-1} \sum_{\substack{l=1 \\ l \neq f}}^{L} \alpha_f^* \cdot \alpha_l \cdot c_{0,k}^* \cdot c_{u,k+(\tau_f-\tau_l)/T_{\text{chip}}} \cdot s_{u,\lfloor (k+(\tau_f-\tau_l)/T_{\text{chip}})/\text{SF} \rfloor} \tag{7.29}
$$

$$
+ \frac{1}{\text{SF}} \sum_{f=1}^{L} \sum_{k=i\cdot\text{SF}}^{(i+1)\cdot\text{SF}-1} \alpha_f^* \cdot c_{0,k}^* \cdot n_{0,k,f}
$$

Nevertheless, the type of spreading sequences put to use must be carefully selected since orthogonal spreading sequences may originate high level of MAI caused by the multipath environment, even in a synchronous environment (downlink).

7.4 ORTHOGONAL FREQUENCY DIVISION MULTIPLEXING

OFDM is a transmission technique adopted by many high data rate communication systems such as IEEE 802.11n, IEEE 802.11ac, and IEEE 802.16e standards, being suitable for frequency selective fading channels [Cimini 1985; Liu and Li 2005]. As opposed to conventional single-carrier transmission techniques, where the information symbols are transmitted in a single stream, OFDM technique splits the symbols into several lower rate streams, which are then transmitted in orthogonal parallel subcarriers. Therefore, the symbol period is increased, making the signal less sensitive to ISI. To avoid interference between subcarriers, the several streams are transmitted in orthogonal subcarriers.

It is known that sinusoids with frequencies spaced by $1/T$ form an orthogonal basis set in a T-duration interval, and a periodic signal with period T can be represented as a linear combination of the orthogonal sinusoids. This means that the orthogonality between subcarriers is assured by using the discrete Fourier transform (DFT) and inverse DFT (IDFT). In practice, OFDM is normally implemented through an efficient technique called fast Fourier transform (FFT) and inverse FFT (IFFT). Therefore, OFDM signals are commonly generated by computing the N-point IFFT, where the input of the IFFT is the frequency domain representation of the OFDM signal. The output of the IFFT is the time domain representation of the OFDM signal, and the N-point IFFT out is defined as a useful OFDM symbol. This *time domain* OFDM signal is composed of N subcarriers, as depicted in Figure 7.9b, as opposed to a single-carrier signal also depicted in the same figure [Marques da Silva et al. 2010].

Although an OFDM signal splits the symbol stream into several parallel substreams (each one associated to a different subcarrier), the ISI can still occur within each substream. To mitigate the effects of ISI caused by channel delay spread, each block of N IDFT coefficients is typically preceded by a CP or a guard interval consisting of N_G samples (N_G stands for the number of samples at the CP), such that the length of the CP is, at least, equal to the time span of the channel (channel length). The CP is simply a repetition of the last N_G time-domain symbols. The prefix insertion operation is illustrated in Figure 7.10.

Although most of the processing of an OFDM transmitter is implicitly performed using IDFT function (as depicted in Figure 7.11a), the elementary processing of an OFDM transmitter is similar to that of an FDM transmitter. Figure 7.12 shows the implicit processing of an OFDM transmitter. As can be seen, such processing includes the FDM processing that consists of modulating different channels with different subcarriers, followed by an adder module (i.e., sum at signal level). Nevertheless, contrary to FDM, the OFDM processing comprises a single symbol stream (instead of multiple independent symbol streams), which is split into lower data rate channels.

Therefore, the basic idea of the OFDM transmission technique consists of splitting a higher rate into a group of parallel lower rate streams, and modulating each lower rate stream with a subcarrier

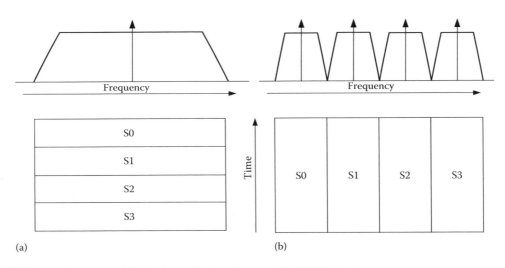

FIGURE 7.9   Spectrum of (a) single-carrier signal versus (b) OFDM signal.

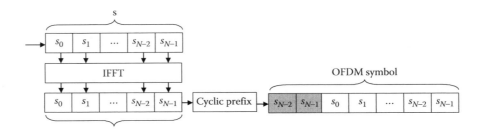

FIGURE 7.10   Cyclic prefix insertion.

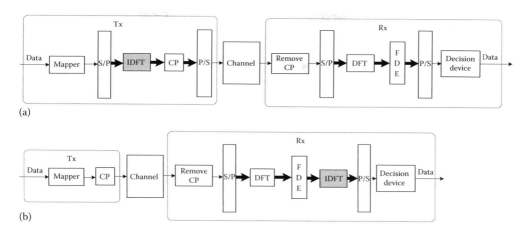

(a)

(b)

FIGURE 7.11 Generic transmission chain for (a) OFDM and (b) SC-FDE.

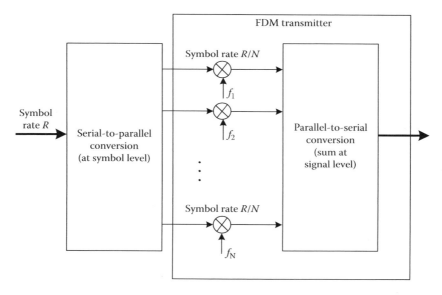

FIGURE 7.12 Implicit baseband processing of an OFDM transmitter.

in such a way that the resulting parallel signals are ideally uncorrelated. Another difference between FDM and OFDM signals relies on how these subcarriers are uncorrelated. While FDM signals use guard bands to assure uncorrelation between adjacent subcarriers,[*] in OFDM, the uncorrelation is implemented by using the IDFT (at the transmitter) and DFT (at the receiver) processing. This results in a much more efficient manner, which typically translates into much higher channel efficiency. As can be seen from Figure 7.13, OFDM signals from adjacent subcarriers present some level of overlapping in the frequency domain. Nevertheless, the mathematical OFDM implementation using the IDFT/DFT assures that the signals are uncorrelated at the receiver. Although these uncorrelated and parallel subcarrier signals are frequency separated, they are summed in time (second box in Figure 7.12), before being transmitted together. It is worth noting that a typical limitation of OFDM signals results from the fact that the spurious free dynamic range of an OFDM amplifier

---

[*] The FDM spectrum in plotted in Chapter 6.

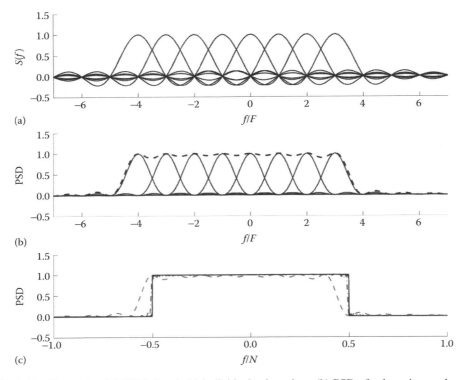

FIGURE 7.13 Transmitted OFDM signal: (a) individual subcarriers; (b) PSD of subcarriers and envelope; and (c) PSD of envelope signal.

is very demanding, in order to respond to this amplitude variations that results from the instantaneous sum of N subcarrier signals (parallel-to-serial conversion). This is more visible for higher number of subcarriers.

Since the symbol rate of each subcarrier signal has a symbol rate N times lower than that of the original signal, the level of ISI that results from the multipath channel is much lower. Another great advantage of OFDM signals results from the fact that, even though if a subcarrier experiences a deep fading or other type of interference, since the symbols are interleaved (serial-to-parallel conversion), these errors can easily be recovered using error correction techniques.

The subcarrier frequencies are known by the receiver, allowing recovering the original symbol stream from the received signal. Nevertheless, there are a number of nonideal effects that degrade the performance of OFDM systems, such as the local oscillator offset,[*] the window location offset,[†] and carrier interference. These impairments results from the inability of the receiver to track the subcarrier frequencies, which results in some level of degradation. Moreover, as previously mentioned, the use of low dynamic range devices in the transmission chain heavily degrades the performance of OFDM signals.

The subcarrier spacing is determined by the IDFT size N and input sampling rate of the IDFT. A subset of active subcarriers are mapped with the data modulation symbols and pilot symbols. The remaining subcarriers are left inactive prior to the IDFT to simplify the transmitter implementation.

---

[*] Local oscillator (also known as *local oscillator frequency offset* of the receiver), resulting in loss of orthogonality between adjacent subcarriers.
[†] The window location offset may place some subcarriers out of the receiver's windows.

If the CP length is greater than the length of the channel, the linear convolution of the transmitted sequence of IDFT coefficients with the discrete-time channel is converted into a circular convolution. As a result, the effects of the ISI and intercarrier interference (ICI) are completely and easily removed. After removal of the guard interval, each block of N received samples is converted back to the frequency domain using a DFT. Each of the N frequency domain samples are processed with a simple one-tap frequency domain equalizer and applied to a decision device to a metric computer.

In conventional time domain signals, the equalization process consists of a series of convolution operations, whose length is proportional to the time span of the channel. This means that for severely time-dispersive channels the receiver can be very complex. In OFDM systems, the equalization is performed as a simple multiplication of the OFDM signal spectrum with the frequency response of the channel. This represents a great advantage in terms of processing requirements and effectiveness, as compared to the equalization process normally employed in time domain signals.

A variant of conventional OFDM schemes is the OFDM access (OFDMA), where the multiple access is achieved by assigning subsets of OFDM subcarriers to individual users, allowing simultaneous low data rate transmission from several users.

From the mathematical point of view, the OFDM signal associated to the lth transmitted block has the form

$$s_l\left(t\right) = \sum_{n=-N_G}^{N-1} s_{n,l} h_T\left(t - nT_S\right) \tag{7.30}$$

where:
  T_S denotes the symbol duration
  N_G denotes the number of samples at the CP
  $h_T\left(t\right)$ is the adopted pulse shaping filter

The block $\left\{s_{n,l}; n = 0, 1, \ldots, N-1\right\}$ is the IDFT of block $\left\{S_{k,l}; n = 0, 1, \ldots, N-1\right\}$, where $S_{k,l}$ denotes the transmitted symbols associated to the kth subcarrier, that is, the kth symbol of the lth transmitted block. The signal $s_l\left(t\right)$ is transmitted over a time-dispersive channel, the received signal is sampled, and the CP is removed. The resulting time-domain block is as follows:

$$\left\{y_{n,l}; n = 0, 1, \ldots, N-1\right\} \tag{7.31}$$

If the length of the CIR is smaller than $N_G T_S$ then the DFT of the block $\left\{y_{n,l}; n = 0, 1, \ldots, N-1\right\}$ is $\left\{Y_{k,l}; k = 0, 1, \ldots, N-1\right\}$, with $Y_{k,l} = S_{k,l} H_k + N_{k,l}$, where H_k denotes the channel frequency response associated to the kth subcarrier and $N_{k,l}$ the channel noise. This means that a frequency-selective channel behaves as a flat fading channel at the subcarrier level. Therefore, we can easily invert the channel effects as follows:

$$\tilde{S}_{k,l} = \frac{Y_{k,l}}{H_{k,l}} = \frac{Y_{k,l} H_{k,l}^*}{\left|H_{k,l}\right|^2} \tag{7.32}$$

For phase shift keying (PSK) constellations the information is on the phase and this equalization process is simply accomplished through $\tilde{S}_{k,l} = Y_{k,l} H_{k,l}^*$.

It is worth noting that the asynchronous digital subscriber line (ADSL) implements a type of OFDM transmission technique entitled discrete multitone (DMT). In order to better optimize the signal to the transmission channel, the DMT presents additional features, namely the ability to remove certain subcarriers and the ability to adjust the modulation order and type, independently for each subcarrier.

7.5   SINGLE-CARRIER–FDE

Although OFDM schemes are the most popular block transmission techniques, the same concept can be used with single-carrier modulations. In fact, single-carrier modulation using FDE is an alternative approach based on this principle. Similar to OFDM, with SC-FDE the data blocks are preceded by a CP, long enough to cope with the overall channel length. Owing to the lower envelope fluctuations of the transmitted signals (and, implicitly a lower PMEPR), SC-FDE schemes are especially interesting when a low-complexity and efficient power amplification is required [Falconer et al. 2002]. As can be seen from Figure 7.11, the OFDM transmitter includes the computation of the IDFT, whereas the SC-FDE transmitter is a regular time-domain one, with the exception that the CP is added. Nevertheless, after reception (and after removing the CP), the SC-FDE receiver computes the DFT, before performing the FDE, followed by the IDFT computation. As the signal is transmitted in blocks (to which the CP is added and to which the equalization is performed), SC-FDE transmission is also considered as a block transmission technique.

From the mathematical point of view, SC-FDE signals are similar to OFDM signals, that is, the lth transmitted block has the form

$$s_l\left(t\right) = \sum_{n=-N_G}^{N-1} s_{n,l} h_T\left(t - nT_S\right) \tag{7.33}$$

Once again, with T_S denoting the symbol duration, N_G denoting the number of samples at the CP and $h_T\left(t\right)$ is the adopted pulse shaping filter. However, the lth time-domain block to be transmitted $\left\{s_{n,l}; n = 0, 1, \ldots, N-1\right\}$ is directly obtained from the data signal, without employing IDFT operation (i.e., the symbols are transmitted in the time domain, not in the frequency domain).

Assuming that the CP is longer than the overall CIR of each channel, the lth frequency-domain block before the FDE block (i.e., the DFT of the lth received time-domain block, after removing the CP) is $\left\{y_{n,l}; n = 0, 1, \ldots, N-1\right\}$ and the corresponding frequency-domain block (i.e., the corresponding DFT) is $\left\{Y_{k,l}; k = 0, 1, \ldots, N-1\right\}$, with

$$Y_{k,l} = S_{k,l} H_{k,l} + N_{k,l} \tag{7.34}$$

Once again, $H_{k,l}$ denotes the channel frequency response for the kth subcarrier and lth time-domain block (the channel is assumed invariant in the frame). $N_{k,l}$ is the frequency-domain block channel noise for that subcarrier and the lth block. The block $\left\{S_{k,l}; n = 0, 1, \ldots, N-1\right\}$ is the DFT of the block $\left\{s_{n,l}; n = 0, 1, \ldots, N-1\right\}$.

Although the situation is similar to the OFDM case, it is not desirable to invert perfectly the channel in the SC-FDE case because this might lead to noise enhancement effects that spread to all data symbols. (In the OFDM case, these noise enhancement effects are restricted to the symbols associated with subcarriers in deep fading and does not spread to other data symbols because the data is transmitted in the frequency domain.) This means FDE should be optimized under the MMSE criterion and the samples at the FDE output are given by [Marques da Silva et al. 2010]

$$\tilde{S}_{k,l} = \frac{Y_{k,l} H_{k,l}^*}{\left(\alpha + \left|H_{k,l}\right|^2\right)} \tag{7.35}$$

with

$$\alpha = \frac{E\left\{\left|N_{k,l}\right|^2\right\}}{E\left\{\left|S_{k,2l-j}\right|^2\right\}} \tag{7.36}$$

7.5.1 IB-DFE Receivers

It is well known that decision feedback equalizers (DFEs) [Proakis 2001] can significantly outperform linear equalizers (in fact, DFEs include as special case linear equalizers). Time-domain DFE have good performance/complexity tradeoffs, provided that the CIR is not too long. However, if the CIRs expand over a large number of symbols (such as in the case of severely time-dispersive channels), conventional time-domain DFEs are too complex. For this reason, a hybrid time-frequency SC-DFE was proposed in Benvenuto and Tomasin [2002], employing a frequency-domain feedforward filter and a time-domain feedback filter. This hybrid time-frequency-domain DFE has a better performance than a linear FDE. However, as with conventional, time-domain DFEs, it can suffer from error propagation, especially when the feedback filters have a large number of taps. A promising iterative block-DFE (IB-DFE) approach for SC transmission was proposed in Benvenuto and Tomasin [2002] and extended to transmit/diversity scenarios in Dinis et al. [2003]. Within these IB-DFE schemes, both the feed-forward and the feedback parts are implemented in the frequency domain, as depicted in Figure 7.14. Marques da Silva et al. [2013] and Montezuma et al. [2014] describe a modified IB-DFE receiver optimized for ultra-wideband signals.

For a given ith iteration, the IB-DFE output samples are given by

$$\tilde{S}_k^{(i)} = F_k^{(i)} Y_k - B_k^{(i)} \hat{S}_k^{(i-1)} \tag{7.37}$$

where:

$\{F_k^{(i)}; k = 0,1, \ldots, N-1\}$ and $\{B_k^{(i)}; k = 0,1, \ldots, N-1\}$ denote the feed-forward and feedback equalizer coefficients, respectively

$\{\hat{S}_k^{(i-1)}; k = 0,1, \ldots, N-1\}$ is the DFT of the hard-decision block $\{\hat{s}_n^{(i-1)}; n = 0,1, \ldots, N-1\}$, of the $(i-1)$th iteration, associated to the transmitted time-domain block $\{s_n; n = 0,1, \ldots, N-1\}$

The forward and backward IB-DFE coefficients $\{F_k^{(i)}; k = 0,1, \ldots, N-1\}$ and $\{B_k^{(i)}; k = 0,1, \ldots, N-1\}$, respectively, are chosen so as to maximize the SNIR. In Dinis et al. [2003], it is shown that the optimum feed-forward and feedback coefficients are given by

$$F_k^{(i)} = \frac{\kappa_F^{(i)} H_k^*}{\alpha + \left\{ 1 - [\rho^{(i-1)}]^2 \right\} |H_k|^2} \tag{7.38}$$

and

$$B_k^{(i)} = \rho^{(i-1)} \left[F_k^{(i)} H_k - 1 \right] \tag{7.39}$$

respectively, where $\kappa_F^{(i)}$ is selected to ensure that $\gamma^{(i)} = 1$ and

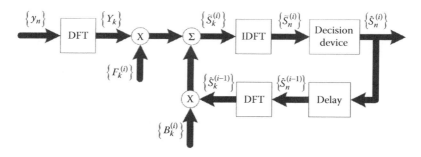

FIGURE 7.14   IB-DFE receiver structure.

$$\rho^{(i)} = \frac{E\left[s_n^* \hat{s}_n^{(i)} \right]}{E\left[|s_n|^2 \right]} \tag{7.40}$$

is a measure of the reliability of the decisions used in the feedback loop. Since the IB-DFE coefficients take into account the overall block reliability, the error propagation problem is significantly reduced. Consequently, the IB-DFE techniques offer much better performances than the non-iterative methods. In fact, the IB-DFE schemes can be regarded as low complexity turbo equalizers [Tuchler et al. 2002] since the feedback loop uses the equalizer outputs instead of the channel decoder outputs. For the first iteration we do not have any information about s_n, meaning that $\rho = 0$, $B_k^{(0)} = 0$, and $F_k^{(0)} = [\kappa_F^{(0)} H_k^*]/(\beta + |H_k|^2)$. Therefore, the IB-DFE reduces to a linear FDE. Clearly, for the first iteration ($i = 0$), no information exists about S_k and the correlation coefficient in Equation 7.40 is zero.

After that first iteration, and if the residual BER is not too high, we can use the feedback coefficients to eliminate a significant part of the residual interference. When $\rho \approx 1$ (after several iterations and/or moderate-to-high SNR), we have an almost full cancellation of the residual ISI through these coefficients, whereas the feed-forward coefficients perform an approximate matched filtering. Clearly, Equation 7.37 could be written as

$$\tilde{S}_k^{(i)} = F_k^{(i)} Y_k - B_k^{(i)} \overline{S}_k^{(i-1)} \tag{7.41}$$

with

$$\overline{S}_k^{(i-1)} = \rho^{(i-1)} \hat{S}_k^{(i-1)} \tag{7.42}$$

Since $\rho^{(i-1)}$ can be regarded as the blockwise reliability of the estimates $\hat{S}_k^{(i-1)}$, $\overline{S}_k^{(i-1)}$ is the overall block average of $S_k^{(i-1)}$ at the FDE output. To improve the performances, we could replace the *blockwise averages* by *symbol averages*, which can be done as described in the following.

If we assume that the transmitted symbols are selected from a quadrature PSK (QPSK) constellation under a Gray mapping rule (the generalization to other cases is straightforward), that is, $s_n = \pm 1 \pm j = s_n^I + j s_n^Q$, with $s_n^I = Re\{s_n\}$ and $s_n^Q = Im\{s_n\}$ (and similar definitions for $\tilde{s}_n$, $\overline{s}_n$, and $\hat{s}_n$), then it can be shown that the log likelihood ratios (LLRs) of the *in-phase bit* and the *quadrature bit*, associated to s_n^I and s_n^Q, respectively, are given by

$$L_n^I = \frac{2\tilde{s}_n^I}{\sigma_p^2} \tag{7.43}$$

and

$$L_n^Q = \frac{2\tilde{s}_n^Q}{\sigma_p^2} \tag{7.44}$$

respectively, where

$$\sigma_p^2 = \frac{1}{2} E\left[|s_n - \tilde{s}_n|^2 \right] \approx \frac{1}{2N} \sum_{n=0}^{N-1} E\left[|\hat{s}_n - \tilde{s}_n|^2 \right] \tag{7.45}$$

Under a Gaussian assumption, it can be shown that the mean value of s_n conditioned to the FDE output $\tilde{s}_n$ is

$$\bar{s}_n = \tanh\left(\frac{L_n^I}{2}\right) + j\tanh\left(\frac{L_n^Q}{2}\right)$$

$$= \rho_n^I \hat{s}_n^I + j\rho_n^Q \hat{s}_n^Q \tag{7.46}$$

where:

$\hat{s}_n^I = \pm 1$ and $\hat{s}_n^Q = \pm 1$ are the hard decisions defined according to the signs of L_n^I and L_n^Q, respectively

ρ_n^I and ρ_n^Q can be regarded as the reliabilities associated to the *in-phase* and *quadrature* bits of the nth symbol

given by

$$\rho_n^I = \frac{E\left[s_n^{I*}\hat{s}_n^I\right]}{E\left[|s_n^I|^2\right]} = \tanh\left(\frac{|L_n^I|}{2}\right) \tag{7.47}$$

and

$$\rho_n^Q = \frac{E\left[s_n^{Q*}\breve{s}_n^Q\right]}{E\left[|s_n^Q|^2\right]} = \tanh\left(\frac{|L_n^Q|}{2}\right) \tag{7.48}$$

(for the first iteration, $\rho_n^I = \rho_n^Q = 0$ and $\bar{s}_n = 0$).

The feed-forward coefficients are still obtained from Equation 7.38, with the blockwise reliability given by

$$\rho^{(i)} = \frac{1}{2N}\sum_{n=0}^{N-1}\left(\rho_n^{I(i)} + \rho_n^{Q(i)}\right) \tag{7.49}$$

Therefore, the receiver with *blockwise reliabilities*, denoted in the following as IB-DFE with hard decisions, and the receiver with *symbol reliabilities*, denoted in the following as IB-DFE with soft decisions, employ the same feed-forward coefficients; however, in the first, the feedback loop uses the *hard-decisions* on each data block, weighted by a common reliability factor, whereas in the second, the reliability factor changes from symbol to symbol (in fact, the reliability factor is different in the real and imaginary component of each symbol).

It is also possible to define a turbo FDE receivers based on IB-DFE receivers that, as conventional turbo equalizers, employs the channel decoder outputs instead of the uncoded *soft decisions* in the feedback loop.

7.6   DIVERSITY COMBINING ALGORITHMS

Before MIMO systems are introduced, it is worth describing diversity combining algorithms, as MIMO systems can be viewed as transmit and receive diversity. Therefore, those signals need to be properly combined to allow the exploitation of diversity.

Multipath fading is an important problem experienced in wireless communications. Fading results in fluctuations in the signal amplitude that degrades the BER performance at the receiver. Diversity schemes attempt to mitigate this problem by finding independently faded paths in the mobile radio channel. Usually this involves providing replicas of the transmitted signal over time, frequency, space, or multipath. Diversity is the most important contributor to reliable wireless communications. It can be processed at the transmitter or at the receiver side (or both).

Note that both WCDMA and OFDM are efficient schemes because they exploit diversity. WCDMA allows the exploitation of multipath diversity (with the RAKE receiver), whereas OFDM transmission technique allows the exploitation of frequency diversity.

The diversity system must optimally combine the receiver-diversified waveforms so as to maximize the resulting signal quality. There are several combining algorithms, namely selection combining, maximal ratio combining, equal gain combining (EGC), or the mean square error (MSE)-based combining. In addition, this combining can either be performed at the received signal level or at the symbol level. These combining schemes are explained in Sections 7.6.1 through 7.6.4.

7.6.1 SELECTION COMBINING

The selection combining algorithm selects the strongest signal from the various signals at its input. This criteria is used by the selective transmit diversity (STD) scheme, where the receiver selects one of the several transmit antennas (instead of one of the several receive antennas).

7.6.2 MAXIMAL RATIO COMBINING

The maximal ratio combining algorithm performs a weighted sum of the various signals at its input. This criterion is optimum only in the presence of noise, as it maximizes the SNR, while it minimizes the noise. The resulting SNR level of the signal at the combiner's output corresponds to the sum of the elementary SNRs.

In the case of WCDMA signals, the MRC is optimum only when interference does not exist. It may correspond to a synchronized network, with flat fading, and with orthogonal spreading sequences.

Figure 7.15 shows a two-branch maximum ratio combiner performed at the receiver's side, where ⊙ stands for the convolution operation. It is assumed that receive diversity is implemented, and that the propagation environment is flat fading. As can be seen, to remove the phase, and consequently, to allow a coherent sum, the signal at each branch of the receiver is convolved with the complex conjugate of the channel. Furthermore, since the signals from different branches are in phase, and weighted by the square of the absolute value of the (sum of the) channel coefficients (after the channel complex conjugation

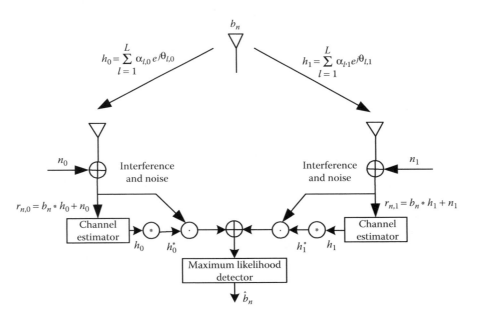

FIGURE 7.15 Two branches maximum ratio combiner (at the receiver).

convolution), they are summed. This corresponds to the maximum ratio combining criteria, as the signals are properly weighted by the square of the absolute values of the channel gains.

7.6.3 EQUAL GAIN COMBINING

The EGC algorithm performs a coherent sum of all signals at its input. The coherent sum requires co-phasing all signals, to avoid signal cancellation. Then, each signal branch is weighted with the same factor, irrespective of the signal amplitude. While providing diversity, this criterion tends, in most scenarios, to achieve a performance worse than that obtained with the MRC. Since the combining weights are equal in the several branch, if the signal from one branch has a deep fade, this effect is felt in the resulting signal. This scheme presents a great advantage, as knowledge about the channel state information (CSI) is not required, and therefore, estimation circuits are not needed, leading to a very simple combiner structure.

7.6.4 MSE-BASED COMBINING

The MSE-based combining criterion tends to lead to better performances in the presence of interference, namely, multipath interference or MAI. The combining weights are calculated based on the information provided by some way. A traditional way consists of using a pilot or a training sequence that allows the receiver to periodically evaluate a coefficient that corresponds to some difference (mean square error) between the transmitted signal and the received one. Let us consider that there are a total of N receive branches that are intended to be combined to provide diversity (e.g., output from the N fingers from L multipaths or output from the N receive antennas).

Let us denote $\hat{b}_n$ as the output data symbol from the nth receive branch and $\hat{\mathbf{c}}_n$ as the output training symbol vector from the nth receive branch. Similarly, $\tilde{\mathbf{c}}$ stands for the known training symbol vector $\mathbf{c}$, needed to implement the MSE-based combining algorithm.

Before proceeding to the MSE-based combining criteria, it is important to normalize the signals involved. Therefore, the normalization coefficient for each of the nth (receive) branch becomes

$$a_n = E\left[\hat{\mathbf{c}}_\mathbf{n}./\tilde{\mathbf{c}}\right] \tag{7.50}$$

where the operand ./ stands for the point-by-point division.

Since it is intended to calculate the expected value of $E\left[\hat{\mathbf{c}}_\mathbf{n}./\tilde{\mathbf{c}}\right]$, $\tilde{\mathbf{c}}$ and $\hat{\mathbf{c}}_n$ are vectors (group of symbols), instead of a single value (symbol), such that an occurrence of a symbol error in the detection within the training sequences vector is averaged. Therefore, the length of the vectors $\hat{\mathbf{c}}_n$ and $\tilde{\mathbf{c}}$ should be as maximum as possible, so as to lead to the expected value. The normalized training vector becomes

$$\hat{\mathbf{c}}_\mathbf{n}^{normalized} = \frac{\hat{\mathbf{c}}_\mathbf{n}}{a_n} \tag{7.51}$$

We are now able to apply the MSE-based combining criteria to each branch of the N independent and identical distributed (i.i.d.) branches, during training sequence, as given below:

$$\begin{aligned}
\mathrm{MSE}(\mathbf{c}_\mathbf{n}) &= E\left[\left|\hat{\mathbf{c}}_\mathbf{n}^{normalized} - \tilde{\mathbf{c}}\right|^2\right] \\
&= E\left[\left|\frac{\hat{\mathbf{c}}_\mathbf{n}}{a_n - \tilde{\mathbf{c}}}\right|^2\right] \\
&= E\left[\left|\frac{\hat{\mathbf{c}}_\mathbf{n}}{E}\left[\hat{\mathbf{c}}_\mathbf{n}./\tilde{\mathbf{c}}\right] - \tilde{\mathbf{c}}\right|^2\right]
\end{aligned} \tag{7.52}$$

After all of the described procedure has been implemented to each one of the N receive branches, during the training sequence phase, to allow the calculation of $\mathrm{MSE}(\mathbf{c_n})$, $n = 1, \ldots, N$, we are now able to weight the received symbols from different branches, before combining them. This makes

$$\hat{b}_n^{\text{weighted}} = \frac{\hat{b}_n^{\text{detected}}}{\mathrm{MSE}(\mathbf{c_n})}$$

$$= \frac{\hat{b}_n^{\text{detected}}}{E\left[\left|\dfrac{\hat{\mathbf{c}}_\mathbf{n}}{E\left[\hat{\mathbf{c}}_\mathbf{n}./\tilde{\mathbf{c}}\right] - \tilde{\mathbf{c}}}\right|^2\right]} \tag{7.53}$$

where:

$\hat{b}_n^{\text{weighted}}$ stands for the new normalized estimated value of the symbol b (data transmitted symbol) obtained from the nth receive branch

$1/\mathrm{MSE}(\mathbf{c_n})$ means the combining weight corresponding to the nth receive branch

From Equation 7.53 it is viewed that a received signal closer to the transmitted one results in a more powerful weight. The opposite also applies. The output of the MSE-based combiner, resulting from the N (receive) branches becomes

$$\hat{b} = \sum_{n=1}^{N} \hat{b}_n^{\text{weighted}}$$

$$= \sum_{n=1}^{N} \frac{\hat{b}_n^{\text{detected}}}{\mathrm{MSE}(\mathbf{c_n})} \tag{7.54}$$

$$= \sum_{n=1}^{N} \frac{\hat{b}_n^{\text{detected}}}{E\left\{\left|\hat{\mathbf{c}}_\mathbf{n}/\left[E\left(\hat{\mathbf{c}}_\mathbf{n}./\tilde{\mathbf{c}}\right)\right] - \tilde{\mathbf{c}}\right|^2\right\}}$$

which corresponds to the estimate of the transmitted symbol b, obtained using diversity from the N receive branches.

7.7 RAKE RECEIVER

In a multipath channel, the original transmitted signal is reflected from obstacles such as buildings and mountains, and the receiver sees several copies of the signal with different delays. Assuming spread spectrum is employed, and if signals arrive shifted in time with more than one chip apart from each other, the RAKE receiver can resolve them. Actually, from each multipath signal's point of view, other multipath signals can be regarded as interference and is partially suppressed by the processing gain. However, a further benefit is obtained if the resolved multipath signals are combined using a RAKE receiver. Hence, the signal waveform of CDMA signals facilitates the utilization of multipath diversity. Expressing the same phenomenon in the frequency domain means that the bandwidth of the transmitted signal is larger than the coherence bandwidth of the channel, and the channel presents frequency selectivity. In the case of single path, multipath diversity provided by the RAKE receiver would not bring any added value.

A RAKE receiver consists of a bank of decorrelators, each receiving a different multipath signal. After despreading by the decorrelators, the signals are combined using, usually, the MRC algorithm, as described in Section 7.6.2. Since the received multipath signals present uncorrelated fading coefficients, diversity order, and thus, performance is improved. Figure 7.16 illustrates the principle of the RAKE receiver. After spreading and modulation in the transmitter, the signal is passed through a multipath channel, which can be modeled by a tapped delay line. In the multipath channel there are L multipath components with different delays $(\tau_1, \tau_2, \ldots, \tau_L)$, attenuation factor $(\alpha_1, \alpha_2, \ldots, \alpha_L)$, and phase components $(\theta_1, \theta_2, \ldots, \theta_L)$, each corresponding to a different propagation path. The RAKE receiver has, ideally, a different finger for each multipath component. In each finger, the received signal is correlated by the spreading sequence, which is previously time-aligned with the delay of the corresponding multipath signal. After despreading, each captured multipath signal (in each jth finger of the RAKE) is weighted, and they are then combined. Moreover, when the MRC algorithm is considered to combine the several multipath signals, each finger aligns the phase of the signal corresponding to that multipath, and weights it by the channel attenuation, in order to allow a coherent sum. Therefore, when the RAKE considers a MRC algorithm, the signal in each finger is weighted by the complex conjugate of the corresponding multipath coefficient $\left(w_j = [\alpha_j e^{j\theta}]^* = \alpha_j e^{-j\theta}\right)$ to allow a coherent sum. Each finger of the RAKE receiver corresponds to a decorrelator chain depicted inside each box of Figure 7.16. Owing to the mobile movement, the scattering environment will change, and thus, the delays and attenuation factors will change as well. Therefore, it is necessary to measure the tapped delay line profile and to reallocate RAKE fingers whenever the delays have changed by a significant amount. Small-scale changes, less than one chip, are taken care of by a code tracking loop, which tracks the time delay of each multipath signal.

The full exploitation of DS-CDMA advantages is only achieved by the use of the RAKE receiver, because this tool takes advantage of the multipath diversity. In CDMA, the utilization of the spread spectrum, with a RAKE receiver allows using the frequency selectivity of the channel as a kind of diversity. The resolution of the RAKE receiver is the chip period. This is the reason why the resolution is improved when the spreading factor is increased. With a lower chip period, it better discriminates different multipaths, allowing higher order of multipath diversity. On the other hand, when the chip period is too long, the several multipaths are received within the same chip period, providing chip interference and not allowing the corresponding discrimination and diversity.

The implicit diversity provided by the RAKE receiver is similar to the explicit diversity provided by the antenna diversity, as well as the corresponding diversity equations.

Although the MRC is usually employed in the RAKE receiver, some other algorithms defined in Section 7.6 may, as well, be employed for multipath combining. The RAKE receiver, depicted in Figure 7.16, consists of a bank of J fingers, each one associated to a different multipath of the received signal.

While the conventional CDMA receiver consists of a single decorrelator, a RAKE receiver considers a different finger (decorrelator) for each multipath that is intended to be captured, and to exploit diversity. The finger outputs are then combined to form a decision statistic. The structure is equivalent to more practical forms, in which the received signal is first filtered with a chip pulse shaping matched filter and despreading is performed using received chip sample and the spreading sequence.

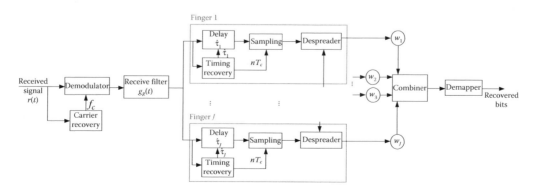

FIGURE 7.16 Scheme of a RAKE receiver.

The jth finger is used to despread the received spread-spectrum signal, resulting in the set of despread values:

$$\hat{b}_{j,k,n} = \sum_{m=\left[(n-1)\cdot S_F\right]+1}^{n\cdot S_F} y_{j,m} \cdot s_{k,m}^*, \quad n = 1\ldots P, \; j = 1\ldots J \tag{7.55}$$

where $y_{j,m}$ stands for the mth sample (sampled at a chip rate) of the output of the receive filter (chip shaping), with a delay corresponding to the jth finger (i.e., to the corresponding multipath) of the RAKE receiver.

Furthermore, $\hat{b}_{j,k,n}$ corresponds to the nth estimated symbol of the kth user, whose estimate is performed using the signal of the jth captured finger (whose value is based on the corresponding delayed version of the received signal $r[t - \tau_j]$). These despread values are combined using combining coefficients $w_1, w_2, \ldots, w_J$ to produce a decision statistic as described in Section 7.6.

$$\hat{b}_{k,n} = \sum_{j=1}^{J} w_j \cdot \hat{b}_{j,k,n} \tag{7.56}$$

where $\hat{b}_{k,n}$ is subject to a block-decision, typically a hard-decision.

Considering that MRC is performed, the impulse response of the RAKE receiver becomes

$$h_R(\tau,t) = \sum_{j=1}^{J} \hat{\alpha}_j(t) e^{-j\hat{\theta}_j(t)} \delta\left(\Delta - t + \hat{\tau}_j\right) \tag{7.57}$$

which is the complex conjugate of the channel impulse. Δ is some nominal delay that is required to ensure physical implementation of the RAKE, whose value should be higher than the delay corresponding to the last multipath captured by the RAKE, corresponding to the last finger. Its frequency response becomes

$$H_r(f,t) = \sum_{j=1}^{J} \hat{\alpha}_j(t) e^{-j\hat{\theta}_j(t)} e^{j2\pi f\left(\hat{\tau}_j - \Delta\right)} \tag{7.58}$$

where $\hat{x}$ stands for the estimated value of x and $\hat{\alpha}_j(t) e^{-j\hat{\theta}_j(t)}$ represents the estimate of the complex conjugate of the multipath coefficient $\alpha_j(t) e^{j\theta_j}$. For this reason, it is needed to make the channel estimate of amplitudes, time delays, and phase shifts of the several multipaths of the channel. The biggest limitation of the RAKE receiver comes from the fact that this estimate is never completely accurate due to channel variation (caused by the movement of the terminal, movement of the environment around terminals, or changes in the atmosphere), or because it may not detect all multipaths.

7.8 MULTIPLE INPUT MULTIPLE OUTPUT

The basic concept behind MIMO techniques relies on exploiting the multiple propagation paths of signals between multiple transmit and multiple receive antennas. The use of multiple antennas at both the transmitter and receiver aims to improve performance or symbol rate of systems, without an increase of the spectrum bandwidth, but it usually requires higher implementation complexity [Rooyen et al. 2000; Marques da Silva and Correia 2001, 2002a,b, 2003b; Hottinen et al. 2003; Marques da Silva et al. 2012a].

In the case of frequency selective fading channel, different symbols suffer interference from each other, whose effect is usually known as *intersymbol interference*. This effect tends to increase with the increase of the symbol rate.

By exploiting diversity, multiantenna systems can be employed to mitigate the effects of ISI. The antenna spacing must be larger than the coherence distance to ensure independent fading across different antennas [Foschini 1996; Foschini and Gans 1998; Rooyen et al. 2000]. Alternatively, different antennas should use orthogonal polarizations to ensure independent fading across different antennas.

The various configurations, shown in Figure 7.17, are referred to as single input single output (SISO), multiple input single output (MISO), single input multiple output (SIMO), or MIMO. SIMO and MISO architectures are a form of receive and transmit diversity (TD) schemes. SIMO comprises a single antenna at the transmitter and multiple antennas at the receiver, whereas MISO consists of the opposite. Moreover, MIMO architectures can be used for combined transmit and receive diversity, as well as for the parallel transmission of data or spatial multiplexing (SM) [Marques da Silva and Monteiro 2014; Marques da Silva and Correia 2014]. When used for SM, MIMO technology promises high bit rates in a narrow bandwidth, therefore, it is of high significance to spectrum users. MIMO systems comprise the transmission of a different signal from each transmit element so that receive antenna array receives a superposition of all the transmitted signals. Figure 7.18 presents a generic diagram of a MIMO system. MIMO systems comprise a pre-processing at the transmitter, or a post-processing at the receiver side, or both.

For M transmit and N receive antennas, the spectral efficiency, expressed in bits per second per Hertz (bit/s/Hz), is defined by [Telatar 1995; Foschini and Gans 1998]

$$C_{EP} = \log_2\left[\det\left(\mathbf{I}_N + \frac{\beta}{M}\mathbf{HH}' \right) \right] \tag{7.59}$$

where:
$\mathbf{I}_N$ is the identity matrix of dimension $N \times N$
$\mathbf{H}$ is the channel matrix
$\mathbf{H}'$ is the transpose-conjugate of $\mathbf{H}$
β is the SNR at any receive antenna

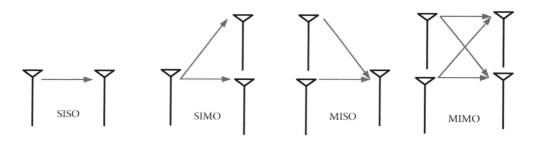

FIGURE 7.17 Multiple antennas configurations.

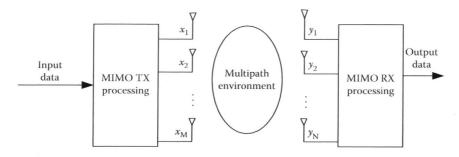

FIGURE 7.18 Generic diagram of a MIMO scheme.

Foschini and Telatar demonstrated that the capacity grows linearly with $m = \min(M, N)$, for uncorrelated channels [Foschini 1996; Telatar 1999].

MIMO schemes are implemented based on multiple-antenna techniques. These multiple-antenna techniques can be of different forms, such as the following:

- Space-time block coding (STBC)
- Multilayer transmission
- Space division multiple access (SDMA)
- Beamforming

STBC is essentially a MISO system. Nevertheless, the use of receive diversity makes it a MIMO, which corresponds to the most common configuration for this type of diversity. STBC-based schemes focus on achieving a performance improvement through the exploitation of additional diversity, while keeping the symbol rate unchanged. Open loop TD schemes achieve diversity without previous knowledge of the channel state at the transmitter side. On the contrary, closed loop TD requires knowledge about the CSI at the transmitter side. STBC, also known as *Alamouti scheme*, is the most known open loop technique [Alamouti 1998] (see Figure 7.19). 3GPP specifications define several TD schemes, such as the standardized closed loop modes 1 and 2 [3GPP 2003a], or the open loop STBC [3GPP 2003b] for two transmit antennas. The STD is a closed loop TD and is not exactly an STBC scheme. Nevertheless, since this is also a TD scheme, this subject is dealt with alongside STBC.

On the other hand, multilayer transmission and SDMA belong to another group, entitled SM, whose principles are similar but whose purposes are quite different. The goal of the MIMO based on multilayer transmission[*] scheme relies on achieving higher symbol rates in a given bandwidth, whose increase rate corresponds to the number of transmit antennas. Using multiple transmit and receive antennas, together with additional post-processing, multilayer transmission allows exploiting multiple and different flows of data, increasing the throughput and the spectral efficiency. In this case, the number of receive antennas must be equal or higher than the number of transmit antennas. The increase of symbol rate is achieved by *steering* the receive antennas to each one (separately) of the transmit antennas, in order to receive the corresponding symbol stream. This can be achieved through the use of the nulling algorithm.

Finally, the beamforming is implemented by antenna array with certain array elements at the transmitter or receiver being closely located to form a beam array (typically separated half

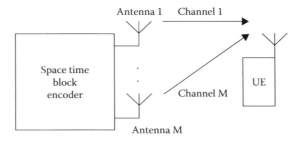

FIGURE 7.19 Generic block diagram of an open loop transmitter scheme.

---

[*] The same principle is applicable to SDMA, but where different transmit antennas correspond to different users, which share the spectrum.

wavelength). Since it steers the transmit (or receive) beam toward the receive (or transmit) antenna, this scheme is an effective solution to maximize the SNR. As a result, an improved performance or coverage can be achieved with beamforming.

7.8.1   SPACE-TIME CODING

Although space-time coding is essentially a MISO system, the use of receiver diversity makes it a MIMO, which corresponds to the most common configuration for this type of diversity. The STBC is a TD technique, being particularly interesting for fading channels. A possible scenario for its application is the downlink transmission of a cellular environment, where the base station uses several transmit antennas and the mobile terminal has typically a single one [Alamouti 1998].

STBC-based schemes focus on achieving a performance improvement through the exploitation of additional diversity, while keeping the symbol rate unchanged [Alamouti 1998; Tarokh et al. 1999]. Symbols are transmitted using an orthogonal block structure, enabling simple decoding algorithm at the receiver [Alamouti 1998; Marques da Silva et al. 2004, 2009].

The Alamouti's TD scheme is the widely known STBC scheme, and requires some processing from the transmitter. It can be implemented either in the time domain or in the frequency domain. In the latter case, it is referred to as *space-frequency block coding.* The current description focuses on the time-domain coding. The extension to frequency-domain coding is straightforward.

Let us consider the *l*th transmitted symbol defined by

$$s_l \left(t \right) = a_l \, h_T \left(t - n T_S \right) \tag{7.60}$$

where:

a_l refers to the symbol selected from a given constellation
T_S denotes the symbol duration
$h_T \left(t \right)$ the adopted pulse shaping filter

The signal $s_l \left(t \right)$ is transmitted over a time-dispersive channel. The received signal is sampled, and the resulting time-domain block is y_l, which is then subjected to FDE.

In the following, we will adopt the notation: lower and upper case signal variables correspond to time and frequency domain variables, respectively. The mapping between one and the other is achieved through DFT and IDFT operations (i.e., $X_l = \text{DFT}\{x_l\}$ and $x_l = \text{IDFT}\{X_l\}$).

The *l*th frequency-domain block before receiver's equalization process is $Y_l = \text{DFT}\left(y_l \right)$, with

$$Y_l = \sum_{m=1}^{M} S_l^{(m)} H_l^{(m)} + N_l \tag{7.61}$$

In the above equation, $H_l^{(m)}$ denotes the channel frequency response for the *l*th time-domain block (the channel is assumed invariant in the frame) and the *m*th transmit antenna. N_l is the frequency-domain block channel noise for the *l*th block.

---

* The Alamouti's transmit diversity scheme can also be applied simultaneously to space, time, and frequency. In this case, it is referred to as the *space-time-frequency block coding* (STFBC).

7.8.1.1 STBC for Two Antennas

Considering the STBC with two transmit antennas (STBC2), the lth time-domain block to be transmitted by the mth antenna ($m = 1$ or 2) is $s_l^{(m)}$, with [Alamouti 1998]

$$
\begin{aligned}
s_{2l-1}^{(1)} &= a_{2l-1} \\
s_{2l-1}^{(2)} &= -a_{2l}^* \\
s_{2l}^{(1)} &= a_{2l} \\
s_{2l}^{(2)} &= a_{2l-1}^*
\end{aligned}
\tag{7.62}
$$

where a_l refers to the symbol selected from a given constellation (e.g., a QPSK constellation) under an appropriate mapping rule, to be transmitted in the lth time-domain block. Considering the matrix-vector representation, we define $\mathbf{s}_l^{[1,2]}$

$$
\mathbf{s}_l^{[1,2]} = \begin{bmatrix} a_{2l-1} & a_{2l} \\ -a_{2l}^* & a_{2l-1}^* \end{bmatrix}
\tag{7.63}
$$

where different rows of the matrix refer to transmit antenna order and different columns refer to symbol period orders.

The Alamouti's post-processing for two antennas becomes [Alamouti 1998; Marques da Silva et al. 2009],

$$
\begin{aligned}
\tilde{A}_{2l-1} &= \left[Y_{2l-1} H_l^{(1)*} + Y_{2l}^* H_l^{(2)} \right] \beta \\
\tilde{A}_{2l} &= \left[Y_{2l} H_l^{(1)*} - Y_{2l-1}^* H_l^{(2)} \right] \beta
\end{aligned}
\tag{7.64}
$$

where $\beta = \left[\displaystyle\sum_{m=1}^{M} \left| H_l^{(m)} \right|^2 \right]^{-1}$.

Defining

$$
\mathbf{Y}_l^{[1,2]} = \begin{bmatrix} Y_{2l-1} & Y_{2l}^* \\ Y_{2l} & -Y_{2l-1}^* \end{bmatrix} \quad \text{and} \quad \mathbf{H}_l^{[1,2]} = \begin{bmatrix} H_l^{(1)*} & H_l^{(2)} \end{bmatrix}^T
$$

Equation 7.64 can be expressed in the matrix-vector representation as $\tilde{\mathbf{A}}_l^{[1,2]} = \left[\mathbf{Y}_l^{[1,2]} \times \mathbf{H}_l^{[1,2]} \right] \times \beta$, where $\tilde{\mathbf{A}}_l^{[1,2]} = \begin{bmatrix} \tilde{A}_{2l-1} & \tilde{A}_{2l} \end{bmatrix}^T$.

Finally, the decoded symbols become

$$
\tilde{A}_{2l-j} = \overbrace{A_{2l-j} \sum_{m=1}^{M} \left| H_l^{(m)} \right|^2 \beta}^{\text{Desired symbol}} + \overbrace{N_{2l-j}^{eq}}^{\text{Noise component}} \qquad j = 0,1
\tag{7.65}
$$

where $N_{k,l}^{eq}$ denotes the equivalent noise for detection purposes.

7.8.1.2 STBC for Four Antennas

Orthogonal codes of rate one, using more than two antennas, does not exist for modulation orders higher than 2. Schemes with 4 and 8 antennas with code rate one only exist in the case of binary transmission. If orthogonality is essential (fully loaded systems with significant interference levels), a code with $R < 1$ should be employed.

The current description focus on a nonorthogonal scheme with $M = 4$ transmit antennas (STBC4) [Marques da Silva et al. 2009], presenting a code rate one.

The symbol construction for the STBC with four antennas can be generally written as [Marques da Silva et al. 2009]

$$\mathbf{s}_l^{[1,4]} = \begin{bmatrix} \mathbf{s}_l^{[3,4]} & \mathbf{s}_l^{[1,2]} \\ \mathbf{s}_l^{[1,2]*} & -\mathbf{s}_l^{[3,4]*} \end{bmatrix} \tag{7.66}$$

where $\mathbf{s}_l^{[3,4]}$ is the same as $\mathbf{s}_l^{[1,2]}$ (see Equation 7.63), by replacing the subscripts 1 by 3 and 2 by 4, as well as by replacing $2l$ by $4l$ (e.g., $a_{2l-1} \rightarrow a_{4l-3}$).

Similar to Equation 7.62, and considering the STBC with four transmit antennas as in Equation 7.66, the time-domain blocks to be transmitted by the mth antenna ($m = 1, 2, 3,$ or 4) are $s_l^{(m)}$, which can be expressed in the matrix-vector representation as

$$\mathbf{s}_l^{[1,4]} = \begin{bmatrix} a_{4l-3} & a_{4l-2} & a_{4l-1} & a_{4l} \\ -a_{4l-2}^* & a_{4l-3}^* & -a_{4l}^* & a_{4l-1}^* \\ a_{4l-1}^* & a_{4l}^* & -a_{4l-3}^* & -a_{4l-2}^* \\ -a_{4l} & a_{4l-1} & a_{4l-2} & -a_{4l-3} \end{bmatrix} \tag{7.67}$$

Note that, in Equation 7.67 different rows of the matrix refer to different transmit antennas and different columns refer to different symbol periods.

The lth frequency-domain block, before the receiver's equalization process, is $y_l = \text{IDFT}\{Y_l\}$, with Y_l as defined by Equation 7.61.

Let us define

$$\tilde{\mathbf{A}}_l^{[1,4]} = \begin{bmatrix} \tilde{A}_{4l-3} & \tilde{A}_{4l-2} & \tilde{A}_{4l-1} & \tilde{A}_{4l} \end{bmatrix}^T \tag{7.68}$$

$$\mathbf{Y}_l^{[1,4]} = \begin{bmatrix} \mathbf{Y}_l^{[3,4]} & -\mathbf{Y}_l^{[1,2]*} \\ \mathbf{Y}_l^{[1,2]} & \mathbf{Y}_l^{[3,4]*} \end{bmatrix} \tag{7.69}$$

and

$$\mathbf{H}_l^{[1,4]} = \begin{bmatrix} H_l^{(1)*} & H_l^{(2)} & H_l^{(3)} & H_l^{(4)*} \end{bmatrix}^T \tag{7.70}$$

with $\mathbf{Y}_l^{[3,4]}$ as defined for $\mathbf{Y}_l^{[1,2]}$ by replacing the subscripts 1 by 3 and 2 by 4, as well as by replacing $2l$ by $4l$ (e.g., $Y_{2l-1} \rightarrow Y_{4l-3}$).

The post-processing STBC for four antennas ($M = 4$) becomes

$$\tilde{\mathbf{A}}_l^{[1,4]} = \begin{bmatrix} \mathbf{Y}_l^{[1,4]} \times \mathbf{H}_l^{[1,4]} \end{bmatrix} \times \beta \tag{7.71}$$

Finally, the decoded symbols become [Marques da Silva and Dinis 2011]

$$\tilde{A}_{4l-j} = \overbrace{A_{4l-j} \sum_{m=1}^{M} \left| H_l^{(m)} \right|^2}^{\text{Desired symbol}} - \overbrace{C \cdot A_{4l-p}}^{\text{Residual interference}} + \overbrace{N_{4l-j}^{eq}}^{\text{Noise component}} \qquad (7.72)$$

with $j = 0,1,2,3$, and with $p = 3 - j$. We also define

$$C = \frac{2 \operatorname{Re} \left\{ H_l^{(1)*} H_l^{(4)} - H_l^{(2)} H_l^{(3)*} \right\}}{\sum_{m=1}^{M} \left| H_l^{(m)} \right|^2} \qquad (7.73)$$

which stands for the residual interference coefficient generated in the STBC decoding process of order 4.

The signal processing for nonorthogonal scheme with eight transmit antennas is defined in Marques da Silva et al. [2009] and in the references therein.

It is worth noting that although the described STBC scheme is a MISO, by adopting receive diversity, this can be viewed as a MIMO system.

7.8.2 Selective Transmit Diversity

The STD is a closed loop TD, where the transmitter selects one out of multiple transmit antennas to send data, based on the one that presents a more beneficial propagation environment.

Assuming the downlink direction, the STD scheme comprises a low rate feedback link from the receiver (MS) telling the transmitter (BS) which antenna should be used in transmission. It is worth noting that by adopting receive diversity, this can also be viewed as a type of MIMO technique.

As depicted in Figure 7.20, the STD comprises a common or dedicated pilot sequence, which is also transmitted. Different antennas with specific pilot patterns/codes enable antenna selection. Then, the transmitter has a link quality information about the M (number of transmit antennas) links. Based on link information, it transmits a single symbol stream over the best antenna. Then, another decision process is made. The receiver is supposed to re-acquire the carrier phase $\theta_k(t)$ after every switch between antennas. The antenna switch (AS) has the capability to switch every slot duration.

The decision for the antenna selection of the STD using WCDMA signals combined with the RAKE receiver should select the signal from the transmit antenna whose multipath diversity order

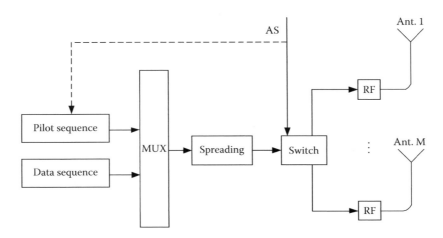

FIGURE 7.20 Scheme of a STD with feedback indication.

is higher. In other words, it should select the transmit antenna m, experiencing a channel with L multipaths that maximizes the quantity $\max\left[\sum_{l=1}^{L}\left|h_{m,l}\right|^2\right]$. This process is similar to the selective combiner that can be employed in receive diversity.

7.8.3 MULTILAYER TRANSMISSION

The goal of the MIMO based on multilayer transmission scheme relies on achieving higher data rates in a given bandwidth, whose increase rate corresponds to the number of transmit antennas [Foschini 1996; Foschini and Gans 1998; Nam and Lee 2002]. In this case, the symbols stream is split in a serial to parallel converter into several transmit antennas, and each transmit antenna sends a lower symbol rate. The reconstruction of the original symbol stream is performed at the receiver side.

Currently, MIMO systems are typically employed to improve the spectral efficiency. This is achieved by increasing the throughput sent in a certain bandwidth. A widely employed technique for this purpose relies on the use of multilayer transmission, as opposed to space-time coding (which aims to provide a performance improvement by exploiting diversity).

In multilayer transmission, the number of receive antennas must be equal to or higher than the number of transmit antennas. The increase of symbol rate is achieved by *steering* the receive antennas to each one (separately) of the transmit antennas, in order to receive the corresponding data stream. This can be achieved by using the nulling algorithm. With the sufficient number of receive antennas it is possible to resolve all data streams, as long as the antennas are sufficiently spaced so as to minimize the correlation [Marques da Silva et al. 2005b].

Let us consider the generic scheme of the multilayer MIMO depicted in Figure 7.21. The $M \times N$ MIMO scheme is spectral efficient and resistant to fading, where the BS uses M transmit antennas and the MS uses N receive antenna [Marques da Silva et al. 2004, 2012a]. As long as the antennas are located sufficiently far apart, the transmitted and received signals from each antenna undergo independent fading.

As depicted in Figure 7.22, two different multilayer MIMO schemes can be considered: scheme 1 and scheme 2 [Marques da Silva et al. 2004]. Scheme 1[*] directly allows an increase of data rate whose increase rate corresponds to the number of transmit antennas. Scheme 2 allows the exploitation of diversity, without increasing the data rate. The diversity combining is performed using any combining algorithm, preferably the MSE-based combining algorithm. In the case of scheme 2 (depicted in Figure 7.22b), antenna switching is performed at a symbol rate, where gray dashed lines represent the signal path at even symbol periods, in case of two transmit antennas. Output signals are then properly delayed and combined to provide diversity using a combining algorithm.

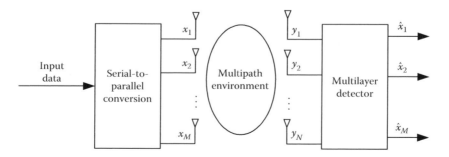

FIGURE 7.21   Generic diagram of the $M \times N$ multilayer MIMO scheme.

---

[*] Note that, in Figure 7.22, *Mod* and *Spr* stand for modulator and spreader, respectively.

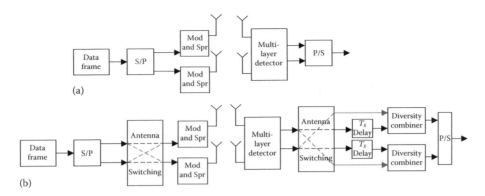

FIGURE 7.22 Diagram of the 2×2 multilayer MIMO alternatives: (a) scheme 1 and (b) scheme 2.

Let us focus on the MIMO based on multilayer transmission, as depicted in Figure 7.21, where the system is equipped with M transmit and N receive antennas (with $N \geq M$). At the transmitter side, the data stream is demultiplexed into M-independent substreams, and then each substream is encoded (modulated) and transmitted by M antennas. Each spread data symbol x_m is sent to the mth transmit antenna. It is considered that the radiated signals are propagated through a multipath frequency selective fading channels. At the receiver, the detector estimates the transmitted symbols from the received signals at the N receive antennas.

The lowpass equivalent transmitted signals at the M antennas are respectively given by

$$
\begin{cases}
x_1(t) = \sqrt{E_S} \cdot s_l^{(1)} \cdot h_T(t - nT_S) \\
\\
\quad \cdot \\
\\
\quad \cdot \\
\\
\quad \cdot \\
\\
x_M(t) = \sqrt{E_S} \cdot s_l^{(M)} \cdot h_T(t - nT_S)
\end{cases}
\tag{7.74}
$$

where s_l refers to the symbol selected from a given constellation (e.g., a QPSK constellation) under an appropriate mapping rule, to be transmitted in the lth time domain block from the mth transmit antenna (with $2 \leq m \leq M$). Moreover, as in Equation 7.60, T_S refers to the symbol interval and $h_T(t)$ is the pulse waveform. $s_l^{(1)}$ and $s_l^{(M)}$ are the data symbols fed to the first and to the Mth transmit antenna. The baseband equivalent of the N-dimensional received vector $\mathbf{y} = \begin{bmatrix} y_1 & y_2 & \dots & y_N \end{bmatrix}^T$ at sampling instants is expressed by

$$
\mathbf{y} = \sum_{l=1}^{L} \mathbf{H}_l \mathbf{x} + \mathbf{z}
\tag{7.75}
$$

where:

$\mathbf{x} = \begin{bmatrix} x_1 & x_2 & \dots & x_M \end{bmatrix}^T$ denotes the transmitted symbol vector, with each element having the unit average power

$\mathbf{H}_l : \mathbf{H}_1 \ \mathbf{H}_2 \ \dots \ \mathbf{H}_L$ denotes the $N \times M$ channel matrix

whose elements $h_{m,n}$ at the nth row and mth column is the channel gain from the mth transmit antenna to the nth receive antenna, that is, $\left(h_{m,n} \right)_l = \left(\alpha_{m,n} e^{j\theta_{m,n}} \right)_l \delta(t - [\tau_{m,n}]_l)$. L refers to the number of multiparts of the channel and the index *l* stands for the multipath order. It is considered a discrete tap-delay-line channel model where the channel from the *m*th transmit antenna to the *n*th receive antenna comprises discrete resolvable paths, expressed through the channel coefficients. The $N \times M$ sets of temporal multipaths corresponding to paths between the multiple transmit antennas and the receive antennas experience i.i.d. Rayleigh fading. The elements of the N-dimensional noise vector $\mathbf{z} = \begin{bmatrix} z_1 & z_2 & \dots & z_N \end{bmatrix}^T$ are also assumed to be i.i.d. complex Gaussian random variables with zero mean and variance σ_n^2.

Different types of detectors can be employed in multilayer MIMO systems. Linear detectors include the decorrelator and the MMSE. Nevertheless, the former presents the disadvantage of introducing noise enhancement. Among the suboptimal detectors, the SIC or the vertical-Bell laboratories layered space-time (V-BLAST) detector are two examples. A detector widely employed in multilayer transmission scheme is the V-BLAST detector, which typically comprises a SIC, alongside a decorrelator or with a MMSE detector. The maximum likelihood sequence estimator (MLSE) detector is also an option. Nevertheless, since the complexity and processing increases exponentially with the increase in the number of antennas, its real implementation tends to be limited to very specific scenarios.[*]

Lately, two types of decoders have been investigated: the lattice and the sphere detector. These types of detectors tend to achieve good performances with a reduced complexity.

7.8.4 SPACE DIVISION MULTIPLE ACCESS

The goal of SDMA relies on improving the capacity,[†] while keeping the spectrum allocation unchanged. SDMA is a technique that allows multiple users exploiting spatial diversity as a multiple access technique, while using the same spectrum.

As previously referred, both SDMA and multilayer transmission belong to the same group, entitled SM. Therefore, the basic concept is common. Nevertheless, while in multilayer transmission an increase of the symbol rate is achieved by considering multiple antennas at the transmitter side, using the SDMA, it is assumed that each transmitter has a single antenna, and the multiple number of receive antennas allow steering the different flows of data corresponding to different users. Consequently, SDMA is commonly employed in the uplink, where the transmitter (UE) has a single antenna, whereas the receiver (BS) has several antennas. Figure 7.23 depicts an SDMA configuration applied to the uplink. As an SM technique, to allow using the nulling algorithm at the receiver side, SDMA assumes that the number of antennas at the receiver is equal or higher than the number of users that share the same spectrum. With such approach, the receiver can decode the signals from each transmitter, while avoiding the signals from the other transmitters (interfering signals). Similar to the decoding performed in multilayer transmission, this can be achieved by using the nulling algorithm.

In the V-BLAST detector, the symbols from the transmit antenna with the highest SNR is first detected using a linear nulling algorithm such as zero forcing (ZF) or MMSE detector. The detected symbol is regenerated, and the corresponding signal portion is subtracted from the received signal vector using typically a SIC detector. This cancellation process results in a modified received signal vector with fewer interfering signal components left. This process is repeated, until all symbols from all antennas are detected. According to the detection-ordering scheme in Foschini [1996], the detection process is organized so that the symbols from the transmit antenna with the highest SNR is detected at each detection stage. Clearly, the processing of SDMA is almost the same as that of the MIMO based on multilayer transmission. Therefore, the reader should refer to Section 7.8.3 for a detailed description of SM detectors.

---

[*] Owing to this reason, the MLSE detector is commonly referred to as *brute force detector*.
[†] Supporting more users per cell.

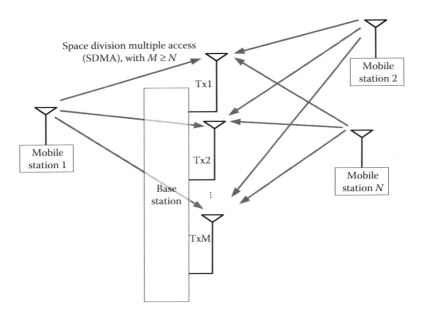

FIGURE 7.23 Example of SDMA scheme applied to the uplink.

7.8.5 BEAMFORMING

In STBC and SM MIMO schemes, the antenna elements that form an array are usually widely separated in order to form a TD array with low correlation between them. On the other hand, the beamforming is implemented by antenna array with array elements at the transmitter or receiver being closely located to form a beam. The beam is generated with the uniform linear antenna array (ULA). The ULA antenna elements spacing is typically half wavelength. Beamforming is an effective solution to maximize the SNR, as it steers the transmit (or receive) beam toward the receive (or transmit) antenna [Marques da Silva et al. 2009]. As a result, an improved performance or coverage is achieved with beamforming. In the cellular environment, this translates in a reduced number of required sites to cover a given area. Figure 7.24 depicts a transmitting station sending signals using the beamforming, generated with the ULA. The beamforming allows transmitting a higher power signal directed toward the desired station, while minimizing the transmitted power toward the other stations. This allows a minimization of the interfering signals.

The beamforming consists of M identical antenna elements with 120° half power beam width (HPBW) [Schacht et al. 2002]. Each antenna element is connected to a complex weight w_m, $1 \leq m \leq M$. An analogy can be made with a receiving array antenna.

The weighted elementary signals are summed together making an output signal as follows [Schacht et al. 2002]:

$$y(t,\theta) = \sum_{m=1}^{M} w_m x \left[t - (m-1) \frac{d}{c} \sin\theta \right] \qquad (7.76)$$

where:
  $x(t)$ is the signal sent by the first antenna element
  d is the distance between elements
  c is the propagation speed
  θ is the direction of arrival (or departure) for the main sector $\left(-60° < \theta < 60° \right)$

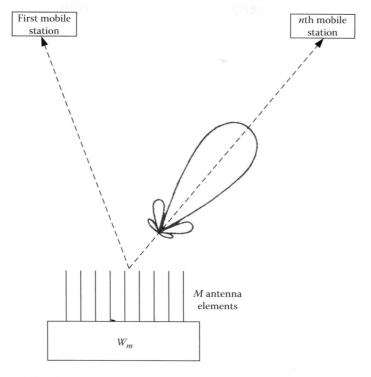

FIGURE 7.24 Simplified diagram of a beamforming transmitter.

In the frequency domain, the output signal can be written as

$$Y(f,\theta) = \sum_{m=1}^{M} w_m X(f) e^{-j2\pi f (m-1)(d/c)\sin\theta} \tag{7.77}$$

which makes the relationship between the output signal and the input signal given by

$$H(f,\theta) = \frac{Y(f,\theta)}{X(f)} = \sum_{m=1}^{M} w_m e^{-j2\pi f (m-1)(d/c)\sin\theta} \tag{7.78}$$

For narrowband beamforming, f is constant and θ is variable. For the beam to be directed toward the desired direction θ_1, we have

$$w_m = e^{j2\pi f (m-1)(d/c)\sin\theta_1} \tag{7.79}$$

in which, for the case of $d = \lambda/2$, Equation 7.79 results in [Schacht et al. 2002]

$$w_m = e^{j\pi(m-1)\sin\theta_1} \tag{7.80}$$

In other words, for $\theta = \theta_1$, $H(f,\theta_1)$ reduces to

$$H(f,\theta_1) = M \tag{7.81}$$

which is the maximum attainable amplitude by beamforming.

7.8.6 Multiuser MIMO

The MIMO techniques previously exposed are typically employed in the concept of single-user MIMO (SU-MIMO). This considers data being transmitted from a single user into another individual user. An alternative concept is the multiuser MIMO (MU-MIMO), where multiple streams of data are simultaneously allocated to different users, using the same frequency bands.

When the aim relies on achieving a performance improvement, a SU-MIMO is normally employed using an algorithm such as STBC. On the other hand, when the aim relies on achieving higher throughputs using a constrained spectrum, we have two options: in the downlink, the MU-MIMO is typically the solution; in the uplink, SM is normally employed (in this case, multilayer transmission).

The approach behind MU-MIMO is similar to SDMA. Nevertheless, while SDMA is typically employed in the uplink,[*] the MU-MIMO is widely implemented in the downlink. This allows sending different data streams into multiple UE. In this case, instead of performing the nulling algorithm at the receiver side, the nulling algorithm needs to be performed using a pre-processing[†] approach at the transmitter side (BS). This is possible because the BS can accommodate a high number of transmit antennas and the UE can only accommodate a single or reduced number (lower) of receive antennas. In the downlink of a MU-MIMO configuration, the number of transmit antennas must be equal to or higher than the number of multiple data streams that are sent to multiple users, at the same time, and occupying the same frequency bands (the opposite of the SDMA approach). In this configuration, the nulling algorithm is implemented, at the transmitter side, using a pre-processing algorithm such as the ZF, MMSE, dirty paper coding, and so on. Alternatively, instead of implementing the above-described SM principle, the MU-MIMO can be performed using the beamforming algorithm. In any case, the MU-MIMO requires accurate downlink CSI at the transmitter side. Obtaining CSI is trivial using TDD mode, being more difficult to be obtained when FDD is employed. In FDD mode, CSI is normally obtained using a feedback link.

It is worth noting that users located at the cell edge, served by the MU-MIMO, may experience SNR degradation due to intercell interference, interuser interference,[‡] additional path loss, or limited BS transmit power (which results from the use of pre-coding). A mechanism that can be implemented to mitigate such limitation relies on employing a dynamic MIMO system, where the MU-MIMO is employed everywhere, except at the cell edge. At the edge of the cell, the BS switches into SU-MIMO (using, e.g., space-time coding), which translates in an improvement of performance. Alternatively, coordinated multipoint (CoMP)[§] transmission is known as an effective mechanism that improves the performance at the cell edge, resulting in a more homogenous service quality, regardless of the users' positions.

7.9 ADVANCED MIMO APPLICATIONS

The challenge facing the mobile telecommunications industry today is how to continually improve the end-user experience, to offer appealing services through a delivery mechanism that offers improved speed, service attractiveness, and service interaction. In order to deliver the required services to the users with the minimum cost, the technology should allow better performances, higher throughputs, improved capacities, and higher spectral efficiencies.

The basic concept of MIMO relies on the transmission of signals through multiple paths, between multiple transmit and multiple receive antennas. Instead of representing an additional interference level, these multiple paths can be used as an advantage. While in MIMO systems, the multiple transmit or the multiple receive antennas are co-located, advanced cellular network architectures

---

[*] Because the nulling algorithm requires a high number of receive antennas.

[†] This is commonly referred to as *pre-coding*.

[‡] Interference among users that share the spectrum and that are separated by the MU-MIMO spatial multiplexing.

[§] Sometimes also referred to as *downlink base station cooperation*.

may also achieve the same level of diversity, but using antennas belonging to different base stations or relay nodes. Nevertheless, a certain level of synchronization or coordination is normally required between those stations. These techniques are described in Sections 7.9.1 through 7.9.4.

7.9.1 BASE STATION COOPERATION

Base station cooperation aims at improving the performance of UE located at the edge of the cells, or aims at increasing the throughput of UE in an area covered by multiple BSs.

An important requirement of 4G systems is the ability to deliver a homogenous service, regardless of the users' positions [Lee et al. 2012]. Users located at the cell edge may experience a degradation of the SNR due to intercell interference, additional path loss or due to limited eNode-B[*] transmit power. In case MU-MIMO is employed, the power constraint becomes more important[†] and the SNR degradation of UE located at the cell edge can even be deeper than in SISO environments. In these scenarios, base station cooperation plays an important role, as it allows the exploitation of additional diversity or the delivery of a high and constant throughput, regardless of the users' positions.

Base station cooperation stands for the ability to send or receive data from/to multiple adjacent BSs to/from UE. Instead of employing the conventional hard handover—with base station cooperation-independent antenna elements of different BSs are grouped together, forming a cluster—the UE may experience a throughput increase or performance improvement (i.e., SNR increase).[‡] Figure 7.25 shows the downlink base station cooperation using a cluster of three BSs. The dashed area corresponds to the area where the UE is able to simultaneously receive signals from three BSs. Using base station cooperation, adjacent BSs can simultaneously transmit the same data. In this case, a combining algorithm must be employed at the UE side, such as the maximum ratio combining, the MSE-based combining, or the selective combining. This technique is employed

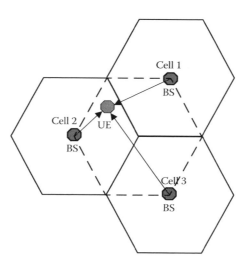

FIGURE 7.25 Downlink base station cooperation using a cluster of three BSs.

---

[*] Evolved Node-B (eNode-B) refers to the base station of LTE and 4G systems. This corresponds to the evolution of the Node-B, considered by UMTS.

[†] In MU-MIMO the power at the transmitter side (i.e., at the BS, in case of downlink) becomes very demanding because the pre-coding may, instantly, require a high level of power. Consequently, using pre-coding, the dynamic range of signals tends to increase.

[‡] A performance improvement can be achieved through the exploitation of additional diversity using, for example, space-time coding.

in macrodiversity, as described in Section 7.9.1.2. Alternatively, some type of pre-processing can be employed at the BSs side, such that the signals that reach the UE do not require any type of post-processing. In this case, this cooperation is commonly referred to as coordinated multipoint transmission [Correia and Marques da Silva 2014], as described in Section 7.9.1.1.

Base station cooperation can also be employed in the uplink, being also referred to as CoMP reception. In this case, traditional receive diversity is employed, using combining algorithms such as the MRC, or the MSE. Alternatively, independent parallel streams of data can be sent out by different UE transmit antennas (typically, a different data stream is sent out by each transmit antenna element), while different BSs decode these different streams using the above-described SDMA algorithm.

7.9.1.1 CoMP Transmission

Using CoMP transmission, independent antenna elements of different BSs are grouped together, forming a cluster, and the UE can experience a throughput increase or performance improvement. A pre-processing is typically employed at the BSs side such that the signals that reach the UE do not require any time of post-processing.

In case each BS uses a MIMO scheme, the resulting MIMO can be viewed as a *giant MIMO*, consisting of the combination of the independent antenna elements from different BSs (see Figure 7.26) [Marques da Silva et al. 2014; Reis et al. 2014].

Coordinated multipoint transmission comprises the coordinated transmission of signals from adjacent BSs, and the corresponding reception from a UE. The signal received at the UE side consists of the sum of independent signals sent by different BSs.

CoMP transmission is an important technique that can mitigate intercell interference, improve the throughput, exploit diversity, and therefore, improve the spectrum efficiency. Note that CoMP transmission allows a spectrum efficiency improvement, even at the cell edge. CoMP transmission can be viewed as a special type of MU-MIMO.

Similar to MU-MIMO, a pre-processing such as beamforming, ZF, MMSE, or dirty-paper is typically employed, in order to assure that the UE receives a signal coming from multiple BSs. In this case, the UE commonly employs a low-complexity and regular detector. Alternatively, noncoherent joint transmission and coordinated scheduling is employed. In this case, CoMP transmission can be associated to carrier aggregation, activating or de-activating some carriers in order to optimize the performance. This technique is employed in LTE-Advanced (3GPP Release 10) [Marques da Silva et al. 2012b].

Similar to the MU-MIMO, accurate downlink CSI is required at BSs side, which consists of an implementation difficulty. In case of TDD, obtaining CSI is trivial, as the uplink and downlink

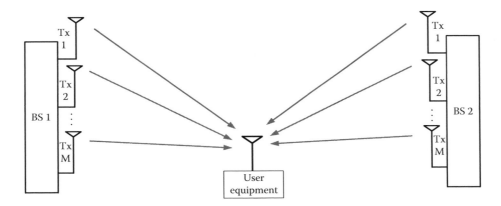

FIGURE 7.26   Base station cooperation (CoMP transmission) combined with downlink MIMO.

channels are almost the same. Nevertheless, in case of FDD, obtaining CSI at the transmitter side is a complex task.

Depending on the way the coordination between different BSs is performed, and the way CSI is obtained (in FDD mode), the following two different architectures can be implemented:

- *Centralized architecture*: In this architecture, there is a central control unit (CU) that decides about the transmission scheme and resources allocation to be used by different BSs. In this case, the CU is connected to different BSs of the cluster. Each UE estimates the downlink CSI of the signals received from each BS. Then, the CSI is sent back to the corresponding BS. At a third stage, CSI is sent from different BSs of the cluster to the CU through backhaul links. Based on CSI, the CU decides about the transmission scheme and resources allocation to be used by each BS, and sends this information to different BSs of the cluster. A major limitation of this architecture relies on the latency, whose factor may result in performance degradation due to fast CSI variations [Diehm et al. 2010].
- *Distributed architecture*: In this architecture, each BS is associated with a different CU, and the decision about the transmission scheme and resources allocation is performed independently at the BS level. Then, this information is exchanged between the cluster's BSs. In this case, each UE estimates the downlink CSI to different BSs and sends this joint data back not only to the BS of reference, but to all BSs. This way the CU associated to each BS has information about different downlink CSI and makes the decision accordingly. An advantage of this architecture relies on the fact that latency is much reduced, and there is no need to use backhaul links for the purpose of exchanging CSI. Nevertheless, this architecture is more subject to errors caused by the uplink transmission [Papadogiannis et al. 2009].

Finally, it is worth noting that the use of CoMP transmission requires performing a cost–benefit analysis, as CoMP allows a performance/network improvement at the cost of a larger amount of overhead on the over-the-air feedback links and in the backhaul links used to interconnect different BSs.

7.9.1.2 Macrodiversity

Macrodiversity refers to the transmission of the same information by different BSs to the UE in the downlink. Therefore, this can be viewed as a particular type of downlink base station cooperation. Macrodiversity aims at supplying additional diversity[*] in situations where the terminal is far from the BSs. This allows compensating the path loss affecting the transmission to a UE located at the edge of the cell. Consequently, this allows a reduction of the amount of transmit power required to reach a distant receiver, thus increasing network capacity. Therefore, macrodiversity can be viewed as a special type of base station cooperation.

There are two types of networks to be considered: the multifrequency networks (MFNs) and the single frequency networks. The BSs to which a terminal is linked to are referred to as the active sets.

In broadcast, the global CIR tends to be longer due to the longer distances between the transmitter (BS) and the farer of the different receivers (UEs). If the cyclic extension is long enough, the global CIR will be the sum of the independent CIRs. This can be seen from Figure 7.27. This enables the SFN concept exploiting the macrodiversity effect.

Macrodiversity is used during soft handover in order to ensure smooth transitions between two cells or two sectors of the same cell, increasing the multipath diversity and reducing the risk of call drop.

Using the same transmitting frequencies (from different BSs), diversity is exploited and deep fading tends to be avoided. In case the channel profile between each of the two BSs and the UE is single path, the resulting signal is viewed as a two-path channel profile, and the diversity is

---

[*] Since it provides transmit diversity, macrodiversity can be viewed as an MISO system.

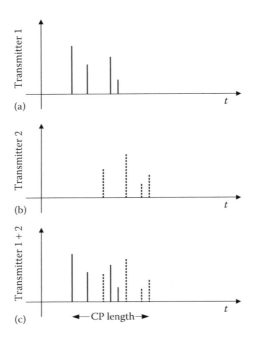

FIGURE 7.27 Global CIR (c) is composed of the sum of the CIR of the several transmitters (a + b).

exploited. Nevertheless, since a single receiver is employed, although diversity is exploited, the frequency selectivity increases. In fact, this effect can be viewed as one signal received from one BS interfered by the signal received from another BS. Nevertheless, this additional frequency selectivity is mitigated because OFDM signals make use of equalizers at subcarrier level, as long as the appropriate CP is employed. Consequently, the macrodiversity presents overall benefits in terms of diversity.

The performance gain brought by macrodiversity depends on the diversity order of the channel. A two-path channel benefits more from macrodiversity than a six-path channel, because the latter already exhibits a high multipath diversity order. Since MBMS in SFN mode is simultaneously broadcasted in several cells, macrodiversity for MBMS in SFN mode does not consume network resources.

In the case of MFN, the UE is required to estimate the carrier of each BS it is linked to. This increases its power consumption. In this case, the signals received from different BSs (especially far ones) may be significantly delayed with regard to those received from near BSs. This requires extramemory at the terminal in order to store the received signals for further combining. Alternatively, an additional synchronization procedure between the BS transmitters is required. In the downlink, the combining takes place at the UE, which has to demodulate and then combine the signals received from the different BSs. The extracomplexity added by macrodiversity then depends on the receiver type. In the case of an equalizer, one has to be set up and operate for each BS the UE is linked to. Moreover, the UE must estimate one transmission channel per BS.

In the special case of OFDM, the following two main cases for macrodiversity can be distinguished:

- BSs are synchronized, at least to allow UE's receiving signals from two or more BSs with a time difference smaller than the CP.
- BSs are not synchronized.

In the first case, the BSs can transmit identical signals to the UE on the same time-frequency resource. This is possible because the signals will superimpose within the CP. In this situation, no

ISI occurs as long as the sum of the time differences plus the maximum delay of the CIR is shorter than the CP. Therefore, the terminal can employ a single receiver to demodulate the superimposed signals. This means that it will perform a unique DFT. In this scenario, macrodiversity behaves just like TD (from a unique transmitter with multiple spaced antennas). When different BSs transmit the same data over the same subcarriers, the resulting propagation channel is equivalent to the allegorical sum of all propagation channels, which increases the diversity gain (deep fading is avoided).

When BSs send the same data over different subcarriers (MFN), the maximum diversity can be achieved since each data symbol benefits from the summation of the propagation channel powers (i.e., frequency diversity).

If the BSs are not synchronized, the terminal will need separated receiver chains to demodulate the signals from the distinct BSs. Moreover, to avoid interference, orthogonal time-frequency resources have to be allocated to different BSs. This is still very complex to fulfill, thus in the general case, interference will occur, and spectral efficiency will decrease.

Fast cell selection is one option for macrodiversity in unicast mode (selective diversity). Intra-BS selection should be able to operate on a subframe basis. An alternative to intra-BS macrodiversity scheme for unicast consists of a simultaneous multicell transmission with soft-combining. The basic idea of multicell transmission is that, instead of avoiding interference at the cell border by means of intercell-interference coordination, both cells are used for the transmission of the same information to a UE, thus, reducing intercell interference as well as improving the overall available transmit power. Another possibility of intra-BS multicell transmission relies on the exploitation of diversity between the cells with space-time processing (i.e., by employing STBC through two cells). Assuming BS-controlled scheduling and that fast/tight coordination between different BS is not feasible, multicell transmission should be limited to cells belonging to the same BS. For multicell broadcast, soft combining of radio links should be supported, assuming a sufficient degree of inter-BS synchronization, at least between a subset of BSs.

7.9.2 MULTIHOP RELAY

Multihop relay is a technique that can improve the coverage and capacity for high quality multimedia broadcast and multicast transmissions in mobile networks by providing a homogenous service, regardless of the users' positions, allowing high data rates for UE even at the cell edge.

UEs at the cell edge suffer from high propagation loss and high intercell interference from neighbor cells. Other UE reside in areas that suffer from strong shadowing effects or require indoor coverage from outdoor BS. These impairments originate a degradation of the SNR, which translates in a reduced service quality. Thus, the overall goal of multihop relay is to bring more power to the cell edge and into shadowed areas, while inducing minimal additional interference for neighbor cells. Moreover, increased cell capacity tends also to be achieved by using multihop relay. In addition, multihop relay may also represent an increased life cycle of UE batteries.

There are several mechanisms that can be used to implement multihop relays [Sydir and Taori 2009]. As defined in the following, one technique employed in multihop relay consists of using TDD [Stencel et al. 2010], where the communication between the BS and the relay node (RN) is performed in a time slot different from the one utilized for the communication between the RN and the UE. On the contrary, Schoene et al. [2008] proposed the use of multihop relay associated to FDD.

Sections 7.9.2.1 through 7.9.2.3 describe the most important methods utilized in multihop relay.

7.9.2.1 Adaptive Relaying

As previously described, UE at the cell edge suffer from high propagation loss and high intercell interference from the neighbor cells. Other UE reside in areas that suffer from strong shadowing effects. The obvious solution for this would be to decrease cell sizes by installing additional base stations that represents an increase of network infrastructure costs. Contrarily, adaptive relay nodes can provide temporary network deployment, avoiding outage of service in the area, while keeping the investment cost at an acceptable level. The positioning of relay nodes is selected in a manner that an improvement of service quality or cell coverage is achieved.

In opposition to conventional repeaters working with the amplify-and-forward strategy, adaptive relays are understood to work in a decode-and-forward style. By performing this, relays amplify and retransmit only the wanted component of the signal they receive and suppress the unwanted portions (i.e., regeneration is performed). The disadvantage of an adaptive relay, compared to a conventional repeater, relies on the additional delay introduced into the transmission path between BS and UE. Moreover, depending on the algorithm, a signaling overhead may also be required. The gain is based on the fact that the transmission path is split up into smaller areas, representing a reduction of the propagation loss.

Fixed relay stations positioned at a specified distance from the BS (see Figure 7.28) tend to increase the probability that a UE receives enough power from the BS. This deployment concept sectorizes the cell in an inner region where the UE (e.g., UE 1 in the figure) receives signals from the BS plus some relay stations, and an outer region where only signals from relay stations are strong enough to be received by the UE (e.g., UE 2 in the figure).

7.9.2.2 Configurable Virtual Cell Sizes

Configurable virtual cell sizes were proposed in Kudoh and Adachi [2003], consisting of a wireless multihop virtual cellular network. This configuration includes the following components:

- *Central port*: This corresponds to a BS acting as a gateway to the core network.
- *Wireless ports*: They correspond to relay stations that may or may not communicate with the UE, and relay the signals from and to the central port.

The wireless ports that are communicating directly with the UE are called end wireless ports. The wireless ports are stationary and can act together with the central port as one virtual base station. The central port and the end wireless ports introduce additional diversity into the cell, such that the transmit power can be reduced. This also results in a reduced interference level for other virtual cells. The differences between conventional cellular networks and virtual cellular networks can be seen from Figures 7.29 and 7.30.

From the perspective of MBMS, configurable virtual cell sizes could be employed to adapt the cell size to the user distribution and service demands. Service needs may present spatial and temporal variations.

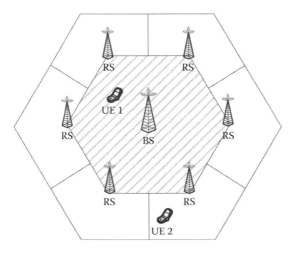

FIGURE 7.28 Two-hop relaying architecture.

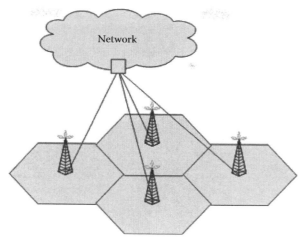

FIGURE 7.29 Conventional cellular network.

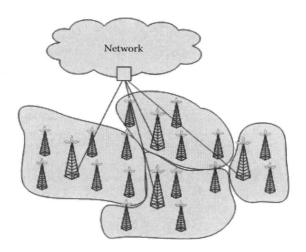

FIGURE 7.30 Virtual cellular network.

7.9.2.3 Multihop Relay in 3GPP

Relay nodes were introduced in 3GPP Release 9 [3GPP 2010a] as a special type of eNB* that is not directly connected to the core network. A RN receives data that was forwarded by an eNB† that is connected to the evolved packet core‡ (EPC). Upon receiving such data, the RN sends it to the UE that are under its area of coverage. This technique can be seen from Figure 7.31. This is a very interesting option for operators because, compared to eNB, RNs present structures less expensive to deploy and to maintain. They can provide temporary network deployment and new services in an area (e.g., when a sporadic event that concentrates a lot of people in the same geographical area, such as a summer festival). The use of a RN allows a fast deploying and inexpensive way to solve the problem, and can also provide coverage in small areas not covered by eNBs.

---

* eNB stands for evolved Node-B (eNode-B), that is, a Node-B employed in LTE. Note that a Node-B is a BS in the UMTS environment.
† This eNB is acting as Donor eNB (DeNB).
‡ Evolved packet core corresponds to the core of the LTE (all-IP) network.

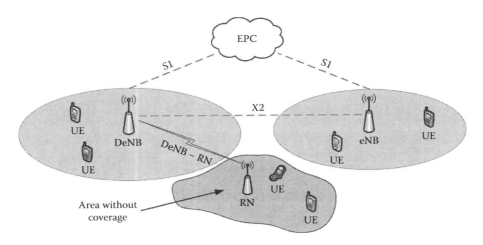

FIGURE 7.31 Example of relay node in E-UTRAN.

In 3GPP [2010b] four relay architectures are proposed and studied. Those architectures differ from each other in terms of expected behavior of the RN/DeNB and how the data is sent within the EPC until it reaches the UE. The 3GPP study concludes that an architecture where a RN acts as a proxy for S1/X2 has the most overall benefits, having been incorporated in 3GPP Release 10 (LTE-Advanced).

The detailed RN architecture chosen for LTE-Advanced can be seen from Figure 7.32. In this architecture, we can see that the RN plays simultaneously two roles:

- From the network's point of view (particularly for the DeNB) the RN acts as an ordinary UE (denoted as Relay-UE)
- From the UE point of view the RN acts as a normal eNB

This way the network can abstract itself from establishing a point-to-point connection with each and every UE. In fact, it only has to establish a connection with the RN as it would normally perform with an ordinary UE, and then forward all data that was destined to UE to the RN (see Figure 7.31). Then, the RN, on its own, will forward all the data received from the network to the respective UE.

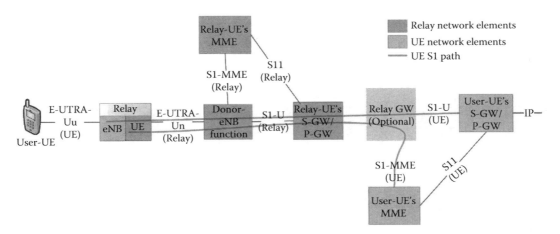

FIGURE 7.32 Relay architecture chosen for LTE-Advanced. (Adapted from 3GPP, TR 36.806-v9.0.0, *Evolved Universal Terrestrial Radio Access [E-UTRA]*; *Relay Architectures for E-UTRA [LTE-Advanced]*, European Telecommunications Standards Institute, France, May 2010b.)

The process of relaying is completely transparent, that is, a UE always assumes to be connected to an eNB because RNs are viewed as eNB from the UE side, and as a consequence no changes to the communication protocol or interfaces used by UE have to be made.

In the context of multiresolution for Evolved-MBMS (E-MBMS),* the use of relaying fits perfectly. With the introduction of RN in E-MBMS we can have different zones with different grades of service. Let us consider again the example depicted in Figure 7.31. In areas covered by eNBs, UE are expected to have high signal-to-interference plus noise ratio (SINR) when they are closer to the center of the cell, reducing as the UE approach the cell edges. If a hierarchic 64-QAM modulation with coding rate 3/4 is being employed, UE at the cell edge will experience low levels of QoS or even service outage due to low SINR. If an RN is strategically positioned at the cell edge of existing eNBs, we can expect that area to have an improved coverage and the RN can adapt its transmitting conditions to provide a different resolution of the service in the area.† With such approach, the use of RNs means that, in E-UTRAN, when providing E-MBMS, UE will experience higher coverage and different service resolutions for the same E-MBMS contents depending on their location and their receiving conditions.

According to [3GPP 2010a], RNs can operate in two different modes:

- *Inband mode*: The link between DeNB and the RN uses the same carrier frequency as the link between RN-UE and eNB-UE.
- *Outband mode*: The link between DeNB and the RN uses a different carrier frequency than that of the RN-UE link.

It is worth noting that the transmit power of RNs is lower than that of eNBs. The objective is to diminish intercell interference caused by the introduction of RNs operating in the same band and carrier frequency as eNBs (inband mode), and to limit the area of coverage of a given RN (since RNs are used to cover small areas that cannot be properly covered by existing eNBs).

In order to reduce intercell interference, RNs can also be employed in E-MBMS subframes to create *transmission gaps*. These transmission gaps are illustrated in Figure 7.33. As can be seen, in the first instance where *Time = t*, the DeNB transmits a segment of data labeled *Data 1* to all the UE inside its area of coverage and also to the RN. At this instance, the RN stores *Data 1* in memory and is not transmitting any data to the users inside its area of coverage, thus not creating any intercell interference. At the next instance *Time = t + 1*, a second set of data labeled *Data 2* is transmitted from DeNB to all the UE within its area of coverage. On its turn, the RN is not receiving *Data 2* but, instead, is transmitting to the UE within its area of coverage the *Data 1* that was stored in memory at *Time = t*, generating intercell interference.

As a result of this intercell interference pattern produced by RNs, the total intercell interference is greatly reduced (RNs are only causing interference half of the total transmission times) with a tradeoff between throughput and intercell interference.

7.9.3 Multiresolution Transmission Schemes

With MBMS, it makes sense to have two or more classes of bits with different error protection, to which different streams of information can be mapped. Depending on the propagation conditions, a given user can attempt to demodulate only the more protected bits or also the other bits that carry

---

* Evolved-MBMS (E-MBMS) framework constitutes the evolutionary successor of MBMS, and is envisaged to play an essential role for the LTE-A proliferation in mobile environments. E-MBMS was initially introduced in 3GPP Release 8 (LTE), with further improvements to be implemented in LTE-A. The basic role of these techniques/mechanisms relies on the power and resources optimization during MBMS transmissions. The objective of E-MBMS is to provide services with different QoS requirements depending on the channel conditions experienced by different users.

† Instead of transmitting video with high frame rate and low resolution, we can half the frame rate and transmit with higher resolution.

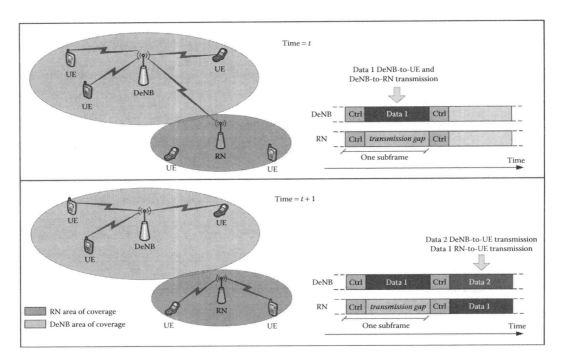

FIGURE 7.33   Example of transmission gaps in E-MBMS.

the additional information. By using nonuniformly spaced signal points in hierarchical modulations, it is possible to modify the different error protection levels [Vitthaladevuni and Alouini 2001, 2004].

Multiresolution techniques are interesting for applications where the data being transmitted is scalable, that is, it can be split into classes of different importance. For example, in the case of video transmission, the data from the video source encoders may not be equally important. The same happens in the transmission of coded voice. The introduction of multiresolution in a broadcast cellular system deals with source coding and transmission of output data streams. In a broadcast cellular system, there is a heterogeneous network with different terminal capabilities and connection speeds. For the particular case of video, scalable video consists of a common strategy presented in the literature that adapts its content within a heterogeneous communications environment [Li 2001; Liu et al. 2003; Vetro et al. 2003; Dogan et al. 2004; Holma and Toskala 2007].

Scalable video, depicted in Figure 7.34, provides a base layer for minimum requirements, and one or more enhancement layers to offer improved quality at increasing bit/frame rates and resolutions [Li 2001]. Therefore, this method significantly decreases the storage costs of the content provider.

Besides being a potential solution for content adaptation, scalable video schemes may also allow an efficient usage of power resources in MBMS, as suggested in Cover [1972]. This is depicted in Figure 7.34, where three separate physical resource blocks (PRBs) are provided for one MBMS service (e.g., @ 384 kbps): one for the base layer, at one-third, the rate of the total bit rate (128 kbps), and with a power allocation that can cover the whole cell range (UE3); another PRB for the first enhanced layer, also at one-third, the rate of the total bit rate (but aggregate rate of 256 kbps), with less power allocation than that of the base layer (UE2); and a third PRB for the second enhanced layer transmitted with small power to be received near the base station (UE1, with aggregate rate of 384 kbps).

The system illustrated in Figure 7.34 consists of three QoS regions, where the first region receives all the information, whereas the second and third regions receive the most important data. The QoS regions are associated to the geometry factor that reflects the distance of the UE from the base station antenna.

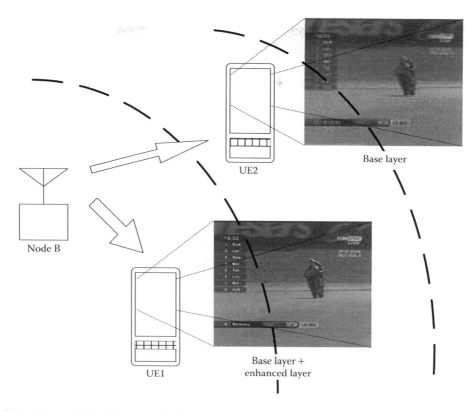

FIGURE 7.34   Scalable video transmission.

Scalable video transmission can be implemented using different techniques. One possibility relies on the use of hierarchical constellations. The system illustrated in Figure 7.35 consists of 64-QAM hierarchical constellation. Strong bit blocks are QPSK demodulated, medium bit blocks are 16-QAM demodulated, and weak bit blocks are 64-QAM demodulated. Furthermore, hierarchical constellations may also be combined with different channel coding rates. This corresponds to the concept of adaptive modulation and coding [Souto et al. 2007].

It is worth noting that the nonuniform QAM constellation concept has already been incorporated in the DVB Terrestrial (DVB-T) standard.

Another possibility to implement scalable video transmission relies on the use of SM MIMO technique, where each transmit antenna sends a different data stream. The first data stream (most powerful) may include the base layer, whereas the enhancement layer may be sent by a second antenna (less powerful data stream). Depending on the power and channel conditions, a certain UE may successfully receive either the two streams or only the base layer.

7.9.4 GREEN RADIO COMMUNICATIONS

The tremendous expansion of mobile network terminals has contributed to the increase of the environmental footprint. Currently, more than four billion subscribers are using mobile phones around the world. The operation of both mobile phones and network infrastructure require huge quantities of electrical energy that currently represents up to 50% of the operational costs.

Energy-efficient wireless transmission techniques, alongside with energy-efficient network architecture and protocols, hardware optimization, and renewable energy sources contribute to the

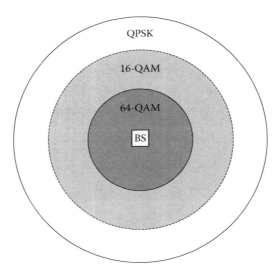

FIGURE 7.35 Example of the type of demodulation used inside a cell for a transmission of a 64-QAM hierarchical constellation.

implementation of the green cell networks concept, as they allow a reduction of the carbon emission footprint. The carbon footprint reduction can be achieved through advancements in the following three main different dimensions:

- *Architecture and protocols*: Improvements in the transmission techniques and network protocols and architectures
- *Components*: More efficient hardware implementation in terms of energy dependency
- *Energy provisioning*: Including a new strategy based on the generalized use of renewable energy sources

The architecture and protocols dimension can be implemented in two different fronts: the network level and the link level.

The network level comprises the dimensioning and adjustment of the network as a whole such that a reduction of the energy consumption is possible without radically compromising the network performance and the spectral efficiency [Correia et al. 2010; Ferling et al. 2010]. This is achieved through the implementation of network protocols, such as base station cooperation, multihop relay (see Figure 7.36), MIMO systems, hierarchical cellular structure,* or interference mitigation schemes (including pre-coding). These techniques may support a reduction of the transmit power by both BSs and UE [Ericsson 2007].

Mobile ad-hoc networking (MANET) is a concept that allows the implementation of networks without making use of BSs, and therefore, achieving the same capability with less energy and with less visual impact and investment. Energy efficiency is also achieved by dynamically allowing an adaption of the cellular network architecture to traffic load fluctuations. In this case, the network architecture should be sufficiently dynamic such that it adapts to variations of the number of UE, variations of the throughputs required by different UE, or to various UE geographic densities. Cognitive radio brings flexibility to the network, as it constitutes a mechanism that allows a dynamic and more efficient use of the spectrum. This translates in higher throughput per user or higher number of users per cell by exploiting spectrum opportunities. Cognitive radio constitutes

---

* Using macrocells, microcells, picocells, femtocells, and so on.

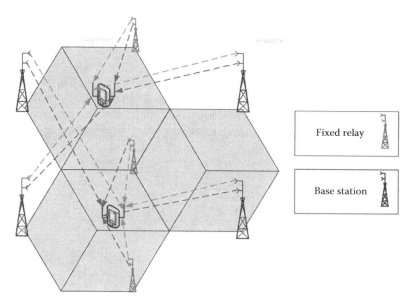

FIGURE 7.36   Cooperative MIMO system with fixed relay station.

an efficient alternative to the traditional approach where the networks were dimensioned to peak traffic scenarios.

The link level front is associated to the individual energy utilization in the interface between an UE and a BS. This includes synchronization and channel estimation techniques. Owing to multiplexing of pilot/training and data symbols, some of the available bandwidth and energy has to be consumed for accomplishing the transmission of the pilot symbols. Since the CIR can be very long, especially for block transmission schemes, the required channel estimation overheads can also be high, namely for fast-varying scenarios. This translates in a reduction of the useful bit rate, decreasing the spectral efficiency and increasing the energy needs. A promising method for overcoming this problem relies on the employment of implicit pilots, also known as pilot embedding or superimposed pilots, which are added to the data block, instead of being multiplexed with it [Marques da Silva and Dinis 2011]. This means that we can significantly increase the pilots' density, while keeping the system capacity. Note that the MIMO implementation constitutes a mechanism to improve the spectral efficiency, throughput, or number of users per cell. Nevertheless, it requires additional energy spent in pilots, as the channel estimation is independently performed for each antenna pair, using a different pilot stream per transmit antenna. This results in an energy efficiency degradation of the MIMO relating to the SISO. The implementation of embedded pilots allows the MIMO system becoming an energy-efficient technique, as the energy spent in pilot symbols is very much reduced, and thus the MIMO can improve the SNR, coverage or throughput, without additional transmit power (or even with a power reduction).

From the components perspective, the basic concept relies on the development of signal processing techniques for smart components in order to obtain a reduction of the energy consumption. Moreover, the relationship between the output power and the consumed power of an equipment or component (e.g., a transmitter) should also be maximized. This can be viewed as power efficiency, and can also be improved by implementing a careful design and advanced signal processing. It is known that the energy consumed in a power amplifier (PA) represents between 50% and 80% of the electrical energy consumed in a BS [Correia et al. 2010]. OFDM transmission technique is characterized by high peak to average power ratio (PAPR)[*] levels, whose implementation requires a PA operating well below the saturation point. This results in poor power efficiency. A signal processing technique commonly employed to improve the energy efficiency of a PA relies on decreasing the PAPR by using a pre-coding technique that reduces the dynamic range of a block transmission signal. This technique allows a PA operation closer to the saturation threshold, which translates in an energetic efficiency gain. Another signal processing technique currently in research relies on the implementation of smart cooling.[†] It is also worth noting that cooling systems tend to be more demanding in high power transmitters. Reducing the required transmit power allows a simpler and less powerful cooling system.

Finally, a new energy provisioning strategy based on the use of renewable energy sources is another dimension that can be exploited in order to achieve a reduction of the carbon emission footprint. This includes the implementation of photovoltaic cells and/or wind generators in base stations. Moreover, photovoltaic cells can also be employed in mobile phones or in human clothes to charge it.

CHAPTER SUMMARY

This chapter provided a view about advanced transmission techniques to support current and emergent multimedia services. Initially, it provided a description about advances in wireless systems and their technical demands. Then, the spread-spectrum communications were introduced, as well as CDMA, OFDM, and SC-FDE.

In another stage, the diversity combining algorithms were introduced, including the selection combining, the maximal ratio combining, the EGC, and the MSE-based combining algorithms. Then, the RAKE receiver was introduced, as employed by CDMA receivers to allow the exploitation of the multipath diversity. The main advantage of the CDMA relies on the ability to exploit multipath diversity, which is implemented using a RAKE receiver. It was described that increasing the CDMA spreading factor allows increasing the level of multipath diversity exploited by the RAKE receiver.

Then, the MIMO systems were introduced, which make use of multiple transmit and receive antennas, and consist of techniques that allow the additional exploitation of diversity, an improvement of the spectral efficiency, or the increase of the symbols rate. It was viewed that the space-time coding aims to provide TD. When space-time coding is combined with receive diversity, this results in a MIMO system, where both transmit and receiver diversities can be exploited. It was also described that multilayer transmission is a type of SM, consisting of a MIMO architecture used for the parallel transmission of data, where each transmit antenna sends a different data stream, supporting an increase of throughput. It was viewed that the multilayer scheme requires that the number of receive antennas be equal or higher than the number of transmit antennas. SDMA was also introduced, consisting also of an SM. Nevertheless, while multilayer transmission sends different data streams, SDMA allows the sharing of spectrum by different users. It was shown that the beamforming allows the creation of an electromagnetic beam that concentrates the power toward the desired user, while minimizing the interfering power to other users.

It was viewed that SU-MIMO considers data being transmitted from a single user, into another individual user. On the other hand, MU-MIMO can be viewed as an alternative, where multiple streams of data are simultaneously allocated to different users, using the same frequency bands. This is equivalent to SDMA. Nevertheless, while SDMA is employed in the uplink, MU-MIMO is used in the downlink.

In another stage, advance MIMO applications were introduced. This included the study of base station cooperation, multihop relay, and multiresolution transmission schemes. These are

---

[*] Such PAPR level increases with the increase of the number of OFDM subcarriers.
[†] Cooling systems represent up to 30% of the consumed BS energy.

applications of the MIMO concept, where the transmit and receive antennas are not co-located, being however properly coordinated to exploit diversity, or to achieve a certain type of gain.

Finally, the concept of green radio communications was introduced, aiming at reducing the carbon footprint by using the energy spent in radio communications in a more efficient manner.

REVIEW QUESTIONS

1. What are the differences between FDMA, TDMA, and CDMA?
2. What are the differences between FDMA and OFDM?
3. What is the advantage of using a RAKE receiver?
4. Which type of transmission technique is employed associated to a RAKE receiver?
5. Which forms of MIMO systems do you know? What are the differences between them?
6. What is the difference between beamforming, multilayer transmission and SDMA?
7. What does multiresolution technique stands for?
8. Which types of multiresolution techniques do you know? How do they work?
9. What is the difference between spread spectrum and DS-CDMA?
10. What is the difference between closed loop and open loop MIMO techniques?
11. What are the differences between OFDMA and SC-FDE?
12. Which types of diversity combining algorithms do you know? What are their differences?
13. What is the difference between Alamouti-like and the multilayer MIMO techniques?

LAB EXERCISES

1. Using the Emona Telecoms Trainer 101 laboratory equipment, and volume 1 of its laboratory manual, perform experiment 19—spread spectrum—DSSS modulation and demodulation.
2. Using the Emona Telecoms Trainer 101 laboratory equipment, and volume 2 of its laboratory manual, perform experiment 14—PN sequence spectra and noise generation.
3. Using the Emona Telecoms Trainer 101 laboratory equipment, and volume 3 of its laboratory manual, perform experiment 1—full (IQ branch) demodulation of a QPSK signal.
4. Using the Emona Telecoms Trainer 101 laboratory equipment, and volume 3 of its laboratory manual, perform experiment 2—line code decoding and decision making.
5. Using the Emona Telecoms Trainer 101 laboratory equipment, and volume 3 of its laboratory manual, perform experiment 5—signal constellation diagrams.

8 Services and Applications

LEARNING OBJECTIVES

- Describe the application layer of TCP/IP.
- Identify and describe the different services provided by the application layer.
- Identify and describe the different application layer protocols.
- Describe the IP telephony and IP videoteleconference, and their implementations.

A data service comprises a functionality that is provided to the end user, making use of a specific program (application program). Examples of data services include IP telephony, web browsing, e-mail, and file transfer. On the other hand, an application program is a program, external to the network architecture, which allows to gain access to a certain service. For example, Microsoft Outlook, Microsoft Outlook Express, and Groupwise are three examples of specific application programs that can be employed to have access to the e-mail service. Web browsing is also a service that can be accessed making use of browsers such as Internet Explorer, Mozilla Firefox, Google Chrome, and so on. This corresponds to a user's program, external to the network architecture.

Finally, as shown in Figure 8.1, the application layer is the layer of the open system interconnection (OSI) or transmission control protocol/Internet protocol (TCP/IP) stack, which provides some network functionalities to the application program, making use of specific application layer protocols, such as the simple mail transfer protocol (SMTP), file transfer protocol (FTP), and telnet. Note that the application program described above has a different meaning than the one for the network architecture.

The different layers below the application layer are responsible for performing functions, which are common to the different service types. Contrarily, the application layer is responsible for implementing tasks that are specific to the data service to be provided to the end user. The application layer defines functionalities in the interaction between ending hosts (e.g., between a client and a server).

The application layer comprises different protocols, one or more per data service. Moreover, these protocols are different for different reference models (e.g., OSI or TCP/IP). While the file transfer access and management (FTAM) is the protocol of the OSI reference model, which allows the file transfer, the TCP/IP defines the FTP and the trivial FTP.

Depending on the protocol stack used to implement the network architecture, the application layer deals with different functionalities. While the application layer of the OSI reference model is limited to the application layer itself, the TCP/IP model includes the functionalities that the OSI reference model assigns to the application layer, presentation layer, and the session layer.

The session layer comprises a group of functionalities that involves the management of a session. They can be split into the following four functionalities:

1. Setup of sessions
2. Synchronization
3. Dialog management
4. Activities management

Setup of sessions: A session consists of a logical link, which is established between end entities. It has a meaning different from a physical link, which is established and managed by the transport layer. In case a physical link, supporting a certain logical link,

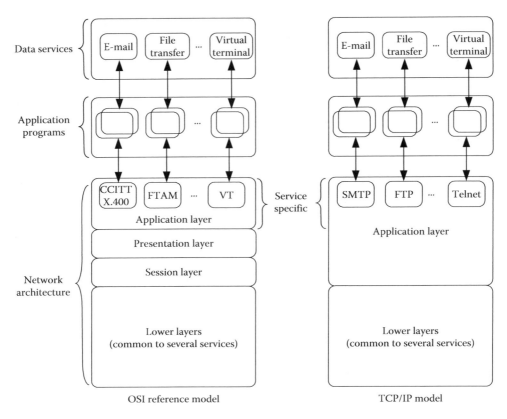

FIGURE 8.1 Difference between application layer (OSI and TCP/IP models), application programs, and services.

fails, a new physical link is established and the session (logical link) continues with its exchange of data (SPDU*).

Synchronization: It considers the periodic transmission of synchronization points, which are properly acknowledged by the receiving entity. In case the transport link fails, the session data does not need to be retransmitted from the beginning, but from the last synchronization point correctly acknowledged. Note that, from the session layer perspective, an acknowledgment means that the received data has been properly processed.[†]

Dialog management: Although the physical link may support full-duplex operation, some user processes only support, from the logical link perspective, half-duplex operation. This layer manages the communication by defining the entity that is allowed to send data. This is performed using a token that is transferred from one entity to another

Activities management: The application data is typically split into a set of elementary activities. Each activity is properly managed and signalized using marks,[‡] in order to register that it has initiated or finalized. These marks are important in order to allow the receiving entity to identify the borders of an activity. An example of a sequence of two elementary activities can be a copy of a file followed by its deletion. The data cannot be deleted

---

* The SPDU is defined in Chapter 2.
† Good reception of data is acknowledged at the transport layer, whereas good processing of data is acknowledged at the session layer.
‡ Exchanged by ending entities.

(second activity) without being sure that the copy (first activity) has been successfully performed. The ISO 8327/CCITT X.225 standard is an example of a standard that defines the session layer.

The presentation layer refers to the way the data is represented and exchanged in the interaction among different ending entities. Moreover, it provides means for the establishment and use of abstract syntax, which allows the exchange of data. An important syntax commonly employed among different entities is the abstract syntax notation 1 (ASN.1). The task of this layer can be viewed as the definition of a language (standard) to use in the communication among different hosts. Instead of allowing the communication being performed with the internal language of a host (e.g., computer), a standardized language is established, and all hosts must use such language while interacting with others. This avoids the need to have successive translators (gateways) among all different possible host languages. This way, it is only necessary to perform the translation between the host's internal language and a common standardized language. Note that the presentation layer also deals with encryption and compression of data, as these two functionalities change the language format used in communications. Such functionalities need to be properly negotiated and accepted before the exchange of data takes place. The ISO 8823/CCITT X.226 standard is an example of a presentation layer standard.

From the ISO reference model perspective, the application layer deals with tasks that are specific to the application program, whereas the session and presentation layers take care of tasks that are common to different application programs. From the TCP/IP model perspective, the application layer takes care of all tasks defined for the application layer, presentation layer, and the session layer of the OSI reference model.

Sections 8.1 through 8.6 describe different data services with a description of the corresponding TCP/IP application layer protocols used to allow the access to such data service.

8.1 WEB BROWSING

Web browsing is the most important Internet service. It is a service that allows a user (client) to gain access to multimedia information stored in a server. The server can be located in an intranet, extranet, or in the Internet. The access to multimedia information through the Internet is referred to as *web browsing*. Moreover, web browsing is performed in a space called *World Wide Web* (www). It consists of a client/server service, where a user retrieves information from a web page. A web page can provide a myriad of media, such as text, tables, graphics, voice, images, and video.

8.1.1 HYPERTEXT TRANSFER PROTOCOL

The hypertext transfer protocol (HTTP) is the TCP/IP application layer protocol, which allows a client to use the www space. The HTTP is a client/server protocol supported by the TCP, and therefore, a certain level of data reliability is assured. The web server can be accessed by using a certain IP address. Moreover, the default port address corresponding to a web server is 80. In HTTP, the different types of media are encoded in text. Since the data exchange is limited to text, the HTTP is simple. Moreover, since HTTP is supported by the TCP transport protocol, the access to each web page (transaction) follows three different elementary phases: connection setup, data exchange, and connection termination. This results in an independent treatment of each transaction (web page access). Once the client has successfully received the desired web page, the TCP connection is terminated.

A web page can be viewed as a file that is stored in a server. A client intending to retrieve a web page sends a GET message to the server with the requested uniform resource locator (URL), and the server responds with the corresponding web page (HTTP response). This can be viewed as a download of an hypertext markup language (HTML) file.

During data exchange between a client and a server, the HTTP may use either 7-bit ASCII format or the multipurpose Internet mail extension (MIME) format. The MIME consists of a message format that allows other non-ASCII characters, as well as binary formats containing images, sounds, and other files. There are different versions of HTTP, namely the HTTP 0.9, 1.0, and 1.1 [RFC 2616], with small differences in the message formats.

The exchange of HTTP data between a client and a server can be established in two basic ways: direct mode or indirect mode. As shown in Figure 8.2, the direct mode comprises the establishment of a direct connection between a client and a server. Contrarily, the indirect mode comprises an intermediate node, where two or more successive TCP connections are established, in order to establish the required end-to-end connection, and to gain access to the web page. As depicted in Figure 8.3, the indirect mode may have to be employed due to the use of a proxy server or a gateway.

A proxy is a server that is located at the edge of an internal network.* All connections, established between an internal host (host located in the intranet) and an external host (e.g., a server located in the Internet), are broken down into two successive connections for the purpose of avoiding direct connectivity, as well as for the purpose of traffic filtering (i.e., network security). From the client perspective, the proxy acts as a server. Contrarily, from the server perspective, the proxy acts as a client. This represents a protective measure from certain types of network attacks.

Another device that may break down a TCP connection is a gateway. A gateway is typically a network device that changes one protocol into another. As an example, a host located in a Microsoft network that needs to have access to a server located in an IP network requires a gateway to perform the translation between the two network protocols. As described in Chapter 16, a gateway may also be employed in the implementation of the IPsec cryptographic protocol. In this case, the hosts from the outside of an organization connect to an IPsec gateway,† located in the interior of an organization, through the Internet using the IPsec cryptographic protocol, while the gateway connects to a server in plaintext (in clear). Contrarily, the connections between internal hosts (intranet) are performed in plaintext.

Note that, in some cases, the access to a web page can be performed without having to establish a connection with a server. A cache is a memory that stores previously accessed web pages, for a certain period of time. Both hosts and proxies have cache memories. In the later case, when an internal network host (client) requests an access to a previously accessed web page (which could have been performed by another internal client), the proxy sends the web page to the client without establishing a TCP connection with the server. This results in a more effective use of the external bandwidth and lower delays.

FIGURE 8.2   Direct HTTP connection.

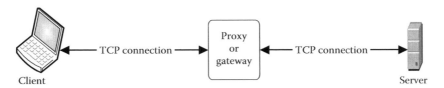

FIGURE 8.3   Indirect HTTP connection.

---

* Typically in a demilitarized zone (DMZ), as defined in Chapter 16.
† An IPsec gateway is typically a firewall or a router that runs the IPsec protocol.

8.2 E-MAIL

Electronic mail (e-mail) consists of a service that allows the exchange of messages between different end users. When a message is sent by a client, such message is uploaded to the message source's server, and is then forwarded to the message recipient's server. Finally, the message is downloaded from the recipient's server into the recipient's host. Depending on the application layer protocol used to provide this network service, the message can either contain a text or contain media files such as video and audio.

8.2.1 SIMPLE MAIL TRANSFER PROTOCOL

The SMTP [RFC 2821] is the TCP/IP application layer protocol that allows a client to upload an e-mail to an e-mail server (user mailbox) or the exchange of an e-mail among different e-mail servers, through an IP-based network, such as the Internet (see Figure 8.4). Moreover, SMTP is also employed by an e-mail server to send e-mails to clients. Note that the connection between the source's server and the recipient's server either can be established directly or may go through one or more intermediate servers.

SMTP comprises two elementary agents: user agents and message transfer agents. User agents allow entities to send messages using domain name system (DNS) names of type mailbox@location. On the other hand, message transfer agents are responsible for forwarding messages to the destination server.

The SMTP is supported by the TCP, and therefore, a certain level of data reliability is assured. Nevertheless, in case a message as a whole (not its TCP segments) fails to be properly delivered to the recipient's host, the source is neither informed about the failure nor message recovery is implemented. As described in Chapter 9, the default SMTP server (TCP) port is 25.

The SMTP message format comprises text type data, consisting of a string of 7-bit ACSII characters. In case a message contains other types of media (pictures, videos, executable files, or other than 7-bit ASCII characters code*), the MIME format is utilized. Note that when a message contains annexes, the message format is automatically switched into the MIME format. The mapping between MIME and SMTP messages is performed using e-mail clients or servers.

In addition to the SMTP, which is used by the server side (to exchange e-mails between successive e-mail servers or to deliver a message to a client) or by a client (to upload an e-mail), there are other two important protocols widely associated with e-mail service:

- *Post office protocol* (*POP*) [RFC 1225]: This allows a client to receive an e-mail from an e-mail server (download) using authentication. The well-known port of POP version 3 (POPv3) is 110.
- *Internet message access protocol* (*IMAP*) [RFC 2045]: This presents some additional features other than those available with the POP, namely the ability to read e-mails located in a server (through downloading or webmail), while leaving a copy in the mail server. It also allows the share of folders among different users, as well as the search of e-mails in a server. IMAP servers listed on well-known TCP port 143.

FIGURE 8.4 Simple mail transfer protocol.

---

* For example, national language characters.

8.3 FILE TRANSFER

The file transfer consists of a service that allows a client downloading or uploading files from/to a server. It consists of a service based on the client–server architecture.

8.3.1 FILE TRANSFER PROTOCOL

The FTP is an IP-based application layer protocol used to allow the transfer of files [RFC 114; 959]. Since file transfer is a service sensitive to errors, the FTP is supported on the TCP transport protocol. The FTP uses two separate connections, respectively for the purpose of control and for data. The control is employed by the client to initiate an FTP session with the server, and for identification and authentication. Note that the default TCP well-known port* used by the FTP server for connection control is 21. Then, in a second stage, the transfer of data is performed, which can be either in active mode or in passive mode. In active mode, the server initiates a connection from the well-known TCP port 20 to a client's high port (a port number above 1023). In passive mode, high ports are used by both the client and the server. An FTP connection can be initiated by a client, using the Windows command line screen (MS-DOS prompt), typing FTP, followed by the alphanumeric identification of the server (e.g., FTP std.au.edu).

The trivial FTP consists of an FTP version with lower capabilities. While the FTP is connection oriented (TCP-based), the trivial FTP is connectionless, and is supported on the user datagram protocol (UDP) transport protocol. Consequently, the trivial FTP results in a nonreliable FTP. Nevertheless, it is typically employed for the transfer of small files, when data transmission is periodically repeated, or in cases where the time involved in the data transfer is an important requirement (Figure 8.5).

8.4 IP TELEPHONY AND IP VIDEOTELECONFERENCE

Some of the drivers that motivated the implementation of the IP telephony and IP videoteleconference protocols are as follows:

- The possibility to integrate voice and data services into a single and common network infrastructure.
- The low cost of packet switching equipment, as compared to circuit switching.
- The high availability and the high available bandwidth of modern deployed packet switching networks.

The exchange of IP telephony or IP video teleconference (VTC) data through the packet switching network comprises the following steps:

- The audio and/or video† acquisition and encoding using a codec (see Chapter 6).
- The obtained data is then encapsulated into packets.

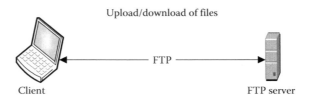

Upload/download of files

FTP

Client   FTP server

FIGURE 8.5 File transfer protocol.

---

* Well-known ports are within the range 0–1023.
† Note that the original audio/video consists of analog information.

- These packets are then transmitted through a packet switching network (e.g., an IP-based network, such as the Internet), using routers as intermediate nodes, instead of PABX.*
- At the destination, the transported data is then de-encapsulated, and the extracted digital data is then converted into analog data,† for human presentation.

Note that, after an initial connection setup, the exchange of voice data is performed in a peer-to-peer mode.

As described in Chapter 1, both IP telephony and VTC are delay intolerant real-time traffic. This means that their performance degrades heavily when media transfer is subject to a certain amount of delay, or delay variation, that is, jitter. 200 ms is a typical maximum acceptable end-to-end delay in signals that allows an acceptable performance.

Chapter 6 described analog and digital encoding standards and techniques employed in both audio (telephony) and video (television). Sections 8.4.1 and 8.4.2 focus on two protocols that are commonly employed for the exchange of both IP telephony, VTC, and application data through packet switching networks such as the Internet. The insertion of these protocols into the TCP/IP stack is depicted in Figure 8.6.

8.4.1 ITU-T H.323

ITU-T defines a set of recommendations for the transfer/exchange of voice and video over different network types. The ITU-T H.323 [2009] is an ITU-T recommendation that consists of an application layer protocol, and specifies the procedures and protocols for the exchange of voice, video, and data conferencing for packet switching networks such as IP-based or IPX-based, for both point-to-point and point-to-multipoint multimedia communications.

Similarly, the ITU-T H.320 [2004] defines similar services for ISDN networks, ITU-T H.324 [2009] for PSTN, and ITU-T H.321 [1998], ITU-T H.310 [1995] specifies it for ATM networks.

In order to implement the above-described services, the ITU-T H.323 [2009] comprises the following four different network component types (see Figure 8.7) [Liu and Mouchtaris 2000].

- *Terminals*: They consist of the equipment (hardware and software) used for human interfacing. This can be a personal computer, a smartphone, or a personal digital assistant running a softphone or a packet-switching-based telephone.

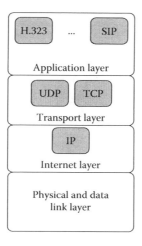

FIGURE 8.6 IP multimedia protocols and the TCP/IP stack.

---

* As considered in conventional circuit switching networks.
† Using the same codec as the one employed for the data encoding, at the transmitter side.

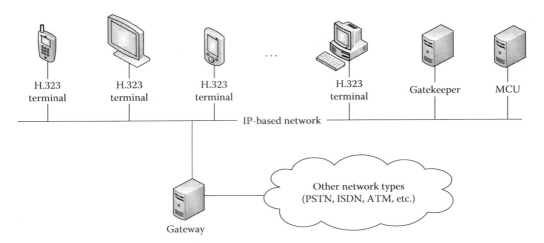

FIGURE 8.7   ITU-T H.323 components.

- *Gatekeeper*: It is responsible for performing address resolution, namely for mapping a telephone number or username/e-mail address into a transport address.* It performs a task similar to the DNS mapping protocol.† In case the terminals have knowledge about the transport addresses of destination endpoints, a gatekeeper may not be necessary. Such mode of operation is referred to as *direct mode*. Nevertheless, a gatekeeper also performs additional functionalities such as admission control‡ and bandwidth management. Therefore, the use of a gatekeeper is essential to avoid congestion, especially in the case of medium-to-large networks.
- *Gateway*: In case interoperability is required between different network types,§ a gateway is employed. Such network device assures voice and video codec translation, data format translation, as well as signaling translation. This network component is required only when translation is necessary.
- *Multipoint control unit*: This component allows H.323 point-to-multipoint communications, that is, conferencing. Without the multipoint control unit, only point-to-point communication is possible.

Figure 8.8 depicts the H.323 protocol stack used in IP-based networks.

As can be seen, depending on the functions that are to be accomplished, different protocols are used by the ITU-T H.323 [Thom 1996; Minoli and Minoli 2002]:

- *Real-time protocol* (*RTP*): It is responsible for controlling a session,¶ and for allowing the exchange of voice/video traffic. Since H.323 voice and video traffic is transferred using the UDP transport protocol, the RTP is an intermediate protocol that adds a header, with a number of fields. This complements the UDP service by adding some level of reliability** to the exchanged data. The RTP fields are as follows:

---

* Concatenation of an IP address with a port number.
† Nevertheless, instead of converting an alphanumeric name into an IP address (DNS), it translates a telephone number or username/e-mail address into a transport address (gatekeeper).
‡ A gatekeeper must authorize a call.
§ For example, between an IP-based and a PSTN-based or an ISDN-based network.
¶ Nevertheless, the TCP/IP stack does not comprise the session layer, whose functionalities are performed by the application layer. Therefore, it can be stated that the RTP is an application layer protocol.
** Note that corrupted packets are not detected and/or corrected by the RTP.

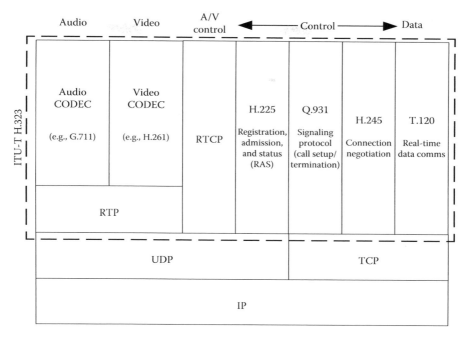

FIGURE 8.8 ITU-T H.323 protocol stack for IP-based networks.

- *Sequencing of packets*: It is implemented through the addition of a sequence number field into the RTP header. This allows the delivery of packets in the correct sequence, the detection of lost packets,[*] and the avoidance of packets duplication.
- *Payload identification*: This field allows the identification of the audio and/or video codec, which can be employed for different purposes, such as bandwidth management.
- *Frame identification*: It is used to signalize the beginning and the end of frames.
- *Source identification*: It is used to identify the originator of packets. This is relevant in the case of point-to-multipoint communications.
- *Intramedia synchronization*: It is used to implement buffering and avoid a certain level of jitter.
- *Real-time control protocol (RTCP)*: It consists of a control protocol that allows the RTP to perform its tasks. It indicates the identity of a caller and provides a variety of network information, which can be useful for keeping the network performing well (e.g., average delay, jitter, and the rate of lost packets).
- *H.225 (RAS)*: It is used by terminals to register with the gatekeeper and to request it a call.
- *Q.931 Signaling protocol*: It is used for call setup and termination between two or more terminals.
- *H.245*: It is used in a communication for the negotiation of audio and/or video codec.
- *T.120*: It consists of a protocol that manages real-time data conferencing such as instant messaging, file transfer, and application sharing. Note that these types of data communications are performed using the TCP, instead of the UDP transport protocol (employed in voice and video communications).

Figure 8.9 plots the different phases of a voice and/or video call, with the indication of the corresponding protocols.

---

[*] Such information is only used for statistical purposes.

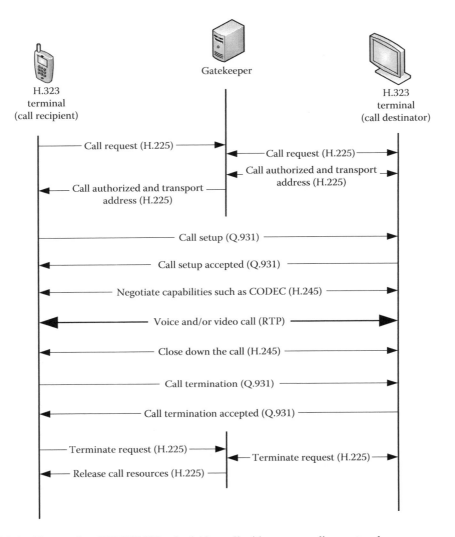

FIGURE 8.9   Phases of an ITU-T H.323 voice/video call with corresponding protocols.

8.4.2 SESSION INITIATION PROTOCOL

Contrary to ITU-T H.323, which defines a set of standards for each specific functionality, the session initiation protocol (SIP) [RFC 326; 2543; 3261] is an application layer protocol, specified by the Internet engineering task force, being limited to a signaling protocol, used for setting up, configuring, and terminating multimedia sessions [Minoli and Minoli 2002]. This is performed using an HTTP-like client–server request/response model, using text type messages,* after which the exchange of voice data is performed in a peer-to-peer mode. Similar to the HTTP, multimedia calls performed with the SIP can make use of authentication and encryption protocols such as IPsec, SSL, or TLS.† Moreover, similar to the ITU-T H.323, the SIP is used for voice, video, and/or data‡ for point-to-point and point-to-multipoint communications over IP-based networks. Nevertheless,

---

* It uses most of the HTTP text-based codes and header fields.

† See Chapter 16 for a detailed description of these cryptographic protocols.

‡ For example, instant messaging, file transfer, and application sharing.

the exchanged traffic is transparent to the SIP [Schulzrinne and Rosenberg 2000], using other protocols such as the RTP, and using the RTCP as a control protocol.* Moreover, the session description protocol is used to specify protocols (e.g., codec) in the data transaction. Note that the SIP comprises the exchange of multimedia traffic supported on either the UDP or the TCP, whereas the ITU-T H.323 specifies that voice and video use UDP and application data uses the TCP.

The SIP signaling protocol comprises different network component types (see Figure 8.10):

- *User agents*: They consist of the equipment (hardware and software) used in the client/server interaction, for the purpose of connection admission and control. The client endpoint is called *user agent client*, and can be a personal computer or a personal digital assistant running a softphone or a packet-based telephone. The server endpoint is referred to as the *user agent server*, being employed to receive session initiation requests and to return the corresponding responses.
- *Registration server (registrar)*: It consists of a server that keeps the location of different SIP users. A certain user with a certain telephone number or username/e-mail address can physically move along an IP-based network. This information is stored in the registrar, as well as the corresponding transport layer address. Similar to the gatekeeper of the ITU-T H.323 protocol, the registrar is responsible for translating telephone numbers or username/ e-mail addresses into transport layer addresses.
- *Proxy server*: It is responsible for forwarding SIP requests and responses to their destination. A LAN implementing a multimedia service using the SIP has its own proxy server.†
- *Redirect server*: It is responsible for returning the location of another SIP agent, to which the call forwarding must be transferred.
- *Gateway*: Similar to the ITU-T H.323 protocol, it assures interoperability among different network types.
- *Multipoint control unit*: Similar to the ITU-T H.323 protocol, it allows point-to-multipoint multimedia communications.

Figure 8.11 plots the different phases of a voice and/or video call using the SIP. By comparing the exchanged SIP messages against those of the ITU-T H.323 protocol (Figure 8.9), it is easy to conclude that the SIP presents a much lower level of complexity.

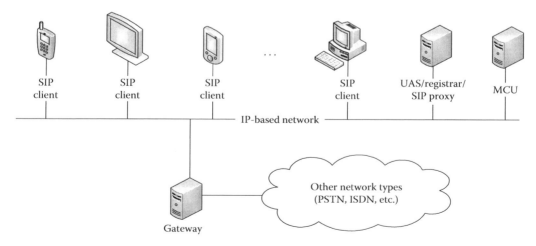

FIGURE 8.10 SIP components.

---

* Both the RTP and the RTCP were already described for ITU-T H.323.
† It is normally a resident of the same server as the registrar.

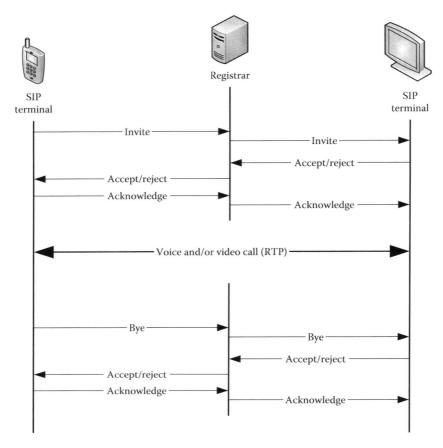

FIGURE 8.11 Phases of a SIP voice/video call.

8.5 NETWORK MANAGEMENT

Several proprietary and standardized protocols were developed in order to allow the remote control and monitoring of network devices, such as routers, printers, workstations, or switches. These protocols were extended in order to incorporate the remote control and monitoring* of other equipment than simply network devices. Such equipment can be, for example, a satellite transceiver or an electrical diesel generator. The remote control capability allows powering on a satellite transceiver and changing its transmitting frequency or its transmitting power through a network. The monitoring capability allows receiving messages in case the satellite antenna is not directed toward the expected position, or in case the transmitter fails. For this to be possible, both the controlled equipment (e.g., the satellite transceiver) and the remote control station need to be interconnected through a network and need to use a network management protocol.

Different network management protocols were standardized. The most commonly known is the IP-based simple network management protocol (SNMP) [RFC 1448], which can be found in three different versions (version 1, 2, and 3). The SNMP v3 is commonly referred to as *enhanced SNMP* due to its improved capabilities. The OSI reference model also standardized a network management protocol entitled *common management information protocol*. Moreover, many proprietary network

---

* In this context, the remote management function through the network can be of two types: control or monitoring. The control function corresponds to an active behavior (e.g., changing the parameters of an equipment), whereas monitoring is a passive behavior (e.g., receiving status data from an equipment).

management protocols have been developed in order to allow the remote control and monitoring of equipment. In case such pieces of equipment are intended to be remotely controlled with a common SNMP, a gateway is required.

8.5.1 SIMPLE NETWORK MANAGEMENT PROTOCOL

The SNMP is an IP-based network management protocol, supported on the UDP (see Figure 8.12). The reason for using the connectionless mode is not because data reliability is not a requirement, but because such data reliability can be assured through the repetition of the transmitted data. The SNMP comprises the periodic retransmission of messages. In case a message is received corrupted, the multiple retransmissions of a message assure that the information is correctly interpreted by the destination.

As depicted in Figure 8.13, in order to allow the remote management of equipment, four different elements need to be implemented[*]:

- *Agent*: It is the device that is intended to be remotely controlled and/or monitored (managed) through the network. For the device to be subject to remote management, it needs to be equipped with an SNMP agent, which comprises a local database with actual and historic variables (object states).

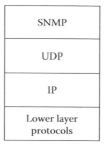

FIGURE 8.12 Location of the SNMP into the TCP/IP stack.

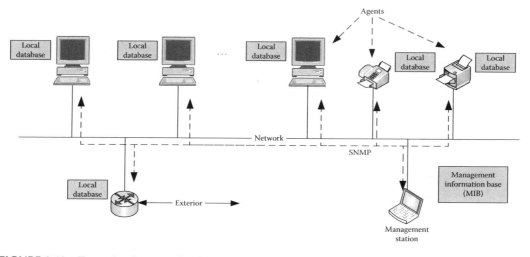

FIGURE 8.13 Example of a network using the SNMP.

---

[*] Note that the remotely managed equipment may be located within a local area network (intranet) or outside it (i.e., in an extranet).

- *Management station*: It corresponds to a computer or workstation that runs specific soft-ware that allows sending and receiving network management messages to/from an agent.
- *Network management protocol*: The communication between an agent and a management station is carried out using an SNMP [RFC 1448].
- *Management information base*: It consists of a hierarchical database, stored in the management station, which compiles the data stored in all different agents (the set of all local databases).

The three most important SNMP messages exchanged between a network management station and the agents are as follows:

- *Get*: It is used for network monitoring purposes, namely for the management station to request the value of an object from an agent.
- *Set*: It is used by the management station to send a value of an object to an agent. Consequently, it is employed for network control purposes.
- *Notify*: It is sent by an agent to the management station to report a failure event (monitor-ing purposes).

8.6 NAMES' RESOLUTION

Names' resolution is the act of translating a numerical address into an alphanumeric name and vice versa. The most used names' resolution protocol is the DNS, as defined in Section 8.6.1.

8.6.1 DOMAIN NAME SYSTEM

The IPv4 protocol comprises 32 bit-size addresses that are used to identify hosts. These addresses are commonly referred to as *four groups of decimal numbers*. Similarly, the IPv6 protocol comprises 128 bit-size addresses, being commonly referred to as *eight groups of four hexadecimal numbers*. These numbers are difficult to memorize. Consequently, it is easier to memorize the address of a host if a name is employed, instead of a sequence of numbers. For the sake of simplicity, the alphanumeric names used to identify the hosts are normally correlated with the organization's name, or with the department's name to which it belongs. These alphanumeric names used in the www space are known as *domain names*.

DNS [RFC 1034; 1035] consists of a mapping protocol that performs the conversion between alphanumeric names into IP addresses. It is an IP-based application layer protocol, which runs over UDP (connectionless).

As depicted in Figure 8.14, the conversion between alphanumeric names and IP addresses can be performed in two different directions:

- *Binding*: It corresponds to the conversion of alphanumeric names[*] (typed by users) into IP addresses (inserted into packet headers, before transmission). This is the most common DNS operation.
- *Reverse mapping*: It corresponds to the translation of IP addresses into alphanumeric names. In some circumstances, this operation may also be required.

The DNS mapping protocol can be viewed as the yellow pages of an IP network.

FIGURE 8.14 Example of DNS operation: (a) binding and (b) reverse mapping.

---

[*] Note that the alphanumeric names used in the World Wide Web are case sensitive.

An example of an alphanumeric IP address is www.std.au.edu. The syntax of alphanumeric names presents different hierarchical levels, with each one being referred to as a *domain*. Higher level domains are located at the right of alphanumeric addresses. The top level domain typically identifies a country or an organization type.[*] This hierarchical organization of alphanumeric names is associated to hierarchical routing,[†] as well as to the corresponding hierarchical organization of IP addresses.

The top level domain of the example above is edu, which stands for education, being followed by the second higher domain (au),[‡] and then by a third level domain (std).[§] The described rules are applicable to both the Internet as well as to private IP-based networks. It is worth noting that the domain names start by the letters www or, alternatively, by the letters http.

The DNS operation comprises the existence of a number of key elements:

- *Domain name space*[¶]: This corresponds to a list of alphanumeric names (which corresponds to a set of IP addresses).
- *DNS database*: This stores the mapping between alphanumeric names and IP addresses.
- *Name resolver*: This is a program that runs in a server, or in a local host, in order to respond to DNS requests.
- *Name server*: This is a server that stores the DNS database and the name resolver, and responds to DNS requests to perform the mapping.

The DNS operation can be carried out in two different modes:

- *Recursive*: A host sends a mapping request to a DNS server. If the interrogated server does not have the response, it searches the response in other servers and returns it to the caller.
- *Iterative*: A host sends a mapping request to a DNS server. If the interrogated server does not have the response, it returns to the host a pointer toward the following DNS server to which the request should be sent.

As can be seen from Figure 8.15, the typical DNS operation comprises a user program that requests the conversion between a domain name into an IP address. Such request is typically sent to the resolver

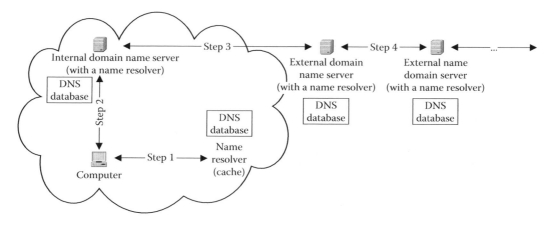

FIGURE 8.15 DNS operation.

---

[*] For example, *uk* stands for the United Kingdom, *edu* refers to an educational organization, *mil* identifies a military organization, and *int* stands for an international organization.
[†] See Chapter 10.
[‡] May refer to a university name.
[§] May refer to a department within the university.
[¶] The Internet Assigned Numbers Authority is globally responsible for managing the domain name space.

module resident in the local host (step 1) or, alternatively, in a local server in the same domain (step 2). In case the typed domain name by the user is not in the local database in cache, the interrogated server searches it in other servers (recursive method, i.e., step 3, 4, ...) or returns a pointer to the following DNS server to which the caller should send the request (iterative method). When the IP address is obtained, such information is stored in cache (in the local host and/or local server) for a certain period of time. Alternatively, if the mapping is not found, the user program receives an error message.

CHAPTER SUMMARY

This chapter provided an overview about the application layer protocols of the TCP/IP stack, including a description of different services supported by the application layer protocols. The difference between a service, the application layer of the network architecture, and an application program was described.

The web browsing service was described, as well as the HTTP, the application layer protocol that supports the web browsing service. It was viewed that the HTTP is a client-server protocol, supported by the TCP, which allows a client using the www space, that is, downloading an HTML file from a server. Moreover, it was described that the HTTP may use either the 7-bit ASCII format or the MIME format. The MIME consists of a message format that allows other non-ASCII characters, as well as binary formats containing images, sounds, and other files.

The e-mail service was described, including the SMTP. It was described that the SMTP is used to upload e-mails from a client into a server, or to support the exchange of e-mails between different servers. Since the SMTP does not support authentication, the download of e-mails is normally performed either using the POP or using the IMAP.

The file transfer service was also described in this chapter, as well as the FTP, the application layer protocol that supports the file transfer service. It was viewed that, since the file transfer is a service sensitive to errors, the FTP is supported on the TCP transport protocol.

Both IP telephony and VTC were also dealt. It was described that, after an initial setup, the exchange of IP telephony traffic is performed in a peer-to-peer mode. Moreover, a description about the H.323 protocol was given, as well as about the SIP, both being utilized to support IP telephony or VTC services. It was viewed that, while the H.323 protocol uses the UDP to support the transfer of multimedia traffic, the SIP may support the transfer of multimedia traffic on either TCP or UDP transport protocol. It was viewed that the gatekeeper is the network component of the H.323 protocol, responsible for performing address resolution, namely for mapping a telephone number, username, or an e-mail address into a transport address. In the SIP, this task is performed by the registrar.

The network management service was described, as well as the SNMP, which is an application layer protocol utilized to provide the network management service. It was viewed that the SNMP is an IP-based network management protocol supported by the UDP.

Finally, a description about the names' resolution service was given, as well as about the DNS protocol, consisting of an application layer protocol used to support the names' resolution service. It was described that the binding corresponds to the conversion of alphanumeric names into IP addresses, whereas reverse mapping corresponds to the reverse operation. It was also viewed that, in the recursive mode, if an interrogated server does not have the response, it searches the requested mapping in other servers and provides the response to the host.

On the contrary, in the iterative mode, if an interrogated server does not have the response, it sends to the host a pointer toward the following DNS server to which the request should be sent.

REVIEW QUESTIONS

1. Which methods can be employed by the DNS protocol to find the mapping between a domain name and its IP address?
2. What are the key elements of the DNS operation?

3. What is the DNS protocol used for? Why is it employed?
4. What are the three most important SNMP messages exchanged between the network management station and the agents?
5. What are the four basic elements that need to be in place, in order to allow the remote management of equipment?
6. Why is the SNMP supported on the UDP?
7. What is the SMTP used for?
8. What is the difference between the POP and the SMTP?
9. What is the difference between the POP and the IMAP?
10. What is the MIME protocol used for?
11. What are the differences between the application layer of the TCP/IP model and of the OSI reference model?
12. What are the tasks carried out by the session layer of the OSI reference model?
13. In the scope of the SNMP, what is the management information base used for?
14. Which kinds of indirect HTTP connections do you know? Define each one.
15. What are the differences between an application layer of a network architecture, an application program, and a data service?
16. What is the application layer of a network architecture used for?
17. What is the presentation layer of the OSI reference model used for?
18. Which are the network components comprised by the ITU-T H.323 standard?
19. Which are the network components comprised by the SIP?
20. What are the differences between ITU-T H.323 and SIPs?
21. What are the functionalities carried out by the RTP?
22. What is the RTCP used for?
23. Which are the transport layer protocols utilized by ITU-T H.323 and SIPs?
24. What are the different protocols that compose the ITU-T H.323 standard used for?
25. In the scope of the ITU-T H.323 standard, what is the T.120 protocol used for?
26. To which TCP/IP layer do ITU-T H.323 and SIPs belong?
27. In the scope of the ITU-T H.323 standard, what are the basic differences between the exchange of voice/video traffic and the exchange of data traffic?
28. In the scope of SIP, what is the registration server (registrar) used for?
29. Which of the two studied IP multimedia protocols present lower level of complexity? Why?

LAB EXERCISES

1. Download and install the free network analyzer Wireshark. Open the application in a PC and select the interface connected to the Internet. You will see the IP datagrams that are being exchanged by the network interface card (NIC) of the PC, with the source IP address, destination IP address, and protocol type, alongside with other information. Use the web browser to access a certain website. In Wireshark, open a datagram that is carrying the HTTP. In a window that appears at the bottom, visualize the content of the frame, the MAC sublayer, the IP datagram, the TCP segment, and the content of the HTTP (corresponds to the payload data, normally a part of a HTML file). Repeat this process for the other application layer protocols, such as the SMTP, POP3, and FTP.

9 Transport Layer

LEARNING OBJECTIVES

- Describe the transport layer of TCP/IP.
- Identify and describe the different transport layer protocols, their properties, and header formats.
- Describe the concept of integrated services (IntServ) and differentiated services (DiffServ).

The transport layer of the transmission control protocol/Internet protocol (TCP/IP) stack is an end-to-end layer that aims to provide, as much as possible, the required quality of service (QoS) to the application layer, at the lowest cost. In order to assure this functionality, the transport layer aims to achieve a trade-off between the QoS parameters requested by the application layer and the QoS available at the network layer (which is a function of the instantaneous traffic).

The QoS parameters dealt with and negotiated by the transport layer includes the following:

- Requested bit rate in each direction
- Bit error rate (BER)
- Maximum end-to-end delay
- Maximum end-to-end jitter
- Priority rules
- Probability of failure in a connection establishment
- Maximum time to establish a transport connection
- Maximum time to terminate a transport connection
- Cost
- Security protection mechanisms

Depending on the service and application layer protocol, the following are the four most important service requirements:

- Bit rate requirements[*]
- Sensitivity to loss of data
- Delay sensitivity
- Variable or constant bit rate service

As an example, the IP telephony (VoIP) service provides low requirements in terms of bit rate, presenting low sensitivity to loss of data (typically, the voice service accepts a BER better than 10^{-3}). Nevertheless, the IP telephony QoS requires a low delay, while presenting a high sensitivity to delay variations (jitter). The IP telephony data exchange is performed over the user datagram protocol (UDP), whereas the real-time transport protocol used by IP telephony at the application layer assures the correct sequence of the exchanged segments. Another example is a download, which

---

[*] Bit rate is commonly referred to as bandwidth. Nevertheless, Chapter 3 describes the relationship between these two factors.

makes use of the FTP, supported on the TCP transport protocol. This requires an approximately high bit rate and a very high sensitivity to loss of data. On the other hand, this service presents a reduced sensitivity to delay and to jitter.

In order to assure that the required bit rate is made available to the application layer, the transport layer may group multiple network layer connections into a single application layer connection. Let us consider the case where the network layer has a 512 kbps connection established, and the application layer is requesting a 1024 kbps connection. In this situation, the transport layer may establish a second 512 kbps network connection and, in a transparent manner, offer a 1024 kbps connection to the application layer. Similar function may be offered when the application layer is requesting a datagram size higher than the maximum transmission unit (MTU). In this situation the transport layer performs the segmentation (at the transmitting side) and reassembling (at the receiving side) of data, such that it fits within the MTU.

The address of the TCP/IP transport protocol is the port number, being composed of a 16-bit address (see Figure 9.1). The Internet Assigned Numbers Authority assigns port numbers between 0 and 1023, called well-known ports. These port numbers are to be used by servers and not to be used by hosts (see Tables 9.1 and 9.2). Port numbers between 1,024 and 49,151 are also registered ports but are used for lesser known services. Finally, port numbers between 49,152 and 65,535 are port numbers to be used by the hosts.

A TCP/IP connection is identified by a source and a destination socket. A socket consists of the concatenation of an IP address with a port number, separated by colons. An example of a socket is the address pair 173.22.83.10:80. The first group of 32 bits is the IP address (173.22.83.10), whereas the second group of 16 bits corresponds to a port number (80). In this case, since 80 is a

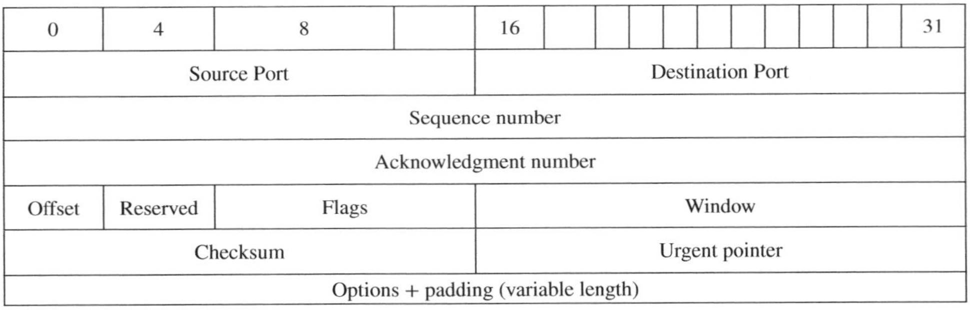

0	4	8		16										31
Source Port				Destination Port										
Sequence number														
Acknowledgment number														
Offset	Reserved	Flags		Window										
Checksum				Urgent pointer										
Options + padding (variable length)														

FIGURE 9.1   TCP segment header.

TABLE 9.1
List of Some Well-Known TCP
Port Numbers

TCP Port Number	Service
21	FTP
23	Telnet
25	SMTP
80	HTTP
110	POP3
143	IMAP
443	HTTPS
161	SNMP

TABLE 9.2
List of Some Well-Known UDP
Port Numbers

UDP Port Number	Service
53	DNS
69	TFTP
161	SNMP

well-known port number, this transport layer address corresponds to an HTTP service provided by a server.

The application layer generates a continuous flow of data that is segmented and transmitted by the transport layer. The message format of the TCP/IP is called a segment, and may be transmitted by using one of the following different protocols:

- User datagram protocol
- Transmission control protocol

These protocols are defined in Sections 9.1 and 9.2.

9.1   TRANSMISSION CONTROL PROTOCOL

The TCP is a transport layer protocol used to provide the application layer with a connection-oriented service between two end entities.

9.1.1   TCP PROPERTIES

Since the TCP is connection oriented, it requires the previous connection establishment before the data is exchanged. Moreover, since the TCP provides reliability,* it assures that the exchanged data, delivered by the transport layer to the application layer, presents the following properties:

- *Data is delivered with reliability*: The TCP assures reliability by using error control (CRC), associated to the PAR procedure.
- *Packets are delivered to the application layer in the correct sequence*: Note that the packets may reach the receiver out of order, due to the use of datagram method. Re-ordering of packets is possible using the sequence number, being one of the fields of the TCP segment header.
- *Packet losses are detected and corrected*: As previously described, one of the various datagram's fields is the time to live. This field consists of the maximum number of nodes that a datagram may cross. If a datagram crosses a higher number of nodes, it is discarded. Consequently, by using the TCP such discard is detected by the transport layer, being requested the repetition of the segment transported in the discarded datagram.
- *Duplication of packets is avoided*: Duplication of packets occurs as a result of flooding in the routing protocol. Since the TCP segments are numbered, the receiver assures that the segments are delivered to the application layer in the correct order.

---

* A connection-oriented service is always a confirmed service.

Error control is provided by the TCP using the following PAR procedures:

- The transmitting entity sends a number of segments corresponding to the length allowed by the sliding window protocol. Further, the transmitting entity activates a timer and waits for an acknowledgment (ACK) from the recipient entity (correct reception of data) within a certain time period.
- If the segments are correctly received by the recipient, its ACK is sent back to the transmitter (service confirmation). The error detection technique is implemented by the TCP using cyclic redundancy check (CRC) codes.* Note that although the data link layer also makes use of error control mechanisms,† some residual‡ errors exist. Moreover, packets can be either discarded by the intermediate routers or duplicated; in the worst case, they can arrive out of order. Consequently, the TCP does not assume a reliable service provided by the lower layers. Consequently, the TCP has mechanisms to deal with such impairments.
- In case the transmitting entity does not receive the ACK within the expected time period, it assumes that the amount of data did not reach the recipient correctly, and repeats its transmission.
- Then the transmitter proceeds with the transmission of the data.

Note that, in addition to error control, this procedure allows flow control. When the receiver informs the transmitter that a certain amount of data was correctly received (error control), it is also informing that it is ready to receive more data (flow control).

In order to make an efficient use of the available transmission medium, while providing flow control, the TCP makes use of the sliding window protocol.§ Therefore, instead of sending a segment at a time (as in the case of the stop-and-wait protocol), a group of segments is sent together. Moreover, signaling data (e.g., ACKs) are often exchanged using piggybacking, where a segment is used simultaneously to send data and to acknowledge data previously received from the opposite direction. The sliding window protocol is implemented by using number of octets as a reference. Before the data is exchanged, the receiver informs the transmitter about the number of octets that can be sent without ACKs. This maximum number of octets is signalized in a TCP header field entitled window (the window size is expressed in number of octets). Such number of octets (window size) is encapsulated into several segments. Moreover, the maximum number of octets that can be transported in a single segment is known as the maximum segment size (MSS).

9.1.2 TCP SEGMENT HEADER FORMAT

The TCP is used to support the following application protocols: FTP, Telnet, SMTP, HTTP, and so on. Figure 9.1 depicts the different fields of a TCP header, with the corresponding indication of each field's size expressed in number of bits. The TCP segment header is composed of two parts: (1) a fixed length part, with a total size of 20 octets, which includes 10 mandatory fields between the source port and the urgent pointer (each line of Figure 9.1 has a total of 32 bits, i.e., four octets) and (2) a variable length part that corresponds to the options plus the padding.

In the following, each of the segment's header fields is defined:

- *Source port*: It consists of a 16-bit transport layer address that identifies the source of a transport layer connection.
- *Destination port*: It consists of a 16-bit transport layer address that identifies the destination of a transport layer connection.

---

* The reader should refer to Chapter 12 for a detailed description of the CRC codes.
† With some exceptions, such as the SLIP.
‡ An error control mechanisms does not assure 100% error-free frames.
§ The sliding window protocol is detailed in Chapter 12.

- *Sequence number*: It consists of a 32-bit field, which identifies the sequence of the first octet within a segment.
- *ACK number*: It consists of a 32-bit field, which indicates the sequence number of the next expected data octet (to be sent within the next segment). In case the previous data was properly received by the receiving entity, this number corresponds to the sequence number immediately after the last octet contained in the last segment. Alternatively, in case of error, this number corresponds to the first octet of the segment received with error, which is meant to be retransmitted.
- *Offset*: It consists of a 4-bit number, representing the header length expressed in groups of 32 bits. This corresponds to the shift between the beginning of a segment and its payload. In fact, this equals the options length (expressed in group of 32 bits) plus five (fixed length header).
- *Reserved*: It is a reserved 4-bit field.
- *Flags*: It consists of eight individual flags, each with one bit (CWR, ECE, URG, ACK, PSH, RST, SYN, and FIN). When a flag is activated, it means that a certain function is in use:
  - When the URG flag is activated, it means that the urgent pointer field is to be considered by the receiving entity of the segment.
  - When the ACK flag is activated, it means that the ACK number field is to be considered by the receiving entity of the segment.
  - The RST flag is used to terminate a confused connection or to refuse a new connection establishment.
  - The SYN flag is used to establish a new connection.
  - The FIN flag is used to terminate a previously established connection.
- *Window*: It is a 16-bit size number, used by the recipient entity to inform the transmitter about the maximum number of octets that can be sent, using the sliding window protocol (flow control). It allows the transmission of several segments without ACK, in total carrying a number of octets lower than the window field.
- *Checksum*: It consists of 16 redundant bits, used to perform error detection of the entire segment, using CRC codes.
- *Urgent pointer*: It consists of a 16-bit number. Adding this number to the sequence number, we obtain the last sequence number of the urgent data.
- *Options*: Due to their rare use, the following fields are optional:
  - *MSS*: It consists of a 32-bit number, representing the largest amount of data, specified in number of octets that can be carried in a single segment. This number is negotiated during the TCP connection establishment. In order to avoid segmentation, the MSS should be smaller than the MTU minus the IP header length (the MTU depends on the data link layer used).
  - *Window scaling*: It consists of a 24-bit number, used to introduce a scale factor to the number expressed in the window field (which is limited to 65,535 octets). This is used in high data rate transmission, to increase the size of the window.
  - *Selective ACK permitted*: It is a 16-bit number, used to implement selective repeat, instead of the pure sliding window protocol. This allows an efficiency gain in terms of bandwidth.
- *Padding*: It is used to ensure that the header length is an integral number of 32-bit words.

Note that the total segment size (LEN; see Figure 9.2) is not explicitly defined in any field. This can be calculated from the total length field of the IP datagram subtracting the Internet header length of the IP datagram and subtracting the segment offset. Naturally, before this sum is completed, all numbers need to be expressed in the same units (e.g., in octets or in bits).

FIGURE 9.2   Example of TCP data exchange.

9.1.3   TCP HANDSHAKING

Figure 9.3 depicts the handshaking used in a TCP connection establishment, where SP stands for source port and DP stands for destination port. The establishment of a TCP connection is a three-way handshaking: the initial step comprises the establishment request, by the client to the server, using the SYN message. The second step corresponds to the response from the server, acknowledging the establishment request, that is, accepting the request to establish the connection. This step is signalized using the SYN and ACK messages, together. Finally, the third step corresponds to the ACK, sent by the client, using the ACK message.

Similarly, Figure 9.2 depicts an example of TCP data exchange. As previously described, the TCP uses the sliding window protocol. From the first message depicted in this figure, it is seen that the receiving entity (client) is authorizing the transmitting entity (server) to send 200 octets using the sliding window protocol (WIN = 200), and that the MSS is 100 (MSS = 100). Let us assume that the transmitting entity (server) has a total of 280 octets to send. This needs to be split into three segments, the two initial segments with maximum load of 100 octets and the third segment with the final 80 octets. The server sends two segments that totalize 200 octets, and waits for the ACK from the receiver (client), because the window size is 200. Since an error occurred in the second segment, the ACK number corresponds to the sequence number of the first octet of the second segment that needs to be repeated. Then, the server retransmits the segment received with errors and the third segment (with only 80 octets). The data exchange is finalized with the ACK sent by the receiving entity (client).

For the case of a segment received free of errors, the ACK number corresponds to the sequence number of the first octet of the previously received segment summed with the number of octets sent in such segment (LEN). In case of error detection, the ACK number corresponds to the sequence number of the first octet of the corrupted segment in the previous window of segments.

Figure 9.4 depicts the procedure to terminate a TCP connection. As can be seen, the client sends a segment to the server with the FIN flag activated, and this entity responds initially with an ACK message, stating that the message was correctly received. Then, in order to accept the request to terminate the connection, a segment is sent by the server with the ACK and FIN flags activated.

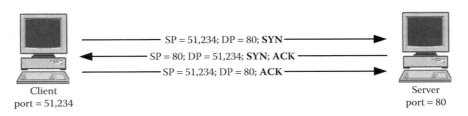

FIGURE 9.3   TCP connection establishment.

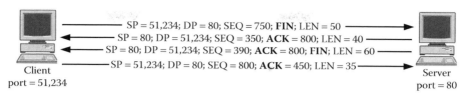

SP = 51,234; DP = 80; SEQ = 750; **FIN**; LEN = 50
SP = 80; DP = 51,234; SEQ = 350; **ACK** = 800; LEN = 40
SP = 80; DP = 51,234; SEQ = 390; **ACK** = 800; **FIN**; LEN = 60
SP = 51,234; DP = 80; SEQ = 800; **ACK** = 450; LEN = 35

Client
port = 51,234

Server
port = 80

FIGURE 9.4   TCP connection termination.

This corresponds to the confirmation from the server. Finally, the client acknowledges the FIN message received from the server.

As previously described, the TCP requires that the receiving entity keeps sending successive ACKs to the transmitting entity, and retransmissions are made whenever ACKs are not received by the transmitting entity. This handshaking represents additional overhead in the network, as well as additional signal delays. This is the reason why delay-sensitive services (e.g., voice and video streaming, etc.) do not make use of the TCP. These services are normally implemented using the UDP, as defined in Section 9.2. Contrarily, services that require data reliability, without presenting delay sensitivity, are normally implemented over the TCP.

9.2   USER DATAGRAM PROTOCOL

Unlike the TCP, the UDP is connectionless and provides an unconfirmed service to the application layer.

9.2.1   UDP PROPERTIES

As a connectionless service, the UDP does not require the previous connection establishment, allowing the rapid exchange of data between entities.

Nevertheless, the data is delivered based on the best effort, that is, without assuring reliability of the exchanged data, and without any kind of flow control. Since the UDP does not introduce delays (in opposition to the TCP), it is normally used by delay-sensitive services (which are normally not demanding in terms of data reliability), or by services whose data is periodically and redundantly retransmitted (such that an error is rapidly corrected by its update). An example of a delay-sensitive but not error-sensitive service is the audio, the video streaming, or the IP telephony. Example of application layer protocols that may be supported on the UDP due to its data redundancy is the SNMP, as well as the DNS protocol.

The UDP only provides a minimum service to its upper layer, acting as a simple interface between the application layer and the network layer.

9.2.2   UDP DATAGRAM HEADER FORMAT

Figure 9.5 depicts the several fields of an UDP datagram header, with the corresponding indication of each field's size expressed in number of bits. As in the case of the TCP segment header, the two initial fields are the source and destination addresses. The length field does not exist in the TCP segment header, having to be calculated from the packet length and from the TCP segment header length (offset). Finally, the checksum is a group of 16 bits used to allow error detection of the UDP header. This field is optional. If not used, its value is set to zero. When an error is detected in the UDP datagram header, the segment is just discarded. Note that, unlike TCP error detection, the UDP error detection is never applied to the payload.

0				16										31
Source port					Destination port									
Length					Checksum									

FIGURE 9.5 UDP datagram header.

9.3 INTEGRATED AND DIFFERENTIATED SERVICES

As previously described, there are a number of QoS requirements that need to be followed in order to assure that a service achieves the expectations. These requirements include the following parameters:

- *Delay*: Services such as audio or video streaming can accommodate a low level of delay (are delay-sensitive).
- *Jitter*: This corresponds to a variation in delay. While a small amount of delay can still be supported by audio or video streaming, its variation rapidly degrades the quality of these services.
- *Throughput*: It is the bit rate that is required to support a specific service. This can be fixed or variable. While the video streaming requires a high and fixed bit rate, the web browsing is not very demanding in terms of throughput, but is a service that requires a variable bit rate.
- *Packet loss*: The increase of the number of lost packets corresponds to a degradation of the BER. There are other factors that may degrade the BER, for example, the errors generated in the physical and data link layers. Nevertheless, the errors generated in physical and data link layers do not rely on the scope of the provision of QoS by the transport layer. While the voice telephony can perform well with a BER up to 10^{-3}, a file transfer typically requires a BER better (lower) than 10^{-6}.

The need to achieve such QoS requirements is the result of the integration of different services into a common network infrastructure.* In order to provide QoS, instead of simply delivering data based on the *best effort*, there are models that can be implemented in TCP/IP networks. These models were normalized by IETF, being named as follows [RFC 1633]:

- Integrated services (IntServ)
- Differentiated services (DiffServ)
- Multiprotocol label switching (MPLS)

MPLS is a mechanism currently used in packet networks for the provision of QoS. Nevertheless, when MPLS is not implemented, DiffServ is currently the solution widely adopted. The reader should refer to Chapter 14 for the description of MPLS.

9.3.1 INTEGRATED SERVICES

The IntServ [RFC 2211; 2212; 2213; 2215] aims to provide QoS in networks taking into account the different service requirements. The IntServ is a model that implements a group of mechanisms on the flow-by-flow basis. The provision of QoS is achieved by applying individual reservations defined

---

* This is also referred to as the convergence of the telecommunications.

by flow specifications,* combined with the resource reservation protocol (RSVP) [RFC 2205] to a group of routers in a network. The packets belonging to an end-to-end connection that pass through a group of routers, benefit from a resource reservation that may avoid latency, or packets discard. Consequently, packets belonging to a service such as audio or video streaming may benefit from higher priority in intermediate routers. To allow the provision of resources, IntServ classify the different traffic into a wide range of levels of granularity.

The exchange of traffic is preceded by a connection establishment phase that requests a certain level of QoS from the network (between two end entities), along a certain route. The intermediate routers receive such request and, based on the previously allocated resources, on the available resources, and on the type of data, decide whether or not it is possible to assign the required resources (buffers, bandwidth, etc.) to the new connection. Such resource reservations are implemented using the RSVP.

The drawback of the IntServ is that, since it rigidly assigns resources to connections, and since the resources are limited, new connections can be refused or existing connections can be dropped, due to lack of resources. These effects of congestion are more visible in high dimension networks. Moreover, the amount of signaling needed to handle the QoS requirements in networks that implement IntServ tends to be too high. Consequently, IntServ is only normally adopted in low dimension networks.

9.3.2 Differentiated Services

DiffServ [RFC 2475] is an alternative protocol that can be implemented to tentatively allow the provision of QoS in high dimension networks, while reducing the required resources (overhead, bandwidth, processing, etc.). While the IntServ operates over individual flows of data, DiffServ is applicable to high volumes of data, reducing the required signaling, reducing the granularity, and, consequently, the complexity. DiffServ handles different connections of the same type using the same principles.† This implies the prior negotiation of resources reservation for the traffic generated by a certain customer. Based on the customer's payment, an ISP offers a certain service-level agreement (SLA), which defines the maximum amount of data for each class of traffic (IP telephony, video streaming, file transfer, web browsing, etc.), as well as the type of warranties that are offered to each traffic class. Different traffic classes are identified by the DSCP field of the datagrams' headers. When a packet enters the ISP cloud, intermediate routers read the DSCP field, and the packet is treated according to the negotiated SLA. If the amount of traffic exceeds that negotiated in the SLA, packets can be dropped or delayed (or an additional fee may be charged to the customer). Moreover, in case of congestion, for the same traffic class (identified by the DSCP field), packets belonging to a customer with higher SLA have precedence. Remind, for a certain SLA, packets with lower DSCP field have priority.

DiffServ is the mechanism currently widely implemented in the Internet for the provision of QoS, when MPLS is not the implemented solution.

CHAPTER SUMMARY

This chapter provided a view about the transport layer of the TCP/IP stack. A description about the importance of the transport layer in the provision of the QoS was given, as well as the definition of the way transport layer addresses, known as *port addresses*, are utilized.

It was described that, in order to provide QoS, the transport layer achieves a trade-off between the QoS parameters requested by the application layer and the QoS available by the network layer.

---

* The flow specification depends on the type of traffic that is being exchanged (e.g., file transfer, web browsing, and video streaming).
† For example, different IP telephony connections or different file transfer connections belonging to customers with the same SLA are handled by the network in the same manner.

A description about the TCP was provided in this chapter. The properties of the TCP was described, as well as the TCP header format, with the description of each of the header fields. Moreover, the TCP handshaking was detailed, including the description of the TCP connection setup, data exchange, and connection termination. It was viewed that a TCP connection setup is a three-way handshaking. Moreover, it was also described that the exchange of data, with the TCP, is performed using the sliding window protocol.

A description about the UDP was provided. This included the description of the properties of the UDP, as well as the description of each of the UDP header fields and format. It was viewed that the UDP header comprises four fields: (1) the source and (2) the destination ports, (3) the length, and (4) the checksum. Moreover, it was viewed that the checksum is only utilized to perform error detection in the header, being an optional functionality.

It was described that both IntServ and DiffServ are techniques that can be employed to provide QoS, alongside with the MPLS. The need to achieve such QoS requirements results from the integration of different services into a common network infrastructure. It was described that IntServ aims to provide QoS by implementing a group of mechanisms on the flow-by-flow basis, that is, by applying individual reservations to a group of routers in a network. It was studied that IntServ comprises the provision of QoS by applying individual reservations defined by flow specifications, combined with the RSVP, to a group of routers in a network. Moreover, it was shown that DiffServ provides QoS in high dimension networks, while reducing the required resources, such as bandwidth, overhead, and processing. While the IntServ operates over individual flows of data, DiffServ is applicable to high volumes of data, reducing the required signaling, reducing the granularity and, consequently, the complexity. Finally, it was described that DiffServ is implemented based on the SLAs previously negotiated by each organization and using the datagram field DSCP.

REVIEW QUESTIONS

1. What are the functionalities of the transport layer of the TCP/IP architecture?
2. What is the difference between the IntServ and the DiffServ protocols?
3. What are the differences between the TCP and the UDP?
4. Give examples of services and application layer protocols supported on the TCP and the UDP.
5. Which kinds of impairments are dealt with by the TCP?
6. How is the QoS provided by the IntServ protocol?
7. What is the reason why the exchange of IP telephony packets, namely, the ITU-T H.323, normally considers the UDP instead of the TCP?
8. What are the phases of a TCP connection establishment?
9. A client intends to terminate a TCP connection with a server. Describe the messages' exchange necessary to terminate such connection.
10. How is the error control and flow control implemented by the TCP?
11. Why do the TCP segments include a sequence number?
12. How is the QoS provided by the DiffServ protocol?
13. What are the disadvantages that may result from the use of the TCP relating to the UDP?
14. What is a well-known port?
15. What is the difference between a jitter and a delay?
16. What are the basic QoS requirements?
17. Knowing that the data link layer normally makes use of error control mechanisms, why does the TCP also implement these mechanisms?
18. Why is the SLA used by the DiffServ protocol?
19. A client intends to establish an HTTP connection with a server. Describe the messages' exchange necessary to establish the connection.

LAB EXERCISES

1. Download and install the free network analyzer Wireshark. Open the application in a PC and select the interface connected to the Internet. You will see the IP datagrams that are being exchanged by the NIC of the PC, with the source IP address, destination IP address, and protocol type, alongside with other information. Select a datagram corresponding to the TCP. In a window that appears at the bottom, visualize the content of the frame, the MAC sublayer, the IP datagram, and the TCP segment. Verify the content of the TCP segment, the header fields, and the way error control and flow control is performed. Now, returning to the upper window, visualize a TCP retransmission that results from a not received or a corrupted received segment. At a third stage, select a UDP message. At the bottom, verify its content and the header fields. Check the difference between the UDP and the TCP segment.

10 Internet Layer
Addressing and Configuration

LEARNING OBJECTIVES

- Describe the addressing format used by the Internet layer of TCP/IP, including in IPv4 and IPv6.
- Configure IPv4 and IPv6 parameters in Cisco equipment using Cisco IOS, namely the line configuration submode and the interface configuration submode.
- Describe and configure the dynamic host configuration protocol (DHCP).
- Describe and configure the network address translation (NAT) and the port address translation (PAT).

The Internet protocol (IP) is the most important network protocol of the transmission control protocol (TCP)/IP stack. It provides a basic service to the delivery of datagrams, enabling a wide list of application and services. The basic functionalities of the IP are as follows:

- Definition of the address space in the Internet
- Routing of datagrams across the network nodes
- Fragmentation and reassembling of datagrams

In order to implement these functionalities, the IP makes use of datagrams, consisting of its elementary message format, used for end-to-end transmission* of data. In order to allow this, actions are to be taken by each node (router) of the network. A router is a device that makes Internet layer switching of datagrams. Normally, different routers exist between the source and the destination of an end-to-end connection. An end-to-end connection is composed of a concatenation of several layer 2 links (data link layer), and the router is the device that is located between the two successive layer 2 links (normally of different types). In the IP, each router has to decide about the best way to forward the datagrams and, to allow such routing, the datagram transports information about the destination address.

As previously described in Chapter 2, a datagram is composed of transport layer data (segment), to which an IP header is added. This is the basic concept of encapsulation, where, at the transmitter side, the upper layer data (segment) is transferred to an adjacent lower layer (datagram), and added with overhead (source and destination IP address, sequence numbers, etc.). The opposite operation, called de-encapsulation, is performed at the destination side, where a lower layer removes its layer's header (e.g., IP header) before the data (IP payload) is transferred to an upper layer (as a segment).

Since the IP is a connectionless protocol, there is no need to previously establish the connection, before the data is exchanged. Moreover, since the IP is datagram-based protocol (instead of virtual circuits), different datagrams may follow different paths, and the datagrams may reach the destination out of order. In case data reliability is necessary for the provision of quality of service, the

---

* End-to-end transmission can be viewed as the transmission between a client and a server, or between two client workstations.

TCP is adopted. The TCP is responsible for ordering datagrams, correcting datagrams with errors, detecting and correcting lost datagrams, as well as for discarding duplicated datagrams.

IP version 4 is the most currently used version. The IP version 6, initially referred to as IP next generation, was standardized by the RFC 2460, and we face, currently, a transition phase for its full migration. Among the different advantages of the IPv6 relating to the IPv4, the former provides a much higher addressing space. Moreover, this protocol presents advanced security capabilities, the ability to provide improved QoS, and the configuration of hosts in an IPv6 network is much simplified (plug and play).

10.1 IP VERSION 4

10.1.1 IPv4 CLASSFUL ADDRESSING

As introduced in Chapter 2, an IPv4 address consists of a 32-bit address, being composed of four groups of eight binary numbers (10101101.00001111.11111100.00000011). For the sake of simplicity, an IPv4 address is normally displayed in four groups of decimal numbers, using the dotted-decimal notation (e.g., 173.15.252.3). The Internet Assigned Numbers Authority (IANA) is the entity responsible for the assignment of IP address ranges to regional network information centers, whereas regional network information centers are responsible for the assignment of IP addresses to domains, such as Internet service providers (ISPs), companies, and so on.

As can be seen from Figure 10.1, IPv4 addresses in classful mode are split into address classes, from A to E. Moreover, an IPv4 address comprise two parts: one or more octets that identify the network part of the IP address and one or more octets that identify the host part of the IP address, within a certain network. A class A address comprises the leftmost octet to identify the network part of the IP address, whereas the remaining three octets identify the host part of the IP address. A class B address comprises the two leftmost octets to identify the network part of the IP address, whereas the remaining two octets identify the host part of the IP address. In the case of class B address, note that the two leftmost bits are fixed to 1, 0. Therefore, the address range of the network part of the IP address can only use the remaining 14 bits, that is, two octets, minus the two static bits used to identify the address class. The address range of the host part of a class B address can use 16 bits, that is, the two full octets.

Class D addresses are used for multicast addresses, which are used to identify all addresses within a certain group of interfaces. The use of multicast addresses, instead of broadcast addresses, allows a more efficient use of the network resources, as traffic is only sent to those interfaces that are intended to receive it.

There are two different possible ways of identifying the class of an IP address. One way relies on the observation of the leftmost octet of an IP address in decimal notation. As can be seen from Table 10.1, the leftmost octet value determines the border between classes.

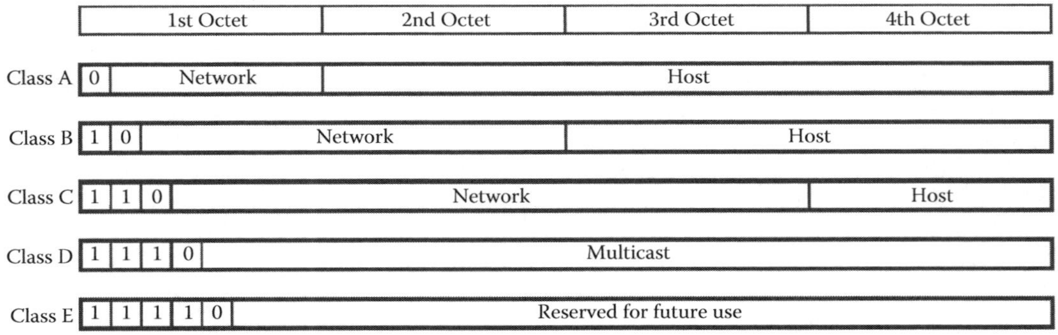

	1st Octet	2nd Octet	3rd Octet	4th Octet
Class A	0 Network	Host		
Class B	1 0 Network		Host	
Class C	1 1 0 Network			Host
Class D	1 1 1 0 Multicast			
Class E	1 1 1 1 0 Reserved for future use			

FIGURE 10.1 Classes of IPv4 addressing.

TABLE 10.1

Mapping between the Address Class and the Leftmost Octet Value

Class	Decimal Range of the Leftmost Octet	Binary Range
Class A	0 ... 127	0XXXXXXX
Class B	128 ... 191	10XXXXXX
Class C	192 ... 223	110XXXXX
Class D	224 ... 239	1110XXXX
Class E	240 ... 255	11110XXX

The second possibility requires the conversion of the leftmost octet from decimal to binary, and the observation of the position of the leftmost zero. Class A has the leftmost zero in the most significant bit (MSB). Class B has the leftmost zero in the second position, and the MSB is 1. Class C has the leftmost zero in the third position, and the two left bits are 1.

Taking the IP address 173.15.252.3 as an example, entering the value 173 into Table 10.1 it is clear that this is a class B address. Using the second method, converting the value 173 into binary we obtain 10101101. Since the leftmost zero is in the second position, we conclude that this IP address belongs to class B. From Figure 10.1, being of class B, we know that the two leftmost octets identify the network, and the two rightmost octets identify the host within a certain network.

There are some addresses that have specific applications and, for this reason, they cannot be used as the others. These addresses are listed in Table 10.2.

As can be seen from Table 10.2, the IP address with all zero bits (0.0.0.0) stands for the host itself (own IP address). This is the default address before having being assigned an IP address to a host (e.g., by the DHCP or manually by a network administrator). Note that the notation included in Table 10.2 is decimal, but an octet with a decimal value zero (0) corresponds to all bits with zero value in the octet (00000000). Similarly, an octet with a decimal value 255 corresponds to all bits with one value in the octet.

TABLE 10.2

Reserved IP Addresses

Reserved IP Addresses	
Own IP address	0.0.0.0
Network address	X.0.0.0 (Class A)
	X.X.0.0 (Class B)
	X.X.X.0 (Class C)
Broadcast in own network	255.255.255.255
Broadcast in a specific network	X.255.255.255 (Class A)
	X.X.255.255 (Class B)
	X.X.X.255 (Class C)
All networks	255.X.X.X (Class A)
	255.255.X.X (Class B)
	255.255.255.X (Class C)
Loopback	127.x.x.x

An IP address with all zero value bits in the position of the host part of the IP address (e.g., X.X.0.0 for class B*) represents the network address. Note that the IP address 173.15.252.3, considered in the example above, corresponds to class B. Therefore, the network IP address becomes 173.15.0.0.

Moreover, the IP address with all one value bits in the position of all octets (255.255.255.255) corresponds to the broadcast address in the own network. The IP address with all one value bits in the position of the octets that represent the host part of the IP address corresponds to the broadcast address in a specific network (e.g., X.X.255.255 is the broadcast address within the network X.X.0.0 [class B]). The IP address with all one value bits in the position of the network part of the address (e.g., 255.255.X.X for class B) represents all hosts with the host part of the IP address X.X, in all class B addresses. Finally, the IP address with 127 decimal value in the leftmost octet represents the loopback address. This is used for testing purposes, when a host sends data to itself.

Taking into account the IP address classes defined in Figure 10.1 and the reserved IP addresses listed in Table 10.2, the number of networks and the number of hosts per IP address class can be quantified (see Table 10.3). The class A has a total of 8 bits assigned to the network, but the leftmost bit is fixed to zero. Therefore, 7 bits are used to identify networks ($2^7 = 128$), but the all zero value bits cannot be used (reserved for the identification of the own IP address) and the decimal value 127 is reserved for loopback, leaving a total of $2^7 - 2 = 126$ existing class A networks. Similarly, 24 bits are used to address hosts within each class A network, but the all zero value bits (own network) and the all one value bits (broadcast address) cannot be used, leaving a total of $2^{24} - 2 = 16,777,214$ hosts within each network. Following the same rational, the number of class B and C networks and hosts listed in Table 10.3 can easily be deduced.

In addition to the above-mentioned reserved IP addresses, each address class has its own reserved address range, used as private addresses. Private addresses are addresses reserved by IANA to be used only inside LANs. Private IP addresses are also referred to as site-local addresses. A network administrator may use these private addresses without any permission from the IANA. Nevertheless, the router should implement a network address translation (NAT) or port address translation (PAT),[†] such that the private IP addresses are not used in the Internet world. The private IP address ranges are listed in Table 10.4, whereas Table 10.5 lists the link-local addresses. Link-local addresses are known as automatic private IP addressing (APIPA) in Windows operating systems. Similar to the private addresses, APIPA addresses are only to be used inside private networks. Moreover, a host using an APIPA address that needs to exchange data with the Internet world must use NAT/PAT.

The link-local addresses (APIPA) [RFC 3927] can be used as an alternative to the DHCP, consisting of a mechanism that can be employed by a network to allow hosts getting IP addresses. In case a host contacts the DHCP server for obtaining an IP address without any response, and if the APIPA is configured in the network interface card (NIC), the operating system automatically assigns an IP address within the APIPA range as listed in Table 10.5. The APIPA protocol assures that this

TABLE 10.3

Quantification of the Number of Networks and Hosts Per Address Class

Class	Number of Networks	Number of Hosts
A	$2^7 - 2 = 126$	$2^{24} - 2 = 16,777,214$
B	$2^{14} = 16,384$	$2^{16} - 2 = 65,534$
C	$2^{21} = 2,097,152$	$2^8 - 2 = 254$

---

* Where X stands for any decimal number between 0 and 255.
† The NAT/PAT is defined later in this section.

TABLE 10.4
Private (Site Local) IP Address Ranges

Private IP Address Ranges	
Class A	10.0.0.0–10.255.255.255
Class B	172.16.0.0–172.31.255.255
Class C	192.168.0.0–192.168.255.255

TABLE 10.5
Link-Local (APIPA) Address Range

Addresses	Mask
169.254.0.1–169.254.255.254	255.255.0.0

address is unique within the LAN. This procedure allows obtaining an IP address without making use of the DHCP server or without having to manually configure it. This procedure is very useful in small size networks.

A router is an Internet layer device that is used to interconnect two or more networks or WAN segments. In addition, these two or more interfaces are normally of different types, that is, the data link layer protocols in the interfaces are typically different. Note that a router has an IP address per NIC installed and configured. As can be seen from Figure 10.2, NIC1 is connected to a class B network and has a class B IP address (171.139.1.2), whereas NIC2 is connected to a class C network and has a class C IP address (191.139.18.1). From this example, it is viewed that a router does not have a single IP address. On the contrary, a router has as many IP addresses as the NICs. The configuration of IP addresses in Cisco 1800 series routers is described in Section 10.3.2.3, and an example is included in Appendix III.

10.1.2 IPv4 Classless Addressing

The IP address space has been described in classful mode. This mode is very inefficient, since the rigid allocation of bits to host or network part of the IP address leaves many IP addresses unused. Therefore, due to this reason, and due to the rapid growth of the number of Internet hosts, the available IPv4 addressing space is depleted. The measure that can mitigate this limitation is making use of the IPv4 address is classless mode.[*]

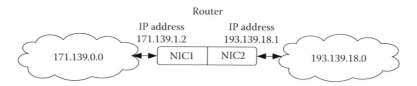

FIGURE 10.2 Example of a router with two NICs.

---

[*] Other mechanisms such as the extensive use of private addresses, DHCP, and NAT/PAT also contribute to optimize the reduced IPv4 address space.

In the classless mode, although bits are left pre-assigned to classes, we add another level of flexibility by using a subnet mask. In classless mode, the border between the network and the host identification is determined by the subnet mask. Mask of a network is an address with logic state 1 bits in the position that identifies the network part of an IP address, and logic state 0 bits in the position that belongs to the host part of the IP address. Using the classless mode, the border between network and host part of the IP address is no longer defined by the complete octets, that is, by 8-bit segments. Classless mode can use any number of bits to address network and host parts of addresses, and the address range is not divided into classes. The IP range previously defined for any of the classes (e.g., class A), can now be utilized without any restriction between the network and host assignments.

A different notation for specifying the network prefix consists of counting the number of bits with state one in the subnet mask and appending that number to the address with a slash (/) separator. This is referred to as bit count notation of the subnet mask. In the classless mode, some bits previously employed in classful mode to address hosts are now used to address subnetworks within a certain network. The remaining bits are then used to address hosts within each subnetwork.

Completing the logic AND operation between the binary version of an IP address and the binary version of the subnet mask lead us to the network and subnetwork parts of an IP address. Similarly, the host part of an IP address can be obtained by performing the logic NAND operation between an IP address and the subnet mask.

Let us examine a host with an IP address 10.20.230.140 and with subnet mask 255.255.192.0 (see Figure 10.3). The subnet mask in binary representation is 11111111.11111111.11000000.00000000. We may refer to this IP address and subnet mask pair as 10.20.230.140/18, since the subnet mask has a total of 18 bits with logic state 1 bits (bit count notation).

In the example of Figure 10.3, performing the AND operation between the IP address and the subnet mask, we conclude that the host is in the network address 10.20.192.0. Similarly, performing the NAND operation, we conclude that the host part of the IP address is 0.0.38.140. This can be seen from Figure 10.4.

IP address—decimal	10	20	230	140
IP address—binary	00001010	00010100	11100110	10001100
Subnet mask—decimal	255	255	192	0
Subnet mask—binary	11111111	11111111	11000000	00000000
Result of AND operation—binary	00001010	00010100	11000000	00000000
Net/subnet address is the result of AND operation—decimal	10	20	192	0

FIGURE 10.3 Calculation of the net or subnet part of the IP address.

IP address—decimal	10	20	230	140
IP address—binary	00001010	00010100	11100110	10001100
Subnet mask—decimal	255	255	192	0
Subnet mask—binary	11111111	11111111	11000000	00000000
Result of NAND operation—binary	00000000	00000000	00100110	10001100
Host part of the IP address is the result of NAND operation—decimal	0	0	38	140

FIGURE 10.4 Calculation of the host part of the IP address.

Since the IP address is to be written and read at Internet layer, workstations (network end stations) and routers (network nodes) are the devices that need to have access to it. Routers are the devices responsible for routing the datagrams between the source and the destination, using the datagrams' destination IP address for this purpose, alongside routing tables. Routers may perform their function only based on the destination address or using the subnet mask as well, in case the classless mode is being considered and the router is compatible with. Traffic routing between IP networks only based on IP address is called classful routing. On the other hand, the methodology of routing taking, as well, subnet mask into consideration is referred to as classless interdomain routing (CIDR) [RFC 1519]. While in the classful mode, routers forward datagrams based on the network part of the destination's IP address, using the CIDR routing, these devices read both the network part and the subnetwork part of the destination's IP address, and use these two components to decide about which output interface to use to forward the datagram.

Due to lack of IPv4 addresses, making use of classless mode, subnetworks can be created with a single IPv4 network address. In this case, only a subnetwork IP address can be assigned to an organization, instead of a whole network IP address. This results in a much more efficient use of the limited IPv4 address space.

Let us consider an example of an organization with the need to have six IP addresses.[*] If the organization is given a class C network IP address in classful mode, from Table 10.3, we know that it has a total of 254 full IP addresses (host IP addresses), while requiring only six addresses. Consequently, the organization is leaving 248 unused host IP addresses, and these unused IP addresses cannot be used by anyone else. With classless mode, the organization can be given, for example, a class C network IP address, but with a subnet mask 255.255.255.248 (this corresponds to /29 in the bit count notation). In this case, since the default (classful) class C subnet mask is 255.255.255.0 (/24) and the classless subnet mask is 255.255.255.248 (/29), a total of $29 - 24 = 5$ bits are used to create subnetworks. This allows a total of $2^5 = 32$ subnetworks, while the remaining 3 bits are left to address hosts within each subnetwork. The number of hosts within each subnetwork comes $2^{32-29} - 2 = 2^3 - 2$, that is, 6 hosts. Therefore, instead of having a full class C address allocated to this company which only requires 6 IP addresses, a subnetwork with the subnet mask 255.255.255.248 is enough, as it allows the six required IP addresses, leaving the remaining 31 subnetworks possible to be allocated to other small companies. From this, we conclude that, using the classless mode, a single class C network IP address can be re-used by many organizations, without spoiling unused addresses, therefore, with a much better efficiency.

The subtraction of the number two in the computation of the number of hosts results from the fact that

- The all 0 bits cannot be used (represents the subnetwork address).
- The all 1 value bits cannot be used (represents the subnetwork broadcast address).

The subnetworks that can be created with 5 bits are 00000, 00001, 00010, 00011, 11110, …, 11111. The six hosts that can be addressed with 3 bits are 001, 010, 011, 100, 110, and 110 (remember that, as described above, the sequences 000 and 111 cannot be used). Tables 10.6 through 10.8 show the subnet masks that can be used with different address classes.

In case of the class C, the default subnet mask is 255.255.255.0, which does not support any subnetwork. This corresponds to /24 in the bit count notation, since it has a total of 24 bits with logic state 1.

The first subnet mask that supports subnetworks is the 255.255.255.128. This corresponds to /25 in the bit count notation. It has a single bit to create subnetworks. Since the number of subnetworks is given by 2^n, where n is the number of bits with logic state 1 beyond the default subnet mask, with this subnet mask a total of 2 subnetworks can be created. With regard to the number of hosts possible to

---

[*] This can be the number of simultaneous IP addresses used dynamically by a higher number of hosts, to have access to the Internet using the NAT.

TABLE 10.6

List of Possible Class C Subnet Masks

(in Decimal and Bit Count Notation)

Class C Subnet Masks

255.255.255.0 (default)	/24
255.255.255.128	/25
255.255.255.192	/26
255.255.255.224	/27
255.255.255.240	/28
255.255.255.248	/29
255.255.255.252	/30

TABLE 10.7

List of Possible Class B Subnet Masks

(in Decimal and Bit Count Notation)

Class B Subnet Masks

255.255.0.0 (default)	/16
255.255.128.0	/17
255.255.192.0	/18
255.255.224.0	/19
255.255.240.0	/20
255.255.248.0	/21
255.255.252.0	/22
255.255.254.0	/23
255.255.255.0	/24
255.255.255.128	/25
255.255.255.192	/26
255.255.255.224	/27
255.255.255.240	/28
255.255.255.248	/29
255.255.255.252	/30

be accommodated within each subnetwork, this is given by $2^m - 2$, where m stands for the number of logic state 0 bits in the subnet mask. Since this mask has $32 - 25 = 7$ bits with the logic state 0, the number of hosts accommodated within each subnetwork comes to $2^7 - 2 = 126$ hosts.

The following subnet mask that supports subnetworks is the 255.255.255.192. This corresponds to /26 in the bit count notation. It has two bits to create subnetworks, which results in a total of four subnetworks possible to be created. The four subnetworks have the following subnetwork addresses: 0 0, 0 1, 1 0, and 1 1, respectively. The same rational applies to the other subnet mask.

Note that the subnet mask with 31 logic state 1 bits, that is, with the mask 255.255.255.254, is not used. This results from the fact that such mask has a single bit with logic state 0, resulting in a number of hosts equal to 0. This subnet mask is not listed in Tables 10.6 through 10.8. The same rational applies to the subnet masks using addresses from other classes.

In order to better understand the calculation of IPv4 addresses using the classless mode, let us focus on the example shown in Figure 10.5. Let us consider the network IP address 195.1.2.0 that is intended to be split into multiple subnetworks. Considering the subnet mask 255.255.255.192, this

TABLE 10.8
List of Possible Class A Subnet Masks
(in Decimal and Bit Count Notation)

Class A Subnet Masks

255.0.0.0 (default)	/8
255.128.0.0	/9
255.192.0.0	/10
255.224.0.0	/11
255.240.0.0	/12
255.248.0.0	/13
255.252.0.0	/14
255.254.0.0	/15
255.255.0.0	/16
255.255.128.0	/17
255.255.192.0	/18
255.255.224.0	/19
255.255.240.0	/20
255.255.248.0	/21
255.255.252.0	/22
255.255.254.0	/23
255.255.255.0	/24
255.255.255.128	/25
255.255.255.192	/26
255.255.255.224	/27
255.255.255.240	/28
255.255.255.248	/29
255.255.255.252	/30

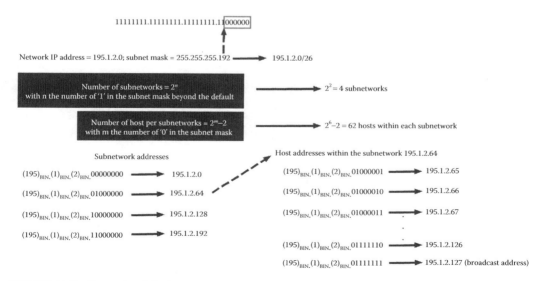

FIGURE 10.5 Example of IPv4 address calculation in classless mode.

results in /26 in bit count notation. With this mask, the number n becomes 2, that is, we have two bits used to address subnetworks. Remember that the number of bits used to address subnetworks include those in the subnet mask with logic state 1, beyond the default subnet mask. This results in a total of four subnetworks possible to be created in this class C network address, with this subnet mask. Moreover, with this mask, the number m becomes 6, that is, we have six bits used to address hosts within each subnetwork. Remember that the number of bits used to address hosts within each subnetwork include those, in the subnet mask, with logic state 0. This results in a total of 62 hosts possible to be allocated within each subnetwork.

The following question is, "What are these four subnetwork addresses?" Since we have two bits to address subnetworks, the subnetwork addresses are those with all possible sequences of 2 bits, in the position of the 2 bits with logic state 1 in the subnet mask, beyond the default class C subnet mask. The first subnetwork is the one with the sequence 00 in the position of the 2 bits with logic state 1 in the subnet mask, beyond the default class C subnet mask, and zeros elsewhere in the same octet. This results in the subnetwork address 195.1.2.0. The second subnetwork is the one with the sequence 01 in the same position. This results in the subnetwork address 195.1.2.64. The third subnetwork is the one with the sequence 10 in the same position. This results in the subnetwork address 195.1.2.128. The fourth subnetwork is the one with the sequence 11 in the same position. This results in the subnetwork address 195.1.2.192.

The following question is, "What are the host addresses within each subnetwork?" Let us focus on the second subnetwork, that is, 195.1.2.64. The bits used to address hosts are those with the logic state 0 in the subnet mask. In our example, these bits are the last six bits. Therefore, the possible host addresses are those corresponding to all possible combination of six bits, except the sequence with logic state 0 in all bits, as this corresponds to the subnetwork address, and except the sequence with logic state 1 in all bits, as this corresponds to the broadcast address within the subnetwork. The first host address comes 01000001 in the rightmost octet. Note that the two leftmost bits 01 corresponds to the subnetwork address, whereas the six rightmost bits correspond to the host address. Using the decimal notation, this corresponds to the IP address 195.1.2.65. Note that this value 65 corresponds to the one after the subnetwork address, which is 64. The second host address comes 01000010 in the rightmost octet. Using the decimal notation, this corresponds to the IP address 195.1.2.66, whose decimal value of the rightmost octet follows the previous one (65). The last host address is the one before the broadcast address in the subnetwork. The broadcast address has the following rightmost octet: 01111111, that is, 127 in the decimal notation. Remember that the broadcast address has logic state 1 in all bits used to identify hosts, which, in our example, corresponds to the rightmost six bits, of the rightmost octet. Note that this value 127, corresponds to the one before the following subnetwork address, which is 128. The last host address within our subnetwork has the following rightmost octet: 01111110. This corresponds to 126 in decimal notation. Note that this value 126, corresponds to the one before the broadcast address within the subnetwork, which is 127.

For high number of bits used to address subnetworks (taken from the bits used to address hosts in the classful mode), the calculation of subnetting becomes too complex to be performed using the binary notation. There is a rule that allows the calculation of subnetting using the decimal notation, which becomes much simpler than using the binary calculation. Such subnetting calculation using the decimal notation is summarized in Table 10.9.

Table 10.10 shows an example of subnet calculation, for a class C address, considering the network IP address 195.1.2.0, with the subnet mask 255.255.255.240. Note that this mask has four bits with logic state 1 and four bits with logic state 0, in the last octet. Consequently, using the previously exposed formulas, the number of subnetworks becomes 16, and the number of hosts per subnetwork becomes 14. Moreover, the base number becomes 16, making the last octet of the subnetwork addresses, in decimal notation: 0, 16, 32, ..., 240. Similarly, since the broadcast addresses are those that precedes the following subnetwork addresses, the last octet of the broadcast addresses, in decimal notation, becomes 15, 31, 47, ..., 255. Finally, the valid host addresses are those between each subnetwork address and the corresponding broadcast address.

TABLE 10.9
Subnet Calculation Using the Decimal Notation

Subnet Calculation

Number of subnetworks (NSN)	$NSN = 2^n$, where n is the number of bits with logic state 1 in the subnet mask, beyond the default subnet mask
Number of hosts per subnetwork (NHPN)	$NHPN = 2^m - 2$, where m is the number of bits with logic state 0 in the subnet mask
Subnetwork addresses	Multiples of the base number. Base number = 256-subnet number. Subnet number is the decimal value of the octet that identifies the subnetwork in the subnet mask. The resulting subnetwork addresses are composed of the network part of the original IP address, concatenated with multiples, from 0 to NSN-1, of the base number (as subnetwork part)
Broadcast address within each subnetwork	Corresponds to the address that precedes the following subnetwork addresses
Valid host addresses within each subnetwork	All valid IP addresses between the subnetwork address and the following subnetwork address, except the subnetwork and the broadcast addresses

TABLE 10.10
Example of Subnet Calculation for Class C (Network IP Address 195.1.2.0/28)

Subnet Calculation (IP Address = 195.1.2.0; Subnet Mask = 255.255.255.240)

Number of subnetworks	$NSN = 2^4 = 16$
Number of hosts per subnetwork	$NHPN = 2^4 - 2 = 14$
Subnetwork addresses	Base number = $256 - 240 = 16 \rightarrow$ subnetworks are 195.1.2.0, 195.1.2.16, 195.1.2.32, …, 195.1.2.224 195.1.2.240
Broadcast in the subnetwork address	Broadcast addresses are 195.1.2.15, 195.1.2.31, 195.1.2.47, …, 195.1.2.239, 195.1.2.255
Valid host addresses within each subnetwork	Address range for subnetwork 195.1.2.0 is 195.1.2.1 $\rightarrow$ 195.1.2.14; address range for subnetwork 195.1.2.16 is 195.1.2.17 $\rightarrow$ 195.1.2.30, …

Table 10.11 corresponds to the breakdown of the example shown in Table 10.10, that is, shows the subnetwork decomposition with most of the addresses, for a class C address, considering the network IP address 195.1.2.0, with the subnet mask 255.255.255.240.

Table 10.12 shows an example of subnet calculation, for a class B address, considering the network IP address 128.1.0.0, and with the subnet mask 255.255.192.0 (11111111.11111111.11000000. 00000000 in binary notation). Note that, using the class B, the two rightmost octets are employed to create subnetworks, instead of a single octet, as considered by class C addresses. This subnet mask has two bits with logic state 1 and $8 + 6 = 14$ bits with logic state 0. Consequently, the number of subnetworks becomes 4, while the number of hosts per subnetwork becomes $2^{14} - 2 = 16,382$. It is worth noting that the value 192, in the subnet mask, is in the position of the second rightmost octet. Consequently, the multiples of the base number, used to calculate the subnetwork addresses, are also placed in the second rightmost octet, while leaving the rightmost octet in zero.

Table 10.13 corresponds to the breakdown of the example shown in Table 10.12, that is, lists the subnetwork decomposition with most of the addresses, for a class B address, considering the network IP address 128.1.0.0, and with the subnet mask 255.255.192.0. Moreover, Table 10.14 lists the subnetwork decomposition with most of the addresses, for a class B address, considering the network IP address 128.1.0.0, and with the subnet mask 255.255.255.192. Comparing these two examples it is worth noting that the multiples of the base number are placed in the position of the

TABLE 10.11

Breakdown of Subnet Calculation for Class C (Network IP Address 195.1.2.0/28)

Subnetwork address	195.1.2.0	195.1.2.16	32	48	64	...	240
First address in subnetwork	195.1.2.1	17	33	47	65	...	241
Last address in subnetwork	195.1.2.14	30	46	62	78	...	254
B/C address in subnetwork	195.1.2.15	31	47	63	79	...	255

TABLE 10.12

Example of Subnet Calculation for Class B (Network IP Address 128.1.0.0/18)

Subnet Calculation (IP Address = 128.1.0.0; Subnet Mask = 255.255.192.0)	
Number of subnetworks	$NSN = 2^2 = 4$
Number of hosts per subnetwork	$NHPN = 2^{14} - 2 = 16,382$
Subnetwork addresses	Base number = $256 - 192 = 64 \rightarrow$ subnetworks are 128.1.0.0, 128.1.64.0, 128.1.128.0, and 128.1.192.0
Broadcast in the subnetwork address	Broadcast addresses are 128.1.63.255, 128.1.127.255, 128.1.191.255, and 128.1.255.255, respectively
Valid host addresses within each subnetwork	Address range for subnetwork 128.1.0.0 is 128.1.0.1 $\rightarrow$ 128.1.63.254; address range for subnetwork 128.1.64.0 is 128.1.64.1 $\rightarrow$ 128.1.127.254

TABLE 10.13

Breakdown of Subnet Calculation for Class B (Network IP Address 128.1.0.0/18)

Subnetwork address	128.1.0.0	128.1.64.0	128.0	192.0
First address in subnetwork	128.1.0.1	64.1	128.1	192.1
Last address in subnetwork	128.1.63.254	127.254	191.254	255.254
B/C address in subnetwork	128.1.63.255	127.55	191.255	128.1.255.255

TABLE 10.14

Breakdown of Subnet Calculation for Class B (Network IP Address 128.1.0.0/26)

Subnetwork address	128.1.0.0	128.1.0.64	0.128	0.192	1.0	1.64	1.128	...	255.192
First address in subnetwork	128.1.0.1	0.65	129	192.1	1.1	1.65	1.129	...	255.193
Last address in subnetwork	128.1.0.62	0.126	0.190	255.254	1.62	1.126	1.190	...	255.254
B/C address in subnetwork	128.1.0.63	0.127	0.191	255.255	1.63	1.127	1.191	...	128.1.255.255

incomplete octet of the subnet mask. The *incomplete* octet is the one that is filled with bits of different logic states, that is, with *0* and *1*, whereas *complete** octets are only filled with either *0* or *1* logic state bits (but not a mix of both). In the case of the example shown in Table 10.14, since the *incomplete*

---

* 255 is a *complete* octet as this corresponds to the sequence 11111111 in binary notation. Similarly, 0 is also a *complete* octet as this corresponds to the sequence 00000000 in binary notation.

octet is the rightmost octet, and since the default class B address has the two leftmost octets to identify the network,* the second rightmost octet varies from 0 to 255, with increments of 1.

Table 10.15 shows an example of subnet calculation, for a class A address, considering the network IP address 11.0.0.0, and with the subnet mask 255.255.192.0 (11111111.11111111.11000000.000 00000 in binary notation). Note that, using the class A, up to three rightmost octets can be employed to create subnetworks. This subnet mask has $2 + 8 = 10$ bits with logic state 1, and $6 + 8 = 14$ bits with logic state 0. Consequently, the number of subnetworks becomes $2^{10} = 1,024$, while the number of hosts per subnetwork becomes $2^{14} - 2 = 16,382$. It is worth noting that the value 192 in the subnet mask (i.e., the *incomplete* octet), is in the position of the second rightmost octet. Consequently, the multiples of the base number, used to calculate the subnetwork addresses, are also placed in the second rightmost octet, while leaving the rightmost octet in 0. Note that the third rightmost octet varies from 0 to 255, with an increment of one.

Table 10.16 corresponds to the breakdown of the example shown in Table 10.15, that is, lists the subnetwork decomposition with most of the addresses, for a class A address, considering the network IP address 11.0.0.0, and with the subnet mask 255.255.192.0. Moreover, Table 10.17 lists the subnetwork decomposition with most of the host addresses, for a class A address, considering the network IP address 11.1.0.0, and with the subnet mask 255.192.0.0. Similar to the class B example, comparing these two examples it is worth noting that the multiples of the base number are placed in the position of the *incomplete* octet of the subnet mask. In the case of the example shown in Table 10.16, since the *incomplete* octet is the second rightmost octet, and since the default class A address has the leftmost octet to identify the network, the second leftmost octet varies from 0 to 255, with increments of 1.

TABLE 10.15
Example of Subnet Calculation for Class A (Network IP Address 11.0.0.0/18)

Subnet Calculation (IP Address = 11.0.0.0; Subnet Mask = 255.255.192.0)	
Number of subnetworks	$NSN = 2^{10} = 1,024$
Number of hosts per subnetwork	$NHPN = 2^{14} - 2 = 16,382$
Subnetwork addresses	Base number = $256 - 192 = 64 \rightarrow$ subnetworks are 11.0.0.0, 11.0.64.0, 11.0.128.0, 11.0.192.0, 11.1.0.0, 11.1.64.0, 11.1.128.0, …, 11.255.192.0
Broadcast address in the subnetwork	Broadcast addresses are 11.0.63.255, 11.0.127.255, 11.0.191.255, 11.0.255.255, 11.1.63.255, 11.1.127.255, 11.1.191.255, …, 11.255.255.255, respectively
Valid host addresses within each subnetwork	Address range for subnetwork 11.0.0.0 is 11.0.0.1 $\rightarrow$ 11.0.63.254; address range for subnetwork 11.0.64.0 is 11.0.64.1 $\rightarrow$ 11.0.127.254; …; address range for subnetwork 11.255.192.0 is 11.255.192.1 $\rightarrow$ 11.255.255.254

TABLE 10.16
Breakdown of Subnet Calculation for Class A (Network IP Address 11.0.0.0/18)

Subnetwork address	11.0.0.0	11.0.64.0	0.128.0	0.192.0	1.0.0	1.64.0	1.128.0	…	255.192.0
First address in subnetwork	11.0.0.1	0.64.1	0.128.1	0.192.1	1.0.1	1.64.1	1.128.1	…	255.192.1
Last address in subnetwork	11.0.63.254	0.127.254	0.191.254	0.255.254	1.63.254	1.127.254	1.191.254	…	255.255.254
B/C address in subnetwork	11.0.63.255	0.127.255	0.191.255	0.255.255	1.63.255	1.127.255	1.191.255	…	11.255.255.255

---

* Consequently, up to the two leftmost octets can be employed to create subnetworks.

TABLE 10.17
Breakdown of Subnet Calculation for Class A (Network IP Address 11.0.0.0/10)

Subnetwork address	11.0.0.0	11.64.0.0	128.0.0	192.0.0
First address in subnetwork	11.0.0.1	64.0.1	128.0.1	192.0.1
Last address in subnetwork	11.63.255.254	127.255.254	191.255.254	255.255.254
B/C address in subnetwork	11.63.255.255	127.255.255	191.255.255	11.255.255.255

10.1.2.1 Variable Length Subnet Mask

We have viewed that the use of IPv4 address spacing in classless mode is more efficient than in classful mode. This results from the fact that, using the classless mode, we may assign a subnetwork to an organization whose required number of IPv4 addresses is lower than the number of addresses made available by the whole network in classful mode. In this scenario, the remaining subnetworks may be assigned to other organizations. Nevertheless, the exposed classless mode considers that different subnetworks generated from a certain network address group have all the same subnet mask.

The variable length subnet mask (VLSM) is a mechanism that further improves the efficiency of the IPv4 address spacing usage by enabling different subnet masks to different subnetworks. Using VLSM, the selection of subnet masks is performed taking into account the number of hosts required for each subnetwork.

The assignment of the IPv4 address space for different subnetworks is performed using the following four steps:

1. Order the number of address requirements for different subnetworks by descending order (include the subnetwork and broadcast addresses).
2. Identify the number of bits in the subnet mask of each subnetwork with the logic state 0, such that the required number of addresses is fulfilled (the remaining bits in the subnet mask have logic state 1).
3. Define the subnet mask for each subnetwork.
4. Define the beginning and end of the IPv4 address space for different subnetworks.

Let us consider an example, where the network address 191.168.1.0 is to be split into multiple subnetworks, and that a company requires three subnetworks: one subnetwork with 9 hosts, another with 2 hosts, and a third one with 30 hosts. Based on the procedure above we have the following:

1. Ordering the required number of hosts by descending order, we obtain 30, 9, and 2 hosts. Taking into account the subnetwork and broadcast addresses we obtain the following address requirements for different subnets, by descending order:
   a. 30 hosts $\rightarrow$ 32 addresses
   b. 9 hosts $\rightarrow$ 11 addresses
   c. 2 hosts $\rightarrow$ 4 addresses
2. Deducting the required number of logic state 0s in the subnet mask that supports such requirements, this comes:
   a. 32 addresses $\rightarrow$ five 0s in the subnet mask that will make a total of 32 available addresses (twenty-seven 1s in the subnet mask) $\rightarrow$ 11111111 11111111 11111111 11100000.
   b. 11 addresses $\rightarrow$ the exponent of two upper layer corresponds to 16, which requires four 0s in the subnet mask which will make a total of 16 available addresses (twenty-eight

1s in the subnet mask). Note that three 0s would not be enough to accommodate 11 addresses → 11111111 11111111 11111111 11110000.

    c. 4 addresses → two 0s in the subnet mask which will make a total of 4 available addresses (thirty 1s in the subnet mask) → 11111111 11111111 11111111 11111100.

3. Defining the subnet mask for each subnetwork, we obtain the following:
    a. 32 addresses → 255.255.255.224
    b. 11 addresses → 255.255.255.240
    c. 4 addresses → 255.255.255.252

4. Defining the IPv4 address space for each subnetwork, we obtain the following:
    a. 32 addresses → address space → 191.168.1.0 up to 191.168.1.31 (base number is $256 - 224 = 32$, making the address range bounded between 0 and $32 - 1 = 31$).
    b. 11 addresses → address space → 191.168.1.32 up to 191.168.1.47 (base number is $256 - 240 = 16$, making the address range bounded between 32 and $32 + 16 - 1 = 47$).
    c. 4 addresses → address space → 191.168.1.48 up to 191.168.1.51 (base number is $256 - 252 = 4$, making the address range bounded between 48 and $48 + 4 - 1 = 51$).

Note that the assignment of the address space for different subnetworks starts with the IPv4 address following the last address (broadcast) assigned to the previous subnetwork. For example, the second subnetwork (11 addresses) starts with the IPv4 address 191.168.1.32 that follows the last address assigned for the previous subnetwork (191.168.1.31).

By comparing VLSM against the conventional classless mode, it is easily understandable that the former is more efficient. With the conventional classless mode, one would assign the *wider* subnet mask that fits the requirements of all subnetworks, in terms of number of addresses. In the case of the above example, the subnet mask would be 255.255.255.224, which would accommodate 32 addresses for each subnetwork. Consequently, considering the three subnetworks, the IPv4 address space that would be reserved using the conventional classless mode would be $3 \times 32 = 96$ addresses, starting from 191.168.1.0 up to 191.168.1.95. For the same situation (example above), the VLSM mechanism simply reserves the address space 191.168.1.0 up to 191.168.1.51. This involves the assignment of almost half of the IPv4 address space, than those reserved for the conventional classless mode, leaving the remaining address space for other utilizations and assignments.

10.1.3 IPv4 Datagram

As previously described, at the transmitting host, a segment is encapsulated into the payload field of an IP datagram, being de-encapsulated at the receiving host. Similarly, an IP datagram is successively encapsulated and de-encapsulated into the payload of different data link layer frames. At intermediate nodes (routers), an IP datagram is taken from the payload frame, the destination IP address is read, and the decision about the output interface to use to forward it is taken, based on the routing table. Afterward, re-encapsulation of the IP datagram into the payload of the new data link layer frame takes place. In order to allow the routing of an IPv4 datagram, hosts[*] need to have access to a myriad of datagram fields. Figure 10.6 depicts the different fields of an IPv4 datagram header.

The different IPv4 header fields are described in the following:

- *Version*: As can be seen from Figure 10.6, it is a 4-bit length field, presenting the value 4 for the IPv4.
- *Internet header length* (*IHL*): It is a 4-bit length field, being composed of a number that expresses the header length (including options and padding), in groups of 32 bits.

---

[*] Workstations, routers, gateways, and so on.

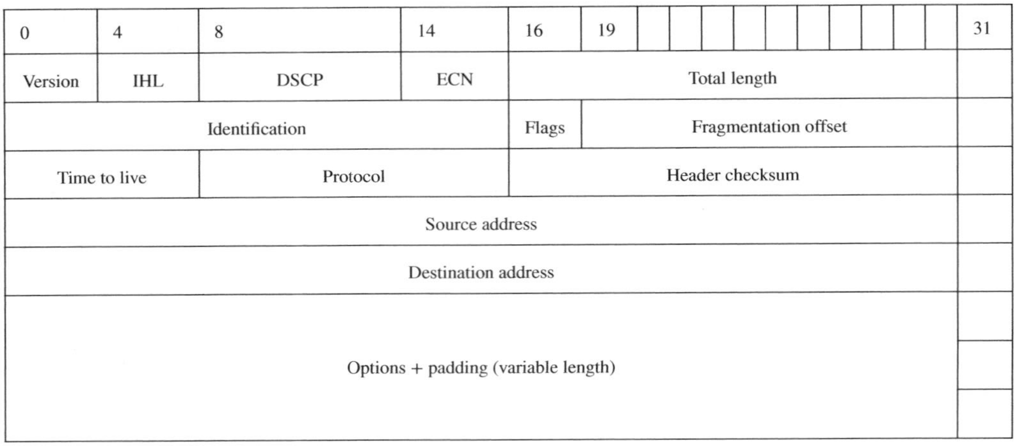

0	4	8	14	16	19									31
Version	IHL	DSCP	ECN		Total length									
Identification				Flags		Fragmentation offset								
Time to live		Protocol			Header checksum									
Source address														
Destination address														
Options + padding (variable length)														

FIGURE 10.6   IPv4 header.

- *Differentiated services code point* (*DSCP*): It is used in differentiated services (DiffServ) [RFC 2474], for the provision of QoS by intermediate routers. These requirements include delay, jitter, packet loss, or throughput.[*] This field is identified by a number between 0 and 15. Using the DiffServ option, datagrams with lower DSCP numbers have priority over those with higher DSCP numbers. In the past, this field was named type of service (ToS).
- *ECN*: It stands for explicit congestion notification [RFC 3168], being an optional feature. It enables network congestion notification.
- *Total length*: It represents the total datagram length, expressed in octets. The minimum datagram length is 20 octets (only the header, without options or payload), and the maximum datagram length is 65,535 octets. In case the maximum data field length[†] of the frame (maximum transmission unit [MTU]) is shorter than the datagram length, fragmentation is carried out.
- *Identification*: It consists of a sequence number used for identifying fragments of an original IP datagram.
- *Flags*: It consists of three bits, which are as follows:
  - *Bit 0*: reserved (must be zero value)
  - *Bit 1*: do not fragment
  - *Bit 2*: more fragments
- *Fragmentation offset*: It gives information about which part of the fragmentation does a certain datagram belong.
- *Time to live* (*TTL*): It specifies the number of hops that a datagram may be subject to. Each router decrements this field by one. When this field reaches zero, the datagram is discarded. TTL is used to prevent IP packets from traveling endlessly over networks due to routing loops or other problems.
- *Protocol*: It gives information about the upper layer protocol encapsulated in a datagram [RFC 790] (e.g., 6 for TCP and 17 for UDP).
- *Header checksum*: It consists of error detection redundant bits only applied to the header of the datagram. Since the Internet layer does not comprise error control in the payload data, in case an error is detected in the header, the datagram is discarded.

---

[*] The audio or video streaming are delay and jitter-sensitive services, whereas the file transfer is a packet loss/bit error rate sensitive service.

[†] Note that the payload length of the data link layer has normally a variable length, depending on the amount of data (packet size) to transport between two adjacent nodes.

- *Source address*: 32-bit address of the source host. In case NAT is in use, this source IP address may not be the original one.
- *Destination address*: 32-bit address of the destination host. In case NAT is in use, this destination IP address may not be the original one.
- *Options*: Due to its rare use, this field is optional, including some prior routing instructions such as the following:
  - *Source routing*: A list of router addresses through where the datagram should follow. The implementation of source routing partially transforms the IP switching from datagram method into virtual circuits method.
  - *Route recording*: It records information about the routers that the datagram passed through.
  - *Stream identification*: It is used for streaming services (e.g., video streaming or voice over IP).
- *Padding*: It is used to fill the rest of the datagram with logic state 0 bits, such that the header length is an integer number of a 32-bit word.

10.2 IP VERSION 6

The massification of the Internet, in combination with the inefficient use of the limited IPv4 address space, drove the need for a new Internet address scheme. The depletion of the IPv4 address space was the most important motivation for the development of the IPv6 [RFC 2460; 1752]. The IPv6 makes use of a 128-bit address, instead of the 32-bit address considered by the IPv4. This makes available a total of $2^{128} = 3.40 \times 10^{38}$ addresses, which allows accommodating the future terminals that result from the massification of the Internet and from the machine-to-machine communications, namely of the Internet of things. Mechanisms previously used to optimize the limited IPv4 address space, such as the NAT, PAT, and DHCP, are not required in IPv6.

Besides the wider address space made available by the IPv6, there were other Internet specifications that were intended to be improved with the new IP version. The IPv6 includes the following improved features:

- *Improved QoS mechanisms*: With the integration of the different services into a common telecommunications infrastructure, there was the need to make routing faster and smarter, with a more efficient and effective way of implementing packets[*] differentiation. This is a characteristic of the IPv6, allowing a better handling of packets belonging to different services. Some mechanisms were improved in order to reduce the resources consumed by intermediate routers. An example is the lower number of fields of an IPv6 header, as compared to the number of IPv4 fields (despite the longer mandatory IPv6 header, namely due to longer IPv6 addresses). This translates in lower resources consumption by intermediate nodes, as well as faster routing. Another example is the IPv6 source routing (optional mechanism), which presents several improvements relating to the IPv4 source routing. This mechanism allows specifying some of the routers that will be visited by packets belonging to a certain flow of data. In addition, the flow label field (mandatory header field) allows prioritized handling by intermediate routers. These mechanisms allow improving the QoS for different types of services.
- *Improved security functions*: While the IPsec is an IPv4 option, the exchange of packets in IPv6 networks is always carried out using the IPsec framework. This mechanism prevents attacks against the confidentiality, integrity, and authenticity. The reader should refer to Chapter 16 for a detailed description of IPsec.

---

[*] Note that the IPv6 uses the term *packet*, instead of datagram (as considered by IPv4).

- *Support for higher user mobility*: The host (interface) identification of an IPv6 address is unique, and can be automatically generated from the MAC address. Consequently, the movement of a user across different networks gives him the possibility to keep connectivity (using the same interface part of the IPv6 address) without administrative intervention.* With such automatic MAC/host address conversion, the ARP is no longer needed. Moreover, the Internet control message protocol version 6 (ICMPv6) used by IPv6 improves the auto-configuration of hosts, allowing an easier and quicker discovery of neighboring nodes.
- *Faster and easier routing*: The IPv6 address is hierarchical, consisting of a global routing prefix, subnet identification, and interface identification. The global routing prefix is normally assigned to sites by geographical regions. Since the IPv4 addresses were not organized by regions, interdomain routers had to consult enormous routing tables to allow the selection of the output interface to forward a datagram. Conversely, since most of the IPv6 routing is performed inside a certain hierarchical level, this results in shorter routing tables that allows a more efficient use of the routers resources (CPU, memory, bandwidth, etc.), enabling a faster routing.

IPv6 considers the same type of dynamic routing protocols as IPv4 to generate routing tables, but only with the minimal modifications to accommodate the changes of the new address space.

10.2.1 IPv6 Addressing

Similar to the IPv4, IANA is the entity responsible for the assignment of groups of IP addresses among different regions. As can be seen from Figure 10.7, an IPv6 address [RFC 2373] is composed of 128 bits (16 octets).† For the sake of simplicity, an IPv6 address is displayed in eight groups of four hexadecimal digits,‡ separated by colons. Each group of four hexadecimal digits is within the range 0000-FFFF.

Within an IPv6 address, zeros can be omitted. As an example, the address 3D0F:00F3:0000:0000: 0000:0000:0000:5A3C can simply be referred to as 3D0F:00F3::5A3C (contiguous zeros may be omitted).§ The position of the omitted zeros is marked with double colons. Nevertheless, this abbreviation can only be used once in an IPv6 address. Otherwise, it would be impossible to know how many zeros had been replaced by double colons. Moreover, within each group of four hexadecimal digits, the leftmost zeros can be omitted. Therefore, the address: 3D0F:00F3::5A3C is equivalent to 3D0F:F3::5A3C (i.e., F3 instead of 00F3), or the address 3D0F:00F3:0000:FE01:0000:01C3:0003: 5A3C is equivalent to 3D0F:F3:0:FE01:0:1C3:3:5A3C.

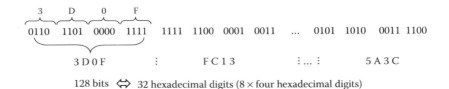

128 bits ⟺ 32 hexadecimal digits (8 × four hexadecimal digits)

FIGURE 10.7 An example of an IPv6 address in both binary and hexadecimal notation.

---

* Only the network/subnetwork part of the IPv6 address changes. This can be discovered by establishing contact with neighboring nodes (without administrative intervention).
† The theoretical limit of the IPv4 address space is 2^{32}, corresponding to 4,294,967,296. Due to the rapid growth of the Internet, the available IPv4 address space is depleted. IPv6 solves this problem, as its address is composed of 128 bits, which makes a wide address space available for the Internet world (2^{128} addresses).
‡ Note that a hexadecimal digit may encode a total of 4 bits.
§ Another example can be the address 3D0F:00F3:0000:0000:0000:0000:0000:0000, which is equivalent to 3D0F:00F3::.

Global unicast IPv6 address presents the generic composition described in Table 10.18. Note that the Interface ID (interface identification) is composed of 64 bits. Moreover, the 64-n bits allocated to the global routing prefix plus the n bits allocated to the subnet ID makes a total of 64 bits. The value of n depends on the address type.

The global routing prefix identifies the network, whereas the subnet ID identifies the subnetwork (these two items identify where the host is connected to). The interface ID is the part of the address that uniquely identifies an interface (this item identifies who you are). This address composition is similar to the composition of an IPv4 address in classful mode. A host that moves across different networks or subnetworks keeps the same interface ID, but changes the global routing prefix and/or the subnet ID. Consequently, it can be stated that a host only has an interface ID, not a whole IPv6 address.

In order to understand the IPv6 address decomposition, it is worth introducing the notion of prefix. A prefix can be used for two different purposes, which are as follows:

- To identify an address range
- To separate the network and subnetwork part of an address from the interface part of an address

In the scope of the identification of an address range, the notation X::/Y stands for the address X:: with the prefix /Y. This consists of an address range starting by the address X000...000 with Y fixed binary digits within the address X, up to ZFFF...FFF. In case the address is in binary notation, since Y stands for the number of fixed binary digits, the fixed bits Z equals the fixed bits X. In case the address is in hexadecimal notation, since the hexadecimal digit X is composed of four bits, if the number of fixed binary digits corresponds to a number lower than four, the hexadecimal Z may differ from X by varying the remaining 4-Y bits. Let us consider the example of the global unicast address range defined by 2000::/3 (the three first binary digits are fixed). This represents the address range starting in the address 2000:: (0010 0000 0000 ...) up to 3FFF:FFFF:FFFF:FFFF:FFFF:FFFF:FFFF:FFFF (0011 1111 1111...).

Using the prefix to separate the network and subnetwork part of an address from the interface part of an address, the concept is the same as used in IPv4 addresses using the bit count notation.

As an example, from the address 31CF:0123:4567:89AB:CDEF:0123:4567:89AB/64 we may conclude that the network and subnetwork part of the address has a total of 64 bits (with the address 31CF:0123:4567:89AB:0000:0000:0000:0000), and that the remaining 64 bits are used to identify the interface within the network (CDEF:0123:4567:89AB).

The IPv6 considers the following three types of addresses:

- *Unicast* identifies a single network interface [RFC 2374]. As described above, the global unicast addresses are located within the address range 2000::/3 (the three leftmost bits of the IPv6 address are fixed to 001).
- *Anycast* identifies any address within a certain address group. For example, an anycast address may refer to the closer node within a certain unicast address range. Anycast addresses use the same address space as unicast, but for different purposes.
- *Multicast* identifies all addresses within a certain a group of interfaces. Note that there are no broadcast addresses in IPv6. The broadcast address is a special case of the multicast address. The multicast addresses are located within the address range FF00::/8 (the eight leftmost bits of the IPv6 address are fixed to 11111111).

TABLE 10.18

Generic Composition of a Unicast IPv6 Address

Global Routing Prefix	Subnet ID	Interface ID
64-n bits	n bits	64 bits

Table 10.19 summarizes the reserved IPv6 address ranges, pre-allocated for different purposes.

As can be seen from Table 10.19, there are three different types of unicast addresses, which are as follows:

- The global unicast address have a global scope, being globally utilized in the Internet.
- The site-local unicast address corresponds to the IPv4 private address. The site-local unicast addresses can be used within a certain site (these addresses are unique within such site).
- The link-local unicast address corresponds to the IPv4 APIPA address. The link-local unicast address can only be used within a specific site. The link-local configuration is used by the operating system in the absence of having another type of IPv6 address. These addresses are used between on-link neighbors and for neighbor discovery on the same link. An IPv6 router never forwards the link-local traffic beyond the link. These addresses are local and not advertised in the Internet world. The procedure assures that these addresses are unique within the site.

The composition of a global unicast address is summarized in Table 10.20. As can be seen from Table 10.20 and Figure 10.8, IPv6 addresses have several hierarchical layers, entitled top-level aggregation (TLA), next-level aggregation (NLA), and site-level aggregation (SLA). The TLA identification (TLA ID) is the highest level in the IPv6 routing hierarchy, being used to identify long-haul providers. Similar to the network addresses, groups of TLA are allocated by IANA to regional Network Information Centers, which then allocates them to global ISP. The NLA identification (NLA ID) is used to identify a regional site within a global ISP. Finally, the SLA identification (SLA ID) is employed internally by a site domain (regional), to identify subnetworks and to create its own local addressing hierarchy [RFC 2374].

These aggregation levels are allocated by geographical regions. This results in shorter routing tables, as most of the routing is performed within a certain level. The upper or lower level prefixes are only used if the address is not within a certain level. Moreover, very often a fixed node/address is used to reach different layers. This also gives a contribution to shortening the routing tables. This organization of the IPv6 addresses space results in shorter routing tables, which translates in faster routing of the packets along the network, facilitating the provision of QoS.

An IPv6 address can be generated using different methodologies, which are as follows:

- Statically assigned a full IPv6 address (e.g., manually assigned using the Cisco IOS in Cisco systems devices, as normally performed in IPv4).
- Statically assigned an interface ID using the EUI-64 notation.
- Dynamically assigned, using, for example, an address provided by a DHCP server, making use of the DHCPv6 protocol.

TABLE 10.19
Reserved IPv6 Address Ranges

Address Type	Binary Prefix	IPv6 Notation
Unspecified	000…0	::/128 or 0:0:0:0:0:0:0:0
Loopback	000…1	::1/128 or 0:0:0:0:0:0:0:1
Link-local unicast	1111111010	FE80::/10
Site-local unicast	1111111011	FEC0::/10
Global unicast	001	2000::/3
Multicast	11111111	FF00:/8

TABLE 10.20

Generic Composition of a Global Unicast Address

001	TLA ID	Res	NLA ID	SLA ID	Interface ID
					Uniquely Identifies an Interface
	Public Domain			**Site Domain**	
3 Bits	13 Bits	8 Bits	24 Bits	16 Bits	64 Bits
Identifies global unicast IPv6 addresses	*Top-Level Aggregation ID (TLA ID)*—Identify long haul providers (the highest level in the IPv6 routing hierarchy) similar to network addresses, groups of TLAs are allocated by IANA to regional network information centers, which then allocates them to global ISP. There are a total of $2^{13} = 8192$ TLA ID	For future TLA or NLA ID expansion	*Next-Level Aggregation ID*—Identifies a regional site within a global ISP (a TLA ID)	*Site-Level Aggregation ID*—Employed internally by a site domain (regional), to identify subnetworks/ create its local addressing hierarchy	Uniquely identifies a network host. It uses the EUI-64 notation and is automatically generated from the MAC address

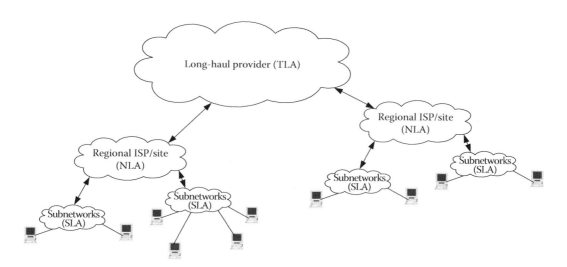

FIGURE 10.8 Hierarchical structure of IPv6 address space and routing.

The interface ID of a unicast address is 64 bit long and can be generated using the EUI-64 notation. It is directly generated from the NIC MAC address. As previously described, a MAC address is composed of 48 bits, grouped into six octets separated by colons, where each octet is denoted by two hexadecimal digits (e.g., 01:23:45:67:89:AB). The resulting EUI-64 address consists of the MAC address subject to the hexadecimal sequence *FFFE* added in the middle of the MAC address. Taking the previous MAC address as an example, the resulting EUI-64 address becomes 0123:45**FF:FE**67:89AB. Note that the interface ID groups four hexadecimal digits, instead of two, as considered by the MAC address. Using this conversion, a host obtains automatically an IPv6 address from the MAC address (being allocated in factory), without the need to obtain an address from a DHCP server.* The host needs to assure that this automatically generated interface ID is not in use by any other host in the network. This is achieved through broadcast queries. Remind that the network prefix needs to be added on the left of the interface ID.

The interface ID of an IPv6 address is kept unchanged, regardless of the network and subnetwork where the host is located. A global unicast IPv6 address within a site is identified by a total of 80 bits, from which 64 bits are used by the interface ID, and 16 bits are used by the site domain ID, that is, to create subnetworks. Moreover, since the complete IPv6 address consists of the concatenation of the network ID, subnet ID, and interface ID, all that a host needs to know in order to find out its complete IPv6 address is to discover the network and subnetwork ID. This can be obtained through advertisement with the neighboring routers. Consequently, a user/host that moves across different networks/subnetworks only changes the corresponding part of the IPv6 address, while keeping the interface ID unchanged. Moreover, since an IPv6 address is automatically obtained from the MAC address and from information obtained from neighboring routers, it is easily concluded that the IPv6 protocol does not need to use the ARP.

Table 10.21 shows the generic decomposition of a multicast address. As can be seen, the leftmost bits consist of a fixed group of eight bits with logic state 1 (ff in hexadecimal), followed by four bits (flags) that gives an indication about whether the multicast address is permanently or transiently assigned. The following four bits identify the scope, namely if the multicast address is only within a subnetwork, a network, a local ISP, a long-haul provider, or global. Finally, the 112 rightmost bits are used to create multicast groups. Note that a multicast group may coincide with the broadcast of a network, subnetwork, site, ISP, or global, but may also consist of a group of individual and specific hosts distributed along different networks, subnetworks, or ISPs.

It is worth noting that a URL that makes use of the IPv6 protocol uses a notation that consists of the IPv6 address inside brackets. As an example, the URL address http://[2123:0123:4567::89AB:CDEF]/test.html corresponds to accessing the file test.html, using the http, available from the IPv6 destination address 2123:0123:4567::89AB:CDEF.

10.2.2 IPv6 Packet

As previously described, in order to reduce the required processing by routers, translating in faster routing, the number of fields included in the IPv6 packet header was reduced, as compared to the IPv4 datagram header. Example of fields existing in the IPv4 that do not exist in the IPv6 mandatory

TABLE 10.21
Generic Composition of a Multicast Addresses

1111 1111	Flags	Scope	Group ID
8 Bits	**4 Bits**	**4 Bits**	**112 Bits**
It means that the IPv6 address is of multicast type	Identifies whether the multicast address is permanently or transiently assigned	Identifies the multicast address scope: interface-local, site-local, organization local, or global	Identifies a multicast group

---

* In some special situations, such autoconfiguration is not suitable. In such cases, the DHCP version 6 (DHCPv6) protocol may be used to configure the hosts.

fields include IHL, identification, flags, fragment offset, and header checksum. Nevertheless, the mandatory part of the IPv6 header is longer than that of the IPv4. This is due to the longer IPv6 addresses (128 bits), as opposed to the 32-bit size IPv4 addresses.

Figure 10.9 depicts the IPv6 header, with the mandatory part and the extension part. The mandatory header has a fixed length of 40 octets, being depicted in Figure 10.10, whereas the extension header is composed of fields necessary for handling special functionalities. The field lengths are depicted in Figure 10.10, expressed in number of octets. Note that some fields were moved from the (fixed part of the) IPv4 header into the extension part of the IPv6 header. This way, the mandatory part only contains the fields that are always to be read by all the intermediate nodes. The aim was

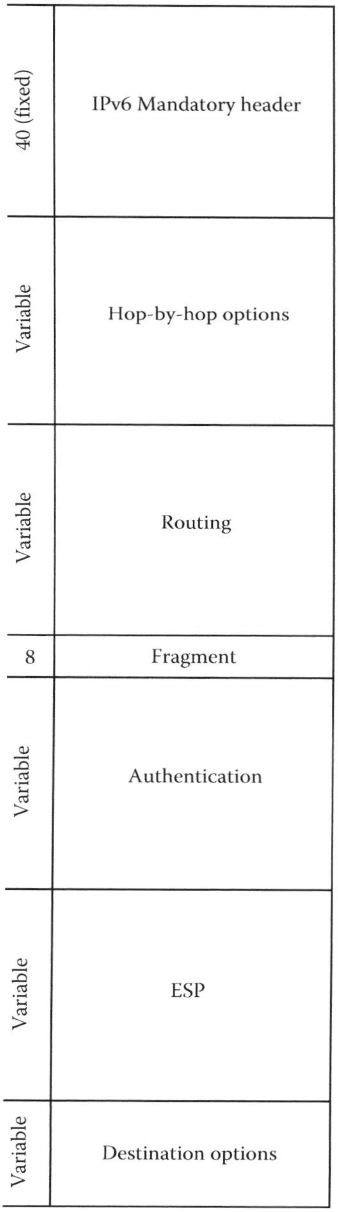

FIGURE 10.9 Generic decomposition of the IPv6 header.

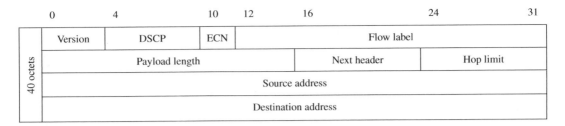

FIGURE 10.10   Mandatory part of IPv6 header.

to reduce the overhead, simplify the processing, and speeding up the routing. The extension header contains the following fields [RFC 4260]:

- *Hop-by-hop options*: It is used to carry special information required to be processed by intermediate routers, such as resource reservation protocol (RSVP) for IPv6 or to implement the jumbo payload. The payload has a size up to 64,000 octets without special options. A larger payload can be implemented using the jumbo payload option, when properly signalized in the hop-by-hop options field.
- *Routing*: It is used to implement source routing. The implementation of source routing partially transforms the IP switching from datagram method into virtual circuits method. The addresses of the intermediate nodes that compose the fixed part of the route are included in this optional field.
- *Fragment*: It is used to implement and manage the fragmentation and reassembling of packets.
- *Authentication*: It consists of a mechanism that provides protection from attacks against the authenticity and integrity of packets, as well as from anti-replay attacks. This is an option of the IPsec framework, as detailed in Chapter 16. Note that while the IPsec was optional in IPv4, the exchange of IPv6 data between routers is always implemented using the IPsec mechanisms.
- *ESP*: It stands for encapsulating security payload, and consists of a mechanism that provides protection from attacks against the confidentiality, authenticity, and integrity, as well as from anti-replay attacks. This is an option of the IPsec framework, as detailed in Chapter 16.
- *Destination options*: It consists of an optional information to be handled by the destination router.

Note that the IPv6 fragmentation is only implemented by end stations of an end-to-end connection, whereas IPv4 fragmentation is carried out by intermediate nodes (routers). In the case of the IPv6, intermediate routers never implement fragmentation. The exchange of IPv6 packets must be preceded by a path MTU discover that dictates the maximum fragment size transmitted. Consequently, as opposed to the IPv4, the fragment offset is not included in the mandatory part of the IPv6 header.

As can be seen from Figure 10.9, the mandatory part of the IPv6 header contains the following fields:

- *Version*: It identifies an IPv6 packet.
- *DSCP*: It is used for the provision of QoS by intermediate routers. These requirements include delay, jitter, packet loss, or throughput. Using the DiffServ option, packets with lower DSCP numbers have priority over those with higher DSCP numbers.

- *ECN*: It stands for explicit congestion notification, being used to explicitly notify nodes of congestion situations. The fields DSCP and ECN are also jointly referred to as traffic class.
- *Flow label*: It is used to signalize the packets belonging to a specific flow of data that require special attention by intermediate routers. This is used to preserve the required QoS. The flow label field allows prioritized handling by intermediate routers (e.g., for voice, audio, or video streaming packets).
- *Payload length*: It consists of a 16-bit size field and corresponds to the length of the extension headers plus the payload. This field varies between 0 and 65,535 octets. Note that the maximum payload is 64,000 octets without special options. The extension header hop-by-hop field allows implementing a larger payload entitled jumbo payload. Conversely, in case the data field length of the frame (MTU) is shorter than the maximum packet length, fragmentation is carried out, which needs to be properly signalized in the fragment field of the extension header.
- *Next header*: This is a pointer to the header of the upper layer protocol (TCP, UDP, and ICMP) carried in the packet payload. In case an extension header is used, this field points toward the first extension header. Note that the next header field also exists in the extension headers. In this case, this field points toward the following extension header or to the upper layer protocol (next header field of the last extension header).
- *Hop limit*: This field corresponds to the time to live field of an IPv4 header. It specifies the number of hops that a packet may be subject to. Each router decrements this field by one. When this field reaches zero, the packet is discarded.
- *Source address*: 128-bit address of the source host.
- *Destination address*: 128-bit address of the destination host.

The configuration of an interface with an IPv6 address is described in Section 10.3.2.3. Moreover, the configuration of a router with an IPv6 routing protocol and the configuration of routers using both IPv4 and IPv6 is described in Chapter 11.

10.3 CISCO INTERNETWORK OPERATING SYSTEM

Most of the Cisco system routers and switches make use of the Cisco IOS in order to implement their functions, as well as to allow the configuration of different parameters. Depending on the device version, there are different versions of Cisco IOS, with some differences among them. The current description considers the Cisco IOS included in Cisco 1800 series routers and in Cisco Catalyst 2900/2960 series switches.

These network devices have typically the following memories:

- Read-only memory (ROM), which is responsible for storing the power-on-self-test (POST), and the bootstrap instructions. Moreover, it also contains a scaled-down version of the Cisco IOS. The POST is used for testing the hardware. Old versions of Cisco IOS store the POST in the ROM.
- Flash memory, which consists of an electrically erasable programmable read-only memory (EEPROM), which is responsible for storing Cisco IOS images.
- Non-volatile RAM (NVRAM), which stores the startup configuration.
- Random access memory (RAM), which is responsible for temporary storing frames or packets (layer 2 or 3 switching, respectively) and for storing the Cisco IOS, as well as for storing the running configuration. The running configuration corresponds to the active

configuration where the changes in the configuration of the device are stored (e.g., the configuration of an interface with an IP address and the name given to the device). Finally, the ARP cache[*] is also stored in the RAM.

When a router or a switch is started up, it executes the following sequence of actions:

- It reads the POST from the ROM and executes the corresponding test, and loads the bootstrap program.
- It reads the Cisco IOS from the flash memory (EEPROM) and loads it into the RAM memory.
- It reads the startup configuration (startup-config) from the NVRAM and loads it into the RAM memory in the form of running configuration (running-config).

10.3.1 Introduction to Cisco IOS

The access or configuration of Cisco systems devices is performed using the command line interface (CLI) or using the setup configuration, through one of the following lines:

- Console port (CON)
- Auxiliary port (AUX)
- Telnet

The console and auxiliary lines are located in the posterior part of the Cisco routers and switches (see Figure 10.11), whereas the telnet is the way a user can remotely connect with the equipment through the network.[†] Moreover, while a router is a device that typically has two or three interfaces, a switch has typically 12, 24, or more interfaces. Moreover, in a switch, the interfaces are normally placed in the frontal part of the equipment. Contrarily, in a router, the interfaces are normally placed in the posterior part of the equipment.

Many communication applications can be used to allow communicating with a Cisco systems device. An important application is the Windows HyperTerminal. Nevertheless, HyperTerminal is no longer available in Windows 7 and 8. Therefore, the application PuTTY is currently widely employed for this purpose, with additional features relating to HyperTerminal, such as the ability to configure a Cisco device through telnet and secure shell (SSH) (in addition to the console port).

FIGURE 10.11   Console and auxiliary lines in a router.

---

[*] The ARP cache contains the IPv4 address to MAC address mappings.

[†] The user must type *telnet*, followed by the destination IP address of the host with whom the telnet is to be established. This is performed using the Windows command line screen (MS-DOS prompt).

The connectivity of a PC with a device through the CON or AUX needs to be configured with the following parameters (see Figure 10.12):

- Bit rate 9600 bps
- 8 Data bits
- No parity
- 1 Stop bit, no flow

The setup configuration only allows basic access and configuration of Cisco systems devices. On the other hand, the CLI* consists of the most important interface through which the operator, administrator, and manager may fully interact with Cisco systems devices. Nevertheless, the user must be familiar with the syntax used in the configuration. This syntax is covered in the remainder of this section.

Depending on the type of interaction with Cisco systems devices, the CLI allows the following three basic command modes:

- *User mode*: This is the default mode after having logged into the device. It allows the user to execute only basic commands relating to device user utilities (not administration utilities). This mode is identified by the symbol > in the prompt, following the router's name, as follows:
  - Router>
- *Privileged mode*: This is the mode that allows viewing the details of the running configuration and introducing minor changes to the running configuration (running-config). Moreover, the access into the configuration mode (defined in the following) is performed from the privileged mode. This mode is associated to the ability to execute device administrative utilities. After having entered into the user mode (default mode), the user must type the *enable* command to enter the privileged mode (see Figure 10.13). Once in the privileged mode, the symbol # appears in the prompt, following the router's name as follows:
  - Router#

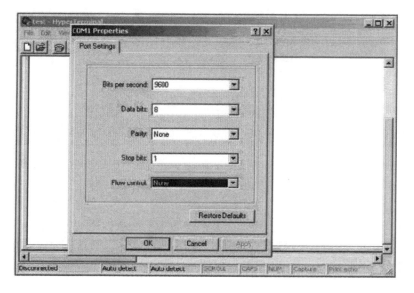

FIGURE 10.12   Connectivity parameters.

---

* In order to enter the CLI, one should respond *no* to the question: *Enter the initial configuration dialog?*

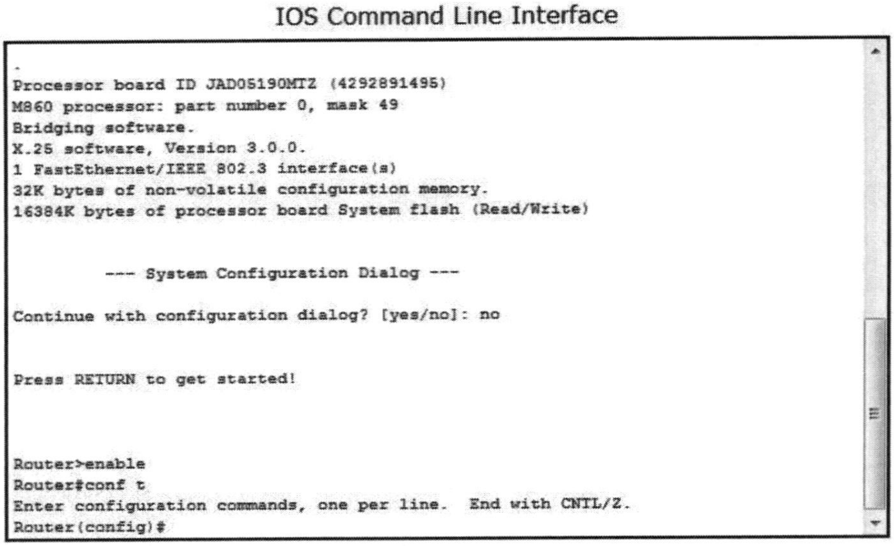

IOS Command Line Interface

```
Processor board ID JAD06190MTZ (4292891495)
M860 processor: part number 0, mask 49
Bridging software.
X.25 software, Version 3.0.0.
1 FastEthernet/IEEE 802.3 interface(s)
32K bytes of non-volatile configuration memory.
16384K bytes of processor board System flash (Read/Write)


        --- System Configuration Dialog ---

Continue with configuration dialog? [yes/no]: no


Press RETURN to get started!



Router>enable
Router#conf t
Enter configuration commands, one per line.  End with CNTL/Z.
Router(config)#
```

FIGURE 10.13   Entering into the privileged and configuration modes.

- In order to restrict the access to the privileged mode, which gives full access to the device, a password protection can be configured. The privileged mode can be accessed through either the CON line or telnet, whereas the auxiliary line only gives access to the user mode. The privileged mode can be left by issuing the command *disable*, and the CLI can be left by issuing the command *logout*. Then, the Cisco IOS syntax becomes, respectively:
  - Router#disable
  - Router#logout
- *Configuration mode*: This mode allows modifying the running configuration. To enter into this mode, one has to first enter the privileged mode, after which one must type the *configure terminal* command (see Figure 10.13). This mode is identified by the characters (config)# in the prompt, following the router's name. The Cisco IOS syntax used to enter the configuration mode is as follows:
  - Router#configure terminal

To save the inserted parameters into the running configuration one needs to enter the sequence of keyboard keys control and Z (CTRL-Z) or to type the *exit* command.

The configuration mode includes two different submodes, namely:

- *Interface configuration*: It allows configuring the interfaces, namely the interface IP address, interface description, and so on. This submode is identified by the characters (config-if)#, following the router's name (see Figure 10.14). The Cisco IOS syntax required to enter the interface configuration submode becomes
  - Router(config)#interface *interface* (*interface* stands for the interface identification)
    - *Example*: Router(config)#interface Ethernet 0/0
- *Line configuration*: It allows configuring the lines used to gain access to the device (console [CON], auxiliary [AUX], telnet [vty]). The configuration of an access password necessary to allow using a certain line is an example of a line configuration parameter (see Figure 10.15). This submode is identified by the characters (config-line)#, following the router's name. The syntax required to enter the line configuration submode becomes

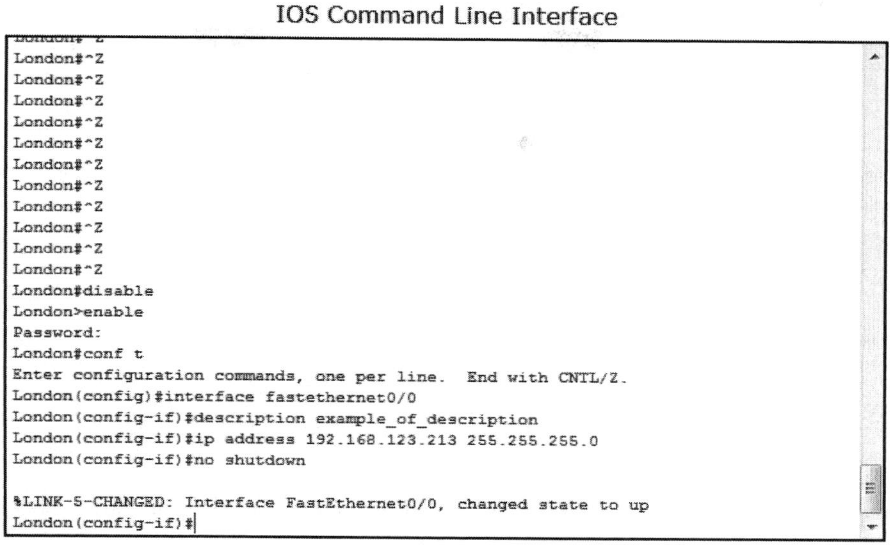

IOS Command Line Interface

```
London#^Z
London#^Z
London#^Z
London#^Z
London#^Z
London#^Z
London#^Z
London#^Z
London#^Z
London#^Z
London#^Z
London#^Z
London#disable
London>enable
Password:
London#conf t
Enter configuration commands, one per line.  End with CNTL/Z.
London(config)#interface fastethernet0/0
London(config-if)#description example_of_description
London(config-if)#ip address 192.168.123.213 255.255.255.0
London(config-if)#no shutdown

%LINK-5-CHANGED: Interface FastEthernet0/0, changed state to up
London(config-if)#
```

FIGURE 10.14 Configuring IP address.

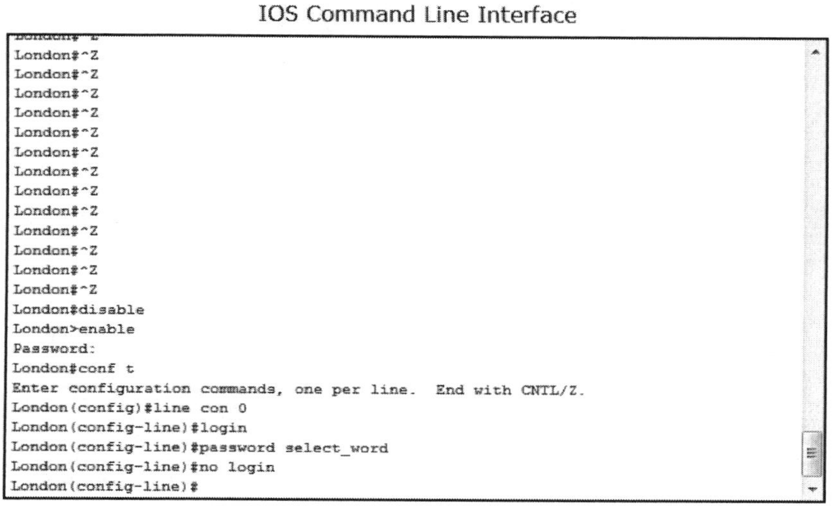

IOS Command Line Interface

```
London#^Z
London#^Z
London#^Z
London#^Z
London#^Z
London#^Z
London#^Z
London#^Z
London#^Z
London#^Z
London#^Z
London#^Z
London#^Z
London#disable
London>enable
Password:
London#conf t
Enter configuration commands, one per line.  End with CNTL/Z.
London(config)#line con 0
London(config-line)#login
London(config-line)#password select_word
London(config-line)#no login
London(config-line)#
```

FIGURE 10.15 Configuring a line.

- Router(config)#line *line* (*line* stands for the line identification, such as con, aux, or vty)
  - *Example*: Router(config)#line console 0

Before we start configuring the router, it is important to make sure that the loaded configuration is not more than the one we really want. Therefore, it is a good rule to *clean* the startup configuration, and then reload the router with the *cleaned* startup configuration, before we start modifying the running configuration. This is performed, from the configuration mode, using the following syntax (two command lines):

- Router(config)#erase startup-config
- Router(config)#reload

The second command makes the router restarting, using the *cleaned* startup configuration, instead of any unknown startup configuration, and loading it as initial running configuration. Moreover, once the router's configuration is finalized, it is a good rule to save such running configuration as startup configuration. This is performed using the following syntax:

- Router(config)#copy running-config startup-config

Alternatively, this can be performed through

- Router(config)#copy run start

A network administrator may also copy the running-configuration (or startup-configuration) into a trivial file transfer protocol (TFTP) server. This is achieved by issuing

- Router(config)#copy running-config TFTP

Then the server asks for the IP address of the TFTP server where the file should be saved, and for the name that should be given to the file. The reverse operation can be performed with the following syntax:

- Router(config)#copy TFTP running-config

The previous command line can also be applied to the startup-configuration, instead of the running configuration, by replacing running-config by startup-config.

It is worth noting that the help command is always available from the CLI, when in the privileged or configuration modes, just by issuing (see Figure 10.16):

- Router(config)#?

Alternatively, one may issue a sequence of characters that corresponds to the initial letters of a command, followed by"?," as follows:

- *Example*: Router(config)#cl?

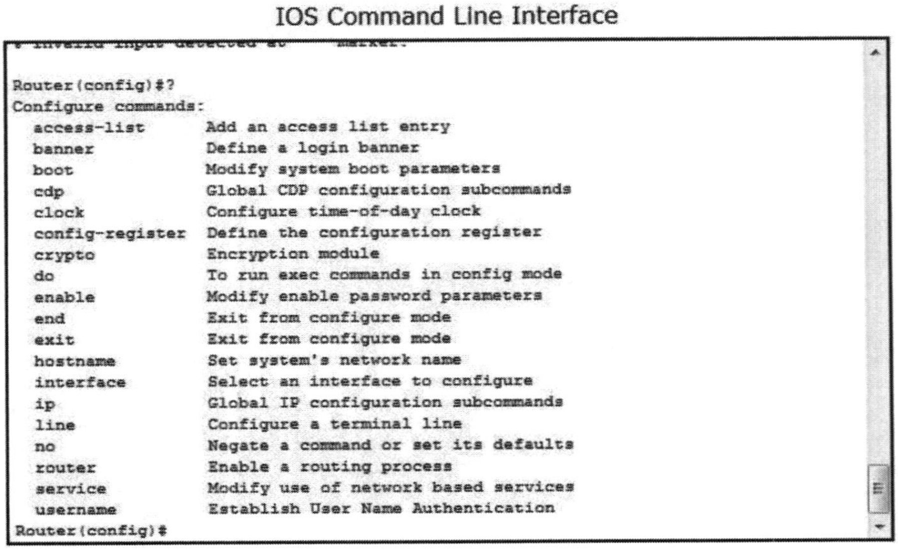

IOS Command Line Interface

```
% Invalid input detected at      marker.

Router(config)#?
Configure commands:
  access-list      Add an access list entry
  banner           Define a login banner
  boot             Modify system boot parameters
  cdp              Global CDP configuration subcommands
  clock            Configure time-of-day clock
  config-register  Define the configuration register
  crypto           Encryption module
  do               To run exec commands in config mode
  enable           Modify enable password parameters
  end              Exit from configure mode
  exit             Exit from configure mode
  hostname         Set system's network name
  interface        Select an interface to configure
  ip               Global IP configuration subcommands
  line             Configure a terminal line
  no               Negate a command or set its defaults
  router           Enable a routing process
  service          Modify use of network based services
  username         Establish User Name Authentication
Router(config)#
```

FIGURE 10.16 Help command.

The prompt lists the commands initiated by the written characters. With these inserted characters, the following list of commands appears on the screen: *clear* and *clock*. Moreover, the command TAB can be used to complete commands (initiated by a sequence of characters).

10.3.2 BASIC CONFIGURATION OF CISCO ROUTERS AND SWITCHES

As previously described, the configuration of the Cisco systems devices involves modifying the running configuration. The modification of the running configuration is possible using the configuration mode, which is accessed by issuing *configure terminal** from the privileged mode.

10.3.2.1 Configuration Mode

Once in the privileged or configuration mode, before entering any of the configuration submodes, some basic parameters can be configured, namely

- From the privileged mode, one can set the time and date using the syntax (see Figure 10.17):
  - Router#clock set HH:MM:SS DD MON YR
    - *Example*: Router#clock set 10:59:59 25 NOV 2014
- From the configuration mode, one can configure an access password to the privileged mode. This can be either a ciphered or a clear text password, and can be configured, respectively, with the syntax (see Figure 10.17):
  - Router(config)#enable **secret** *password* (*password* stands for the ciphered password)
  - Router(config)#enable **password** *password* (*password* stands for the clear text password)
- From the configuration mode, one can configure the router's name by issuing (see Figure 10.17):
  - Router(config)#hostname *hostname* (*hostname* stands for the router's name)

IOS Command Line Interface

```
Mode processor: part number 0, mask 49
Bridging software.
X.25 software, Version 3.0.0.
1 FastEthernet/IEEE 802.3 interface(s)
32K bytes of non-volatile configuration memory.
16384K bytes of processor board System flash (Read/Write)


        --- System Configuration Dialog ---

Continue with configuration dialog? [yes/no]: no


Press RETURN to get started!



Router>enable
Router#clock set 10:08:00 30 JAN 2013
Router#configure terminal
Enter configuration commands, one per line.  End with CNTL/Z.
Router(config)#enable secret secret_word
Router(config)#hostname London
London(config)#
```

FIGURE 10.17 Setting the time and date.

---

* Note that the Cisco IOS allows using abbreviated commands. In this context, instead of issuing *configure terminal*, the user may just issue *conf t* to enter the configuration mode from the privileged mode. The abbreviated command should be such that there is no other command started with the inserted abbreviation.

- From the configuration mode, one can configure a banner message of the day (MOTD), using the syntax:
  - Router(config)#banner motd #*message*#
    - *Example*: Router(config)#banner motd #Only authorized personnel can have access to the router's privileged or configuration mode#
- From the configuration mode of a router, one can disable DNS lookup table by issuing the syntax:
  - Router(config)#no ip domain-lookup

10.3.2.2   Line Configuration Submode

From the configuration mode, one can enter the line configuration submode, using the syntax:

- Router(config)#line *line* (*line* stands for the line identification namely console line, auxiliary line, or telnet [in the case latter case, five telnet lines are normally configured, from 0 to 4])
  - *Example*: Router(config)#line console 0

The most frequent configured item, in any of the line submodes, is the access password. In order to configure an access password to enter the line *line*, the following sequence of syntaxes must be issued (see Figure 10.15):

- Router(config)#line *line*
- Router(config-line)#login (makes the system requesting a line login password)
- Router(config-line)#password *password* (*password* stands for the password)
  - *Example*:
    - Router(config)#line con 0
    - Router(config)#login
    - Router(config)#password secret_word

In order to not request a login password, one must issue:

- Router(config-line)#no login

Finally, if one intends to disable a certain line access, the following syntax must be used:

- Router(config)#line *line*
- Router(config-line)#login
- Router(config-line)#no password
  - *Example*:
    - Router(config)#line vty 0 4 (disables the telnet access using lines 0–4)
    - Router(config)#login
    - Router(config)#no password

10.3.2.3   Interface Configuration Submode (IPv4 and IPv6)

From the configuration mode, one can enter the interface configuration submode using the syntax:

- Router(config)#interface *interface* (*interface* stands for the interface identification, such as Ethernet, fast Ethernet, and serial, with the slot and port where the interface is installed in the router)
  - *Example*: Router(config)#interface ethernet0/1

Some of the parameters that can be configured in the interface submode include:

- Configuring an interface description (see Figure 10.18):
  - Router(config-if)#description *description* (*description* is a description)
    - *Example*: Router(config-if)#description Interface to Amsterdam

- Configuring an interface with a static IP address* is performed with the following syntax (see Figures 10.18 and 10.19)[†]:
  - Router(config-if)#ip address *int_IP_address subnet_mask* (*int_IP_address* and *subnet_mask* stand for the IP address to assign to the interface and its subnet mask)

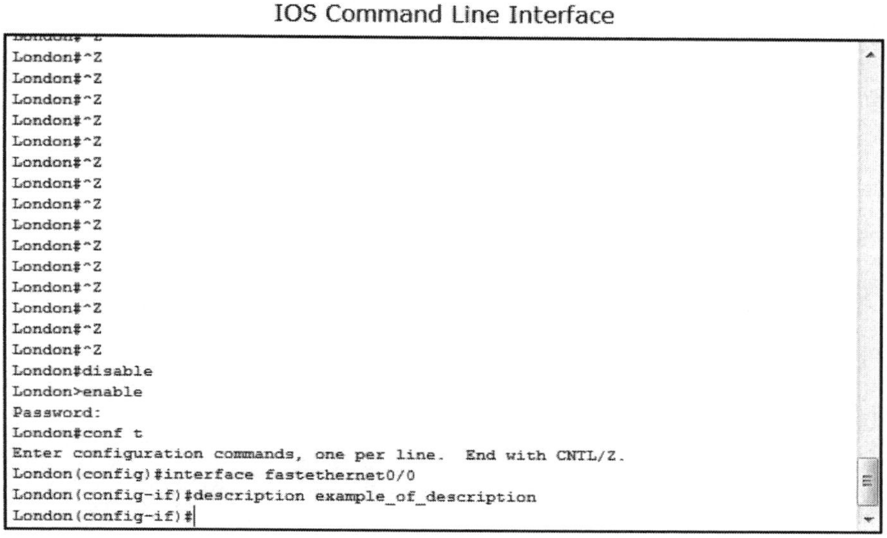

IOS Command Line Interface

```
London#^Z
London#^Z
London#^Z
London#^Z
London#^Z
London#^Z
London#^Z
London#^Z
London#^Z
London#^Z
London#^Z
London#^Z
London#^Z
London#^Z
London#^Z
London#disable
London>enable
Password:
London#conf t
Enter configuration commands, one per line.  End with CNTL/Z.
London(config)#interface fastethernet0/0
London(config-if)#description example_of_description
London(config-if)#
```

FIGURE 10.18 Configuring an interface description.

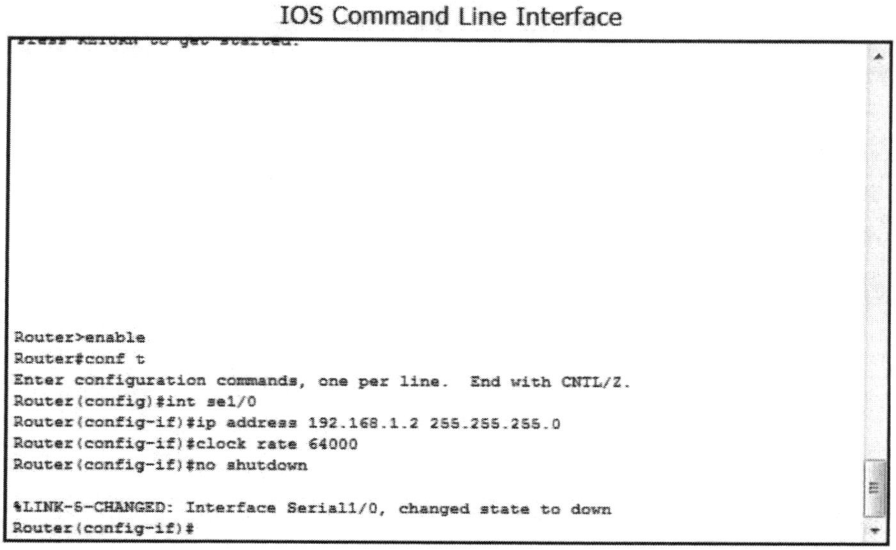

IOS Command Line Interface

```
Press RETURN to get started.




Router>enable
Router#conf t
Enter configuration commands, one per line.  End with CNTL/Z.
Router(config)#int se1/0
Router(config-if)#ip address 192.168.1.2 255.255.255.0
Router(config-if)#clock rate 64000
Router(config-if)#no shutdown

%LINK-5-CHANGED: Interface Serial1/0, changed state to down
Router(config-if)#
```

FIGURE 10.19 Configuring an interface with IP address and clock rate.

---

* This configuration does not apply to switches.
[†] The configuration of a dynamic IP address in an interface with the DHCP procedure is described in the following.

- Router(config-if)#no shutdown (to activate the interface)
  - *Example*:
    - Router(config-if)#ip address 191.123.213.10 255.255.0.0
    - Router(config-if)#no shutdown
- When a serial cable* is used to interconnect two routers, the configuration must consider one router acting as data terminal equipment (DTE) and the other acting as data communication equipment (DCE).† The DCE is the device responsible for sending a synchronism clock to the DTE, required for the communication between these devices. For the communication to be possible, the router acting as DCE needs to send to synchronism clock to the router acting as DTE. This is achieved using the syntax (see Figure 10.19):
  - Router(config-if)#clock rate *clock_rate* (*clock_rate* stands for the clock rate expressed in bps)
    - Router(config-if)#clock rate 32000
- Finally, the interface configuration can be checked by using one of the syntaxes‡:
  - Router#show interface (allows verifying configuration of all interfaces)
  - Router#show interface *interface* (allows verifying configuration of the interface *interface*)
    - Router#show IP interface etherhet0/0
  - Router#show IP interface brief (shows a brief configuration of all interfaces, as shown in Figure 10.20)
  - Router#show running-configuration (or simply as *show run*, which shows the full configuration of the running-configuration)
  - Router#show running-configuration | section interface (or simply as *show run | s interface*, which shows the section interface of the running-configuration; naturally the word *interface* is an example and any other configuration word can be used, such as *access-list*, *line*, etc.; note that the command | does not work in the packet tracer simulator, only in real Cisco equipment)
  - Router#show running-configuration | include interface (or simply as *show run | i interface*, which shows the part of the running-configuration that includes interface)

IOS Command Line Interface

```
Router#^Z
Router#^Z
Router#^Z
Router#^Z
Router#^Z
Router#^Z
Router#show ip interface fastethernet0/0
FastEthernet0/0 is administratively down, line protocol is down (disabled)
   Internet protocol processing disabled
Router#show ip interface brief
Interface              IP-Address      OK? Method Status                    Protocol

FastEthernet0/0        unassigned      YES manual administratively down down

Serial1/0              192.168.1.2     YES manual down                      down

Serial1/1              unassigned      YES manual administratively down down

Serial1/2              unassigned      YES manual administratively down down

Serial1/3              unassigned      YES manual administratively down down
Router#
```

FIGURE 10.20   Command showing IP interface brief.

---

* The HDLC protocol is widely used as the data link layer protocol over serial cables to interconnect two routers.
† An example of a DCE is a modem, whereas a workstation acts as a DTE.
‡ Note that the show commands are always issued from the privileged mode, not from the configuration mode.

Finally, in order to verify an ARP table on a router (mapping between IP addresses and MAC addresses), the following syntax can be used:

- Router#show arp

Similarly, the command *arp -a*, typed in the Windows command line screen (MS-DOS prompt) of a workstation, allows verifying the workstation's ARP table.

When configuring a Cisco router with IPv6, the first action relies on activating IPv6 traffic-forwarding. This is performed in configuration mode by issuing the following syntax:

- Router(config)#ipv6 unicast-routing

Then, the router's interfaces must be configured with IPv6 addresses. As described above, this can be statically configured (manually configured) or using the DHCP autoconfiguration. In the former case, an administrator may manually insert a full IPv6 address, using the following syntax:

- Router(config)#interface *interface* (*interface* stands for the interface identification; this is used to enter the interface configuration submode)
- Router(config-if)#ipv6 address *ipv6_address/prefix*
  - *Example*:
    - Router(config)#interface ethernet0/1
    - Router(config-if)#ipv6 address 2002:123:1234:2767::8CAF/64

Alternatively, one may simply insert the global routing prefix and subnetwork ID, and declare that the interface ID is to be automatically computed from the MAC address using the EUI-64 notation, as follows:

- Router(config)#interface *interface*
- Router(config-if)#ipv6 address *routing_prefix/prefix* eui-64
  - *Example*:
    - Router(config)#interface ethernet0/1
    - Router(config-if)#ipv6 address 2002:123:1234:2767::/64 eui-64

When issuing EUI 64, the router knows that it has to compute the full IPv6 address from both the network prefix (network part of the address) and from the MAC address, that is, the router knows that the part of the IPv6 address that corresponds to the interface ID is to be computed from the MAC address, using the EUI 64 notation. Assuming that the MAC address is 01:23:45:67:89:AB, the resulting interface ID becomes 0123:45**FF:FE**67:89AB. This can be verified with the command *show ip interface et0 / 1*.

Finally, an IPv6 interface configuration can be checked by using the command *show ipv6 interface, show ipv6 interface* interface, or *show ipv6 interface brief*. Note the similarity between these commands and those utilized in IPv6. The difference relies on modifying the command *IP* by *IPv6*.

10.3.3 DYNAMIC HOST CONFIGURATION PROTOCOL

The assignment of an IP address to a host can be either static or dynamic. In the former case, the network administrator uses the network settings of a host to assign an IP address. Nevertheless, assigning IP addresses statically does not allow using such addresses when those hosts are switched off, or disconnected from the network. A more efficient mechanism relies on the use of DHCP [RFC 2131]. The DHCP is another example of a protocol that contributes to achieve a better usage of the depleted IPv4 address space. The DHCP assigns dynamically IP addresses to hosts. Most of the home routers run the DHCP server, but a computer server can also be employed for this task. The DHCP allows a DHCP server (router or computer) dynamically assigning IP addresses to different LAN workstations. The assigned IP addresses can be either private or public, depending on the specific environment and use. The DHCP is also normally used by ISPs, such that every time a client powers on a router, it receives a new IP address. Note that, similar to the NAT and the PAT, the dynamic modification of IP addresses performed by the DHCP also represents an added value from the security point of view.

When a host connects to a network, by default it assigns to itself the IP address 0.0.0.0. Then, it contacts the DHCP server, requesting the assignment of an IP address. The communication between a client and a DHCP server comprises UDP messages (see Figure 10.21):

1. The client broadcasts the message DHCPDISCOVER to port 67, using 0.0.0.0 as its own IP address, and meaning that it is requesting the assignment of an IP address.
2. The DHCP server responds with the unicast message DHCPOFFER to port 68 that includes the following fields (the DHCP server creates a new entry into its ARP table with the new assigned IP address and the MAC address of the client):
   a. The IP address assigned to the host and the corresponding subnet mask.
   b. The address of the DNS server that should be used by the host.
   c. The IP address of the default gateway that should be used by the host.
   d. The lease period (the period over which the client can use the assigned IP address).
3. The client broadcasts the message DHCPREQUEST to port 67, asking the DHCP server to verify that the received IP address is correct, that is, not corrupted by errors, and that it had really been assigned by the DHCP server.
4. The DHCP server responds with the unicast DHCPACK message to port 68, confirming that the received data is correct, or correcting it in case of error.

The IP address of the default gateway corresponds to the IP address of the router to which all traffic whose recipient is not an internal host should be forwarded to.

It is finally worth noting that some type of hosts should use static addressing (not dynamic). This is the case of servers, network printers, or routers. If the server IP address were dynamic, how could an external client know its IP address to access it? Naturally, the DNS server could map the different IP addresses into the domain name. But what if the DNS server used dynamic addressing? As a rule of thumb, the network administrator should assign static IP addresses to servers, network printers, routers, and switch management IP addresses, as these types of hosts do not change either physically or logically.

Before DHCP was employed, there was another protocol entitled bootstrap protocol (BOOTP), but that present few differences related to the DHCP. The BOOTP assigns addresses statically, while the DHCP assignment is dynamic. The BOOTP assignment is permanent, while the DHCP

FIGURE 10.21   Four-stage handshaking procedure of DHCP.

assignment is periodic. Finally, the number of BOOTP configuration fields is limited to 4, while the DHCP fields include 20 parameters.

10.3.3.1   DHCP Configuration

The configuration of a Cisco router as a DHCP server is performed with the following syntax:

- Router(config)#ip dhcp excluded-address *lowest_IP_address* [*highest_IP_address*] (addresses to be excluded from the DHCP pool)
- Router(config)#ip dhcp pool *pool_name*
- Router(dhcp-config)#network *IP_address* [mask | /prefix-length] (identifies a pool of addresses started from *IP_address* plus [mask | /prefix-length], except the excluded IP addresses declared above)
- Router(dhcp-config)#default-router default_router_*IP_address*
- Router(dhcp-config)#dns-server *dns_server_IP_address*
  - *Example*:
    - Router(config)#ip dhcp excluded-address 192.168.1.1
    - Router(config)#ip dhcp excluded-address 192.168.1.253 192.168.1.254
    - Router(config)#ip dhcp pool IP_POOL
    - Router(dhcp-config)#network 192.168.1.0 255.255.255.0
    - Router(dhcp-config)#default-router 192.168.1.254
    - Router(dhcp-config)#dns-server 192.168.1.253
    - Router(dhcp-config)#end

The DHCP server is activated by default. In case one wants to disable it, this can be performed with *no service dhcp*, being re-activated with *service dhcp*.

In case a host requesting a DHCP address is not in the same network as the DHCP server, the intermediate routers need to forward DHCP request and responses. This is configured by inserting the IP address of the DHCP server after the command *ip helper-address* in the interface from where the DHCP messages are initially received, followed by the IP address of the DHCP server, as follows (see Figure 10.22, where router 1 is the Router_intermediate and where *interface* is Fa0/0):

- Router_intermediate(config)#interface *interface*
- Router_intermediate(config-if)#ip helper-address *dhcp_server_IP_address*
  - *Example*:
    - Router 1(config)#interface fa0/0
    - Router 1(config-if)#ip helper-address 10.1.1.2

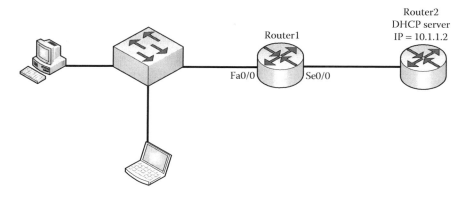

FIGURE 10.22   DHCP forwarding.

The addresses assigned by the DHCP server, and the corresponding MAC addresses, can be verified using the following syntax:

- Router#show ip dhcp binding

The addresses that belong to a DHCP pool can be verified using the syntax:

- Router#show ip dhcp pool

A client host of a Cisco systems device using Cisco IOS can also be configured with the DHCP address, instead of with a static IP address. This is performed using the following syntax:

- Router(config)#interface *interface*
- Router(config-if)#ip address dhcp
- Router(config-if)#no shutdown
  - *Example*:
    - Router(config)#int fa0/0
    - Router(config-if)#ip address dhcp
    - Router(config-if)#no shutdown

In case a wireless access point is utilized, the access point DHCP (used to assign IP addresses to wireless hosts) should be disabled. With the wireless access point DHCP disabled, even the hosts connected through wireless medium obtain IP addresses directly from a DHCP server, in accordance with the procedure previously described, instead of using the DHCP of the wireless access point.

10.3.4 Network and Port Address Translation

Due to the rapid depletion of the IPv4 address space, several mechanisms had to be developed in order to optimize the unallocated address space. The classless mode of IPv4 address scheme was one of these mechanisms. Other mechanisms include the NAT, the PAT, as well as the DHCP.

It was already described that private address ranges can freely be used within local area networks, without any permission from the Network Information Center, as long as these IP addresses are not used in the Internet world. Table 10.4 shows the private address ranges for different IPv4 address classes. A workstation, located in a LAN with Internet access, may have either a public IP address or a private IP address, as long as the router performs the address translation from a private into a public (for outgoing datagrams).* This address translation can be implemented using one of the two following methodologies:

- *NAT*: The router that interconnects a LAN with a MAN or a WAN, for Internet access, is running the NAT application, which is responsible for mapping private into public addresses. Depending on the number of simultaneous workstations that need to have access to the Internet, a single or multiple public IP addresses are used by the NAT. The NAT changes the source IP address in every outgoing datagrams, whereas the destination IP address are changed in every incoming datagrams. The NAT procedure is depicted in Figure 10.23.
- *PAT*: The router that interconnects the LAN with a MAN or WAN for Internet access, is running the PAT application, being responsible for mapping private IP addresses into a single public IP address. The PAT changes the source IP address of a datagram from a private address into a unique public IP address. Moreover, the PAT changes the outgoing segment's source port into one out of a range of ports used by the PAT router in the Internet world. Consequently, although the outgoing datagrams from different workstations present

---

* And from a public into a private (for incoming datagrams).

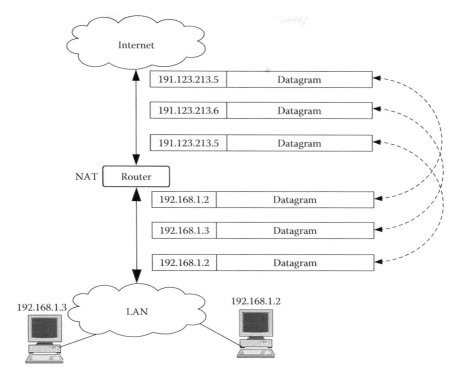

FIGURE 10.23 Network address translation.

the same source IP address, they are differentiated by different source ports. The opposite operation is performed by the PAT for incoming datagrams. Consequently, for incoming traffic, the PAT converts different destination ports in segments (and a single public IP address) into the different private IP addresses used by different LAN workstations.[*] The PAT procedure is depicted in Figure 10.24. The PAT is also referred to as the *NAT overload*.

It is worth noting that, very often, routers run simultaneously the NAT and the PAT protocols. In this case, instead of having a single public IPv4 address, the router has two or more addresses, as well as a range of port addresses to use in the Internet world. This concept is commonly referred to as the NAT overload.

Although allowing a better address space usage, these protocols also prevent the IP addresses from being known from the Internet world.[†] This represents an added value from the security point of view.

10.3.4.1 Dynamic NAT and PAT Configuration

For the sake of saving addresses, hosts located inside private domains may use private (local) addresses. In this case, the address translation between a private and a public address needs to be carried out by the router (see Figure 10.25) for packets that travel to outside of the private domain (and the reverse operation for packets that travel in the opposite direction).

---

[*] After this conversion, since the LAN works at layer 2, the router consults the ARP table to find the MAC address that corresponds to the destination's private address. In case such MAC address is not part of the ARP table, a broadcast ARP packet is sent to the LAN in order to identify the MAC address that corresponds to a certain IP address (this procedure is known as ARP protocol). In this case, such new entry is added to the ARP table.

[†] This is normally used to avoid a server's IP address (e.g., a Web server) from being known from the Internet.

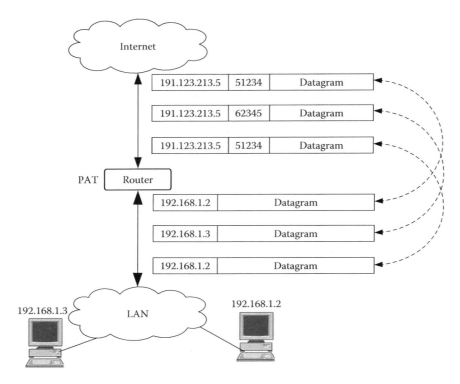

FIGURE 10.24   Port address translation.

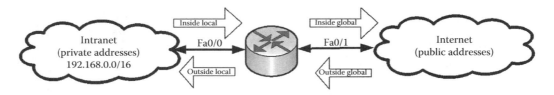

FIGURE 10.25   Two possible address translations: private (local) to public (global) or public to private.

In case of clients, the traditional address translation is the dynamic NAT/PAT. In this situation, there is no rigid translation between a private and a public address. Such translation varies from time to time. In this case, the public address that a certain host presents in the Internet world varies. Note that the private IP addresses are also normally assigned dynamically to clients using DHCP.

Regardless of whether it uses static or dynamic address translation, an intranet host that accesses the Internet has typically a private address (inside local) and a public address (inside global). Similarly, an Internet host has a public address (outside global) and, optionally, may also be identified inside the intranet by a private address (outside local).

The configuration of dynamic address translation is carried out in four different steps, which are as follows:

1. Define a pool of addresses to use in the NAT translation (not used in case of static address translation).
2. Create an access control list (ACL) to identify which packets are subject to address translation (in case of static address translation, this can be omitted).

3. Activate the address translation.
4. In the router that performs the address translation, identify the interface that is linked to the internal area (inside) and the one linked to the external area (outside).

The definition of the pool of addresses to use in the NAT translation is performed using the following syntax:

- Router(config)#ip nat pool *pool_name pool_start pool_end* [netmask *mask* | prefix-length *bits*]
  - *Examples*:
    - Router(config)#ip nat pool public_addresses 193.1.2.3 193.1.2.10 netmask 255.255.255.0
    - Router(config)#ip nat pool public_addresses 193.1.2.3 193.1.2.10 prefix-length 24

The creation of an ACL used to identify the packets that are subject to address translation can be defined as a standard or extended ACL. In this chapter we briefly describe standard ACL. The reader should refer to Chapter 16 for a detailed description of standard and extended ACLs. A standard ACL simply allows packet filtering based on the source IP address. The syntax of a standard ACL is as follows:

- Access-list *list_number* [**permit** | deny] *source_address* [*wild_card_mask*]
  - *Example*: Access-list 10 permit 192.168.0.0 0.0.255.255
  - *Example*: Access-list 20 permit host 11.20.1.2

Naturally that, in the scope of NAT/PAT, the deny option of a standard ACL is not employed, as the aim of the ACL creation is to define the range of source IP addresses that are subject to address translation. Note that the standard ACLs use a *list_number* within the range 1–99. Nevertheless, extended ACLs may also be employed (see Chapter 16).

The third step comprises the activation of the address translation. This is performed using the following syntax:

- Router(config)#ip nat [inside | outside] source list [*acl_name* | *acl_number*] pool *pool_name* [overload]
  - *Examples*:
    - Router(config)#ip nat inside source list outgoing_acl_nat pool public_addresses
    - Router(config)#ip nat inside source list outgoing_acl_nat pool public_addresses overload

Finally, the identification of the area where the router's interfaces are connected to is configured using the following syntax:

- Router(config)#interface *interface*
- Router(config-if)#ip nat [inside | outside]
  - *Example*:
    - Router(config)#interface FastEthernet 0/0
    - Router(config-if)#ip nat inside
    - Router(config-if)#interface FastEthernet 0/1
    - Router(config-if)#ip nat outside

It is worth noting that inside address translation in the outgoing direction is always performed by the router's interface connected to the internal network (defined by the command *ip nat inside* applied to the internal interface). On the other hand, inside incoming address translation is always performed by the router's interface connected to the Internet (defined by the command *ip nat outside* applied to the external interface).

Focusing on the network depicted in Figure 10.25, assuming inside translation, the complete dynamic address translation configuration, for the NAT case, becomes

> Router(config)#ip nat pool public_addresses 193.1.2.3 193.1.2.10 netmask 255.255.255.0
>   (step 1 of NAT)
> Router(config)#ip access-list extended outgoing_acl_nat (step 2 of NAT)
> Router(config-ext-nacl)#permit ip 192.168.0.0 0.0.255.255 any (this example uses an extended
>   ACL)
> Router(config-ext-nacl)#exit
> Router(config)#ip nat inside source list outgoing_acl_nat pool public_addresses (step 3 of
>   NAT)
> Router(config)#interface fastEthernet 0/0 (step 4 of NAT)
> Router(config-if)#ip nat inside
> Router(config-if)#interface fastEthernet 0/1
> Router(config-if)#ip nat outside
> Router(config)#Ctrl-Z

The overload command that can be added at the end of the command activation line allows the address translation using the NAT overload, that is, using multiple global ports, while overlapping one or more global IP addresses (corresponding to multiple local IP addresses).

The above examples focus on the inside translation, that is, from the intranet to the Internet. As previously described, the outside address translation is facultative. Remind that outside translation corresponds to the translation between a public address (outside global) of an Internet host into a private address (outside local).

It is worth noting that, by simply configuring the inside address translation (outgoing translation), the source IP address (and eventually the source port address, in case PAT is being performed) of outgoing packets is subject to a translation from a private into a public address, while the destination IP address (and eventually the destination port address) of incoming packets is also subject to a translation, in this case from a public into a private address.

In case the NAT overload is implemented with a single global IP address, this operation is referred to as PAT. This situation can also be applied to the same scenario of Figure 10.25. Nevertheless, a single inside global IP address is utilized, instead of a pool. Commonly, instead of assigning an inside global IP address, the PAT simply uses the IP address configured in the external interface. In the scenario depicted in Figure 10.25, the external interface corresponds to Fastethernet 0/1.

The activation of the address translation using PAT is performed with a different syntax, and the definition of a pool of addresses is not required.

- Router(config)#ip nat [inside | outside] source list [*acl_name* | *acl_number*] interface *int_type_name* overload
  - *Example*:
    - Router(config)#ip nat inside source list outgoing_acl_nat interface fa0/1 overload

For the same scenario of the network depicted in Figure 10.25, but assuming dynamic address translation using PAT (with a single inside global address), the complete configuration becomes

> Router(config)#access-list 9 permit ip 192.168.0.0 0.0.255.255 (a standard and numbered ACL
>   can be utilized, instead of an extended and named ACL)

Router(config-ext-nacl)#exit
Router(config)#ip nat inside source list 9 interface fastEthernet 0/1 overload
Router(config)#interface fastEthernet 0/0
Router(config-if)#ip nat inside
Router(config)#interface fastEthernet 0/1
Router(config-if)#ip nat outside
Router(config)#Ctrl-Z

10.3.4.2   Static NAT Configuration

When a private domain (e.g., an intranet) hosts a server that provides a service to the Internet world, the public IP address of such server needs to be fixed and known from the Internet. The solution typically relies on allocating a fixed private IP address to the server, and using static address translation (between a fixed private and a fixed public address). Note that, in this scenario, PAT may also be implemented (also referred to as *port forwarding*).

In the configuration of static address translation, step 1 (pool definition) and step 2 (ACL definition) of dynamic NAT procedure described above are omitted, being defined when activating the address translation. Concerning the activation of the static address translation process, the following syntax is employed:

- Router(config)#ip nat [inside | outside] source static [*protocol_type*] inside_local [port_local] inside_global [port_ global]
  - *Example*: Router(config)#ip nat inside source static 192.168.1.1 193.12.13.14
  - *Example*: Router(config)#ip nat inside source static **tcp** 192.168.1.1 **8080** 193.12.13.14 **80** (this is also referred to as port forwarding because it forwards traffic from one port to another)

Note that the insertion of the protocol type is facultative. This is only required when static PAT is to be performed. This can be seen from the second example above.

Focusing on the network depicted in Figure 10.26, assuming that static address translation is to be applied to the intranet server with the private IP address 192.168.1.1 and port address 8080 into the global IP address 193.12.13.14, and port address 80, the complete configuration becomes

Router(config)#ip nat inside source static tcp 192.168.1.1/8080 193.12.13.14/80 (activates address translation)
Router(config)#interface fastEthernet 0/0
Router(config-if)#ip nat inside

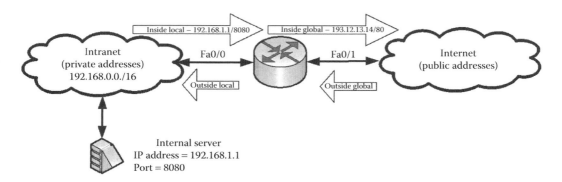

FIGURE 10.26   Example of a network where static address translation should be employed.

Router(config-if)#interface fastEthernet 0/1
Router(config-if)#ip nat outside
Router(config)#Ctrl-Z

Once the static and dynamic address translations have been configured, one can verify the state of address translations with the following syntax (from the privileged mode):

• Router#show ip nat translations

Moreover, one can verify the address translations statistics using the following syntax:

• Router#show ip nat statistics

CHAPTER SUMMARY

This chapter provided a view about the addressing and configuration of the Internet layer.

The IPv4 was described. The classful addressing of IPv4 was introduced, consisting of the basic addressing IPv4 format. Since the classful addressing of IPv4 is very inefficient, and due to the rapid growth of the number of Internet hosts, the available address space is depleted. The measure that can mitigate this limitation relies on making use of the IP address in classless mode, having also been studied in this chapter.

Then, the VLSM was also introduced, consisting of a type of classless addressing where, depending on the required number of hosts, different subnetworks can make use of different subnet masks.

The IPv4 datagram format was studied, including the description of each of the fields. The depletion of the IPv4 address spacing was the most important motivation for the development of the IPv6, having been studied in this chapter, comprising a 128-bit address, instead of the 32-bit address considered by the IPv4. The IPv6 packet format was studied, including the description of each of the fields.

The Cisco IOS was also studied in this chapter, including an introduction to configuration of Cisco routers and switches, namely the different modes of operation. The line configuration and interface configuration submodes were described, including the configuration of an IP address in an interface, for both IPv4 and IPv6.

The DHCP was described, consisting of a mechanism that also contributes to the use of the depleted IPv4 address spacing in a more efficient way. Moreover, the DHCP configuration using Cisco IOS was also studied in this chapter.

Finally, the address translation was studied, including the dynamic NAT, the static NAT, and the PAT (also referred to as the NAT overload). The address translation also contributes to the use of the depleted IPv4 address spacing in a more efficient way. Finally, the configuration of static and dynamic NAT and PAT was studied in this chapter.

REVIEW QUESTIONS

1. What is the broadcast address of the subnetwork where the host with the IP address 10.96.2.33/12 is?
2. What are the advantages of the IPv6, relating to the IPv4?
3. What does an APIPA address stands for? In which cases can it be used?
4. What is the network address where the host with the IP address 194.123.3.4 is located (classful mode)?
5. Why are IPv6 routing tables shorter than IPv4?

6. What is the NAT/PAT used for?

7. What is the DHCP used for?

8. What is the complete IPv6 address corresponding to 3D0F:00F3::5A3C?

9. What are the advantages of the IPv4 address space in classless mode relating to the classful mode?

10. For which purposes is Cisco IOS *enable secret* command (followed by a given password) used for?

11. What are the mechanisms that can be used to maximize the limited address space made available by the IPv4 protocol?

12. Using EUI-64 notation, what is the interface ID of the IPv6 address corresponding to an NIC with the MAC address 02:13:45:67:98:BA?

13. In which cases must the clock rate be configured in a router?

14. Given the network address 10.0.0.0, with subnet mask 255.255.192.0, present the subnetwork addresses, as well as the broadcast addresses.

15. What is the Cisco IOS command *no shutdown* used for?

16. What is the most important reason that motivated the evolution of the IP from IPv4 to IPv6?

17. What is the address of the second subnetwork, considering the network IP address 126.12.0.0, with the subnet mask 255.255.255.192 (classless mode)?

18. Given the network address 193.123.10.0/27, present the subnetwork addresses, as well as the broadcast addresses.

19. What is the class of the IP address 193.136.235.1 (classful mode)?

20. Given the network address 10.0.0.0, with subnet mask 255.255.255.192, present the subnetwork addresses, as well as the broadcast addresses.

21. What is the network and subnetwork part of the IPv6 address 31CF:0123:4567:89AB:CDEF:0123:4567:89AB/64?

22. How many networks can the class A IPv4 address accommodate (classful mode)?

23. Given the network address 10.0.0.0, with subnet mask 255.192.0.0, present the subnetwork addresses, as well as the broadcast addresses.

24. How many hosts can be allocated in a classful class B network address?

25. What are the subnet masks possible to be used with a class B network address (express the result using the bit count notation)?

26. Consider the class C network with the address 193.136.235.0/27. Choose the correct statement:
    a. Eight subnetworks can be created.
    b. The address of the first subnetwork is 193.136.235.32.
    c. The number of hosts per subnetwork is 62.
    d. All previous responses are correct.

27. Given the EUI-64 address (interface ID) 01:23:45:FF:FE:67:89:AB, of an IPv6 address, what is the corresponding MAC address?

28. What are the functionalities of the IP?

29. What is the subnetwork address to which the host with the IP address 10.88.1.2/11 belongs?

30. Consider a host with the MAC address 01:23:45:67:89:AB. What is the corresponding interface ID of the IPv6 address, assuming it is generated using EUI-64 notation?

31. How many hosts can be allocated in a subnetwork where the host with the IP address 10.76.1.2/11 is located?

32. How many bits compose an IPv6 address?

33. Why is IPv6 routing faster than IPv4?

34. How in an IPv6 address composed?

35. What is the DHCP used for?

36. Consider the network address 192.168.1.0, from where three subnetworks are intended to be created, respectively, with 16, 2, and 12 hosts. Determine the address spacing, taking into account the VLSM methodology.
37. Consider the network address 10.20.0.0, from where three subnetworks are intended to be created, respectively, with 12, 2, and 250 hosts. Determine the address spacing, taking into account the VLSM methodology.

LAB EXERCISES

1. Download and install the free network analyzer Wireshark. Open the application in a PC and select the interface connected to the Internet. You will see the IP datagrams that are being exchanged by the NIC of the PC, with the source IP address, destination IP address, and protocol type, alongside other information. Select an outgoing datagram. What is the source IP address? Is it a private or a public IP address? And what is the destination IP address? In a window that appears at the bottom, visualize the content of the IP, including the different header fields. In case you find IPv6 packets being exchanged by the NIC, select one of these and inspect the header field. Check the difference between the IPv4 and IPv6 packets.
2. Considering the network depicted in the figure below, assign addresses to hosts, using the classless mode.

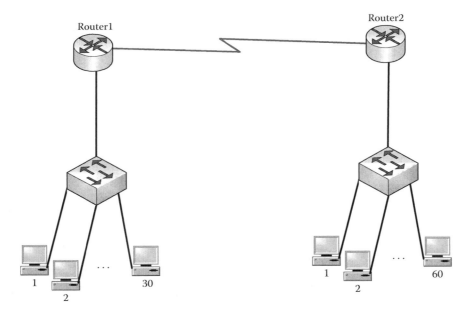

3. Considering the network depicted in the previous figure, assign addresses to hosts, using the address 11.12.13.0/24, and using VLSM.
4. Consider the network depicted above. After having assigned addresses to hosts using VLSM (as requested by the previous exercise), configure interfaces with IP addresses using the Packet Tracer simulator (use Cisco IOS in routers).
5. Repeat the previous exercise with real network equipment (reduce the number of computers).
6. Consider the network depicted in the figure below. Using the Packet Tracer simulator, configure interfaces with IP addresses, as plotted. Before starting, *clean* the startup configuration, and then re-load the routers with the *cleaned* startup configuration. In both routers, configure a ciphered access password to the privileged mode and another to the console port, a hostname, and a MOTD.

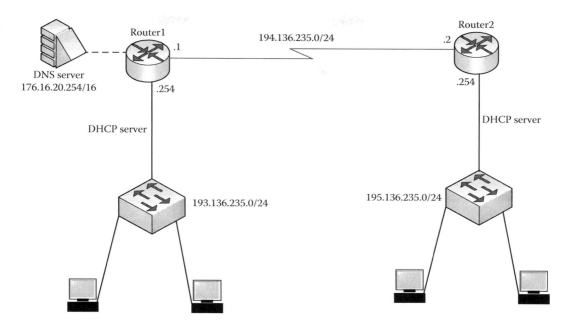

7. Repeat the previous exercise with real network equipment (reduce the number of computers).
8. Considering the network depicted in the figure below, configure dynamic NAT and static translations for the intranet and DMZ. Use the inside global addresses in the range 11.0.0.0/24 for the intranet and 12.0.0.0/24 for the DMZ.

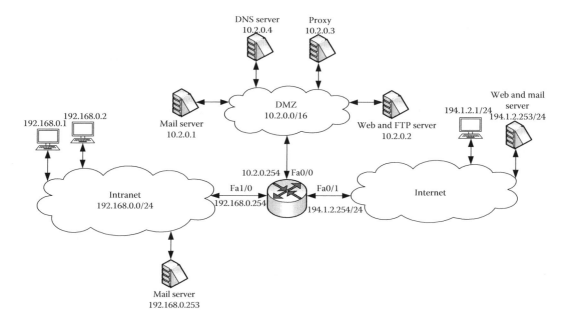

9. Repeat the previous exercise using NAT overload, instead of regular dynamic NAT.

11 Internet Layer
Routing and Configuration

LEARNING OBJECTIVES

- Describe the routing algorithms and protocols used in TCP/IP.
- Configure IPv4 and IPv6 routing protocols in Cisco equipment using Cisco IOS.
- Identify the difference between a distance vector routing protocol and a link-state routing protocol.
- Describe the Internet control message protocol (ICMP).
- Define fragmentation and reassembling.
- Describe the possible migration mechanisms used during IPv4 to IPv6 transitions.
- Describe the Cisco discovery protocol (CDP), and its utilization with the Cisco IOS.

Routing is used to make datagrams reach their destinations, after passing through several intermediate routers (nodes). Routers perform layer 3 switching (selection of the output interface to forward a datagram) based on the destination IP address and based on information contained in routing tables. A routing table gives information about the output interface to use in order to forward datagrams to a certain destination network.

The Internet is organized in clouds that are interconnected. As can be seen from Figure 11.1, an autonomous system (AS) can be viewed as a cloud and is composed of several LANs, MANs, and WANs, belonging to an organization (e.g., an ISP or a system of corporate LANs). An AS comprises a number of networks under the administrative control of a single entity that presents a common routing policy to the Internet.

Within an AS, routers forward datagrams among different LANs using interior gateway protocols (IGP), which consist of a group of protocols used to generate entries into routing tables in order to allow forwarding datagrams inside the organization's network (AS). Similarly, exterior gateway protocols (EGPs)[*] consist of a group of protocols used to generate entries into routing tables in order to allow forwarding datagrams between different ASs. A router without connection with the exterior of an AS only uses one or more IGPs. A router that connects with other routers of the same AS and with routers of different ASs[†] needs to use both IGP and EGP.

The interconnection between ASs has similarities with the interconnection of different LANs. Both can be viewed as clouds. Nevertheless, while LANs within an organization are interconnected using IGP-type protocols, the EGP-type protocols are used to interconnect different ASs. Moreover, while switches are used within LANs to switch traffic between different hosts (layer 2 switching), interior gateway routers (without connection with the exterior of an AS) are used to switch traffic between different interior LANs (layer 3 switching).

Very often, in order to make a datagram reach a certain destination, there is the need to cross multiply different ASs. Therefore, the different ASs need to communicate, such that they update their routing table entries.

---

[*] The border gateway protocol (BGP) is an example of an EGP. Nevertheless, EGP also refers to a protocol itself, not only to a family of (external) protocols.
[†] A router that connects to other routers of different ASs is referred to as a border router.

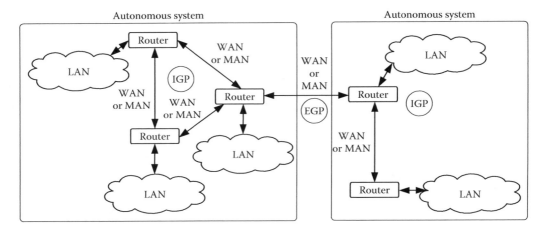

FIGURE 11.1 AS and interior and exterior gateway protocols.

Figure 11.2 shows an example of an IPv4 routing table. The different IP addresses shown in Figure 11.2 are class B, but the prefix /24 corresponds to class C. Consequently, we may conclude that the device (a host or a router) that has this routing table is using the classless addressing mode. From the second line, we see that the network 174.168.235.0, with subnet mask 255.255.255.0 (prefix /24), is directly connected (C on the left) through a serial interface with the device. The network 175.168.235.0/24 is directly connected through a fast Ethernet interface, whereas the network 176.168.235.0/24 is directly connected through a serial interface. From the first line, it is shown that the network 173.168.235.0/24 can be reached via other device (with the IP address 174.168.235.1, which belongs to the network 174.168.235.0 and which, from the second line, is directly connected to the router that has this routing table). The first line also gives us information that the remote device with IP address 174.168.235.1 can be reached through a serial interface (Serial1/0). Therefore, we conclude that this is a remote network (not directly connected). Moreover, the letter O means that the routing protocol used to generate this entry of the routing table was the open shortest path first (OSPF). Routing protocols are used to allow devices to create routing table. Finally, from the last line, it is shown that the remote network 177.168.235.0/24 can be reached via other device (176.168.235.2), and that the protocol OSPF is being used to generate this routing table entry. The information inside brackets [110/1562] is as follows: 110 is the administrative distance (AD), and it corresponds to the OSPF protocol[*]; 1562 is the metric distance (or cost), which corresponds to a value used by routers to make decisions about the path to use in order to make the datagrams reach the destination. AD is a value allocated to each routing protocol or mechanism that translates its reliability. The lower an AD is, the more reliable it is. In case there is more than one path to reach a certain destination, the router

O   173.168.235.0/24 [110/782] via 174.168.235.1, 00:00:00, Serial1/0

C   174.168.235.0/24 is directly connected, Serial1/0

C   175.168.235.0/24 is directly connected, FastEthernet0/0

C   176.168.235.0/24 is directly connected, Serial1/1

O   177.168.235.0/24 [110/1562] via 176.168.235.2, 00:00:00, Serial1/1

FIGURE 11.2 An example of an IPv4 routing table.

---

[*] Each routing protocol has a different AD.

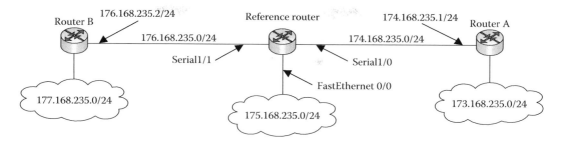

FIGURE 11.3 The network topology deducted from the routing table depicted in Figure 11.2.

TABLE 11.1

Longest Match Operation Performed during Routing Table Lookup (Refers to Routing Table in Figure 11.2)

Longest Match Performed in Routing Operation

Destination address of packet	177.168.235.2	**10110001.10101000.11101011**.00000010
Route 1	173.168.235.0/24	**1010**1101.10101000.11101011.00000000
Route 2	174.168.235.0/24	**1010**1110.10101000.11101011.00000000
Route 3	175.168.235.0/24	**1010**1111.10101000.11101011.00000000
Route 4	176.168.235.0/24	**101100**00.10101000.11101011.00000000
Route 5	177.168.235.0/24	**10110001.10101000.11101011**.00000000

Note: The bold bits correspond to those that match with the destination IP address.

---

selects the route with the lower administrative and/or metric distance, as defined in the following. Figure 11.3 shows the network topology that can be deducted from this routing table.

Table 11.1 shows the way routers perform routing table lookup, which is based on the longest match operation. A router compares the destination IP address of a packet against the multiple destination networks existing in the routing table (only the network part of the destination addresses is considered). The selected route corresponds to the one from which the destination network achieves a longest match with the destination IP address of the packet. Note that this comparison is performed in binary notation. From the example shown in Table 11.1, it is viewed that route 5 is the one that has the longest match, as there are a total match of 24 bits between the destination IP address of the packet and the destination network of the routing table. Then, according to Figure 11.2, the selected route is via 176.168.235.2 (IP address of the next router), being accessible through the router's output interface Serial1/1.

Routing algorithms are normally grouped into the following two main categories:

- Nonadaptive algorithms
- Adaptive algorithms

11.1 ADMINISTRATIVE AND METRIC DISTANCES

In case there is more than one path to a certain destination, the router selects the path (by this order)

1. With the lower AD* (listed in Table 11.2).
2. With the lower metric distance (in case there is more than one route for the same destination with the same AD that is calculated with the same protocol).

---

* A lower AD corresponds to a more reliable method used to calculate the route.

TABLE 11.2
List of Administrative Distances

Protocol	Administrative Distance
Directly connected Interface	0
Static route	1
EIGRP (internal routes)	90
IGRP	100
OSPF	110
IS-IS	115
RIP	120
EGP	140
EIGRP (imported routes)	170
Unknown source	255

The metric distance (also referred to as *cost*) consists of a value that ranks each path to a certain destination. The preferable path (preferred route) has the lowest metric distance and is included in the routing table, whereas the undesirable path has the highest metric distance. Finally, in case there is more than one path with the same lowest metric distance, equal cost load balancing is implemented, that is, the traffic is split into these routes with equal metric distances. The parameters taken into account in the calculation of the metric distance depends on the routing protocol.

Figure 11.4 depicts an example with two paths between the nodes A and D. There is one path through node B, whose calculation is performed with the OSPF protocol, and another path through node C, whose calculation is performed with the routing information protocol (RIP). According to the rule (1), the node A forwards the data through the node B because the OSPF protocol presents a lower AD (110) than the (RIP) (120).

In the example of Figure 11.5, both paths are calculated with the OSPF protocol, and the route through node B presents a cost (or metric distance) of 20, whereas the route through node C presents a cost of 11. Then, according to rule (2), the selected route is through node C, because this node presents the lowest metric distance.

It is worth noting that all dynamic routing protocols (e.g., OSPF, RIP, interior gateway routing protocol [IGRP], enhanced interior gateway routing protocol [EIGRP], intermediate system to intermediate system [IS-IS]) allow equal cost load balancing. This means that in case there are

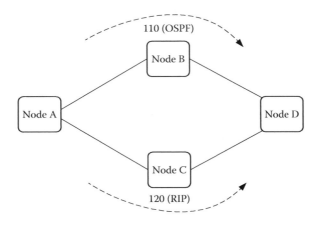

FIGURE 11.4 Administrative distances and routing priorities.

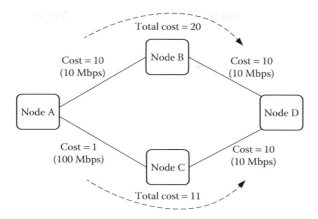

FIGURE 11.5   Metric distances (costs) and routing priorities using the OSPF protocol.

multiple paths to a certain destination network whose administrative and metric distance are the same, the traffic is equally split between the multiple paths. This means that odd packets follow one path, whereas even paths follow the other path. In this case, the routing table contains the single destination network, but shows multiple exit interfaces, one for each equal cost path. The protocols OSPF and RIP support up to four equal cost load balancing. Moreover, EIGRP and IGRP allow unequal cost load balancing. In this scenario, in case there are multiple paths to the same destination network, but with different metric distances, load balancing is still performed. In this case, the split of traffic takes into account the metric distance, that is, it sends more packets through the distance with lower metric distance. Load balancing leads to an improvement of network performance, and is automatically activated if the routing table has multiple paths to a destination.

11.2   ROUTE SUMMARIZATION

Using adaptive routing protocols, a router advertises to its neighbors the network addresses directly connected to it and those advertised by neighbors of neighbors. Note that a router does not advertise host addresses, but network addresses. In case classless mode is employed, a router may also advertise subnetwork addresses, instead of full network addresses. Without route summarization, a router that receives such advertisements inserts a different entry into its routing table corresponding to each network (or subnetwork) advertised by its neighbor.

Let us focus on Figure 11.6. Without router summarization, Router1 would have four different entries into its routing table corresponding to the remote networks 171.10.129.0/24, 171.10.130.0/24, 171.10.131.0/24, and 171.10.132.0/24. These entries correspond to the routes advertised by Router2. In order to reduce the length of routing tables, and faster routing table lookup, Router2 may perform route summarization, and simply advertise the result of such route summarization, instead of the four independent networks. As can be seen from Figure 11.6, the summary route is 171.10.128.0/21. Consequently, Router1 reduces the size of its routing table and enables a faster routing processing. As can be seen from Figure 11.6, route summarization consists of route aggregation, and of advertising a contiguous set of addresses as a single address with shorter subnet mask. Using classless adaptive routing protocols (e.g., RIPv2, EIGRP, OSPFv2), the classful boundaries are ignored, and summarization is permitted with masks that are less than that of the default classful mask.

As can be seen from Figure 11.6, to compute a summary route, one should initially list the networks in binary format (highlighted bits correspond to those that are common and accessible through the route that performs the advertisement). Then, count the number of leftmost bits that are the same. From the depicted example, we see that there are a total of 21 common leftmost bits, corresponding to the subnet mask for the summary route /21, or 255.255.248.0. Then, add zero value

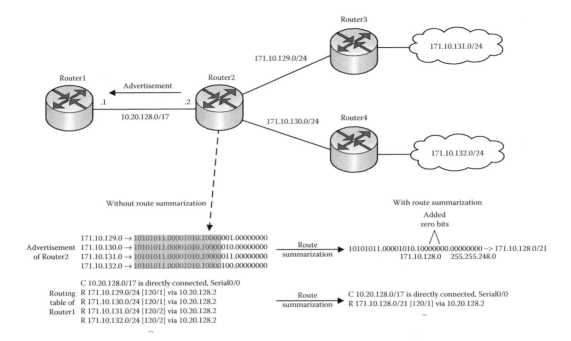

FIGURE 11.6 Route summarization.

bits to the position of the remaining rightmost bits. The resultant summary route is 171.10.128.0/21, which should be advertised by Router2 to Router1, and thus, reducing the length of the routing table of Router1. It is worth noting that the summary route 171.10.128.0/21 also includes other routes beyond those desired, namely the route 171.10.133.0/24, 171.10.134.0/24, and 171.10.135.0/24, but this is not a problem, as those networks are not in the neighborhood.

Another term widely used in networking is supernet. A supernet is a summary route whose length is less than the corresponding classful. An example of a supernet is the address 171.128.0.0/9. Note that the corresponding classful address has a subnet mask /16, whereas this supernet has a subnet mask /9.

11.3 STATIC ROUTING AND FLOODING

It was described that routing algorithms can be grouped into two categories: adaptive and nonadaptive routing algorithms. Moreover, in the case of nonadaptive algorithms, these can be of two types: static or flooding algorithm.

With nonadaptive algorithms the entries into the routing tables are not calculated using network topologies such as traffic or bandwidth metrics. The selection of the path to utilize is previously computed, and pre-loaded into the router.

A static entry into the routing table consists of an action manually performed by the network administrator. Routers have knowledge about the networks connected to it. Therefore, one must only configure static routes to remote networks, that is, to networks not directly connected to the router that is being configured. Although different paths may exist for datagrams to reach a certain destination, using a static entry, the network administrator is forcing traffic to follow a certain fixed path. One disadvantage results from the fact that, in case such path is suddenly interrupted (or subject to bad conditions), the traffic is not automatically switched over to another path. Using static route, such re-routing of the traffic can only be performed through human intervention, that is, changing the required entry into the routing table. In this case, the traffic is interrupted during a certain time period. The reader should refer to Sections 11.3.1 and 11.3.3 for a detailed description of the configuration of static routing into Cisco routers.

The default route, also referred to as gateway of last resort, comprises a single static route to where a router should forward all packets whose destination IP address is outside its own internal network. When a router is only connected to another router, the default route configuration is typically utilized. A common situation is when the router only has two interfaces: one connected to a LAN and another interface connected to another router. In this case, all packets received from the LAN, whose destination is not another host outside the LAN, should be forwarded to the other router (default route).

It is worth noting that routing decision is performed based on the longest match. This means that the destination IP address of a packet is compared against the different network addresses existing in the router's routing table. In case a static route is configured, and there is not any other network in the routing table that presents a longer match, the packet is forwarded to the static route. In this scenario, if the static route was not configured, the packet would be dropped, as there is no existing network that matches the destination IP address of the packet.

Flooding is another nonadaptive algorithm, through which a router forwards an input datagram into all different outputs. Consequently, the datagrams will reach the destination with redundancy (the same datagram will be repeated), and through all possible paths. Using the flooding algorithm, it is assured that one of the paths is the optimum one. An important disadvantage results from the fact that the network becomes overloaded with many replicas of each datagram, which may degrade its performance.

Besides the configuration of lines and interfaces, there is the need to configure static or dynamic routing protocols in routers. This is necessary for a router to be able to create and update routing tables and, consequently, to be able to forward packets. Moreover, workstations and switches also require some level of configuration.

The configuration of a workstation is relatively simple and normally consists of configuring the IP address (which can be static or using a DHCP server), as well as the IP address of the default gateway.*

With regard to the switch, which is normally the central node of a LAN, a configuration is required in case VLANs are to be created. Moreover, if a switch is to be remotely controlled, an IP address must be assigned to its management VLAN. Otherwise, only basic parameters, such as the device's name and description are configured, if desired.

11.3.1 STATIC ROUTE CONFIGURATION

Routers have knowledge about the networks connected to them. Therefore, one must only configure static routes to remote networks, that is, to networks not directly connected to the router that is being configured.

The example of configuration provided in this subsection refers to the configuration of Router1, of the internetwork example depicted in Figure 11.7.

As can be seen from Figure 11.8, the configuration of a static route is performed using the following syntax:

- Router(config)#ip route *dest_address subnet_mask next_router_address* (*dest_address* is the network IP address of the destination to be reached, *subnet_mask* is the corresponding subnet mask, and *next_router_address* is the IP address of the interface of the next router to where the packets should be forwarded to, in order to reach the destination network).
  - *Example*: Router(config)#ip route 193.123.213.0 255.255.255.0 192.125.215.4

---

* The default gateway consists of a router to where all the packets generated in the workstation should be forwarded to.

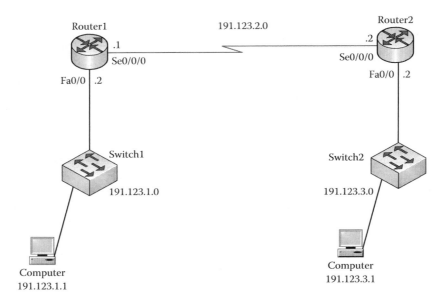

FIGURE 11.7 Example of an internetwork.

```
                  IOS Command Line Interface
Router#^Z
Router#^Z
Router#^Z
Router#^Z
Router#^Z
Router#^Z
Router#^Z
Router#disable
Router>enable
Router#conf t
Enter configuration commands, one per line.  End with CNTL/Z.
Router(config)#int fa0/0
Router(config-if)#ip address 191.123.1.2 255.255.255.0
Router(config-if)#no shutdown
Router(config-if)#exit
Router(config)#int se0/1/0
Router(config-if)#ip address 191.123.2.1 255.255.255.0
Router(config-if)#no shutdown
Router(config-if)#exit
Router(config)#ip route 191.123.3.0 255.255.255.0 191.123.2.2
Router(config)#^Z
%SYS-5-CONFIG_I: Configured from console by console
Router#
```

FIGURE 11.8 Configuring a static route.

Naturally, these commands must be inserted as many times as the number of remote networks to be statically configured.

It is worth noting that the router's decision about to which interface should a packet be forwarded to requires consulting twice the routing table: with the first, the IP address of the next router is extracted; then the router again consults the routing table to verify which exit interface gives access to the network where the IP address of the next router is located. This can be viewed from the routing table plotted in Figure 11.2 where, from the last row, it is viewed that the network 177.168.235.0/24 is accessible via 176.168.235.2. Then, at a second stage, from the penultimate row, it is viewed that the network 176.168.235.2/24 (where the next router with the IP address 176.168.235.2 is located) is accessible via the exit interface serial1/1. In this particular case, the exit interface could also be extracted from the last row. Nevertheless, in many situations, this is not the case.

In case of point-to-point connections, a static route can also be configured, in a more efficient manner, by using the following syntax:

- Router(config)#ip route *dest_address subnet_mask interface* (*interface* is the identification of the exit interface of the router being configured that should be used to forward packets to the destination remote network).
  - *Example*: Router(config)#ip route 193.123.213.0 255.255.255.0 Serial0/0/0

In this case, we have replaced the IP address of the next router, by the exit interface of the router under configuration that should be used to send packets. With this procedure, we have avoided to consult the routing table a second time to verify which exit interface gives access to the network where the next router is placed. Nevertheless, it is important to keep in mind that the procedure of inserting static routes with the exit interface is only possible in case of serial interfaces (point-to-point), while not possible in case of multiaccess networks (e.g., Ethernet). In case of multiaccess networks, the router needs to obtain the address of the next router, because the ARP procedure needs to be implemented in order to obtain the MAC address corresponding to the address of the next router.[*] In an Ethernet network, if the static route were configured with the exit interface (e.g., Ethernet 0/0) without the next hop address, the router would not have any way to obtain the MAC address (the IP packet is encapsulated into an Ethernet frame, with the destination MAC address of the next hop router). In the case of a multiaccess network, one can configure simultaneously with the IP address of the next hop and with the exit interface. This also speeds up the routing table lookup process, being configured with the following syntax:

- Router(config)#ip route *dest_address subnet_mask next_router_address interface*
  - *Example*: Router(config)#ip route 193.123.213.0 255.255.255.0 191.123.213.2 et0/1

An entry into a routing table can be removed by using the following syntax:

- Router(config)#**no** ip route *dest_address subnet_mask next_router_address*

11.3.2 FLOATING ROUTE CONFIGURATION

Let us now focus on the internetwork depicted in Figure 11.9, where there exists a main route between Router1 and Router2, and a floating route (i.e., a backup route) between these two routers. The configuration of a static floating route is performed with the following syntax:

- Router(config)#ip route *dest_address subnet_mask* [*interface* | *next_host_address*] **floating_admin_dist**
  - *Example*: Router(config)#ip route 193.123.213.0 255.255.255.0 Serial0/0/1 **254**

For a route to be kept floating (static), the AD of the floating route needs to be higher than the main route. Note that routes with lower AD are chosen by precedence. The AD of a static route is 1. Therefore, one can configure a floating route with an AD as *floating_admin_dist* 254 (255 corresponds to an unreachable route). Configuring a floating route with 254 allows it being used as a backup route to all dynamic routing protocols, as their ADs are always lower than 254 (see Table 11.2).

Alternatively, Router1 and Router2 could be configured with two dynamic routing protocols. In such situation, the route calculated using the dynamic routing protocol with the lower AD would be

---

[*] Note that a packet is encapsulated into a MAC frame with a source and destination MAC address.

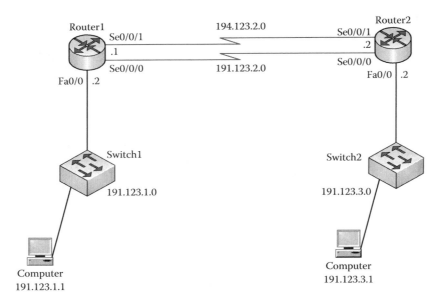

FIGURE 11.9 Example of an internetwork with a floating route.

utilized as the main route, whereas the route calculated with the dynamic routing protocols that has a higher AD would be left as a floating route.

11.3.3 DEFAULT ROUTE CONFIGURATION

In case a router should forward all the packets to a single other router, the required syntax is as follows (see Figure 11.10):

- Router(config)#ip route 0.0.0.0 0.0.0.0 *next_router_address*
  - *Example*: Router(config)# ip route 0.0.0.0 0.0.0.0 192.125.215.4

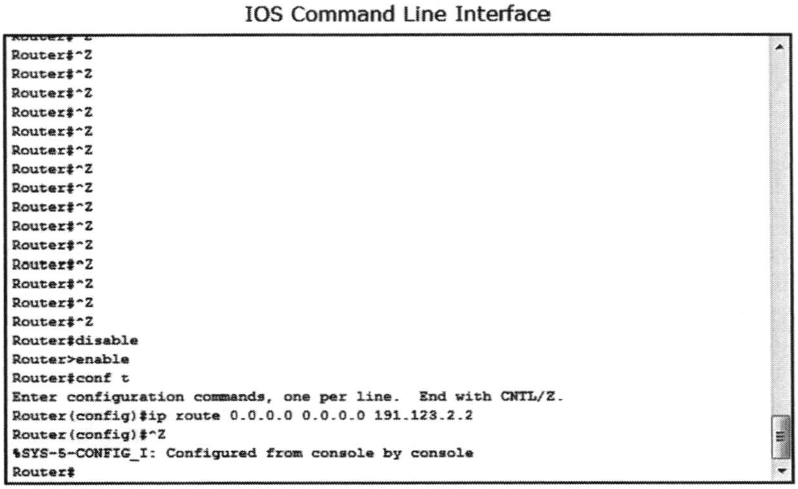

```
               IOS Command Line Interface
Router#^Z
Router#^Z
Router#^Z
Router#^Z
Router#^Z
Router#^Z
Router#^Z
Router#^Z
Router#^Z
Router#^Z
Router#^Z
Router#^Z
Router#^Z
Router#^Z
Router#^Z
Router#disable
Router>enable
Router#conf t
Enter configuration commands, one per line.  End with CNTL/Z.
Router(config)#ip route 0.0.0.0 0.0.0.0 191.123.2.2
Router(config)#^Z
%SYS-5-CONFIG_I: Configured from console by console
Router#
```

FIGURE 11.10 Configuring a default static route.

In this case, the first group of 0.0.0.0 stands for all networks, whereas the second group of 0.0.0.0 stands for all subnet masks. This is known as the default route, or as gateway of last resort, or even as *quad-zero* route. When a router is only connected to another router, the default route configuration is typically utilized. A common situation is when the router only has two interfaces: one connected to the LAN and another interface connected to another router. In this case, all the packets received from the LAN, whose destination is not another host within the LAN, should be forwarded to the other router (default route). Similar to the static routes described above, in case of serial interfaces, the routing table look up process can be accelerated by using the description of the output exit interface, instead of the address of the next hop router. Then, the following syntax comes:

- Router(config)#ip route 0.0.0.0 0.0.0.0 *interface*
  - *Example*: Router(config)# ip route 0.0.0.0 0.0.0.0 se0/1

After having configured all routers, one can verify the routing tables by issuing the following command*:

- Router#show ip route

Moreover, if one intends to follow the sequence of nodes of a packet until they reach a certain destination IP address *dest_ address*, the Cisco IOS syntax must be used†:

- Router#traceroute *dest_ address*
  - *Example*: Router#traceroute 193.123.212.3

In case one intends to verify when routes are added, modified, or deleted, in real time, even during router's configuration, the following syntax should be applied to the router:

- Router#debug ip routing

This functionality can be disabled by issuing the following Cisco IOS syntax:

- Router#undebug ip routing
- Router#undebug all

11.4 ADAPTIVE ROUTING ALGORITHMS AND PROTOCOLS

Adaptive routing algorithms are used by different routing protocols to compute routing tables. These routing tables are used to determine the output interface to use in order to make the packet reach a certain destination network.

Routing protocols are used by routers to exchange known information about the network. This information is then used to calculate the entries into the routing tables, that is, to calculate paths.

---

* In a workstation, the command "route print" or "netstat -r" (typed in the Windows command line screen [MS-DOS prompt]) allows verifying the configured routing table. A workstation's routing table can be modified by issuing "route" followed by one of the options that can be viewed from "route /?" (issued in the Windows command line screen [MS-DOS prompt]). Moreover, the command "ipconfig" (typed in the Windows command line screen [MS-DOS prompt]) allows the visualization of the IP configuration in workstations' interfaces, the subnet mask, and the default gateway. In a workstation, the command "netstat" alone allows viewing all active TCP connections.

† Similarly, the network administrator may use the commands "ping [dest_IP_address]" and "tracert [dest_IP_address]" in workstations to check the connectivity or to follow the sequence of nodes of a packet until they reach a certain destination IP address.

Although the exchange of data generated by some routing protocols to allow the creation of routing tables is carried out over UDP datagrams, routing protocols such as the RIP, the OSPF, the legacy IGRP, the EIGRP, and the IS-IS belong to the Internet layer (layer 3). As can be seen from Figure 11.11, in case of the RIP, there is a lower layer protocol (routing protocol) invoking a higher layer protocol (transport protocol—UDP). This is possible in TCP/IP stack, while the OSI reference model only allows an upper layer to make use of the services provided by a lower layer (and not the reverse).

Adaptive algorithms support decision making of routers, which translates in routing tables entries, taking into account one or a combination of metrics, such as the network topology, the traffic load, the bandwidth, and so on. In addition, different adaptive algorithms use different mechanisms to obtain information about the used metric. Since these metrics suffer time variations, the routing tables' entries are permanently being updated and changing. As a result of this metrics variation, a certain traffic flow (e.g., a download), may initially follow a certain path (a sequence of routers) but, after a certain time period, the remaining datagrams may be routed through different intermediate nodes. Note that, naturally, adaptive protocols compute the entries into routing tables without human intervention.

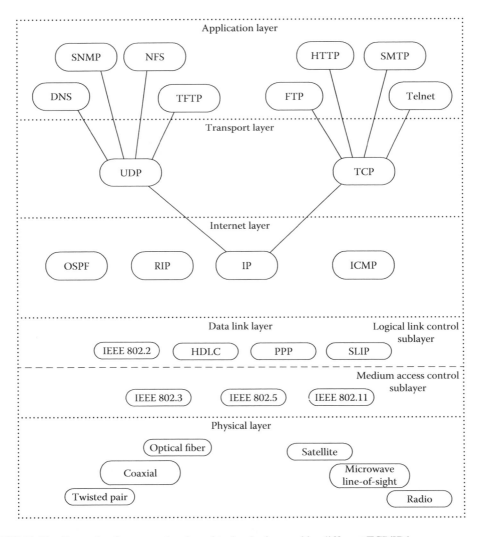

FIGURE 11.11   Example of some protocols and technologies used by different TCP/IP layers.

There are two families of routing protocols that make use of adaptive algorithms:

- Distance vector
- Link state

11.4.1 Classification of Adaptive Routing Protocols

Routing protocols can be distance vector or link state based. Depending on whether the information in a router is simply the distance or the full network topology, the adaptive routing protocol is classified, respectively, as a distance vector or a link state. As can be seen from Table 11.3, another classification for routing protocols relies on whether they are classful or classless.

Adaptive routing protocols allow the automatic construction of routing tables by exchanging routing information with neighbor routers.

At the beginning of this chapter, it was described that IP addresses were initially employed and distributed using the classful mode. Later on, it was viewed that the explosion of Internet users would rapidly lead to a depletion of the limited address spacing available to the classful mode. Then, in order to use the limited address spacing in a more efficient manner, the classless mode will be utilized. A classful routing protocol simply exchanges the network information (updates) in classful mode, without exchanging subnet masks, and the built routing table is also presented in classful mode. On the other hand, a classless routing protocol exchanges the network information in classless mode, including the corresponding subnet masks in this exchange, and the built routing table is presented in classless mode. Moreover, CIDR routing and VLSM are used by the family of classless routing protocols. By default, automatic summarization at major network boundaries (classful) is applied by both RIP versions. Nevertheless, RIPv2 also summarize routes with a subnet mask that is smaller than classful subnet mask.

11.4.2 Distance Vector Protocols and Their Configuration

Distance vector routing protocols compute the distance to each destination node within an AS (for IGP-type protocols) or in remote ASs (for EGP-type protocols) based on a metric. The used metric* can be the number of hops, the bandwidth, the traffic, the delay, and so on. Some distance vector protocols make use of a combination of several of these elementary metrics. The path selection to a network is based on the shortest distance. The computation of the shortest distance (best path) to a network using a distance vector protocol is typically performed using the Bellman–Ford algorithm,† which is as follows:

- Set the cost for itself as 0, and a certain value for adjacent nodes (1 when the metric is the number of hops).
- The preliminary routing table is sent to adjacent nodes with this cost information.

TABLE 11.3
Classification of Routing Protocols

Classification	Classful	Classless	IPv6
Distance vector	RIPv1; IGRP	RIPv2; EIGRP	RIPng; EIGRP for IPv6
Link state	EGP	OSPFv2; IS-IS; BGPv4	OSPFv3; IS-IS for IPv6; BGPv4 for IPv6

---

* The metric is also referred to as *cost* or *metric distance*.
† An exception is the EIGRP protocol, which uses the diffusing update algorithm (DUAL).

- Each adjacent node saves the most recent received routing information and uses this information to compute its own routing table by adding its cost/metric information to those included in the received routing table. Then, the lowest distance to each destination network is the one included in the new routing table. Finally, the node resends this new routing table to other neighbors (neighbors of neighbors).

Observing a routing table, each node knows which output interface should be used to forward packets to another network. Nevertheless, distance vector routing protocols do not compute the overall network topology. Distance vector protocols simply compute the distances to other networks. The mostly known distance vector routing protocols are the RIP, the IGRP, and the EIGRP.

11.4.2.1 Routing Information Protocol

The RIP is a dynamic routing protocol used inside ASs, and therefore, it consists of an IGP-type protocol. Using the RIP, the network nodes send periodically the whole routing tables to the other neighboring nodes. This is the traditional way nodes use this protocol family to update their routing tables. In case routing table updates are not accurate, such inaccuracy is propagated along the whole network. Moreover, the amount of overhead generated in long networks is high, and the convergence time* is also high. In addition to periodical updates, the RIP also sends updates when a change in the network is detected, or when a network modification is informed through an update received from another router (triggered updates). The most important advantage of the RIP relies on its simple configuration.

The AD of the RIP is 120, and the metric used by the RIP is the hop count, being exchanged between neighbors every 30 s (the default hello time). Using the RIP, a maximum of 15 hops is allowed for a datagram to be delivered to a destination network. If the metric distance is higher than this value, the packet is considered as unreachable, being discarded. Moreover, similar to other routing protocols, the RIP supports equal cost load balancing, up to four different routes. This means that when a router has up to four different routes to the same destination network with the same cost (metric distance), each packet of each group of four consecutive packets is sent through a different route.

RIP messages are encapsulated in a UDP datagram, with 520 as both source and destination ports.

In order to improve the stability, and to avoid routing loops, RIP uses the following three different mechanisms in the exchange of updates:

- *Hold-down times*: If an update is not received to refresh an existing route within the hold-down time, the route is marked as invalid (metric = 16), but kept in the routing table until the flush time. If an update is not received until the flush time, the route is removed from the routing table. Then, the router queries the neighboring routers to check if any of those have routes to the lost remote network. By default, the hold-down time corresponds to six times the hello time (180 s). By default, the flush time corresponds to eight times the hello time (240 s). Hold-down times prevents routing loops, when the network connectivity is not stable, or during convergence.
- *Split horizon*: A routing table update is never advertised through the interface where the change have occurred. In case the main route had gone down, and the neighbor had received the update (i.e., split horizon not in use), such neighbor could use such (invalid) route to send packets, creating a loop.
- *Poison reverse*: A routing table update is also advertised through the interface where the change has occurred (reverse advertisement), but with an infinity distance (poison), that is, as an invalid route. Advertising an infinity distance prevents other routers from sending packets over an invalid route.

---

* Time necessary to build in the routing table.

Two important versions of the RIP exist: the RIP version 1 (RIPv1) [RFC 1058] and the RIP version 2 (RIPv2) [RFC 1723]. As previously described, RIPv1 is classful, whereas RIPv2 is classless (see Table 11.4). This means that RIPv2 advertises both network addresses and the corresponding subnet masks, whereas RIPv1 simply advertises the network addresses (without subnet masks). RIPv2 supports VLSM, whereas RIPv1 does not.

By default, automatic summarization at major network boundaries (classful) is applicable to both RIP versions. Nevertheless, RIPv2 also summarizes routes with a subnet mask that is smaller than classful subnet mask. This way, depending on whether RIPv1 or RIPv2 is employed, the subnet mask applied to the updates can also be different. In case of RIPv2, the subnet masks are sent alongside the network addresses in the updates, and therefore, the procedure is simple. On the contrary, in the case of RIPv1, the procedure differs as follows:

- If the interface from which an update arrived belongs to the same network as the routing update, then the subnet mask of the interface is applied to the network in the routing update.
- If the interface from which an update arrived belongs to a different network as the routing update, then the classful subnet mask of the network is applied to the network in the routing update.

While RIPv1 does not support discontiguous subnetworks, RIPv2 does support discontiguous subnetworks. Discontiguous subnetworks comprise a network that separates another major network into multiple subnetworks. As can be seen from Figure 11.12, the network 171.123.0.0 is a discontiguous network, as its subnetworks (171.123.1.0/24 and 171.123.2.0/24) are separated by the (sub)network 10.20.30.0/24. In the scope of RIP, a boundary router is a router that has multiple interfaces in more than one major classful network.

Note that RIPv2 supports discontiguous subnetworks because the subnet mask is sent alongside with the subnetwork information. Nevertheless, the auto-summary needs to be manually disabled,

TABLE 11.4

Main Differences between RIPv1 and RIPv2

Protocol	Class Mode	Sends Subnet Mask in Updates	VLSM	Auto-Summary	Discontiguous Subnetworks	Updates	Authentication
RIPv1	Classful	No	No	Yes	No	Broadcast	No
RIPV2	Classless	Yes	Yes	Yes	Yes	Multicast	Yes

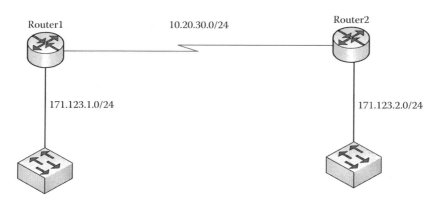

FIGURE 11.12 Discontiguous subnetworks.

once it is enabled by default. Otherwise, the route to 171.123.1.0/24 would be summarized as 171.123.0.0/16, and the route to 171.123.2.0/24 would also be summarized as 171.123.0.0/16. This would make the router not to forward traffic to the two subnetworks of 171.123.0.0/16.

With regard to the transmission method, RIPv1 routing updates are sent in broadcast mode, while RIPv2 uses multicast for this purposes. From the security point of view, while RIPv1 does not support authentication, RIPv2 may use authentication when sending route updates.

IPv6 considers the same type of routing protocols as IPv4, but with minimal modifications to accommodate the changes. The IPv6 uses an improved version of the RIPv2, entitled RIPng (RIP of the next generation).

The configuration of the RIP using the Cisco IOS is quite simple. After having entered into the configuration mode, one uses the following Cisco IOS syntax:

- Router(config)#router rip
- Router(config-router)#network *net_IP_address*
    - *Example* (*see Figure 11.12*):
        - Router1(config)#router rip
        - Router1(config-router)#network 10.0.0.0
        - Router1(config-router)#network 192.168.2.0

Note that *net_IP_address* stands for the network IP address to where each router's interfaces are connected to. The second command line must be inserted as many times as the number of interfaces that a router has connected to different networks (Figure 11.13).

The networks declared using these commands are those that will be advertised to the neighbors. Moreover, updates are also sent through the interfaces of these declared networks. In case of a LAN, since typically no other routers exist, the overhead bandwidth can be reduced or the security improved by disabling such route updates. This is performed issuing the command line:

- Router(config-router)#passive-interface *interface*
    - *Example*: Router1(config-router)#passive-interface Fastethernet 0/0

In the above syntax, *interface* stands for the interface type and number (interface ID). Note that although the passive mode disables the updates from being sent through this interface, the network connected to such interface is still advertised in the updates sent to other interfaces (as the network has been declared). Instead of declaring each of the individual passive interfaces, it is a good configuration rule to, initially, declare all interfaces as passive using the command "passive-interface default," and then declare each of the nonpassive interfaces individually. This presents the advantage that one does

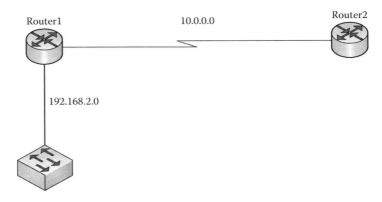

FIGURE 11.13 Example of internetwork for RIP configuration.

not forget interfaces in active mode. In case one interface is left in passive mode when it should be in active mode, then the routing protocol does not work properly. Then, the following syntax comes:

- Router(config-router)#passive-interface default
- Router(config-router)#no passive-interface *interface*
  - *Example*:
    - Router(config-router)#passive-interface default
    - Router(config-router)#no passive-interface se0/0
    - Router(config-router)#no passive-interface se0/1

It has been described that there are two versions of the RIP: version 1 and version 2. By default, the RIP configuration corresponds to the version 1 (RIPv1). As can be seen from Table 11.3, if one intends to use the RIP in classless mode, then the version 2 (RIPv2) must be employed. One can switch from RIPv1 into RIPv2 by issuing the following syntax:

- Router(config-router)#version 2

It is worth noting that, by default, both RIPv1 and RIPv2 perform route summarization. In case multiple subnetworks corresponding to a common route summary are reachable through different interfaces, then route summarization must be disabled. This is performed using the following Cisco IOS syntax:

- Router(config-router)#no auto-summary

Let us now suppose that a certain router is simultaneously configured with a static route and with the RIP. In this case, we may want to send the static route information in the RIP advertisements. This is achieved using the following syntax:

- Router(config-router)#default-information originate

Let us consider the internetwork depicted in Figure 11.14. Moreover, Figure 11.15 shows the configuration of Router1, assuming the configuration of a static route, alongside with the RIPv2. Moreover, since different subnetworks that correspond to a certain route summary are reachable through different interfaces, the auto-summary was disabled by issuing the syntax *no auto-summary*.

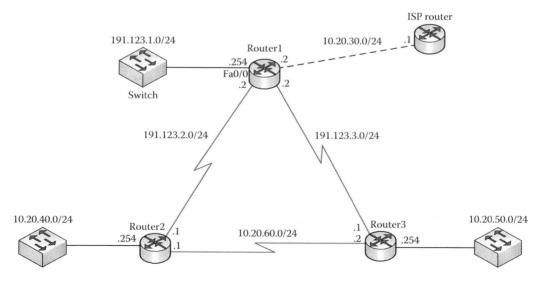

FIGURE 11.14   Example of an internetwork.

IOS Command Line Interface

```
Router>enable
Router#conf t
Enter configuration commands, one per line.  End with CNTL/Z.
Router(config)#ip route 0.0.0.0 0.0.0.0 10.20.30.1
Router(config)#router rip
Router(config-router)#version 2
Router(config-router)#network 191.123.1.0
Router(config-router)#network 191.123.2.0
Router(config-router)#network 191.123.3.0
Router(config-router)#passive-interface fa0/0
Router(config-router)#no auto-summary
Router(config-router)#default-information originate
Router(config-router)#
```

FIGURE 11.15 Configuring the protocol RIPv2 in a router.

Figure 11.16 shows the routing table of Router1 from Figure 11.14, after the configuration with RIPv2 and static route to the ISP. As can be seen, this internetwork presents multiple discontiguous networks. Since RIPv2 is employed, which comprises sending the subnet masks in the routing updates, and since automatic summarization was disabled (using the "no auto-summary" command), the routing table of Router1 has enough and correct routing information to the multiple remote networks.

Regarding the routing behavior, both RIPv1 and RIPv2 may route in one of following two modes:

- *Classless behavior*: Using this behavior, in case a default route is configured, the router uses the default route if the destination network of the packet does not match any of the routes in the routing table (remember that the routing decision is performed based on the longest match, and that the default route is used in case there is no other with longer match). In this mode, the packet is dropped only if there is not any destination network

IOS Command Line Interface

```
       D - EIGRP, EX - EIGRP external, O - OSPF, IA - OSPF inter area
       N1 - OSPF NSSA external type 1, N2 - OSPF NSSA external type 2
       E1 - OSPF external type 1, E2 - OSPF external type 2, E - EGP
       i - IS-IS, L1 - IS-IS level-1, L2 - IS-IS level-2, ia - IS-IS inte
       * - candidate default, U - per-user static route, o - ODR
       P - periodic downloaded static route

Gateway of last resort is 10.20.30.1 to network 0.0.0.0

     10.0.0.0/8 is variably subnetted, 5 subnets, 2 masks
R       10.0.0.0/8 [120/1] via 191.123.2.1, 00:00:47, Serial0/1/0
                   [120/1] via 191.123.3.1, 00:01:03, Serial0/1/1
C       10.20.30.0/24 is directly connected, FastEthernet0/1
R       10.20.40.0/24 [120/1] via 191.123.2.1, 00:00:17, Serial0/1/0
R       10.20.50.0/24 [120/1] via 191.123.3.1, 00:00:06, Serial0/1/1
R       10.20.60.0/24 [120/1] via 191.123.2.1, 00:00:17, Serial0/1/0
     191.123.0.0/24 is subnetted, 3 subnets
C       191.123.1.0 is directly connected, FastEthernet0/0
C       191.123.2.0 is directly connected, Serial0/1/0
C       191.123.3.0 is directly connected, Serial0/1/1
S*   0.0.0.0/0 [1/0] via 10.20.30.1
Router#
```

FIGURE 11.16 Routing table of Router1 after the RIPv2 configuration.

in the routing table that matches the destination network of the packet and if the default route is not configured. The classless behavior is configured using the following Cisco IOS syntax:

- Router(config-router)#ip classless
- *Classful behavior*: Using this behavior, in case a default route is configured, the router will use the default route *only* in case the class of the destination network of the packet (class A, B, or C) is not in the routing table. Otherwise, the packet is forwarded to the interface where the classful network is located, or the packet is dropped. Let us suppose that classful behavior is configured in Router1 of Figure 11.16, and that a packet is received, whose destination IP address is 10.0.0.1. Observing the routing table depicted in Figure 11.16 we do not find the network 10.0.0.0, making the router to drop the packet because there are subnetworks of the network 10.0.0.0, but these do not match with the destination IP address of the packet (10.0.0.1). Still in the same scenario, but assuming that the destination IP address of the packet is 173.123.213.1, since there is not any network or subnetwork belonging to the network 173.123.0.0, the packet is sent to the default route. The classful behavior is configured using the following Cisco IOS syntax:
- Router(config-router)#no ip classless

The reader should not confuse between routing behavior and class mode used by the RIPv1 and RIPv2. Note that RIPv1 is always classful and RIPv2 is always classless. The routing behavior refers to the way routing decision is performed, and not to how the routing table is built (or whether or not subnet masks are sent in the updates).

In addition to the command "show ip route" previously described, there are other two syntaxes that are widely employed in troubleshooting, which are as follows:

- Router#show ip protocols
- Router#show run | include route (command | only works in Cisco equipment, not in packet tracer simulator)
- Router#debug ip rip

The command "show ip protocols" is normally used to check the timers information configured in RIP, namely the hello time, hold-down time, and flush time (see Figure 11.17). In addition, using this

IOS Command Line Interface

```
Router#show ip protocols
Routing Protocol is "rip"
Sending updates every 30 seconds, next due in 28 seconds
Invalid after 180 seconds, hold down 180, flushed after 240
Outgoing update filter list for all interfaces is not set
Incoming update filter list for all interfaces is not set
Redistributing: rip
Default version control: send version 2, receive 2
  Interface              Send  Recv  Triggered RIP  Key-chain
  Serial0/1/1            2     2
  Serial0/1/0            2     2
Automatic network summarization is not in effect
Maximum path: 4
Routing for Networks:
        191.123.0.0
Passive Interface(s):
        FastEthernet0/0
Routing Information Sources:
        Gateway         Distance      Last Update
        191.123.2.1       120         00:00:11
        191.123.3.1       120         00:00:01
Distance: (default is 120)
```

FIGURE 11.17 The result of the command "show ip protocols" applied to Router1.

command one can verify many other details about the route configuration, such as whether or not automatic route summarization is in effect, or whether or not any interface is in passive mode. Note that this command is valid for all adaptive routing protocols, not being limited to the RIP.

Finally, the following Cisco IOS syntax is utilized for viewing real-time RIP routing updates and other network parameters as they modify state:

• Router#debug ip rip

11.4.2.2 Enhanced Interior Gateway Routing Protocol

EIGRP consists of a Cisco proprietary distance vector routing protocol. The EIGRP is derived from the legacy IGRP. While IGRP is a classful routing protocol, EIGRP is classless. With EIGRP, the subnet mask is transmitted in a route update, VLSM are supported, and discontiguous networks are permitted. Although EIGRP is incorporated into the group of distance vector protocols, its updates are only sent when the network changes. Note that both IGRP and EIGRP are Cisco proprietary protocols and only operate on Cisco routers.

It has been shown that RIP is limited to a maximum of 15 hops. Beyond such network size, another routing protocol has to be utilized. This was the main motivation for the development, in 1985, of the IGRP. Later in 1992, the EIGRP was released, having incorporated a number of improvements to the IGRP.

The EIGRP registers and computes the distances to destination networks using protocol-dependent modules (PDMs). The EIGRP may work with different packet switching protocols such as IP, IPX, and Apple Talk. Consequently, a different set of PDM modules exists for each packet switching protocol. Moreover, within each protocol, the PDM of EIGRP uses the following three different tables:

• Neighbor table
• Topology table
• Routing table

The neighbor table of a certain router includes the identification of the neighboring nodes (alongside with their IP addresses), whereas the topology table comprises the identification of all known networks, even those beyond the neighboring networks, alongside their metric distances to each network. Note that a certain destination network can be reached through different paths. The topology table includes the multiple paths to each destination network, alongside the different metric distances. Finally, the routing table includes the list of destination networks, even those beyond neighboring networks, with the corresponding metric distances. Note that while the topology table includes multiple routes (and the corresponding metrics), the routing table is limited to the route to each network with the lowest metric distance. Using the Cisco IOS, one can verify each of these three tables with the following command lines (from the privileged mode):

• Router#show ip eigrp neighbors
• Router#show ip eigrp topology
• Router#show ip route

In addition to the equal cost load balancing supported by other routing protocols, the EIGRP is the only one that supports *unequal* cost load balancing. This means that when a router running the EIGRP has multiple routes to the same destination network, even with different costs (metric distances), the packets of a data exchange are distributed along the different available routes based on the route's cost. In this case, more packets are sent through the route with the lowest cost, whereas fewer packets are sent through the route with the highest cost. By default, load balancing using four paths is enabled in routers configured with EIGRP. The number of paths used for load balancing can be checked under the designation *maximum path* using the Cisco IOS syntax (see Figure 11.18):

- Router#show ip protocols

Unlike the RIP whose packet exchange is performed using UDP datagrams, the EIGRP uses the reliable transport protocol (RTP) for sending and receiving EIGRP packets. The need for the use of the RTP results from the fact that the EIGRP is not limited to IP packets, being also utilized with IPX and Apple Talk, and therefore, this cannot be constrained to the TCP/IP stack. The RTP may use both reliable and unreliable deliveries. Reliable deliveries require the use of confirmation messages, using acknowledgment packets.

In EIGRP, the construction of different tables is taken care by the DUAL. This makes another difference of the EIGRP, relating to other distance vector protocols, which normally use the Bellman–Ford algorithm. The DUAL algorithm enables a convergence time shorter than that of RIP, avoids loops, and incorporates a number of backup routes that are very useful in case a connection is lost.

The DUAL algorithm comprises the following types of packets:

- *Hello packets*: It consists of the first phase and aims to discover and establish adjacencies with neighboring routers, being exchanged using the RTP in unreliable mode.
- *Update packets*: It propagates routing updates, when a route is lost or when a new route exists. Update packets are exchanged using the RTP in reliable mode, using multicast.
- *Query packets*: It is sent by a router when a route to a network is lost, and the topology table has no other backup route to such lost network. Query packets are exchanged using the RTP in reliable mode, using multicast or unicast.
- *Reply packets*: It is sent by a router as a response to a query packet. Reply packets are exchanged using the RTP in reliable mode, using unicast.
- *Acknowledgment packets*: It acknowledges correct receipt of EIGRP packets that had been sent using the RTP in reliable mode (update, query, and reply packets).

IOS Command Line Interface

```
Router#show ip protocols

Routing Protocol is "eigrp  123 "
   Outgoing update filter list for all interfaces is not set
   Incoming update filter list for all interfaces is not set
   Default networks flagged in outgoing updates
   Default networks accepted from incoming updates
   EIGRP metric weight K1=1, K2=0, K3=1, K4=0, K5=0
   EIGRP maximum hopcount 100
   EIGRP maximum metric variance 1
Redistributing: eigrp 123, static
   Automatic network summarization is not in effect
   Maximum path: 4
   Routing for Networks:
      191.123.0.0
   Passive Interface(s):
      FastEthernet0/0
   Routing Information Sources:
      Gateway         Distance      Last Update
      191.123.3.1      90            6890
      191.123.2.1      90            7625
   Distance: internal 90 external 170

Router#
```

FIGURE 11.18 Verifying the values for K metric weights with the command "show ip protocols" in Router1 of Figure 11.14.

Unlike RIP that periodically sends routing updates, the EIGRP updates are only sent when there is a change in the route status. This reduces the amount of bandwidth and overhead utilized by the adaptive routing protocol. Moreover, it is commonly stated that the EIGRP updates are partial and bounded. They are partial because update packets only include route modifications, not the exchange of the whole routing table. They are bounded because update packets are only sent to the affected routers. Those routers not impacted by the route modification are not informed about the network change.

Let us focus again on the internetwork depicted in Figure 11.14. As can be seen from the routing table of Router1, depicted in Figure 11.19, the letter D is shown in some EIGRP entries of a routing table, standing for DUAL algorithm, used to compute the metric to remote networks. As can be seen from Table 11.2, the AD of 90 for EIGRP refers to internal routes. The remaining parameters shown in the EIGRP routing table are the same as those described for the RIP routing table.

In case a route is imported from another routing protocol and advertised by EIGRP, the AD becomes 170, instead of 90. One example of an imported route is a default route advertised by EIGRP to its neighbors. This is configured in the router with the default route using the following Cisco IOS syntax[*]:

- Router(config-router)#redistribute static

The AD = 170 can be seen from the routing table of Router2, plotted in Figure 11.20.

Note that a routing table should, in principle, show a single route to each remote network. Nevertheless, from Figure 11.20, the remote network 191.123.3.0 can be reached through two different routes: (1) via the router 10.20.60.2 or (2) via the router 191.123.2.2. In principle, the multiple links to a certain destination should not be shown in a routing table, but in a topology table. However, in this particular case, the metric distances of these two routes are the same (9849856). This is the only case when a routing table shows more than one route to a destination. In this case, equal cost load balancing is implemented.

IOS Command Line Interface

```
Router#show ip route
Codes: C - connected, S - static, I - IGRP, R - RIP, M - mobile, B - BGP
       D - EIGRP, EX - EIGRP external, O - OSPF, IA - OSPF inter area
       N1 - OSPF NSSA external type 1, N2 - OSPF NSSA external type 2
       E1 - OSPF external type 1, E2 - OSPF external type 2, E - EGP
       i - IS-IS, L1 - IS-IS level-1, L2 - IS-IS level-2, ia - IS-IS inte
       * - candidate default, U - per-user static route, o - ODR
       P - periodic downloaded static route

Gateway of last resort is 10.20.30.1 to network 0.0.0.0

     10.0.0.0/24 is subnetted, 4 subnets
C       10.20.30.0 is directly connected, FastEthernet0/1
D       10.20.40.0 [90/2172416] via 191.123.2.1, 00:12:52, Serial0/1/0
D       10.20.50.0 [90/2684416] via 191.123.2.1, 00:00:06, Serial0/1/0
D       10.20.60.0 [90/2681856] via 191.123.2.1, 00:12:52, Serial0/1/0
     191.123.0.0/24 is subnetted, 3 subnets
C       191.123.1.0 is directly connected, FastEthernet0/0
C       191.123.2.0 is directly connected, Serial0/1/0
C       191.123.3.0 is directly connected, Serial0/1/1
S*   0.0.0.0/0 [1/0] via 10.20.30.1
Router#
```

FIGURE 11.19 Routing table of Router1 after EIGRP configuration (refers to internetwork depicted in Figure 11.14).

---

[*] Note that the syntax utilized by EIGRP is different from the RIP that uses *Router(config-router)#default-information originate*.

IOS Command Line Interface

```
Router#show ip route
Codes: C - connected, S - static, I - IGRP, R - RIP, M - mobile, B - BGP
       D - EIGRP, EX - EIGRP external, O - OSPF, IA - OSPF inter area
       N1 - OSPF NSSA external type 1, N2 - OSPF NSSA external type 2
       E1 - OSPF external type 1, E2 - OSPF external type 2, E - EGP
       i - IS-IS, L1 - IS-IS level-1, L2 - IS-IS level-2, ia - IS-IS inte
       * - candidate default, U - per-user static route, o - ODR
       P - periodic downloaded static route

Gateway of last resort is 191.123.2.2 to network 0.0.0.0

     10.0.0.0/24 is subnetted, 3 subnets
C       10.20.40.0 is directly connected, FastEthernet0/0
D       10.20.50.0 [90/2172416] via 10.20.60.2, 00:24:00, Serial0/1/1
C       10.20.60.0 is directly connected, Serial0/1/1
     191.123.0.0/24 is subnetted, 3 subnets
D       191.123.1.0 [90/2172416] via 191.123.2.2, 00:23:58, Serial0/1/0
C       191.123.2.0 is directly connected, Serial0/1/0
D       191.123.3.0 [90/9849856] via 10.20.60.2, 00:11:13, Serial0/1/1
                    [90/9849856] via 191.123.2.2, 00:11:13, Serial0/1/0
D*EX 0.0.0.0/0 [170/2195456] via 191.123.2.2, 00:23:58, Serial0/1/0
Router#
```

FIGURE 11.20 Routing table of Router2 after EIGRP configuration with an imported route (AD = 170) (refers to internetwork depicted in Figure 11.14).

The configuration of the EIGRP using the Cisco IOS is similar to the RIP, with few differences. Naturally, before starting with the EIGRP configuration, one must configure the interfaces. As can be seen from Figure 11.21, the EIGRP configuration starts by issuing from the following configuration mode:

- Router(config)#router eigrp *process_id*
  - *Example*: Router(config)#router eigrp 123

In the syntax above, *process_id* stands for the process identification (ID) number, and represents an instance of the routing protocol running on a router. In EIGRP, the AS number is automatically assigned to process ID number, consisting of a 16-bit length number. Note that all routers in an

IOS Command Line Interface

```
Router>
Router>
Router>
Router>
Router>
Router>
Router>enable
Router#conf t
Enter configuration commands, one per line.  End with CNTL/Z.
Router(config)#ip route 0.0.0.0 0.0.0.0 10.20.30.1
Router(config)#router eigrp 123
Router(config-router)#network 191.123.1.0
Router(config-router)#network 191.123.2.0
Router(config-router)#network 191.123.3.0
Router(config-router)#passive-interface fa0/0
Router(config-router)#no auto-summary
Router(config-router)#redistribute static
Router(config-router)#^Z
Router#
%SYS-5-CONFIG_I: Configured from console by console

Router#
```

FIGURE 11.21 Cisco IOS configuration of Router1 with EIGRP (refers to internetwork depicted in Figure 11.14).

EIGRP routing domain must use the same process ID number. Afterward, EIGRP routing configuration proceeds with the configuration of the networks connected to each of the router's interfaces, using the following syntax:

- Router(config-router)#network *net_IP_address*
  - *Example*:
    - Router(config-router)#network 191.123.213.0
    - Router(config-router)#network 11.213.15.0

In the syntax above *net_IP_address* stands for the network or subnetwork address connected to each router's interface. This command line must be inserted as many times as the number of the router's interfaces, allowing the router's interfaces to send and receive EIGRP packets. Similar to the RIP configuration, in case one interface is not connected to a router (which is typically the situation of a router's interface connected to a LAN), one may configure the router to not send updates through such interface. This is performed by issuing the following syntax:

- Router(config-router)#passive-interface *interface*
  - *Example*: Router(config-router)#passive-interface fa0/0

As described for the RIP, it is always a better configuration methodology to, initially, configure all interfaces in passive mode, and then disable the passive mode in the required interfaces. This is performed issuing the following syntaxes:

- Router(config-router)#passive-interface default
- Router(config-router)#no passive-interface *interface*
  - *Example*:
    - Router(config-router)#passive-interface default
    - Router(config-router)#no passive-interface se0/0
    - Router(config-router)#no passive-interface se0/1

In the syntax above *interface* stands for the interface type and number.

Since the EIGRP automatically summarize routes at classful boundaries, one may want to disable that function. Disabling automatic summarization in a router is performed using the syntax:

- Router(config-router)#no auto-summary

Finally, in case a default route is to be included in the EIGRP updates, the following syntax should be used:

- Router(config-router)#redistribute static

Figure 11.22 shows the neighbor table for Router1 of the internetwork depicted in Figure 11.14. This table shows the different hosts connected to the router, with their addresses, and a number of other parameters, such as the interface, hold time, and so on.

Figure 11.23 shows the topology table for Router1 of the internetwork depicted in Figure 11.14. This table shows the multiple networks known by the router, including those directly connected and remote networks, alongside the feasibility distance (FD) and the interface. As can be seen from Table 11.5, FD corresponds to the metric distance in the EIGRP environment.

In the topology table, the letter P on the left of the network address means that the interface is in passive mode, meaning that the convergence time has terminated and that the route is stable. In the case of remote networks, the FD number appears twice or more in the topology table: the first

IOS Command Line Interface

```
Router#
Router#
Router#
Router#
Router#
Router#
Router#
Router#
Router#
Router#
Router#
Router#
Router#show ip eigrp neighbors
IP-EIGRP neighbors for process 123
H    Address         Interface      Hold Uptime    SRTT   RTO   Q   Seq
                                    (sec)          (ms)         Cnt Num
0    191.123.2.1     Se0/1/0        12   00:22:25  40     1000  0   25
1    191.123.3.1     Se0/1/1        10   00:09:37  40     1000  0   24

Router#
```

FIGURE 11.22 Neighbor table of Router1 after EIGRP configuration (refers to internetwork depicted in Figure 11.14).

IOS Command Line Interface

```
Router#show ip eigrp topology
IP-EIGRP Topology Table for AS 123

Codes: P - Passive, A - Active, U - Update, Q - Query, R - Reply,
       r - Reply status

P 191.123.1.0/24, 1 successors, FD is 28160
         via Connected, FastEthernet0/0
P 0.0.0.0/0, 1 successors, FD is 51200
         via Rstatic (51200/0)
P 191.123.2.0/24, 1 successors, FD is 2169856
         via Connected, Serial0/1/0
P 191.123.3.0/24, 1 successors, FD is 9337856
         via Connected, Serial0/1/1
P 10.20.50.0/24, 1 successors, FD is 2684416
           via 191.123.2.1 (2684416/2172416), Serial0/1/0
           via 191.123.3.1 (9340416/28160), Serial0/1/1
P 10.20.60.0/24, 1 successors, FD is 2681856
           via 191.123.2.1 (2681856/2169856), Serial0/1/0
           via 191.123.3.1 (9849856/2169856), Serial0/1/1
P 10.20.40.0/24, 1 successors, FD is 2172416
           via 191.123.2.1 (2172416/28160), Serial0/1/0
Router#
```

FIGURE 11.23 Topology table of Router1 after EIGRP configuration (refers to internetwork depicted in Figure 11.14).

TABLE 11.5
EIGRP Topology Designations

EIGRP Topology Nomenclature	
EIGRP Nomenclature	**Meaning**
Successor	Next hop router
Feasibility distance (FD)	Metric distance to the destination
Successor's reported distance (SRD)	Neighbor's metric distance
Feasible successor (FS)	Backup next hop router
Feasibility condition	SRD < FD

time in the same row as the address of the remote network, and the second time inside brackets, after the next hop address. Let us focus on the remote network 10.20.40.0/24, whose FD is 2172416, and the next hop router is 191.123.2.1, achieved through the router's interface Serial0/1/0. In the EIGRP environment, the next hop router is referred to as successor router. Inside brackets we have (2172416/28160). In this case, 28160 represents the successor's reported distance (SRD), which stands for the neighbor's metric distance. For a certain route to be selected, the feasibility condition must be achieved: this states that the SRD must be smaller than the FD (SRD < FD). In other words, this means that the metric distance of the neighbor must be smaller than the own router's metric distance. Otherwise, such route would not be the best. From Figure 11.23 we confirm that the feasibility condition is achieved for all the routes to remote networks. Finally, from Table 11.5, we see that the feasible successor (FS) corresponds to the backup next hop router. The FS is the next hop router that should be used in case the connection to the successor (primary next hop router) fails. The FD via the FS should be higher than the FD via the successor, as the route is selected based on the lowest cost.

The topology table depicted in Figure 11.23 shows multiple routes to some of the destinations. In the case of the destination 10.20.50.0/24, we see that FD is 2684416, and that there are two routes: one main route via 191.123.2.1 (successor) with a FD 2684416, and a backup route via 191.123.3.1 (FS) with FD of 9340416. This can be seen more clearly from Figure 11.24. For both routes, the feasibility condition is achieved, as their successor's reported distances are lower than the FDs. Finally, it is worth to refer that the main route (route via the successor) is always the one that appears first, and it corresponds to the one with the lowest FD.

It has been described that the DUAL algorithm is utilized to compute the EIGRP routes. We have shown different values for the FD and for successor's reported distance. Nevertheless, the computation of these parameters has still not been described. The EIGRP may use the following parameters as metrics:

- Bandwidth
- Delay
- Reliability
- Load

The complete composite formula utilized by the EIGRP as metric is

$$\text{Metric} = \left[\frac{K_1 \times \text{BW} + (K_2 \times \text{BW})}{(256 - \text{LD}) + K_3 \times \text{DL}} \right] \times \left[\frac{K_5}{(\text{RLBT} + K_4)} \right] \tag{11.1}$$

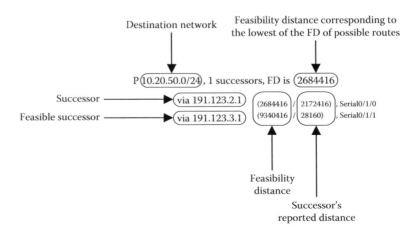

FIGURE 11.24   Description of the routes to 10.20.50.0/24 (shown in topology table of Figure 11.23).

where:
    BW is the bandwidth factor
    LD is the load factor
    DL is the delay factor
    RLBT is the reliability factor

By default, only bandwidth and delay is taken into account. Therefore, the K metric weights become $K_1 = 1$ (bandwidth); $K_2 = 0$ (load); $K_3 = 1$ (delay); $K_4 = 0$ (reliability); and $K_5 = 0$ (reliability). In this case, the metric of Equation 12.1 becomes

$$\text{Metric} = (\text{BW} + \text{DL}) \tag{11.2}$$

The values for K metric weights can be modified entering into the route configuration submode and using the following syntax:

- Router(config-router)#metric weights *tos* $K_1\, K_2\, K_3\, K_4\, K_5$

where 1 or 0 can be assigned to these parameters, depending on whether or not such parameters are to be taken into account, respectively. As can be seen from Figure 11.18, if one intends to verify the values for the K metric weights, then the following syntax should be employed (from the privileged mode):

- Router#show ip protocols

When computing the metric to a destination, the value for BW corresponds to

$$\text{BW} = \frac{10^7}{\min[\text{bandwidth (kbps)}]} \times 256 \tag{11.3}$$

where $\min(\text{bandwidth [kbps]})$ stands for the minimum of the bandwidths, expressed in kbps, configured in the multiple outgoing interfaces used by the route. In case a route comprises multiple hops, the value for *bandwidth* is the least of these parameters configured in the interfaces of the routers utilized by the route. Moreover, the value for DL corresponds to the sum of the delays of the multiple outgoing interfaces in the several hops expressed in microseconds $\left(\sum_{n=1}^{\text{no. of hops}}[\text{delay }(\mu s)]\right)$, divided by 10 and multiplied by 256

$$\text{DL} = \frac{\sum_{n=1}^{\text{no. of hops}}(\text{delay }[\mu s])}{10} \times 256 \tag{11.4}$$

The value for $RLBT$ corresponds to a dynamic measure of the likelihood that a link will fail. Finally, the value for LD is a dynamic measure that reflects the load traffic utilized by a link. Both $RLBT$ and LD are expressed as a fraction of 255 (see Figure 11.25). A better route is achieved with a higher $RLBT$ and with a lower LD.

Note that the value for bandwidth used in the calculation of BW does not correspond to the link bandwidth, but consists of a static bandwidth value configured in the router's interface, in the outgoing direction. Both sides of a serial link should be configured with the same bandwidth. By default, most serial interfaces use a bandwidth value of 1.544 Mbps, corresponding to a T1 interface. Moreover, the default bandwidth value for a fast Ethernet interface is 100 Mbps, and the default bandwidth value for a Gigabit Ethernet interface is 1 Gbps, and so on. One can modify the

IOS Command Line Interface

```
Serial0/1/0 is up, line protocol is up (connected)
  Hardware is HD64570
  Internet address is 191.123.2.2/24
  MTU 1500 bytes, BW 1544 Kbit, DLY 20000 usec,
       reliability 255/255, txload 1/255, rxload 1/255
  Encapsulation HDLC, loopback not set, keepalive set (10 sec)
  Last input never, output never, output hang never
  Last clearing of "show interface" counters never
  Input queue: 0/75/0 (size/max/drops); Total output drops: 0
  Queueing strategy: weighted fair
  Output queue: 0/1000/64/0 (size/max total/threshold/drops)
     Conversations  0/0/256 (active/max active/max total)
     Reserved Conversations 0/0 (allocated/max allocated)
     Available Bandwidth 1158 kilobits/sec
  5 minute input rate 6 bits/sec, 0 packets/sec
  5 minute output rate 10 bits/sec, 0 packets/sec
     13 packets input, 841 bytes, 0 no buffer
     Received 0 broadcasts, 0 runts, 0 giants, 0 throttles
     0 input errors, 0 CRC, 0 frame, 0 overrun, 0 ignored, 0 abort
     12 packets output, 879 bytes, 0 underruns
     0 output errors, 0 collisions, 1 interface resets
     0 output buffer failures, 0 output buffers swapped out
     0 carrier transitions
     DCD=up  DSR=up  DTR=up  RTS=up  CTS=up
Serial0/1/1 is up, line protocol is up (connected)
  Hardware is HD64570
  Internet address is 191.123.3.2/24
  MTU 1500 bytes, BW 1544 Kbit, DLY 300000 usec,
       reliability 255/255, txload 1/255, rxload 1/255
  Encapsulation HDLC, loopback not set, keepalive set (10 sec)
  Last input never, output never, output hang never
  Last clearing of "show interface" counters never
--More--
```

FIGURE 11.25 Verifying the metric parameters with the command show interfaces in Router1 of Figure 11.14.

value for this parameter by entering into the interface configuration submode and issuing the following syntax:

- Router(config-if)#bandwidth *value_in_kbps*
  - *Example*: Router(config-if)#bandwidth 2048

In the above syntax, *value_in_kbps* stands for the bandwidth value we want to assign to the interface, expressed in kbps. Note that this command does not modify the physical bandwidth of the link.

Similarly, the delay parameter can also be modified from the interface configuration submode by issuing the following syntax:

- Router(config-if)#delay *value_in_micro_seconds*
  - *Example*: Router(config-if)#delay 25000

As can be seen from Figure 11.25, one can verify the values for bandwidth, delay, load, and reliability by issuing the syntax:

- Router#show interfaces

This shows the values for these parameters in all router's interfaces. Alternatively, the following syntax can be issued to verify the parameters in a certain router's interface:

- Router#show interface *interface*
  - *Example*: Router#show interface fa0/0

Let us calculate the metric for the route from Router1 of Figure 11.14 to network 10.20.60.0/24. As can be seen from the topology table, there are two different routes to this network, namely through Router2 and through Router3, with $FD = 2681856$ and 9849856, respectively. We will calculate the FDs (metric distances) for both cases and confirm these values, and verify that the route with the lowest FD ($FD = 2681856$) is the one that appears in the routing table of Router1. Using either Router2 or Router3, a total of two hops exists to reach the destination network 10.20.60.0/24. Let us assume $K_1 = K_3 = 1$ and $K_2 = K_4 = K_5 = 0$ in Router1, Router2, and Router3, respectively.

First, we focus on the route through Router2. Let us use the highlighted metric values shown in Figure 11.25 for the interface Serial0/1/0 of Router1 (link to Router2) and the same values in Router2 (to connect with 10.20.60.0/24). The bandwidth in the two interfaces of the two routers are the same, therefore, the least is 1544 kbps, equal to each of the two. Then, from (12.3), BW becomes $BW = 10^7/1544 * 256 = 1658031$. From Equation 12.4, we compute DL as $DL = (20000 + 20000)/10 * 256 = 1024000$, where 20000 corresponds to the delay, expressed in microseconds, configured in the two serial interfaces of the two routers. Then, from Equation 12.2 we obtain metric $= (1658031 + 1024000) = 2682031$. From the topology table of Router1 depicted in Figure 11.24 we obtain a value of 2681856, which is very similar to the metric obtained from our calculations. The difference relies on the fact that the bandwidth of T1 interface is not exactly 1.544 Mbps, as this is an approximate value.

Let us now focus on the route through Router3. The connection between Router1 and Router3 is established using the interface Serial0/1/1 of Router1 (with parameter values shown at the bottom of the Figure 11.25). Note that the delay of this interface has been previously modified, as it shows DLY = 300000 (delay = 300,000 μs, when the default is delay = 20,000 μs). Then, for the interface from Router3 to the destination network, we use the default values. From Equation 12.3, BW becomes $BW = 10^7/1544 * 256 = 1658031$. Then, from Equation 12.4, we compute DL as $DL = (300000 + 20000)/10 * 256 = 8192000$. Finally, from Equation 12.2 we obtain metric $= (1658031 + 8192000) = 9850031$. From the topology table of Router1 we obtain a value of 9849856 for the FD of the route to network 10.20.60.0/24, through Router3 (via 191.123.3.1), which is also similar to the calculated value. Naturally, the selected route, which is the one shown in the routing table, is the one with the lowest FD, corresponding to the route through Router2 ($FD = 2681856 < FD = 9849856$). This can be viewed from the routing table of Router1, depicted in Figure 11.19.

Similar to RIP, the timers information configured in EIGRP (e.g., hello time and hold time) can be verified using the following syntax (see Figure 11.18):

- Router#show ip protocols

In addition, with this command, one can verify many other details about the route configuration, such as whether or not automatic route summarization is in effect, or whether or not any interface is in passive mode. Finally, if one intends to view real-time EIGRP routing updates and other network parameters as they modify state, the following Cisco IOS syntax must be issued:

- Router#debug eigrp fsm

The reader should refer to Appendix III for an example of an internetwork configured with EIGRP.

11.4.3 LINK-STATE PROTOCOLS AND THEIR CONFIGURATION

Link-state routing protocols compute a topologic database in each router, with the overall map of network connectivity, whereas distance vectors protocols simply compute the distances to each remote node. The map has the form of a graph, showing all network nodes and their interconnection. Link-state protocols can use configured bandwidth in the interfaces or delay (among other parameters), as metric parameters.

The map of topology that dictates the construction of the routing table is built as follows:

- Each node finds the neighboring nodes by sending a *hello* packet. A node that receives a *hello* packet responds with a *hello* packet as well. Using this procedure each node measures the metric to its neighbors.
- Then, each node sends to the neighboring nodes link-state advertisement packets (LSPs). Such packets include the information about adjacent neighboring nodes as well as the corresponding metrics. Then, the neighboring nodes re-send the LSP to the neighbors of neighbors, following the flooding procedure. From the received LSPs, the nodes can then build the network topology, including the indication of the cost between adjacent nodes.

Once such network map is built, each node computes the shortest (best) path to each destination node within its AS (for IGP-type protocols) or to remote ASs (for EGP-type protocols). Such shortest path computation is performed using the Dijkstra algorithm [Dijkstra 1959]. The entries into the routing tables consist of the calculated shortest paths using this algorithm. The shortest path consists of the path between a reference router and each destination node that has the lowest metric distance.

Unlike the distance vector protocols family where the whole routing tables are propagated, nodes using link-state protocols only send to the neighboring nodes the changes in the topology,* and this changes are only sent whenever a change in the topology is detected (instead of periodically). The most widely known link-state protocols are the OSPF and the IS-IS.

The Dijkstra algorithm is used to compute the shortest path, that is, the minimum distance, between any two nodes that use link-state protocols. This is the reason why this algorithm is also referred to as shortest path. This is the tool utilized by link-state protocols for the calculation of routing tables.

This procedure needs to be implemented independently for all nodes. Each node is considered, at a time, as the reference node (initial node), and the minimum distance is calculated to all other network nodes. Then, another node is used as the reference node.

We assume that the distance between adjacent nodes is already known (calculated using the *hello* packets and propagated using the LSPs). Note that the distances can be different in different directions. The Dijkstra algorithm considers the following steps:

1. Set all nodes as unvisited, and the initial node as the reference one.
2. Assign an initial distance zero to the reference node (node A, in Figure 11.26) and infinity to all other nodes. The distance of a node is the distance between the initial node and such remote node.
3. Calculate the distance from the initial node to each one of the adjacent nodes through the several different paths (routes). Assume the selected path between the initial node and each adjacent node as the path that presents the lowest distance (shortest path). Note that the shorter distance may have a higher number of hops, whereas a longer distance may present a lower number of hops. Mark these adjacent nodes as visited, and register the nodes' distance as lowest computed distance (instead of infinity). Note that the path to the visited nodes will not be computed again (the best path has already been selected).

---

* For example, new nodes that enter the network, nodes that leave the network, links that become out of order, a cost/metric to another node that suffers a variation, and so on.

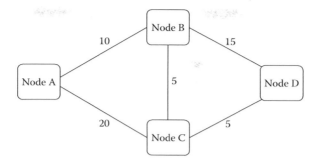

FIGURE 11.26 Example of a network whose shortest path is calculated with the Dijkstra algorithm.

4. Repeat the distance calculation to adjacent of adjacent nodes, using the procedure previously described for the adjacent nodes. Increase the distance from the initial node successively by one, until the distance between the initial node. At the end of this procedure, all network nodes have been calculated.

Let us consider the example of the network plotted in Figure 11.26 to describe the Dijkstra algorithm. Using the node A as the initial node, we set the distance to itself as 0 and infinity to all others. According to procedure 3, the minimum distance to node B is 10 (direct path) and the minimum distance to node C is 15 (two hop path, through node B).[*] Set these the node's distances. Then, according to procedure 4, the minimum distance to node D is to be computed. The distance through node B is 25, whereas the distance through node C is 20.[†] Consequently, the node D distance is set to 20, and the path between node A and node D is selected through nodes B and C. As a conclusion, the calculated distances are as follows:

- Node A: 0
- Node B: 10 (direct path)
- Node C: 15 (through node B)
- Node D: 20 (through node B and node C)

From this data, the routing table of node A can be constructed with the minimum distances to all remote nodes. Consequently, the initial node (node A) has knowledge that it has to send all the packets to node B (the default interface to use). Then, this procedure needs to be repeated to other network nodes, using the others as the reference nodes.

11.4.3.1 Open Shortest Path First

The OSPFv1 was initially released in 1989, but was never deployed. Later on, in 1991, OSPFv2 was released by RFC 1247, consisting of an updated version of the previous OSPFv1. In 1998, an update to the OSPFv2 was standardized by RFC 2328. The IPv6 uses the OSPFv3, which is the OSPFv2 properly modified to accommodate the IPv6 address space [RFC 2740], as well as to integrate the link-local address of the IPv6.

The OSPF protocol is a link-state protocol, and therefore, it computes the whole network topology, as opposed to distance vector protocols that normally limits to compute the distance to multiple destinations.[‡] Moreover, as can be seen from Table 11.3, OSPFv2 consists of a classless routing

---

[*] Note that the direct path between nodes A and C does not correspond to the lower distance.

[†] Note that the distance to node C was already set as 15 (mark as visited). Consequently, the route between nodes A and D through node C was not recomputed.

[‡] The EIGRP is an exception, as it is a distance vector protocol, but computes the whole network topology.

protocol, supporting classless and VLSM addressing, discontiguous networks, and comprising the transmission of the subnet masks in route updates, alongside the destination address. Note that the OSPF protocol does not automatically summarize routes at major network boundaries. Similar to EIGRP, the OSPF route updates are only sent when the network changes. This represents a difference to the RIP, which comprises the periodic exchange of the whole network information. Comparing the OSPF against other routing protocols, the following additional advantages are presented:

1. The OSPF supports type of service.
2. The OSPF allows the subdivision of an AS into subgroups, by using different areas.

Contrary to RIP whose network information is exchanged using UDP datagrams, OSPF does not use UDP or TCP. OSPF updates are encapsulated directly into IP datagrams with protocol number 89. The AD of the OSPF protocol is 110. This means that the OSPF protocol is less reliable than EIGRP, whose AD is 90 for internal routes, but more reliable than RIP, whose AD is 120.

With the purpose of computing the whole network topology and routing tables, the OSPF routing protocol comprises a number of different packet types:

- *Hello packets*: It used to discover neighbors and to establish adjacencies, consisting of the initial stage of this routing protocol. By default, hello packets are exchanged in multicast mode, every 30 s (hello interval). Moreover, by default, the dead interval corresponds to four times the hello interval. In case the dead time is reached without having received a hello packet, the adjacent network in cause is considered down. It is worth noting that the dead interval must be higher than the hello interval, or the network will not work properly. Moreover, the hello interval and the dead interval must be the same between neighbors. Hello and dead intervals can be modified using the following Cisco IOS syntax applied to a certain interface (interface configuration submode):
  - Router(config-if)#ip ospf hello-interval *hello_interval*
  - Router(config-if)#ip ospf dead-interval *dead_interval*
- *Database description*: These packet types are exchanged by neighboring routers, after the establishment of adjacencies, for the purpose of synchronizing the link-state databases of multiple routers with information about the network.
- *Link-state request (LSR)*: It is used to request link-state updates.
- *Link-state update (LSU)*: It is used to exchange link-state advertisements (LSAs), that is, to exchange information about neighbors, and neighbor of neighbors, and the corresponding path costs. This information is stored in the link-state database. The exchange of LSUs is performed using the flooding algorithm, with confirmations. This exchange of information becomes problematic in case of multiaccess networks with a high number of routers, leading to enormous consumptions of bandwidth and traffic. This problem can be overcome by using a designated router (DR), as described in the following.
- *Link-state ACKnowledgment (LSAck)*: It is used to acknowledge correct receipt of the other type of OSPF packets. Note that the OSPF protocol has its own confirmation protocol.

Let us focus again on the internetwork depicted in Figure 11.14. As can be seen from the routing table of Router1, depicted in Figure 11.27, the letter O is shown in an OSPF entry of a routing table, standing for OSPF routing protocol. As can be seen from Table 11.2, the AD of the OSPF routing protocol corresponds to 110, whose value is also shown in the routing table, alongside the cost, inside brackets. The remaining parameters shown in the OSPF routing table are the same as those described for the RIP and for the EIGRP routing table.

Figure 11.28 shows the routing table of Router2. Since there are two equal cost routes to reach the network 191.123.3.0, those two routes appear in the routing table, namely via 10.20.60.2 and 191.123.2.2. This is the only case when a routing table shows more than one route to a destination.

IOS Command Line Interface

```
Router>enable
Router#show ip route
Codes: C - connected, S - static, I - IGRP, R - RIP, M - mobile, B - BGP
       D - EIGRP, EX - EIGRP external, O - OSPF, IA - OSPF inter area
       N1 - OSPF NSSA external type 1, N2 - OSPF NSSA external type 2
       E1 - OSPF external type 1, E2 - OSPF external type 2, E - EGP
       i - IS-IS, L1 - IS-IS level-1, L2 - IS-IS level-2, ia - IS-IS inte
       * - candidate default, U - per-user static route, o - ODR
       P - periodic downloaded static route

Gateway of last resort is 10.20.30.1 to network 0.0.0.0

     10.0.0.0/24 is subnetted, 4 subnets
C       10.20.30.0 is directly connected, FastEthernet0/1
O       10.20.40.0 [110/65] via 191.123.2.1, 00:02:13, Serial0/1/0
O       10.20.50.0 [110/65] via 191.123.3.1, 00:00:25, Serial0/1/1
O       10.20.60.0 [110/128] via 191.123.2.1, 00:00:14, Serial0/1/0
                   [110/128] via 191.123.3.1, 00:00:14, Serial0/1/1
     191.123.0.0/24 is subnetted, 3 subnets
C       191.123.1.0 is directly connected, FastEthernet0/0
C       191.123.2.0 is directly connected, Serial0/1/0
C       191.123.3.0 is directly connected, Serial0/1/1
S*   0.0.0.0/0 [1/0] via 10.20.30.1
Router#
```

FIGURE 11.27 Routing table of Router1 after OSPF configuration (refers to internetwork depicted in Figure 11.14).

IOS Command Line Interface

```
Router>enable
Router#show ip route
Codes: C - connected, S - static, I - IGRP, R - RIP, M - mobile, B - BGP
       D - EIGRP, EX - EIGRP external, O - OSPF, IA - OSPF inter area
       N1 - OSPF NSSA external type 1, N2 - OSPF NSSA external type 2
       E1 - OSPF external type 1, E2 - OSPF external type 2, E - EGP
       i - IS-IS, L1 - IS-IS level-1, L2 - IS-IS level-2, ia - IS-IS inte
       * - candidate default, U - per-user static route, o - ODR
       P - periodic downloaded static route

Gateway of last resort is 191.123.2.2 to network 0.0.0.0

     10.0.0.0/24 is subnetted, 3 subnets
C       10.20.40.0 is directly connected, FastEthernet0/0
O       10.20.50.0 [110/65] via 10.20.60.2, 00:02:00, Serial0/1/1
C       10.20.60.0 is directly connected, Serial0/1/1
     191.123.0.0/24 is subnetted, 3 subnets
O       191.123.1.0 [110/65] via 191.123.2.2, 00:03:57, Serial0/1/0
C       191.123.2.0 is directly connected, Serial0/1/0
O       191.123.3.0 [110/128] via 10.20.60.2, 00:02:00, Serial0/1/1
                    [110/128] via 191.123.2.2, 00:02:00, Serial0/1/0
O*E2 0.0.0.0/0 [110/1] via 191.123.2.2, 00:03:57, Serial0/1/0
Router#
```

FIGURE 11.28 Routing table of Router2 after OSPF configuration with an imported route (refers to internetwork depicted in Figure 11.14).

In this case, equal cost load balancing is implemented. In all other situations, the routing table simply shows a single route, corresponding to the one with the lowest cost. Note that, unlike EIGRP, in case a route is imported from another routing protocol (e.g., default route), and advertised in OSPF, the AD keeps as 110 (see Figure 11.28).

The configuration of the OSPF routing protocol using the Cisco IOS is similar to the EIGRP, with few differences. Naturally, before starting with the OSPF configuration, one must configure the interfaces. As can be seen from Figure 11.21, the OSPF configuration starts by issuing, from the configuration mode:

- Router(config)#router ospf *process_id*

where *process_id* stands for process identification (ID) number, which represents an instance of the routing protocol running on a router. In OSPF, the process ID number consists of a 16-bit length number, and may differ to other OSPF routers in the routing domain. The OSPF configuration is followed by the syntax:

- Router(config-router)#network *net_IP_address wild_card* area *area_number*
  - *Example*:
    - Router(config-router)#network 191.123.213.0 0.0.0.255 area 0
    - Router(config-router)#network 11.213.241.0 0.0.0.255 area 0

where *net_IP_address* stands for the network or subnetwork address connected to each router's interface, *wild_card* stands for the wild card mask (instead of subnet mask), and *area_number* stands for the area number to which this router belongs.[*] This command line must be repeated as many times as the number of different networks or subnetworks connected to the router. As described above, an advantage of the OSPF protocol results from the fact that an AS can be split into different areas, facilitating the management and simplifying routing tables. The area number (*area_number*) identifies an area within an AS.

In case a router's interface is not connected to another router (which is typically the situation of a router's interface connected to a LAN), one may configure the router not to send updates through such interface. This is possible using the following Cisco IOS syntax:

- Router(config-router)#passive-interface *interface*
  - *Example*: Router(config-router)#passive-interface fa0/0

where *interface* stands for the interface identification. Then, if a default route is to be included in the OSPF updates, similar to RIP, the following syntax is used[†]:

- Router(config-router)#default-information originate

Since the OSPF routing protocol does not automatically summarize routes at classful boundaries, disabling automatic summarization in a router is not required[‡] (Figure 11.30).

After having configured all routers, one can verify the routing tables by issuing the following syntax (see Figure 11.29)[§]:

- Router#show ip route

---

[*] The first area must be 0.
[†] Note that the command line utilized by the OSPF differs from that of EIGRP (*redistribute static*).
[‡] The command line "no auto-summary" is not required in OSPF.
[§] In a workstation, the command "route print" or "netstat -r" (typed in the Windows command line screen [MS-DOS prompt]) allows verifying the configured routing table. A workstation's routing table can be modified by issuing "route" followed by one of the options that can be viewed from "route /?" (issued in the Windows command line screen [MS-DOS prompt]). Moreover, the command "ipconfig" (typed in the Windows command line screen [MS-DOS prompt]) allows the visualization of the IP configuration in workstations' interfaces, the subnet mask, and the default gateway. In a workstation, the command "netstat" alone allows viewing all active TCP connections.

IOS Command Line Interface

```
Router>enable
Router#conf t
Enter configuration commands, one per line.  End with CNTL/Z.
Router(config)#ip route 0.0.0.0 0.0.0.0 10.20.30.1
Router(config)#router ospf 321
Router(config-router)#network 191.123.1.0 0.0.0.255 area 0
Router(config-router)#network 191.123.2.0 0.0.0.255 area 0
Router(config-router)#network 191.123.3.0 0.0.0.255 area 0
Router(config-router)#passive-interface fa0/0
Router(config-router)#default-information originate
Router(config-router)#^Z
Router#
%SYS-5-CONFIG_I: Configured from console by console

Router#
```

FIGURE 11.29 Cisco IOS configuration of Router1 with OSPF (refers to internetwork depicted in Figure 11.14).

Moreover, if one intends to verify the sequence of nodes followed by a packet until it reaches a certain destination IP address *dest_IP_address*, the following Cisco IOS syntax must be issued[*]:

- Router#traceroute *dest_IP_address*
  - *Example*: Router#traceroute 193.123.212.3

Figure 11.30 shows the neighbor table for Router1 of the internetwork depicted in Figure 11.14. This table can be viewed by issuing the following syntax:

- Router#show ip ospf neighbor

IOS Command Line Interface

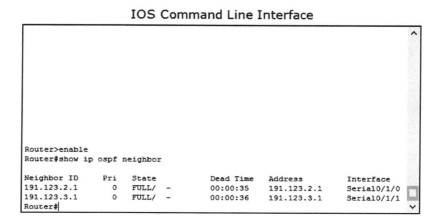

```
Router>enable
Router#show ip ospf neighbor

Neighbor ID     Pri   State         Dead Time   Address       Interface
191.123.2.1      0    FULL/  -      00:00:35    191.123.2.1   Serial0/1/0
191.123.3.1      0    FULL/  -      00:00:36    191.123.3.1   Serial0/1/1
Router#
```

FIGURE 11.30 Neighbor table of Router1 after OSPF configuration (refers to internetwork depicted in Figure 11.14).

---

[*] Similarly, the network administrator may use the commands "ping [dest_IP_address]" and "tracert [dest_IP_address]" in workstations to check the connectivity or to follow the sequence of nodes of a packet until they reach a certain destination IP address.

This syntax shows the different hosts connected to the router; with their addresses; and a number of other parameters, such as the interface, dead time, priority, and so on.

The information contained in the link-state database is utilized by the Dijkstra algorithm to calculate the shortest path to each destination node. Then, the result of this calculation is included in the routing table. The shortest path route (lowest cost) is selected first as path for the route. This is the origin of the name *open shortest path first*. The metric utilized by the OSPF protocol is as follows:

$$\text{Metric} = \sum_{n=1}^{\text{no of hops}} \frac{10^8}{(\text{bandwidth [bps]})} \tag{11.5}$$

where bandwidth (bps) stands for the bandwidth configured in the router's interface, expressed in bits per second. Note that the value for bandwidth used in the metric calculation does not correspond to the link bandwidth, but consists of a static bandwidth value configured in the router's interfaces, in the outgoing direction. Both sides of a serial link should be configured with the same bandwidth. Moreover, while in EIGRP the metric is calculated taking into account the lowest bandwidth of the multiple outgoing routers' interfaces in a route, using the OSPF routing protocol, one must calculate the metric for each hop and, according to Equation 12.5, the metric route is the cumulative sum of the independent metrics for different hops. Figure 11.5 depicts an example of an internetwork, with bandwidths and with the resulting costs (metrics).

By default, most serial interfaces use a bandwidth value of 1.544 Mbps, corresponding to a T1 interface. Moreover, the default bandwidth value for a fast Ethernet interface is 100 Mbps, and the default bandwidth value for a Gigabit Ethernet interface is 1 Gbps, and so on. One can modify the value for this parameter by entering into the interface configuration submode and issuing the following syntax:

- Router (config-if)#bandwidth *value_in_kbps*
  - *Example*: Router (config-if)#bandwidth 1544

where *value_in_kbps* stands for the bandwidth value we want to assign to the interface, expressed in kbps. Note that this command does not modify the physical bandwidth of the link. Moreover, instead of modifying the static bandwidth of a certain interface, which results in a changed cost, one can directly modify the cost. The cost can be modified in a certain interface by issuing the following syntax (interface configuration submode):

- Router(config-if)#ip ospf cost *cost_value*
  - *Example*: Router(config-if)#ip ospf cost 1234

where *cost_value* stands for the cost value to assign to the interface under configuration.

As can be seen from Figure 11.25 (same as EIGRP), one can verify the bandwidth values using the following syntaxes:

- Router#show interfaces (for checking the bandwidth values configured in all router's interfaces)
- Router#show interface *interface* (for checking the bandwidth value configured in a certain router's interface)
  - *Example*: Router#show interface fa0/0

As can be seen from Figure 11.31, the cost of an interface can also be verified in a router's interface by issuing one of the following syntaxes:

- Router#show ip ospf interface (for all interfaces)
- Router#show ip ospf interface *interface* (for a specific interface)

IOS Command Line Interface

```
Router#
Router#
Router#
Router#show ip ospf interface se0/1/0
Serial0/1/0 is up, line protocol is up
  Internet address is 191.123.2.2/24, Area 0
  Process ID 321, Router ID 191.123.3.2, Network Type POINT-TO-POINT, Cost: 64
  Transmit Delay is 1 sec, State POINT-TO-POINT, Priority 0
  No designated router on this network
  No backup designated router on this network
  Timer intervals configured, Hello 10, Dead 40, Wait 40, Retransmit 5
    Hello due in 00:00:01
  Index 2/2, flood queue length 0
  Next 0x0(0)/0x0(0)
  Last flood scan length is 1, maximum is 1
  Last flood scan time is 0 msec, maximum is 0 msec
  Neighbor Count is 1 , Adjacent neighbor count is 1
    Adjacent with neighbor 191.123.2.1
  Suppress hello for 0 neighbor(s)
Router#
```

FIGURE 11.31 Verifying the cost parameter with the command "show ip ospf interface serial0/1/0" in Router1 of Figure 11.14.

In the OSPF environment, a router is known by its router identifier (router ID). A router ID is assigned to a router using one of the following options, by descending order:

1. By statically assigning a certain IP address as router ID using the syntax:
   a. Router(config-router)#router-id *ip_address* (*ip_address* is an IP address to be used as router ID; this command can be issued when configuring a router with the OSPF).
2. If router ID is not explicitly configured, as described above, the highest IP address of any of the loopback interfaces is used as router ID.
3. If router ID is not explicitly configured and loopback interfaces are not configured, the highest IP address of any of the active interfaces is used as router ID.

The advantage of using the IP address of a loopback interface over an active interface relies on the fact that a loopback interface does not fail, whereas an active interface may fail. This results in an improved OSPF stability. As can be seen from Figure 11.31, the router ID can be verified using the following syntax:

- Router#show ip ospf interface

Moreover, the following syntax also allows verifying the router ID:

- Router#show ip protocols

As described above, the exchange of LSU packets is performed using the flooding algorithm, with confirmations. This becomes problematic in case of multiaccess networks with multiple routers (see Figure 11.32), leading to enormous consumption of bandwidths and traffics. This problem can be overcome by using a DR and a backup DR (BDR). Let us consider a multiaccess network, with a central switch and with n routers connected to it. The number of adjacencies comes $n(n-1)/2$, that is, the number of adjacencies increases exponentially with the number of routers n. In this case, the number of LSU packets would be enormous, as well as their corresponding confirmations. This would consume a high amount of bandwidth. Therefore, instead of establishing $n(n-1)/2$ adjacencies, within a multiaccess network (not in point-to-point networks) an elected DR is responsible for establishing adjacencies, that is, to send and receive LSAs. Moreover, there is a BDR that takes up on duty, in case the DR becomes unavailable. In this case, other routers of the multiaccess network (DRothers) send LSAs to DR and BDR using the multicast address 224.0.0.6. Then, the DR forwards LSAs to DR others using the multicast address 224.0.0.5.

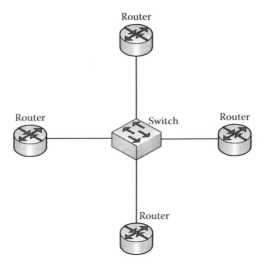

FIGURE 11.32 Example of a multiaccess network.

Within the multiaccess network, the router with the highest OSPF interface priority becomes the DR, whereas the router with the second highest OSPF interface priority is elected as the BDR. In case interface priorities are the same, the router with the highest router ID is elected as the DR/BDR.

The priority of an interface can be modified issuing the following syntax (in the interface configuration submode):

- Router(config-if)#ip ospf priority *priority_number*
  - *Example*: Router(config-if)#ip ospf priority 3

where *priority_number* is the priority number to assign to the interface, within 0–255.

As can be seen from Figure 11.30, in case one intends to view the interface priority of the neighboring routers, the following syntax must be issued:

- Router#show ip ospf neighbor

Moreover, for verifying the identification of the DR and BDR (in case there is one), the following syntax is used:

- Router#show ip ospf interface

Similar to RIP and EIGRP, timer information configured in OSPF can be verified using the following syntax (e.g., hello time and dead time):

- Router#show ip protocols

In addition, with this syntax one can also verify many other details about the route configuration, such as whether or not any interface is in passive mode. Finally, if one wants to view real-time OSPF routing updates and other network parameters as they modify state, the following syntax must be issued:

- Router#debug ip ospf events

Appendix III contains an example of a network configuration, including the configuration of the different interfaces.

11.5 INTERNET CONTROL MESSAGE PROTOCOL

The ICMP is a protocol used by operating systems to allow the exchange of error messages, control messages, for diagnostic purposes, or for flow control purposes of a network. This protocol is used by workstations and routers.

As can be seen from Figure 11.11, we see that ICMP* belongs to the Internet layer, being defined by RFC 792. Nevertheless, the ICMP is implemented over the IP (which is also an Internet layer protocol). In fact, an ICMP message is directly encapsulated into an IP datagram, which provides a connectionless and nonconfirmed service.

Examples of ICMP messages include the following:

- *Time to live exceeded message*: This message is sent back to a sender of a datagram, when it is discarded. This is signalized to the user with the signal . .
- *Destination unreachable message*: This message is sent back to a sender of a datagram when the destination's IP address could not be found, that is, when the routing table contained in a router has no information about the destination IP address contained in a packet. This is signalized to the user with the signal *U*.
- *Echo request/reply message*: The echo request message is sent to a certain destination's IP address to check whether or not it exists. The response to the echo request message is the reply message. These two messages are used by the *ping* command.† This is signalized to the user with the signal !.
- *Redirect message*: This message is used when a path of a datagram is different from that considered by previous datagrams.

11.6 FRAGMENTATION AND REASSEMBLING

A datagram is encapsulated into a frame at the transmitter side, and de-encapsulated at the receiver side. Therefore, the maximum datagram/packet‡ size depends on the frame type used to encapsulate it.

A router is a device that is responsible for performing switching of packets between different networks, or between a network and a network segment. A router normally makes use of different access technologies (data link layer) in different interfaces, which translates in different frame types and maximum payload sizes. The MTU is the largest number of bytes that can be carried in a frame payload.§ Note that the payload length of the data link layer frame is normally variable, depending on the amount of data to transport between two adjacent nodes.

If a router is switching a packet from a data link layer with a longer MTU into a data link layer with a shorter MTU, and if the datagram length is longer than the shorter MTU, fragmentation is implemented. In this case, each packet is decomposed into a group of smaller pieces. Conversely, when a router is switching a packet from a data link layer with a shorter MTU into a longer MTU, reassembling occurs, that is, the small pieces are re-grouped so as to re-create the packet with the original size.

---

* The ICMP used with the IPv4 is also referred to as ICMPv4, whereas the ICMP protocol used with IPv6 is called ICMPv6.
† The *ping* command is used to test for network connectivity. The "ping" command (as well as the "tracert" command) makes use of the ICMP to send messages between devices [RFC 792].
‡ The term *datagram* is used by IPv4, whereas the IPv6 uses the term *packet*. Since many of the functions described in this chapter are applicable to both IPv4 and IPv6, these terms are used interchangeably.
§ Note that frame header is not included.

Unlike to the IPv4 where fragmentation is carried out by intermediate nodes (routers), IPv6 fragmentation is only implemented by end stations of an end-to-end connection. In this case, intermediate routers never implement fragmentation. Consequently, the IPv6 protocol must discover the MTU path, in advance, before data exchange is initiated. Table 11.6 shows the MTU of different data link layer technologies used in LAN.

It is worth noting that different packet fragments transport the maximum load (allowed by the MTU), whereas the last packet fragment transports the rest of the fragmented packet. Figure 11.33 depicts a router responsible for interconnecting a IEEE 802.11 LAN with an ISDN WAN segment. Since the LAN has a MTU = 2312 bytes, and since the WAN has a MTU = 1500 bytes, the router

TABLE 11.6
Default MTU for Different Data Link Layer Technologies

Data Link Layer	MTU (Bytes)
PPP	296
SLIP	296
Ethernet	1500
ISDN	1500
HDLC	1600
IEEE 802.11	2312
IEEE 802.5	4464

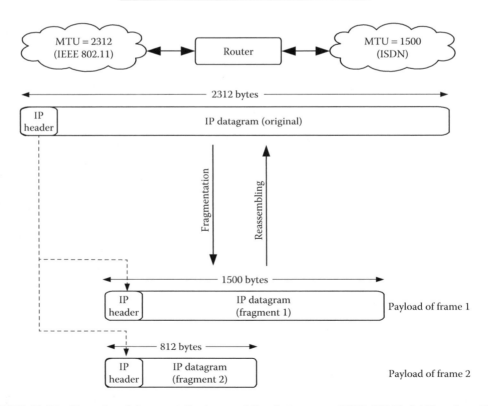

FIGURE 11.33 Example of fragmentation/reassembling between an IEEE 802.11 LAN and an ISDN segment.

needs to implement fragmentation in one direction (LAN to WAN) and reassembling in the opposite direction (WAN to LAN). As can be seen, the first fragment transports the maximum load (1500 bytes), whereas the second fragment transports the rest of the packet (2312 − 1500 = 812 bytes).

11.7 IPv6 TRANSITION AND CONFIGURATION

11.7.1 TRANSITION FROM IPv4 INTO IPv6

The transition from IPv4 into IPv6 is normally implemented in several steps. This migration should be smooth and normally includes a mapping between IPv4 and IPv6 addresses.

Dual stacking: When a router is simultaneously connected to IPv4 and IPv6 networks, the preferred option involves dual stacking. In this case, a router or a multilayer switch is simultaneously running IPv4 and IPv6 protocols either in the same interface or in different interfaces. In the former case, the interface is considered as dual-stacked. The dual stacking is the preferred solution to migrate from IPv4 into IPv6, as it allows a gradual transition.

6to4 Tunnel: When the traffic is mainly IPv6, whereas part of the network is still IPv4, a common procedure relies on the establishment of an implicit IPv6 tunnel in the IPv4 network to allow the exchange of IPv6 packets encapsulated into IPv4 datagrams (see Figure 11.34). In this case, the IPv6 protocol makes use of the IPv4 network as a data link layer. This requires the communication between IPv4 and IPv6 nodes. In this case, the protocol field of the IPv4 header presents the value 41.

The source host (computer) sends IPv6 packets, as it is connected to an IPv6 network, but uses an IPv4/IPv6 address. The need to use IPv4/IPv6 address results from the fact that the host is connected to a hybrid network composed of IPv6 and IPv4 segments. Therefore, the source host encapsulates IPv4 addresses into IPv6 addresses (source and destination) using 6to4 unicast address mapping.

The 6to4 unicast address uses the 16 leftmost bits fixed to 2002 (2002::/16) of type 2002:XXYY:ZZWW::/48, defined as follows [Blanchet 2006; Davies 2008]:

- XX is the hexadecimal notation of the leftmost IPv4 octet
- YY is the hexadecimal notation of the second IPv4 octet
- ZZ is the hexadecimal notation of the third IPv4 octet
- WW is the hexadecimal notation of the rightmost IPv4 octet

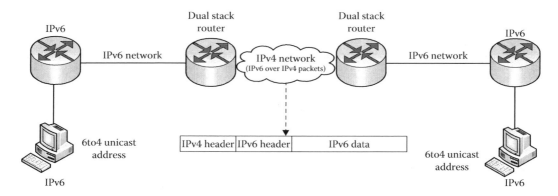

FIGURE 11.34 6to4 tunnel.

As an example, the IPv4 address 193.123.213.25 can be translated into the following IPv6 address: 2002:C17B:D519::/48. Note that the 32 bits of the IPv4 address uses the fields Res and NLA ID (see Chapter 10). Following this approach, the IPv4 nodes can be used to allow the exchange of IPv6 packets. An IPv4 router of a tunnel reads the destination IPv6 address to find the output interface to forward the packet, but only considers a part of the address (bits 17th to 48th).

Manually configured IPv6 tunnel: This is an alternative to 6to4 tunnel, used in a similar scenario, but with the difference that the IPv4 network utilized by IPv6 packets is static. In this case, IPv6 packets are also encapsulated into IPv4 datagrams when they are transported within the IPv4 network, and the edge routers utilized in the border between the IPv6 and IPv4 networks are dual stack routers. The source and destination hosts use normally IPv6 addresses (source and destination). As can be seen from Figure 11.35, the ingress IPv4 router encapsulates the IPv6 packet into the IPv4 datagram data field, and uses the ingress IPv4 address as IPv4 source host, while the IPv4 destination host corresponds to the IPv4 address of the egress router of the IPv4 tunnel.

Hybrid configuration: Another possibility is a hybrid configuration, whereas some nodes are IPv4 and others are IPv6. In this case, the approach may rely on encapsulating an IPv4 address into an IPv6 address. This is normally referred to as *IPv4-mapped-IPv6 address*, being defined as 0:0:0:0:FFFF:<ipv4 address> [Blanchet 2006].* In this case, only the 32 rightmost bits out of the 64 bits allocated to the interface ID are used (as described in Chapter 10).

11.7.2 IPv6—RIPng Configuration Using Cisco IOS

In Chapter 10, the configuration of IPv6 addresses in the routers' interfaces have been described. In this chapter, we focus on the configuration of the routing protocols. In this particular case, we focus on IPv6, and on the configuration of the RIPng, consisting of the RIP version specifically modified for use with IPv6.

As described in Chapter 10, when configuring a Cisco router with IPv6, the first action relies on activating IPv6 traffic forwarding. This is performed from the configuration mode by typing the following syntax:

* Router(config)#ipv6 unicast-routing

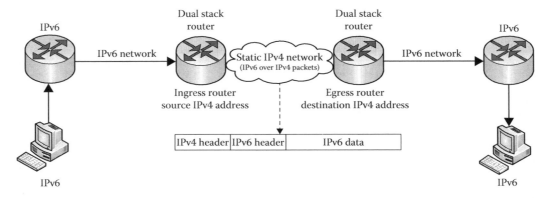

FIGURE 11.35   Manually configured IPv6 tunnel.

---

* Naturally, this is equivalent to ::FFFF:<ipv4 address>.

Then, the router's interfaces must be configured with IPv6 addresses. The reader should refer to Chapter 10 (interface configuration submode) for a detailed description of an interface configuration with IPv6 addresses.

The configuration of the RIPng protocol in Cisco routers using the Cisco IOS is performed using the following syntaxes:

- Router(config)#ipv6 router rip *process_ID* (*process_ID* identifies a RIPng process)
- Router(config)#interface *interface*
- Router(config-if)#ipv6 rip *process_ID* enable
    - *Example*:
        - Router(config)#ipv6 router rip rip_process
        - Router(config)#interface ethernet0/1
        - Router(config-if)#ipv6 rip rip_process enable
        - Router(config)#interface serial0/0
        - Router(config-if)#ipv6 rip rip_process enable

Finally, it is worth to refer that the commands utilized for verifying IPv6 configuration are similar to those utilized in IPv4, but replacing the command IP by IPv6 (e.g., "show ipv6 interface," "show ipv6 interface *interface*," "show ipv6 interface brief," and "show ipv6 route").

11.8   CISCO DISCOVERY PROTOCOL

CDP is a Cisco proprietary protocol used to detect neighbor Cisco equipment (e.g., routers, switches, and firewalls), and to verify their parameters, including the following:

1. Device identifiers (e.g., hostname)
2. Address list (e.g., addresses of network interfaces)
3. Interface identifier (e.g., Ethernet 0/0, serial0/1)
4. Capabilities list of the equipment
5. Type of platform (e.g., Cisco Router 1841)
6. Hold time

CDP is a layer 2 protocol, which allows detecting neighbor Cisco equipment, not equipment beyond neighbors. The action of gathering information about neighbor Cisco equipment is possible by entering the following syntax:

- Router#show cdp neighbors

Moreover, more detailed information can even be obtained by issuing the following syntax:

- Router#show cdp neighbors detail

Finally, after getting information about a neighbor, one can establish a telnet session with it by issuing the following syntax:

- Router#telnet *ip_address*

where *ip_address* is the IP address of the neighbor equipment, previously obtained using the CDP. This allows gathering information about the equipment with which a telnet session is established, and then, gathering information about neighbors of neighbors using the CDP. This process can be repeatedly performed, such that a network administrator gathers information about all the Cisco devices present in its network.

Note that the CDP is typically enabled by default. This may represent a vulnerability from the security point of view, as an intruder may gather information about the network devices using the CDP. For the sake of security protection, it is a good rule to disable the CDP by issuing the following syntax from the configuration mode:

- Router(config)#no cdp run

CHAPTER SUMMARY

This chapter provided a view about the routing and configuration of the Internet layer, including the description of the routing algorithms and protocols. Routing protocols are used by routers to build routing tables. It is based on routing tables that the routers decide the output interface that should be used to forward a datagram, and to make it reach the recipient's address.

The concept of administrative and metric distance is utilized by routers to build the routing tables. Route table lookup consists of a mechanism used by routers to make routing decisions. It was viewed that the route table lookup is based on the longest match, between the destination IP address and the different entries of the routing table. Route summarization consists of a mechanism that reduces the length of routing tables and fastens routing tables look-up. Static route, default route, and flooding were introduced. It was viewed that the default route is a specific type of static route.

This chapter described the distance vector protocols, which comprise the computation of the distance to each destination network based on a certain metric. The RIP and the EIGRP are two distance vector protocols, whose configuration in routers was described in this chapter, using the Cisco IOS.

This chapter also described the link-state protocols, which comprise the computation of a topologic database in each router, with the overall map of network connectivity. The Dijkstra algorithm was described, consisting of the algorithm utilized by link-state protocols to compute the metric to different destination networks. The OSPF is an example of a link-state protocol that has been studied in this chapter, including its configuration in routers using the Cisco IOS.

The ICMP was also described, being used by the network for management purposes. Different transition architectures used to implement the migration from IPv4 into IPv6 were described. The configuration of the RIPng was addressed, consisting of the RIP used in IPv6.

Finally, the CDP was described, as well as its configuration in Cisco devices using the Cisco IOS.

REVIEW QUESTIONS

1. In RIP, what is the administrative distance?
2. Consider a router with the following routing table:
   O 173.168.235.0/24 [110/782] via 174.168.235.1, 00:00:00, Serial1/0
   C 174.168.235.0/24 is directly connected, Serial1/0
   C 175.168.235.0/24 is directly connected, FastEthernet0/0
   C 176.168.235.0/24 is directly connected, Serial1/1
   a. Draw the network diagram corresponding to the information obtained from the routing table.
   b. According to the routing rules, what is the address of the next router to which the datagrams should be forward in order to reach a host with the destination IP address 173.168.235.3?
3. What is the meaning of the Cisco IOS command "IP route 0.0.0.0 0.0.0.0 194.136.235.1"?
4. What does MTU stand for? What is the relationship between MTU and frame size?
5. Which categories of routing algorithms do you know?

6. What are the most important differences between a link-state routing protocol and a distance vector routing protocol?

7. What is the default maximum number of hops allowed by the RIP before a datagram is discarded?

8. What is the relationship between the MTU and the datagram length? What measures can be implemented to adjust these two parameters?

9. What is the meaning of the Cisco IOS command "IP route 193.123.213.0 255.255.255.0 194.136.235.1"?

10. Given the IPv4 address 194.136.235.1, what is the corresponding IPv4-mapped-IPv6 address?

11. What does fragmentation stand for?

12. What does administrative distance stand for?

13. What does metric distance stand for?

14. Consider a router with the following routing table:
    C 175.168.235.0/24 is directly connected, FastEthernet0/0
    C 176.168.235.0/24 is directly connected, Serial1/1
    C 177.168.235.0/24 is directly connected, Serial1/0
    R 178.168.235.0/24 [120/2] Via 176.168.235.2
    R 179.168.235.0/24 [120/3] Via 177.168.235.1

    a. Draw the network diagram corresponding to the information obtained from the routing table.

    b. According to the routing rules, what is the address of the next router to which the datagrams should be forward in order to reach a host with the destination IP address 178.168.235.3?

15. In the OSPF protocol, how do you quantify the metric distance?

16. Which type of routing protocols compute a topologic database in each router, with the overall map of connectivity?

LAB EXERCISES

1. Consider the network diagram depicted in the figure below. Using the Packet Tracer simulator, configure the whole network, assuming static route. Verify the rooting tables in each of the routers and hosts. Before starting, *clean* the startup configuration, and then re-load the routers with the *cleaned* startup configuration. In the three routers, configure a ciphered access password to the privileged mode and another to the console port, a hostname, and a motd.

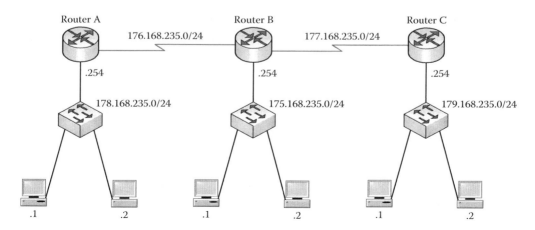

2. Repeat the previous exercise, but with the RIPv2 routing protocol, instead of static route.
3. Repeat the previous exercise, but with the EIGRP.
4. Repeat the previous exercise, but with the OSPF routing protocol.
5. Repeat the previous exercise, but using real network equipment.
6. Consider the network diagram depicted in the figure below. Using the Packet Tracer simulator, configure the whole network, using the OSPF routing protocol. Keep the network 194.123.2.0/24 as a floating root. Verify the rooting tables in each of the routers and hosts.

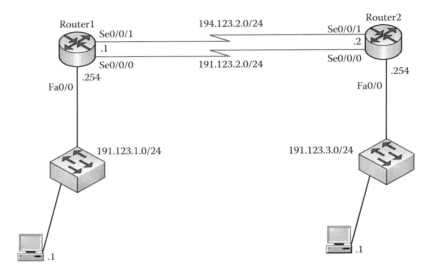

7. Repeat the previous exercise, but with static route, instead of OSPF.
8. Repeat the previous exercise, but using real network equipment.
9. Consider the network depicted in the figure below. Using the Packet Tracer simulator, configure the whole network, using OSPF. Before starting, *clean* the startup configuration, and then re-load the routers with the *cleaned* startup configuration. In both routers, configure a ciphered access password to the privileged mode and another to the console port, a hostname, and a motd.

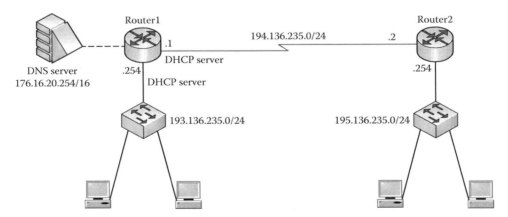

10. Consider the network diagram depicted in the figure below. Using the Packet Tracer simulator, configure the whole network, assuming EIGRP. Verify the rooting tables in each of the routers and hosts.

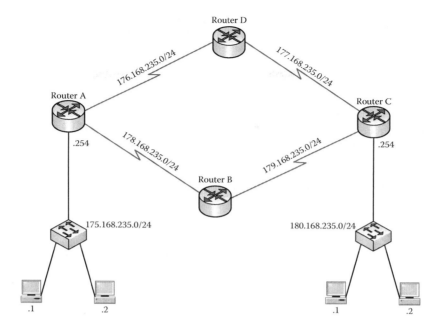

11. Consider the network diagram depicted in the figure below. Using the Packet Tracer simulator, configure the whole network, assuming static route. Verify the routing tables in each of the routers and hosts. What is the route selected from Router C to reach the host located in the network with IP address 178.168.235.0/24.

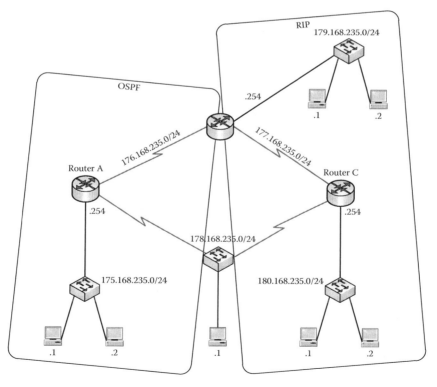

12. Repeat the previous exercise, but using real network equipment.

12 Data Link Layer

LEARNING OBJECTIVES

- Define the data link layer (DLL) of TCP/IP.
- Describe the DLL protocols used in TCP/IP.
- Identify and describe the different sublayers of the DLL.
- Identify the different types of LAN devices and equipment.
- Describe and configure the spanning tree protocol (STPr) using Cisco IOS.
- Describe and configure virtual local area networks (VLANs) using Cisco IOS.
- Describe and configure the VLAN trunking protocol (VTP) using Cisco IOS.

The physical layer is responsible for allowing the exchange of bits between two adjacent nodes[*] of a network. However, these bits are subject to channel impairments, such as noise, interference, or distortion. All of these channel impairments may originate corrupted bits, which degrades the performance. As described in Chapter 3, the bit error probability increases with

- The increase in the power of noise, interference, and distortion.
- The increase in distance without regeneration.
- The decrease in the transmitted power.
- The decrease in the reliability of the transmission medium. A radio transmission medium is typically less reliable than an optical fiber. Consequently, the bit error probability of the former transmission medium tends to be worse than the latter.

The DLL makes use of error control techniques to keep the errors at an acceptable level. Depending on the medium that is being used to exchange data, error control can be performed using either error detection or error correction techniques. In the case of error detection, codes such as cyclic redundancy checks (CRCs) or parity bits are used to allow errors being detected at the receiver side, and the receiver may request the retransmission of the frame. However, if the medium is highly subject to noise and interferences, the choice is normally the use of error correction. In the latter case, the level of overhead per frame is higher, but it avoids successive repetitions, which also translates in a decrease of overhead. In both cases, the DLL handles blocks of bits to which the corresponding overhead is added. These blocks of bits are referred to as frame. Moreover, error correction can also be used when the transmission is unidirectional or when latency must be avoided (the handshaking associated with error detection and retransmission introduces delays).

As can be seen from Figure 12.1, the DLL is directly above the physical layer, which is responsible for allowing a reliable exchange of data between two adjacent nodes. The DLL is composed of the following two sublayers:

- Logical link control (LLC) that deals with flow control and error control. These functionalities are implemented using protocols such as IEEE 802.2, HDLC, PPP, or SLIP.
- Medium access control (MAC) that determines when a station is allowed to transmit within a LAN. Note that this sublayer exists only in the case of a LAN (and some types of MAN),

---

[*] For example, between adjacent routers, or between a workstation and a router.

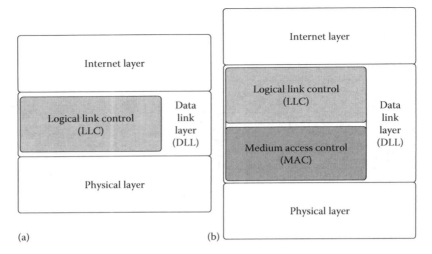

FIGURE 12.1   Data link layer of (a) WAN and (b) LAN/MAN.

that is, in multiuser networks. When stations share the transmission medium in a LAN, it is said that the access method is with collisions (e.g., Ethernet). In this case, the MAC sublayer is responsible for defining when a station is allowed to transmit in such a way that collisions among transmissions from different stations are avoided (which originate errors). On the other hand, when stations of a LAN do not share the transmission medium, it is said that the method is without collisions (e.g., Token Ring). This sublayer is implemented using protocols such as IEEE 802.3, IEEE 802.4, IEEE 802.5, or IEEE 802.11.

The Internet layer makes use of the services provided by the DLL. In fact, the Internet layer is an extreme-to-extreme layer, which is responsible for allowing data being routed along the several network nodes, such that it reaches the final destination. Therefore, the Internet layer makes use of a concatenation of DLLs, a different one for each different network hop.[*] A hop can be a satellite link, using a certain DLL protocol, whereas another link can be an optical fiber link, using the same or another DLL protocol.

12.1   LAN DEVICES

Before the description of error control and flow control is given, it is worth describing the devices employed within a LAN or within a MAN.

12.1.1   Hub

A hub is a network device that retransmits in all output interfaces[†] the bits present at one of its input (see Figure 12.2). In addition, it performs regeneration of signals, which mitigates the channel impairments (noise, interference, distortion, etc.). The hub also acts as a repeater, which is an important functionality when the network is longer than the maximum segment size (MSS). Nevertheless, the hub is not able to detect errors.[‡] The star is the most common physical topology of a network that employs a hub. Using a hub (repeater) as a central node, the logical topology of the network

---

[*] A link between two adjacent nodes.
[†] Contrary to the switch.
[‡] Contrary to some types of bridges and switches.

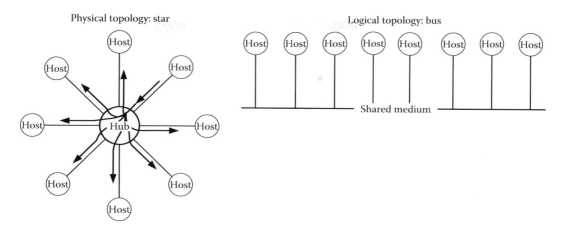

FIGURE 12.2 Physical and logical topology of a network employing a hub.

becomes a bus topology. By definition, a bus topology consists of a network that makes use of a common transmission medium. It is worth noting that the hub deals with bits, and therefore it works at layer 1 of the OSI model.

12.1.2 BRIDGE

Contrary to a hub, which is normally employed as the central node of a network, a bridge is typically employed to interconnect two network segments. Moreover, these two segments may use the same MAC sublayer protocol or different MAC sublayer protocols,* as long as the LLC sublayer protocol is the same. As defined below, both the IEEE 802.3 and IEEE 802.5 MAC sublayer protocols make use of the common IEEE 802.2 LLC sublayer protocol, and therefore a bridge may be used to interconnect these two types of networks.

Another common application of the bridge relies on its ability to achieve a segmentation of a collision domain. The bridge breaks a unique collision domain into two smaller collision domains (see Figure 12.3). Naturally, this reduces the number of hosts that share the medium, and to where the CSMA-CD mechanism is applied. Consequently, the network performance tends to be improved. Moreover, because a certain network host stops receiving all the network frames, this represents an advantage from the security point of view.

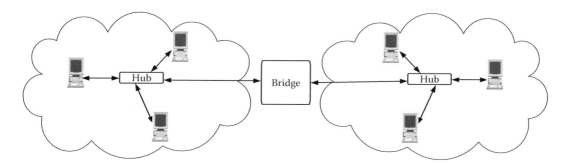

FIGURE 12.3 Network segmentation with a bridge.

---

* As long as the bridge interfaces are compatible with these two specific types of MAC sublayer protocols.

In any of the above-described bridge applications, when a frame arrives a bridge port, it must decide whether or not the frame has to be forwarded to the other network segment. This decision is taken based on the destination MAC address of the frame, that is, using layer 2 of the OSI model.

The mapping between the output port and the MAC address is performed based on a MAC address table,* which is stored in the bridge. Initially, this table is empty. Consequently, when a switch receives a frame, it forwards the frame to all output interfaces (ports). As the bridge receives frames, it registers the port and the corresponding source MAC address of the frame, and registers this mapping in the MAC address table. After having exchanged a certain number of frames, the table stored in the bridge has full knowledge about the network segments where each of the MAC addresses of the network nodes is located.

There may be multiple paths between two network points. Multiple paths exist in the case of a network with a mesh topology (see Chapter 1). In this case, the bridge has to decide which one to use. The discovery and decision of the paths is performed making use of the STP (IEEE 802.1D). The obtained information is utilized to remove loops, as well as to keep information about redundant paths. In case a path is interrupted, the bridge switches to the backup path. Moreover, there are situations where a bridge uses two or more paths simultaneously to perform load balancing.

Contrary to the hub, which performs the repetition of bits by hardware, the bridge operation is performed by software. Consequently, the delay introduced by this network device is typically much higher.

The bridge works using the store-and-forward mode, defined as follows: it accepts a whole received frame and stores it in memory. Then the CRC calculation is performed and the output interface is calculated using the MAC address table. It still verifies the frame delay between the origin and destination, before it forwards it to the output interface.

12.1.3 SWITCH

Similar to a bridge, a switch performs the segmentation of a network into smaller collision domain segments. In fact, in case each segment connects a single host, the collision domains are avoided, and consequently, the CSMA-CD mechanism described below is not applicable. However, while the bridge is typically equipped with only two interfaces (which can be of different MAC sublayer types), the number of interfaces of the switch is typically high. Moreover, these interfaces are typically of the same type (e.g., IEEE 802.3/Ethernet protocol). Nowadays, the switch tends to be the central node of the network, instead of the previously used hub. Moreover, the bridge has been replaced by the switch.

It is possible to have different switches linked in cascading† (see Figure 12.5). Typically, an IEEE 802.3 network is implemented using a star as the physical topology. Nevertheless, while the corresponding logical topology of a network employing a hub is the bus, the star is the logical topology that results from the use of the switch as the central node. This results in a significant improvement in the network performance. A network using a hub as a central node only allows a host transmitting at a time. In contrast, a switch allows up to half of the network hosts transmitting to the other half of the network (half duplex).‡ Assuming that a LAN works at 10 Mbps in half-duplex mode, and that it has a total of ten hosts, then we may have up to five hosts transmitting at 10 Mbps to the other five hosts (see Figure 12.4). This results in a cumulative network throughput of 50 Mbps.

---

* A MAC address table is similar to a routing table. While the routing table performs mapping between a destination IP address of a packet and an interface, the MAC address table performs mapping between a destination MAC address of the frame and a port.
† For example, a main switch may be used to serve a whole organization, while a second layer of switches may be employed for different departments.
‡ In case the network is full duplex, it is possible to have all network hosts transmitting to all network hosts, simultaneously, at the maximum data rate.

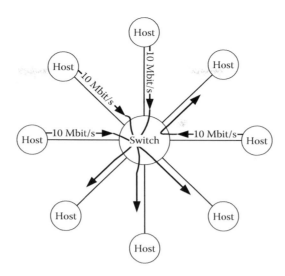

FIGURE 12.4 Example of half-duplex transmission in a LAN with a switch.

In case the network works in full-duplex mode, then the maximum cumulative network throughput becomes 100 Mbps.

It is worth noting that most of the switches currently available in the market present autosensing. This is a capability that allows a switch detecting and adapting to different throughputs in different interfaces, as well as to half or full duplex. Note that the switch may connect with different devices at different speeds. It may connect with hosts at 100 Mbps, whereas the connectivity with a server may be at 10 Gbps.

Similar to the bridge, the switch performs the forwarding of frames based on the destination MAC address. Therefore, this device works at layer 2 of the OSI model.* The forwarding of frames is performed based on a MAC address table, which maps destination MAC addresses into output ports. Similar to the bridge, the MAC address table is filled in as the switch receives frames from the corresponding interfaces (ports). Moreover, there may be more than one path between two network nodes. The STP (IEEE 802.1D) normally resolves that by using a metric based on the lower number of hops, while the other paths are kept as redundancy. Load balancing is also possible in high-dimensional LANs.

Contrary to the bridge, which performs the switching of frames by software, the switch performs its functionality through hardware. Consequently, the latency introduced by a switch is typically much lower than that of a bridge.

Depending on the type of switch, the forwarding of frames is performed using one of the following modes:

- *Store-and-forward*: It accepts a whole received frame and stores it in memory. Then the CRC is computed and the output interface is calculated using the MAC address table. It also verifies the frame delay between the origin and destination, before it forwards it into the calculated output. This is the mode that achieves the best level of integrity. Nevertheless, the delay introduced with this mode tends to be higher than in the case of the other modes.
- *Cut-through*: It only reads the initial octets of the frame header, up to the destination address field (to be able to compute the output port), before the frame is forwarded. The advantage

---

* In any case, it is currently possible to find network devices that perform both layer 2 (switch) and 3 (router) switching.

of this mode relies on the maximization of the throughput, while the disadvantage relies on the risk of forwarding corrupted frames.

- *Fragment free*: It only accepts the initial 64 octets of a frame before the selection of the output port is made and the switching path is established. This is used to allow the detection of a collision, as established by the CSMA-CD mechanism. Switches, using this mode, are normally connected to a hub,* and, therefore, their current application is limited.

12.1.4 SPANNING TREE PROTOCOL

As previously described, the switch and the bridge perform layer 2 switching based on the destination MAC address contained in a frame and taking into account the information present in a MAC address table.[†] A MAC address table maps MAC addresses into output ports. Initially, this table is empty. Consequently, the equipment forwards a frame received in one of the ports to all ports, except the one from where the frame was received. When a switch or a bridge receives frames from different ports, the corresponding source MAC addresses are registered and associated with certain ports (switch interfaces), resulting in the construction of the MAC address table. Most of the Cisco switches (e.g., Catalyst 2960 series switches) that make use of Cisco IOS allow checking MAC address tables, from the privileged mode, using the syntax:

- Switch#show mac-address-table

Small-size networks have typically a physical topology consisting of a single layer of switches (e.g., IEEE 802.3 networks). Medium- to high-dimensional networks have typically a topology with several layers of switches (see Figure 12.5). To improve the resistance from failures, it is a good choice to keep redundant paths in a network. This results in a network with a mesh configuration, where there exist multiple paths between different network points. In this case, a bridge or a switch has to decide which output port to use, among several possibilities. These decisions are taken by switches and bridges using the spanning tree protocol (STPr), which results in an efficient way of building MAC address tables. In this case, the output ports of a switch that belong to a selected route are placed in the forwarding state, whereas the others can be placed in the blocking state. IEEE 802.1D is a widely known STPr standard.

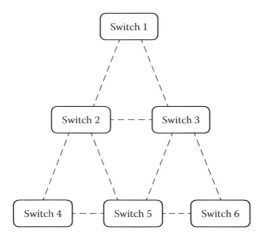

FIGURE 12.5   Example of a network with loops.

---

* This is the reason why the CSMA-CD mechanism is applicable.
† The MAC address table is also referred to as the switching table.

Figure 12.5 shows an example of a mesh network with several layers of switches, namely a main switch (corporate switch), second layer of switches (e.g., branch switches), and third layer of switches (e.g., division switches). As can be seen, the communication between switch 1 and switch 5[*] can be established through switch 2 or through switch 3. In addition, it can also be established through switch 4 or 6, although with a higher number of hops.

The STPr avoids loops by spanning the mesh network in a tree. When a switch or a bridge is established, an exchange of data among these devices is performed within the network, and a sequence of opening and closing of ports is executed. During this phase, the port LEDs of the equipment show the orange color. In the steady state, the loops are avoided by disabling those links between those two points that are not part of the selected path between any two points. Once the STPr finalizes its transient phase, the port LEDs of the equipment become green.

The STPr avoids loops in the network, while keeping information about redundant paths. The existence of loops results in flooding and network overload, which represents a performance degradation of the network, or even network breakdown. The STPr uses the spanning tree algorithm (STA) to determine whether switch ports are placed in the forwarding state or in the blocking state and to select a path, among different options.

There are two types of problems that can occur as a result of a frame loop:

- *Broadcast storms*: A broadcast frame is sent through all output ports except the one from where it was received. Let us suppose that switch 4 of Figure 12.5 forwards a broadcast frame (received from any connected PC) to switches 2 and 5. In a second stage, switch 2 forwards the broadcast frame to switches 1, 3, and 5, while switch 5 forwards the broadcast frame to switches 2, 3, and 6. Then, in a third stage, the switches that received the broadcast frame send the frame through all output ports except those from where they were received. This means that, for example, switch 5 forwards the broadcast frames to switches 3, 4, and 6. Noting that switch 4 is the one from where the broadcast frame was initially sent, the process is restarted inside the same loop. Contrary to the IP packet, the IEEE 802.3 frame does not include any field to discard it when it enters a loop.
- *Duplicate unicast frames*: Let us consider that switch 4 has a unicast frame to send to switch 1, and that its MAC address table does not have any entry with the output interface to use (the MAC address table is still empty). In this case, switch 4 forwards the unicast frame through all output interfaces except the one from where the unicast frame was received (it is assumed that it was received from a PC connected to switch 4). Therefore, switch 4 sends the frame to switches 2 and 5, duplicating the unicast frame. Then, these two switches forward the frames through only one output interface (in case they have the MAC address table properly filled) or through all other output interfaces. In the latter case, the unicast frame is further duplicated, originating more duplicated frames in the network. At the end of the path, switch 1 will receive several replicas of the unicast frame initially sent by switch 4.

Based on the previous description, we can summarize the aim of the STPr as follows:

1. Avoid broadcast storms
2. Avoid duplicate unicast frames
3. Define the best path to forward frames, based on the shortest path to the root bridge
4. Build the MAC address table

Note that items 1 and 2 above are obtained by avoiding the loops.

---

[*] In fact, the purpose is to establish a communication between switch 1 and a host connected to switch 5, not to switch 5 itself. Switch 5 is only an interim network device.

The STA starts by electing a switch within a broadcast domain that becomes a root bridge. The root bridge is the switch where the metric distances from all other network switches are computed. The root bridge of a network is the switch with the lowest bridge identification (BID). The BID is a switch identifier generated using the following switch parameters (see Figure 12.6):

- Priority value
- MAC address of the switch
- Optional extended system identification

The root bridge is the switch with the lowest priority (see Figure 12.6). In the case of two switches with the lowest priority, the one with the lowest MAC address is elected as the root bridge. Finally, in case priority values and MAC addresses are the same, the root bridge is the switch with the lowest optional extended system identification.

One can modify the priority value of Catalyst 2960 series switches using Cisco IOS, from the configuration mode, using the syntax:

- Switch(config)#spanning-tree vlan *vlan_number* priority *priority_value*

Note that this action is normally performed repeatedly for each different VLAN configured in the switch. From the above description, one can conclude that a network may have a certain switch acting as the root bridge for a certain VLAN, whereas the root bridge of another VLAN can be a different switch.

The root bridge election process is performed through the exchange of frames entitled bridge protocol data units (BPDUs) between switches. Initially, each switch sends BPDUs to its neighbors, with its BID and identifying itself and the root bridge. After converging, the root bridge is found and this information is propagated along the network.

Once the root bridge has been found, the STA makes each network switch to determine the shortest path to the root bridge. This shortest path becomes the selected one and the corresponding switch ports are placed in the forwarding state. The other switch ports can be placed in the blocking

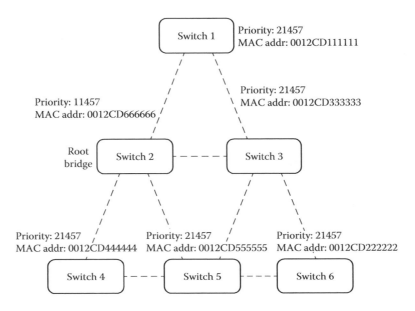

FIGURE 12.6 Root bridge election process.

state to block redundant paths,* that is, to avoid loops. A path cost is calculated using the port cost of switches, and these depend on port speeds. Table 12.1 lists the port costs for different link speeds, according to the revised IEEE specification. Let us suppose that one wants to find the shortest path from switch 4 to 2 (the root bridge), and that 1 Gbps trunk links are used between all switches. In this case, the direct path between these two switches has a cost of 4, whereas the path through switch 5 presents an overall path cost of $2 \times 4 = 8$. Therefore, because the direct path presents the lowest cost, this becomes the selected path, whereas the path through switch 5 is blocked. This can be seen from Figure 12.7.

One can modify the default IEEE port costs in Catalyst 2960 series switches using Cisco IOS by entering the interface configuration submode using the syntax:

- Switch(config-if)#spanning-tree cost *cost_value*

TABLE 12.1

Port Costs According to IEEE Specification

Link Speed	Port Cost
10 Mbps	100
100 Mbps	19
1 Gbps	4
10 Gbps	2

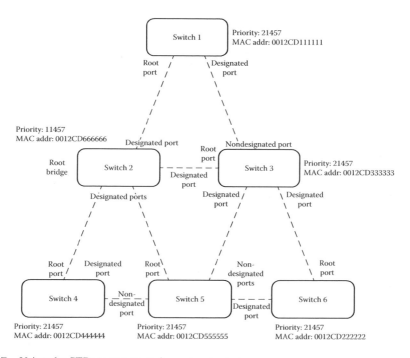

FIGURE 12.7 Using the STPr to overcome loops (port roles).

---

* As can be seen from Figure 12.7, the switch ports that belong to a redundant path can either be in the designated role (forwarding state) or in the nondesignated role (blocking state). When a switch port that belongs to a redundant path is in designated role, the port of the other switch connected directly to it is always in the nondesignated role (blocking state).

Note that the root bridge is a single point of failure. Therefore, the STA determines a primary root bridge and a secondary root bridge (the switch with the second lowest BID), for each VLAN. In case the primary root bridge is placed out of order, the secondary root bridge assumes these functions and the traffic is reestablished.

Depending on whether or not a switch port forwards traffic, and based on the switch proximity to the root bridge (based on the shortest path), switch ports have different roles (see Figure 12.7):

- *Root port*: A switch port in the forwarding state that is closest to the root bridge.
- *Designated port*: All switch ports placed in the forwarding state, except the one selected as the root port.
- *Nondesignated port*: All ports that were placed in the blocking state to prevent loops.

From the above description, one can summarize the three steps of the STPr:

- Elect the root bridge
- Elect the root ports
- Elect the designated and nondesignated ports

One can verify port and path costs, the state of ports, the BID of the switch, as well as find out which switch is the root bridge, from the privileged mode, using the following syntax:

- Switch#show spanning-tree

12.2 LLC SUBLAYER

Similar to the upper layers that make use of the services made available by the lower layers, the DLL makes use of the service provided by the physical layer. Because the physical layer is responsible for a nonreliable exchange of bits, the DLL adds the required reliability. Consequently, the LLC sublayer is responsible for allowing a reliable exchange of data between two adjacent nodes of a network. This is achieved by implementing the following functionalities:

- Error control
- Flow control
- Grouping isolated bits into frames

Note that the LLC sublayer is established using one of the following modes:

- Connectionless and nonconfirmed
- Connectionless and confirmed
- Connection-oriented

As described in Chapter 1, the connection-oriented mode requires the previous setup of the connection before data is exchanged. In addition, it considers the connection termination after the exchange of data. Because the connection-oriented mode is always confirmed, it uses either error detection with retransmission of frames or error correction. Moreover, for error control, and for allowing the delivery of frames to the Internet layer in the correct sequence, the frames are numbered.

The nonconfirmed connectionless mode is utilized in scenarios where the error probability is reduced (e.g., the transmission of bits in an optical fiber), or in scenarios where an upper layer is responsible for performing error control. This mode does not assure reliability of data, that is, there is no feedback from the receiver to the transmitter about whether or not it was correctly received.

In case confirmation is used, error control can be implemented using either error detection with retransmission or error correction. In the case of error detection with retransmission, there is a feedback link informing the transmitter about whether or not the data was received free of errors. Using error detection or error correction,* there is an additional level of processing, overhead, and delay in signals.

Normally, the transmitter can send data faster than the receiving entity is able to receive. To avoid loss of bits, the receiver needs to send feedback (control data) to the transmitter about whether or not it is ready to receive more data. This is achieved through flow control.

The DLL creates groups of bits to which the corresponding overhead (redundant bits to allow error control and flow control) is added. This group of bits, with a specific format depending on the DLL protocol, is referred to as a frame. A frame consists of a group of bits necessary to allow the implementation of error control and flow control.

Figure 12.8 shows the decomposition of a frame, as composed of the payload data (data received from the Internet layer for transmission) and this layer overhead.

The frame overhead consists of the start of frame, the destination and source address, the control bits, the redundant bits for error control, and the end of frame. The start and end of frame are used to allow frame synchronization (i.e., layer 2 synchronization), that is, for the receiver to understand when the frame starts and when it terminates. In addition, in case the link is asynchronous, the start of frame may also be utilized to allow bit synchronization (i.e., layer 1 synchronization). The control bits are used for the management of flow control and error control. The redundant bits are used to allow the detection of errors in a frame, or to implement error correction.

The start and end of frame can be signalized using different procedures, namely:

- Delimiting character string
- Flag
- Violation of the line coding mechanism

Delimiting character string: In this case, the receiver detects the start of the frame through the reception of a sequence of two predefined characters. As can be seen from Figure 12.9, these two characters are the data link escape (DLE) and the start of text (STX). Moreover, the receiver detects the end of a frame by receiving another group of two predefined characters, namely the DLE and the end of text (ETX). A limitation of this procedure comes from the fact that the bits corresponding to the DLE character may be part of the payload data. In this case, to avoid that the receiver becomes confused, instead of sending a single DLE in the payload data, the transmitter sends twice the DLE character, and the receiver removes one of these characters.

Flag: This is the most common frame synchronization method. In this case, the receiver detects the start and end of frame through the reception of a predefined sequence of bits (see Figure 12.10). It is composed of a sequence of bits with low probability of occurrence within the payload data part of a frame. The PPP and the HDLC protocols use the sequence

Start of frame	Address	Control	Payload data	Redundant bits for error control	End of frame

FIGURE 12.8 Generic frame format.

DLE	STX	...	DLE	ETX

FIGURE 12.9 Delimiting character string.

---

* Connection-oriented or connectionless and confirmed modes.

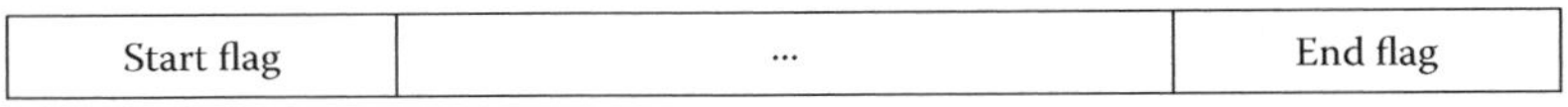

Start flag	...	End flag

FIGURE 12.10 Flag.

01111110 as the flag. A common limitation of this procedure comes from the fact that a part of the payload data may have a sequence of bits equal to the flag. In this case, the receiver could interpret it as the start or end of a frame. To avoid this, a procedure known as bit stuffing is implemented, defined as follows for the HDLC protocol: every time the transmitter detects a sequence of five "1" logic state bits in the data, it inserts a "0" logic state bit as the sixth bit. The receiver performs the reverse operation. With the bit stuffing procedure, it is assured that a sequence of six "1" logic state bits is never transmitted within the data field.

Violation of the line coding mechanism: In this case, the receiver detects the start and end of a frame through the reception of a predefined sequence of signal levels that is not allowed to occur within the normal transmission of data. As an example, the biphase Manchester line coding technique always considers a transition (from 0 to 1 or from 1 to 0) at the middle of the bit period (see Chapter 6). The absence of such transition may be used as a signaling for the start and/or end of a frame.

12.2.1 ERROR CONTROL TECHNIQUES

Depending on the transmission medium and DLL protocol that is employed to exchange data, error control can be performed using either error detection or error correction techniques [Benedetto et al. 1997]. In the case of error detection, codes such as CRC or parity bits can be used to allow errors being detected at the receiver. In case an error is detected, the receiver requests a retransmission of a corrupted frame. On the other hand, in case the transmission medium is highly subject to channel impairments (e.g., wireless medium), the choice goes normally to the use of error correction, instead of error detection.[*] In the latter case, the level of overhead per frame is higher, but it avoids successive repetitions, which also translates in a decrease of overhead. Moreover, error correction can also be used when the transmission is unidirectional or when latency must be avoided (the handshaking associated with error detection and retransmission introduces delays).

With regard to error detection and error correction codes, the code rate is an important parameter that is worth defining. The code rate R_C is defined as [Benedetto et al. 1997]

$$R_C = \frac{n}{k} \tag{12.1}$$

where n stands for the number of information bits and k for the number of transmitted bits.

As can be seen from Figure 12.11, an error control algorithm considers a block of n information bits to which m redundant bits are added, to allow the error detection or error correction capability. These redundant bits are calculated from the original n information bits. The output of the error control algorithm comprises a total of $k = n + m$ bits (codeword), which are transmitted.

n information bits ⟶ Error control algorithm ⟶ *k* encoded bits
(*n* information bits + *m* redundant bits)

FIGURE 12.11 Error control encoder.

---

[*] In some cases, these two techniques are used together.

Error control techniques are commonly employed by modems, as well as by many protocols such as the HDLC, PPP, and TCP.

12.2.1.1 Error Detection Codes

Once a frame is received, the host checks for errors using the redundant bits (requires a bidirectional communication). In case an error is detected, the following two possibilities exist:

- The corrupted frame is requested for retransmission.
- The receiver discards the corrupted frame, and an upper layer implements an error control mechanism.

In case an error is detected and the retransmission procedure is utilized, the frame is repeated by the transmitting host. When a frame is transmitted, the transmitting host starts a timer (chronometer). Depending on the type of confirmation,* the procedure used by the receiving host for signalizing the transmitting one is different, as defined in the following.

In the case of positive confirmation, the receiving entity sends a feedback message in case the frame is free of errors. This procedure is commonly referred to as positive acknowledgment with retransmission (PAR) and the feedback message is known as acknowledgment (ACK). The positive confirmation procedure can be resumed as follows (see Figure 12.12):

- *With errors*: If the ACK is not received by the transmitting entity within a certain time period (controlled by the transmitter's timer), it assumes an error and retransmits the frame.
- *Without errors*: If the ACK is received by the transmitting entity within the expected time period, it assumes correct reception of the frame and proceeds with the transmission of the following frame.

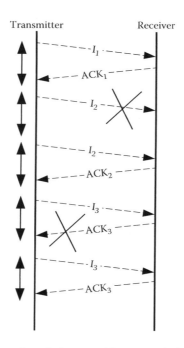

FIGURE 12.12   Example of positive acknowledgment with retransmission handshaking.

---

* As described in Chapter 1, the confirmation is employed in connection-oriented services, or in confirmed connectionless services.

In the case of negative confirmation, the receiving entity only sends a feedback message in case of errors. This procedure is commonly referred to as negative acknowledgment (NAK). The negative confirmation procedure can be resumed as follows:

- *With errors*: If a feedback message is received by the transmitting entity within a certain time period informing that the frame was in the presence of errors, this entity repeats the frame.
- *Without errors*: If a feedback message is not received within a certain time period informing the transmitting entity about the presence of errors, this entity assumes correct reception of the frame and proceeds with the transmission of the following frame.

The advantage of negative confirmation relies on the lower amount of data exchanged, as opposed to the positive confirmation. Note that, in both cases, the frames are numbered for use by the positive or negative confirmation handshaking.

Figure 12.12 shows an example of the PAR procedure. The transmitter sends a first information frame (*I1*) to the receiver. This frame is received free of errors, the receiving host acknowledges (*ACK1*) the correct reception of this frame, and this confirmation is received within the expected time period. In the case of the second frame, an error is detected at the receiving side. Consequently, the latter host does not send the corresponding acknowledgment. The emitter timer reaches the timeout and the second frame is retransmitted. In the case of the third frame (*I3*), it is correctly received, but it is the corresponding acknowledgment (*ACK3*) that does not reach the transmitter. Consequently, the transmitter assumes that the third frame was received corrupted, and retransmits the frame. In the last situation, although the receiving host expects the fourth frame, it knows that the last received frame is a repetition of the previously sent because the frame is numbered, and discards it. Otherwise, the receiving host would have assumed this frame as the fourth, and would incorrectly deliver it to the network (upper) layer.

As previously described, when the error control mechanism is based on positive confirmation, the receiver only sends a feedback signal to the transmitter when the message is correctly received. In this case, flow control is associated with error control in the sense that the feedback signal informs the transmitter that

- The previously received message is free of errors (error control).
- The receiver is able to receive more data (flow control).

In contrast, the negative confirmation considers a feedback signal sent by the receiver only in case an error is detected in the received message. Consequently, using the negative confirmation, flow control is not automatically associated with error control.

12.2.1.1.1 Hamming Codes

Let us consider mapping between information bits and encoded bits (codewords), as depicted in Figure 12.13.

The Hamming distance is the minimum number of bits between any two codewords (blocks of encoded bits). Figure 12.14 shows the distance between any two codewords. As can be seen, the minimum distance between any two codewords is 5, which is the Hamming distance.

The error correction and detection capabilities of *Hamming codes* are deducted from the Hamming distance, as follows [Benedetto et al. 1997]:

- The maximum number of corrupted bits r that can be corrected is obtained from the Hamming distance d as $r = (d-1)/2$.
- The maximum number of corrupted bits s that can be detected is obtained from the Hamming distance d as $s = d-1$.

Information   Encoded bits
    bits   (codeword)

00 → 00000 00000

01 → 01010 10101

10 → 10101 01010

11 → 11111 11111

FIGURE 12.13   Example of mapping between information bits and encoded bits.

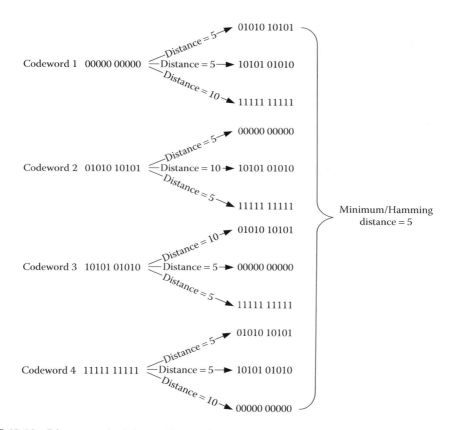

FIGURE 12.14   Distance and minimum distance between codewords.

Focusing again on the example of Figure 12.14 and from the above equivalences, we conclude that:

- The maximum number of corrupted bits that can be corrected is obtained as $r = (5-1)/2 = 2$. Let us assume that the transmitted block is 00000 00000. If the received block is 00000 01010 (two corrupted bits), the closest codeword is 00000 00000. If the receiver uses this code as an error correction code, the receiver deducts the latter block as the estimated transmitted block. In this case, the receiver's decision is correct. On the other hand, if the received block is 00001 01010 (three corrupted bits), then the closest codeword is 10101 01010. If the receiver opts by the latter sequence as the estimated transmitted one, then the receiver makes a wrong estimate. Consequently, it is confirmed that using the proposed code as an error correction code, a maximum of two bits can be corrected.

- The maximum number of corrupted bits that can be detected is obtained as $s = 5 - 1 = 4$. Assuming that the transmitted block is 00000 00000, if the received block is 00101 01010 (four corrupted bits), then the receiver concludes that such codeword does not correspond to any of the valid codewords and requests the retransmission of data or just discards it. If the received block is 10101 01010 (five corrupted bits), because this sequence of bits corresponds to a valid block, the receiver assumes this as the estimated transmitted one. In the latter case, the receiver makes the wrong estimate. Consequently, it is confirmed that using the proposed code as an error detection code, a maximum of four bits can be detected.

12.2.1.1.2 Parity Bits

A basic error detection mechanism relies on the use of parity bits. These parity bits are redundant bits used by the receiver to check for errors. The parity bits, which result from a predefined operation, are added to a frame at the transmitting side. The receiver performs the same operation and observes these redundant bits. The parity can be even or odd.

The transmitting entity applies the eXclusive-OR (XOR) operation to the frame, and adds the resulting parity bit to the sequence (one additional bit). For odd parity, if the number of "1" logic state bits in the frame is odd, the parity bit is set to "1." For even parity, if the number of "1" logic state bits in the frame is even, the parity bit is set to "1." Otherwise, the parity bit is set to "0." Finally, the frame and the parity bits are transmitted together. Figure 12.15 shows two examples of odd parity, while Figure 12.16 shows two examples of even parity.

The receiving entity performs the same XOR operation to the information bits and checks for the received parity bit. In case the frame is received free of errors, the result of this processing originates the same parity bit as the one received. In case an odd number of bits have been received corrupted, the result of this processing originates a parity bit different from the one received. This can be seen from Figure 12.17, where the second bit has been received corrupted. Note that with this method, there is no way to find out which one is the corrupted bit.[*]

	Original message	Odd parity	Resulting data for transmission
Example 1	1 1 0 0 1	1	1 1 0 0 1 1
Example 2	1 0 0 1 0	0	1 0 0 1 0 0

FIGURE 12.15 Examples of odd parity.

	Original message	Even parity	Resulting data for transmission
Example 1	1 1 0 0 1	1	1 1 0 0 1 0
Example 2	1 0 0 1 0	0	1 0 0 1 0 1

FIGURE 12.16 Examples of even parity.

---

[*] Otherwise, such bit would be corrected, and this would be an error correction code.

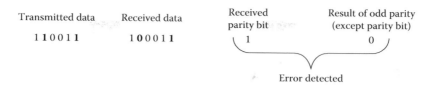

FIGURE 12.17   Example of an error detected by the receiver using odd parity.

In case an even number of corrupted bits are received, the result of the parity check operation corresponds to the received parity bit. In this case, the parity operation is not able to detect the errors.

In the case of the example of Figure 12.15, which consists of error detection, the number of information bits is 5 and the number of transmitted bits is 6. From Equation 12.1, the code rate becomes $R_C = 5/6 = 0.8333$.

12.2.1.1.3 Cyclic Redundancy Check

The CRC is the most common error detection technique used nowadays. Examples of protocols that use CRC codes include the PPP, the HDLC, the Ethernet/IEEE 802.3, and the TCP. CRC codes are also used to check the integrity of stored data.

The CRC encoding considers a block of n information bits to which m redundant bits are added, to allow the error detection capability. These redundant bits are calculated from the information bits using a certain predefined generator polynomial. Note that the generator polynomial is known by the transmitting and receiving entity. Finally, a total of $k = n + m$ bits are transmitted.

The processing of the transmitting entity is as follows:

- Add* a total of m zeros to the block of n information bits, where m corresponds to the degree of the generator polynomial.
- Divide the resulting block by the bits that result from the generator polynomial (see Figure 12.18). The rest of this division originates a total of m bits.
- A total of k bits, consisting of the concatenation of the n information bits with the rest of the division (m redundant bits), are transmitted.

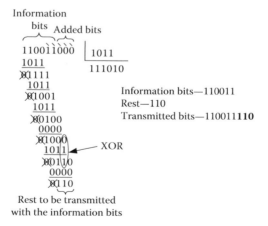

FIGURE 12.18   Example of CRC encoding.

---

* Concatenate.

Based on the received block of k bits, the receiving entity performs the following processing to check the integrity of the data:

- The block composed of k received bits is divided by the bits that result from the generator polynomial. If the rest of this division is zero, then the received block is assumed free of errors. If the rest of this division is different from zero, then the received block is assumed corrupted.
- Alternatively, the receiver may extract the initial n bits of the received block composed of k bits, followed by a similar processing as that performed by the transmitting entity (add zeros and divide by the bits that result from the generator polynomial). If the obtained rest equals the received one, then the received block is free of errors. Otherwise, the received block is corrupted.

Note that the transmitted block includes the original information bits, to which some redundant bits (the rest of the division) are added. In this case, the CRC is referred to as a systematic error control code. On the other hand, error control codes whose generated codeword does not directly include the original information bits are called nonsystematic.

Let us consider an example of CRC use, where the information bits consist of the sequence 110011 and the generator polynomial is $P(x) = x^3 + x + 1$. Based on the above description, the processing of the transmitting entity becomes (see Figure 12.18):

- Because the degree of the generator polynomial is 3, the number of zeros added to the information sequence is also 3. The resulting sequence becomes 110011**000**.
- The generator polynomial $P(x) = x^3 + x + 1$ is mathematically equivalent to $P(x) = 1 \times x^3 + 0 \times x^2 + 1 \times x + 1 \times x^0$. Then, the sequence of bits that result from the generator polynomial is 1011. The division is depicted in Figure 12.18, and the rest of this operation becomes **110** (the number of bits taken from the rest of the division corresponds to the degree of the generator polynomial). Note that the binary difference operation performed to calculate the rest consists of the module 2 adder without carry.[*]
- The transmitting sequence that consists of the concatenation of the information bits with the rest becomes 110011**110**.

Let us suppose that the block was received free of errors. Based on the above description, the processing of the receiving entity becomes (see Figure 12.19):

- The received block of bits (110011110) is divided by the bits that result from the generator polynomial (1011). As can be seen from Figure 12.19, the rest is zero. Consequently, we may conclude that the received block is free of errors.

We may now consider the example depicted in Figure 12.20, where we assume that the received block is corrupted, once the received bit sequence is 110011100, instead of 110011110.

Based on the above description, the processing of the receiving entity becomes:

- The received block of bits (110011100) is divided by the bits that result from the generator polynomial (1011). As can be seen from Figure 12.20, the rest is not zero (010). Consequently, we may conclude that the received block is corrupted.

---

[*] A modulo 2 adder is implemented with an OR-exclusive gate (XOR), whose relationship between the input and the output bits is $0 \oplus 0 = 0$; $0 \oplus 1 = 1$; $1 \oplus 0 = 1$; $1 \oplus 1 = 0$.

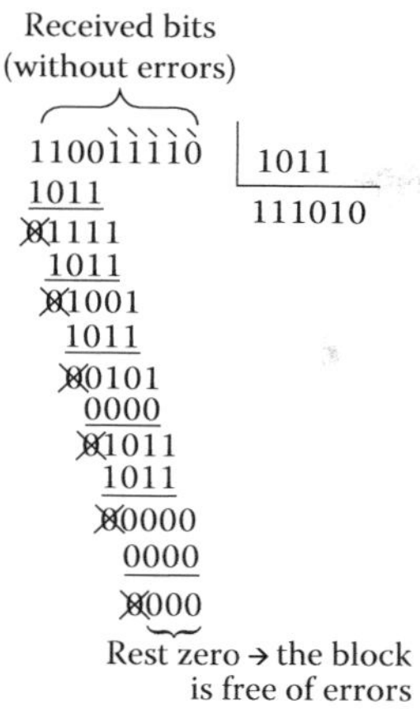

Received bits
(without errors)

$$
\begin{array}{r}
11001\overset{\text{\tiny\\\\\\}}{1110} \\
1011 \\
\hline
\text{\char'170}1111 \\
1011 \\
\hline
\text{\char'170}1001 \\
1011 \\
\hline
\text{\char'170}0101 \\
0000 \\
\hline
\text{\char'170}1011 \\
1011 \\
\hline
\text{\char'170}0000 \\
0000 \\
\hline
\text{\char'170}000
\end{array}
\qquad
\begin{array}{l}
1011 \\
\hline
111010
\end{array}
$$

Rest zero → the block
is free of errors

FIGURE 12.19 Example of CRC decoding without errors in the received block.

Received bits
(corrupted)

$$
\begin{array}{r}
11001\overset{\text{\tiny\\\\\\}}{1100} \\
1011 \\
\hline
\text{\char'170}1111 \\
1011 \\
\hline
\text{\char'170}1001 \\
1011 \\
\hline
\text{\char'170}0101 \\
0000 \\
\hline
\text{\char'170}1010 \\
1011 \\
\hline
\text{\char'170}0010 \\
0000 \\
\hline
\text{\char'170}010
\end{array}
\qquad
\begin{array}{l}
1011 \\
\hline
111010
\end{array}
$$

Rest is not zero → the received
block is corrupted

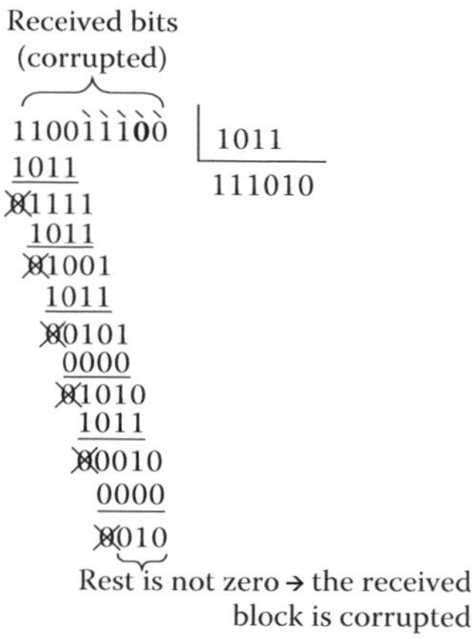

FIGURE 12.20 Example of CRC decoding with errors in the received block.

12.2.1.2 Error Correction Codes

Error correction codes allow the receiving entity to correct one or more corrupted bits within a received block, without having to request a retransmission. As previously described, error correction codes are typically employed in transmission mediums highly subject to impairments, that is, where the BER is degraded. In such situations, the use of error detection and retransmission would lead to successive retransmissions of data, which would translate in excessive additional bandwidth and delay. Error correction codes are also used in channels where retransmission is not possible, such as data broadcast. This capability is normally implemented in a modem, but can also be included in a protocol.

To start the description of error correction codes, we return to the previous description of the parity bits. We have seen that a traditional implementation of error detection relies on the use of parity bits. We may now consider the situation where the parity operation is performed in two dimensions. Let us consider the example depicted in Figure 12.21. In this case, the parity operation is applied independently to each frame (in rows) and, simultaneously, it is applied in columns, to a group of frames. The transmitting data consists of each frame with the corresponding parity bit, by ascending order, followed by the odd parity bits applied in columns.

		Odd parity
Frame 1	1 0 1 0 1	1
Frame 2	1 1 0 0 1	1
Frame 3	1 1 0 0 0	0
Frame 4	0 1 0 1 0	0
Odd parity	1 1 1 1 0	0

FIGURE 12.21   Example of odd parity applied in two dimensions.

Considering that the third bit in frame 3 is received corrupted, the receiver is now able to identify the location of such corrupted bit. This is done using triangulation of the row and column parity bits. This example is depicted in Figure 12.22. In this case, the receiver is now able to perform error correction, instead of only detecting the presence of a corrupted bit.

The price to pay for this additional capability is the reduced code rate R_C^* of the error correction as compared to the code rate of error detection. We have seen that the code rate of the example of Figure 12.15, which consists of error detection, is $R_C = 5/6 = 0.8333$. On the other hand, the code rate corresponding to the error correction depicted in the example of Figure 12.22 becomes $R_C = 25/36 = 0.69444$, which translates in a higher overhead associated with error correction than that of error detection.

The error correction codes are generically referred to as forward error correction (FEC) codes. There are two main types of FEC codes:

- Convolutional codes
- Block codes

Note that in some applications, these two types of error correction codes are combined, originating the so-called concatenated codes.

12.2.1.2.1   Convolutional Codes

Convolutional codes encode an arbitrary group of input bits, using a certain logic function. The parity operation depicted in Figure 12.21 can be seen as a convolutional code. Figure 12.23 shows another example of a convolutional encoder. Because there is no output bit that fed back to the input, this belongs to the group of nonrecursive codes [Benedetto et al. 1997].

This example implements the following operations:

$$k_1 = n_1 \oplus n_0 \oplus n_{-1}$$
$$k_2 = n_0 \oplus n_{-1}$$
$$k_3 = n_1 \oplus n_0$$

(12.2)

		Received odd parity	Processed odd parity
Frame 1	1 0 1 0 1	1	1
Frame 2	1 1 0 0 1	1	1
Frame 3	1 1 1 0 0	0	1
Frame 4	0 1 0 1 0	0	0
Received odd parity	1 1 1 1 0	0	0
Processed odd parity	1 1 0 1 0	0	0

FIGURE 12.22   Correction of a corrupted bit using odd parity in two dimensions.

---

* Obtained from Equation 12.1.

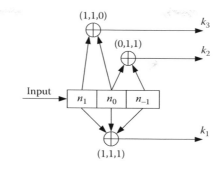

FIGURE 12.23 Example of a nonrecursive convolutional encoder.

Note that the symbol ⊕ in Equation 12.2 stands for modulo 2 adder (XOR). The encoder is normally initialized by filling all registers of the shift register with zero logic state bits. Then, for each data bit that is fed, the shift register shifts once to the right. An important parameter of a convolutional code is its constraint length. This corresponds to the number of previous input bits that a certain output bit depends on. This constraint length equals the number of memory registers.

In the example depicted in Figure 12.23, for each input bit, three output bits are generated using the logic operations. The corresponding code rate is $R_C = 1/3$. The generator polynomial of this convolutional encoder is $G_1 = (1,1,1)$, $G_1 = (0,1,1)$, $G_3 = (1,1,0)$, and its constraint length is 3. Moreover, because the output sequence does not include the input sequence, this convolutional code is referred to as nonsystematic.

Assuming that the first input bit is 1 and that, initially, the registers are fed with 0, the state of the shift registers is (1,0,0), respectively for (n_1, n_2, n_3), and the output bits are (1,0,1), respectively for (k_1, k_2, k_3). Assuming that the second input bit is 0, then the corresponding state of the shift registers is (0,1,0), respectively for (n_1, n_2, n_3), and the output bits are (1,1,1), respectively for (k_1, k_2, k_3).

Another way of expressing this convolutional code is using the Z transfer function. The Z transfer function of the convolutional encoder depicted in Figure 12.23 is as follows:

$$H_1(z) = 1 + z^{-1} + z^{-2}$$

$$H_2(z) = z^{-1} + z^{-2} \tag{12.3}$$

$$H_3(z) = 1 + z^{-1}$$

Figure 12.24 shows another example of a convolutional encoder. Because there is an output bit (k_1) that is fed back to the input, this convolutional code is considered as recursive. Moreover, because the output bit k_3 corresponds to the input bit, the output sequence includes a replica of the input sequence (among other bits). Consequently, the convolutional encoder depicted in Figure 12.24 is systematic [Benedetto et al. 1997].

The convolutional codes are normally decoded using the Viterbi algorithm.

12.2.1.2.2 Block Codes

Block codes encode a fixed group of input bits to generate another fixed group of output bits. They use n input bits to generate a codeword of length k from a certain alphabet. The encoding depicted in Figure 12.13 is an example of a block code. Reed–Solomon codes are among the most used block codes [Benedetto et al. 1997]. The Hamming code is another type of block code.

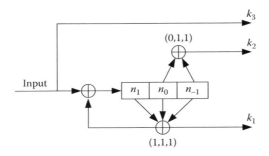

FIGURE 12.24   Example of a recursive convolutional encoder.

12.2.1.2.3   Adaptive Modulation and Coding

Adaptive modulation and channel coding rate considers changes to the modulation and coding rate as a function of the link conditions. If a user experiences poor link conditions, his modulation order can be reduced (e.g., from 16QAM to QPSK), reducing the required SNR level to achieve an acceptable BER performance or, alternatively, decreasing the coding rate. The opposite can happen when a user has very good link conditions, increasing the modulation order and/or increasing the coding rate, to achieve a higher throughput.

12.2.1.2.4   Interleaving

It is known that the bits are normally corrupted in bursts. This is the result of channel impairments such as deep fading, impulsive noise, or even an instantaneous strong interference. As previously described for the Hamming distance, the error correction codes are able to correct up to a certain number of bits. Beyond this number, error correction codes are not able to correct those bits. To improve the error correction capability, the error correction is normally associated with interleaving. Interleaving is used to remove the sequential properties of errors and allow the error correction codes to perform better [Benedetto et al. 1997].

An interleaver is somewhat similar to the scrambler described in Chapter 6. Nevertheless, while the interleaver simply changes the sequential position of the bits, splitting a sequence of corrupted bits into several frames, the scrambler performs a mathematical operation using shift registers, at the transmitting side. The de-interleaver, located at the receiver side, performs the opposite operation, repositioning the bits into the original sequence. This way, the number of corrupted bits that appear in each frame becomes possible to be corrected by an error correction code.

As can be seen from Figure 12.25, the interleaving operation is performed after the error correction encoding algorithm is applied at the transmitting side, and the de-interleaving is performed before the error correction decoding algorithm is applied at the receiving side.

Let us focus on a certain group of encoded bits[*] before interleaving. After the interleaving operation, this group of encoded bits is split among different transmitted frames. The effect of instantaneous channel impairment will affect a certain transmitted frame, that is, a certain group of interleaved bits. Because the receiver performs the deinterleaving operation, the bits of the affected transmitted frame are split among different codewords, which will then be present at the input of the error correction decoder. Because this group of bits does not consist of a long sequence of corrupted bits anymore, error correction can easily perform its functionality.

12.2.1.2.5   Puncturing

Puncturing is an operation that can be applied to the output of the error correction encoder to artificially increase its code rate (see Figure 12.26). This operation is achieved by periodically removing

---

[*] Block at the output of the error correction encoder.

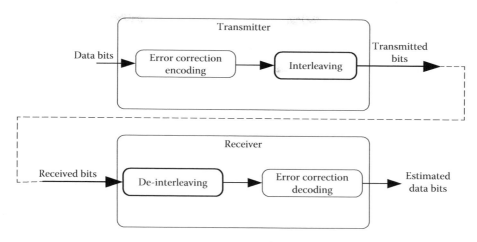

FIGURE 12.25 Error correction and interleaving.

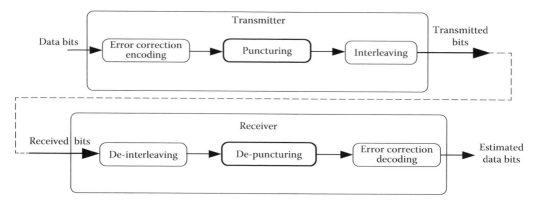

FIGURE 12.26 Location of the puncturing and de-puncturing blocks within the transmitting and receiving chain.

some bits from a codeword at the transmitting side, whereas the receiver inserts zero value bits in the corresponding predefined positions [Benedetto et al. 1997]. The puncturing operation should be performed in such a way that the error correction code should be able to correct the inserted zero value bits to the correct logic states.

As can be seen from the example depicted in Figure 12.27, the puncturing operation consists of removing the bits b_1 and b_6 within each 10-bit codeword. The resulting punctured codeword is only composed of eight bits, instead of ten. The punctured code rate is increased from $R_C = 1/2$ (before puncturing) to $R_C = 5/8$ (after puncturing). The depuncturing operation, performed by the receiver, adds zero value bits in the positions of the removed bits, while it is expected that the error correction coding is able to correct those inserted bits into the correct logic state.

12.2.2 AUTOMATIC REPEAT REQUEST

As previously described, error detection is normally the selected error control mechanism for the transmission of data services (e.g., file transfer and web-browsing) through most of the reliable transmission mediums (e.g., optical fiber and twisted pair). Moreover, when a frame is detected as corrupted, two possibilities exist: the frame is discarded and the error is handled by a higher

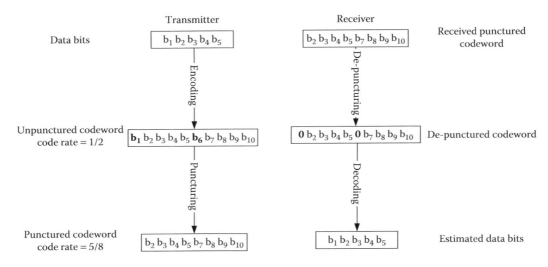

FIGURE 12.27   Puncturing.

layer or a request for frame retransmission is sent back by the receiving entity to the transmitting entity. The latter procedure is employed when the DLL uses confirmed services.[*] As previously described, the retransmission procedure may use positive (PAR) or negative (NAK) confirmation, and the frames are numbered to allow accurate acknowledgments and handling of the repeated frames.

The most common DLL retransmission protocol is the automatic repeat request (ARQ) which consists of a positive confirmation technique. Three different versions of the ARQ protocol exist, as defined in the following subsections.

12.2.2.1   Stop-and-Wait ARQ

Using Stop-and-Wait ARQ, each frame is separately acknowledged by the receiver, using positive confirmation. When a frame is transmitted, the transmitting entity starts a timer and waits for the corresponding acknowledgment from the receiving entity. When a frame is received, the receiving entity uses the redundant bits to check for errors. In case the frame has been received free of errors, the entity sends the corresponding acknowledgment using a feedback channel (positive confirmation). Otherwise, in case the frame has been received corrupted, no signal is sent back to the transmitting entity. In this case, the timer of the transmitting entity reaches the timeout and it considers the frame as lost, proceeding with its repetition. In case the acknowledgment is received, the transmitting entity proceeds with the transmission of the following frame. Note that the acknowledgment message sent back by the receiving entity to the transmitting entity may use a dedicated frame or may be piggybacked.[†] In both cases, the acknowledgment (control information) is sent within the control field.

The example depicted in Figure 12.12 corresponds to the Stop-and-Wait ARQ. As previously described, the frames are numbered to allow the frames and the acknowledgments being correctly identified. Let us suppose an example where an acknowledgment is lost. In such situation, the transmitting entity assumes that the frame was received with errors and repeats its retransmission. If the frames had no numbering, the retransmission would be processed by the receiver as a new one.

---

[*] Connection-oriented or connectionless services.
[†] Sent in the control field of an information frame.

Nevertheless, because the frame is numbered, the receiver is able to identify that this frame is a repetition of a frame previously correctly received, and discards it.

A limitation of this protocol results from the fact that every frame is acknowledged, and the following frame is not transmitted before the previous frame has been correctly acknowledged. This represents an additional overhead and delay. Let us suppose a link using a GEO satellite. In this case, although the transmission rate can be high, the bottleneck is the propagation time which corresponds to approximately 260 ms in each direction. Consequently, even though if a frame consisting of 32 octets is transmitted at a rate of 256 kbps, in 1 ms, the transmitting entity stops its transmission and waits for the acknowledgment. The acknowledgment arrives only at the instant $t = 2 \times 260 + 1 = 521$ ms.[*] From this example, we verify that this version of the ARQ protocol corresponds to a very inefficient use of the expensive satellite space segment. Some advancements are achieved by other versions of the ARQ protocol described in the following.

12.2.2.2 Go-Back-N ARQ

Go-Back-N ARQ considers the acknowledgment of a group of frames,[†] instead of each frame independently. This allows a more efficient use of the transmission medium. An important parameter is the selection of the number of successive frames that are transmitted together in a window. In the example of the GEO satellite link, by choosing 521 frames in a group (window) to be transmitted simultaneously, it is assured that the acknowledgment[‡] of the first frame is received at time $t = 2 \times 260 + 1 = 521$ ms. This corresponds to the transmission instant of the 521th frame. Afterward, the transmitting entity may send another frame (522th frame) and, simultaneously, the acknowledgment of the second frame is received. Consequently, by using the Go-Back-N ARQ the transmitter is avoided to stop and wait for the reception of the acknowledgment. This corresponds to a more efficient use of the satellite resources.

In the above description, all the frames were assumed free of errors. Nevertheless, in case one frame (within the whole group) is received corrupted, the receiving entity does not send the corresponding acknowledgment and the timer of the transmitting entity reaches the timeout. Consequently, the corrupted frame is retransmitted followed by all other frames included in the window, including those transmitted frames that were correctly received. Returning to the example of the GEO satellite link, let us suppose that the first frame is received corrupted. Consequently, because the corresponding acknowledgment is not received at the time instant $t = 2 \times 260 + 1 = 521$ ms, the timer of the transmitter reaches the timeout and the first frame is retransmitted followed by all other 520 previously transmitted frames. Note that there is a variation of this protocol where the receiving entity sends a reject message (REJ) in case the frame is received corrupted. Therefore, using the Go-Back-N ARQ, the transmitter always sends frames in sequence and the retransmissions are signalized using the timeout or the reject message procedure. Moreover, the receiver discards all frames received after a corrupted one. Consequently, the transmitter must resend the first frame, followed by all other 520 frames that have already been transmitted.[§] In this case, while the transmitter has a window of 521 frames,[¶] because the receiver discards all frames after a certain corrupted one, it can be stated that the receiving window is 1, whereas the transmitting window is N. The number of N frames in the window is such that it allows the transmitting entity not having to stop the transmission and wait for the reception of an acknowledgment. This is defined by

$$N = \frac{2 \times T_P}{T_F} + 1 \tag{12.4}$$

---

[*] 260 ms corresponds to the frame propagation time, and another 260 ms corresponds to the acknowledgment propagation time.
[†] Some authors refer to this group of frames as window of frames.
[‡] The acknowledgment is also referred to as "Receive Ready."
[§] Even though if only the first one out of the 521 frames had been received corrupted.
[¶] It is allowed to send a group of 521 frames.

where T_P stands for the propagation time,[*] and T_F stands for the frame duration[†] (i.e., frame transmission time).

Figure 12.28 depicts an example of the Go-Back-N ARQ protocol using the reject message. Note that instead of explicitly using the acknowledgment n (ACK) message, the receiver sends a receive ready $n + 1$ (RR) message. The meaning is similar, but instead of explicitly signalizing that the nth-order frame was correctly received, it signalizes that it is ready to receive the $(N + 1)$th frame. By sending the RR $n + 1$, the receiver is, implicitly, informing the transmitting entity that all previous frames were correctly received. The use of the receive ready message is directly related to flow control functionality, as defined in the sliding window protocol.

Even though if the reject message is employed, the timer needs to keep being employed because the RR message may be lost or corrupted. In case the timeout is reached without having received the RR message, the transmitting entity sends a RR message with the P bit active (RR $P = 1$). This corresponds to an interrogation from the transmitting into the receiving entity about which frame is it ready to receive. Then, the receiving entity responds as appropriate. This is depicted in Figure 12.28.

12.2.2.3 Selective Reject ARQ

The Selective Reject ARQ is similar to the Go-Back-N ARQ. In both cases, the transmission window is higher than 1, consisting of a value that is a function of the propagation time. Nevertheless, while in the case of the Go-Back-N ARQ the receiving window is 1 (the receiver discards all frames after a corrupted received frame), in the Selective Reject ARQ the receiving window is such that the received frames after a corrupted one are stored in a memory. When the receiver rejects a certain corrupted frame (REJ n), the transmitting entity only resends a certain corrupted frame, returning

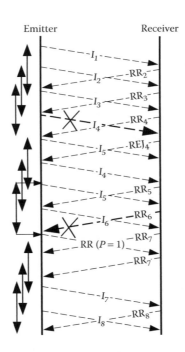

FIGURE 12.28 Example of the Go-Back-N ARQ protocol for a transmit window of $N = 3$ using the reject message.

---

[*] Approximately 260 ms in the case of the GEO satellite link.
[†] 1 ms for a frame composed of 32 octets and a transmission rate of 256 kbps.

to the normal transmission sequence (i.e., the transmitting entity does not retransmit all frames after a corrupted one).

The transmit window of the selective reject ARQ is $2 \times N + 1$, and the receive window is N. As in the case of the Go-Back-N ARQ, the value of N is such that it allows the transmitting entity not having to stop the transmission and to wait for the reception of an acknowledgment. The value N is defined by

$$N = \frac{T_P}{T_F} \qquad (12.5)$$

Figure 12.29 shows an example of the Selective Reject N protocol. The only difference relating to the Go-Back-N protocol (Figure 12.28) relies on the fact that the former only requires that a certain corrupted frame is resent, whereas the Go-Back-N protocol requires that all frames after a certain rejected one are retransmitted.

12.2.3 FLOW CONTROL TECHNIQUES

The transmitter is normally able to transmit more data than the receiver is able to receive and process. To avoid loss of data, the receiver needs to send feedback (control data) to the transmitter informing about whether or not it is ready to receive more data. This handshaking is known as flow control.

When error control is based on error detection with positive confirmation (PAR), flow control is automatically associated and performed. In this case, when the receiving entity sends a feedback signal stating that the previously received frame was correctly received (error control), it is also informing the transmitting entity that it is ready to receive another frame. Otherwise, the receiver does not send the acknowledgment until the moment it is ready to receive more data. Although flow control can be jointly performed with error control procedure, it consists of a different functionality.

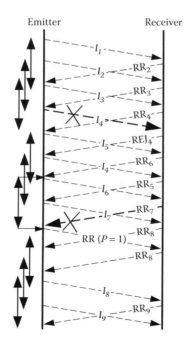

FIGURE 12.29 Example of the Selective Reject ARQ protocol for a transmit window of $(2 * N + 1) = 3$.

Naturally, when error detection is implemented using negative confirmation, because a feedback is only sent by the receiving into the transmitting entity in case of errors, the flow control functionality is not implicitly performed. The same applies when error correction is employed.

Depending on the adopted ARQ procedure, there are two basic versions of the flow control protocol that can be implemented, namely:

- Stop-and-wait
- Sliding window

12.2.3.1 Stop and Wait

This type of flow control is implicitly and automatically performed when the Stop-and-Wait ARQ protocol is implemented. It is worth remembering that the ARQ uses positive confirmation. Similar to the corresponding ARQ protocol, this procedure is performed independently for every frame. The transmitter sends a single frame and, when the receiver is ready to receive more data, it sends the receive ready message, as described for the Stop-and-Wait ARQ. This can be seen from Figure 12.28, where the acknowledgment is achieved using the receive ready message.

12.2.3.2 Sliding Window

The sliding window flow control is associated with the Go-Back-N ARQ or Selective Reject ARQ protocols. When any of these ARQ protocols are implemented, the sliding window flow control is implicitly implemented.

As previously described, the objective of the transmitting window is to allow that the transmitter keeps sending data (successive frames), without having to stop and wait for the reception of confirmations. Ideally, if everything occurs as expected, the transmitting entity receives confirmations while still transmitting data. This allows the maximization of the transmission medium usage.

An important aspect of this protocol relies on the fact that the frames need to be numbered. The transmitting window N is calculated by for Go-Back-N ARQ or by for Selective Reject ARQ, and is a function of the relationship between the propagation time and the frame duration. The window size N corresponds to the maximum number of frames that can be transmitted without confirmation. If the confirmations are delayed, the transmitter may also have to delay the transmission of the following frames. Nevertheless, if this delay is within the window size, the transmitter has permission to proceed with the transmission of more data. This concept is plotted in Figure 12.30, where the lower window edge (LWE) corresponds to the position of the last confirmed frame, and the upper window edge (UWE) is an upper bound for the last frame that can be transmitted. These two parameters are related by $UWE = LWE + N$.

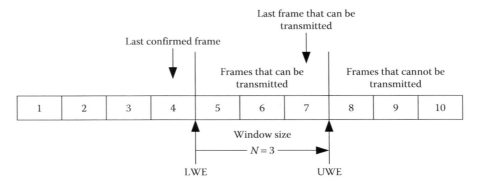

FIGURE 12.30   Example of sliding window flow control with $N = 3$.

12.3 LLC PROTOCOLS

The primary function of a LLC protocol is to provide error control and function control. LLC protocols are used to either interconnect different networks (WAN and MAN) or allow the exchange of data within a LAN. While the most common LLC protocol in LAN is IEEE 802.2, the choice of the type of LLC protocol used to interconnect different networks depends on different parameters, such as the required bandwidth, the quality of service, the cost, the reliability, and the availability.

The previous sections described which functionalities are implemented by the LLC sublayer, and how these functionalities are implemented. The following subsection describes the most used LLC employed in LANs: IEEE 802.2, whereas LLC protocols utilized in MAN and WAN networks are detailed in Chapter 14.

12.3.1 IEEE 802.2 Protocol

The Institute of Electrical and Electronics Engineers created the IEEE 802 committee, whose objective relied on the creation of LAN and MAN standards to assure interoperability among different existing technologies (Token Ring, Ethernet, etc.). To achieve this goal, the IEEE 802 committee created several subcommittees. As can be seen from Figure 12.31, the IEEE 802.2 subcommittee was one of these, whose objective was the standardization of a protocol for the LLC sublayer. This resulting protocol has the name of the subcommittee that created it, that is, IEEE 802.2 1998 protocol, having been adopted by ISO/IEC and renamed as ISO/IEC 8802 - 2:1998. Moreover, other subcommittees were created, for the standardization of different MAC sublayer technologies employed in LAN or MAN networks. It is important to refer to that while the protocols used by the TCP/IP stack above the DLL are standardized by IETF using request for comments, the protocols of the DLL and physical layer are typically standardized by multiple standardization organizations, such as IEEE, ISO, ANSI, ITU, EIA/TIA, and FCC.

As can be seen from Figure 12.31, important LAN/MAN sublayer standards include:

- IEEE 802.3—CSMA-CD
- IEEE 802.4—Token Bus
- IEEE 802.5—Token Ring
- IEEE 802.11—CSMA-CA

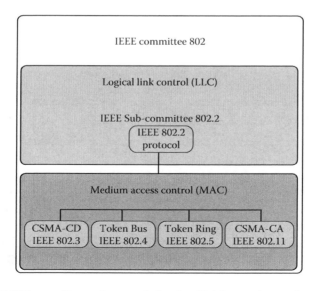

FIGURE 12.31   IEEE 802 committee and some relating data link layer subcommittees.

DSAP	SSAP	Control	Payload data
1 octet	1	1–2	Variable (minimum 46 octets)

FIGURE 12.32   IEEE 802.2 PDU format.

Note that different LAN/MAN subcommittees were dependent on the LLC subcommittee (IEEE 802.2 subcommittee). Moreover, in addition to the subcommittees described here, others were created. Nevertheless, we focus only on those with interest for the current description.

An important note that is worth mentioning is the fact that all different IEEE LAN/MAN protocols (IEEE 802.3, 802.4, 802.5, 802.11, etc.) make use of the same LLC protocol, which is responsible for performing error control and flow control: the IEEE 802.2 protocol.

The LLC sublayer based on the IEEE 802.2 protocol data unit (PDU*) transports, in the payload data field, the packets generated by the Internet layer (upper layer). Moreover, the PDUs are transported in the frame payload data fields of the MAC sublayers (IEEE 802.3, 802.4, 802.5, 802.11, etc.). As can be seen from Figure 12.32, the IEEE 802.2 PDU is similar to that of HDLC, being composed of four different fields:

- *Destination service access point*: It consist of a one-octet address that identifies the LLC destination.
- *Source service access point*: It consist of a one-octet address that identifies the LLC source.
- *Control*: It consist of one- or two-octet field used to allow the handshaking associated with error control and flow control. It is worth noting that the redundant bits for error control using CRC are included as part of the MAC sublayer (not part of the LLC PDU).
- *Payload data*: It consists of a variable length field, carrying upper layer packets, with a minimum size of 46 octets. In case the packet size is lower than that, the payload data is zero padded to this value.

Depending on the type of handshaking performed by the IEEE 802.2 LLC, it may provide the following types of services to the Internet layer:

- *Type 1*: Connectionless and nonconfirmed service. In this case, error control and flow control are not provided.
- *Type 2*: Connection-oriented service. In this case, the protocol sliding window is employed to implement flow control, whereas the protocol Go-Back-N ARQ assures that the frames are delivered free of errors. Moreover, the duplication of frames is avoided, and the correct sequence of frames is guaranteed.
- *Type 3*: Connectionless but confirmed service. In this case, the protocol stop and wait is used to provide error control and flow control. Nevertheless, this protocol is not as efficient as the connection-oriented one (type 2).

12.4   MAC SUBLAYER

As previously described, this sublayer is responsible for managing and controlling the access of hosts into a common transmission medium, within a LAN or a MAN. Depending on the type of management, the access control can be of synchronous or asynchronous types.

Synchronous type of access control mechanisms uses rigid allocation of bandwidth to hosts, regardless of whether or not these hosts use such bandwidth. Examples of synchronous access control

---

* A PDU is the message format of the LLC sublayer.

mechanisms include FDMA, TDMA, CDMA, and so on. GSM networks make use of the TDMA access control mechanism. In this case, a frequency carrier is split into eight time slots, one for carrying data relating to a user's communication in one direction.

In contrast, asynchronous type of access control mechanisms allows the bandwidth being allocated as a function of the users' needs and of the available network capacity. The asynchronous allocation of resources can be performed using different methods:

- *Demand assignment multiple access*: Similar to synchronous mechanisms, it is also based on FDMA, TDMA, or CDMA. However, the number of frequency carriers, time slots, or code sequences is dynamically allocated, as a function of the users' need and of the network available capacity. This requires a centralized management of the resources.
- *Round robin*: This access control mechanism is based on a token that circulates in the network in a certain direction. When a host receives the token, it is allowed to send data to the shared medium for a certain maximum period of time. After this time period, the host must stop the transmission and the token is forwarded to its adjacent host. In the next round, the host may continue its transmission of data. After having received the token, in case the host has no data to transmit, it passes the token immediately to the adjacent host, without consuming a rigid time period. The advantage of the round robin mechanism relating to the TDMA relies on the fact that if a host does not need to use the resources (has no data to transmit), it does not consume a fixed bandwidth. Example of protocol that makes use of this access control mechanism includes the Token Ring and the Token Bus.
- *Contention*: This distributed access control mechanism is based on the ability of any station to start transmitting in case the shared medium is idle (free). In case the channel is busy (being used by another station), it waits until the channel becomes idle, after which it starts transmitting. This mechanism is very simple and works well with low and medium traffic conditions. Because the probability of collision increases exponentially with the increase of the traffic level, this mechanism tends to collapse under high traffic conditions. Examples of contention access control mechanisms include the CSMA-CD[*] or the CSMA-CA.[†]

12.5 MAC PROTOCOLS

At the beginning of computer networks, there were only proprietary protocols developed by companies, and the shared market was the metric of success. Among the existing important LAN/MAN protocols were the Token Ring developed by IBM and the Ethernet created by the DIX consortium (Digital, Intel and Xerox). Because these protocols were not standardized and were incompatible, the IEEE 802 committee created several subcommittees (see Figure 12.31). Each subcommittee became responsible for the standardization of each relevant LAN/MAN protocol. This was the solution to accommodate all important proprietary protocols, each with some advantages and disadvantages, instead of standardizing a single protocol, while leaving all the others out.

As can be seen from Figure 12.31, IEEE created a group of subcommittees for the creation of different MAC sublayer standards. However, these standards are directly related to the physical layer used to allow its implementation (10BASET, 10GBASEFX, etc.). An IEEE 802.3 network includes the definition of the access control to the shared medium, the frame format, etc. Moreover, the corresponding physical layer includes the transmission rate, the type of cable, the MSS, the type of line

---

[*] The CSMA-CD is the contention mechanism used by the IEEE 802.3 protocol.
[†] The CSMA-CA is the contention mechanism used by the IEEE 802.11 protocol.

encoding technique (e.g., Manchester, and nonreturn-to-zero), the type of connectors, and so on. As can be seen from Figure 12.33, these different physical parameters vary depending on the IEEE 802.3 version. 10BASET and 10GBASEFX are two examples of physical layer technologies used to implement an IEEE 802.3 network.

The following subsections describe the most important LAN/MAN IEEE standards. Although IEEE 802.11 consists of a MAC protocol utilized in LAN, because it is wireless, this is dealt with in Chapter 15.

12.5.1 IEEE 802.3 PROTOCOL

The Ethernet technology, initially developed and commercialized by the DIX consortium, was the basis for the development of the IEEE 802.3 standard. Nevertheless, there are small differences between these two, namely small variations of the content of different frame fields. The current description focuses on IEEE 802.3, which is currently the most implemented standard [IEEE 802.3 2008].

IEEE 802.3 is a standard responsible for the access control to the shared medium based on a contention mechanism. In addition, the physical layer is also defined. It receives data from the upper sublayer (IEEE 802.2 LLC) and encapsulates it into the payload data field of the IEEE 802.3 MAC frame, followed by the addition of the corresponding frame header. Then, the frame is broadcasted in the network, which consists of a shared transmission medium.[*]

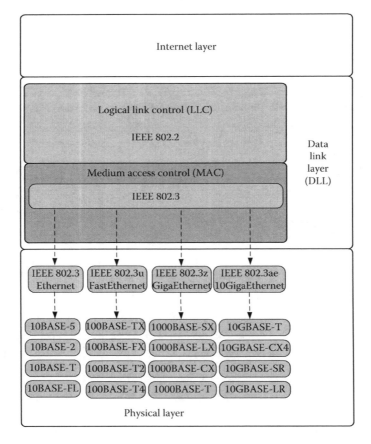

FIGURE 12.33 IEEE 802.3 standard and its versions and technologies.

---

[*] The IEEE 802.3 standard assumes a hub as the central node of the network. This device makes the transmission medium a shared resource.

Figure 12.34 depicts the IEEE 802.3 frame format, with the indication of different fields and corresponding sizes. The frame includes the following fields:

- *Preamble*: It consists of a flag used to allow the synchronization of the receiver with the transmitter at both physical and data link layer (i.e., at bit and frame synchronization). It is composed of 56 bits (7 octets) consisting of alternating 0 and 1 logic state bits.
- *Start of frame delimiter* (*SFD*): It consists of eight bits (1 octet) with the sequence "10101011." It is used to signalize the start of a frame. Moreover, the SFD, together with the preamble, is used to allow the receiver to synchronize with the transmitter.
- *Destination address*: It consists of the physical address of the destination NIC. The physical address is also referred to as hardware address or as MAC address. It is composed of six octets (48 bits), with the following decomposition:
  - The initial 24 bits are called organizationally unique identifier, being globally allocated by IEEE to the organization or hardware manufacturer (see http://www.neotechcc.org/forum/macid.htm for a list of vendor codes).
  - The final 24 bits are internally managed and allocated by the organization or manufacturer in a way to avoid duplications (in case the destination MAC address is a broadcast, this contains all 1s in the 48 bits of the MAC address, i.e., ff:ff:ff:ff:ff:ff).
- *Source address*: It consists of the physical address of the source NIC with the structure defined for the destination address.
- *Type/size*: It consists of a two-octet field used to identify either the frame size or the frame type:
  - If the field content is a decimal number lower than 1500, that number represents the number of octets transported in the payload data field of the frame.
  - If the field content is a decimal number higher than 1536, the field is used to represent the type of layer 3 data transported in the payload data field of the frame.
- *Payload data*: It is used to transport LLC data (and the LLC data is used to transport layer 3 data, i.e., packets[*]). In case its content has a size lower than 46 octets, the upper sublayer (IEEE 802.2 1998) performs zero padding of the PDU up to this value. Zero padding is the action of adding zeros to assure that a field presents a certain minimum size.
- *CRC*: Also referred to as FCS, it consists of redundant bits transmitted with the data to allow the receiver to check the integrity of the frame using the CRC. The error verification is applicable to the whole frame except the preamble, SFD and CRC. Note that the LLC protocol (upper sublayer) does not make use of any error control technique. It let the MAC sublayer verify the errors. In case the MAC sublayer detects an error, it informs the LLC sublayer which is then responsible for requesting the repetition of the PDU. The LLC sublayer implements the handshaking necessary to allow the flow control and error control (receive ready, reject, etc.). The generator polynomial used by the IEEE 802.3 standard is[†]:

$$P(x) = x^{32} + x^{26} + x^{23} + x^{22} + x^{16} + x^{12} + x^{11} + x^{10} + x^8 + x^7 + x^5 + x^4 + x^2 + x + 1 \quad (12.6)$$

Preamble	S F D	Destination address	Source address	Type size	Payload data	CRC
7 octets	1	6	6	2	46–1500 (variable)	4

FIGURE 12.34 IEEE 802.3 frame format.

---

[*] In case of the IEEE 802.3 standard, the maximum transmission unit (MTU) is 1500 octets. This corresponds to the maximum payload size of a frame, which also corresponds to the maximum size of a packet (layer 3 data), added to the LLC overhead. As previously described, in case the packet size is higher than 1500 octets, fragmentation is implemented by a router at the border of an IEEE 802.3 network.

[†] This is the same generator polynomial used by the HDLC protocol in 32 bits mode.

The contention mechanism employed by the IEEE 802.3 standard is the carrier sense multiple access–collision detection (CSMA-CD), which is defined as follows (see Figure 12.35):

- A host that intends to initiate a frame transmission within the network starts by listening to the channel (carrier sense).
- If the shared medium is idle (i.e., if there is not any host transmitting), the host may initiate the transmission. Otherwise, if another host is transmitting (medium is busy), it waits a backoff period of time. After this period, the host rechecks the channel availability, to start the transmission.
- Albeit the channel could be idle, and a host could have started the transmission, another host could have started the transmission in a very close moment. This originates a collision.* To assure that such collisions are detected, a host needs to keep listening to the

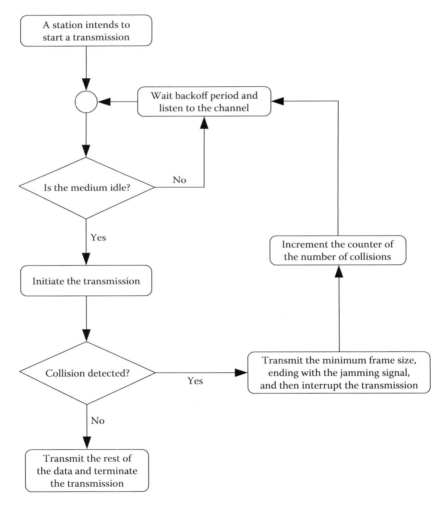

FIGURE 12.35 CSMA-CD contention mechanism.

---

* The reason for the hosts not to, initially, detect each other results from the fact that the network has a certain length, which corresponds a certain period of time for the propagation of the electric signals. When a host starts the transmission, the generated electric signals need to propagate along the transmission line to achieve another space location within the physical network.

medium (receiving), while transmitting. If a host detects a collision, the host must proceed with the transmission such that the minimum frame size (512 bit) is transmitted and such that the transmission is finalized with the jamming signal.* Then, the host must interrupt its transmission. This is plotted in Figure 12.36 for the case where a collision has been detected at the start of the frame (within the initial 44 payload octets). Otherwise the host transmits immediately the jamming signal, and then interrupts the transmission. This procedure ensures that all receivers detect a collision by checking the received corrupted CRC, including those located in the other physical extreme of the network. After the jamming signal, both hosts must interrupt their transmissions and try to restart after a listening strategy defined as follows:

- *Nonpersistent*: If the medium is idle, transmit immediately. If the medium is busy, wait a backoff period† and listen again to the medium. Note that it may happen that the medium is already idle, but the host is still waiting a random period.
- *1-Persistent*: If the medium is idle, transmit immediately. If the medium is busy, keep waiting and listening to the medium. Then, when it becomes idle, transmit immediately. If more than a single host is waiting for the medium to be idle, the transmissions of these waiting hosts will certainly collide. This represents a disadvantage of the 1-persistent scheme.
- *P-Persistent*: If the medium is idle, the transmit probability is p, while the probability to delay one time unit‡ is $(1-p)$. If the medium is busy, wait for being idle, and then use the previous procedure. If the transmission is delayed, after the waiting time use the previous procedure. The value of p varies, but typical values are between 0.5 and 0.75.

The backoff time used in IEEE 802.3 is referred to as binary exponential backoff. It consists of a random waiting period, whose mean value is doubled during the initial 10 retransmission attempts. The mean value of the random waiting period is kept unchanged for more than six additional retransmission attempts. In case 16 unsuccessful collisions occur, the host gives up attempting and sends an error to the upper layer. Then, the cycle restarts after a larger backoff (this value increases with the increase in the number of groups of 16 attempts).

12.5.1.1 Maximum Collision Domain Diameter

The maximum collision domain diameter (MCDD)§ is defined as the maximum distance between the two furthest network nodes. From this parameter, we may calculate the total time it takes for the

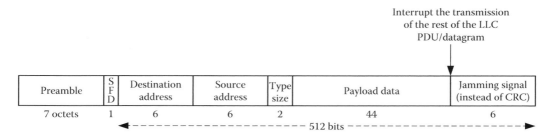

FIGURE 12.36 Transmitted frame in case a collision is detected within the initial 44 payload octets.

---

shortest frame (minimum frame size) to travel round trip between the two furthest network nodes in that domain. Note that this time takes into consideration the round trip time (double) to allow the collision (interfering) signal to return to the transmitting source.

The minimum frame size of an IEEE 802.3 network (at 10 and 100 Mbps) is 512 bit.[*] The period of time necessary to transmit this minimum frame is referred to as slot time. The slot time corresponds to the maximum period of time where a collision can be detected. Beyond such slot time, a collision never occurs, as there is enough time for a certain transmission to reach and being detected by all other network nodes. If the minimum frame size were shorter, the MAC sublayer could not detect a collision, and the collision/error would have to be handled by another layer (e.g., by the TCP), or the data would be corrupted. This is avoided by defining the MCDD.

The MCDD consists of the maximum length (diameter) of a domain, where a collision can be handled by the MAC sublayer. It is dimensioned for half of the slot time period (round trip), resulting in a 256-bit duration. Nevertheless, to take into account additional delay in network devices (e.g., repeaters) the one-way calculations are performed for a total of 232 bits.

Considering a 10 Mbps IEEE 802.3 network, the time for transmitting 232 bits becomes

$$T = \frac{\text{number of bits}}{\text{bit rate}} = \frac{232}{10^7} = 23.2 \ \mu\text{s} \tag{12.7}$$

From the physics, the MCDD becomes

$$\text{MCDD} = v \cdot T \tag{12.8}$$

where:
  v is the propagation speed of signals in the transmission medium (coaxial cable, twisted pair, and optical fiber)
  T is the result of Equation 12.7

Taking into account the approximate propagation speed in the coaxial cable of $v = 1.22 \times 10^8$ m/s (note that the light speed in the vacuum is $v = 3 \times 10^8$ m/s), from Equation 12.8 the MCDD becomes 2800 m.

Using the same principle, the time for the transmission of 232 bits in a 100 Mbps IEEE 802.3 network becomes $T = 2.32 \ \mu\text{s}$, and the MCDD results in approximately 280 m for a coaxial cable (205 m for a twisted pair).

At low speeds, the maximum network size is normally constrained by the MSS,[†] instead of the MCDD. This is the case of a 10BASET network, where the MSS is limited to 100 m, while the MCDD is 280 m. Nevertheless, for speeds of 1 Gbps or above, the MCDD becomes the constraint. In this case, a mechanism called carrier extension is employed, to overcome such limitation. The carrier extension consists of a number of bits transmitted after and together with the regular frame, when its length is shorter than 4096 bits (see Figure 12.37). The objective is to assure that the

FIGURE 12.37   Carrier extension.

---

resulting frame length (original frame plus carrier extension) is a minimum of 4096 bits long (used in the calculation of the 1 Gbps IEEE 802.3 slot time). Naturally, in case the frame length is longer than 4096 bits, the carrier extension is not added. Note that the minimum frame size is kept as 512 bits long, and thus, interoperability with 10 and 100 Mbps IEEE 802.3 networks is assured. The resulting extension of the slot time from 512 bits into 4096 bits extends the MCDD of the 1 Gbps IEEE 802.3 standard from around 20 m up to around 200 m.

An alternative solution that can be adopted by 1 Gbps or faster IEEE 802.3 networks is the frame bursting procedure. Instead of spending bandwidth at transmitting the carrier extension, if the transmitting host has several short frames to transmit, these frames can be sent together, linked with the interframe gap (IFG). The IFG consists of a predefined bit pattern. This improves the network performance, relating to the carrier extension use. Figure 12.38 depicts the frame bursting procedure. Note that the initial frame is always sent using the carrier extension, whereas the others can be transmitted using the frame bursting procedure. The maximum length of the linked short frames is limited to 56,536 bits. The use of the frame bursting results in a higher minimum frame size, which translates in a higher MCDD.

The above-described CSMA-CD mechanism is applicable to a network that makes use of a linear hub, which translates in a shared transmission medium. The hub is a repeater, which repeats in every output the bits present at one of the inputs. Consequently, the medium becomes shared and the CSMA-CD is applicable. Nevertheless, in case the central node of the network is a switch (or a bridge), instead of a hub, because this device only switches the frames to the output where the host with the frame's destination MAC address is located, the effect of the CSMA-CD is neglected. In fact, the CSMA-CD is also applicable but, as the use of the switch makes the transmission medium not shared, the effect of the CSMA-CD is neglected.

In any case, note that IEEE 802.3 is a standard defined for the worst case scenario, which is the case of a network with a hub as a central node. The use of a switch is a modification of the IEEE 802.3 standard, where the MAC contention mechanism based on the CSMA-CD is not followed. Moreover, in case a switch is employed in a 1 Gbps or faster IEEE 802.3 network, the carrier extension and frame bursting do not need to be employed.

The CSMA-CD facilitates the access to the shared medium as long as the traffic rate is below 40%–50%. Beyond this threshold, the performance of the CSMA-CD degrades heavily.

12.5.1.2 Physical Layer Employed in IEEE 802.3 Networks

As can be seen from Figure 12.33, there are several physical layer implementations of an IEEE 802.3 network. An example of a physical implementation is 100BASE-FX, whose meaning is described in Figure 12.39. The first group of digits (100) refers to the transmission rate, while the second group (BASE or BROAD) refers to the signaling type (baseband or broadband). Finally, the last group of digits or letters refers to the MSS divided by 100 (e.g., 5 from 10BASE-5 stands for the MSS of 500 m) or to the transmission medium (FX from 100BASE-FX stands for optical fiber).

Tables 12.2 through 12.5 present the physical characteristics of IEEE 802.3 at 10 Mbps, 100 Mbps, 1 Gbps, and 10 Gbps, including parameters such as the type of cabling employed, the physical topology, the line encoding technique, the MSS, and the type of connectors employed.

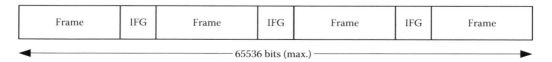

FIGURE 12.38   Frame bursting.

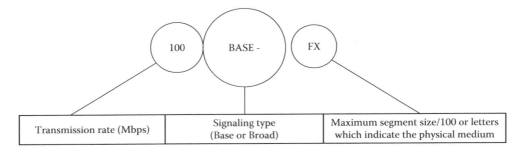

FIGURE 12.39   Identification of IEEE 802.3 technologies employed in the physical layer.

TABLE 12.2
Physical Characteristics Used in IEEE 802.3 (10 Mbps)

	IEEE 802.3—10 Mbps (Ethernet)			
	10BASE-5	**10BASE-2**	**10BASE-T**	**10BASE-FX**
Cabling	12 mm coaxial cable (RG8/RG11)	6 mm coaxial cable (RG58)	UTP Cat.3 4 or 5 (4 pairs)	Multimode fiber (62.5/125 mm) (two fibers for full duplex)
Physical topology	Bus	Bus	Star	Star
Encoding	Manchester	Manchester	Manchester	Manchester
MSS (m)	500 (2500 maximum with up to 4 repeaters)	185 (300 if repeaters are not employed)	100	2000
Type of connectors	BNC	BNC	RJ45	ST or SC

TABLE 12.3
Physical Characteristics Used in IEEE 802.3u (100 Mbps)

	IEEE 802.3u—100 Mbps (FastEthernet)			
	100BASE-T		**100BASE-X**	
Designation	**100BASE-T2**	**100BASE-T4**	**100BASE-TX**	**100BASE-FX**
Cabling	UTP Cat.3, 4, or 5 (two pairs)	UTP Cat.3 (four pairs: one for Tx, another for Rx, two pairs are negotiated)	STP or UTP Cat.5 (two pairs)	Multimode fiber (62.5/125 mm) (two fibers for full duplex)
Physical topology	Star	Star	Star	Star
Encoding	Manchester	8B/6T	4B/5B with scrambling	4B/5B
MSS (m)	100	100	100	412
Type of connectors	RJ45	RJ45	RJ45	ST or SC

Naturally, increased speeds have been achieved with latest standards. Optical fiber cables have played an important contribution to achieving higher speeds, and higher ranges without the need to use repeaters/regenerators (i.e., increased MSS). From the two types of optical fiber cables, the single mode is the one that achieves the best performance. Another important parameter that allows achieving higher transmission rates is the use of more efficient

TABLE 12.4

Physical Characteristics Used in IEEE 802.3z (1 Gbps)

	IEEE 802.3z—1000 Mbps (GigabitEthernet)			
	1000BASE-X			
Designation	1000BASE-CX	1000BASE-SX	1000BASE-LX	1000BASE-T
Cabling	STP (two pairs)	Multimode fiber (62.5/125 mm) (two fibers)	Single-mode optical fiber (two fibers)	UTP Cat.5 (four pairs negotiated for Tx or Rx)
Physical topology	Star	Star	Star	Star
Encoding	8B/10B	8B/10B	8B/10B	4D-PAM5 with scrambling and FEC
MSS (m)	25	220	5000	100 (200 with repeater)
Type of connectors	DB9 or HSSDC	ST or SC	SC	RJ45

TABLE 12.5

Physical Characteristics Used in IEEE 802.3ae (10 Gbps)

	IEEE 802.3ae—10 Gbps		
	10GBASE-T	10GBASE-SR	10GBASE-ER
Cabling	UTP Cat.5 or better (four pairs)	Multimode fiber (62.5/125 mm) (two fibers)	Single-mode optical fiber (two fibers)
Physical topology	Star	Star	Star
Encoding	4D-PAM10 with scrambling and FEC	64B/66B	64B/66B
MSS (m)	100	300	40,000
Type of connectors	RJ45	ST or SC	SC

line coding techniques, as well as the combination of the multiple pairs for transmission or receiving data.[*]

It is important to refer to that the channel impairments tend to increase with the increase in the transmission rates. This tends to originate a degradation of the BER performance. The use of error detection mechanisms makes the receiver to keep requesting repetitions of frames. To avoid the resulting increased overhead and delay, the high transmission rates made over transmission mediums of low performance (e.g., twisted pairs) is normally associated with FEC (error correction) and scrambling (to improve the signal quality).

Finally, it is worth noting that IEEE 802.3ae (10 Gbps) can be used in both LAN and MAN networks. 10GBASE-ER[†] can be employed as a MAN, as an alternative to SDH, ATM, or even MPLS.

12.5.2 IEEE 802.5 Protocol

The Token Ring protocol, initially developed and commercialized by IBM, was the foundation for the standardization of IEEE 802.5 [1998], having been adopted by ISO/IEC and renamed as

---

[*] As an example, 1000BASE-T allows transmitting at a speed of 1 Gbps using four UTP Cat. 5 pairs. These pairs are all used for transmit, all for receive, or some pairs for transmit and others for receive.
[†] ER stands for Extended Range.

ISO/IEC 8802 - 5:1998. Nevertheless, there are small differences between the Token Ring and the IEEE 802.5 standard, namely small variations on the content of different frame fields. The current description focuses on the IEEE 802.5 standard, as more adopted in LAN than the Token Ring.

IEEE 802.5 consists of a MAC sublayer standard that defines a MAC mechanism based on the above-described round robin, as well as the physical layer. This sublayer receives data from the upper sublayer (IEEE 802.2 LLC), and encapsulates it into the payload of the IEEE 802.5 MAC frame, followed by the addition of the corresponding frame header. Then, the frame is transmitted over the LAN.

The IEEE 802.5 standard considers the transmission speed of 4 Mbps using UTP cabling, or 16 Mbps with STP cabling. Currently, the speed of 100 Mbps is already possible. The differential Manchester line encoding technique is employed in the IEEE 802.5 standard.

The logical topology comprised by the IEEE 802.5 standard is the ring, where a token circulates in a certain direction. When a host receives the token, it is allowed to send data to the common transmission medium for a certain maximum period of time. After such period, the host must interrupt the transmission and the token is forwarded to the adjacent host. In the next round, the host may continue its transmission. After having received the token, in case the host has no data to transmit, it immediately forwards the token to the following adjacent host, without consuming a rigid time period. This represents an advantage relating to the TDMA, where fixed time slots are allocated to different users, regardless of whether or not they have data to transmit.

Advantages of IEEE 802.5 relating to IEEE 802.3 are as follows:

- IEEE 802.5 assures a maximum delay for a host to transmit a frame, whereas IEEE 802.3 does not. This maximum delay corresponds to the maximum time necessary for a token to circulate in the ring. This characteristic is especially important in case of high traffic load, where the successive collisions experienced in IEEE 802.3 networks tend to collapse it or to insert a delay higher than that acceptable for the provision of QoS (e.g., for IP telephony or video streaming).
- IEEE 802.5 networks allow prioritization of traffic, which is another advantage as compared to IEEE 802.3 networks.

Disadvantages of IEEE 802.5 relating to IEEE 802.3 are as follows:

- IEEE 802.3 is much simpler than IEEE 802.5, without the need to have a management station responsible for the control of the token.
- IEEE 802.3 network does not present a long minimum waiting time as that of IEEE 802.5 (corresponding to time required for the token to go around the ring).
- IEEE 802.3 tends to be the preferable standard because of its simplicity and good performance under low traffic load conditions.

Albeit the ring is the considered logical topology of IEEE 802.5, the physical topology is the star, where a hub or a switch is employed as a central node. This can be seen from Figure 12.40. In case the central node is a hub, tokens and frames are sent to all stations (common transmission medium) but only the next adjacent host processes the data. Note that it was previously described that a network that makes use of a hub as a central node has a logical bus topology. Nevertheless, because the IEEE 802.5 standard uses a token to control the access to the shared medium, the logical topology is transformed into a ring. In this case, although a token is used to implement the round robin access mechanism, a hub makes the medium a diffusion channel.* The delay inserted in the signal by a hub corresponds to one bit period. This is the time necessary to implement the regeneration of bits.

---

* Similar to the IEEE 802.3 standard, with a hub as a central node.

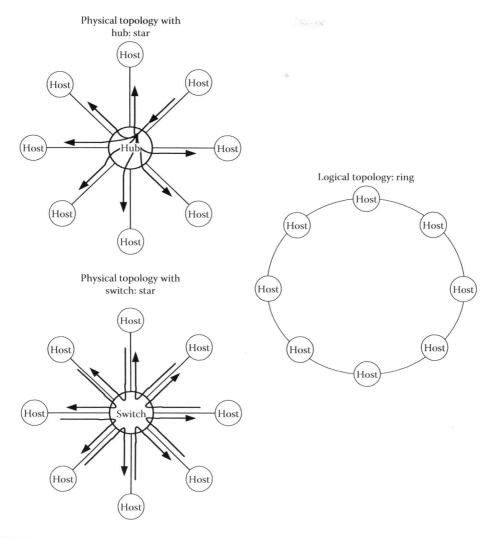

FIGURE 12.40 Physical and logical topologies of the IEEE 802.5 standard.

In contrast, in case a switch is employed, the transmission medium is not shared, and this device forwards only frames and tokens from a host to its adjacent, following a certain direction of the logical ring. In this case, the channel is not of diffusion type. The delay inserted by a switch in the signal depends on its type (store-and-forward, cut-through, etc.).

The operation of the MAC sublayer defined by the IEEE 802.5 standard is as follows:

- A token circulates in the network, from host to host, in a certain direction of the ring.
- If a certain host has data to transmit, when the token is received, the host keeps it and transmits data for a maximum period of time corresponding to the token holding period (THP). The default THP value is 10 ms.[*] During this period, the host can transmit one or more frames. Note that a host is only allowed to transmit data, in case the priority of the

---

[*] IEEE 802.5 networks working at 4 Mbps may transmit approximately one frame with the maximum load (4500 octets), or several lower length frames, in one THP period. At a speed of 16 Mbps, more frames are allowed in each THP period.

frame is higher than the priority of the token.* This priority management represents a great advantage of IEEE 802.5 relating to IEEE 802.3 networks.

- A host listens to the channel, accepting the frames whose destination MAC address is its own address. Otherwise, the frames are forwarded to the adjacent host.
- The frames circulate from host to host, around the ring, being copied by the destination, and removed by the host that generated it, after one round.
- Once a host terminates the transmission of data (or in case it has no data to transmit), it immediately forwards the token to its adjacent host of the ring.

The IEEE 802.5 standard requires a host to act as a management station. Any host of the network may act as a management station. Its functions include the following:

- Frame removal, in case the host that generated the frame did not remove it after one round.
- Generation and insertion of the token into the ring, as well as its monitoring. Moreover, the management station is also responsible for defining the priority level of the token.
- Detection of lost tokens. Due to channel impairments, a token may lose its sense. In this case, a new token has to be inserted, while the old one must be removed.

Figure 12.41 depicts the format of an IEEE 802.5 frame, while Figure 12.42 depicts the format of an IEEE 802.5 token.

The content of the frame fields is described in the following:

- *Start delimiter (SD)*: It delimits the start of the frame, using a violation of the differential Manchester line coding technique.
- *Access control (AC)*: It contains priority bits to distinguish between a frame and a token. Moreover, it also contains priority bits as well as other bits used for monitoring purposes.
- *Frame control (FC)*: It indicates the type of data transported in the payload data field.
- *Destination address (DA)*: It consists of the address of the destination host.
- *Source address (SA)*: It consists of the address of the source host.
- *Payload data*: It transports IEEE 802.2 PDU. The maximum length of the frame payload data is 4500 octets.
- *CRC*: Also referred to as FCS, it uses the same CRC function as the one defined for IEEE 802.3 (defined by Equation 12.6), applied to all frame fields, except the SD, CRC, ED, and FS.
- *Ending delimiter (ED)*: It indicates whether a frame is the last one of a group of frames, or an interim frame within such group. Moreover, it gives an indication about when a CRC error is detected in a frame.

S D	A C	F C	Destination address	Source address	Payload data	CRC	E D	F S
1	1	1	2–6 octets	2–6	46–4500 (variable)	4	1	1

FIGURE 12.41   IEEE 802.5 frame format.

S D	A C	E D
1	1	1 octet

FIGURE 12.42   IEEE 802.5 token format.

---

* The priority of the token is defined by the management station, based on the priority of the traffic that the different hosts have to transmit.

- *Frame status (FS)*: It consists of a group of bits that are changed by the destination host when it is copied. It is based on these bits that either the transmitting host or, alternatively, the management station removes the frame from the ring.

Note that the three fields contained in the token are also present in the frame, presenting the same meanings.

12.5.3 Fiber Distribution Data Interface Protocol

The fiber distribution data interface (FDDI) is another MAC sublayer standard based on the round robin access mechanism, with MAN applications. The basic operation of the FDDI standard is very similar to that of IEEE 802.5. While IEEE 802.5 is an IEEE standard and used in LAN, the FDDI was standardized by American National Standards Institute (ANSI) as ANSI X3.139 [1987], as well as by ISO as ISO 9314–2 [1989] for the MAC sublayer. The corresponding physical layer was standardized as ANSI X3.148 [1988] and ISO 9314–1 [1989].

The FDDI considers two rings based on optical fiber, an operational and a backup ring. Switching from one into another is automatically performed in case of failure. It allows data rates of 100 Mbps, covering an area of up to 100 km, and up to 500 network hosts. The regeneration distance is 2 km for multimode fiber, and 60 km for single-mode fiber. Moreover, the digital coding technique is the 4B/5B, which achieves a better efficiency than the differential Manchester employed in the IEEE 802.5 standard.

Similar to the IEEE 802.5 standard, the FDDI was designed to work with the IEEE 802.2 LLC,[*] as an upper sublayer. It receives data from the upper sublayer, and then encapsulates it into the payload data of the FDDI MAC frame, followed by the addition of the corresponding frame header, as depicted in Figure 12.43. Moreover, the content of the token is depicted in Figure 12.44, whose fields are the same as those described for the frame content.

The content of the frame fields is defined in the following:

- *Start of frame sequence (SFS)*: It comprises these subfields:
  - *Preamble (PA)*: A sequence of predefined bits used for synchronism purposes.
  - *Start delimiter (SD)*: A sequence of predefined bits that delimits the start of a frame.
  - *Frame control (FC)*: It indicates the type of data transported in the payload data field.
- *Destination address (DA)*: It consists of the address of the destination host.
- *Source address (SA)*: It consists of the address of the source host.
- *Payload data*: It transports IEEE 802.2 PDU. The maximum length of the frame payload data is 4500 octets.

◄———SFS———►						◄EFS►
PA	SD FC	Destination address	Source address	Payload data	CRC	ED FS
8 octets	1 1	6	6	0–4500 (variable)	4	1 1,5

FIGURE 12.43 FDDI frame format.

PA	SD FC ED
8 octets	1 1 1

FIGURE 12.44 FDDI token format.

---

[*] Although the FDDI standard is not an IEEE standard, it interfaces with the IEEE 802.2 LLC sublayer.

- *CRC*: It uses the same CRC function as the one defined for IEEE 802.3, as defined by, applied to the FC, DA, SA, and payload data.
- *End of frame sequence* (*EFS*): It comprises two subfields:
  - *Ending delimiter* (*ED*): It consists of a group of bits used to signalize the end of a frame. Moreover, these bits give indication about whether a frame is the last one or an interim.
  - *Frame status* (*FS*): It consists of a group of bits that are changed by the destination host when it is copied. It is based on these bits that either the transmitting host or, alternatively, the management station removes the frame from the ring. Moreover, this field allows the signalization when an interim node detects an error in a frame.

12.5.4 Digital Video Broadcast Standard

The digital video broadcast (DVB) is a suite of standards that define the physical layer and DLL of digital video distribution [Watkinson 2001]. While the MPEG consists of a set of codec standards, defining the way the analog video is digitized and compressed, the DVB standards define the way the digital video is transmitted (e.g., modulation scheme, error correction type and associated code rate, and frame structure). It is worth noting that the digital video transmitted using a DVB standard is encoded and compressed using an MPEG standard (DVB-MPEG). An exception is the DVB for handheld devices, entitled DVB-H, which comprises its own specific encoding and compression algorithm.

The DVB system includes a variety of standards for different transmission mediums, namely

- *Terrestrial television* (*VHF/UHF*): DVB-T and DVB-T2
- *Cable*: DVB-C and DVB-C2
- *Satellite*: DVB-S, DVB-S2 and DVB-SH
- *Digital terrestrial television for handheld terminals*: DVB-H, DVB-SH

Different modulation schemes and error correction codes are employed in different DVB standards. Transmission mediums more subject to channel impairments make use of lower order modulation schemes, as well as lower error correction code rates.

Note that the mobile TV, using the LTE infrastructure, is expected to be a competitor to DVB-H-based TV broadcast.

12.6 VIRTUAL LOCAL AREA NETWORKS

VLANs can be created in switches to allow different logical networks with a single physical network. VLANs are used for different purposes, namely

- For providing access to different networks with a single switch
- For creating different broadcast domains with a single switch
- For achieving logical isolation among different hosts (security)

Providing access to different networks with a single switch: The access to different networks (or subnetworks) is normally granted by the use of a router. Creating multiple VLANs in switches allows gaining access to multiple networks without having to make use of a router to separate the networks. In addition, a single switch can be employed, instead of multiple switches for multiple networks, and thus, VLAN allows a reduction in the amount of hardware. This can be seen from Figure 12.45. Moreover, it is possible to use more than a single switch within a physical network, to have access to multiple logical networks.

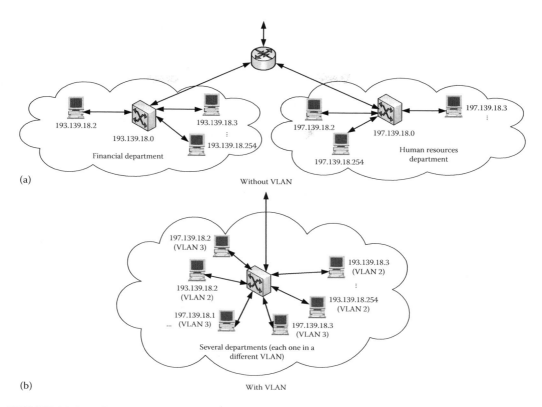

FIGURE 12.45 Getting access to multiple networks with (a) conventional configuration or (b) VLAN configured in a switch.

This is advantageous to reduce the amount of cabling (a single cable is used between switches, instead of multiple cables from a single switch into multiple hosts) or when more switch ports are needed than those made available by a single switch. In this case, different switches must be connected in trunk mode and an encapsulation protocol must be configured. The most used encapsulation protocol is IEEE 802.1Q. An example of a VLAN configuration in multiple switches is provided in Appendix III.

Creating different broadcast domains with a single switch: Normally, a broadcast domain corresponds to a whole organization. However, it is known that most of the broadcast traffic is limited to a department, while the rest of the organization's hosts are overloaded with such undesired traffic. The creation of different VLANs allows reducing the amount of undesired traffic that the hosts receive. This is achieved by creating different VLANs, a different one corresponding to each different broadcast domain (e.g., for different departments). This represents an improvement in network performance. Figure 12.45b depicts an example of two VLANs created in a switch, each one to be used by a different organization department (e.g., VLAN 2 for the financial department and VLAN 3 for the human resources department[*]). Consequently, each department will benefit from a different broadcast domain. It is also worth noting that with the use of a switch configured with multiple VLANs, the flexibility is greatly improved. Let us suppose the case where each

---

[*] Note that the VLAN 1 is the default.

department is physically located in different rooms, and a new employee that belongs to the human resources department is placed in the room that physically belongs to the financial department. With the use of VLAN, the network manager only has to reconfigure the switch port that serves this new employee from the financial VLAN into the human resources VLAN.

Logical isolation among different hosts (*security*): Let us suppose that the employees of an organization need to have access to three different networks: intranet, extranet (or Internet), and IP telephony. Using the conventional approach and assuming that isolation needs to be assured among these different networks (for security purposes), three different physical networks are needed. Nevertheless, if three VLANs are created, a single physical network can be employed, whereas three different logical networks are implemented over it. In this case, the traffic cannot be exchanged between different virtual networks (VLANs) (except if such capability is explicitly configured using a router). This can be seen from Figure 12.46b, where the network 193.139.18.0 (data) is isolated from the network 197.139.18.0 (IP telephony), using a common physical network infrastructure. The only way to enable the exchange of data among different networks is making use of a router. In this case, the switch port connecting the router is configured in trunk mode, whereas the router's interface needs to be configured using as many subinterfaces as the number of VLANs to be interfaced with and using an encapsulation protocol (e.g., IEEE 802.1Q). Note that there can be multiple switches configured with multiple VLANs, where the interconnection between them is performed in trunk mode (using, e.g., the IEEE 802.1Q protocol), whereas the switch's interfaces used to provide access to workstations are configured in access mode.

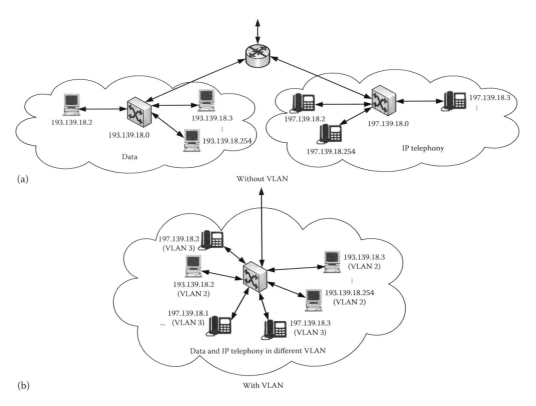

FIGURE 12.46 Isolating networks with VLAN using a single physical infrastructure (a) creating broadcast domains using a router (without VLAN), and (b) creating broadcast domains using VLANs.

12.6.1 Configuration of Virtual Local Area Networks

In case multiple VLANs are to be created, this configuration needs to be programmed in switches. Different VLANs are used to reduce the broadcast domain of a network, for security reasons or to improve the management flexibility level of a network.

Traditionally, a LAN corresponds to a single physical and logical network, which is associated with a single network or subnetwork IP address.[*]

A switch is responsible for forwarding frames to the destination host based on the MAC address. A switch keeps knowledge about the MAC address in each interface using MAC address tables. A MAC address table performs mapping between an interface and the corresponding MAC address. Initially, while the MAC address table is empty, a received frame is forwarded to all other output interfaces of the switch. As frames are being received, the switch registers the source MAC address of the received frame, and maps it to the corresponding interface, building the MAC address table. The Cisco IOS allows checking MAC address tables using the syntax:

- Switch#show mac-address-table

If one intends to send data to all hosts within, a LAN uses the corresponding broadcast IP address. Nevertheless, it is known that an organization (such as a company) is typically composed of different departments,[†] and most of the broadcasts are restricted to the department. If we create different VLANs, and associate each department with a different VLAN, the broadcast traffic can then be sent only to the own department, without having to overload the rest of the LAN with traffic. Naturally, each VLAN should be split into different subnetworks, each with a certain broadcast address. Without this capability, within the same organization, different physical LANs[‡] would be required, each with a different network or subnetwork IP address, and a router to assure the interconnection between different subnetworks. In addition, the use of VLANs brings an additional level of flexibility, as a workstation physically located in a room traditionally used by a certain department may be used by a person belonging to another department. The only action that is required is to assign the interface of the switch from one into another VLAN, or to move the patch cord from one into another interface of the switch.

An important characteristic of the VLAN implementation relies on the inability of a switch to allow the exchange of data between different VLANs. This can only be achieved by using a router properly configured. This characteristic can be used to logically isolate different networks or subnetworks used for different purposes within the same physical network infrastructure. Consequently, a common network infrastructure can be used for different purposes: a VLAN can be used, for example, to provide Internet access, another VLAN to give access to the voice over IP, another VLAN to provide access to the intranet, and so on. Depending on the service that is intended to be given to a host intends, such host is configured with the corresponding network or subnetwork IP address. This represents great savings in terms of cabling and network devices, while gaining an additional level of flexibility.

The configuration of switches typically follows two different steps: (1) the creation of VLANs and (2) the assignment of interfaces to VLANs, as follows:

- Creation of a VLAN: this is configured from the privileged mode, using the following sequence of syntaxes:
  - Switch#configure terminal
  - Switch(config)#vlan *vlan_number*
  - Switch(config-vlan)#name *vlan_name*

---

[*] In case NAT/PAT is considered, a LAN may use a single external IP address, whereas internally, different private or public addresses are considered.

[†] For example, financial department, human resources department, etc.

[‡] One LAN per department.

Here *vlan_number*[*] is the number assigned to the VLAN and *vlan_name* stands for the name assigned to the VLAN.[†] This sequence of two command lines should be issued as many times as the number of VLANs to create. The created VLANs can be saved into the running configuration by issuing (see Figure 12.47):

- Switch(config-vlan)#CTRL-Z

- *Assignment of an interface to a VLAN*: As can be seen from Figure 12.48, this is configured from the configuration mode,[‡] from where each interface of the switch is configured,

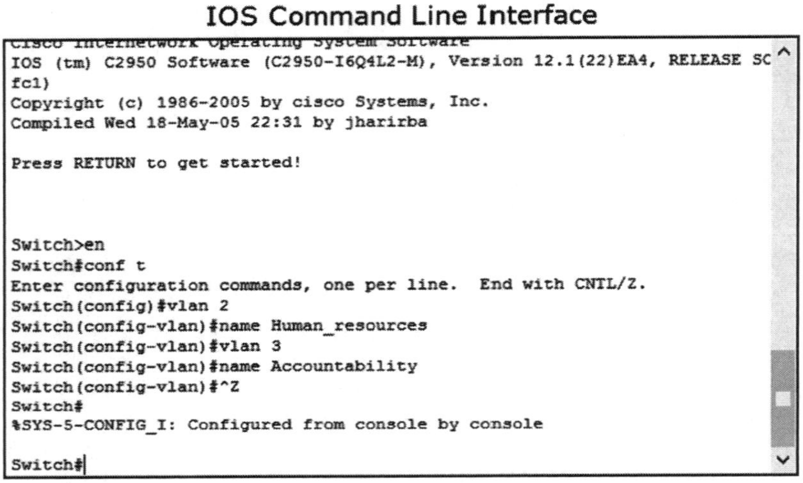

IOS Command Line Interface

```
CISCO Internecwork Operating System Software
IOS (tm) C2950 Software (C2950-I6Q4L2-M), Version 12.1(22)EA4, RELEASE SC
fc1)
Copyright (c) 1986-2005 by cisco Systems, Inc.
Compiled Wed 18-May-05 22:31 by jharirba

Press RETURN to get started!




Switch>en
Switch#conf t
Enter configuration commands, one per line.  End with CNTL/Z.
Switch(config)#vlan 2
Switch(config-vlan)#name Human_resources
Switch(config-vlan)#vlan 3
Switch(config-vlan)#name Accountability
Switch(config-vlan)#^Z
Switch#
%SYS-5-CONFIG_I: Configured from console by console

Switch#
```

FIGURE 12.47   Creating VLAN.

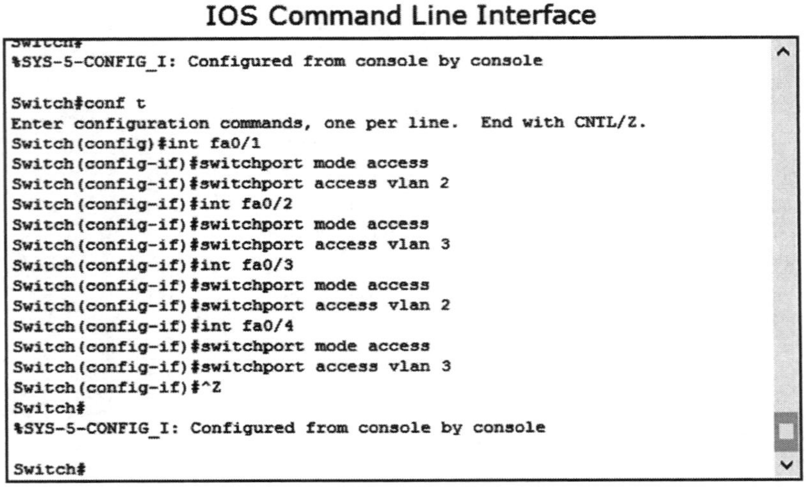

IOS Command Line Interface

```
Switch#
%SYS-5-CONFIG_I: Configured from console by console

Switch#conf t
Enter configuration commands, one per line.  End with CNTL/Z.
Switch(config)#int fa0/1
Switch(config-if)#switchport mode access
Switch(config-if)#switchport access vlan 2
Switch(config-if)#int fa0/2
Switch(config-if)#switchport mode access
Switch(config-if)#switchport access vlan 3
Switch(config-if)#int fa0/3
Switch(config-if)#switchport mode access
Switch(config-if)#switchport access vlan 2
Switch(config-if)#int fa0/4
Switch(config-if)#switchport mode access
Switch(config-if)#switchport access vlan 3
Switch(config-if)#^Z
Switch#
%SYS-5-CONFIG_I: Configured from console by console

Switch#
```

FIGURE 12.48   Assigning interfaces to VLANs.

---

[*] Note that the default VLAN is the VLAN 1. This VLAN is not configurable.
[†] For example, human resources department.
[‡] As previously described, this is accessed from the privileged mode by issuing *configure terminal*.

followed by the declaration of the VLAN number to which the interface belongs. This is configured, for each of the interfaces of the switch (or the interfaces remain in the default VLAN [VLAN 1]), using the sequence of syntaxes:

- Switch(config)#[interface *interface* | interface range *interface_range*]
- Switch(config-if-range)#switchport mode access
- Switch(config-if-range)#switchport access vlan *vlan_number* (*vlan_number* corresponds to the VLAN number)
  - *Example*:
    - Switch(config)#interface range et0/1–4
    - Switch(config-if-range)#switchport mode access
    - Switch(config-if-range)#switchport access vlan 2
    - Switch(config)#interface et0/5
    - Switch(config-if-range)#switchport mode access
    - Switch(config-if-range)#switchport access vlan 3

This configuration can be saved into the running configuration by issuing CTRL-Z keyboard keys. Finally, the user can verify the assignment of the interfaces to different VLANs by issuing one of the following syntaxes:

- Switch#show vlan
- Switch#show vlan brief (see Figure 12.49)
- Switch#show running-configuration
- Switch#show interfaces vlan *vlan_number* (see Figure 12.50)
  - *Example*: Switch#show interfaces vlan 3
- Switch#show interfaces *interface* switchport (see Figure 12.51)
  - *Example*: Switch#show interfaces et0/1 switchport

Voice over IP is an important type of traffic that is handled by networks. An IP telephone is normally connected to a switch, whereas a workstation connects to the IP telephone for the exchange of data traffic. In this scenario, the IP telephone acts as a switch and the connection between the telephone and the switch carries both VoIP traffic and data traffic (trunk mode).

IOS Command Line Interface

```
Switch(config-if)#switchport access vlan 2
Switch(config-if)#int fa0/8
Switch(config-if)#switchport access vlan 3
Switch(config-if)#int fa0/9
Switch(config-if)#switchport access vlan 2
Switch(config-if)#^Z
%SYS-5-CONFIG_I: Configured from console by console
Switch#show vlan brief

VLAN Name                             Status    Ports
---- -------------------------------- --------- -------------------------------
1    default                          active    Fa0/10, Fa0/11, Fa0/12, Fa0/13
                                                Fa0/14, Fa0/15, Fa0/16, Fa0/17
                                                Fa0/18, Fa0/19, Fa0/20, Fa0/21
                                                Fa0/22, Fa0/23, Fa0/24
2    Human_resources                  active    Fa0/1, Fa0/3, Fa0/5, Fa0/7
                                                Fa0/9
3    Accountability                   active    Fa0/2, Fa0/4, Fa0/6, Fa0/8
1002 fddi-default                     active
1003 token-ring-default               active
1004 fddinet-default                  active
1005 trnet-default                    active
Switch#
```

FIGURE 12.49 Command shows VLAN brief.

IOS Command Line Interface

```
Switch#show interfaces vlan 1
Vlan1 is administratively down, line protocol is down
    Hardware is CPU Interface, address is 0000.0ca2.c238 (bia 0000.0ca2.c23
    MTU 1500 bytes, BW 100000 Kbit, DLY 1000000 usec,
        reliability 255/255, txload 1/255, rxload 1/255
    Encapsulation ARPA, loopback not set
    ARP type: ARPA, ARP Timeout 04:00:00
    Last input 21:40:21, output never, output hang never
    Last clearing of "show interface" counters never
    Input queue: 0/75/0/0 (size/max/drops/flushes); Total output drops: 0
    Queueing strategy: fifo
    Output queue: 0/40 (size/max)
    5 minute input rate 0 bits/sec, 0 packets/sec
    5 minute output rate 0 bits/sec, 0 packets/sec
        1682 packets input, 530955 bytes, 0 no buffer
        Received 0 broadcasts (0 IP multicast)
        0 runts, 0 giants, 0 throttles
        0 input errors, 0 CRC, 0 frame, 0 overrun, 0 ignored
        563859 packets output, 0 bytes, 0 underruns
        0 output errors, 23 interface resets
        0 output buffer failures, 0 output buffers swapped out
Switch#
```

FIGURE 12.50   Command shows interfaces vlan *vlan_number.*

IOS Command Line Interface

```
Switch#show interfaces fa0/3 switchport
Name: Fa0/3
Switchport: Enabled
Administrative Mode: dynamic auto
Operational Mode: down
Administrative Trunking Encapsulation: dot1q
Operational Trunking Encapsulation: native
Negotiation of Trunking: On
Access Mode VLAN: 1 (default)
Trunking Native Mode VLAN: 1 (default)
Voice VLAN: none
Administrative private-vlan host-association: none
Administrative private-vlan mapping: none
Administrative private-vlan trunk native VLAN: none
Administrative private-vlan trunk encapsulation: dot1q
Administrative private-vlan trunk normal VLANs: none
Administrative private-vlan trunk private VLANs: none
Operational private-vlan: none
Trunking VLANs Enabled: ALL
Pruning VLANs Enabled: 2-1001
Capture Mode Disabled
Capture VLANs Allowed: ALL
Protected: false
Appliance trust: none
Switch#
```

FIGURE 12.51   Command shows interfaces *interface* switchport.

The configuration of the switch interface connecting to the IP telephone under the above-described typical scenario is as follows:

- Switch(config)#interface *interface*
- Switch(config-if)# mls qos trust cos (allows the switch interface to give priority to voice traffic, but the whole network must be configured to give priority)
- Switch(config)#switchport voice vlan *voice_vlan_number* (*voice_vlan_number*[*] corresponds to the voice VLAN number)

---

[*] Typically, voice VLAN is assigned to VLAN 150.

- Switch(config-if)#switchport mode access (by default, switch ports are in access mode; if not previously modified, this step can be omitted)
- Switch(config-if)#switchport access vlan *data_vlan_number* (*data_vlan_number* corresponds to the data VLAN number)
  - *Example*:
    - Switch(config)#interface et0/1
    - Switch(config-if)#mls qos trust cos
    - Switch(config-if)#switchport voice vlan 150
    - Switch(config-if)#switchport mode access
    - Switch(config-if)#switchport access vlan 2
    - Switch(config-if)#end

12.6.1.1   Configuration of the Management VLAN

It was described that a Cisco router can be remotely configured through telnet or through secure shell (SSH). An advantage of using SSH over telnet relies on the use of authentication and encryption. Moreover, remote access and remote configuration of some Cisco devices are also possible through SNMP and HTTP. In the case of a switch, this can be accessed through a management VLAN. By default, the management VLAN is VLAN 1, which enables any user connected to the switch to manage it. To avoid this default configuration that represents a vulnerability from the security point of view, a VLAN number different from a data VLAN should be assigned as the management VLAN. Moreover, an IP address must be assigned to the management VLAN to allow the switch to be remotely configured through telnet or SSH. It is worth noting that the ports assigned to the management VLAN should be secured.* Otherwise, any device could get remote access to the switch, and could remotely configure it, which might represent a security violation. The creation of a management VLAN is performed using the following steps:

- Switch(config)#vlan *management_vlan_number* (*management_vlan_number* is the number of the new VLAN to be used as the management VLAN)
- Switch(config-vlan)#name management
- Switch(config-vlan)#exit
- Switch(config)#interface vlan *management_vlan_number*
- Switch(config-if)#IP address *ip_address subnet_mask* (the IP address and subnet mask of the management VLAN that is then utilized to remotely access the switch)
- Switch(config-if)#no shutdown
  - *Example*:
    - Switch(config)#vlan 99
    - Switch(config-vlan)#name management
    - Switch(config-vlan)#exit
    - Switch(config)#interface vlan 99
    - Switch(config-if)#ip address 193.123.213.5 255.255.255.0
    - Switch(config-if)#no shutdown
    - Switch(config-if)#exit
    - Switch(config)#interface range et0/3–5 (securing the switch ports)
    - Switch(config-if-range)#switchport mode access
    - Switch(config-if-range)#switchport access vlan 99
    - Switch(config-if-range)#switchport port-security[†]

---

* For example, using the command "switchport port-security" and creating a secret password to access the telnet line.
[†] The reader should refer to Chapter 16 for the description of Cisco switch security.

– Switch(config-if-range)#switchport port-security mac-address *static_mac_address*
(where *static_mac_address* is the mac-address of the host(s) that is allowed to
access the switch port)
– Switch(config-if-range)#switchport port-security violation shutdown (if the mac-
address of the host connected to et03–5 is not allowed [the one above configured],
the port is shutdown)
– Switch (config-if-range)#line vty 0 15
– Switch (config-line)#password secret_word
– Switch (config-line)#login
– Switch (config-line)#CTRL-Z

12.6.1.2   Configuration of the VLAN Default Gateway

As described above, if a switch is to be remotely controlled, an IP address must be assigned to
the management VLAN. Moreover, in case one intends to get access, from the command line
interface of a switch to devices located in remote networks, the default gateway also needs to
be configured in a switch. The default gateway is the IP address of the router where all traffic
destined to a remote network must be forwarded to. Note that a switch with different VLANs has
a different default gateway IP address for each different VLAN. This is performed using the following syntax:

• Switch(config)#interface vlan *vlan_number* (done for each switch VLAN)
• Switch(config-if)#ip default-gateway *ip_address_of_default_gateway_for_the_vlan*
  • *Example*:
    – Switch(config)#interface vlan 10
    – Switch(config-if)#ip default-gateway 193.123.10.254
    – Switch(config)#interface vlan 20
    – Switch(config-if)#ip default-gateway 195.123.10.254

12.6.2   Inter-VLAN Routing

It was described that an important characteristic of VLAN relies on the inability to exchange
traffic between the switch's interfaces connected to different VLANs. This advantage can be
used for security purposes (e.g., to separate voice and data traffic). Let us suppose that each
corporate department corresponds to a different VLAN. In this case, inter-VLAN connectivity is
normally required. Then, a layer 3 device needs to be connected to the switch, while the switch
interface connected to the layer 3 device needs to be configured in trunk mode. A trunk is an
interconnection between two switches, or between a router and a switch, used for the transportation of data traffic of more than one VLAN and control traffic. Trunks enable the extension of
VLANs over the whole network. Alternatively, one could configure the interconnection between
two switches or between a router and a switch in access mode but using as many cables as the
number of VLANs.

The IEEE 802.1Q protocol [IEEE 802.1Q 2011] is used for coordinating trunks on IEEE 802.3
interfaces. Note that the IEEE 802.3 frame header does not provide any information about the
VLAN. This simply transports a packet, encapsulated into the frame. When in trunk mode, IEEE
802.3 frames need to be tagged with information about which VLAN does the transported frame
refers to. Subsequently, when Ethernet frames are placed on a trunk they need additional information about the VLANs they belong to. This is accomplished by using the 802.1Q encapsulation
header. This header adds a tag to the original Ethernet frame specifying the VLAN to which the
frame belongs to.

As can be seen from Figure 12.34, this is performed by the IEEE 802.1Q encapsulation header, comprising (Figure 12.52):

- Addition of a header field entitled tag control information (TCI) after the type/length field of the IEEE 802.3 frame.
- Inclusion of the hexadecimal value 0×8100 (33024 in decimal) in the type/length field. This makes the network device to look for the content of the TCI, when a frame is received with this value in the type/length field.
- Recalculation of the CRC.

Moreover, the TCI contains the following subfields:

- User priority (UP), 3 bits long, consisting of a method used to give priority to certain frames with respect to others (useful in VoIP), whose functionality is implemented using the IEEE 802.1p protocol.
- Canonical format identifier (CFI), 1 bit long, which allows IEEE 802.5 frames to be carried over IEEE 802.3 links.
- VLAN identifier (VLAN ID [VID]), comprising 12 bits used for identifying which VLAN the frame in the trunk belongs to.

Note that a trunk is used to carry data and control frames belonging to different VLANs, properly tagged.[*]

In terms of the switch configuration, an interface can be placed in trunk mode by issuing the following sequence of syntaxes:

- Switch(config)#[interface *interface* | interface range *interface_range*]
- Switch(config-if-range)#**switchport mode trunk**
  - *Example*:
    - Switch(config)#interface fa0/2
    - Switch(config-if-range)#switchport mode access (configure access mode)
    - Switch(config-if-range)#switchport access vlan 2
    - Switch(config)#interface fa0/3
    - Switch(config-if-range)#switchport mode access (configure access mode)
    - Switch(config-if-range)#switchport access vlan 3
    - Switch(config)#interface fa0/1
    - Switch(config-if-range)#switchport mode trunk (configure trunk mode)
    - Switch(config)#interface fa0/4
    - Switch(config-if-range)#switchport mode trunk (configure trunk mode)

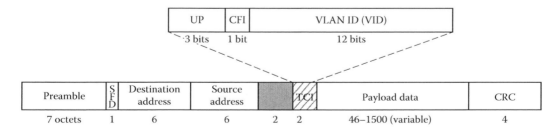

FIGURE 12.52 IEEE 802.3 frame format with IEEE 802.1Q encapsulation.

---

[*] Marked with the identification of the VLAN to which the frame belongs.

12.6.2.1 Configuration of the Native VLAN

In addition to data VLAN frames, there are the native VLAN frames that are used to carry control traffic, such as traffic used by the STP (this is not data traffic, but control traffic). This also needs to be explicitly configured in a trunk; otherwise, only data traffic is allowed in a trunk.

By default, the native VLAN corresponds to VLAN 1. This is modified by creating a new VLAN as previously described, using a number different from 1, and assigning the name native to the VLAN (see creating new VLANs in Figure 12.47).

Afterwards, one must configure a trunk to allow control traffic (native VLAN) to flow over it. This is performed by issuing the sequence of syntaxes (see Figure 12.53):

- Switch(config)#interface *interface*
- Switch(config-if)#switchport mode trunk
- Switch(config-if)#switchport trunk native vlan *vlan_number* (this command line is only applicable for control traffic [native VLAN]; *vlan_number* is the identification of the VLAN used as the native VLAN).
  - *Example*:
    - Switch(config)#interface et0/1
    - Switch(config-if)#switchport mode trunk
    - Switch(config-if)#switchport trunk native vlan 99
    - Switch(config-if)#CTRL-Z

Inter-VLAN connectivity can be implemented with an external router to forward traffic between different VLANs. Alternatively, a multilayer switch can be used to perform this function. The following subsections provide a description of the configuration of a router or a multilayer switch necessary to allow inter-VLAN connectivity.

IOS Command Line Interface

```
Switch>en
Switch#conf t
Enter configuration commands, one per line.  End with CNTL/Z.
Switch(config)#hostname L2Switch
L2Switch(config)#vlan 2
L2Switch(config-vlan)#name Accountability
L2Switch(config-vlan)#vlan 3
L2Switch(config-vlan)#name Human_resources
L2Switch(config-vlan)#vlan 99
L2Switch(config-vlan)#name Native
L2Switch(config-vlan)#exit
L2Switch(config)#int fa0/1
L2Switch(config-if)#switchport mode access
L2Switch(config-if)#switchport access vlan 2
L2Switch(config-if)#int fa0/2
L2Switch(config-if)#switchport mode access
L2Switch(config-if)#switchport access vlan 2
L2Switch(config-if)#int fa0/3
L2Switch(config-if)#switchport mode access
L2Switch(config-if)#switchport access vlan 3
L2Switch(config-if)#int fa0/4
L2Switch(config-if)#switchport mode access
L2Switch(config-if)#switchport access vlan 3
L2Switch(config-if)#int fa0/5
L2Switch(config-if)#switchport mode trunk
L2Switch(config-if)#switchport trunk native vlan 99
L2Switch(config-if)#^Z
L2Switch#
%SYS-5-CONFIG_I: Configured from console by console

L2Switch#
```

FIGURE 12.53   Configuring a switch with an interface in trunk mode and the others in access mode.

12.6.2.2 Inter-VLAN Connectivity with a Router

Inter-VLAN can be implemented with an external router connected to the switch that has multiple VLANs, whereas the router forwards the traffic between different VLANs. The interconnection between the switch and the router can be performed in two different ways:

- *Using multiple cabling connections between the router and the switch, one per VLAN (see Figure 12.54)*: In this case, the switch and router ports are configured in access mode, using the conventional router and switch configuration. Using this configuration, each router interface is configured in a different subnet. A drawback of this configuration relies on the fact that adding more VLANs corresponds to adding more cabling, which corresponds to an additional hardware complexity, cost, and interfaces usage.
- *Using a single cabling connection between the router and the switch*: In this case, the switch and router interfaces are configured in trunk mode. This configuration is referred to as router-on-a-stick (see Figure 12.55). This requires the creation of a different subinterface for each VLAN to be interconnected, and each subinterface should be configured in a different subnet. The advantage of this configuration relies on the fact that a large number of VLANs can be added without impacting the cabling and interface usage. Nevertheless, it becomes more difficult to troubleshoot.

The configuration of inter-VLAN connectivity using multiple cables is performed using the previously described configuration in access mode, for both the switch and the router. With regard to the configuration in trunk mode, the interface of the switch that connects to the router is configured in trunk mode, following the steps previously described and, if existing, allowing the native VLAN

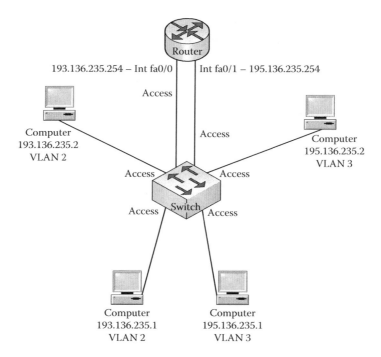

FIGURE 12.54 Inter-VLAN routing with multiple cables (two, in this case) between the router and the switch (access mode).

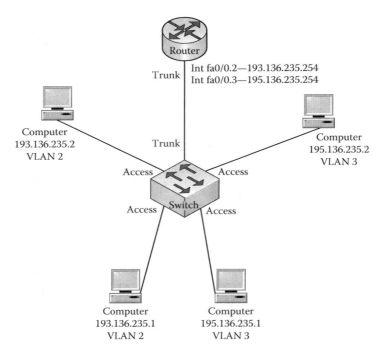

FIGURE 12.55   Inter-VLAN routing with a router (router-on-a-stick).

to be interconnected. In terms of the router's configuration, this is performed using the following sequence of syntaxes from the configuration mode (see Figure 12.56):

    a. Router(config)#interface *interface* (the identification of the router's interface connected to the switch).

    b. Router(config-if)#no ip address (to remove any IP address from this router's interface,[*] otherwise packets are not received by the local subinterface).

    c. Router(config-if)#no shutdown (to activate this router's interface[†]).

    d. Router(config-if)#interface *interface. sub_interface*[‡] (the router must be configured with multiple subinterfaces, where each subinterface is to be connected to a different VLAN).

    e. Router(config-subif)#encapsulation dot1q *vlan_number* [native][§] (uses the encapsulation protocol IEEE 802.1Q).

    f. Router(config-subif)#IP address *int_ip_address subnet_mask* (the IP address and subnet mask of the subinterface).

    g. Router(config-subif)#CTRL-Z (save and exit).

The commands d, e, and f above are to be issued as many times as the number of subinterfaces to create and configure.

---

[*] Only the subinterfaces will have IP addresses.
[†] Only the interface is activated, not the subinterfaces.
[‡] For example, Router(config)#int fa0/0.2 (2 is the subinterface to configure).
[§] When configuring the native VLAN, the command "native" must be added to allow the router's trunk to carry control traffic.

IOS Command Line Interface

```
Router>en
Router#conf t
Enter configuration commands, one per line.  End with CNTL/Z.
Router(config)#int fa0/0
Router(config-if)#no shut

Router(config-if)#
%LINK-5-CHANGED: Interface FastEthernet0/0, changed state to up

Router(config-if)#no ip address
Router(config-if)#int fa0/0.2
Router(config-subif)#
%LINK-5-CHANGED: Interface FastEthernet0/0.2, changed state to up

Router(config-subif)#encapsulation dot1q 2
Router(config-subif)#ip address 193.136.235.1 255.255.255.0
Router(config-subif)#int fa0/0.3
Router(config-subif)#
%LINK-5-CHANGED: Interface FastEthernet0/0.3, changed state to up

Router(config-subif)#encapsulation dot1q 3
Router(config-subif)#ip address 195.136.235.1 255.255.255.0
Router(config-subif)#int fa0/0.99
Router(config-subif)#
%LINK-5-CHANGED: Interface FastEthernet0/0.99, changed state to up
Router(config-subif)#encapsulation dot1q 99 native
Router(config-subif)#ip address 194.136.235.1 255.255.255.0
Router(config-subif)#^Z
Router#
```

FIGURE 12.56 Configuring a router with subinterfaces for VLAN interconnectivity (router-on-a-stick).

12.6.2.3 Inter-VLAN Connectivity with a Multilayer Switch

Inter-VLAN communication is also possible by using a multilayer switch (layer 2 and 3 switch), instead of a router. This topology is depicted in Figure 12.57, where an external router is included but is not part of the inter-VLAN connectivity (it is only utilized to describe the configuration of the multilayer switch interface that connects to it). The layer 2 switch is configured as before, with the interfaces connected to computers in access mode and with the interface connected to the multilayer switch in trunk mode. The interface of the external router connected to the multilayer switch is configured as in a regular router-to-router connection (configure the interface with an IP address, configure a routing protocol or a static route, etc.). With regard to the multilayer switch, layer 3 switching needs to be enabled in the multilayer switch, and an IP address needs to be assigned to each of the VLANs to which routing is to be allowed, followed by the configuration of the routing (dynamic routing protocol, default static route to external router, etc.). This is performed using the following syntax:

- L3switch(config)#ip routing (enables layer 3 switching)
- L3switch(config)#int vlan *vlan_number*
- L3switch(config-if)#ip address *ip_address subnet_mask* (assign the IP address to VLAN)
- L3switch(config-if)#no shutdown
- L3switch(config-if)#ip route 0.0.0.0 0.0.0.0 [*next_hop_address* | *exit_interface*] (alternatively, a dynamic routing protocol can be configured)
  - *Example*:
    - L3switch(config)#ip routing
    - L3switch(config)#int vlan 2 (after the creation of vlan2)
    - L3switch(config-if)#ip address 193.136.235.254 255.255.255.0
    - L3switch(config-if)#no shutdown
    - L3switch(config-if)#int vlan 3
    - L3switch(config-if)#ip address 195.136.235. 254 255.255.255.0
    - L3switch(config-if)#no shutdown

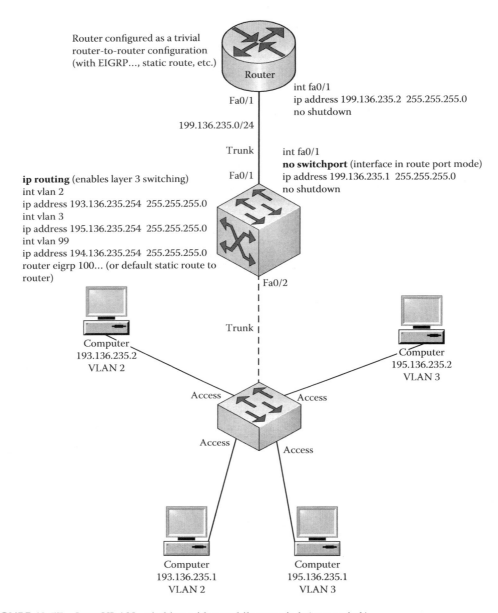

Router configured as a trivial
router-to-router configuration
(with EIGRP..., static route, etc.)

Router

int fa0/1
ip address 199.136.235.2 255.255.255.0
no shutdown

Fa0/1

199.136.235.0/24

Trunk

int fa0/1
no switchport (interface in route port mode)
ip address 199.136.235.1 255.255.255.0
no shutdown

Fa0/1

ip routing (enables layer 3 switching)
int vlan 2
ip address 193.136.235.254 255.255.255.0
int vlan 3
ip address 195.136.235.254 255.255.255.0
int vlan 99
ip address 194.136.235.254 255.255.255.0
router eigrp 100... (or default static route to
router)

Fa0/2

Trunk

Computer
193.136.235.2
VLAN 2

Computer
195.136.235.2
VLAN 3

Access   Access

Access   Access

Computer
193.136.235.1
VLAN 2

Computer
195.136.235.1
VLAN 3

FIGURE 12.57   Inter-VLAN switching with a multilayer switch (core switch).

- L3switch(config-if)#int vlan 99
- L3switch(config-if)#ip address 194.136.235. 254 255.255.255.0
- L3switch(config-if)#no shutdown
- L3switch(config-if)#exit
- L3switch(config)#ip route 0.0.0.0 0.0.0.0 199.136.235.2
- L3switch(config)#CTRL-Z

With regard to the interfaces, as can be seen from Figure 12.57, both are configured in trunk mode, whereas the interface connected to the router needs to be configured in route-port mode and with an IP address assigned, as follows:

- L3switch(config)#interface *interface*
- L3switch(config-if)#switchport mode trunk (trunk mode)
- L3switch(config-if)#no switchport (to activate the route-port mode)
- L3switch(config-if)#ip address *ip_address subnet_mask* (assign the IP address to interface)
- L3switch(config-if)#no shutdown
  - *Example*:
    - L3switch(config)#interface fa0/1 (interface connected to the router)
    - L3switch(config-if)#switchport mode trunk
    - L3switch(config-if)#no switchport
    - L3switch(config-if)#ip address 199.136.235.1 255.255.255.0
    - L3switch(config-if)#no shutdown
    - L3switch(config)#interface fa0/2 (interface connected to L2switch)
    - L3switch(config-if)#switchport mode trunk

Appendix III contains an example of an internetwork configuration with inter-VLAN routing, using both an external router (router-on-a-stick) and a multilayer switch.

Chapter 16 (under the topic firewalls) describes the configuration of access control lists in Cisco routers.

12.6.3 VLAN TRUNKING PROTOCOL

The VTP consists of a protocol used to automatically propagate the VLAN configuration to neighbor switches within a network (bounded by a router or a layer 3 switch). The use of the VTP protocol results in a simplified management of the VLAN database across multiple switches. Using the VTP, and configuring a switch as a VTP server and the others as VTP clients, a network manager simply has to create VLANs in the VTP server. These created VLANs are replicated to switches configured as VTP clients, simplifying the configuration effort and avoiding configuration inconsistencies. Note that the VTP only propagates normal range VLANs (1–1005), that is, extended range VLANs are not propagated. Moreover, to allow VTP propagation, VTP switches must be interconnected in trunk mode. The propagation of the VTP configuration is carried out using VTP advertisement packets within a VTP domain. A VTP domain comprises a number of switches within a network that share the VLAN configuration using the VTP.

A switch can be configured, within a VTP domain, in one of three modes:

- *VTP server*: A switch that can create, delete, or rename VLANs for the whole network. Instead of having to create, delete, or rename VLANs in each of the network switches, a network manager can simply perform these functions in a switch configured as a VTP server. These configurations are then replicated to switches configured as VTP clients or VTP servers within the VTP domain using VTP advertisements, as long as the configuration revision number of the manipulated switch is higher than the configuration revision number of the other switches. A switch can be placed in VTP server mode from the configuration mode of the switch using the syntax "vtp mode server." Note that all the switches of a network must have the same domain name (case sensitive), being configured using the syntax "vtp domain *domain_name*." As a rule of thumb, to keep a backup, a network should have two switches configured as VTP servers (see Figure 12.58). Note that the VLAN information is stored in the nonvolatile random access memory (NVRAM) of the VTP server. By default, a switch is configured as a VTP server.
- *VTP client*: A switch that replicates the VLAN configuration received from a VTP server (using VTP advertisements) with a configuration revision number higher than the one previously stored in the VTP client. Moreover, in case a network manager modifies the VLAN configuration locally in a VTP client switch, this modification is locally implemented, but

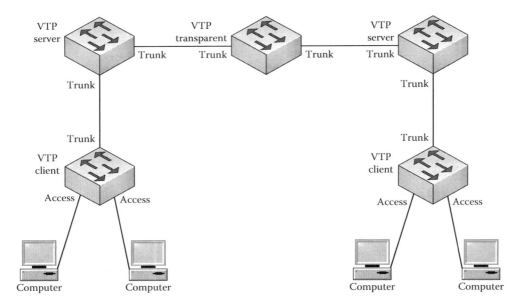

FIGURE 12.58 VTP protocol.

not replicated to the other switches of the VTP domain. Note that the VTP client stores VLAN information in the random access memory (RAM), whereas the VTP server does it in the NVRAM. A switch can be placed in VTP client mode from the configuration mode of the switch using the syntax "vtp mode client."

• *VTP transparent*: A switch configured as VTP transparent simply forwards received VTP advertisements to neighboring switches, but its own configuration is not modified according to these VTP advertisements. Moreover, in case a network manager modifies the VLAN configuration in a VTP transparent switch, this modification is locally implemented, but not replicated to other switches of a VTP domain. A switch can be placed in VTP transparent mode from the configuration mode of a switch using the syntax "vtp mode transparent."

In addition to VTP mode, a number of other VTP parameters can be configured in a switch. Such parameters can be viewed in a switch from the priviledge mode using the syntax "show vtp status," and include the following parameters:

• *VTP version*: A switch can be configured with VTP version 1 or 2. By default, a switch uses version 1. The version 2 is configured using the syntax "vtp version 2."
• *Configuration revision*: It displays the configuration revision number. Everytime the VLAN information is modified, such as adding or deleting a VLAN, the configuration revision number is increased by one unity. In contrast, in case the VTP domain name is modified, the configuration revision number is reset (returns to zero). By default, the configuration revision number is zero.
• *Maximum VLANs supported locally*: It displays the maximum number of VLANs supported by the switch. This number depends on the type of switch, but a common value for this parameter is 255.
• *Number of existing VLANs*: It displays the number of VLANs configured in the switch, including those configured locally (if applicable) and those synchronized from a VTP server. The default number of existing VLANs is 5.
• *VTP operating mode*: It displays the VTP mode of operation of a switch, being server, client, or transparent. By default this parameter is set to the VTP server.

- *VTP domain name*: It displays the domain name of a switch. All switches that belong to the same domain must use the same VTP domain name. By default, the VTP domain name is *null*.
- *VTP pruning mode*: It displays whether or not pruning is configured in the switch (enabled or disabled). When pruning mode is enabled, a switch only forwards frames with a certain destination MAC address through a trunk link in case the corresponding host is accessible through that trunk (a MAC address table is kept associated with each interface connected to a trunk). This results is a more efficient use of the network bandwidth, as well as improved network security. When pruning mode is disabled, all frames are always sent through trunk links. By default, the VTP pruning mode is set to disabled.
- *VTP V2 mode*: It displays whether or not VTP version 2 is enabled (enabled or disabled). By default, VTP V2 mode is in disabled mode.
- *VTP traps generation*: When enabled, VTP traps are sent to a network management station.
- *MD5 digest*: It displays whether or not MD5 is enabled, being utilized for checking integrity and authenticity in VTP advertisements.
- *Configuration last modified*: It displays the date and time of the last VTP advertisement that materialized a VLAN modification, including the IP address of the switch responsible for such modification.

It was described that VLAN information is synchronized within a VTP domain using VTP advertisements. Depending on the type of information that is carried out, these advertisements can be of the following three types:

- *Summary advertisements*: They are sent immediately by a switch configured as either a VTP client or a VTP server when its configuration is modified, or when a request advertisement is received. Moreover, summary advertisements are also periodically transmitted, by a VTP server, in every 5 min. The aim of a summary advertisement relies on sending the VTP configuration revision number, and domain name modifications to neighboring switches.
- *Subset advertisements*: A subset advertisement relies on sending information, for synchronization purposes, about adding, deleting, or renaming VLANs, or to modify the MTU size. A subset advertisement is only sent by a VTP server when a request advertisement is received.
- *Request advertisements*: It is sent by a switch when: (1) a summary advertisement arrives with a higher configuration revision number; (2) the switch is reset or the VTP domain name is modified; and (3) a subset advertisement is missing. When a request advertisement is received, the VTP server responds with a summary advertisement, followed by a subset advertisement.

When initiating the VTP configuration of switches that belong to a network, the network manager must assure that all switches are set to their default settings. This can be performed by resetting the switches. Moreover, when adding a new switch to an existing VTP domain, the network manager must assure that this switch has a configuration revision number lower than the others already working in the VTP domain. Otherwise, the VLAN configuration of the new switch is propagated to the other switches, which can disrupt the whole network. One procedure that can be implemented by a network manager to reset the configuration revision number relies on modifying twice the domain name (back and forward). Remember that modifying the domain name results in resetting the configuration revision number.

A password can be configured in a switch participating in a VTP domain. In this case, the authentication is carried out during VTP advertisements. A VTP password can be configured using the syntax "vtp password *password*." Note that passwords and domain names are case sensitive. This means that if, by mistake, passwords or domain names are not exactly the same in different switches, they do not synchronize. A common reason for VTP switches to not synchronize results also from the use of different VTP versions, or because the interconnection between two switches is not established using trunk links, or even because all switches are configured as VTP clients (none is acting as a VTP server).

The most common troubleshooting command used in the VTP configuration is "show vtp status," typed from the privileged mode.

CHAPTER SUMMARY

This chapter provided a view about the DLL. A description of the devices employed in LAN networks was given, namely a hub, a bridge, and a switch.

The STP was described, which is employed to avoid loops within broadcast domains, to calculate paths between network switches, and to build MAC address tables.

A description of the LLC sublayer was provided, including the error control techniques, namely error detection associated with retransmission, and error correction; the ARQ techniques, namely the stop and wait, the go back N, and the selective reject; and the flow control techniques, namely the stop and wait, and the sliding window. It was viewed that the stop-and-wait protocol becomes very ineffective when the propagation time of signals is high. In this case, it is preferable the use of the Go-Back-N ARQ or the Selective Reject ARQ, associated with the flow control sliding window.

It was viewed that the LLC protocols are responsible for providing error control and flow control, that is, the LLC is responsible for providing reliability to the bits exchanged by the physical layer. For this to be possible, the bits are grouped into frames. Because this chapter focuses on LAN networks, the IEEE 802.2 protocol was described here, leaving the HDLC and PPP protocols for another chapter.

It was viewed that the MAC sublayer is employed to regulate the access of hosts to the transmission medium. Several MAC protocols were described, namely IEEE 802.3 (commonly referred to as Ethernet), IEEE 802.5 (commonly referred to as token ring), FDDI, and DVB.

The VLAN concept was introduced, as employed for reducing the broadcast domain, or for performing isolation of devices within the network, which can be an important advantage for security purposes.

Finally, the VTP protocol was introduced, consisting of a protocol used to automatically propagate VLAN configuration to neighbor switches within a network.

REVIEW QUESTIONS

1. What are VLANs used for?
2. What are the functionalities provided by the LLC sublayer? What is the mostly known LLC sublayer protocol used in LAN?
3. What does the MAC sublayer is used for? In which type of networks do the MAC sublayer protocols exist?
4. What are the advantages and disadvantages of the IEEE 802.3 protocol relating to IEEE 802.5?
5. Define the CSMA-CD mechanism.
6. In which layer and sublayer is the error control capability implemented?
7. Consider the CRC generator polynomial $P(x) = x^2 + 1$ with the information source 11011.
8. Consider that the received bits sequence is 1101101 and that the CRC generator polynomial is $P(x) = x^2 + 1$. Verify whether or not the received sequence of bits is free of errors.

9. Which types of flow control protocols do you know? How do they work?
10. Which types of error control protocols do you know?
11. What is the difference between the Go-Back-N ARQ and the Selective Reject ARQ mechanisms?
12. Having a code rate of 1/2, how can we obtain a code rate of 3/4?
13. In which scenarios is it preferable to use error correction mechanisms, instead of error detection?
14. What is the difference between the 10BASE-T and the 1000BASE-T technologies?
15. What does puncturing stands for? How is it implemented?
16. What does code interleaving stand for? What is its advantage and why is it employed?
17. What is the relationship between the Hamming distance and the error correction capability?
18. What is the encapsulation protocol normally employed in VLANs?
19. Which types of mechanisms can be employed for signalizing the start and end of frames?
20. What is the difference between the 1000BASE-CX and the 1000BASE-SX technologies?
21. What is the relationship between the Hamming distance and the error detection capability?
22. Which physical layer mechanisms are employed by 1000BASE-T to allow a throughput of 1 Gbps?
23. What is the IEEE 802.1D protocol used for?
24. What does the MCDD stand for?
25. What does the carrier extension stand for and for which purpose is it employed?
26. In CSMA-CD what does the backoff period stand for? What values may it take?
27. What is the MCDD of an IEEE 802.3 network, working at 10 Mbps?
28. Which kind of listening strategies can be employed in the CSMA-CD mechanism?
29. What is the difference between a recursive and a nonrecursive convolutional code?
30. What is the logical topology of an IEEE 802.5 network?
31. What is the difference between a bridge and a switch?
32. What is the difference between a hub and a switch?
33. What is the difference between a switch and a router?
34. Consider a convolutional code whose outputs are defined by $k_1 = n_1 \oplus n_0 \oplus n_{-1}$, $k_2 = n_1 \oplus n_0$, $k_3 = n_1 \oplus n_{-1}$. Considering that the input sequence is 101, what is the output sequence?
35. Having a single physical network, how can we create several logical networks?
36. What does frame bursting stand for?
37. How does the sliding window protocol work?
38. Which type of error correction codes do you know? What are the differences between them?
39. What is the difference between a systematic and a nonsystematic error control technique?
40. What is the difference between the positive acknowledgment with retransmission (PAR) and the negative acknowledgment (NAK)? What are their relative advantages and disadvantages?
41. How is the code rate of an error code defined?
42. What is the aim of the IEEE 802.1Q protocol?
43. Which topologies can be used to implement inter-VLAN connectivity?

LAB EXERCISES

1. Download and install the free network analyzer Wireshark. Open the application in a PC and select the interface connected to the Internet. You will see the IP datagrams that are being exchanged by the NIC of the PC, with the source IP address, destination IP address, and protocol type, alongside with other information. Select a datagram for inspection in the lower window. In a window that appears at the bottom, visualize the content of the

frame and Ethernet message, including the different header fields. In the outgoing frames, what are the source MAC address and the destination MAC address?

2. Consider the network depicted in the figure below. Assuming the subnet mask 255.255.255.0, configure the plotted network using the Cisco Packet Tracer simulator. Use the ping and tracert commands in hosts. Check the results.

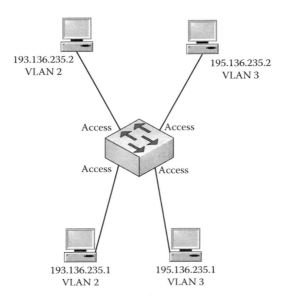

3. Consider the network depicted in the figure below. Assuming the subnet mask 255.255.255.0, configure the plotted network using the Cisco Packet Tracer simulator. Configure routing between different VLANs.

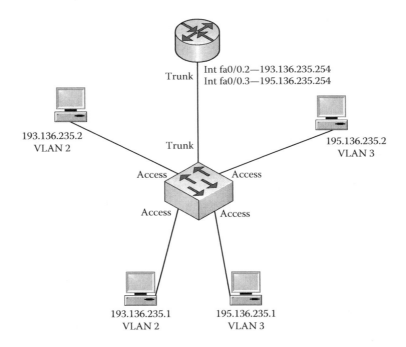

4. Consider the network depicted in the figure below. Assuming the subnet mask 255.255.255.0, configure the plotted network using the Cisco Packet Tracer simulator. Use the OSPF protocol in routers. Configure routing between different VLANs. Use the VTP protocol when configuring the two switches directly connected.

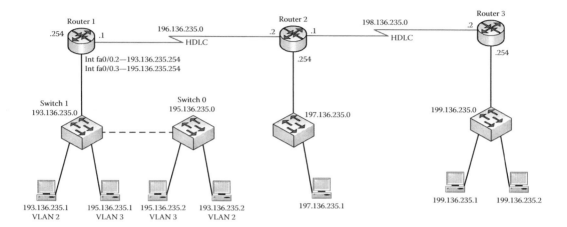

5. Replicate the previous exercise using the EIGRP protocol, instead of the OSPF.
6. Replicate the previous exercise using the static routes, instead of the EIGRP.
7. Replicate exercise 1 using real network equipment.
8. Consider the network depicted in the figure below. Configure the plotted network using the Cisco Packet Tracer simulator. Configure routing between different VLANs.

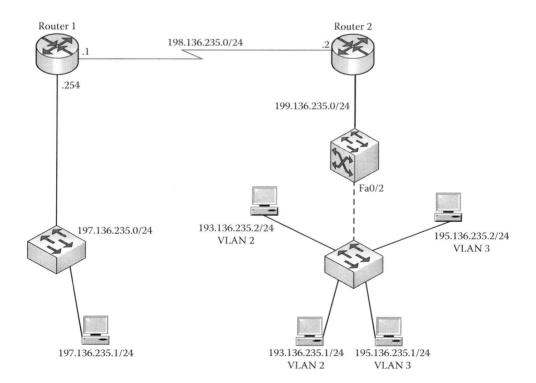

13 Structured Cabling System

LEARNING OBJECTIVES

- Define the concept of structured cabling.
- Identify and describe the subsystems and elements of a structured cabling system.
- Identify and describe the standards and specifications used in structured cabling systems.
- Describe the hardware and accessories used in structured cabling systems.

Legacy cabling systems were dedicated to specific services, such as telephony, computer networks, security systems, video teleconference (VTC), and telemetry. This originated a duplication of cabling that translated in an increase cost, as well as in an inability to adapt to new service requirements and technologies. Moreover, because of lack of standardization, the technical support of these legacy cabling systems was dependent on each manufacturer.

The basic idea behind the structured cabling concept relies on overcoming the above-described drawbacks, by using a modular cabling system specified by standards and independent of the manufacturers, and for use in a wide variety of application services. Moreover, the structured cabling concept is oriented by performances, taking redundancy as an important requirement, and keeping flexibility to support future additional network requirements. Structured cabling consists of a way networks are organized, with the aim of facilitating installation, maintenance, and administration.

Therefore, structured cabling must be generic to support a wide range of communication technologies and contents (data, voice, VTC, etc.), and sufficiently flexible to accommodate the normal technologies' evolution of communication systems and to support an eventual growth of the organization, without the need to radically modify the existing cabling infrastructure.

In order to facilitate the operation and management procedures of cabling, to reduce the amount of traffic, and to make it generic and flexible, cabling should be hierarchically organized in modules (subsystems). These modules comprise the interconnection between buildings (campus backbone), the interconnection between floors within a building (building backbone), and the interconnection between different terminals (horizontal cabling). Moreover, cabling can also be functionally organized in branches, departments, services, and so on, as most of the traffic generated by a certain layer refers to its own.

Moreover, structured cabling is a concept that aims to be independent of the manufacturers. The basic idea relies on making possible mixing a certain cabling element from one manufacturer with another cabling element from another manufacturer. Furthermore, it must be possible to modify the active* equipment (hub, switch, router, etc.) without implication on the passive components (cabling, connectors, adapters, etc.). To make this possible, cabling must be defined by standards, instead of proprietary technologies. These standards can be of different levels, namely international, regional, national, or from associations (e.g., the Telecommunications Industry Association). The aim is to assure the independence of cabling from the manufacturers, as well as to allow a size or technological evolution. Naturally, the structured cabling concept tends to achieve a cost reduction relating to legacy cabling, especially when the whole life cycle is taken into account (i.e., considering installation, operation, administration, upgrades, etc.).

---

* Active equipment corresponds to equipment that performs regeneration of signals, in addition to any other function.

13.1 HIERARCHICAL NETWORK DESIGN

In order to meet the organizational requirements, a common approach relies on splitting the design into multiple layers, where each layer performs a different group of functions (see Figure 13.1). This modular concept presents the following advantages:

- Improves the system reliability as it explores redundancy
- Accommodates changes and scalability without major physical modifications
- Facilitates operation, administration, maintenance, and troubleshooting
- Reduces costs

With this modular approach, the equipment of each layer is kept as simple as possible, just to meet the layer functional requirements. The hierarchical structure of a network design comprises the following discrete layers:

- *Core layer*: This corresponds to the top layer, consisting of high-speed routers[*] and switches optimized for performance and availability. This comprises a high-speed backbone, assuring interconnection between multiple core layer equipment and the Internet. The core layer aggregates traffic from multiple distribution layer equipment, and therefore, the high speed is of major importance.
- *Distribution layer*: This corresponds to the mid-layer, consisting of high-speed routers[†] and switches, focusing on implementing security mechanisms, such as access control lists, and for allowing inter-virtual local area networks (VLAN) routing. The distribution layer aggregates traffic from multiple-access-layer equipment, and therefore, the speed available at this layer is also an important issue.
- *Access layer*: This corresponds to the lower layer, consisting of layer 2 switching, focusing on the connection with user equipment, such as workstations, printers, and

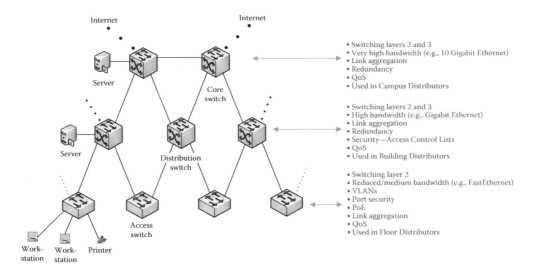

FIGURE 13.1 Hierarchical network design.

---

[*] Routing and switching functionalities can also be implemented by a route switch, where layer 2 and 3 switching is performed.

[†] Routing and switching functionalities can also be implemented by a route switch (also simply referred to as *switch*), where layer 2 and 3 switching is performed.

wireless access points. This is the layer where VLANs are created for a segmentation of the broadcast domain and for security purposes. The other security mechanism implemented at the access layer relies on implementing switch port security. When required, power over Ethernet (PoE) can also be provided by the access switch.* Note that, because of their simplicity, access layer switches are typically less expensive than other layer equipment.

The number of pieces of equipment required for each layer depends on the implementation requirements. A possibility relies on quantifying the number of user devices, and then defining the number of access layer switches required to support the number of user devices. Alternatively, an access switch can be placed at each floor of a building, or dedicated to a corporate department. Then, one must define the number of access switches per distribution switch. Finally, the number of distribution switches per core switch has also to be quantified. Assuming that one distribution switch supports eight access switches, one can scale the network until the number of eight access switches is reached.

An important advantage of a hierarchal network design relies on the reliability that results from the path redundancy. As can be seen from Figure 13.1, this is implemented by connecting an access switch to two different distribution switches, and by connecting a distribution switch to two or more core switches. Naturally, this results in the existence of loops, but these loops are resolved by the spanning tree protocol. In case one cable is interrupted or if a core or distribution switch fails, there is always another path available. Nevertheless, there is no redundancy in case an access layer switch fails or in case a cable that connects to the user device is interrupted.

The choice of the equipment and bandwidth to install is a function of the user community requirements. Moreover, future needs should also be taken into account, such as an increase in the number of corporate employees, future applications that require higher bandwidth, and addition of more servers.

As can be seen from Figure 13.1, a server normally utilized by a single department can be located at the distribution layer, while a server massively utilized by the whole organization should be placed at the core layer.

When specifying a switch for each of the three layers, one should specify the port density, forwarding rate, and link aggregation. The port density consists of the number of ports in a switch, whereas the forwarding rate corresponds to the amount of data a switch can process per second. Finally, when the ports' bandwidth of a switch is lower than required, some switches support link aggregation. In this case, one can interconnect a distribution switch with a core switch using, for example, four parallel Gigabit Ethernet links, instead of a single one.

It is finally worth referring that small environments may implement a hierarchical network design with two layers, instead of three, comprising a core layer and an access layer. In this scenario, the core layer and the distribution layer are merged together, as well as their functionalities.

13.2 STRUCTURED CABLING ELEMENTS AND SUBSYSTEMS

The adoption of the structured cabling concept translates in an increased life cycle. This results from the fact that structured cabling can easily accommodate variations in throughputs, in contents (voice, data video), in size, in topology, or even in technology. Therefore, upgrading structured cabling is not a difficult task, as it is modular, with open interfaces, and defined by open standards, rather than by proprietary technological solutions.

---

* PoE is normally provided as an option, and tends to increase very much the cost of a switch. Therefore, the choice for using PoE must be carefully made.

Structured cabling comprises the following functional elements (see Figure 13.2) [ISO/IEC 11801ed2]:

- *Campus distributor* (*CD*): It comprises a rack with equipment from where the campus backbone cabling diverges. Note that the CD exists only when structured cabling comprises multiple buildings, and typically includes a core layer switch. Therefore, in case a single building is considered, this can be omitted.
- *Campus backbone cabling*: It is used to allow the interconnection between the CD and different building distributors (BDs).
- *BD*: It comprises a rack with equipment that performs the cross-connection between campus backbone cabling and building backbone cabling. The BD may include both the core layer and the distribution layer switches, or simply the core layer switch (leaving the remaining two layers of switches for the floor distributor [FD]). Alternatively, in small environments, the core and distribution layers can be merged into a single layer, called the *core layer*.
- *Building backbone cabling* (*or vertical backbone cabling*): It is used to allow the interconnection between the BD and the FD.
- *FD*: It comprises a rack with equipment that performs the cross-connection between building backbone cabling and horizontal cabling. Depending on the number of telecommunication outlets and their physical distribution, one FD may serve more than one floor* (such that

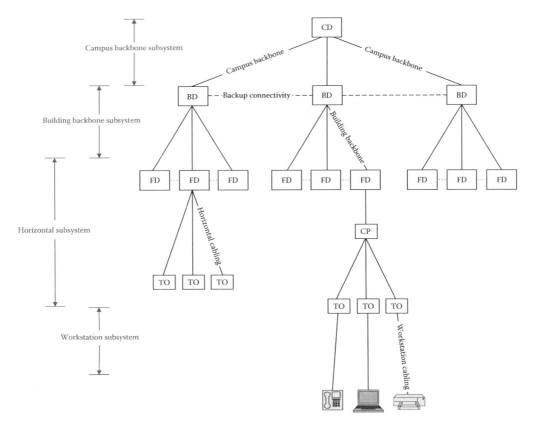

FIGURE 13.2 Subsystems of structured cabling.

---

* For the sake of cabling minimization, an FD should be placed as close as possible to the geographic center of the floor.

the amount of cabling and/or hardware is minimized). On the other hand, an FD should be limited to a maximum of 250 telecommunication outlets. Moreover, a BD can simultaneously be used as a BD and as an FD. The FD may include an access layer switch, or may include both an access layer switch and a distribution layer switch.

- *Horizontal cabling*: It is used to allow the interconnection between the FD and the telecommunication outlets.
- *Consolidation point* (*CP*): Between the FD and the TO, there could be a consolidation point that acts as a small patch panel.
- *Telecommunication outlet* (*TO*): It comprises a socket where the workstation cable connects to (using, e.g., ISO 8877, SC or ST connector). Note that a TO can be single or double. Double TOs are typically employed to connect a telephone and a workstation. Currently, the most common configuration comprises an IP telephone that connects to the TO, while the workstation connects to the IP telephone. Therefore, in this scenario, a single TO is enough per workstation.
- *Workstation cabling*: It is used to allow the interconnection between a TO and terminal equipment (workstation, printer, server, IP telephone, etc.).

A distributor rack comprises multiple pieces of equipment, namely one or more switches or hubs, eventually a router, one or more patch panels, an overvoltage protection, an uninterrupted power supply (UPS), and so on. Section 13.4 describes each of these elements.

Structured cabling is also split into four subsystems [ISO/IEC 11801ed2], where each subsystem comprises a number of functional elements. These subsystems are (see Figure 13.2) as follows:

- *Campus backbone subsystem*: It is used to interconnect different buildings within a campus (such as in a university campus or airport). It comprises a CD and campus backbone cabling, as well as the corresponding terminations. Note that this subsystem may not exist, as a private network may be limited to a building. When present, the campus backbone corresponds typically to a metropolitan area network.
- *Building backbone subsystem*: It is used to interconnect a BD with an FD. It includes a BD, building backbone cabling, as well as the corresponding terminations.
- *Horizontal subsystem*: It is used to interconnect the FDs with the TOs, comprising these two types of functional elements, as well as horizontal cabling.
- *Workstation subsystem*: It is used to interconnect the TOs with the terminal equipment, including devices such as patch cords, adapters, and connectors. It also includes workstation cabling.

13.2.1 Centralized Optical Architecture

The second amendment to the ISO/IEC 11801 standard is entitled ISO/IEC 11801amend2, which introduced the centralized optical architecture concept. When structured cabling is composed of only optical fibers (i.e., twisted pairs are not employed), the horizontal subsystem can be omitted. In this case, optical fibers connect the BD directly with TOs. One or more cables, composed of multiple optical fibers, leave the BD toward multiple TOs. Because optical fibers commonly support the distance between the BD and TOs, regeneration performed by the FD is not required. Moreover, although the cable that leaves the BD carries multiple optical fibers, because the width of optical fibers is much thinner than twisted pairs, the width of the cable composed of a high number of optical fibers is still acceptable. Figure 13.3 depicts the generic diagram of structured cabling with the centralized optical architecture. Because of the lower number of distributors, an installation and maintenance cost reduction is achieved with this architecture. It is finally worth noting that

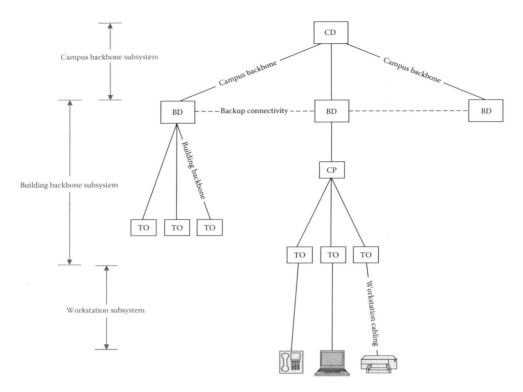

FIGURE 13.3   Subsystems of structured cabling with the centralized optical architecture.

the optical fiber length employed in successive connections depends on the type of optical fiber. In the case of multimode optical fibers, attention should be paid to the modal bandwidth factor, which relates the maximum bandwidth to the maximum length of an optical link.

13.3   STRUCTURED CABLING STANDARDS AND SPECIFICATIONS

Different levels of institutions aimed to release standards on structured cabling. This subject was initially standardized in North America by a joint venture between three institutions: the American National Standards Institute (ANSI), which is a national standardization institution, the Electronic Industries Alliance (EIA), and the Telecommunications Industry Association (TIA), corresponding to manufacturers' associations.

Afterward, the International Organization for Standardization (ISO), in association with the International Electrotechnical Commission (IEC), took the work already performed in North America, introduced some additional modifications, and released international standards on structured cabling.

In Europe, the regional standardization institution entitled *Comité Européen de Normalisation Electrotechnique* (CENELEC) also released standards on this subject.

13.3.1   NORTH AMERICAN STANDARDS

The structured cabling concept was initially standardized by ANSI/TIA/EIA, in 1991, as ANSI/TIA/EIA-568. The aim was to define a generic and flexible cabling concept, independent of the manufacturer's technologies, able to support different traffic types and data rates, and easy to be modified [Stallings 2010]. The definition of subsystems and functional elements of structured cabling, the type of transmission medium to be used in each module, the corresponding maximum length and bandwidth, as well as the type of connectors to use was established in this standard.

This standard was updated in 1995 by a new edition entitled ANSI/TIA/EIA-568-A, where more recent transmission mediums were introduced, as well as additional details previously not taken into account. ANSI/TIA/EIA-568-A does not impose any electromagnetic interference protection to twisted pair cables, such as shielding or screening. Moreover, contrary to the international standard that derived from this [ISO/IEC 11801], no special requirements were considered in terms of fire protection (see Table 13.1).

Figure 13.4 shows an example of the implementation of structured cabling following the standard ANSI/TIA/EIA-568-A. This example considers a private network composed of only a single building; therefore, the campus subsystem does not exist (there is neither CD nor campus cabling). In this figure, various functional elements are depicted. Moreover, the maximum cable distances are also depicted, which are a function of the employed transmission medium. As can be seen from this example, the BD also acts as an FD, which results in a hardware reduction. Moreover, because the BD connects with the exterior (WAN or MAN), it makes use of a router, in addition to the main switch (core switch) and patch panel existing in the same rack. In contrast, the FD does not require a router, simply has an access switch (which is connected to the main switch, located in the BD), as well as a patch panel. A router could be employed in case it is intended to perform a segmentation of the broadcast domain or, alternatively, VLANs can be created in the access switch. Note that a CD has typically a core switch, a BD commonly makes use of a distributor switch, while an FD uses an access switch. In this case, because the CD is not used, a core switch is considered in the BD, and the distributor switch is not employed.

The BD and the FDs are interconnected through the building backbone, and where redundant cabling also exists to prevent from failures. The FD connects to TOs through horizontal cabling that, according to ANSI/TIA/EIA-568-A, can have a maximum length of 90 m (100 m between the FD and the terminal equipment, including patch cords and workstation cabling). Finally, the patch panel allows the flexible connectivity between the TOs and the required positions of the switch, hub, or PABX (Figure 13.4).

In 2001, the new update entitled ANSI/TIA/EIA-568-B introduced some modifications to the previous standard version, namely to connection termination layout and to twisted pair categories (see Chapter 4).

Finally, the 2009 update, entitled ANSI/TIA/EIA-568-C, comprises the state of the art in transmission mediums, and the corresponding update in the definition of maximum lengths and bandwidths. Moreover, the concept of centralized optical architecture was introduced, where cabling comprises only optical fibers and the FD can be removed from the architecture. Naturally, this

TABLE 13.1
Twisted Pair Types of Different Standards

Standard	Sheathing Type	Jacket Type
ANSI/EIA/TIA-568-A	Thermoplastic	UTP
ISO/IEC 11801	Thermoplastic optionally with LSZH[a]	Twisted pair optionally with shielding or screening (UTP, STP, FTP)
EN 50173	Thermoplastic with LSZH (mandatory)	Twisted pair with screening (mandatory) and shielding (optional) (FTP or S/FTP)

[a] LSZH stands for low smoke zero halogen.

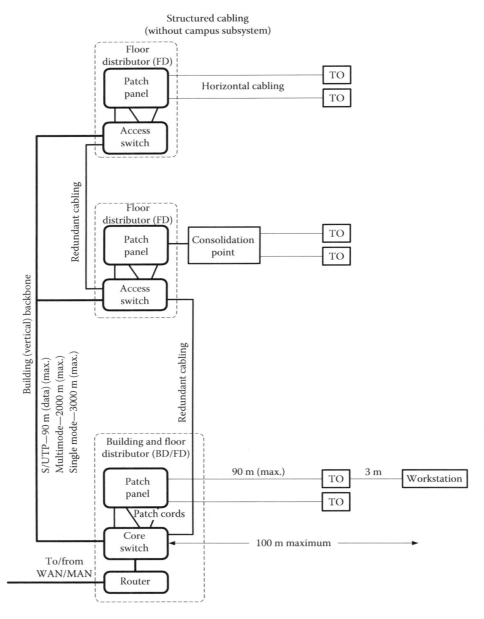

FIGURE 13.4 Example of a structured cabling implementation with functional elements in accordance with ANSI/TIA/EIA-568-A.

architecture comprises one less subsystem (see Figure 13.3), relating to the conventional architecture where twisted pair cabling can also be considered.

13.3.2 INTERNATIONAL STANDARDS

As a follow-up to the standardization that took place in North America, the ISO, jointly with the IEC, released, in 1995, ISO/IEC 11801. This initial structured cabling standard by ISO/IEC corresponds approximately to ANSI/TIA/EIA-568-A, but with some additional requirements, namely the cabling that may optionally present shielding[*] or screening,[†] while North American standard comprises unshielded twisted pairs without any shielding. Moreover, this standard also states that the cable sheathing may present the low smoke zero halogen (LSZH) property as an option. In case of fire, a cable with LSZH characteristics burns with low emissions of smoke, and with zero emissions of halogen. Note that this requirement was not considered by ANSI/TIA/EIA-568-A. Table 13.1 lists the most important differences between the equivalent standards of difference scopes. Moreover, ISO/IEC 11801 also imposes some coverage and the number of users' upper bounds.[‡]

In 2002, the ISO/IEC released a new standard edition, entitled ISO/IEC 11801ed2, where some additional requirements taken into account by ANSI/TIA/EIA-568-B were introduced. Finally, an amendment introduced in 2010 [ISO/IEC 11801amend2] already took into account some of the requirements considered by ANSI/TIA/EIA-568-C, namely the centralized optical architecture. Moreover, the limitations in terms of the number of users and coverage, previously considered, were removed in this standard version [Monteiro and Boavida 2011]. Therefore, with the exception of some minor modifications, one can state that the international standard [ISO/IEC 11801amend2] corresponds approximately to the North American standard [ANSI/TIA/EIA-568-C].

The maximum distances allowed by ISO/IEC 11801amend2 are depicted in Figure 13.5. Note that this standard specifies a maximum link distance of 500 m for the building backbone, regardless of the transmission medium employed. Naturally, these values are maximum figures. Therefore, in the case of twisted pair cabling, one must perform the calculations taking into account the bandwidth under consideration, the attenuation coefficient curve, the ACR, the ELFEXT, and so on. Moreover, in the case of multimode optical fibers, the modal bandwidth factor should be taken into account, together with the used bandwidth, to perform the calculation of the maximum cable length. Note that the interconnection between the CD and the BD and the interconnection between the BD and the FD are recommended to be performed using single-mode optical fiber (for data services). In case structured cabling is only utilized for voice, shielded or unshielded twisted pairs can be employed. The connection between the FD and the TO can be implemented using any type of transmission medium. The decision relies on the required bandwidth, that is, on the application class type.

Table 4.3 shows the maximum bandwidth that a certain twisted pair category can accommodate, but for a distance of 100 m (40 m in the case of category 7). However, the maximum bandwidth that can be supported by a certain twisted pair cable, with a specific length, is quantified in ISO/IEC 11801amend2 as a function of the application class type, that is, as a function of the used bandwidth.

Table 13.2 shows different application class types, from A to F_A. These bandwidths refer to those consumed by network terminals, that is, by hosts or servers. Note that above 1 GHz, optical fiber is normally the preferable transmission medium.

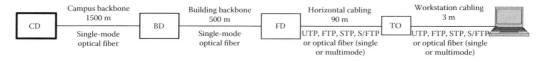

FIGURE 13.5 Maximum distances allowed by ISO/IEC 11801amend2 for data services (recommended cablings).

---

[*] Shielding comprises a metallic braid or sheathing, applied to each individual pair of wires, which protects wires from noise and interferences (namely crosstalk).

[†] When a shielding is applied to multiple pairs, that is, to the whole cable, instead of a single pair of wires, it is referred to as *screening*.

[‡] A maximum distance of 3000 m, maximum coverage area of 1,000,000 m², and supporting a maximum of 50,000 terminals.

TABLE 13.2

Application Class Types [ISO/IEC 11801amend2]

Class	Bandwidth (MHz)	Application
A	100 kHz	Telephony and other low rate applications (e.g., PABX, X.21)
B	1 MHz	Data applications at a low rate (e.g., ISDN)
C	16 MHz	Data applications at a medium rate (e.g., Fast Ethernet, Token Ring)
D	100 MHz	Data applications at a high rate (e.g., Fast Ethernet, Gigabit Ethernet)
E	250 MHz	Data applications at a high rate (e.g., Gigabit Ethernet, 10 Gigabit Ethernet)
E_A	500 MHz	Data applications at a very high rate (e.g., 10 Gigabit Ethernet)
F	600 MHz	Data applications at a very high rate (e.g., 10 Gigabit Ethernet or higher)
F_A	1 GHz	Data applications at a very high rate (e.g., 10 Gigabit Ethernet or higher) and HDTV

As can be seen from Figure 13.6, one can calculate the aggregate bandwidth in building (vertical) backbone cabling by summing the individual bandwidths consumed by terminal stations, where each of these parcels should be properly weighted by a simultaneousness coefficient. This can be expressed mathematically by

$$\text{Aggregate\_BW} = \sum_{i=1}^{N} \left[\text{Indiv\_BW}_i^* \text{simultan\_coef}_i \right] \qquad (13.1)$$

where:
Aggregate_BW stands for the aggregate bandwidth, also referred to as the *total throughput*
N is the number of individual trunks under consideration in the aggregate bandwidth
Indiv_BW$_i$ refers to the bandwidth of the ith trunk
simultan_coef$_i$ stands for the simultaneousness coefficient of the ith trunk

The same rational applies to the calculation of the aggregate bandwidth in a campus backbone. The simultaneousness coefficient translates the way different hosts use the bandwidth. It is known that, within an organization (LAN), the own organization is the recipient of most of the traffic generated

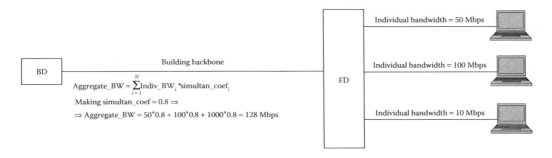

FIGURE 13.6 Calculation of the aggregate bandwidth.

here. Therefore, the simultaneousness coefficient that can be considered in a LAN can be of the order of 0.8. On the other hand, when it comes to WAN interconnection calculation, because of the same reason, the simultaneousness coefficient can be of the order of 0.3.

The maximum distance and bandwidth supported by a certain twisted pair depend on multiple factors. The most important factors are normally the attenuation coefficient and the bandwidth (per 100 m). Other factors such as NEXT, ACR, or ELFEXT may also come into play in order to determine the maximum distance and the corresponding bandwidth supported by a twisted pair cable.

ISO/IEC 11801amend2 shows the maximum length that can be supported by a certain twisted pair category, for various bandwidths (i.e., for different application class types). Some of these values are listed in Table 13.3.

The maximum length of an optical fiber is a function of the used bandwidth, which is calculated from the modal bandwidth factor for multimode optical fibers. In case the available modal bandwidth factor (of multimode fibers) does not allow using multimode optical fibers for a certain bandwidth/distance, then single-mode optical fibers should be employed.

Figure 13.7 depicts an example of structured cabling in accordance with ISO/IEC 11801amend2 specifications, using copper or fiber. This figure corresponds to Figure 13.4 [ANSI/TIA/EIA-568-A] but where the maximum distances and recommended transmission medium types have been modified. The increased distance comprised by ISO/IEC 11801amend2 relating to ANSI/TIA/EIA-568-A results from the fact that, among other factors, more recent transmission mediums are recommended by the former standard. This example considers a private network composed of only a single building; therefore, the campus subsystem does not exist. Note that the BD also acts as an FD, which results in a hardware reduction. Because the BD connects with the exterior (WAN or MAN), it makes use of a router, in addition to the main switch and patch panel existing in the same rack. In contrast, the FD does not require a router,[*] simply has an access switch, as well as a patch panel.[†] The BD and the FDs are interconnected through the building backbone, and where redundant cabling also exists to prevent from failures. The FD connects to TOs through horizontal cabling that, according to ISO/IEC 11801amend2, can have a maximum length of 90 m (100 m between the FD and the terminal equipment, including the patch cords and workstation cabling). Finally, the patch panel allows the flexible connectivity between the TOs and the required position of the switch, hub, or PABX.

The centralized optical architecture, standardized by ISO/IEC 11801amend2, presents some differences relating to the architecture with copper and fiber. As can be seen from Figure 13.8, in

TABLE 13.3
Maximum Distance versus Bandwidth of Twisted Pairs

Class	Category 3 (16 MHz)	Category 4 (20 MHz)	Category 5 (100 MHz)	Category 6 (250 MHz)	Category 7 (600 MHz)
A (100 kHz)	2 km	3 km	3 km	3 km	3 km
B (1 MHz)	500 m	600 m	700 m	TBDa	TBD
C (16 MHz)	100 m	150 m	160 m	TBD	TBD
D (100 MHz)	—	—	100 m	TBD	TBD
E (250 MHz)	—	—	—	100 m	TBD
F (600 MHz)	—	—	—	—	40 m

a TBD stands for to be defined.

---

[*] Nevertheless, an additional router can be employed in case a broadcast domain segmentation is required.
[†] In fact, two patch panels are normally used: one with copper interfaces to connect to TOs, whereas another with optical interfaces is commonly employed to connect to the BD.

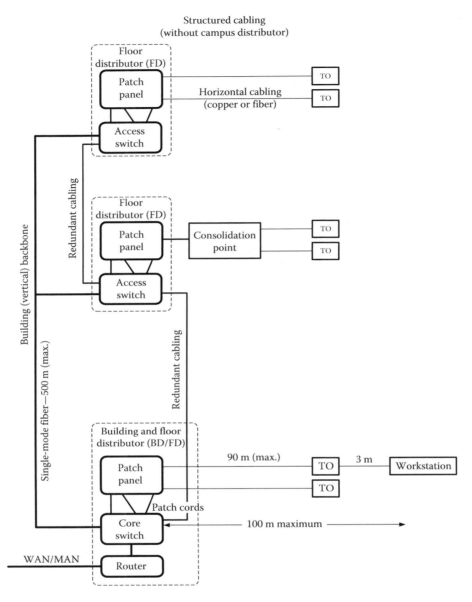

FIGURE 13.7 Example of a structured cabling implementation with functional elements in accordance with ISO/IEC 11801amend2 (recommended cablings).

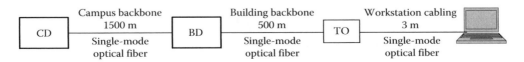

FIGURE 13.8 Maximum distances allowed by ISO/IEC 11801amend2 for the centralized optical architecture.

this case the maximum distance of 500 m supported by the building backbone considers the link between the BD and TOs. The impact of removing the FD from the architecture is reduced for two reasons:

- Typically, optical fibers support the distance of 500 m without having to perform regeneration of signals (in the conventional architecture, this was performed by the switch of the FD). Note that the maximum distance of 500 m is applicable between the BD and the TO.
- The capacity* of an optical fiber cable is extremely higher than that of a twisted pair cable. Therefore, the building backbone cable can accommodate a large number of optical fiber pairs† without impacting the width of the cable.

Figure 13.9 depicts an example of structured cabling in accordance with ISO/IEC 11801amend2 specifications, for the centralized optical architecture. This figure corresponds to Figure 13.4 [ANSI/TIA/EIA-568-A] and to Figure 13.7 (copper and fiber) but where the maximum distances and transmission

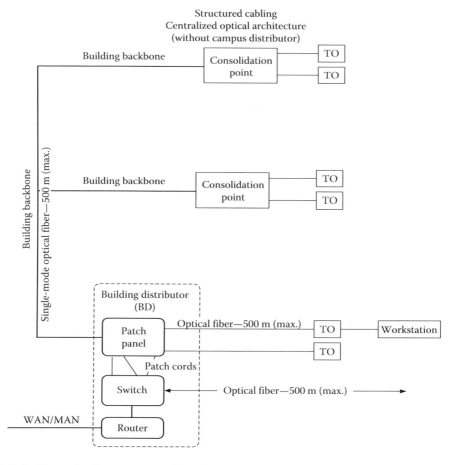

FIGURE 13.9 Example of a structured cabling implementation with the centralized optical architecture in accordance with ISO/IEC 11801amend2.

---

* In this sense, capacity refers to the number of fibers in a cable.
† A pair of optical fibers (i.e., two fibers) is required to allow full-duplex operation, one for transmission and another for reception.

medium types have been modified, alongside without the need to use an FD (as shown in Figure 13.3). Note that the number of required ports in the BD switch is higher than those required for the conventional (noncentralized optical) architecture. With these exceptions, the remaining comments performed in Figure 13.7 are also applicable in Figure 13.9. A cost–benefit analysis should be performed when a choice about the architecture is required. One of the requirements that should be taken into account is the flexibility to support additional requirements, in terms of network size and bandwidth.

13.3.3 EUROPEAN STANDARDS

At the European level, the CENELEC released in 1995 the EN 50173, consisting of a framework for the standardization of the structured cabling concept, taking ISO/IEC 11801 as a baseline. EN 50173 considers the LSZH property as a requirement for the external cable jacket, and requires screening as mandatory, while the shielding is kept as an option. Table 13.1 lists the most important differences between EN 50173 and the corresponding North American and international standards.

In 1997, a new version of this standard was released, entitled [EN 50173-1], where some modifications to the cabling categories and traffic classes were introduced. Finally, in 2007, the CENELEC released the current standard, entitled [EN 50173-2], which is based on ISO/IEC 11801ed2.

13.4 HARDWARE AND ACCESSORIES

Different distributers of structured cabling (CD, BD, and FD) are materialized by a rack that accommodates different equipment pieces and accessories. Routers and switches were already dealt with in previous chapters. Therefore, the following subsections describe the remaining equipment pieces and accessories existing in a typical distributor.

13.4.1 RACK

A rack is used to accommodate different equipment pieces and accessories of a distributor, namely a router (if required), one or more switches (or hubs), a PABX panel (where the PABX connects to), one or more patch panels, a power strip, an uninterrupted power supply, a key and video monitor (KVM), a workstation, a monitor, and so on.

When projecting structured cabling, an engineer should define the list of hardware required per distributor, including some details such as the number of routers and switch ports, and the type of interfaces (e.g., ISO 8877, SC, ST). Naturally, an FD does not require a router, as it is connected to the BD,[*] except if it intends to perform a segmentation of the broadcast domain (e.g., for creating separate subnetworks).

The most common racks employed in structured cabling present a width of 19 in.,[†] as defined by IEC 60297-3-100. Similarly, most of the pieces of network equipment also present a width of 19 in. Exceptionally, some pieces present half of this size. In this case, one can place two (or four) of these pieces side by side, such that the end-to-end width becomes 19 in.

The rack height is mostly expressed in rack units (U); one rack unit (1 U) corresponds to a height of 44.45 mm. The same applies to network equipment. Depending on the number of interfaces, a switch can be 1 U high or higher. The same applies to routers and other hardware pieces (patch panel, UPS, etc.).

Racks may present a wide range of dimensions. A 12-U rack is normally installed on the wall, whereas a 45-U rack can be employed in floor installation. Moreover, the depth of the racks available in the market may also span from 500[‡] up to 800 mm.[§]

---

[*] As both distributors are within the same network. Remember that a router is used to perform the interconnection between different networks.
[†] This corresponds to 482.6 mm.
[‡] Normally employed in wall installations.
[§] Normally employed in floor installations.

When the list of required hardware is completed, a market survey should be performed. This allows making a cost estimate of the distributor and getting knowledge about the size of each piece of equipment. With these figures, an engineer can calculate the required space in a rack, with special attention to the required height. Then, a rack choice must be performed, from those available in the market. For future expansion, it is important to leave some empty space in a rack. Table 13.4 shows an example of rack space estimation.*

In case the distributor presents simultaneously fiber and copper interfaces, the optical fiber patch panel (and other equipment with optical fiber interfaces) should be placed in the upper part of the rack, whereas the copper patch panel is installed in the lower part of the rack. In this case, the central area of the rack is typically reserved for active equipment. Moreover, in an FD, two different patch panels can be employed: one to support the connectivity to TOs (through horizontal cabling), and another to support the connectivity to the BD (through building backbone cabling). The building backbone patch panel should be installed in the upper part of the rack (normally presents optical interfaces).

Distributors should be installed in areas without sun exposition, and in a controlled environment, including temperature and humidity. Moreover, the environment should be clean of dust and corrosive gases.

A rack is built in a way that fresh air enters from the bottom, allows the necessary upstream air circulation while it refrigerates the installed equipment, and, finally, warm air leaves the rack from the top. Similarly, equipment rooms should be equipped with an air conditioning system that, conversely to the rack, collects (warm) air from the top, refrigerates it, and projects (cold) air to the bottom. Note that, when required, a rack may be supplied with a ventilation kit, which reinforces refrigeration. This is an accessory that is installed at the top of the rack for reinforcing the aspiration of warm air from the equipment toward the top of the room.

Figure 13.10 depicts an example of a 19-in. rack, with different hardware installed.

13.4.2 PATCH PANEL

A patch panel aims to provide flexibility between the positions of the switch, hub, or PABX panel and the conductors of horizontal cabling that come from TOs.

The conductors of horizontal cabling that come from TOs connect rigidly to the posterior part of the patch panel (in an FD). Then, the patch cords, which connect in the front part of the patch

---

TABLE 13.4
Estimating the Required Space in a Rack (Example)

Equipment	Height (in Rack Units [U])
Fiber patch panel (connectivity to the BD)	1
4 Patching guides (1/4 U each)	1
Router	1
Switch	1
PABX panel	1
Power strip	1
Copper patch panel (connectivity to TOs)	1
UPS	1
Total height (in rack units [U])	8

---

* This corresponds to the distributor plotted in Figure 13.10.

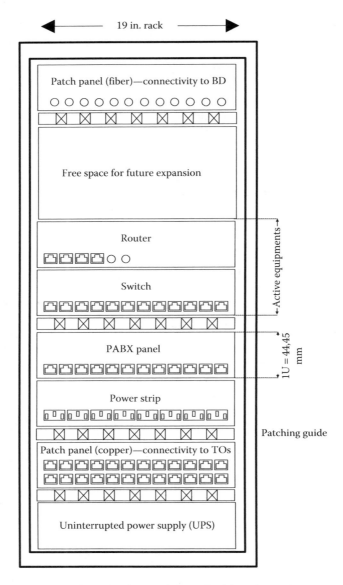

FIGURE 13.10   An example of a floor distributor (19-in. rack with equipment).

panel, allow the corresponding flexible interconnectivity with the required position of the switch, hub, or PABX panel.

Figure 13.11 shows an example of a patch panel, connected to a double TO (for data and telephony). A possible solution may rely on using the upper sockets of the patch panel for data services, whereas the lower sockets are employed for telephony. A different patch cord is utilized to connect to the required device, that is, one patch cord is used to connect to the switch for data services, whereas a second patch cord is employed to connect to the PABX panel for telephony services. Because this switch connects directly to TOs, it is likely a FD. In this case, this switch corresponds

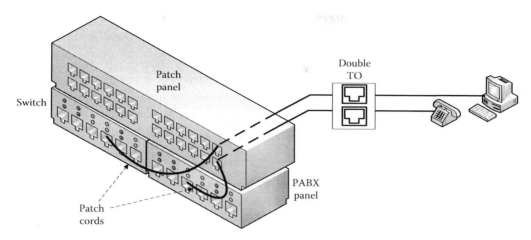

FIGURE 13.11 The patch panel.

to the second hierarchy of switches, which is then connected to the main switch located in the BD, using building backbone cabling.

It is worth noting that the patch panel can have ISO 8877 sockets (twisted pairs) or fiber sockets. In the latter case, this can be of different types, namely SC or ST. In case the centralized optical architecture is adopted, this patch panel is normally located in the BD, and all connectivity is performed using optical fiber.

A mechanism that can be employed to improve the organization of a rack relies on using patching guides (see Figure 13.10). This allows aligning the patch cords.

Note that the example depicted in Figure 13.11 considers the 19-in. patch panel, whereas the switch and the PABX panel present half of this width. For saving vertical space, these two pieces of equipment are placed side by side, such that the end-to-end width becomes 19 in.

13.4.3 ELECTRICAL SUPPLY

The active equipment as well as computers and monitors needs to be supplied with alternate current[*] (AC). A distributor rack is normally supplied with a number of Schucko-type electrical sockets in the power strip, where the communication equipment connects to (see Figure 13.10). Moreover, the supplied voltage should present ground and should be protected from overvoltage, using a voltage regulator and a circuit breaker. Commonly, most of the power strips available in the market present an embedded overvoltage protection.

The distributor rack should also have the capability to work even in case of electrical power failure. This can be achieved either using an emergency electrical source for the whole building or using an individual UPS[†] only for the distributor. In the former case, the emergency energy can either be provided using only a UPS system (for, e.g., the whole building), or the UPS system can simply be employed to assure the provision of electrical power only for an interim time period, until the moment an emergency generator is switched on.

PoE is a very common solution that reduces the need to install electrical sockets close to terminal equipment. PoE, standardized by IEEE 802.3af, allows equipment such as IP telephones,

---

[*] In Europe the AC voltage is 230 V, whereas 115 V is typically employed in the United States.
[†] A UPS also acts as a voltage stabilizer.

security network cameras, wireless LAN access points, and other IP-based terminals to receive power, in parallel to data, over the existing twisted pair infrastructure, without the need to make any modifications in it. The IP-based terminal must be capable of receiving a direct current (DC) power on either the data pairs or the unused pairs of the copper cable. PoE can be directly provided by a switch (when it presents this capability) or by using a PoE injector that is connected in the middle between a switch and the terminal equipment. The maximum supported DC voltage depends on the distance. Naturally, a longer distance supports a lower maximum DC voltage.

13.4.4 Labeling Distributors and Cables

Each distributor should have an identification, which is properly labeled in the cables, as well. As a rule of thumb, a BD is identified by letter A, whereas FDs are identified by letters B, C, D, …, as a function of the distance from the BD [Monteiro and Boavida 2011]. Similarly, each TO should also have an identification which depends on the FD where it connects to. Different sockets of a patch panel should have labels to identify the cable where it connects to. Moreover, the cable that connects in each position of the patch panel is also labeled with the same identification. This identification corresponds to the identification of the TO.

It is finally important to refer that the network manager should have a book register where all designations allocated to different functional elements are kept.

CHAPTER SUMMARY

This chapter provided a view about the structured cabling system concept, including the study of the functional elements and subsystems, existing standards, and implementation strategies. It was viewed that the aim of structure cabling is to define a generic and flexible cabling concept, independent of the manufacturer's technologies, able to support different traffic types and data rates, and easy to be modified.

The conventional architecture for the structured cabling concept was introduced, including the description of the subsystems and their corresponding functional elements. It was viewed that structured cabling considers the campus subsystem, the backbone subsystem, the horizontal subsystem, and the workstation subsystem. Similarly, the centralized optical architecture was also described, including the description of the subsystems, as well as the differences between the centralized optical architecture and the conventional one. It was viewed that, in the centralized optical architecture, the FDs were removed from the architecture. Moreover, because the connectivity is directly established between the BD and multiple TOs, the number of switch ports needs to be sufficiently high.

A view about the existing structured cabling standards was provided, as well as a description of the specifications of each of these standards. It was described that the standardization of structured cabling was initiated in North America by a joint venture between three institutions: ANSI, EIA, and TIA.

It was also described that two international standardization organizations also released standards on structured cabling. These international standards are mainly a follow-up to the standardization that took place in North America. These standards were described, including the definition of the architecture, the cabling technologies employed in each subsystem, and their corresponding maximum lengths and bandwidths.

It was described that a European standardization organization also released standards on this subject. These standards are mainly derived from the international ones, with minimal additional features.

It was described that a market survey facilitates a cost estimate, as well as an estimate of the size of the rack required to accommodate the distributor's equipment.

It was viewed that different distributers of structured cabling are materialized by a rack that accommodates different equipment pieces and accessories. The optical patch panels should be placed at the top of the rack, whereas the active equipment pieces are placed in the middle part. The patch panel is responsible for allowing the flexible connectivity between the terminal equipment and the required positions of the switch, hub, or PABX.

The supply of electrical power was also dealt with in this chapter, including its accommodation into the rack. It was viewed that the power strip accommodates multiple Shucko-type sockets for the supply of AC power. Moreover, the power strip is normally supplied with a voltage regulator and overvoltage protection. A distributor can be provided with an individual UPS, or, alternatively, the emergency electrical power can be externally provided, using only a UPS system, or a UPS system plus a power generator.

PoE was introduced, allowing IP-based terminals to receive electrical power, in parallel to data, over the existing twisted pair infrastructure. This capability can be directly provided by a switch, or using a PoE injector that is connected in the middle between a switch and the terminal equipment.

Finally, the way distributors and cablings are labeled was also dealt with in this chapter.

REVIEW QUESTIONS

1. What are the advantages of a structured cabling system?
2. How can we calculate the aggregate bandwidth in a building backbone?
3. What are the differences between the conventional structured cabling architecture and the centralized optical architecture of a structured cabling system?
4. Which functional elements are included in the horizontal subsystem?
5. Which functional elements are included in the workstation subsystem?
6. Which North American standards on structured cabling did you study?
7. Which international standards on structured cabling did you study?
8. What is the maximum length comprised by ISO/IEC 11801amend2 for building backbone cabling?
9. What is the maximum length comprised by ISO/IEC 11801amend2 for horizontal cabling?
10. What is the simultaneousness coefficient normally employed in LAN and WAN environments?
11. With regard to fire protection, what is the difference between the cabling standardized by ANSI/EIA/TIA-568-A and that standardized by EN 50173?
12. With regard to electromagnetic isolation, what is the difference between the cabling standardized by ISO/IEC 11801 and that standardized by EN 50173?
13. In a multimode optical fiber, what is the factor that determines the maximum bandwidth supported in a certain link distance?
14. In which part of a distributor rack should an optical patch panel be placed?
15. What is the procedure normally used to estimate the height of a rack?
16. Which types of emergency electrical power can be used to support the operation of a distributor?
17. What is the aim of a patch panel?
18. In what ways can PoE [IEEE 802.3af] be provided?
19. With regard to electromagnetic isolation, what is the difference between the cabling standardized by ISO/IEC 11801 and that standardized by ANSI/EIA/TIA-568-A?

LAB EXERCISES

1. Consider the network depicted in the figure below. Configure the plotted network using the Cisco Packet Tracer simulator. Assign addresses as desired using the network 10.0.0.0/16 to create subnetwork using VLSM. Configure the EIGRP routing protocol. Test the access to different web servers using the workstations' browsers.

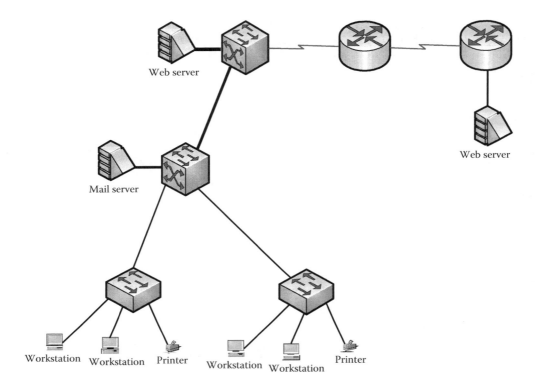

14 Transport Networks and Protocols

LEARNING OBJECTIVES

- Describe the use of transport networks.
- Identify the protocols used in transport networks.
- Describe different types of circuit switching transport networks.
- Describe different types of packet switching transport networks.
- Configure HDLC, PPP, and frame-relay protocols in Cisco routers, using Cisco IOS.

Transport networks consist of metropolitan area networks (MANs) or wide area networks (WANs), being commonly employed to interconnect different local area networks (LANs), MANs, or telephone private automatic branch exchange (PABX). These network types may either belong to a telecommunications operator or be propriety of a certain organization.

Figure 14.1 depicts an example of a transport network (WAN or MAN type), using a ring topology that is employed to interconnect different LANs. Depending on the type of technology, it may present different topologies. An important advantage of the ring topology relies on its redundancy, which translates in resistance from failures.

The transport networks can be grouped into the following categories:

- *Permanent circuit*: The main advantage of this type of connectivity relies on its dedicated and guaranteed bandwidth, without sharing with other users. In addition, the available bandwidth is typically high. The disadvantage is the high cost that is normally associated to this type of connectivity. A permanent circuit is normally leased from a telecommunications operator. Alternatively, a circuit may be implemented and established by an operator or institution as a request to a specific need (using, for example, terrestrial microwave systems or optical fibers). Logical link control (LLC) protocols that can be ran over this type of connectivity include the synchronous data link control (SDLC), high-level data link control (HDLC), and the point-to-point protocol (PPP).
- *Circuit switching*[*] (*integrated services digital network* [*ISDN*] *or public switching transport network* [*PSTN*]): This type of connectivity is the traditional way of sharing circuits between different users. Thus, it leads to a cost reduction, as compared to permanent circuits. Nevertheless, the available bandwidth is typically limited. In case it is used for data communications, the PPP and the link access procedures over D channel (LAPD) are among the most used LLC protocols.
- *Packet switching* (*X.25, frame relay, multiprotocol label switching [MPLS], etc.*): As previously defined for the packet switching (see Chapter 1), its main advantage relies on the statistical multiplexing performed as a function of the instantaneous available resources and of the users need. In addition, while the available bandwidth is typically higher than that of circuit switching,[†] the cost is typically kept within reduced limits, as opposed to

---

[*] In case this type of circuits are used for data communications, dial-up or xDSL modems can be employed.

[†] But typically based on the best effort, that is, not guaranteed.

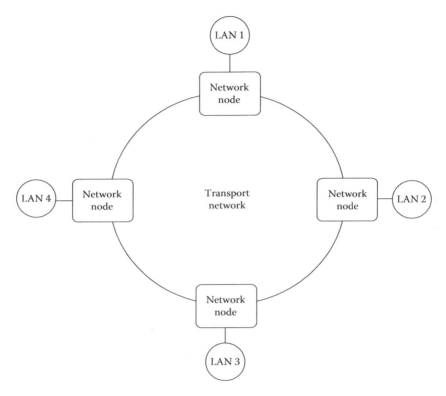

FIGURE 14.1 Example of a transport network (ring topology) used to interconnect different LAN.

the high cost associated to permanent circuits. The most used LLC protocols employed in packet switching include the link access procedures balanced (LAPB) and the Internet Engineering Task Force (IETF).

- *Cell switching (ATM)*: This corresponds to a variation of the packet switching. The reason for calling it cell switching, instead of packet switching, relies on the use of fixed size packets (cells*), instead of variable size packets (MPLS, Internet protocol [IP], etc.).

When the service consists of analog or digital telephony, the traditionally used transport network is of circuit switching type. Contrarily, packet switching is normally employed for data communications.

Initially, transport networks consisted of point-to-point connections between different PABX, using coaxial cables. Currently, single-mode optical fibers are mostly employed, using a ring topology (with two or four optical fibers). The synchronous digital hierarchy (SDH) is normally implemented over optical fibers. Moreover, the MPLS is currently widely employed by telecommunication operators as a packet switching transport network. Most often, MPLS configuration is implemented over an SDH network.

14.1 CIRCUIT SWITCHING TRANSPORT NETWORKS

As described in Chapter 1, this type of switching considers the establishment of a physical path between the origin and the destination of a communication. Although it requires the previous establishment of a connection, it allows a synchronous exchange of data, with a delay only due to the

---

* The ATM uses fixed size packets of 53 octets, where 48 octets correspond to the payload data and 5 octets to the header.

propagation of signals. Consequently, this mode is ideal for delay sensitive media, such as telephony or videoteleconference.

14.1.1 Frequency Division Multiplexing Hierarchy

Frequency division multiplexing (FDM) was previously used in transport networks to support the exchange of analog voice channels between different PABX.* As can be seen from Table 14.1, different hierarchies comprise different tributary orders, with the multiplexing of different number of voice channels. Coaxial cables and terrestrial microwave systems are among the most used transmission media. Each voice channel comprises a bandwidth of 300–3400 kHz, while the frequency carriers are 4 kHz, spaced apart to avoid adjacent channel interference.

Modern telephone networks do not comprise FDM transport networks. Nevertheless, FDM is currently being used in cable television distribution using hybrid networks† (some parts are composed of optical fiber, whereas the final distribution is typically implemented using coaxial cable).

14.1.2 Plesiochronous Digital Hierarchy

The plesiochronous digital hierarchy (PDH) is a family of circuit switching transport networks implemented using TDM. The perfect synchronism between different tributaries is not achieved in PDH because, although they present the same nominal rate, they do not have exactly the same phase and data rate. This results from the fact that different tributaries are synchronized by different and independent clocks, and their accuracy is not sufficient. Consequently, the maximum throughput possible to be exchanged in PDH networks is limited. PDH tributaries are referred to as *plesiochronous* and the corresponding hierarchy is entitled PDH.

As described in Chapter 6, the Nyquist sampling theorem states that the sampling frequency is, at least, the double of the highest frequency component present in the signal, that is, the sampling period is given by $T_a \leq 1/(2B)$. Moreover, the TDM requires that the demultiplexer be synchronized with the multiplexer. Such synchronization is achieved using a synchronism signal that is transmitted in one of the time slots, in parallel with the transported signals. The whole signal is designated as frame, and the synchronism signal is referred to as *framing*.

Similar to the FDM hierarchy, the transported signals consist typically of 300–3400 kHz voice channels. Using TDM, the analog channels need to be encoded with an analog coding system such as pulse amplitude modulation (PAM), pulse density modulation (PDM), and pulse position modulation (PPM) (see Chapter 6). Nevertheless, due to the inability to perform regeneration of signals, analog voice is not well fitted for long-range transmission. Consequently, the TDM is normally employed to transport digital voice, using pulse code modulation (TDM-PCM). In this case, since

TABLE 14.1
European FDM Carrier Standards

| Designation | FDM Hierarchy | |
	Number of Channels	Bandwidth
Group	12	60–108 kHz
Supergroup	60	313–552 kHz
Master group	300	812–2,044 kHz
Supergroup master	900	8,516–12,338 MHz

---

* The reader should refer to Chapter 6 for a description of FDM multiplexing and demultiplexing.
† In some cases, cable television distribution is already being implemented using only optical fiber cables.

the sampling frequency is 8 ksamples/s, the sampling period results in 125 µs, which also corresponds to the period of the frame repetition. Note that the PCM considers that each sample is quantified and encoded with 8 bits, resulting in a throughput of 64 kbps.

The European PDH system is entitled European Conference of Postal and Telecommunications (CEPT), having been normalized by the International Telecommunications Union (ITU) as ITU-T G.732. On the other hand, the PDH system used in the North America is entitled *digital signal* (DS), having also been normalized by the ITU as ITU-T G.733. Moreover, the Japanese PDH system consists of a variation of the DS used in North America. Figure 14.2 depicts the generation of different PDHs. As can be seen, all of the hierarchies are composed of a number of ITU-T G.711 voice channels (TDM-PCM) properly multiplexed. A higher hierarchy is generated using a multiplexer. The generation of a nth order hierarchy requires a stack of n multiplexer. Similarly, the extraction of an nth order tributary requires a stack of n (de)multiplexers.

The primary European multiplexing hierarchy is called CEPT-1 [ITU-T G.732], whereas the North America and Japanese versions are referred to as DS-1 [ITU-T G.733]. Note that the interface provided by a CEPT-1 is referred to as E1, whereas the interface provided by a DS-1 is named as T1.* From the transportation of data point of view, the nomenclature CEPT-x or DS-x is employed, whereas the nomenclature employed in interfaces is E-x, T-x, or J-x.

Both CEPT-1 and DS-1 use word interposition, with 8 bits per word, and a rate of 8000 frames per second (see Table 14.2). The voice codec comprised by the CEPT-1 is the A-law and its frame has 32 time slots. Similarly, the DS-1 considers the µ-law and its frame has a total of 24 time slots.

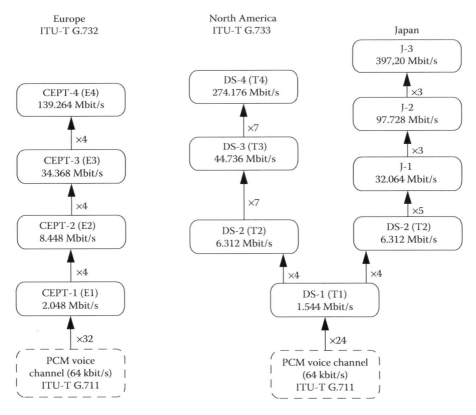

FIGURE 14.2 Generation of PDHs.

---

* The same concept applies to other hierarchies.

TABLE 14.2

Generic Characteristics of the Primary PDH

	PDH	
Designation	CEPT-1 [ITU-T G.732]	DS-1 [ITU.T G.733]
PCM Law	A ($A = 87.6$)	μ ($\mu = 255$)
Number of time slots	32	24
Number of voice channels	30	24
Number of bits per frame	$32 \times 8 = 256$	$24 \times 8 + 1 = 193$
Frame throughput (Mbps)	$256 \times 8 = 2.048$	$193 \times 8 = 1.544$
Framing	In block, using 7-bit words in the time slots 0 of odd frames	Distributed, using the sequence 101010..., composed of the 193rd bit in odd frames
Signaling	In time slot 16, at a rate of 4 bits per channel, split in 16 frames (multiframe)	In 8 bits of each channel in one frame out of six

From the 32 time slots[*] of the CEPT-1, 30 are used to transport 64 kbps telephone channels. In addition, the time slot 0 of odd frames is employed for framing, and the time slot 16 is used for signaling.[†] The CEPT-1 comprises a cumulative throughput of $32 \times 8 \times 8000 = 2.048$ Mbps.

The DS-1 frame comprises a total of $24 \times 8 + 1 = 193$ bits, corresponding to 24 telephone channels of 64 kbps plus an additional bit (per frame) for framing purposes.[‡] The cumulative throughput of the DS-1 frame is $193 \times 8000 = 1.544$ Mbps. The signaling is transported in the data time slots. Specifically, the signaling bits are transmitted in the 6th–12th frame of each multiframe,[§] using the eighth bit of each time slot.[¶] This results in a PCM word of 7 bits, leading to a small degradation of the PCM voice channels.

Higher order tributaries are obtained from the multiplexing of a number of immediately lower order tributaries. With the exception of the primary multiplexing tributary order,[**] all upper multiplexing tributary orders are multiplexed using bit interposition.

It is worth noting that the throughput of a higher hierarchy is higher than the number of lower order tributaries multiplied by their elementary throughputs.[††] This results from the fact that an additional overhead is necessary for framing and for frame justification.

Frame justification consists of the addition or removal of some bits, in order to allow the correct operation of multiplexers and demultiplexers when the different tributaries rate is subject to fluctuations, and therefore, it differs from the nominal rate. When the rate is higher than the nominal, the justification comprises the addition of a bit without information (from time to time), in order to adjust the rates. Some bits are preallocated in frames, for justification purposes. In addition, when bit justification is employed, this needs to be properly signalized using a justification indication bit, which is also a preallocated bit in the frame composition.

---

[*] Numbered from 0 up to 31.

[†] At a rate of 4 bits per channel, split over 16 frames.

[‡] The F-bit is used in odd frames for framing purposes. It presents the pattern 101010.... This corresponds to a distributed framing instead of block framing employed in the CEPT-1.

[§] A multiframe is composed of 12 frames.

[¶] The signaling information refers to the corresponding transported channel in the time slot.

[**] Which uses the word *interposition* (of PCM channels).

[††] As an example, while the throughput of the CEPT-2 is 8.448 Mbps, the cumulative throughput of four CEPT-1 tributaries is $4 \times 2.048 = 8.192$ Mbps (which is lower than 8.448 Mbps).

14.1.3　SYNCHRONOUS DIGITAL HIERARCHIES

The inaccuracy of independent clocks employed in PDH did not allow throughputs higher than 140 Mbps. These throughputs were not enough to face the new information exchange requirements. This issue was the main motivation for the development of a synchronous hierarchy, optimized for optical fiber transmission media.* Moreover, the level operation, administration, and maintenance (OA&M), as well as the level standardization in PDH systems were very low. These capabilities were highly improved in synchronous systems.

Two different synchronous circuit switching transport network systems were developed [Sexton and Reid 1992]: The SDH, employed in Europe and standardized by ITU-T as ITU-T G.707, and the synchronous optical network (SONET), used in North America and standardized by ANSI as ANSI T1.105. These systems consider atomic† clocks that allow the exchange of throughputs at a rate much higher than those possible with PDH systems. Moreover, the higher level of standardization comprised by SDH/SONET‡ between equipment of different manufacturers lead to improved interoperability.

Another important innovation of synchronous transport network systems relies on the ability to insert or extract a tributary from any other tributary of any other order, without the need to have a stack of multiplexers or demultiplexers, as required for the PDH.

Figure 14.3 shows different tributaries of both SDH and SONET hierarchies. As can be seen, different SDH tributaries are referred to as *synchronous transport module* (STM), whereas SONET

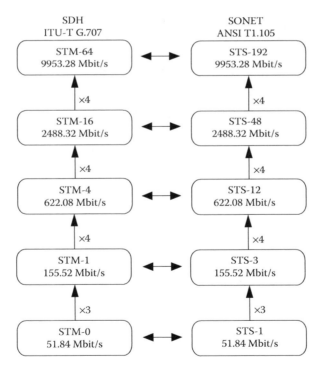

FIGURE 14.3　Tributaries of both SDH and SONET.

---

* This represents an important improvement, as compared to PDH. Nevertheless, it is possible to find pieces of equipment that implement SDH/SONET over other transmission technologies, such as terrestrial microwave systems.

† The central (master) atomic clocks (cesium or rubidium) are employed to achieve a high level of accuracy. The other slave clocks are periodically synchronized making use of the information contained in the SDH frame.

‡ Including the standardization of new OA&M capabilities.

tributaries are named as *synchronous transport signal* (STS). Moreover, there is equivalence between the tributaries' rates of different hierarchies. It is also worth noting that, above STM-1/ STS-3, the multiplexing of higher hierarchical tributaries is performed with four lower order hierarchical tributaries.

The different SDH/SONET tributaries can be used to transport different traffic type, such as multiplexed ITU-T G.711 voice channels, PDH tributaries, and packets belonging to packet switching networks (such as IP, asynchronous transfer mode [ATM], and MPLS). Figure 14.4 shows examples of PDH encapsulation into SDH/SONET. In this figure, boxes with straight lines refer to SDH/ SONET hierarchy and boxes with dash lines refer to PDH hierarchy. Note that Figure 14.4 only depicts some possible encapsulations. Nevertheless, a wide variety of PDH tributaries mixes can be jointly used to generate any SDH/SONET tributary.

14.1.3.1 SDH/SONET Network

An SDH/SONET network comprises a number of elementary devices that allow multiplexing/ demultiplexing and transport of data. Such devices are as follows:

- *Line terminal multiplexer* (*LTM*): It accepts lower order tributaries to generate a higher order tributary. This type of device is employed at the beginning and at the end of a path. This means that all multiplexed tributaries end in this device.
- *Add and drop multiplexer* (*ADM*): It accepts lower order tributaries to generate a higher order tributary. Nevertheless, contrary to the LTM, the ADM can be used in the middle of a path, in order to insert or remove some of the lower order tributaries.

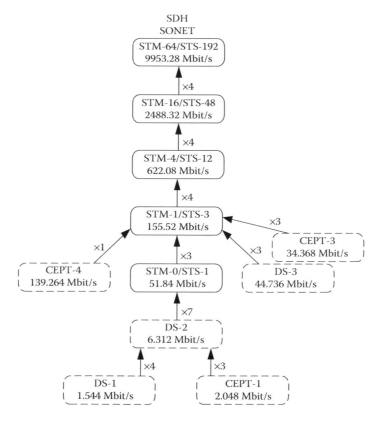

FIGURE 14.4 PDH encapsulation into SDH/SONET.

- *Synchronous digital cross connect (SDXC)*: It is used to perform semipermanent switching along the SDH/SONET network, in order to provide *permanent circuits* to a customer who requires it.
- *Regenerator (REG)*: It is used to perform the regeneration of signals, mitigating the effects of channel impairments. SDH/SONET networks make use of regenerators, typically, every 60 km of optical fibers.

Figure 14.5 shows an example of two SDH networks with interconnection. As can be seen, an STM-16 hierarchy uses ADM-16 to multiplex up to four STM-4 tributaries. In the example depicted in Figure 14.5, an STM-4 tributary can be imported from an STM-4 network. The STM-16 ring and the STM-4 ring are interconnected in two points in order to achieve interconnection redundancy. This results in an overall mesh topology.

It is worth noting that the elementary physical topology employed in SDH/SONET networks is the ring. Since optical fibers are unidirectional, a cable composed of, at least, two optical fibers (one pair) is employed to allow simultaneous bidirectional communications. As depicted in Figure 14.6, in case of two optical fibers, both are employed for operation and backup using different time slots. The external fiber allows the exchange of data in the clockwise direction, whereas the internal fiber performs the same in the anticlockwise direction. It is important to note that each node (ADM)

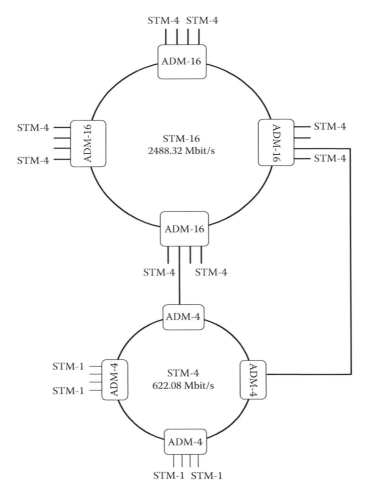

FIGURE 14.5   Example of SDH transport networks.

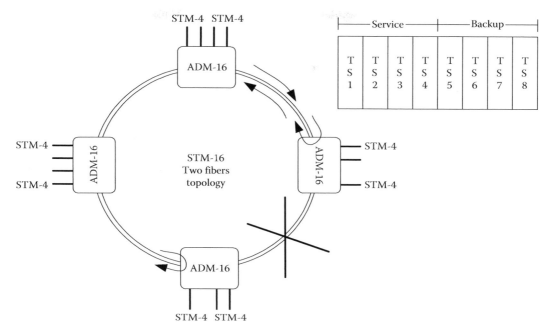

FIGURE 14.6 Failure procedure in a cable with two optical fibers.

receives the same data twice, coming, respectively, by both the external and the internal fiber. The destination node (ADM) combines the data coming from the two directions in order to provide diversity.* The combining technique is the selective combining. This is performed using the path overhead (see Figure 14.10). The resultant signal presents better performance than the two independent signals.

In case the optical fiber cable is cut, such impairment is detected by the two closest ADM and bridges are introduced in those ADM. This is performed using the multiplexing overhead. Consequently, a switch to the backup time slots is performed by the network. Note that this self-recovery capability represents a great advantage of the synchronous systems, as compared to plesiochronous systems.† The plotted example refers to a STM-16 ring. Nevertheless, the same principle applies to other hierarchies. The same concept is also applicable to SONET networks.

As can be seen from Figure 14.7, in case the SDH/SONET ring is implemented with four optical fibers, all time slots of two optical fibers are allocated for operation, while the other two optical fibers (one pair) are reserved for backup. In this case, redundancy is not assured with TDM, but with an extra pair of optical fiber. Similar to the previous case, the closest ADM insert bridges, but those bridges are responsible for forwarding the signals to the redundant fibers in the opposite direction.

Both SDH and SONET networks comprise different layers, as plotted in Figure 14.8. Each layer refers to the communication between certain types of devices that includes a number of functionalities.

The description of different layers and their functionalities is included in Table 14.3, whereas Figure 14.9 plots an example of a network with the identification of different layers using both SDH and SONET terminologies.

---

* The reader should refer to Chapter 7 for the description of combining techniques/diversity schemes.
† Manual recovery employed in plesiochronous systems lead to recovery time of several hours.

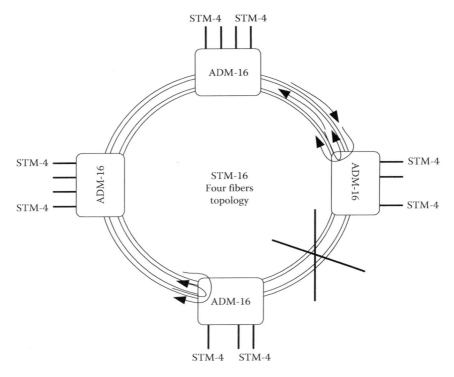

FIGURE 14.7 Failure procedure in a cable with four optical fibers.

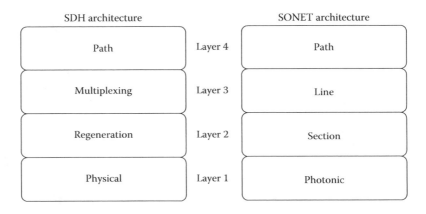

FIGURE 14.8 SDH/SONET architectures.

Each layer has its own overhead type, being processed by the corresponding device. As can be seen from Figure 14.9, the path is considered end-to-end. Consequently, the path overhead is inserted by the initial multiplexer (LTM or ADM), being removed by the final multiplexer (LTM or ADM). The multiplexing section consists of the parts of the path between adjacent multiplexers, including those intermediates (ADM or SDXC). Consequently, the path of the example plotted in Figure 14.9 is composed of two multiplexing sections. In this case, the intermediate multiplexer (ADM or SDXC) removes the initial multiplexing section overhead at its input, and inserts another multiplexing section overhead at its output, corresponding to the second multiplexing section. Identical rational applies to the regeneration section overhead. Note that a regenerator only processes the regeneration section overhead (removed at the input of the device, and inserted at its

TABLE 14.3
SDH/SONET Multi-layer Architecture

Layer	SDH Terminology	SONET Terminology	Functionalities
4	Path	Path	Definition of the end-to-end transported tributary, throughput, etc.
3	Multiplexing	Line	Synchronism, multiplexing, switching, OA&M, type of protection from failures, etc.
2	Regeneration	Section	Regeneration distance, electrical to optical conversion, etc.
1	Physical	Photonic	Definition of the type of optical fiber, light wavelength, transmit power, etc.

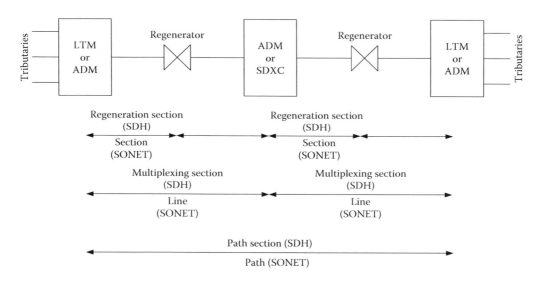

FIGURE 14.9   Example of a SDH/SONET network, with the identification of the architecture.

output), whereas an intermediate multiplexer processes both the regeneration section overhead and the multiplexing section overhead.

14.1.3.2   SDH/SONET Frame Format

The SDH/SONET frame comprises a different header type for each different layer (except for the physical layer) as follows:

- *Path overhead*: It is used to manage the end-to-end path,[*] being inserted at the beginning and removed at the end of the path. This is used to manage the end-to-end path between extreme devices (LTM or ADM). The end-to-end transport of a PDH tributary into SDH/SONET is performed by the path layer and managed by the path overhead. In addition, this layer also implements error protection mechanisms and provides engineering orderwire communication channels[†] (at path level).

---

[*] End-to-end path from the SDH communication point of view.
[†] Note that error protection and engineering orderwire communication channels are provided by all different layers, and managed by the corresponding overheads.

- *Multiplexing section* (*line*) *overhead*: It includes functions such as multiplexing of lower order tributaries, frame synchronization, switching information, error protection, and engineering orderwire communication channels, being processed and used to manage multiplexers.
- *Regeneration section* (*section*) *overhead*: It allows the alignment of the frame, assures error protection, and provides engineering orderwire communication channels. This overhead is processed and used to manage the communication between different regenerators and multiplexers.

Different SDH and SONET hierarchies are formed by octet interposition of the lower hierarchy tributaries. In addition, an STM-1 frame can be formed by octet interposition of three STS-1 frames (STM-0). Due to this reason, the following frame description focus on STS-1, whereas the composition of an STM-1 frame can be deducted in a straightforward manner.

The basic structure of an STS-1 frame is depicted in Figure 14.10. It is composed of a total of 810 octets, split into 9 rows and 90 columns. The frame transmission is performed starting from the 1st up to the 90th octet of each row, from row to row. This structure is transmitted every 125 μs,[*] while another frame with the same structure is transmitted in an equal period. The transmission of $810 \times 8 = 6480$ bits in 125 μs results in a throughput of 51.84 Mbps. Since each octet corresponds to 8 bits, which is repeated every 125 μs, this results in a throughput of 64 kbps.[†]

The frame comprises two main blocks: the header and the synchronous payload envelope (SPE).[‡] The section header[§] comprises three columns and three rows, the pointers correspond to three

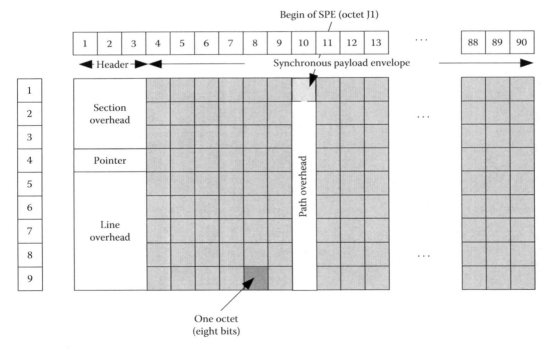

FIGURE 14.10   STS-1 frame format.

---

[*] This is the sampling period, which corresponds to the inverse of the 8000 samples/s sampling rate employed in PCM voice.

[†] 64 kbps corresponds to the throughput necessary to support one PCM voice channel. Nevertheless, one STS-1 octet can be used to transport different type of data, such as IP packets and MPLS packets.

[‡] In the SDH terminology, the SPE is referred to as the *virtual container*.

[§] In the SDH terminology, the section header is referred to as the *regeneration header*.

columns and one row, and the line overhead* is composed of three columns and four rows. The SPE comprises the octets used to transport the payload data and the path overhead (transported in the 10th column).

The pointer is employed to identify the beginning of the SPE within the payload area and to accommodate justification bits. It is worth noting that both SDH and SONET comprise a concept where a SPE/VC does not occupy a rigid position within the payload area. On the contrary, it may fluctuate within the payload area. The initial octet of the path overhead is entitled J1, and marks the beginning of the SPE/VC. Although Figure 14.10 depicts the octet J1 at the top of the path overhead column, due to load fluctuations, this octet may be located anywhere in the payload area.

This transportation concept can be viewed as the payload of a truck. We could have the furniture of 10 houses to be transported at the same time between two different cities. Each family's furniture requires one standard truck. In principle, one truck would be employed for each family's furniture. Nevertheless, a different method could be adopted: each truck could be used to transport part of a family's furniture, while another truck could transport part of another family's furniture. This concept is employed in SDH and SONET. To implement such concept, there is the need to have a pointer toward the beginning of the SPE/VC.

As depicted in Figure 14.11, the STS-1 pointer uses three octets (24 bits). The logic states of the initial six bits are fixed, whereas the 10-bit pointer is used to quantify a value between 0 and 782 octets.[†] This value corresponds to the octet order where the SPE/VC begins, within the payload area.

Due to successive insertions and removals of headers in different devices (regenerators, multiplexors, etc.), the payload (SPE/VC) may travel faster than the corresponding frame.[‡] Another problem that originates with similar effects is when the incoming clock rate is lower than the outgoing clock rate. In order to overcome such fluctuation, an extra octet is inserted into the last octet of the pointer.[§] This is known as *negative justification*. In this case, the value of the pointer needs to be decreased by one. The negative justification is depicted in Figure 14.12.

Contrarily, when the incoming clock rate is higher than the outgoing clock rate, the frame flows faster than the payload. In order to overcome such fluctuation, a stuff octet[¶] is transmitted after the pointer. This is known as *positive justification*. This can be seen from Figure 14.13.

As can be seen from Figure 14.14, the basic structure of an STM-1 frame is composed of a total of 2430 octets, split into 9 rows and 270 columns, formed by octet interposition. Consequently, the STM-1 header comprises 9 columns (instead of 3 columns, as considered for the STS-1).

14.1.4 DIGITAL SUBSCRIBER LINE

The digital subscriber line (DSL) is not exactly a transport network. Nevertheless, it consists of a transmission technique that allows the transport of data over conventional installed copper twisted pairs (analog transmission medium). In addition, the PPP is normally implemented over the DSL. Most common uses of DSL modems are made over circuit switching networks.

1	2	3	4	5	6	7	8	9	10	11	12	13	14	15	16	17	18	19	20	21	22	23	24
0	1	1	0	1	0	Pointer (number between 0 and 782)											Space used for negative justification						

FIGURE 14.11 STS-1 pointer.

---

* In the SDH terminology, the line header is referred to as the *multiplexing header*.
† In case of STM-1 (SDH), nine octets are employed for similar purposes, instead of three.
‡ The frame is delayed by successive overhead processing.
§ In case of STM-1 (SDH), instead of using a single octet for negative justification (STS-1/SONET), a total of three octets are employed.
¶ An octet without information.

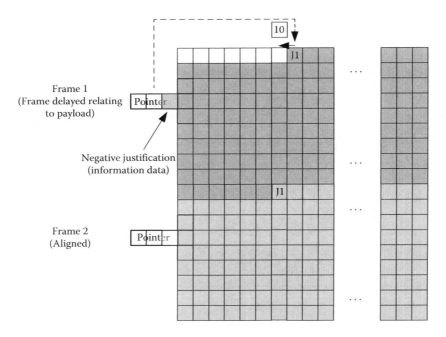

FIGURE 14.12   Frame delayed relating to the SPE (STS-1).

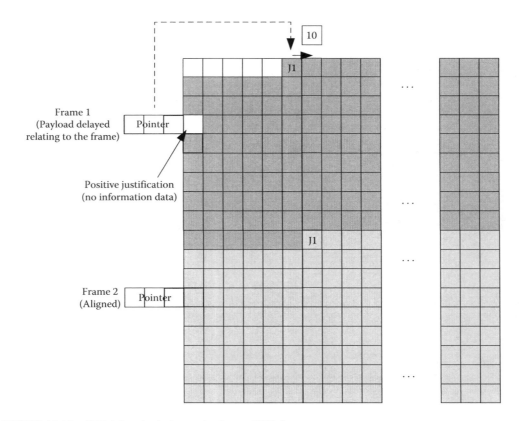

FIGURE 14.13   SPE delayed relating to the frame (STS-1).

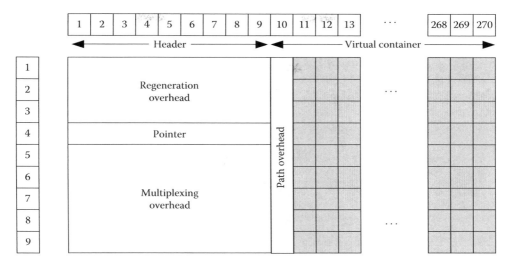

| 1 | 2 | 3 | 4 | 5 | 6 | 7 | 8 | 9 | 10 | 11 | 12 | 13 | ⋯ | 268 | 269 | 270 |

◄──────── Header ────────►◄──────── Virtual container ────────►

FIGURE 14.14 STM-1 frame format.

Different versions of DSL exist (generically referred to as xDSL*), and each version makes use of different transmission techniques to allow the exchange of digital data with different characteristics (throughput, reliability, bandwidth, modulation scheme, etc.). The xDSL transmission technique is implemented by making use of modems at both ends of the transmission medium (e.g., at home and at the ISP site). The transmission medium comprises a copper wire used to connect the customer premises equipment and the DSL access multiplexer. The DSL access multiplexer consists of a multiplexer located at the service provided premises,† used to concentrate the connections from different DSL customers.

As previously described, a modem consists of a device that allows the exchange of digital data over analog transmission media. It implements the modulation at the transmitter side, whereas the demodulation process is carried out at the receiver side. The modulation involves the process of encoding one or more source bits into a modulated carrier wave. Moreover, the modem also implements an error control technique. An important advantage of using a modem, instead of line coding, relies on the fact that the transmitted signal is carrier modulated (bandpass), instead of transmitted in the baseband. This allows selecting the transmission bandwidth where the channel impairments (e.g., attenuation and distortion) are less destructive.

Figure 14.15 depicts a block diagram of a communication chain using xDSL modems.

Note that the communication implemented by the modems is bidirectional. In addition, today's xDSL communications are full duplex. This is possible because modems use, typically, two different

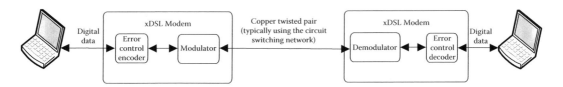

FIGURE 14.15 Block diagram of a communication chain using DSL modems.

---

* For example, for $x = A$, the xDSL becomes asymmetric digital subscriber line (ADSL).
† The service provider premise where the DSL access multiplexer is placed is commonly referred to as the *central office*.

frequency bands (employing the FDD technique), one for transmission and another for reception of signals. As depicted in Figure 14.16, a lower bandwidth is normally utilized for uplink, whereas an upper bandwidth is employed for downlink. Since data communications tend to require a higher throughput in the downlink than in the uplink, most xDSL modems allocate a higher bandwidth for the downlink. In this case, the communication established by the modem is referred to as *asymmetric*. Moreover, a requirement for xDSL modem is ability to keep the analog telephony in simultaneous with the exchange of data. This is possible by keeping a baseband bandwidth allocated for conventional analog telephony, with a typical reserved band of 0–20 kHz. Typical uplink frequency band is between 25 and 140 kHz, whereas the typical downlink band is between 150 kHz and 1 MHz. Latest xDSL standards utilize higher frequencies. An example is the very high data rate DSL (VDSL) whose frequency band utilized for data goes from 25 kHz up to 12 MHz.

In order to mitigate the effects of the channel impairments, most xDSL modems have an equalization module embedded, which is an effective measure to minimize the effects of intersymbol interference. This kind of interference is also mitigated using high-order modulations (e.g., M-QAM) or by employing the OFDM transmission technique. Some type of xDSL modems use a variation of the OFDM transmission technique in the downlink (more demanding direction), entitled *discrete multi-tone* (DMT). Figure 14.17 depicts the generic spectrum of a modem employing DMT transmission technique, where the downlink spectrum is split into multiple subcarriers. Similar to OFDM, the aim is the mitigation of the negative effects of intersymbol interference. The flow of data is split into a lower rate transmission, and each lower rate transmission is independently transmitted in a different subcarrier. The DMT is very similar to the OFDM technique described in Chapter 7, with the difference that some subcarriers are removed and the modulation order of different subcarriers can be different (as a function of the channel impairments experienced by each subcarrier). Similar to the OFDM technique, each DMT subcarrier tends to suffer from flat fading (i.e., nonfrequency selective fading), instead of frequency selective fading. In order to mitigate the remaining fading effects, each subcarrier is subject to an equalization process at the receiver side. This results in a better signal quality, which translates in a more efficient use of the spectrum (expressed in bit/s/Hz).

An alternative to the pre-allocation of different uplink and downlink bandwidths (FDD) consists of using a bandwidth simultaneously for the uplink and downlink. In such cases, the downlink bandwidth includes the uplink bandwidth, and an echo cancellation mechanism[*] is employed to avoid interferences between the uplink and the downlink spectrum. Moreover, some xDSL modems use a single frequency band for both uplink and downlink. In this case, the full-duplex operation is possible by employing TDD. Therefore, one can say that the DSL modem is responsible for

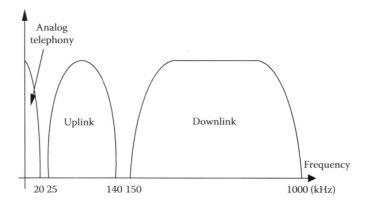

FIGURE 14.16 Generic use of spectrum by an xDSL modem.

---

[*] Implemented using an adaptive equalizer to reject the nondesired signal.

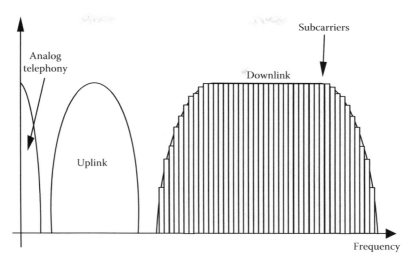

FIGURE 14.17 Generic spectrum of an xDSL modem that uses DMT.

implementing layer 1 and layer 2 connections, where layer 2 link comprises a deterministic medium access control mechanism (TDD).

Table 14.4 shows the most common xDSL standards alongside with their important characteristics. As can be seen, the initial asymmetric DSL (ADSL) does not employ the DMT transmission technique. Instead, the FDD is utilized, using carrierless amplitude phase modulation (CAP) associated to QAM modulation scheme. Since the CAP suppresses the carrier, the required power for the signal transmission is reduced. The ADSL2 allows higher uplink and downlink data rates, as well as an extended range. This results from the use of the DMT transmission technique. On the other side, the high bit rate DSL (HDSL) transmits signals in baseband employing the 2B1Q line coding technique. Since digital transmission is performed in baseband, cohabitation with analog telephony is not possible. Finally, the VDSL supports higher data rates at the cost of lower distances.

TABLE 14.4
xDSL Standards and Characteristics

Designation	XDSL Modem				
	ADSL	**ADSL2**	**HDSL**	**VDSL**	**VDSL2**
Standard	**ANSI-T1.413—1998**	**ITU-T G.992.3**	**ITU-T G.991.1**	**ITU-T G.993.1**	**ITU-T G.993.2**
Transmission technique	CAP—QAM	DMT—M-QAM	Digital baseband transmission using line coding 2B1Q	DMT—M-QAM	DMT
Maximum uplink data rate (Mbps)	1	1.3	1.544 (T1) or 2.048 (E1)	16	100
Maximum downlink data rate (Mbps)	8	12	2.048	52	100
Range	4 km	5 km	4 km	1.5 km	300 m
Observations	One twisted pair (TP)	One TP	HDSL uses two TP. HDSL2 have same performance with single TP	One TP. supports HDTV	One TP. supports HDTV

The extreme data rates provided by both VDSL and VDSL2 allow their use for the provision of television. This makes the twisted pair acting as a competitor to coaxial cables, typically employed in cable TV.

14.1.5 Data Over Cable Service Interface Specification

Data over cable service interface specification (DOCSIS) consists of a set of standards that allows high-speed data communications over coaxial cables, typically installed for the provision of television service. The DOCSIS can be viewed as an alternative to xDSL, but using coaxial cables as transmission medium (instead of twisted pairs). Alternatively, hybrid fiber coaxial, or just optical fiber, may also be utilized as transmission medium.

Figure 14.18 depicts the block diagram of a communication chain using DOCSIS modems.

DOCSIS is currently widely employed by ISPs to provide Internet access to domestic users. This service is normally provided together with the cable television. When telephony service is also provided, the three services are commonly known as *triple play*. In this case, the telephony service is provided using voice over IP over DOCSIS modems (employed between the subscriber home and the ISP).

As can be seen from Table 14.5, the DOCSIS modem shares the spectrum with regular analog or digital video distribution, using FDMA (downlink). The uplink channels are implemented using TDMA (DOCSIS 1.0) or CDMA (DOCSIS 2.0 and 3.0). In the uplink, the higher modulation order and bandwidth of DOCSIS 2.0 and 3.0 allows a higher data rate per channel, as compared to DOCSIS 1.0. In the downlink, all three DOCSIS versions allow the same data rate per channel. Nevertheless, an important difference of the DOCSIS 3.0 as compared to DOCSIS 2.0 relies on its ability to aggregate multiple channels (in both uplink and downlink), which results in a data rate that is N times higher. Typically, four channels are aggregated in DOCSIS 3.0, resulting in a data rate of 122.88 Mbps in the uplink and 171.52 Mbps in the downlink.

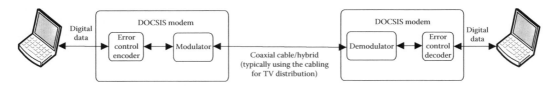

FIGURE 14.18   Block diagram of a communication chain using DOCSIS modems.

TABLE 14.5
DOCSIS Standards and Characteristics

Designation	DOCSIS Modem		
	DOCSIS 1.0	**DOCSIS 2.0**	**DOCSIS 3.0**
Standard	**ITU-T J.112 (Appendix II)**	**ITU-T J.122**	**ITU-T J.222**
Transmission technique	FDMA (TDMA in uplink)	FDMA (CDMA in uplink)	FDMA (CDMA in uplink)
Uplink channel bandwidth/ modulation	200 kHz—3.2 MHz using QPSK or 16-QAM	200 kHz—6.4 MHz using 8 to 128-QAM	200 kHz—6.4 MHz using 8 to 128-QAM
Downlink channel bandwidth/ modulation	6 MHz channels using 64 or 256-QAM	6 MHz channels using 64 or 256-QAM	6 MHz channels using 64 or 256-QAM
Maximum uplink data rate (Mbps)	10.24	30.72	122.88 (in four channels [typical value])
Maximum downlink data rate (Mbps)	42.88	42.88	171.52 (in four channels [typical value])

14.2 PACKET SWITCHING TRANSPORT NETWORKS AND PROTOCOLS

Packet switching networks are of lower costs than circuit switching, and are ideal for the exchange of data. Packet switching networks intend to provide a higher bandwidth (based on the best effort), but at costs close to those of circuit switching networks. The network resources are made available as a function of the users' needs and as a function of the instantaneous network traffic. Therefore, it is normally stated that a packet switching network performs a statistical multiplexing, as opposed to synchronous access control mechanisms (frequency or time multiplexing) typically employed in circuit switching networks.

Packet switching involves the segmentation of a message into small pieces of data, and each piece is switched independently by the network nodes. Each piece of data is referred to as a *packet*. In order to allow its routing by the network nodes, each packet contains additional information (overhead) in a header. Moreover, each node of a packet switching network is able to store packets, in case it is not possible to send it due to temporary congestion. In this case, the time for message transmission is not guaranteed here, but this value is kept within reasonable limits, especially if quality of service (QoS) is offered.

There are different packet switching protocols, such as IP, ATM, MPLS, frame relay (FR), and X.25. While the IP protocol is typically employed by the network to interact directly with users (providing services), the ATM, MPLS, FR, X.25, and PPP are mostly employed in transport networks (WAN and MAN networks).

14.2.1 ASYNCHRONOUS TRANSFER MODE

The ISDN [Prycker 1991] was developed and standardized in the 1980s in order to become the most important domestic and commercial data network, providing both voice and data services. ISDN was also expected to be used as a circuit switching transport network (using, e.g., a primary access of 2.048 Mbps). Nevertheless, its implementation required, in most cases, the replacement of the existing voice-graded UTP cables, which was an important limitation. On the other hand, the rapid development of the Ethernet technologies based on installed UTP cables, as well as the requirements of the new services, became important obstacles for the development of the ISDN standard. The ATM is a protocol intended to fit these new requirements, while providing much higher throughputs necessary to support new multimedia services, support of services that require variable throughputs, delay sensitive services, and support of services sensitive to errors [Handel and Hubber 1991; Prycker 1991]. Note that the ATM is associated to a concept entitled broadband-ISDN (B-ISDN). It is important referring that while the ATM is a protocol, the B-ISDN is a concept that is materialized with a reference model, as depicted in Figure 14.19.

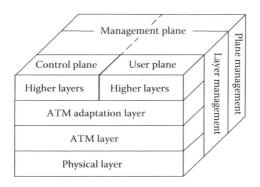

FIGURE 14.19 B-ISDN reference model.

Although the ATM can be employed to connect end users, its most common applicability is as a transport network. In addition, although the ATM protocol can be directly implemented as a baseline protocol of a transport network, its most common implementation is above an SDH or a SONET circuit switching transport network. In this case, the SDH/SONET network is only used to transport the ATM cells, whereas the data of different services is typically encapsulated into the ATM cells.

The ATM is a special type of packet switching protocol. In fact, some authors refer to it as a cell switching or as cell relay. The reason comes from the fact that the ATM packets are of fixed size (cells), as opposed to variable size packets adopted by most of the packet switching protocols. The ATM protocol is based on virtual circuits, instead of datagrams. There is the need to previously establish the circuit,* and the cells are always routed by the same intermediate nodes, until the circuit is terminated.

An important advantage of the ATM relies on its faster switching that results from the use of fixed routes, and from the reduced level of processing and decision by intermediate nodes. When a network node receives a cell, it checks the route to which it belongs (such information is contained in a label, which is part of the ATM cell header) and, consulting the routing table, it verifies which is the output interface to use in order to forward the cell.

Due to both fast switching and small size of cells, the ATM is very well fitted to support delay sensitive services. This makes the ATM protocol well fitted to support voice and video services.

The cell label has a local meaning. This means that the label is removed by each node, and another label is re-inserted. Moreover, since the multiplexing performed by ATM does not follow any time synchronous mechanism (such as time multiplexing), the ATM is referred to as an *asynchronous mode*.

As can be seen from Figure 14.22, an ATM cell has a fixed length of 53 octets. Its header has a length of 5 octets, and the payload data is 48 octets long.

An important characteristic of the ATM relies on the ability to support different classes of traffic, while compatible with their different QoS requirements. The ATM protocol presents the following advantages:

- Support constant or variable data rate services
- Support services from very low up to high data rate requirements
- Support symmetric and asymmetric communication services
- Support different services in a single network (such as voice, video, and multimedia)
- Support delay sensitive services (e.g., telephony and audio or video streaming)
- Support errors sensitive services (e.g., file transfer)

14.2.1.1 B-ISDN Reference Model

According to the ITU-T I.121 recommendation, the ATM is the protocol used to implement the B-ISDN concept. The B-ISDN reference model comprises different layers and planes, as depicted in Figure 14.19.

The B-ISDN reference model includes different classes of services, grouped as follows:

- *Class A*: Circuit switching emulation and audio and video of constant bit rate
- *Class B*: Compressed audio and video of variable bit rate
- *Class C*: Connection oriented data (such as file transfer of web browsing)
- *Class D*: Connectionless data (such as network management protocols or DNS)

---

* Consequently, the ATM protocol is connection oriented.

In order to provide different services, the B-ISDN reference model is split into different layers[*] with the following functionalities [Costa 1997]:

- *Physical layer*: It includes two different groups of functionalities:
  - *Physical medium*: It includes functions that are dependent of the physical medium.
  - *Transmission convergence*: It takes care of the generation and recovery of frames, header error control, cell rate decoupling, and so on.
- *ATM layer*: It is responsible for the multiplexing and demultiplexing of cells, flow control, insertion and removal of headers, and so on. It encapsulates blocks of data with different sizes, generated by the ATM adaptation layer, into cells (fixed size).
- *ATM adaptation layer (AAL)*: It adapts the requirements of the different classes of services provided by the higher layers to the ATM layer. Consequently, the AAL is service dependent. The rate of generated blocks of data and their corresponding sizes is service dependent, being performed by the AAL. Then, these blocks of different size are encapsulated into cells by the ATM layer. In order to support different services, with different requirements, some service-dependent and service-independent functions have to be provided. Consequently, the AAL is split into two sublayers (see Figure 14.20): convergence sublayer (CS), which provides services to the upper layer, and segmentation and reassembling sublayer (SAR), which segments the data received by the CS, in order to allow its encapsulation into ATM cells. To allow these functionalities, the AAL provides four different type of service:
  - *Type 1 AAL*: It supports constant bit rate services (e.g., circuit switching emulation) from the higher layer (class A), keeping the synchronism information between the source and the destination. In addition, it manages errors (lost cells, corrupted cells, cells in wrong sequence, duplication of cells, etc.).
  - *Type 2 AAL*: Multiplexes low data rate channels, such as mobile communications.
  - *Type 3/4 AAL*: It supports classes C and D variable bit rate services with error detection.
  - *Type 5 AAL*: While type 3/4 AAL considers the multiplexing of small blocks of data, type 5 AAL groups the data into large blocks, allowing a more efficient multiplexing (reducing the overhead). Consequently, type 5 AAL is more efficient for data communications.
- *Higher layers*: This layer makes the interface with different classes of service, providing them the required services.

As can be seen from Figure 14.19, the management plane comprises the independent management of each layer (management layer), as well as the management of the B-ISDN reference model as a whole.

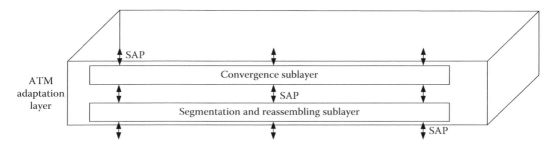

FIGURE 14.20   AAL sublayers.

---

[*] Different layers are interfaced using service access points.

14.2.1.2 ATM Network

As previously described, the ATM network can be employed either as a transport network or be used to provide different services to the end user. In the former case, there are only interfaces between two adjacent network nodes (network–network interface), whereas the latter case comprises interfaces between users and network nodes (user–network interface). The B-ISDN reference model comprises the adaptation of the ATM network to the different services generated by the users.

The transport of data is performed by the ATM network using two different basic layers:

- *Physical layer*: It comprises the exchange of bits within different types of devices.
- *ATM layer*: It comprises layer 3 switching performed by the network nodes.

Similar to the SDH layers, depending on the types of devices, the physical layer of the ATM protocol is split into different sublayers, each one with its own overhead:

- *Transmission path sublayer*: It corresponds to the SDH path layer, and comprises the ATM end-to-end exchange of data, from the local where the data is encapsulated into cells up to the local where those bits are removed from cells.
- *Digital section sublayer*: It corresponds to the exchange of data between, for example, two adjacent SDH multiplexers.
- *Regeneration section sublayer*: It corresponds to the exchange of data between two adjacent regenerators.

Similarly, the ATM layer also comprises two hierarchical sublayers defined as [ITU-T I.113; Costa 1997] (see Figure 14.21):

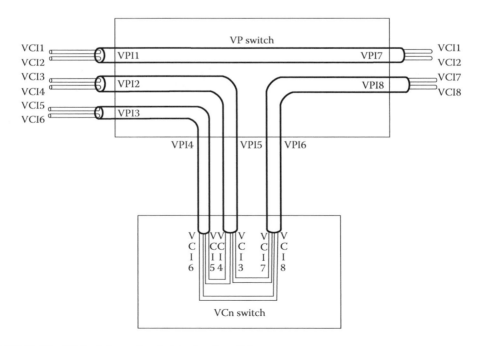

FIGURE 14.21 Virtual channel and virtual path switch.

- *Virtual channel (VCn) sublayer*: It comprises layer 3 switching of a group of different cells that are identified by a common VCn identifier.
- *Virtual path (VP) sublayer*: It comprises layer 3 switching of a group of different virtual channels that are identified by a common VP identifier (VPI), and that are equally switched in a certain path of the route.

A VP switch performs the switching of different VPs, translating the incoming VPIs into other outgoing VPIs. Note that a VP switch performs the switching of multiple VCns, whose inputs and outputs are common. Similarly, a VCn switch performs the switching of different VCns, translating the incoming VCIs into other outgoing VCIs.

14.2.1.3 ATM Cell Format

As can be seen from Figure 14.22, the ATM cells are 53 octets long, being composed of a 48-octet long payload data field and of a 5-octet long header field.

Depending on the type of interface where cells are employed, the headers present two different formats. Those different cells format are depicted in Figure 14.23, where UNI stands for user–network interface and NNI stands for network–network interface. The contents of different header's fields are as follows:

- *Generic control field*: With the default value of 0000.
- *VPI*: It is composed of 8 bits (UNI) or 12 bits (NNI), as above described.
- *Virtual channel identifier (VCI)*: It is composed of 16 bits, as above described.
- *Payload type (PT)*: To signalize the type of data transported in the cell.
- *Reserved (RES)*: For future use.
- *Cell loss priority (CLP)*: In case of congestion, cells with this bit active are first discarded.
- *Header error control (HEC)*: It is composed of 8 bits that are used to check the presence of errors in the cell header.

14.2.2 MULTIPROTOCOL LABEL SWITCHING

The MPLS is employed in packet switching transport networks (MAN/WAN) as well as to implement virtual private networks (VPN), that is, to allow the interconnection of different corporate subnetworks (e.g., intranet).

Header	Payload data
5 octets	48 octets

FIGURE 14.22 ATM cell format.

GPC	VPI	VCI	P T	R E S	CLP	HEC
4 bits	8 bits	16 bits	1	1	2	8 bits

(a)

VPI	VCI	P T	R E S	CLP	HEC
12 bits	16 bits	1	1	2	8 bits

(b)

FIGURE 14.23 Cells header format of (a) UNI and (b) NNI.

The MPLS was designed to support any layer 3 protocol (IPv4, IPv6, IPX, ATM, etc.) [RFC 3031; 3270]. Moreover, it can be implemented over any type of layer 2 protocol, such as SDH, SONET, Ethernet, and PPP. Consequently, it is normally stated that the MPLS belongs to layer 2,5 of the OSI reference model (see Figure 14.24). It is worth noting that the MPLS implementations can either be implemented over SDH/SONET circuit switching network or directly over a physical circuit. The advantage of using a packet switching protocol (MPLS) relies on the ability to reach a better usage of the network resources. Moreover, the MPLS is known as a *packet switching transport protocol* that provides QoS, feature not present in circuit switching networks (SDH/SONET).

Note that different VPNs provided to customers are identified by VPN routing and forwarding (VRF). In other words, a VRF corresponds to a VPN implemented using the MPLS technology.

14.2.2.1 MPLS Network

The MPLS is a connection-oriented protocol, whose forwarding of packets is performed using the virtual circuit method (similar to the ATM protocol).

The virtual circuit method considers the prior establishment of a circuit, through where all packets are forward. Contrary to the datagram method where each router has to decide about the output interface to use in order to forward the packet to the final destination address (based on long routing tables), with the virtual circuits method, the output interface of intermediate nodes are identified by the virtual circuit identification. This facilitates the processing of intermediate routers, enabling an increased speed routing and improved QoS, as compared to the datagram method. Moreover, since all packets are transmitted through the same intermediate routers, the delivery of packets to the final destination is performed in the correct sequence, which also contributes to the provision of QoS.

The MPLS was developed taking into account the weaknesses of the ATM protocol, while maximizing the QoS capabilities in transport networks. An important limitation of the ATM protocol relies on the fact that the overhead is high (5-octets), as compared to the cells size (48-octets), which also results in a low network efficiency.

The current existing physical layer infrastructure allows the exchange of data at a much higher speed than before. This allows exchanging longer frames without introducing delays to the supported services. Consequently, contrary to the ATM, the MPLS allows the exchange of connection-oriented packets with different sizes.

The routing of an MPLS packet is performed making use of a label[*] contained in the packet header. The label identifies the virtual circuit to which the packet belongs. Due to this reason, the MPLS is referred to, by some authors, as a *label switching protocol*.

An example of an MPLS network is depicted in Figure 14.25. As can be seen, the MPLS consists of a WAN or MAN, used to transport large amount of data between multiple LANs. The different MPLS routers are called *label switching router* (LSR). Nevertheless, the LSRs that are located

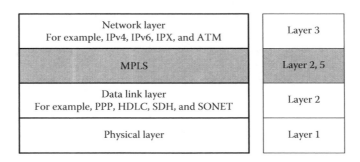

Network layer For example, IPv4, IPv6, IPX, and ATM	Layer 3
MPLS	Layer 2, 5
Data link layer For example, PPP, HDLC, SDH, and SONET	Layer 2
Physical layer	Layer 1

FIGURE 14.24   Location of the MPLS protocol into the OSI reference model.

---

[*] The label is similar to VPI/VCI identifiers employed in the ATM protocol.

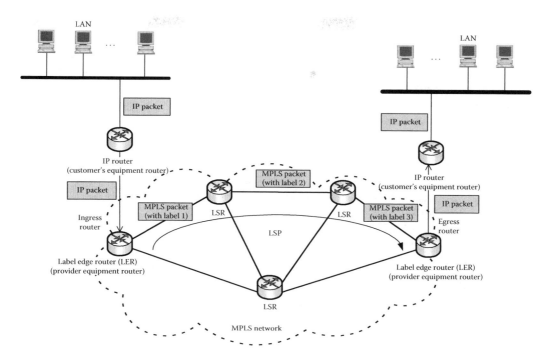

FIGURE 14.25 Example of an MPLS network.

at the edge of the MPLS cloud (network), interfacing with the customer's equipment router[*], are referred to as *label edge router* (LER), or simply as *Edge LSR*. The LER is a provider equipment router, as it is owned by the transport service provider.

Since the MPLS uses the virtual circuit method, the routing path is calculated in advance, before data is transmitted through the network. Such path calculation is performed taking into account the required QoS and the traffic conditions. The MPLS virtual circuit is designated, in the MPLS world, as *label switching path* (LSP), as depicted in Figure 14.25. Moreover, as a preventive measure to avoid congestion, the MPLS performs traffic engineering [RFC 2702]. This consists of load balancing of different traffic flows over different paths.

The IP packet is received by the ingress LER as a regular IP packet. This router translates the IP packet into the MPLS format by simply adding a shim header. The shim header consists of a 32-bit header, composed of four fields. The most important field is the label (20 bits), being employed to identify the virtual circuit and the class of service of the IP packet. It is important noting that, similar to the VPI/VCI identifiers employed in ATM networks, the label has a local meaning. This means that each LSR performs the removal of the label corresponding to the previous point-to-point MPLS connection, adding a new label corresponding to the following point-to-point connection. This function is referred to as *label swap* (performed in LSRs), whereas the addition or removal of a label performed by LERs is referred to as *push* (in ingress LER) or *pull* (in egress LER), respectively. The calculation of the output interface to be used by an MPLS router in order to forward a packet is performed taking into account the incoming label (i.e., the virtual circuit), as well as the information contained in a routing table. Routing tables are built in the same manner as the IP routing table, using static or dynamic protocols. In the MPLS world, the routing tables, that is, the lookup tables, are referred to as *label forwarding information base* (LFIB) or as *forwarding information base*. While an LSR makes use of the LFIB table to route packets (MPLS to MPLS), an

---

[*] The customer's equipment router consists of, for example, an IP router.

ingress LER makes use of the forwarding information base table (IP to MPLS) and an egress LER makes use of the LFIB table as well (MPLS to IP).

When a new LSP is created, the corresponding labels are added to the LFIB. The label distribution protocol (LDP) is employed to perform the discovery of paths (LSPs) and to perform the resulting distribution of labels between different routers. In order to assure the reliability of data, the LDP runs over the TCP protocol. Note that the LDP is accompanied by bandwidth reservation for specific LSPs, and therefore, also contributes for the MPLS to be a QoS enabler protocol.*

A certain LSP can be used to transport different types of traffic (voice, video streaming, file transfer, etc.). For the sake of QoS provisioning, different traffic types must be properly identified with different forward equivalent classes (FECs). As can be seen from Figure 14.26, a label identifies a pair of LSP/FEC. When a new FEC is created, it is assigned to a certain LSP. Note that different FECs of a communication between the same two end points may belong to the same LSP or to different LSPs. In the example depicted in Figure 14.26, the FEC1 and FEC2 belong to LSP1, whereas the FEC3 and FEC4 belong to LSP2.

The MPLS routers process differently the packets with different FECs, that is, packets with different priorities or with different sensitivities to loss of data.

14.2.2.2 MPLS Packet Format

As previously described, the MPLS is a protocol used to transport layer 3 packets. Contrary to the ATM protocol that comprises a certain packet (cell) format, the MPLS limits to add a certain header to the transported layer 3 packets (ATM, IPv4, IPv6, IPX, etc.). The MPLS header (shim header)

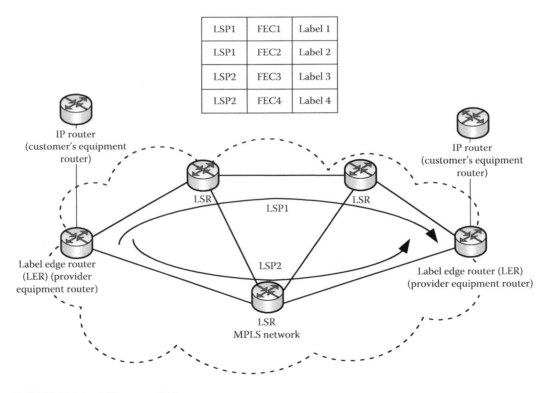

FIGURE 14.26   LSP versus FEC.

---

* Together with the low latency introduced by this protocol and with the use of different processing as a function of the service carried out in a packet.

is typically added between layer 2 and 3 headers. The shim header comprises the following fields (see Figure 14.27):

- *Label*: It is employed to identify the LSP (virtual path), as well as the FEC (type of traffic being transported in the packet).
- *Class of service*: It is used for QoS provisioning and for explicit congestion notification purposes.
- *Stack (S)*: In case multiple labels are inserted into a frame, it indicates a hierarchy. The value one means that the label is the last of the stack. Optionally, the MPLS may implement a hierarchy of virtual paths,[*] requiring multiple shim headers in an MPLS packet.
- *Time to live*: It corresponds to the time-to-live field of IPv4 packets, being assigned to the last label of a stack.

When the transported protocol has fields that identify virtual circuits, the MPLS shim header is directly inserted into the corresponding fields. This is the case of ATM and FR protocols (see Figure 14.28):

- *ATM*: The shim header is inserted into the VPI/VCI fields.
- *Frame relay*: The shim header is inserted into the data link channel identifier (DLCI) field.

Alternatively, the shim header is placed between layer 2 and layer 3 headers. This can be seen from Figure 14.28c. This latter procedure is employed in protocols such as Ethernet, PPP, and token ring.

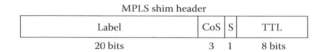

FIGURE 14.27   MPLS shim header format.

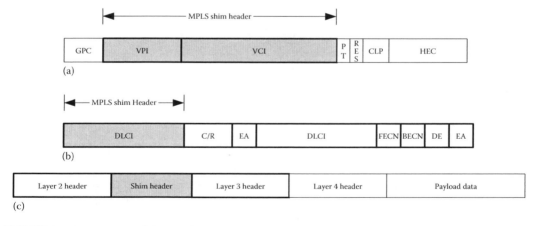

FIGURE 14.28   Location of the MPLS shim header: (a) ATM cell, (b) frame-relay packet, and (c) other layer 3 protocols.

---

[*] The use of multiple labels is referred to as *label stacking*. In this case, the top label (rightmost label) refers to the current real hop, whereas the second label corresponds to the virtual hop. Note that in this scenario consisting of a hierarchy of virtual paths, a virtual hop corresponds to a number of real hops.

14.2.3 HDLC Protocol

The HDLC is a connection-oriented LLC protocol employed in permanent circuits, having been standardized by ISO 3309 and ISO 4335. This is a data link layer (LLC) protocol utilized in transport networks.

HDLC may establish links in either half- or full-duplex modes, and the links may be point-to-point or point-to-multipoint (one to several hosts). Moreover, it considers a synchronous transfer of data, with assured physical layer synchronism (i.e., clock synchronism). Consequently, a flag is only employed at the start and at the end of the HDLC frame to allow the data link layer synchronism, that is, to achieve the frame synchronization.*

The HDLC protocol considers three different types of frames:

- *Information (I)*: It is used to transport upper layer (Internet layer) information data, with error control† data piggybacked (in the control field).
- *Supervisory (S)*: It is employed for error control purposes when piggybacking is not employed, that is, when the receiving entity has no information data to transmit.
- *Unnumbered (U)*: It consists of frames not numbered, being used for several functions such as connection establishment or connection termination, as well as to define the mode of the link to be established, as defined in the following.

The HDLC protocol considers three different types of stations:

- *Primary*: It consists of a station with the ability to control the link. It can keep one or more simultaneous links, and can send commands.
- *Secondary*: It cannot send commands. This kind of station is limited to receive commands and to send appropriate response to these commands.
- *Combined (or mixed)*: It can send and received commands. It can also send the appropriate response to the received commands.

HDLC links can be established in two forms:

- *Unbalanced*: It is composed of a primary station and one or more secondary stations.
- *Balanced*: It is composed of two combined stations.

In addition, HDLC links can transfer data in three different modes:

- *Normal response mode*: It considers an unbalanced link, where a primary station sends commands and the secondary station sends the corresponding responses.
- *Asynchronous response mode*: It is similar to the normal response mode, but where a secondary station may also initiate the data transfer.
- *Asynchronous balanced mode*: Contrary to the two previous modes, it is composed of a balanced link, where any station may initiate the data transfer. This is the most used HDLC mode of data transfer.

Figure 14.29 shows the frame format employed by the HDLC protocol. This frame includes the following fields:

---

* To signalize the start and the end of a frame.
† This includes control data such as receive ready, receive not ready, and selective reject.

Flag	Address	Control	Payload data	CRC	Flag
8 bit	8 or more	8 or 16	Variable	16 or 32	8

FIGURE 14.29   HDLC frame format.

- *Flag*: It consists of the bit pattern 01111110, which is used to signalize the start and the end of a frame. In order to prevent the receiver to incorrectly consider a start or an end of a frame, when a sequence of six or more logic state 1 bits is detected by the transmitting entity in any part of the frame other than the flag, such entity inserts an additional logic state 0 bit after the fifth bit.[*] The receiving entity performs the reverse procedure, that is, when it receives a sequence of five logic state 1 bits, followed by a logic state 0 bit, it removes logic state 0 bit. This procedure is known as *bit stuffing*.
- *Address*: It contains the destination address, being composed of one or several groups of 8 bits. In case the MSB of each group is a logic state 1 bit, it indicates that this is the last group of 8 bits used by the address field. Alternatively, if the MSB of the group is a logic state 0 bit, it indicates that another group of 8 bits is sent as the address field. In addition, the bit pattern 11111111 corresponds to the broadcast address.
- *Control*: It is used to control the connection, and can be 8 or 16 bits long. Depending on the type of frame (I, S, or U), and of the control field length, its content varies, as depicted in Figure 14.30. The use of 8 or 16 bits in the control field needs to be negotiated during setup phase (setup extended mode corresponds to employing 16 bits in the control field). The 16-bit control field format is used when the sequence of the frame numbering is high (e.g., long window size in the sliding window mechanism). Note that only I and S frames can be 16-bits long (U frames are always 8-bits long). As can be seen from Figure 14.30, some bits are variable, whose meanings are defined as follows:
  - $N(S)$: Sending sequence number. It is used to identify the sequence number of the current transmitted frame.
  - $N(R)$: Receiving sequence number. It is used for control purposes[†] to identify the sequence number following a previously correctly received frame. This number corresponds to the sequence number of the frame that the receiving entity is ready to receive.

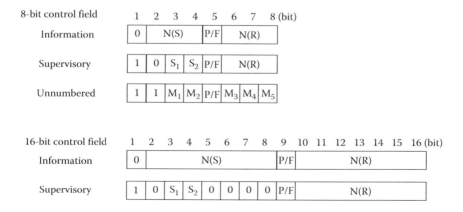

FIGURE 14.30   Format of the control field of an HDLC frame (8 or 16 bits format).

---

[*] As an example, instead of transmitting 111111111111, the sequence 11111011111011 is transmitted.
[†] In case of I frames, the control is performed in piggybacking mode.

- S_n: Supervisory bits, used for flow control and error control, with the following meanings for S_1 and S_2 bits:
    - *00*: Receive ready (RR). This corresponds to the acknowledgment (RR followed by the number of the frame N(R) that the entity is ready to receive).
    - *01*: Reject (REJ). This is used to reject a frame with errors.
    - *10*: Receive not ready (RNR). This is used by the receiving entity when it is not ready to receive more data.
    - *11*: Selective reject (SR). This is used by the receiving entity to reject a specific frame using the selective reject automatic repeat request (ARQ) protocol.
- M_n: Unnumbered bits, used for multipurpose functions, such as setting up and finalizing a connection (see Table 14.6).
- *P/F* (Pool/final bit): When the *P* bit is active (1), the transmitting entity is querying the receiving entity about which frame is it ready to receive. The *F* bit active means that it is a response to a previous query (*P*).
- *Payload data*: It is only used in information (I) frames and in some U frames. In case of I frames, it is used to transport Internet layer data (upper layer).
- *Cyclic redundancy check (CRC)*: It also known as frame check sum, it consists of 16 or 32 redundant bits used for error detection. The generator polynomials used by the HDLC protocol are one of the following two[*]:

$$P_{16}(x) = x^{16} + x^{15} + x^2 + 1$$
$$P_{32}(x) = x^{32} + x^{26} + x^{23} + x^{22} + x^{16} + x^{12} + x^{11} + x^{10} + x^8 + x^7 + x^5 + x^4 + x^2 + x + 1$$
(14.1)

Figure 14.31 shows an example of a handshaking performed between node A and node B, using the three types of frames (I, S, and U). The shown messages are those that are properly encoded using the frame control fields. The message SABME sent using a U frame stands for setup asynchronous balanced mode extended (extended corresponds to 16 bit control field). In addition, DISC stands for disconnect and UA for unnumbered acknowledgement.

14.2.3.1 HDLC Configuration Using Cisco IOS

The HDLC protocol is the default protocol utilized by Cisco routers in WAN interconnections. Therefore, when configuring a serial interface, the encapsulation protocol automatically selected by the Cisco router is the HDLC. There are not special requirements for this. One only has to configure the IP addresses in the interfaces, the clock rate (in data communication equipment [DCE] side of a connection), activate the interface, and so on. In case a different encapsulation protocol is

---

TABLE 14.6
Some HDLC Codes E for M1–M5 Bits and Their Descriptions

Code (Bits M1–M5)	Name	Description
00–001	SNRM	Set normal response mode
11–100	SABM	Set asynchronous balanced mode
00–010	DISC	Disconnect
11–110	SABME	Set asynchronous balanced mode extended
00–110	UA	Unnumbered acknowledgment

---

[*] $P_{16}(X)$ is the generator polynomial used in CRC16 (16 redundant bits), whereas $P_{32}(X)$ is the generator polynomial used in CRC 32.

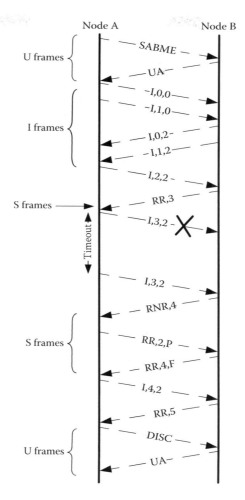

FIGURE 14.31   An example of HDLC handshaking using the control fields.

configured in a serial interface, the following command can be issued in the serial interface to revert to HDLC encapsulation protocol:

- Router(config-if)#encapsulation hdlc (to enable HDLC encapsulation on the interface)

14.2.4   Point-to-Point Protocol

The PPP is an LLC connection-oriented protocol employed in either permanent circuits or circuit switching (e.g., dial-up or xDSL connection), having been standardized by RFC 1663. It can establish links in either half- or full-duplex modes, using synchronous or asynchronous mode, being, however, limited to point-to-point connections.

The PPP is the successor of the serial line IP (SLIP), consisting of a very simple protocol whose functionalities were almost limited to a basic framing, without any advanced functionalities. Conversely, the PPP includes error control, authentication,[*] compression, encryption, link control protocol (LCP), and network control protocol (NCP) capabilities.

---

[*] Namely the PAP and the CHAP. The reader should refer to Chapter 16 for the description of these protocols.

An NCP runs directly over the data link layer protocol, being used to negotiate options and specific parameters of the network layer (which also runs over the data link layer). Examples of NCP are the PPP IP control protocol,[*] the IPv6 control protocol over PPP (IPv6CP), the PPP Apple Talk control protocol, and so on.

For connection-oriented protocols, the NCP is responsible for performing the connection establishment, as well as the connection termination. In addition, it is responsible for configuring and supporting the operation of the network layer.[†] As can be seen from Figure 14.32, the NCP is only initiated after the LCP of the data link layer had been successfully established. Similarly, the LCP is responsible for establishing and configuring a data link layer, as well as for operating and terminating a link. Different data link layers use different LCP protocols, with different message formats. Note that the authentication[‡] is performed after the establishment and configuration of the LCP, being invoked by this protocol. When authentication is required, the NCP link establishment does not start before authentication process succeeds.

The LCP does not have knowledge about the network protocol used to allow the exchange of end-to-end data. This is known, configured, and managed by the NCP. It is important to note that each different data link layer protocol or network layer protocol has as its own NCP. As an example, the IP control protocol consists of the NCP used by the PPP to configure and manage the IPv4 protocol. Changing either the data link layer (e.g., from PPP into HDLC) or the Internet layer protocol (e.g., from IPv4 into IPv6) results in a different NCP, with a different message format.

Several network protocols can be employed[§] (even simultaneously) over a specific data link layer protocol. Consequently, a different NCP is required for each different network protocol being supported.

The following are examples of messages exchanged by both LCP and NCP:

- *Configure request*: It requests the establishment and connection configuration.
- *Configure ACK*: It accepts a configure request, accepting all proposed options.

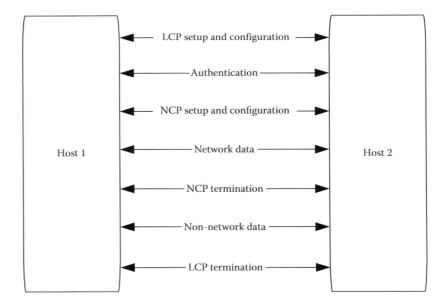

FIGURE 14.32   LCP and NCP protocols.

---

[*] Also referred to as IPv4CP.

[†] Always necessary for both connection-oriented and connectionless protocols.

[‡] For example, the CHAP or the PAP.

[§] For example, IPv4, IPv6, and Apple Talk.

- *Configure NAK*: Some proposed options in the configure request have unacceptable values.
- *Configure reject*: It rejects a configure request because some options are not recognized.
- *Terminate request*: It requests to terminate a connection.
- *Terminate ACK*: It accepts to terminate a connection.
- *Code reject*: Means that invalid or not recognized code was received.
- *Echo request*: A message testing the connection. It requests the counterpart node to respond with an echo reply message for testing a connection.
- *Echo reply*: The response of an echo request message, being used to test the connection.

The frame format of the PPP is depicted in Figure 14.33.

The PPP frame format was developed taking the HDLC format as a baseline. Some similarities were kept, such as the content of the flag field and the generator polynomials used by CRC. The PPP frame includes the following fields:

- *Flag*: It consists of the bits pattern 01111110, which is used to signalize the start and the end of a frame.
- *Address*: Since the PPP is only used in point to point, the address field is fixed to the bit pattern 11111111 (FF in hexadecimal notation). When compression is employed, this field is omitted.
- *Control*: This field is fixed to the bit pattern 00000011 (03 in hexadecimal notation). When compression is used, this field may be omitted.
- *Protocol*: It identifies the type of protocol being handled by the payload data of the frame. Possible protocols being handled by the frame includes IPv4 data, IPv6 data, IPX data, LCP, NCP, authentication, encryption control protocol (ECP), and compression control protocol (CCP).
- *Payload data*: It used to allow the exchange of upper layer data or the exchange of variables relating to the protocol being handled by the frame (e.g., LCP, NCP, ECP, and CCP). In the latter case, some of the payload data is used to transport the following fields (see Figure 14.34):
  - *Code (type)*: It identifies the type of message being exchanged.
  - *Identifier*: A response includes a copy of the identifier employed in the corresponding query.
  - *Length*: It consists of the length of the code, identifier, length field, and data field, expressed in octets.
  - *Data*: Specific information related to the message that is being exchanged, such as options being negotiated.
- *Padding*: It is used to ensure that the PPP frame size is an integer number of 32 bits.
- *CRC*: Also known as *frame check sum*, it consists of 16 or 32 redundant bits used for error detection. The PPP uses the same generator polynomials as those considered by the HDLC protocol (defined by Equation 14.1).

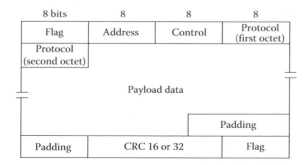

FIGURE 14.33 PPP frame format.

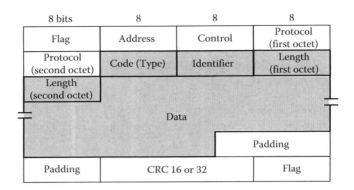

FIGURE 14.34 Payload field used by the protocols being managed by frames.

It is finally worth referring that, similar to HDLC, a router is normally configured with PPP encapsulation as data terminal equipment (DTE). In laboratory, back-to-back connection of two routers using a serial cable is possible. In this case, one side is configured as a DTE and the other as a DCE. The DCE side requires the clock rate configured in the serial interface, as the synchronism signal is sent by this router to the DTE.

14.2.4.1 PPP Configuration Using Cisco IOS

Some Cisco routers can be interconnected using the PPP. As a data link layer protocol, the PPP can be supported, at the physical layer, by a dedicated circuit (either a leased circuit from an ISP or a circuit installed by the owner), by an ADSL circuit, a DOCSIS circuit, and so on.

The PAP is a possible authentication protocol, consisting of a two-way handshaking procedure. The authentication initiator sends the username and password in a clear mode, whereas the authenticator accepts or rejects the authentication process. Note that the passwords configured by both parties must be the same. The PPP configuration with the PAP starts in a configuration mode with the following syntax:

- Router(config)#hostname *Routers_name*
- *Routers_name* (config)#username *other_routers_username* password *password*

Note that, by default, the *Routers_name* must coincide with the username configured by the other party, that is, the hostname must agree with the username sent by the other router (*other_routers_username*).

The rest of the PPP configuration is performed in the interface configuration submode as follows:

- *Routers_name* (config)#interface *interface* (where *interface* stands for the identification of the serial interface to configure [e.g., se0/1])
- *Routers_name* (config-if)#ip address *ip_address subnet_mask* (IP address and subnet mask of the interface, as required for other data link layer protocols)
- *Routers_name* (config-if)#clock rate *clock_rate* (only in case the router's interface is acting as a DCE)
- *Routers_name* (config-if)#no shutdown (to activate the interface)
- *Routers_name* (config-if)#encapsulation ppp (to enable PPP encapsulation on the interface)
- *Routers_name* (config-if)#ppp authentication pap (to enable PAP authentication with PPP)
- *Routers_name* (config-if)#ppp pap sent-username *own_username* password *password*

Note that *own_username* corresponds to the username that should be received from the other party, and should coincide with the hostname configured in this router.

The CHAP is a three-way handshaking procedure, whose authentication process is periodically repeated.

The authentication process starts from the authenticator by sending a random text to the client (challenge). The initiator ciphers the authentication credentials with a function composed of the random text received from the authenticator and a hash function (message digest 5 [MD5]). Then, the result of this function is sent to the authenticator. The authenticator performs the same function and compares the received authentication credentials against the result of its own processing. Note that, when authentication is applied in the two directions, both parties act as initiator and authenticator. The reader should refer to Chapter 16 for a detailed description of the CHAP.

The PPP configuration with the CHAP is similar to PAP, with the difference that the command line "ppp authentication pap" is replaced by "ppp authentication chap" and the command line "ppp pap sent-username *other_routers_username* password *password*" is not applicable. Moreover, the username initially configured from the configuration mode is the username of the other party (whereas in the PAP, it is the own username). Similar to PAP, the configuration of the CHAP starts in configuration mode:

- Router(config)#hostname *Routers_name*
- Router(config)#username *other_routers_username* password *password*

Note that, by default, the *Routers_name* must coincide with the username configured by the other party, that is, the hostname must agree with the username sent by the other router (*other_routers_ username*). Then, the configuration proceeds in interface configuration submode:

- Routers_name(config)#interface *interface*
- Routers_name(config-if)#ip address *ip_address subnet_mask*
- Router(config-if)#clock rate *clock_rate* (only in case the router's interface is acting as a DCE)
- Routers_name(config-if)#no shutdown
- Routers_name(config-if)#encapsulation ppp
- Routers_name(config-if)#ppp authentication chap

The PPP troubleshooting can be performed using the command "debug ppp authentication" or with the "show interface *interface*." The latter command line allows verifying the encapsulation in use by the interface. Moreover, the command "show running-config" allows verifying the type of authentication and encryption, the username, as well as other PPP configuration parameters.

14.2.5 FRAME RELAY

The FR is a data link layer protocol that simply encapsulates packets (IP, IPX, SNA, Apple Talk, etc.) into the FR frame, and forwards it along the FR network. This is a protocol widely used to interconnect a branch office LAN with a headquarters office LAN, or to interconnect two branch office LANs. Therefore, FR circuits can be viewed as WAN or MAN networks. The FR protocol was implemented taking X.25 protocol as a baseline. However, while the FR is a layer 2 protocol, the X.25 protocol is a network layer protocol. Moreover, the bandwidth made available by FR circuits is typically of the order of an E1, a T1, or multiples channels of 64 kbps, and for carrying digital data. On the contrary, X.25 circuits were of very low bandwidth, and for carrying analog data.

When contracting an FR circuit, the service provider guarantees the exchange of data up to a data rate referred to as the *committed information rate*. Nevertheless, some bursts can be allowed by the service provider if the instantaneous traffic allows it, up to a data rate known as *extended information rate*.

As can be seen from Figure 14.35, the customer router is regarded as DTE, whereas the FR switch connected to the DTE is referred to as DCE. The synchronism signal is sent by the DCE, and therefore, it is not required to configure the clock rate in the FR router. The FR cloud is typically the propriety of a telecommunications service provider. The interconnection between the DTE and the DCE is normally implemented using leased circuits with the same bandwidth as the contracted FR circuit. This leased circuit is connected to a serial interface of the router, and the FR encapsulation protocol is configured. An advantage of FR relies on the fact that the internal FR network is meshed, providing alternative paths. Moreover, when comparing an FR circuit against an end-to-end leased circuit (dedicated), the FR tends to be less expensive, as this protocol allows a more efficient use of the equipment resources (bandwidth sharing by multiple users).

An internal circuit between two DTEs is referred to as the *virtual circuit*. This is known as virtual because this circuit is not physically established. Let us focus on the topology depicted in Figure 14.36 where, on the left, a hub-and-spoke topology is depicted, and where Router A is the hub. In this case, there is one virtual circuit between Router A and Router B, and a second virtual circuit between Router A and Router C (alternatively, the full-meshed topology comprises virtual circuits between all routers). Although there is a single physical circuit between Router A (DTE) and the closer FR switch (DCE), there are two virtual circuits implemented over it. Moreover, a virtual circuit is identified in each hop by a DLCI, the 10-bit FR address. Let us focus on the virtual circuit between Router A and Router B. The connection between Router A and the DCE has a specific DLCI, whereas the

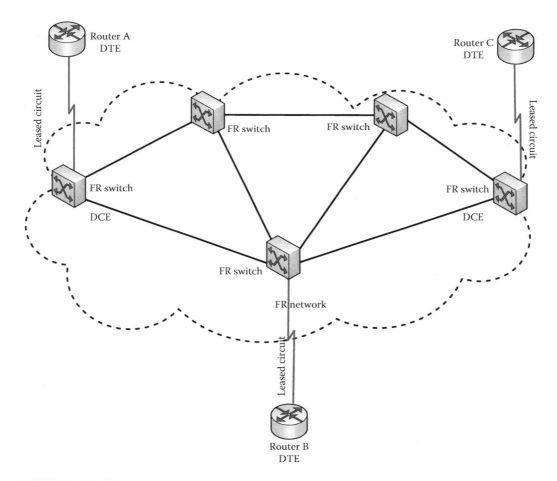

FIGURE 14.35   FR network.

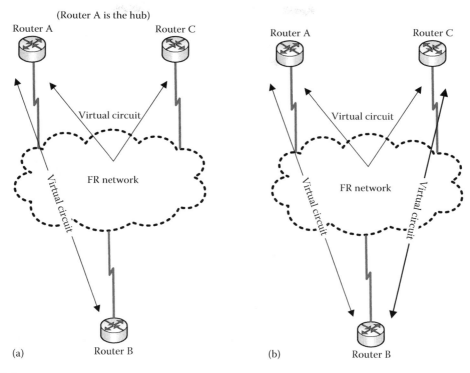

FIGURE 14.36 Frame relay (a) hub-and-spoke topology versus (b) full-meshed topology.

link between DCE and the following FR switch has another DLCI, and so on. Note that the DLCI is unidirectional. This means that a certain hop of a virtual circuit has a certain DLCI and another DLCI in the opposite direction. The DLCIs are assigned by the FR service provider, typically within the range 16–1007. An *FR map* is a database that stores the mapping between the destination IP address and the DLCI. Naturally, since the DLCI has a local validity, the same occurs with the FR map. When a router (DTE) has a packet to send to a remote host through the FR cloud, it encapsulates the packet into the FR frame and searches in the FR map for the DLCI corresponding to the destination IP address. Conversely, when an FR switch receives a frame with a certain DLCI, it may need to find the corresponding IP address. This is obtained using the inverse address resolution protocol (*inverse ARP*). This is typically required when a switch requests a new DLCI using signaling.

As can be seen from Figure 14.37, the FR frame includes a starting and an ending flag for the purpose of synchronization and frame delimiting. The payload data is used to transport packets, such as IP, IPX, SNA, or Apple Talk packets. The frame check sequence[*] is utilized to check errors in the FR header, not in the whole frame. If an error is detected in the header, the frame is discarded.

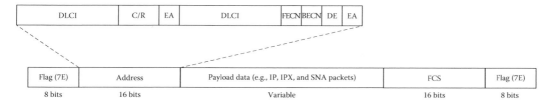

FIGURE 14.37 FR frame format.

---

[*] The FR protocol uses CRC codes as frame check sequence.

Note that the FR does not provide reliability to the transported data. When reliability is required, an upper layer protocol of the end points must implement the functionality (e.g., TCP). In any case, it is worth noting that the level of reliability provided by existing circuits already assure the provision of data with a very low bit error rate and packet loss rate.

The FR address field includes the following subfields:

- *DLCI*: It is used to identify a certain hop of a virtual circuit. The DLCI is the 10-bit address of the FR.
- *Extended address* (*EA*) *bit*: A bit set to 1 in the last octet of the DLCI.
- *C/R*: Not used.
- *Forward explicit congestion notification bit*: A bit set to 1 in a frame received by an FR switch, from a congested link.
- *Backward explicit congestion notification bit*: A bit set to 1 in a frame transmitted by an FR switch, to a congested link.
- *Discard eligibility bit*: When the discard eligibility bit is set to 1, in case of congestion, the frame can be dropped.

Although the internal FR cloud is meshed, the end-to-end FR circuits provided to clients are normally implemented using a hub-and-spoke topology (see Figure 14.36). Alternatively, an institution may always contract a full-meshed topology to the FR service provider to improve the reliability, but this represents an increased cost.

The hub-and-spoke topology comprises a point-to-multipoint link between Router A and the other two routers. Router A (hub) is normally the headquarter's office, whereas the other routers (spokes) are typically part of branches' offices. The point-to-multipoint link can be problematic when dynamic routing protocols are being used. Recall that the split-horizon avoids loops by not allowing packets received in a certain interface to be forward through the same interface. Nevertheless, routing protocol updates require that a certain update packet received by Router A from Router B, be forward to Router C. This is not possible with the hub-and-spoke topology, while Router A is in point-to-multipoint mode. As can be seen from Figure 14.38, there are two ways to solve this problem. One relies on implementing a full-meshed topology. In this case, there is no need to forward routing updates since an update is directly received by all frame-relay nodes, without retransmissions. An alternate method that can be used to solve the split-horizon problem in FR networks relies on implementing subinterfaces in the hub (Router A). As can be seen from Figure 14.38, in this case, two subinterfaces are created in the Router A (hub), where each subinterface corresponds to a different virtual circuit. In this scenario, a routing update received by Router A from Router B can be forwarded to Router C without problem because it uses different subinterfaces, and therefore, the split horizon is not applicable.

The creation of subinterfaces in the hub-and-spoke topology to mitigate the negative effect of split horizon in FR networks modifies the Router A connection from point-to-multipoint into a two (multiple) point-to-point connections. Moreover, this modifies the topology from a hub-and-spoke into a partial-meshed topology.

14.2.5.1 Frame-Relay Configuration Using Cisco IOS

The conventional FR configuration using Cisco IOS is performed in a serial interface using the following syntax:

a. Router(config)#interface *interface*
b. Router(config-if)#ip address *ip_address subnet_mask*
c. Router(config-if)#encapsulation frame-relay [cisco | ietf] (default is cisco)
d. Router(config-if)#frame-relay map ip *dest_ip_address dlci_output* [broadcast] [cisco | ietf]
e. Router(config-if)#frame-relay lmi-type [cisco | ansi | q933a]
f. Router(config-if)#no shutdown (activate only after the frame-relay configuration)

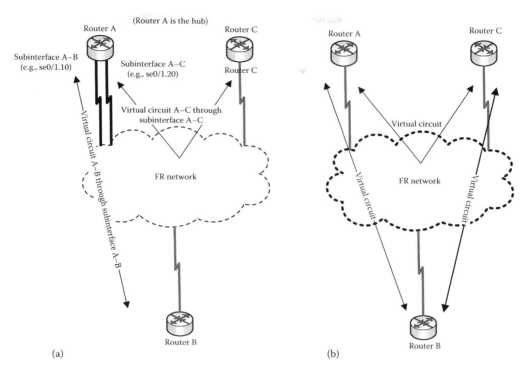

FIGURE 14.38 FR configurations without split-horizon problem: (a) hub-and-spoke topology with subinterfaces and (b) full-meshed topology.

In the above configuration, *dest_ip_address* stands for the destination IP address, that is, the IP address of the remote host of a certain virtual circuit. Moreover, *dlci_output* stands for the DLCI number that is to be configured in the serial interface. Step d must be repeated as many times as the number of virtual circuits to configure. Let us focus on Router B in the full-meshed topology of Figure 14.38. Since Router B has two virtual circuits (one to Router A and another to Router C), the command line creating a mapping between an IP address and the DLCI needs to be typed twice.

Both Cisco and IETF protocols are possible FR encapsulation protocols. Cisco is the default FR encapsulation protocol. In this case, the "cisco" command does not need to be typed. Note that the optional command [cisco | ietf] can be applied after the command line "encapsulation frame-relay," when this is applicable to all FR interfaces. Alternatively, when different virtual circuits use different encapsulation protocols, this should be applied after the command "frame-relay map ip *dest_ip_ address dlci_output* [broadcast]." In the full-meshed topology, Router B may use the FR encapsulation protocol Cisco to connect to Router A, and the IETF protocol to connect to Router C. In this case, the FR encapsulation protocol should be declared after the command "frame-relay map ip *dest_ ip_address dlci_output* [broadcast]," instead of after the command "encapsulation frame-relay."

FR networks are nonbroadcast multiple access (NBMA). This means that the data can only be transmitted from one host to another one, that is, only unicast is allowed. In other words, by default, FR nodes do not allow either multicast or broadcast. Let us consider again that Router A wants to send a routing update. With either topology of Figure 14.38, since the frame-relay network is NBMA, such routing update cannot be sent simultaneously to Router B and Router C. This limitation can be overcome by replicating packets that need to be sent to multiple destinations. This is implemented by typing the command "broadcast," as described above.

While the FR protocol described above consists of a protocol between DTEs, the local management interface (LMI) protocol is configured after the command "frame-relay lmi-type" and comprises the configuration of a protocol utilized in the interface between a DTE (router) and a DCE

(FR switch) or between two adjacent FR switches. This way, a virtual circuit may have a certain LMI protocol in a hop, and another protocol in another hop.

It was described above that in the hub-and-spoke topology, subinterfaces can be created in the hub, to overcome the negative effects of the split-horizon (routing updates) in FR networks. The creation of subinterfaces and their configuration with the FR encapsulation protocol is performed using the following syntax:

a. Router(config)#interface *interface*
b. Router(config)#encapsulation frame-relay [cisco I ietf]
c. Router(config)#interface *interface. sub_interface* point-to-point
d. Router(config-subif)#ip address *ip_address subnet_mask*
e. Router(config-subif)#frame-relay *interface-dlci dlci_output*
f. Router(config-subif)#exit
g. Router(config-if)#frame-relay lmi-type [cisco I ansi I q933a]
h. Router(config-if)#no shutdown (activate only after the frame-relay configuration)

Steps c to e must be repeated as many times as the number of subinterfaces to configure.

In order to better understand the configuration of an FR network, let us configure Router A and Router B of the network depicted in Figure 14.39.

FR configuration in Router B:
 • Router(config)#interface se0/0
 • Router(config-if)#ip address 193.168.1.2 255.255.255.0
 • Router(config-if)#encapsulation frame-relay ietf
 • Router(config-if)#frame-relay map ip 193.168.1.1 61
 • Router(config-if)#frame-relay lmi-type q933a

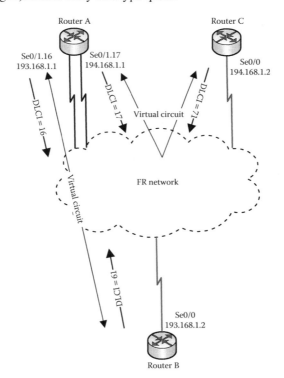

FIGURE 14.39   FR network to configure.

- Router(config-if)#no shutdown
- Router(config-if)#CTRL-Z

FR configuration in Router A:

- Router(config)#interface se0/1
- Router(config)#encapsulation frame-relay ietf
- Router(config)#interface se0/1.16 point-to-point
- Router(config-subif)#ip address 193.168.1.1 255.255.255.0
- Router(config-subif)#frame-relay interface-dlci 16
- Router(config-subif)#interface se0/1.17 point-to-point
- Router(config-subif)#ip address 194.168.1.1 255.255.255.0
- Router(config-subif)#frame-relay interface-dlci 17
- Router(config-subif)#exit
- Router(config-if)#frame-relay lmi-type q933a
- Router(config-if)#no shutdown
- Router(config-if)#CTRL-Z

Finally, the following troubleshooting commands can be utilized in FR:

- Show frame-relay map (shows the FR maps between interfaces and DLCIs)
- Show frame-relay lmi (shows the LMI statistics for the interface)
- Show frame-relay pvc (shows permanent virtual circuit for the interface)
- Debug frame-relay lmi (shows the LMI data exchanged over the time)
- Show interface *interface* (shows interface configuration, including encapsulation parameters)

CHAPTER SUMMARY

This chapter performed a description about transport networks and protocols. It was viewed that transport networks are employed by telecommunication operators to transport large amount of end user data, such as IP packets or voice channels. Transport networks can be grouped into three categories: permanent circuits, circuit switching, and packet switching networks.

It was viewed that transport networks consist of WAN or MAN networks, being commonly employed to interconnect different LAN, MAN, or telephone PABX.

It was viewed that circuit switching transport networks are ideal for delay-sensitive media, such as telephony or VTC. It was described that FDM hierarchy was widely employed in the past to transport multiple analog voice channels between PABX.

The PDH was also described, comprising a quasi-synchronous digital network, used to transport digital voice and data, using TDM. Nevertheless, the frequency and phase fluctuations in the synchronism signals did not allow PDH supporting high rates transmissions. This motivated the migration into the SDH. The SDH is used to transport digital voice and data at high speeds, being optimized for optical fiber transmission media.

The DSL specifications were introduced in this chapter, namely the ADSL, HDSL, and VDSL. It was viewed that the DSL consists of a physical layer specification, employed in modems, to allow the exchange of data through the existing twisted pairs. It was also described that, similar to DSL specifications, the DOCSIS consists of a physical layer specification, employed in modems to allow the exchange of data through the existing cable, used for television broadcast. This cabling can be coaxial, optical fiber, or a mix of these two types.

Packet switching transport networks were also described. It was viewed that transport networks based on packet switching present the typical advantages of packet switching networks, namely the ability to reach a better usage of the network resources as compared to circuit switching, where the resources are preallocated to specific users.

It was described that ATM protocol can be employed by network operators to transport user voice and data traffic. An advantage of ATM relies on the ability to provide QoS.

The MPLS protocol was also introduced, consisting of an alternative to ATM, aiming to mitigate the weaknesses of the ATM, namely the high overhead introduced by ATM, and the inability of ATM to use variable size packets. The MPLS is currently the most used type of packet switching transport networks. A great advantage of MPLS relies on its ability to provide QoS.

The HDLC protocol was also described, consisting of a data link layer employed in WAN. The PPP was also studied, consisting of a data link layer employed in WAN, with additional functionalities relating to HDLC, such as authentication, encryption, and compression. The configuration of the PPP in Cisco routers using Cisco IOS was also studied.

Finally, the FR encapsulation protocol was described in this chapter, consisting of a protocol used by network operators to provide virtual circuits to corporate customers. The configuration of the FR in serial interfaces of routers, using Cisco IOS, was also studied.

REVIEW QUESTIONS

1. What is the ATM adaptation layer used for?
2. What is the difference between ATM and B-ISDN?
3. What are the layers of the B-ISDN reference model and their functions?
4. Why is ATM referred to as *cell switching*, instead of packet switching network?
5. In the scope of the ATM protocol, what is the difference between a virtual path and a virtual channel?
6. In the scope of the ATM protocol, which sublayers are included in the physical layer? What are their functionalities?
7. Which types of ATM cell headers exist? Define the different header fields.
8. What is the difference between an ATM virtual circuit switch and an ATM virtual path switch?
9. Which types of B-ISDN classes of services exist?
10. In the scope of MPLS, what does a FEC consist of?
11. What are the advantages of the MPLS protocol relating to the ATM protocol?
12. What is the MPLS packet format?
13. In which location is the MPLS shim header introduced in the packets/frames?
14. In the scope of MPLS protocol, what does LSP stands for?
15. Which mechanisms are implemented by the MPLS protocol that makes it well fitted for providing QoS?
16. Which type of information is contained in an MPLS shim header?
17. Which mechanisms are employed by different LSRs in order to calculate the output interface to forward a certain packet?
18. What is the difference between an LSR and an LER?
19. What is the difference between a customer's equipment router and a provider equipment router?
20. Describe the working functionalities of an MPLS network.
21. To which OSI reference model layer does the MPLS protocol belong?
22. Define the SDH architecture.
23. What are the reasons that may originate fluctuations between a SPE/VC and a frame?
24. Which mechanisms can be implemented to counteract the fluctuations between a SPE/VC and a frame?
25. What is the difference between the ITU-T G.732 and the ITU-T G.733 standard?
26. Which type of interposition is employed in SDH and SONET hierarchies?
27. Which type of interposition is employed in PDH hierarchy?
28. What are the advantages of the synchronous hierarchies (SDH/SONET) relating to plesiochronous hierarchies?

29. Which type of protection from failures can be employed in an SDH ring using two optical fibers? How does it work?
30. Which type of protection from failures can be employed in an SDH ring using four optical fibers? How does it work?
31. How are different SDH headers managed by different SDH devices?
32. How does a DOCSIS 1.0 modem allow the exchange of digital data over coaxial cables?
33. Which type of transmission techniques can be employed in xDSL modems?
34. What is the difference between DOCSIS 1.0, 2.0, and 3.0?
35. What is an NCP used for?
36. Which studied xDSL version performs the transmission in baseband? What is its line coding technique?
37. What is the difference between an xDSL modem and a DOCSIS modem?
38. What is the difference between DMT and OFDM?
39. Why is echo cancellation used in xDSL modems?
40. What is the typical spectrum utilized by DSL modems?
41. Which types of LLC protocols can be implemented in a permanent circuit for data communications?
42. What is the control field of the HDLC protocol used for?
43. Which kind of HDLC frames do you know?
44. What is the purpose of the protocol field of the PPP header?
45. What is the PPP LCP used for?
46. What are the S frames of the HDLC protocol used for?
47. What is the difference between the PPP and the SLIP?

LAB EXERCISES

1. Consider the network depicted in the figure below. Assuming the subnet mask 255.255.255.0, configure the plotted network using the Cisco Packet Tracer simulator. Consider static routing. Use the PPP with PAP. Test the result using traceroute/ping in routers and tracert/ping in hosts. Configure RIPv2 routing protocol.

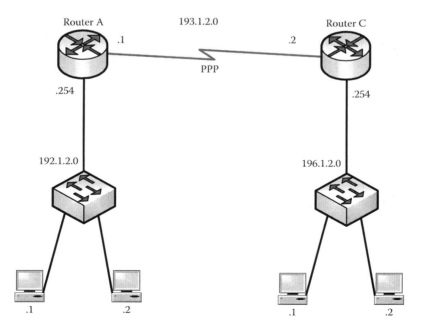

2. Consider the network depicted in the figure below. Assuming the subnet mask 255.255.255.0, configure the plotted network using the Cisco Packet Tracer simulator. Consider RIPv2 routing protocol. Use the PPP with CHAP.

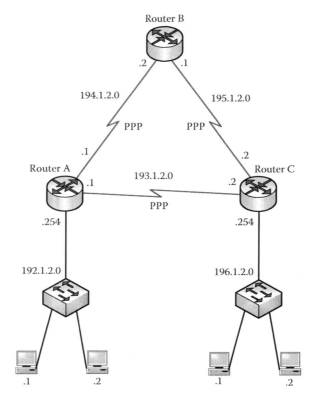

3. Consider the network depicted in the figure below. Assuming the subnet mask 255.255.255.0, configure the plotted network using the Cisco Packet Tracer simulator. Use the PPP in serial links, with the PAP. Use the VTP when configuring the two switches directly connected. Use the OSPF protocol in routers.

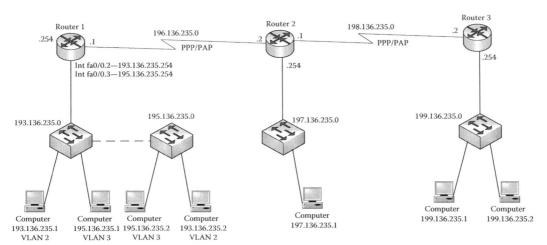

4. Replicate the previous exercise using the EIGRP, instead of the OSPF.
5. Replicate the previous exercise using the static routes, instead of the EIGRP.
6. Replicate the previous exercise using real network equipment.
7. Consider the network depicted in the figure below. Configure the plotted network using the Cisco Packet Tracer simulator. Configure a PPP serial link between the two routers, using the CHAP. Use the RIP.

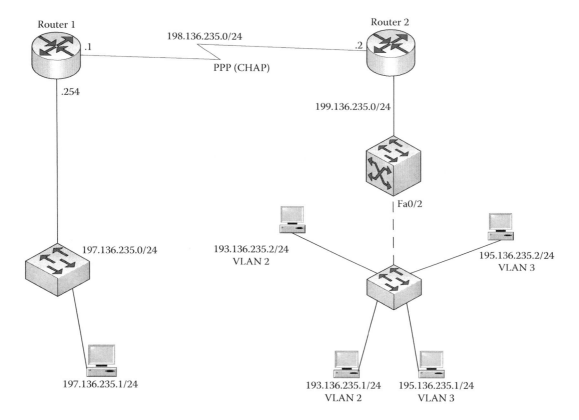

8. Consider the network depicted in the figure below. Assuming the subnet mask 255.255.255.0, configure the plotted network using the Cisco Packet Tracer simulator. Use the OSPF protocol in routers.

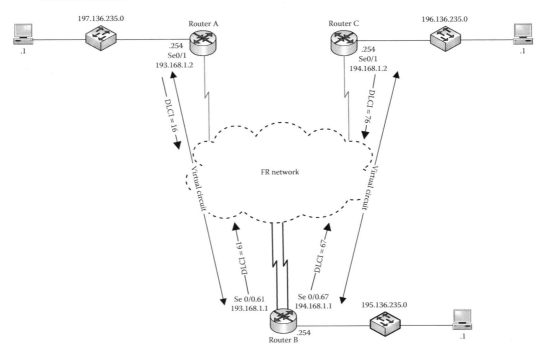

9. Consider the network depicted in the figure below. Configure the plotted network using the Cisco Packet Tracer simulator. Use the EIGRP.

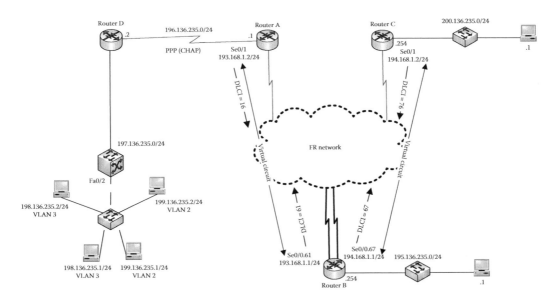

15 Cellular Communications and Wireless Standards

LEARNING OBJECTIVES

- Describe the concept of cellular communications and its hierarchical structure.
- Describe the evolution of cellular systems.
- Identify and describe wireless communication standards and protocols.

The recent need to be in permanent touch with others facilitated an enormous growth of the wireless communications, both in terms of offer and demand of services. In the past, cellular phones were mainly used for voice communications. A constant evolution of services allowed the massification of services such as short message services, multimedia messaging service, and video call. Recently, Internet access has become possible using cellular phones, allowing a new myriad of services such as e-mail, video streaming, and web browsing. In fact, the evolution from the second generation of cellular systems to the third generation was the main driver to allow a sudden increase of traffic due to the new services.

Cellular coverage and capacity, as well as electrical consumption, are key aspects in cellular systems. Currently, there are still some geographic zones where, due to difficult propagation conditions or low population density, the coverage is weak or even inexistent.

15.1 CELLULAR CONCEPT

The cellular network concept was the result of the need to develop a higher capacity for mobile service. It consists of the reutilization, in adjacent locations, of several low-power transmitters, typically with less than 100 W. Figure 15.1 shows a typical cellular network structure, where a base station (BS) is placed at the center of each cell. This BS is used by mobile stations (MSs) located in the corresponding cell, in order to allow them to gain access to the network. Moreover, different BSs are typically interconnected with guided transmission systems (e.g., optical fibers and terrestrial microwave system).

As depicted in Figure 15.2, in order to allow a connection establishment between MS1, located in the left cell, with MS2, located in another cell, the network performs the following steps:

1. MS1 establishes a link over the air with the corresponding BS (BS1).
2. Then, the existing fixed connection between the source BS (BS1) and destination BS (BS2) allows the call being forwarded to the destination's BS.
3. Finally, BS2 establishes a link over the air with the destination mobile station (MS2).

Using these three independent links, a connection can be established between two different users located in different cells.

The areas to be covered by a cellular network are distributed by multiple cells where, to each cell, a set of frequency bands are allocated to BS and MS. The use of the same frequency bands by adjacent cells results in co-channel interference (see Chapter 3). This type of interference can be avoided by not using the same set of frequency bands in adjacent cells, as the propagation losses normally assure the necessary isolation. Following this approach, the resulting geometric pattern utilized by cellular configuration is depicted in Figure 15.3.

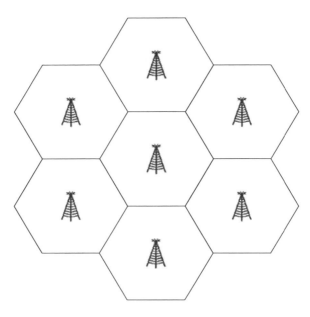

FIGURE 15.1   Cellular network structure.

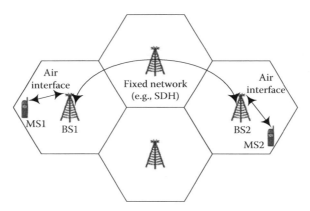

FIGURE 15.2   Example of interconnection between two mobile stations using the corresponding base stations.

The cell geometry consists of a hexagon. Instead of considering a cell consisting of a circle around the center (BS), the hexagon is adopted. With such configuration, the antennas from different BSs are equidistant. In Figure 15.3, each letter represents a different set of frequency bands assigned to each cell. As can be seen, in order to avoid co-channel interference, the same set of bands are not assigned to adjacent cells. This leads to the reuse factor 7, where each group of seven cells use a group of seven set of frequency bands, and the repetition of frequency bands only occur in groups of seven cells. In fact, as described in Chapter 3 for the definition of co-channel interference, code division multiple access (CDMA) networks[*] make use of all sets of frequency bands in all cells, leading to reuse factor 1. This is depicted in Figure 15.4. The resulting residual interference is mitigated by employing multiuser detectors.

---

[*] For example, in the downlink of Universal Mobile Telecommunications System (UMTS).

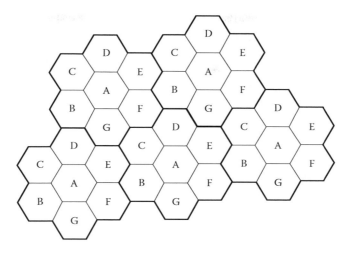

FIGURE 15.3 Typical cellular network structure with reuse factor 7.

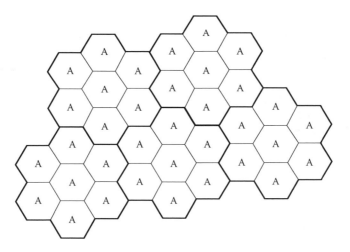

FIGURE 15.4 Typical cellular network structure with reuse factor 1 (normally adopted by CDMA networks).

Cellular coverage is related to capacity. The cellular capacity is viewed as the number of simultaneous calls within a cell. Regardless of the cell dimension, a cell accommodates a certain number of simultaneous calls. Increasing the cell dimension (e.g., macrocell, instead of a microcell) accommodates a lower number of calls per square meter while corresponding to a higher cellular coverage (see Figure 15.5).

The decision on whether or not to implement a lower hierarchy cell relies on the rate of expected calls. Allocating M carriers per cell allows accommodating $M \times N$ calls, with N the number of calls per carrier.[*] Note that the numbers M and $M \times N$ are independent on whether it is a macrocell,

---

[*] GSM accommodates eight TDMA calls per carrier. The number of calls accommodated in each wideband UMTS carrier depends on the CDMA spreading factor.

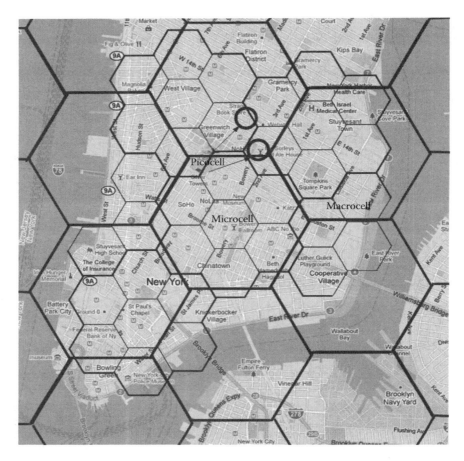

FIGURE 15.5   Hierarchical cellular structure.

a microcell, or a picocell. The above equivalence assumes time division duplexing (TDD).[*] In the case of frequency division duplexing (FDD),[†] the number of calls per cell is halved.[‡]

In 3G networks, the coverage is also related to the throughput available to each user. A lower dimension cell typically corresponds to a better signal quality due to the lower path loss. A better signal quality (higher signal strength and lower noise and interferences) allows typically a higher throughput.

Although this is not a problem inherent of cellular telecommunications, electrical consumption has been, lately, an issue to which the technological research is looking at. In the telecommunications field, this problem can be mitigated by using smart and adaptive antennas, efficient power control (see Section 15.1.5) or with the implementation of hierarchical cellular structures (e.g., macro-, micro-, pico-, and femtocells). In this sense, a lower dimension cell, where an MS is at a shorter distances from a BS, corresponds to a reduced energy consumption.

---

[*] TDD allows full-duplex operation by allocating uplink and downlink channels in a single-carrier frequency, but using different time slots.

[†] FDD allows full-duplex operation by allocating uplink and downlink channels in a two different carrier frequencies.

[‡] As one half of the carriers is used for the uplink, and the other half for the downlink.

15.1.1 MACROCELL

A macrocell is the initially designed type of cell. In order to maximize the coverage, the antennas' BSs are placed in geographically high locations (e.g., on the top of buildings or on the top of hills). The area of a macrocell varies from a few kilometers up to around 30 km (rural environments). The dimension of the cell to be covered depends on the population density and on the propagation environment. In the case of rural environment, the coverage area of a cell tends to be higher than in urban scenarios, as the expected rate of calls is lower in rural areas. Note that the fast fading typically experienced in rural areas is modeled by a Rice distribution. As detailed in Chapter 5, it consists of a Rayleigh fading model (Rayleigh distribution) to which a line-of-sight (LOS) component is summed. Due to the presence of LOS, the shadowing effect is normally not experienced, resulting in a lower rate of path loss* in rural areas, as compared to urban environments. Consequently, a rural environment is much easier to cover and the cell size is typically higher than in urban areas.

On the other hand, an urban environment is typically characterized by the absence of line-of-sight component (Rayleigh distribution), where the shadowing effect is normally experienced (log-normal distribution), and whose path loss rate is higher than in the rural environments.

While in urban environments, a hierarchical cellular structure can be adopted to face the high demand of calls, implementing microcells or even picocells within a macrocell, in rural areas, due to the expected low rate of calls, other lower size cells are normally not implemented.

15.1.2 MICROCELL

A microcell is normally implemented in parallel with a macrocell, following a hierarchical cellular structure (see Figure 15.5). This cell corresponds to an area from few hundreds of meters up to around 2 km. Depending on the signal's quality and desired service, an MS may switch between a macrocell and microcell. The shadowing effect is normally not present in microcells, as the LOS propagation component normally exists between transmit and receive antennas. When shadowing is present, its effect is typically of reduced consequences. Note that a macrocell is more subject to a weak coverage (e.g., due to shadowing effects and to a wider area) than lower dimension cells. Highly populated areas, and the corresponding rate of calls, may justify the implementation and investment in microcells. The signal-to-noise ratio (SNR) made available by microcells is typically better than that of macrocells. Consequently, in the case of 3G (CDMA networks), the available user throughput in microcells is normally higher than in macrocells.

15.1.3 PICOCELL

Picocells are designed for highly populated indoor or outdoor environments, such as hotels, offices, and shopping centers. In addition, the area to be covered varies from few meters up to around 200 m. Since the LOS component is normally present in picocells, the type of fading experienced is typically the multipath Rice fading.

In picocells, the SNR is normally high and the available throughput is maximized. Moreover, the number of calls per square area is also maximized. The implementation of a picocell requires a high investment in infrastructure, being only justified in highly populated areas. Since walls originate high attenuation levels, the implementation of an indoor picocell solves the problem of covering indoor environments from outdoor BS. In order to avoid co-channel interference, the spectrum made available in lower dimension cells must be different from that in higher dimension cells.

---

* The free space path loss has a distance power decay rate 2, whereas in real scenarios, this value varies between 3 and 5. Rural scenarios present typically a distance power decay rate of the order of 3, whereas in urban scenarios, this distance power decay rate varies typically between 4 and 5.

Finally, due to the low distances involved, the battery used in picocells is much lower than that used in higher order cells, thereby translating into saving of energy.

15.1.4 Femtocell

Femtocells make use of the existing cabled infrastructure to provide indoor and small-size environment cellular access. It makes use of small size BSs that typically interconnect with the cellular network through xDSL or cable modem. Femtocells solve coverage problems and enable a reduction in the electrical consumption [Zhang and de la Roche 2010]. It is viewed as economically more effective than picocells. While being considered as a low-cost solution, it enables a high SNR in the interior of domestic houses, offices, or other indoor buildings, resulting in a high throughput available to end users, providing a wide range of services.

Using femtocells, the services of a cellular operator can be viewed as an alternative to the fixed telephone operator as well as to the Internet service provider. Femtocells were not included in the third generation partnership project (3GPP) specifications for universal mobile telecommunications system (UMTS), but are considered for the long-term evolution (LTE). Similar to a picocell, since the penetration rate of electromagnetic waves over walls is limited, a femtocell solves the problem of covering indoor environments from outdoor BS. Due to the low distances involved, the battery used in femtocells is minimized, representing another important advantage of this cell configuration.

15.1.5 Power Control

Power control is the adjustment in the transmitting power in order to optimize the performance of communications. In cellular communications, power control is used to maximize the SNR of signals at the receiver side and the life battery. This is performed using fast and slow power control.

Fast power control intends to mitigate the effects of fast fading caused (multipath channel), in the uplink as well as in the downlink. On the other hand, slow power control intends to compensate for the received power from stations far from the receiver (e.g., at the edge of a cell), in order to compensate for the near-far problem. In addition, slow power control also intends to mitigate the effects of slow fading (shadowing).

The purpose of power control is to make the received SNR as constant as possible, and consequently, to keep the bit error probability approximately stationary. Power control uses the minimum transmit power such that the SNR has the desired level at the receiver side, but not higher than that, as it would translate in interference to other users (e.g., co-channel interference). Note that a user's signal represents interference to the others, especially in a cellular network whose air interface is based on CDMA technology.

Uplink power control improves the performance due to the following reasons: it equalizes the power received from multiple MSs (avoiding near-far problem) and compensates for fading. Moreover, besides increase in life of batteries, interferences from adjacent cells (co-channel interference) are also decreased. Note that co-channel interference is more important in CDMA networks because the reuses factor one is normally adopted. Power control is always employed in CDMA networks, as this technology is highly sensitive to variations in the received power.* This is valid even when the CDMA network adopts multiuser detectors at receivers, whose purpose is the mitigation of the level of multiple access interference.

In the downlink, as the BS sends all the signals to all mobiles in a synchronized manner, all signals are received by all MSs with the same power. Therefore, in this case, the use of power control

---

* A user's signal received with a higher power represents a higher level of multiple access interference to the others.

is only considered for minimizing interferences caused in adjacent cells[*] (especially in CDMA networks), and to compensate for interferences received from adjacent cells.[†]

Power control is performed in a dynamic way, in an open or a closed loop. Open loop power control measures the interference conditions, adjusting the transmitted power to avoid these effects. However, since in FDD networks, the transmit and receive frequency bands are different, fast fading does not present correlation between uplink and downlink, and open loop power control is not an effective mechanism. In this case, closed loop power control is normally adopted. With closed loop power control, the receiver measures the SNR and sends a command to the transmitter, to adjust its power, in order to keep the destination's SNR at a desired level, thereby avoiding fluctuations.

As detailed in Chapter 7, multiple input multiple output (MIMO) systems is another technique that can be employed to maximize the SNR, without having to increase the transmitting power.

15.2 EVOLUTION OF CELLULAR SYSTEMS AND THE NEW PARADIGM OF 4G

The first generation (1G) of cellular networks was analog, having been deployed between 1980 and 1992. 1G included a myriad of cellular systems, namely the total access communication system, the advanced mobile phone system, and the Nordic mobile telephony, among others. These systems were of low reliability, low capacity, low performance, and without roaming capability between different networks and countries. The multiple access technique adopted was frequency division multiple access (FDMA), where signals of different users are transmitted in different (orthogonal) frequency bands.

The second generation (2G) of cellular networks, like the global system for mobile communications (GSM), was widely used between 1992 and 2003. This introduced the digital technology in the cellular environment, with a much better performance, better reliability, higher capacity, and even with the roaming capability between operators, due to its high level of standardization and technological advancements. The multiple access technique used by GSM was time division multiple access (TDMA), where signals generated by different users were transmitted in different (orthogonal) time slots. Narrowband CDMA system was adopted in the 1990s by IS-95 standard, in the United States. IS-95 was also a 2G system.

Afterward, the UMTS, standardized in 1999 by 3GPP[‡] Release 99 (see Table 15.1), proceeded with the utilization of CDMA, in this particular case, using the wideband CDMA (WCDMA). The UMTS consists of a third generation (3G) cellular system.

The CDMA concept relies on different spread spectrum transmissions, each one associated with a different user's transmission, using a different (ideally orthogonal) spreading sequence [Marques da Silva et al. 2010].

LTE can be viewed as the natural evolution of 3G,[§] using a completely new air interface, as specified by 3GPP Release 8, and enhanced in its release 9. Its initial deployment took place in 2010. The LTE comprises an air interface based on orthogonal FDMA (OFDMA)[¶] in the downlink and single-carrier FDMA (SC-FDMA) in the uplink. This allows a spectral efficiency improvement by a factor of 2 to 4, as compared to the high speed packet access (HSPA),[**] making use of new spectrum, different

---

[*] Transmit with the minimum power to accommodate each user, without generating a high level of interference in adjacent cells.

[†] Increase the power in order to increase the SNR level. This is normally necessary for users located farther from the reference BS, that is, at a lower distance from an adjacent BS.

[‡] 3GPP is responsible for specifying and defining the architecture of the European 3G and 4G evolution.

[§] In fact, LTE is sometimes referred to as 3.9G.

[¶] In opposition to WCDMA utilized in UMTS.

[**] Standardized in 3GPP Releases 5–7 (see Table 15.1).

TABLE 15.1

Comparison between Several Different 3GPP Releases

FDD	WCDMA	HSPA	HSPA+	LTE	
TDD	TD-SCDMA[a]	TD-HSDPA	TD-HSUPA	TD-LTE	LTE/IMT Advanced
Deployment	2003	2006/8	2008/9	2010	2014
3GPP Release	99	5/6	7	8/9	10/11/12
Downlink Data Rate	384 kbps	14.4 Mbps[b]	28 Mbps[b]	>160 Mbps[c]	1 Gbps nomadic; 100 Mbps mobile
Uplink Data Rate	128 kbps	5.76 Mbps[b]	11 Mbps[b]	>60 Mbps[c]	500 Mbps nomadic; 50 Mbps mobile
Switching	Circuit + packet switching	Circuit + packet switching	Circuit + packet switching	IP based (packet switching)	IP based (packet switching)
Transmission Technique	WCDMA/ TD-SCDMA	WCDMA/ TD-SCDMA	WCDMA/ TD-SCDMA	Downlink: OFDMA Uplink: SC-FDMA	Downlink: OFDMA Uplink: SC-FDMA
MIMO	No	No	Yes	Yes	Yes
Multihop Relay	No	No	No	No	Yes
Adaptive Modulation and Coding	No	Yes	Yes	Yes	Yes
Cooperative Systems	No	No	No	No	Yes
Carrier Aggregation	No	No	No	No	Yes

[a] Synchronous CDMA.

[b] Peak data rates.

[c] Assuming 20 MHz bandwidth and 2×2 MIMO.

transmission bandwidths from 1.4 up to 20 MHz, alongside with MIMO systems and the all-over IP[*] architecture [Marques da Silva et al. 2012].

In order to fully implement the concept of *anywhere* and *anytime*, as well as to support new and emergent services, users are demanding more and more from the cellular communication systems. New requirements include increasing throughputs and bandwidths, enhanced spectrum efficiency, lower delays, and network capacity, made available by the air interface.[†] These are the key issues necessary to deliver the new and emergent broadband data services. In order to face these requirements, the LTE-Advanced (LTE-A) was initially specified in release 10 of 3GPP, and improved in its release 11 and 12. The LTE-A consists of a fourth generation (4G) cellular system, having being deployed in 2014. It supports peak data rates in the range of 100 Mbps for vehicular mobility to 1 Gbps for nomadic access (in both indoor and outdoor environments). 4G aims to support current and emergent multimedia services, such as social networks and gaming, mobile TV, high-definition television (HDTV), digital video broadcast, multimedia messaging service, and video chat, using the all-over IP concept and with improved quality of service (QoS).[‡]

The specifications for international mobile telecommunications-Advanced (IMT-Advanced[§]) were agreed at the International Telecommunications Union-Radio communications (ITU-R) in [ITU-R 2008]. ITU has determined that *LTE-A* should be accorded the official designation of

---

[*] Internet protocol.

[†] Where the bottleneck is typically located.

[‡] Important QoS parameters include the definition of the required throughput, bit error rate (BER), end to end packet loss, delay, and jitter.

[§] IMT-Advanced is commonly referred to as IMT-A.

IMT-Advanced. IMT-Advanced is meant to be an international standard of the next generation cellular systems.[*]

New topological approaches like cooperative systems, carrier aggregation, multihop relay, advanced MIMO systems, as well as block transmission techniques allow an improved coverage of high rate transmission, improved system performance and capabilities, necessary to fit the advanced requirements of IMT-A [ITU-R 2008].

15.2.1 EVOLUTION FROM 3G SYSTEMS INTO LONG-TERM EVOLUTION

The third generation of cellular system is composed of different evolutions. The initial version, specified by 3GPP Release 99, marked a sudden change in the multiple access technique (see Table 15.1). While the GSM was based on TDMA, 3G makes use of WCDMA to achieve an improved spectrum efficiency and cell capacity. This evolution allowed an improvement rate from few dozens of kbps up to 384 kbps for the downlink and 128 kbps for the uplink. These rates were improved in the following updates, achieving 28 Mbps in the downlink of HSPA+ (3GPP release 7). In order to respond to the increased speed demands of the emergent services, higher speeds became possible with the already deployed LTE, supporting 160 Mbps in the downlink (as defined by 3GPP release 8), and even higher speeds with some additional improvements to LTE baseline introduced in 3GPP Release 9 (e.g., advanced MIMO systems). The LTE air interface was the result of a study item launched by 3GPP named Evolved UTRAN (E-UTRAN). The goal was to face the latest demands for voice, data, and multimedia services, improving spectral efficiency by a factor of 2–4, as compared to HSPA Release 7. The LTE can be viewed as a cellular standard for 3.9G (3.9 generation).

The LTE air interface relies on a completely new concept that introduced a number of technological evolutions as a mean to support the performance requirements of this new standard. This includes block transmission technique using multicarriers, multiantenna systems (MIMO), BS cooperation, as well as the all-over IP concept.

The air interface of LTE considers the OFDMA transmission technique in the downlink and SC-FDMA in the uplink. Depending on the purpose, different types of MIMO systems are considered in 3GPP Release 8. The modulation employed in LTE comprises quadrature phase shift keying (QPSK), 16-QAM or 64-QAM (quadrature amplitude modulation), using adaptive modulation and coding (AMC). When in the presence of noisy channels, the modulation order is reduced and the code rate is increased. The opposite occurs, when the channel presents better conditions.

The LTE comprises high spectrum flexibility, with different spectrum allocations of 1.4, 3, 5, 10, 15, and 20 MHz. This allows a more efficient spectrum usage and a dynamic spectrum allocation based on the bandwidths/data rates required by the users [Astely et al. 2009].

Intra-cell interference is avoided in LTE by allocating the proper orthogonal time slots and carrier frequencies between users in both uplink and downlink. However, inter-cell interference is a problem higher than in the case of UMTS,[†] especially for users at the cell edge. Inter-cell interference can be mitigated by implementing mechanisms such as interference cancellation schemes, reuse partitioning, and advanced BS cooperation.

Another important modification of the LTE, as compared to UMTS, is the all-IP architecture (i.e., all services are carried out on top of IP), instead of the circuit[‡] plus packet[§] switching network adopted by UMTS.

---

[*] Similarly, IMT2000 corresponds to a set of third generation cellular system standards, namely IEEE 802.16e, CDMA2000, and WCDMA.

[†] Due to the lower power spectral density of WCDMA signals, the level of interferences generated in UMTS tends to be lower.

[‡] Circuit switching is employed in UMTS for voice service.

[§] Packet switching is employed in UMTS for data service.

An important improvement of the LTE, compared to the UMTS, relies on its improved capability to support multimedia services.

The multimedia broadcast and multicast service (MBMS), already introduced in 3GPP Release 6 (HSPA), aims to use spectrum-efficient multimedia services, by transmitting data over a common radio channel. MBMS is a system that allows multiple mobile network users to efficiently receive data from a single content provider source by sharing radio and transport network resources. While conventional mobile communications are performed in unicast[*] mode, multimedia services are normally delivered in either broadcast or multicast mode. In broadcast mode, data is transmitted in a specific area (MBMS service area) and all users in the specific MBMS service area are able to receive the transmitted MBMS data. Very often, broadcast communications are established in a single direction (i.e., there is no feedback from the receiver into the transmitter). In multicast mode, data is transmitted in a specific area but only registered users in the specific MBMS service area are able to receive the transmitted MBMS data.

The LTE introduced a new generation of MBMS, entitled *evolved MBMS* (eMBMS). This is implemented in LTE in two types of transmission scenarios [Astely et al. 2009]:

- *Multicell transmission*: Multimedia broadcast over a single frequency network (MBSFN)[†] on a dedicated frequency layer or on a shared frequency layer. The group of cells that receive the same MBSFN multicast data service is referred to as *MBSFN area*.[‡]
- *Single-cell transmission*: Single cell-point to multipoint on a shared frequency layer.

Multicell transmission in single frequency network (SFN) area is a way to improve the overall network spectral efficiency. In MBSFN, when different cells transmit the same eMBMS multimedia data service, the signals are combined, in order to provide diversity for a user equipment (UE) located at a cell boundary. This results in an improved performance and better service quality.

15.2.2 IEEE 802.16 Protocol (WiMAX)

WiMAX stands for Worldwide Interoperability for Microwave Access and allows fixed and mobile wireless access. WiMAX, standardized by the Institute of Electrical and Electronics Engineers (IEEE) as IEEE 802.16, was initially created in 2001 and updated by several newer versions. It consists of a technology that implements a wireless metropolitan area network [Eklund et al. 2002; Andrews et al. 2007; Peters and Heath 2009]. The basic idea of WiMAX relies on providing wireless Internet access to the last mile, with a range of up to 10 km. Therefore, it can be viewed as a complement or competitor of the existing asynchronous digital subscriber line (ADSL) or cable modem, providing the service with the minimum effort in terms of required infrastructures. On the other hand, fixed WiMAX can also be viewed as a backhaul for Wi-Fi [IEEE 802.11], cellular BS, or mobile WiMAX. Since the standard only defines the physical layer and the medium access control (MAC) sublayer, it can be used associated to either IPv4 or IPv6.

In order to allow the operation of WiMAX in different regulatory spectrum constraints faced by operators in different geographies, this standard specifies channel sizes ranging from 1.75 up to 20 MHz, using either TDD or FDD, with many options in between [Yarali and Rahman 2008].

---

[*] Unicast stands for communication whose data destination is a single station.

[†] The MBSFN allows delivering services such as mobile television.

[‡] Within a MBSFN area, if one or more cells are not required to broadcast the multimedia data service, the transmission can be switched off, and the corresponding resources can be released to regular unicast or other services.

The initial version of WiMAX was updated by several newer versions:

- IEEE 802.16-2004, also referred to as *IEEE 802.16d*. This version only specified the fixed interface of WiMAX, without providing any support for mobility [IEEE 802.16–2004]. This version of the standard was adopted by European Telecommunications Standards Institute (ETSI) as a base for the HiperMAN.[*]
- IEEE 802.16-2005, also referred to as *IEEE 802.16e*. It consists of an amendment to the previous version. It introduced support for mobility, handover, and roaming, among other new capabilities [IEEE 802.16e-2005]. In addition, in order to achieve better performances, MIMO schemes were introduced.
- Relay specifications are included in IEEE 802.16j amendment. The incorporation of multihop relay capability in the foundation of mobile IEEE 802.16-2005 is a way to increase both the available throughput by a factor of 3 to 5 and/or coverage (and higher channel reuse factor), or even to fill the *coverage hole* of indoor coverage [IEEE 802.16-2004; Oyman et al. 2007; Peters and Heath 2009]. Multihop relay capability was included in IMT-Advanced [Astely et al. 2009].

In addition to these versions, requirements for the next version Mobile WiMAX entitled IEEE 802.16m [2009] were completed. The goal of IEEE 802.16m version is to reach all the IMT-Advanced requirements as proposed by ITU-R in [ITU-R 2008], making this standard a candidate for the IMT-A. Advances in IEEE 802.16m include wider bandwidths (up to 100 MHz, shared between uplink and downlink), adaptive and advanced TDMA/OFDMA access schemes, advanced relaying techniques (already incorporated in IEEE 802.16j), advanced multiple-antenna systems, adaptive modulation schemes such as hierarchical constellations and AMC, and frequency adaptive scheduling, among other advanced techniques.

The original version of the standard specified a physical layer operating in the range of 10 to 66 GHz, based on OFDM and TDMA technology. IEEE 802.16-2004 added specifications for the 2 to 11 GHz range (licensed and unlicensed), whereas IEEE 802.16-2005 introduced the scalable OFDMA (SOFDMA) with MIMO (space-time coding based, spatial multiplexing based, or beamforming) or advanced antenna systems [IEEE 802.16e], instead of the simple OFDM with 256 subcarriers considered by the previous version.

In terms of throughputs and coverage, these two parameters are subject to a trade-off [IEEE 802.16e]: typically, mobile WiMAX provides up to 10 Mbps per channel (symmetric), over a range of 10 km in rural areas (LOS environment) or over a range of 2 km in urban areas (non-LOS environment) [Ohrtman 2008]. With the fixed WiMAX, this range can normally be extended. Mobile version considers an omni-directional antenna, whereas fixed WiMAX uses a high gain antenna (directional). Throughput and ranges may always change. Nevertheless, by enlarging one parameter, the other has to reduce, otherwise the BER would degrade. In the limit, fixed WiMAX can deliver up to 70 Mbps per channel (in LOS, short distance and fixed access), and may cover up to 50 km (in LOS for fixed access), with a high gain antenna [Ohrtman 2008], but not both parameters simultaneously. Contrary to UMTS where handover is detailed specified, mobile WiMAX has three possibilities but only the first one is mandatory: hard handover, fast BS switching, and macrodiversity handover. Fast BS switching and macro diversity handover are optional, as it is up to the manufacturers to decide about their implementation specifications. Therefore, there is the risk that handover is not possible for these advanced handover schemes between two BS from different manufacturers. Another drawback on the use of WiMAX is the maximum speed allowed in mobility, which is limited to 60 km/h. For higher speeds, the user experiences a high degradation in performance.

The WiMAX version currently available [IEEE 802.16-2005] incorporates most of the techniques also adopted by LTE (from 3GPP) such as OFDMA, MIMO, and advanced turbo coding.

---

[*] High-performance metropolitan area network.

In addition, the inclusion of multihop relay capabilities (IEEE 802.16j) aims to improve the speed of service delivery and coverage by a factor of 3 to 5. Moreover, IEEE 802.16m integrates and incorporates several advancements in transmission techniques that meet the IMT-Advanced requirements, including 100 Mbps mobile and 1 Gbps nomadic access, as defined by ITU-R [2008].

15.2.3 LTE-A AND IMT-ADVANCED

4G aims to support the emergent multimedia and collaborative services, with the concept of *anywhere* and *anytime*, facing the latest bandwidth demands. The LTE-A (standardized by 3GPP) consists of a 4G system. Based on LTE, the LTE-A presents an architecture using the all-over IP concept [Bhat et al. 2012]. The support for 100 Mbps in vehicular and 1 Gbps for nomadic access[*] and a latency lower than 5 ms is achieved with the following mechanisms:

- Carrier aggregation, composed of multiple bandwidth components (up to 20 MHz) in order to support transmission bandwidths of up to 100 MHz.
- Advanced antenna systems, increasing the number of downlink transmission layers to eight and uplink transmission layers to four. Moreover, LTE-A introduced the concept of multiuser MIMO, in addition to the single-user MIMO previously considered by the LTE.
- Multihop relay (adaptive relay, fixed relay stations, configurable cell sizes, hierarchical cell structures, etc.), in order to achieve a coverage improvement and/or an increased data rate.
- Advanced inter-cell interference cancellation (ICIC) schemes.
- Advanced BS cooperation, including macrodiversity.
- Multiresolution techniques (hierarchical constellations, MIMO systems, OFDMA multiple access technique, etc.).

Standardization of LTE-A is part of 3GPP Release 10 (completed in June 2011), and enhanced in its release 11 (December 2012) and release 12 (March 2013).

The IMT-Advanced refers to the international 4G system, as defined by the ITU-R [2008]. Moreover, the LTE-A was ratified by the ITU as an IMT-Advanced technology in October 2010 [ITU 2010].

Within 4G, voice, data, and streamed multimedia are delivered to the user based on an all-over IP packet switched platform, using IPv6. The goal is to reach the necessary QoS and data rates in order to accommodate the emergent services.

Due to the improvements in address spacing with the 128 bits made available by IPv6, multicast and broadcast applications will be easily improved, as well as the additional security, reliability, intersystem mobility, and interoperability capabilities. Moreover, since 4G system relies on a pool of wireless standards, this can be efficiently implemented using the software defined radio platform, being currently an interesting research and development (R&D) area by many industries worldwide.

15.3 IEEE 802.11 PROTOCOL (WI-FI)

The IEEE 802.11 consists of a technology dedicated to interconnect and allow Internet connection of wireless devices in a local area network (LAN) environment. *Wireless Fi*delity, often known as Wi-Fi, is a trademark of Wi-Fi Alliance for certified products based on IEEE 802.11 standard [IEEE 802.11]. This standard specifies the physical layer and the MAC sublayer, offering services to a common 802.2 Logical Link Control.

---

[*] 1 Gbps as a peak data rate in the downlink, whereas 500 Mbps is required for the uplink.

Originally developed for cable replacement in companies, Wi-Fi quickly became very popular in providing IP connectivity in environments such as offices, restaurants, airports, residences, and campuses, covering typically ranges of the order of 100 m outdoors and 30 m indoors [IEEE 802.11].

The initial version of Wi-Fi, invented in 1991 by NCR Corporation/AT&T, in the Netherlands, was standardized as IEEE 802.11, supporting 1 or 2 Mbps in the 2.4 GHz band, using either frequency hopping spread spectrum (FHSS) or direct sequence spread spectrum (DSSS) [Geier 2002]. This version was upgraded by the following newer versions:

- IEEE 802.11a is a standard version that consists of an extension to IEEE 802.11 that allows up to 54 Mbps in the 5 GHz band using OFDM transmission technique, with an approximated range of the order of 35 m.
- IEEE 802.11b allows a data rate of 11 Mbps in the 2.4 GHz band, using DSSS due to its relative immunity to interference (instead of OFDM), with an approximated range of the order of 35 m.
- IEEE 802.11g is an extension of IEEE 802.11b that allows up to 54 Mbps in the 2.4 GHz band, using either OFDM or DSSS transmission techniques, with an approximated range of the order of 35 m.
- IEEE 802.11n is an upgrade in order to allow over 100 Mbps in the 5 GHz band, by using both OFDM transmission technique and the multistreaming MIMO scheme, with an approximated range of the order of 70 m.
- IEEE 802.11ac is an upgrade to IEEE 802.11n, in order to support up to 500 Mbps in the 5 GHz band, using a wider bandwidth of up to 160 MHz, higher number of parallel MIMO streams and higher modulation orders, with an approximated range of the order of 70 m.

The IEEE 802.11 standard uses the carrier sense multiple access-collision avoidance (CSMA-CA) algorithm as the MAC sublayer protocol. This medium is similar to CSMA-CD but using the request to send and clear to send messages, sent by the frame sender and destination, respectively. These additional messages allow the CSMA-CA reaching a performance improvement, as compared to the CSMA-CD, by alerting the other stations that a frame transmission is going to take place [Ohrtman and Roeder 2003]. In addition, this solves the hidden terminal problem,* which can be experienced in an ad hoc mode (making use of an access point) and in infrastructure network mode (with an access point). Just as in an Ethernet LAN, having more users results in a reduction of throughput (within the coverage area). Therefore, its efficiency is limited to a reduced number of users and/or reduced traffic.

Wi-Fi-based products require at least 20MHz for each channel (22MHz in the 2.4GHz band for IEEE 802.11b), and have specified only the license exempt bands 2.4GHz ISM (Industrial, Scientific, Medical), 5GHz ISM and 5GHz unlicensed national information infrastructure (UNII) for operation [Ferro and Potorti 2005].

With the IEEE 802.11 family of standards, a wireless access point (WAP) connects a group of wireless devices into a single cable device (normally a router). A WAP is similar to a network cable switch, performing frame switching based on the MAC address.

Besides allowing connectivity in infrastructure network mode (using a WAP), Wi-Fi also allows ad hoc networks (peer-to-peer interconnection). This means that wireless devices can interconnect directly, without using an IEEE 802.11 WAP. In addition, by using two wireless bridges, an IEEE 802.11 link can be established to interconnect two cable LANs, as long as the two bridges are within Wi-Fi wireless range.

---

* The hidden terminal problem refers to the situation where a terminal's transmitting coverage area may not be within the receiving coverage area of another terminal, which belongs to the same network, whereas an intermediate node (WAP or terminal working in an *ad hoc* mode) may be in both areas of coverage.

The IEEE 802.11 networks are formed by cells known as *basic service set* (BSS). The BSS are created using a WAP (in infrastructure network mode) or without a WAP (in an ad hoc mode). An extended service set (ESS) consists of multiple interconnected WAP (via wireless or via cable) that allows the interconnection of different BSS. Each BSS has a service set identifier (SSID). The SSID is 32 octets long, being used to identify the name of the BSS.

The IEEE 802.11 frames can be of three different types: data, control, and management. Figure 15.6 shows the generic format of all IEEE 802.11 frames. The preamble consists of the synchronism signal (80 bits) composed of alternating 0 and 1 logic state bits, followed by the start of frame delimiter consisting of 16 bits with the following pattern: 0000 1100 1011 1101. The physical layer convergence procedure (PLCP) header is always transmitted at 1 Mbps and contains information about the protocol data unit (PDU) length, as well as information about the transmission rate that will be used to transmit the frame. The MAC data is described below, whereas the final CRC field consists of 16 CRC bits used for error detection of the frame header.

Figure 15.7 shows the format of the IEEE 802.11 MAC data, whose fields have the following meanings:

- *Control*: This is the initial field, whose breakdown of subfields is depicted in Figure 15.8. It is used for control of the MAC sublayer. This field includes the following subfields:
  - *Protocol version*: This identifies the protocol version.
  - *Type*: This contributes to the identification of the type of IEEE 802.11 frame (control, management, or data).
  - *Subtype*: Together with the type, it identifies the type of IEEE 802.11 frame.
  - *To DS*: This bit is set to 1 when the frame is sent to the WAP (for forwarding).
  - *From DS*: This bit is set to 1 when the frame is received from the WAP.
  - *More fragments*: This bit is set to 1 in case the frame is carrying a fragment of a packet. The first fragment has this bit set to 0.
  - *Retry*: This bit is set to 1 in case the frame is a retransmission of a previously sent one.
  - *Power management*: This indicates the power management of the sender after the transmission of the current frame.
  - *More data*: This means that the station has more frames to transmit. This is useful to prevent receiving stations to enter power save mode.

Preamble	PLCP header	MAC data	CRC

FIGURE 15.6   IEEE 802.11 frame format.

		← MAC header →						
Control	Duration ID	Address 1	Address 2	Address 3	Sequence	Address 4	Payload data	CRC
2 octets	6	2	6	6	2	6	0–2312 (variable)	4

FIGURE 15.7   IEEE 802.11 MAC data format.

Protocol version	Type	SubType	To DS	Fm DS	More fragment	Retry	Pwr manag	More data	WEP	Order
2 bits	2	4	1	1	1	1	1	1	1	1

FIGURE 15.8   Decomposition of the control field. (Data from IEEE 802.11, *Wireless LAN MAC and Physical Layer [PHY] Specifications*, ANSI/IEEE Std 802.11; 1999 [E] Part 11, ISO/IEC 8802-11, 1999.)

- *WEP*: This bit is modified when the frame has been processed.
- *Order*: This bit is set to 1 when the frames are sent in sequence (frames and fragments are commonly not sent in sequence).
- *Duration ID*: This may have different meanings. In power-save pool frames, this is the station ID. In other frames, this field identifies the duration used for network allocation vector calculation.[*]
- *Address 1*: This is the recipient MAC address.
- *Address 2*: This is the MAC address of the station that sent this frame. It can be a wireless station or a WAP.
- *Address 3*: If the To DS subfield of the control field is set to 1, the address 3 is the original source MAC address. If the From DS subfield of the control field is set to 1, the address 3 is the destination MAC address.
- *Sequence*: It includes two subfields:
  - *Fragment number field*: It identifies the sequence number of each frame (12 bits).
  - *Sequence number field*: It identifies the number of each fragment (4 bits).
- *Address 4*: This is the source MAC address used in an ad hoc mode.
- *Payload data*: This contains the LLC PDU data received from the upper sublayer (IEEE 802.2 LLC sublayer). Its size varies from 0 up to a total of 2312 octets.
- *CRC*: It is also referred to as frame check sum (FCS), consists of a 32 redundant bits used for error detection. It uses the same CRC generator polynomial as the one defined for IEEE 802.3 (defined Chapter 12).

As a rule of thumb, the IEEE 802.11 WAP should be distributed in an area with a configuration that minimizes uncovered areas. Different WAPs are interconnected using an Ethernet cable. In order to avoid uncovered areas, WAPs should be placed such that there is a coverage overlap of the order of 10% to 15% between two adjacent BSSs.[†] In case these two adjacent WAPs use the same channels, interference would exist in the overlap area. Therefore, adjacent WAPs should use channels sufficiently spaced apart such that interference is avoided. Each one of the 11 channels existing in the 2.4 GHz band (North America) occupies a total of 22 MHz, and has a center frequency separation of 5 MHz. Therefore, in order to avoid interference, the channels selected by two adjacent WAPs should present a minimum of 5 channels separation (e.g., channels 1 and 6, 6 and 11, or 2 and 7).

The authentication and encryption methods utilized in IEEE 802.11 are described in Chapter 16.

CHAPTER SUMMARY

This chapter provided a view about cellular communications and wireless standards, having been split into multiple areas.

A description about the cellular concept was provided, and its hierarchical structure, namely macro-, micro-, pico-, and femtocells. It was viewed that the cellular coverage and capacity is subject to a tradeoff. Regardless whether it is a macro-, micro-, or a picocells, its capacity is approximately constant and is only dependent on the number of time slots and carriers. Reducing the cell size results in a higher capacity that becomes available per square meter of surface.

Moreover, power control was also addressed, as employed to mitigate the near-far problem, and the multiple types of fading. It was viewed that slow power control aims to mitigate the near-far

---

[*] The NAV refers to the duration used by a transmitting station to send a frame and by the receiving station to send the corresponding ACK. This is measured either after the transmission of the RTS or of the CTS.

[†] As described above, depending on the version, the range of IEEE 802.11 varies between 35 and 70 m.

problem and of the shadowing, whereas fast power control mitigates the effects of the multipath fading, that is, fast fading. Due to a higher expected rate of calls, a hierarchical cellular structure can be implemented in an urban environment, composed of macro- and microcells, or even picocells, within the macrocells. On the contrary, in rural environments, due to the lower rate of calls, such hierarchical cellular structure is normally not implemented.

At the second stage, the evolution of cellular systems and the new paradigm of 4G were addressed. This comprised the study of the evolution from 3G systems into LTE. It was viewed that while 3G systems are implemented based on WCDMA transmission technique, both LTE and LTE-A make use of OFDMA. eMBMS was introduced in LTE and aims to use spectrum efficient multimedia services by transmitting data over a common radio channel.

The IEEE 802.16 protocol was also studied, which is commonly referred to as *WiMAX*. It was viewed that the WiMAX allows fixed and mobile wireless access, having been implemented in several versions. The latest is referred to as *IEEE 802.16m*, which aimed to reach all the IMT-Advanced requirements, including wider bandwidths, adaptive and advanced OFDMA access, advanced relaying techniques, advanced MIMO, adaptive modulation schemes, as well as advanced scheduling.

Then, the LTE-A and the IMT-Advanced were described. It was viewed that the LTE-A aims to support the emergent multimedia and collaborative services, with a throughput of 1 Gbps nomadic access and 100 Mbps in mobile environment, having been standardized by 3GPP releases 10, 11, and 12. Moreover, the LTE-A was selected by ITU as IMT-Advanced.

Finally, the IEEE 802.11 protocol was addressed, including the frame fields, as well as the CSMA-CA procedure. It was described that the IEEE 802.11 protocol defines both the physical layer and the MAC sublayer, and interfaces with the IEEE 802.2 protocol, at the LLC sublayer. The IEEE 802.11 protocol uses the CSMA-CA, which solves the hidden terminal problem, as opposed to CSMA-CD, utilized by the IEEE 802.3. To minimize uncovered areas, IEEE 802.11 WAPs should be placed, such that there is a coverage overlap of the order of 10% to 15% between two adjacent BSSs. Finally, it was described that, to avoid interference, the channels utilized by adjacent IEEE 802.11 WAPs should present a minimum of five-channel separation.

REVIEW QUESTIONS

1. What is the difference between a macrocell, a microcell, and a picocell?
2. What are the advantages of using a hierarchical cellular structure?
3. What is the typical propagation environment experienced in rural environments?
4. What is the typical propagation environment experienced in urban environments?
5. What is the purpose of a picocell?
6. What is the difference between a picocell and a femtocell?
7. What is the relationship between a cell size and the cell capacity? Why?
8. What is the reuse factor adopted in different types of cellular networks?
9. Assuming M carriers in the cell and N calls per carrier, what is the maximum number of simultaneous calls in a cell using FDD?
10. What are the advantages of the UMTS, relating to the GSM?
11. What are the differences between LTE and UMTS?
12. What are the differences between LTE and 4G?
13. In which cellular generations is the OFDM transmission technique employed?
14. What are the differences between LTE and WiMAX?
15. Which cellular generations consider IP-based network (i.e., all over IP)?
16. Which cellular generations consider MIMO systems?
17. Which potential technologies are expected to be implemented by 4G in order to provide 100 Mbps in mobility and 1 Gbps nomadic?
18. In which modes can the IEEE 802.11 operate?

19. What is the throughput made available by the different versions of IEEE 802.11?
20. What is the difference between the CSMA-CD and the CSMA-CA?
21. In which known LAN standard is the CSMA-CA mechanism employed?

LAB EXERCISES

1. Consider the network depicted in the figure below. Configure the plotted network using the Cisco Packet Tracer simulator. Consider static routing. Use the point-to-point protocol with password authentication protocol between router A and router B. Using the enhanced interior gateway routing protocol (EIGRP) routing protocol. Configure the PC connected through the WAP WRT300N to receive an IP address from the DHCP server (router A).

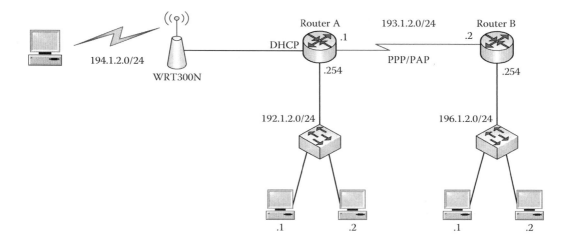

2. Repeat the previous exercise using real network equipment.

16 Network Security

LEARNING OBJECTIVES

- Identify the problems and solutions involved in network security.
- Describe the security services and attack types.
- Describe the types of malware.
- Describe the physical security issues.
- Define the INFOSEC risk management, and quantify the risk.
- Describe different protective measures used to protect from threats and to minimize vulnerabilities.
- Implement a corporate security plan.
- Describe the security mechanisms used in wireless networks.
- Define different security network architectures, and the way they are implemented using firewalls, including virtual private networks (VPNs).
- Describe and configure access control lists (ACLs) in Cisco routers using Cisco IOS.

16.1 OVERVIEW OF NETWORK SECURITY

Globalization has emerged as a result of widespread use of communications and information systems (CIS) by states, organizations, companies, and individuals. The widespread network of networks, called *Internet*, is an important example of what it takes to globalization, and entry into the *information society* or the *information age*.

In order to assist decision making, these technological advancements are demanding deep changes in organizations, emphasizing the human dependence of information (and hence of data). This dependence on technology has reduced the asymmetries in access to information, enabling criminal and terrorist organizations to gain access to almost as much information as states, thereby reducing their competitive disadvantages [Arquilla 1997]. In this new context, new players started using CIS and cyber space to carry out attacks, in order to obtain information and gain knowledge, to neutralize and control systems (e.g., dams, power plants, and telecommunication networks) [Klein 2003]. As an example, in Australia, a hacker managed to shed three million gallons of sewage from a sewage treatment plant [Klein 2003]. In addition, criminals who use these new technologies to carry out their activities benefits from a cover-up, because they are difficult to be identified and even, sometimes, the detection of the attacks can be avoided.

In a broad sense, the information systems security (INFOSEC) can be viewed as a set of technical measures and procedures adopted to prevent unauthorized observation, modification, or denial of the illegitimate use of knowledge, facts, information, skills, or resources. The information that is intended to be preserved can be stored,[*] being processed[†] or in transit[‡] [Maiwald 2003; McClure and Scambray 2005]. The INFOSEC is decomposed into the following subareas:

- Computers security (COMPUSEC)
- Communications security (COMSEC)

---

[*] For example, paper, disk, tape, and diskette.
[†] For example, CPU, RAM, and virtual memory.
[‡] Using a channel for establishing a communication by electronic, electromagnetic, and optical means.

- Network security (NETSEC)
- Emanations security (EMSEC)
- Physical security

Each of these subareas needs to be independently protected. It is worth noting that INFOSEC is a process, not a product or technology. It may use various human means, procedures, and/or technologies to enable its implementation, including antivirus, access control, firewalls, authentication mechanisms (passwords, smart cards, tokens, biometrics, etc.), intrusion detection systems (IDSs), policy management, vulnerability scanning, encryption, and physical security.

Computers security relies on protecting the processing and storage devices. The ISO/IEC 15408 standard defines a common criteria, whose purpose relies on establishing a group of procedures and mechanisms that assure a certain level of confidence. These mechanisms include the whole information system life cycle, from the project phase, operation, up to destruction. A group of computer mechanisms, such as access control or antivirus system, aims to ensure that security policies are followed, and therefore, can be viewed as COMPUSEC mechanisms.

Communications security refers to the protection applied to data in transit between a transmitter and a receiver. The most common type of COMSEC protection relies on using cryptography.

Network security refers to the security of the network infrastructure, except the channel protection (covered by COMSEC). NETSEC refers to the protection of routing, dynamic host configuration protocol (DHCP), domain name server (DNS), and so on. A common type of NETSEC protection relies on ciphering the exchange of routing data.

Emanations security aims to protect the electromagnetic and acoustic radiations that may reveal the original message or information that is being processed or transmitted. A common type of EMSEC protection relies on using tempest or separating the cabling that carries classified data, from those that carries unclassified. Moreover, since an optical fiber is not a metallic medium and therefore it does not radiate electromagnetic signals, it can be used as an EMSEC protection.

Finally, physical security refers to the physical protection of resources. This aims to keep the physical survivability of resources, covering areas such as the electrical power supply, air conditioning, or the physical access control to areas where the systems and equipment are located.

The INFOSEC involves the protection of the information systems from attacks against the following attributes (commonly referred to as *CIAA*) [RFC 2196]:

- Confidentiality (C)
- Integrity (I)
- Availability (A)
- Authenticity (A) and nonrepudiation

Table 16.1 summarizes the various security services* and attack types, which are possible to be carried out through the network. These security services and attacks are characterized in the following sections.

16.2 INFORMATION GATHERING

Before an attack is executed against the confidentiality (access), integrity (modification), authenticity (nonrepudiation), or availability (denial of service [DoS]), a hacker or a cracker starts performing *information gathering*, also referred to as *reconnaissance attack*. This phase relies on determining

---

* The accountability is a complementary security service, which adds protection to the following attributes of information: confidentiality, integrity, and authenticity.

TABLE 16.1

Summary Table of Security Services and Attack Types

Attack Types and Services	Confidentiality	Integrity	Availability	Authenticity	Accountability
Access	X				X
Modification		X			X
Denial of Service			X		
Nonrepudiation				X	X

who has what, and with which vulnerabilities that can be exploited. Moreover, after an attack is carried out, the attacker should destroy evidence of the attack (e.g., delete logs in a network log host).

Note that a typical definition for a hacker is someone who exploits information systems of others as an intellectual challenge, without a malicious intention. On the other hand, a cracker starts acting as a hacker in the exploratory phase, but proceeds with a malicious activity, such as the robbery of data, information destruction, and making a system out of service.

The reconnaissance is the initial phase of an attack and aims to discover and map systems and services, their level of exposition, and the vulnerabilities associated to each one. A reconnaissance attack (information gathering) is normally implemented using the following sequence of procedures:

- *Internet queries*: The aim of an Internet query relies on gathering knowledge about the address space assigned to a certain institution. This can be performed from the exterior using a tool such as nslookup or whois.
- *Ping sweeps*: Once the address space assigned to an institution is known, a hacker pings such address range to determine which addresses are active and reachable from the exterior. This can be automatically performed using a tool such as gping or fping.
- *Port scans*: Once the addresses active from the exterior are known, a port scanner allows determining the ports that are active in each IP address, as well as the type and version of the application in use. From this information, a hacker may extract the vulnerabilities that can be exploited (e.g., SQL injection, buffer overflow, or SYN attack). A common tool widely used for this function is the Nmap.
- *Eavesdropping*: A hacker with internal access to a network, or with a physical access to a network from the exterior, may run a packet sniffer (e.g., Wireshark) to gather useful information that is exchanged in the network (such as username, passwords, and PINs). In case a hub is the central node of the network, this is trivial. Nevertheless, even with a switch, packet sniffing can also be implemented (see MAC address flooding in Section 16.3.1.1). The reader should refer to Section 16.3.1.1 for a detailed description of eavesdropping.

16.3 SECURITY SERVICES AND ATTACK TYPES

Information security services aim to establish protection against different information security attacks. In the context of information, security services include confidentiality, integrity, availability, authenticity, and accountability. All these services implement functions to prevent, mitigate, or detect information security attacks. Note the difference between security services and the mechanisms used to implement these services. Cryptography, digital signature, or digital certificates are few examples of mechanisms that implement these security services, referred to in this book as *protective measures*.

Information security attributes is another concept similar to information services. The attributes include confidentiality, integrity, availability, and authenticity, whereas security services corresponds to a wider concept that also includes accountability, and a set of mechanisms, tools, systems, and procedures to provide protection from attacks against each information attribute.

16.3.1 CONFIDENTIALITY

Confidentiality is the protection from unauthorized observation or access of information or data. A traditional procedure relies on restricting the access to information and data, such that its access is limited to those entities (user or process) who

- Have a security clearance equal to the highest security level of the information or data.
- Need to have access to the information or data (need to know principle).

The information, which is the potential target of an attack, can be in transit, stored, or in processing.

The most common type of protection from attacks against the confidentiality is cryptography. Two basic types of cryptography are normally adopted: symmetric and asymmetric. In the symmetric type of cryptography, the same key is used by both the source and the destination, whereas the asymmetric scheme comprises two different keys: a public and a private key. A message can be ciphered with the private key and deciphered with the public key or vice versa. Moreover, encryption can be applied to the channel (bulk encryption) or to the message (end-to-end encryption). Note that even when encryption is applied, an intruder may extract information from the volume of traffic that is exchanged through a certain channel. The solution to this problem is to use a type of encryption that *fills in* the channel with *empty bits*, such that the traffic is kept approximately constant, independent of the volume of messages exchanged. This is known as *traffic flow security*.

In addition to encryption, authentication (e.g., use of a password, token, smart cards, and biometrics), and access control mechanisms needs also to be included to ensure confidentiality. Access control and authentication methods are part of accountability, whose purpose is to complement the basic services (confidentiality, integrity, and authenticity), ensuring that INFOSEC is properly established.

Sections 16.3.1.1 through 16.3.1.4 describe the most common type of attacks against the confidentiality.

16.3.1.1 Eavesdropping

The most common type of attack against confidentiality is eavesdropping. It consists of listening to a private communication, being applicable to phone calls, e-mails, instant messages, and so on. It is an attack against the information in transit, without interfering with it. Eavesdropping is depicted in Figure 16.1.

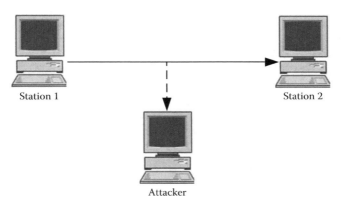

Station 1   Station 2

Attacker

FIGURE 16.1   Eavesdropping.

Since the eavesdropping action does not disable the legitimate destination of the message to have access to it (it is passive), it is difficult to detect. In case of data communications, it is also referred to as *packet sniffing* or only *sniffing*. The execution of such type of attack requires that the network interface card (NIC) allows receiving data (packets) whose destination is a different host. This is possible when a LAN is based on a hub and a host's NIC is configured in promiscuous mode. In case of a hub, all hosts' NIC receive all packets, and the NIC only processes the frames whose destination MAC address is the own MAC address (in such a case, the data is transferred to the host). Note that the execution of packet sniffing in a LAN using a switch becomes more difficult, as this device only forwards frames to the output port where the destination host is located. Note that eavesdropping can also be used to gather information that is useful to carry out a more complex network attack. For example, an eavesdropper may identify usernames and passwords, which is useful to permit access to other resources. Moreover, an eavesdropper may also rubber other important data such as credit card numbers. Finally, since simple network management protocol (SNMP) version 1 (SNMPv1) comprises the exchange of clear text strings, an eavesdropper may easily get control of network devices. This can be mitigated by using SNMPv3, which comprises the exchange of ciphered strings, instead of clear text.

Another method that can be used to execute an eavesdropping attack is the MAC address flooding. This aims to make a switch acting as a hub. Then, one simply has to make the NIC functioning in promiscuous mode and a trivial packet sniffer allows getting access to the characters exchanged through the network. The MAC address flooding requires that an attacker starts flooding the switch with frames whose multiple different source MAC addresses are fictitious. This makes the switch creating entries into the MAC address table for each different source MAC address. Since this table has a finite length, after some time, the switch is performing as a hub, that is, it starts sending frames through all output interfaces, except the one from where the frame came from. A mechanism that can be implemented to mitigate the MAC address flooding relies on assigning static MAC addresses to the different switch interfaces.

Another scenario where eavesdropping can be exploited is the wireless environment. A wireless access point transmission can be received by any NIC that has joined a certain wireless LAN. Note that a wireless access point acts as a link to an Ethernet hub, and the wireless devices can monitor the clear mode frames as long as a cipher is not utilized (open) or when the cipher key is common to all wireless devices (wired equivalent privacy [WEP]).* At this stage, a user can modify the NIC configuration to promiscuous mode and, with the aid of a network monitoring software (packet sniffer), the exchanged data can be intercepted. Then, username, passwords, and other important data can be obtained by a hacker.

Routers, switches, firewalls, and other network devices are commonly remotely controlled through telnet. In this case, the use of a network monitoring software gives access to the clear text configuration characters that are exchanged through the network. This may allow a user to enter a piece of equipment, modify its configuration, and control the whole network. This is commonly referred to as *telnet attack*. While consisting of an internal attack, this may allow proceeding with an attack from an external network. Naturally, the network equipment is even more vulnerable in case the privileged password of a Cisco systems device is sent in clear text (enable password *password*), instead of ciphered (enable secret *password*). A commonly adopted solution to mitigate this vulnerability relies on using a secure connection with the network equipment based on secure shell (SSH), instead of telnet (clear text). Naturally, this requires that the equipment supports SSH, either version 1 (SSHv1) or version 2 (SSHv2). SSHv1 ciphers the link using the Rivest–Shamir–Adleman (RSA) asymmetric protocol, whereas SSHv2 ciphers the link with a combined cryptography, using a session (symmetric) protocol (DES, 3DES, or AES), and an asymmetric cryptography that is simply employed to send the symmetric session key to the

---

* As can be seen from Section 16.9, in case Wi-Fi protected access (WPA) is in use, since the cipher key is different for each wireless device, the clear mode frames cannot be monitored by other users.

receiving party. In case of a Cisco router, this can be configured as an SSH server (using version 2) with the following syntax:

- Router(config)#ip domain-name example.com
- Router(config)#crypto key generate rsa
- Router(config)#ip ssh version 2
- Router(config)#line vty 0 4
- Router(config)#transport input ssh[*]

The DHCP starvation is another widely used method to execute eavesdropping. In this case, an attacker activates a fake DHCP server in an internal network segment. When a client makes a DHCP request in this network segment, the fake DHCP servers respond before the legitimate one, assigning an IP address, subnet mask, DNS server, and a default gateway. In this case, the default gateway address corresponds to the IP address of an attacking host, making all traffic to be received by an attacker.

The Cisco discovery protocol attack can also be performed to execute eavesdropping. It is a layer 2 Cisco proprietary protocol that allows Cisco equipment to meet the neighbors, thereby receiving data such as IP address, type, and equipment model, software version, and native VLAN. This data can be used to facilitate the entry into the neighboring equipment by modifying its configuration. This type of attack can be mitigated by deactivating the Cisco discovery protocol.

Eavesdropping can also be performed by other means. It can be executed by making use of a special equipment (e.g., a directional antenna and receiver) that allows receiving the electromagnetic waves generated by local devices (e.g., by a monitor or a keyboard, as the cable acts as an antenna that emits electromagnetic waves). Eavesdropping can also be implemented using a laser pointed toward the glass of a window that allows listening of a conversation taking place in a room. Eavesdropping can also be executed by listening to a WLAN.

The most important mechanism employed to protect from an eavesdropping attack is the use of cryptography and ciphered access passwords (e.g., challenge handshaking authentication protocol [CHAP]).

16.3.1.2 Snooping

It consists of prying into the private affairs of others, especially by prowling about. The purpose is to read an e-mail content, listen to a telephone talk, or observing a password of a PIN code of others, without authorization. It differs from eavesdropping as the information is not in transit.

More advanced snooping capabilities include the remote monitoring of activities of a resource (e.g., a workstation), or reading of passwords. This can be performed by using a keylogger program installed in the target workstation that automatically sends the captured information through the network to the attacking host (e.g., using an automatic e-mail). Protective measures such as firewalls, IDSs, and auditing tools are normally very effective to protect from snooping.

16.3.1.3 Interception

It consists of the ability to gain access to a private communication, but assuming an active behavior, acting as the source or the destination of a message. The most commonly known type of interception is the *spoofing*, where an attacker acts as the source of a data communication (source spoofing) or as the destination of the data communication (destination spoofing). Both source spoofing and destination spoofing are depicted in Figure 16.2.

This type of attack requires that the attacker is able to authenticate himself as a third party. The robbery of the authentication (e.g., password) can be initiated by other means such as phishing.[†]

---

[*] Alternatively, the command line "transport input all" allows both SSH and telnet, whereas the command line "transport input telnet" only supports telnet.

[†] Phishing consists of inducing a user to contact the wrong entity. This can be performed by sending an e-mail asking for username and passwords.

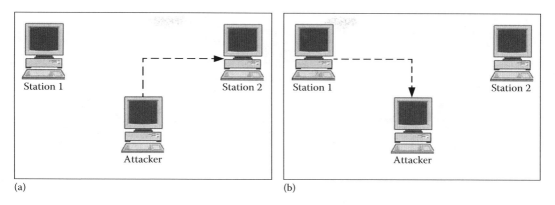

(a)   (b)

FIGURE 16.2   Spoofing: (a) source spoofing and (b) destination spoofing.

The illegal authentication can also be achieved by performing an attack to a DNS server, through the modification of the name's resolution. The illegitimate access to a third party bank account using the e-banking can be viewed as a type of source spoofing.

16.3.1.4   Trust Exploitation

The trust exploitation aims to gain access to a corporate internal network by compromising a host in the demilitarized zone (DMZ)* (see Figure 16.3). This type of attack can be implemented using port redirection. In this case, the compromised host located in the DMZ modifies the destination port of segments and redirects traffic to the internal network, crossing the firewall.

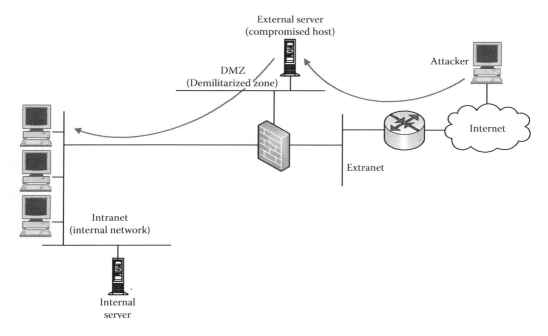

FIGURE 16.3   Trust exploitation.

---

* The reader should refer to Section 16.11.3 for a detailed description of a DMZ.

16.3.2 INTEGRITY

Integrity aims to avoid accidental or malicious modification of information or data without knowledge of its rightful owner. This service has two distinct objectives:

- To ensure that the information or data cannot be added, modified, or erased by unauthorized subjects.
- In case information or data has been added, modified, or erased, the service includes the necessary measures to alert the owner that it is not in the original state.

The key to implement a protective system from attacks against the modification of information or data relies on adding to the content, mechanisms that reveal its modification. This is typically achieved by using a Hash function. The Hash function performs a known processing to the content of a message, and the result of this Hash processing (i.e., a Hash sum that consists of, for example, 128 bit string) is added to the message. Once the message is received, the destination applies the same Hash function to the received message. If the message had been received unchanged, the result of this Hash processing is the same as the one received alongside with the message. Note that the Hash function is a one-way processing. This means that it is not mathematically possible to obtain the message from the Hash sum.

An example of an open source Hash function is the md5sum function. Note that an attacker may be smart enough such that he modifies both the message and the corresponding Hash sum transmitted together with a message. In order to improve the resistance from this advanced attack against the integrity, either the message or the Hash sum should be ciphered. In order to save processing (and time), the Hash sum is normally ciphered (due to its lower length). If the sender ciphers (encrypts) the Hash sum with its asymmetric private key,* the receiver is assured that the message has been sent by its legitimate sender (as the private key is only known by the legitimate sender†). This is the concept of digital signature, which comprises the use of a Hash function and its encryption with the sender's private key. The digital signature provides two levels of security protection [Cross et al. 2003]:

- Ensures the identification of the producer of the message (authenticity)
- Ensures that the signed message was not modified (integrity)

The digital signature is the digital equivalent of a handwritten signature. Note that a handwritten signature does not provide any information to the receiver about the integrity of the message, whereas the digital signature does.

Similar to confidentiality, the effectiveness of the protective measures from attacks against the integrity depends on the combination of a Hash function with the accountability service (access control, authentication, identification, etc.), namely to prevent the creation, modification, or deletion of information and data (e.g., permissions to write data).

16.3.2.1 Man-in-the-Middle

The most common attack against integrity is the man-in-the-middle. It comprises the modification of the content, such as a financial transaction. In this case, the attacker captures the bits in transit and modifies it. Such modification can be the destination bank account of a financial transaction. This type of attack is depicted in Figure 16.4.

---

* Note that an asymmetric cipher comprises a private key and a public key. A message can be ciphered with one of these keys, being deciphered with the other. Nevertheless, for the sake of confidentiality protection, a message is normally ciphered with the destination's public key, being deciphered with the private key. The public key is broadcasted to all potential participants in a communication, whereas the private key is only known by its owner.
† There is no possible processing that allows obtaining a private key from the public key.

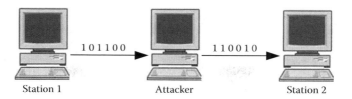

FIGURE 16.4 Man-in-the-middle attack.

This attack is difficult to execute, as the attacker has to be in line between the source and the recipient. Moreover, the attacker must have software that allows him to take charge of the session, such that the legitimate parties do not detect.

An attack against the integrity may also consist of adding or removing words from a message. Sometimes, adding or removing part of a message may change dramatically the sense of a message.

It was above described that since a wireless access point acts like an Ethernet hub, the wireless access point transmissions can be received by any computer NIC that has joined a certain wireless LAN. Note that a wireless access point acts as a link to the Ethernet hub, and the wireless devices can monitor the clear mode frames as long as a cipher is not utilized (open) or when the cipher key is common to all wireless devices (WEP).* With the NIC in promiscuous mode and with the aid of a network monitoring software (e.g., Wireshark) the exchanged data can be intercepted. Moreover, a computer can also be modified to act as a wireless access point for the legitimate computer, and to act as the legitimate computer for the wireless access point. Then, the exchanged data can be modified as desired.

16.3.3 Availability

Availability aims to ensure authorized access to information or data in a timely manner. The effect of attacks against the availability can be mitigated by employing redundancy that is activated in case an attack is detected against the primary system. This may include redundant storage, redundant communications, or redundant servers, as well as disaster recovery plans.

The attacks against the availability are often carried out with the transmission of viruses (spread with human intervention) or worms (which self-propagate through the network). The target of an attack can be an application, a service, a system, a network, bandwidth, and so on. This type of attack may consist of bandwidth consumption, resources consumption (e.g., server), or even the interruption of support infrastructure.

Depending on the target, there are different protective measures that can be undertaken. In 1999, during the North Atlantic Treaty Organization (NATO) intervention in Serbia (due to the province of Kosovo), both parties used the Internet to carry out computer attacks [Klein 2003]. In this conflict, the Serbs performed a DoS attack against the NATO site (an attack on service availability). This attack stopped the NATO web and mail servers, by filling it in with empty messages. The aim of this campaign was to prevent NATO to send out messages to the world. Meanwhile, the Serbian air defense systems were crippled by a cyber attack. These actions often arise spontaneously, by citizens of states or directly executed by the organizations involved in a conflict.

16.3.3.1 Denial of Service

The most common type of attack against availability is the DoS. The purpose is to deny the target the ability to use the applications, services, resources, and so on. The DoS concept is depicted in Figure 16.5.

---

* As can be seen from Section 16.9, in case WPA is in use, since the cipher key is different for each wireless device, the clear mode frames cannot be monitored by other users.

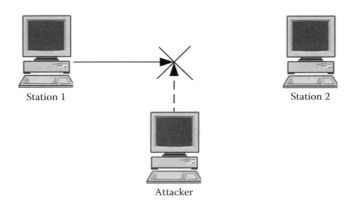

FIGURE 16.5 Generic diagram of a denial of service attack.

A common type of DoS attack is the SYN flooding attack. As can be seen from Figure 16.6, it consists of the emission of consecutive requests to establish transmission control protocol (TCP) connections (SYN messages), forcing the target to keep responding with SYN ACK messages. In this case, the target is flooded with such handshaking procedure, and reaches the maximum number of simultaneous TCP connections possible to be handled by a server. As described in Chapter 9, the client should finalize the three-way handshaking of the TCP connection establishment by sending an ACK message. In this case, such message is not sent by the attacker, leaving the connection semi-opened. This makes the server becoming confused.

Another type of DoS is the smurf attack. As can be seen from Figure 16.7, the attacker pings a remote network* composed of a wide range of hosts, while showing the IP address of the target device as a source address in the ping packet (instead of its own IP address). Therefore, it forces each one of the hosts in the pinged network to respond to the ping request, with ping response packets to the target device. This target host is flooded and overloaded with packets, being placed out of service.

A DoS attack can also be of distributed mode (distributed DoS [DDoS]). The most common DDoS is the botnet. In a botnet attack, an attacker typically spreads out a worm file along different hosts. The worm is typically propagated in chain along the network, and the target is typically a server. These intermediate hosts are known as *zombies*, and can comprise as much as hundreds or even thousands of hosts.

The attack is normally triggered (preprogrammed by the worm) to take place at a certain time, being executed against the target server. The attack can be of SYN type, smurf attack, or any other (e.g., buffer overflow). In the botnet attack, the attacker is difficult to be identified

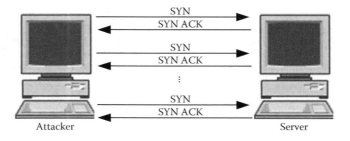

FIGURE 16.6 SYN flooding attack.

---

* Using the broadcast IP address as the destination IP address of the ping packet.

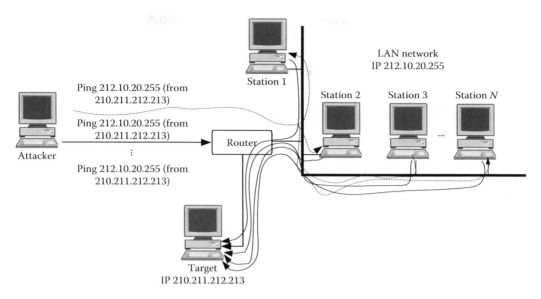

FIGURE 16.7 Smurf attack.

(see Figure 16.8). Note that, most of the time, these third-party hosts do not even have knowledge that they are materializing such attacks.

A botnet attack against the White House web server took place on July 31, 2001 (entitled *Code Red*). The final attack was executed by sending successive ping packets to the White House web server [Klein 2003].

There are other types of DDoS attacks. The peer-to-peer attack can be viewed as a DDoS that uses a peer-to-peer service, which presents vulnerabilities. Such vulnerabilities facilitate the use of malicious tools, which allows implementing the DDoS in the peer-to-peer mode. A popular peer-to-peer service used in DDoS relies on the file sharing. The attacker takes advantage of conception vulnerabilities to control the clients, forcing them to disconnect from the legitimate masters, and trying to connect to the target machine.

In addition to the previously mentioned redundancy, a preventive protection also consists of keeping antivirus and other software* up to date.

16.3.4 AUTHENTICITY

Authenticity aims to ensure that the author of information or data is the declared author. On the other hand, nonrepudiation aims to ensure that the author does not come in due course to deny the authorship of an action. The mechanisms that better ensure authenticity and nonrepudiation consist of using the digital signature. Therefore, the same protection mechanism (digital signature) can be applied for both protection from attacks against integrity and authenticity. This is achieved by ciphering the Hash sum (normally used for integrity) with the sender's asymmetric private key. The receiver decodes the Hash sum with the sender's public key and compares it against the result of the Hash function applied to the received message. If the decoded received Hash sum equals the Hash sum obtained from the received message, then the receiver may conclude that the message was not modified (integrity) and that the sender is legitimate (authenticity). If the two Hash sums are not equal, then one or both of the attributes (integrity or authenticity) are illegal.

---

* For example, operating systems.

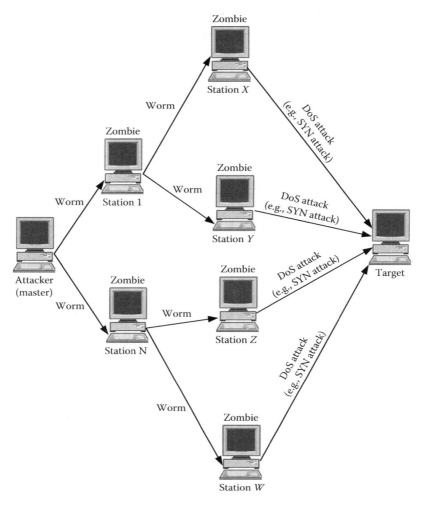

FIGURE 16.8 Botnet attack.

Note that a variety of mechanisms is available that can be used to protect from attacks against authenticity. These mechanisms include the use of authentication mechanisms, such as username and passwords, smart cards, biometrics, access control techniques, as well as several variations and combinations of these mechanisms. A widely used authentication protocol is the challenge handshaking authentication protocol (CHAP) that is adopted by the point-to-point protocol, being executed during the connection establishment and periodically (from a server request). Once the authenticator server has received a request to connect from the client, the CHAP is executed as follows (three-way handshaking):

- *Challenge*: The server sends a variable length random stream.*
- *Response*: The client sends to the server the Hash function applied to the digest. The digest corresponds to the concatenation of the challenge (previously received from the server) with the password. This Hash sum is sent to the server, together with the username.

---

* This is called the *challenge*, being also generically referred to as *token*. The challenge provides protection from a replay attack.

- *Success or failure*: The authenticator server verifies the response from the client against its own calculation. If the response fits with its own calculation, the server sends a success message to the client, otherwise a failure message is sent.

Note that the CHAP requires that both the client and the server know the plaintext password, although it is never explicitly exchanged. The procedure of encrypting the password with a Hash function is widely used. An example of its use is the authentication process of computers.

While not perfect, the CHAP is much better than its predecessor password authentication protocol (PAP), where usernames and passwords are exchanged in clear. Note that the CHAP considers periodic authentication, whereas the PAP only requires authentication during the setup phase.

Other authentication procedures exist where the need for the server to have knowledge about the clients' passwords is avoided.[*] These procedures provide protection from robbery of a passwords file, which is a common type of attack. It is worth noting that most of the attacks come from the interior of the organizations. Therefore, not keeping clear passwords in an authentication server is one important principle of authentication process.[†]

Nonrepudiation is normally associated with authentication measures combined with registration, such that the user cannot, in the future, deny that a certain action was performed by him. Let us consider a user that places a bank transfer order. With measures such as authentication combined with registration (accountability), the user cannot reject the authorship of such bank movement, as the bank keeps a proof, in a record, that such movement was undertaken by the subject.

It is worth noting that preventive measures from attacks against authenticity should also include mechanisms to prevent from using a nonlegitimate server. This attack is normally achieved with a phishing attack (typically an e-mail asking the user to provide username and passwords) or an attack to a DNS server.

16.3.4.1 Replay Attack

In the replay attack, the attacker has access to the bits in transit and records the initial handshaking, which comprises the authentication procedure. Later on, the attacker sends the same sequence of bits to the authenticator machine (e.g., to a server). This can be seen from Figure 16.9.

If the password is sent in clear mode, the replay attack is easy to be achieved. On the other hand, since the CHAP sends the ciphered digest (concatenation of the password and the challenge,

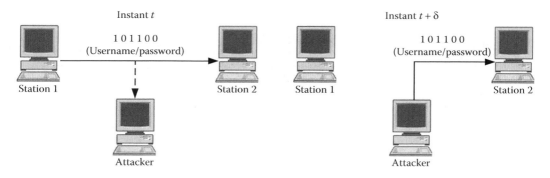

FIGURE 16.9 Replay attack.

---

[*] Instead of ciphering the digest with a Hash sum (as used by the CHAP), a possibility may rely on ciphering the challenge with the result of the Hash function applied to the password. In this case, the authenticator server only needs to store such Hash sum (not the plaintext password), and the authentication computation relies on ciphering the challenge with such Hash sum, followed by a comparison, at the receiver side, with the received sequence.

[†] An authentication server should only keep ciphered passwords (with Hash function).

whereas the challenge consists of a random stream that varies with time), replicating the ciphered digest does not lead to a successful authentication process.

The digital signature by itself does not give protection from a replay attack, as it does not comprise authentication. It only presents authenticity mechanisms,[*] that is, it only ensures that a certain message was generated by a subject. In the context of a replay attack, if such message is resent over the network, the destination party does not have any mechanism to identify this action.

Nevertheless, the authentication protocols of security architecture for IP (IPsec), secure sockets layer (SSL), and transport layer security (TLS) include mechanisms that enable authentication and protection from a replay attack.

16.3.5 Accountability

These security services complement the basic security services (see Table 16.1). The accountability services do not bring any direct added value, but its inexistence represents a great reduction in the effectiveness of the confidentiality, integrity, and authenticity services. The main purpose of this service is to ensure that the entity (user or process) is the legitimate one, and that the events log revel the truth of the facts. These complementary services require using additional resources such as processing, memory, and bandwidth. Using these complementary services as protective mechanisms, an attack becomes more difficult to be implemented, and possible to be registered and traced *a posteriori*. Therefore, a successful attack against the confidentiality or integrity needs also to include an attack against the authentication process, as well as against the registration (events log). A lack in one of these complementary functions represents a lack of the basic security services.

Accountability can be broken down into two groups:

* I3A—Identification, authentication, authorization, and access control (see Figure 16.10)
* MRA—Monitoring, registering, and auditing

I3A intends to find out who is an entity, to prove its authenticity, check its permissions, and allow or deny access to resources. The MRA aims to register events, errors, and accesses; to identify who performed those actions; and to register the moment when those actions were executed. To be effective, it requires an exact log of events, as well as a precise timing reference used in registration.

These individual services are detailed in the following subsections.

16.3.5.1 Identification

The identification function intends to know who the entity (user or process) is that wants to perform a certain action. The identification is normally associated to the authentication (defined in the following). The combination of these two complementary security services are referred to as I&A.

16.3.5.2 Authentication

The authentication function intends to prove that the entity (user or process) who wants to perform a certain action is legitimate, and effectively corresponds to the claimed identification. Note that

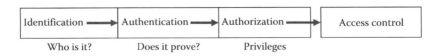

FIGURE 16.10 I3A functions.

[*] Note the difference between authentication and authenticity. The former is a process, whereas the latter is an attribute.

authentication differs from authenticity, as the former is a process, whereas the latter is a security service. The following are different mechanisms used to implement I&A:

- Something that can be known (password and PIN)
- Something that can be possessed (smart card and token*)
- Something that can be seen, that is, biometrics (speech, eye, and footprint)

A more effective I&A process includes a combination of two or more of these mechanisms.

Authentication may use the ciphered password with a Hash function. This procedure prevents the authenticator machine from having to store passwords in clear. This way, an intrusion attack against such machine does not enable him to obtain such password(s). It is also worth noting that it is widely known that most of the attacks come from the interior of the organizations. Therefore, preventing employees from having access to passwords is a basic principle. Note that computers normally store passwords ciphered with Hash codes. The user inserts the password, which is then ciphered and compared against the ciphered password (which is stored in the computer). In this case, if someone wants to forge the authentication, even though if he can reach the file with ciphered passwords, he cannot succeed to authenticate itself. With such mechanism, an attacker may employ one of the following possibilities, in order to forge an authentication using passwords:

- Attack using words from a dictionary
- Attack with variations in the user's names
- Brute force attack

Depending on the system, an attack based on a dictionary may take few minutes. An attack with variations of the user's name may also be quick and easy to achieve. A brute force attack consists of trying all different possible characters, with different lengths. The success of such attack is warranted, as long as there is enough time to perform such attack. Depending on the processing speed, a brute force attack against an authentication system using a password with eight characters may take as much as several months.† Naturally, reducing the password's length leads to a successful brute force attack in much shorter time frame.

16.3.5.3 Authorization

It consists of a complementary security service that is used after I&A. It establishes the privileges for each different entity. It depends on the organization's plan and policy, and defines who may have access to what. Similar to I&A, compromising the authorization represents a compromise of a basic security service.

16.3.5.4 Access Control

It is a complementary security service used after the authorization service. Whenever an entity intends to have access to a resource, it verifies if the entity is in the access list, allowing or denying access to resources. The most important network device that performs access control is the firewall. The access control can be implemented in mandatory mode (mandatory access control) or in discretionary mode (discretionary access control).

16.3.5.5 Monitoring

Monitoring consists of an activity that keeps track of the actions that are being undertaken, including the identification of who does what, and when.

---

* A token consists of an authentication mechanism, such as a hardware device (e.g., a card) which, when connected to, for example, a computer, allows the user to utilize such resource.

† Note that the time for a brute force attack increases exponentially with an increase in the password's length.

16.3.5.6 Registration

Registration consists of an activity that registers the result of the monitoring activity.

16.3.5.7 Auditing

Auditing consists of an activity that should be implemented periodically to evaluate the effectiveness of the security plans and policies in the network. In addition, an audit can also be performed after the detection of security violations, attacks, intrusion detections, and so on. The purpose of this activity is to identify the authorship of an attack, its consequences, which mechanisms or procedures could have been undertaken to prevent such attack, to identify the procedures that failed, and so on.

16.4 MALWARE

The three most important types of malware are virus, worm, and Trojan horse.

A virus is a piece of software that aims to execute a certain malicious function on a host. When a virus is executed, it infects the host by running a certain unwanted code. Note that a virus requires an action to propagate.

A worm is a malware that performs the same functions of a virus, but whose main difference relies on its inherent ability to auto-propagate. Therefore, when the unwanted code is run, this piece of malware not only infects the host but also automatically propagates to other hosts (e.g., using the list of address stored in the host).

Finally, a Trojan horse is a virus or a worm-like malware, with the difference that it looks like any other application. For example, a PowerPoint presentation may have an embedded Trojan horse. When a user runs the presentation, the unwanted code is executed.

It is finally worth noting that a piece of malware can be built to either execute a certain type of attack (against confidentiality, integrity, authenticity, or availability) or gather information (reconnaissance attack).

16.5 PHYSICAL AND ENVIRONMENTAL SECURITY

In addition to conventional attacks against confidentiality, integrity, authenticity, or availability, an attack against the physical network infrastructure or against the environmental control subsystem may deny its utilization. Depending on the method utilized to compromise the network device, the physical and environmental threats can be classified into four classes:

- *Hardware*: It corresponds to denying the utilization of a certain equipment or group of equipment (system), by physically damaging it. This can be mitigated by monitoring and registering the accesses to areas where equipment are located (physical access control and video cameras). Note that an access to an area can also be perpetrated through nonconventional accesses, such as windows, roofs, and pipes. Blocking these nonconventional accesses are important actions.
- *Electrical system*: An equipment or system can be placed out of order by affecting the supply of electrical power. This can be mitigated by remotely monitoring the electrical supply, with alarms. Moreover, the utilization of redundant electrical circuits, uninterrupted power supplies, and emergency generators is important to keep the survivability of equipment and systems.
- *Environment control*: It corresponds to denying the utilization of a certain equipment or group of equipment (system), by affecting the necessary environmental conditions, such as temperature, humidity, and air flow. This can be mitigated by keeping systems of physical access control and video cameras, not only to the equipment room but also to the room

from where the environmental conditions are controlled. Keeping a log of the personnel that accessed the rooms is important to identify the perpetrator, in case an incident occurs. Moreover, it is important to keep a remote monitoring of the environmental operating conditions, with alarms in case of failure.

*   *Installation and maintenance*: The use of an equipment or subsystem can be denied by interrupting the cabling that interconnects different equipment or by physically accessing the console port of a router, switch, or firewall (e.g., modifying the privileges). Moreover, the access to a rack where the patch panel is installed may also deny the utilization of terminals. When installing and maintaining equipment and cablings, it is important to ensure that these assets are not accessed by unauthorized personnel. Finally, keeping cablings and connectors with the proper identification (labels), clean of dust, and in accordance with the installation rules are important actions to keep the survivability of systems.

16.6   RISK MANAGEMENT

In order to maintain access to own information systems, while protecting them from attacks, there is the need to execute the appropriate INFOSEC measures and to maintain an effective control of CIS systems. Therefore, it is first necessary to identify the goods to protect (software, router, firewall, server, etc.), as well as their values (impact in case such good is lost). Second, it is necessary to determine the threats that are likely to occur, namely threats against confidentiality, integrity, availability, and authenticity of information. A threat consists of a potential action or an event that might violate the security of information systems. Third, it is necessary to determine the vulnerabilities and the risks. Vulnerability consists of a potential route of exploitation to the threats. Malicious software can be viewed as a threat, whereas an operating system without the appropriate patches can be viewed as a vulnerability.

A risk is generally defined as a potential loss of assets and values, which are subject to threats that exploit vulnerabilities in a system, organization, or persons. Risk management comprises a process, for the risk monitoring. A risk is often mathematically quantified as

$$\text{Risk} = \text{threat} \times \text{vulnerability} \times \text{value of goods} \tag{16.1}$$

With regard to the value of a good, this does not refer to its monetary value, but consists of the impact to the organization, in case such good fails as a result of an attack.

Risk management should also include the identification of the mitigation measures that can be applied to minimize the risk, as well as the establishment of foundations for a security plan (see Figure 16.11). Each risk is associated to a certain probability and impact, which needs to be noted at this stage. The residual risk corresponds to the initial risk, to which the countermeasures are subtracted (patches, firewall, IDSs, redundancy of systems, cryptography, procedures, policies, etc.). There are measures that aim to minimize the vulnerabilities of information systems (patching, hardening, training, etc.) and measures to protect from threats (firewall, cryptography, backups, etc.). As can be seen from Figure 16.11, geometrically, the risk corresponds to the intersection area between vulnerabilities, threats, and value of good.* Moreover, risk mitigation is achieved through the minimization of vulnerabilities and threats. The residual risk corresponds to the acceptable risk, after the application of the selected mitigation measures, being a top management decision of the organization.

---

* A mathematical multiplication corresponds to a geometrical intersection.

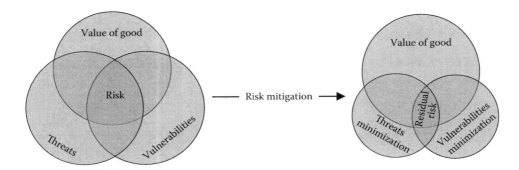

FIGURE 16.11   Information security risk and its mitigation.

16.7   SECURITY PLAN

The security plan consists of a document where the security policy is defined. The security policy comprises a number of procedures, rules, behaviors, and technological mechanisms that must be utilized or followed in order to ensure that the INFOSEC is assured, and that the technological infrastructure is kept in the appropriate form. The security plan aims to protect from attacks against confidentiality, authenticity, and availability.

The security evaluation phase precedes the establishment of a security policy. It comprises the identification of the risks, through the quantification of the value of the goods, the vulnerabilities, and the threats (see Section 16.6). A security policy must include the protective mechanisms and procedures adopted to mitigate the security risk.

A security policy is implemented making use of a security plan, a number of technological mechanisms (firewalls, IDSs, or intrusion preventive systems [IPSs], antivirus, physical access controls, etc.), procedures, rules, and behaviors.

A security plan should be developed by an external consulting company specialized in INFOSEC, with the involvement of the information technologies (IT) department, and of the top-level administration. Note that a security plan should not be developed by the IT department itself, but this department should be included in the development of a security plan. The mitigation of vulnerabilities and threats involves a financial cost and a number of actions and behaviors. This also translates in reducing the ease of use of information systems, that is, functionality. Therefore, a tradeoff needs to be achieved, in terms of security and functionality. This decision must be made by the top-level administration, after an initial analysis by the IT department, and under proposal of an external consulting company in charge of developing the security plan. Therefore, a security plan includes the acceptance of a number of vulnerabilities and threats (those left after the application of the mitigation measures), whose decision is made by the corporate top-level administration. This is the reason why the top-level administration needs to be involved in the development of a security plan.

The above description refers to the security policy in a broad sense, where different policies are included (information policy, remote access policy, e-mail policy, etc.). Using a restrict approach, the security plan comprises a document where multiple policies are defined:

*   *Information policy*: The establishment of an information policy starts by classifying the information in different security levels, and defining the procedures that must be followed in processing, storing, transmission, and destruction of the different security levels of information. Then, the information policy defines the level of security that each corporate employee can have access to, keeping in mind that such access to information must be based on the *need to know* principle. This means that someone authorized to have access

to, for example, confidential information only accesses it in case he needs it for a certain specific task.

- *Security policy*: The establishment of a security policy defines a number of actions, mechanisms, equipment, and procedures utilized for the minimization of the vulnerabilities and the threats against confidentiality, integrity, availability, and authenticity. This includes the establishment of logs (network logging system), identification, authorization, and access control mechanisms (e.g., card reader or fingerprint reader), the interconnection of networks (e.g., between an intranet and the Internet using a firewall), the protective measures against malware (e.g., antivirus), the cryptography, IDS/IPS, and the remote access to an internal corporate network.
- *Computer usage policy*: This policy defines the procedures associated with computers' passwords (e.g., length and characteristics of passwords, modification intervals, and time for a computer without action to request the password from the user), the privileges given to a user to install new software, the level of privacy relating to information stored or processed in the computer, and the definition of the propriety of the information stored or processed in the computer. Moreover, it defines the procedure and locations where a user is authorized to use a corporate laptop, and whether or not the information stored in a laptop must be encrypted.
- *Internet policy*: It defines whether or not a certain host is authorized to access the Internet, and to which types of sites and for which purposes (social networks sites, etc.).
- *E-mail policy*: This policy defines the type of information that can be sent via e-mails (e.g., security level), the level of privacy, and the propriety of information sent in e-mails, the type of attachments that can be sent via e-mails, the action required by a user in case a spam or phishing e-mail is received, and so on.
- *Printer policy*: It establishes the need to keep a log of the pages printed by each user, the number of pages a certain user is authorized to send to the printer, the print resolution and color definitions that must be utilized in printings, the printer that must be utilized for printing different classification levels of information, the consequences of leaving printings with classified information not picked up by the owner, and so on.
- *Backup policy*: It defines the type of information that requires being backed up, as well as its periodicity. Moreover, it defines where backups are kept and if backups of backups are implemented.
- *Remote access policy*: It defines whether or not a user can have full remote access to an internal corporate network, to corporate e-mails, and so on. Moreover, it defines the procedures and technological requirements implemented to allow such remote access, such as the type of authentication and encryption mechanisms to implement in the organization.
- *Information systems projects implementation policy*: It defines the procedures associated to processing, transmission, and storing of information systems projects information, such as requirements definition, risk analysis, testing, configuration management, and handover of systems. A corporate DMZ can be created, with a partner's VPN, to facilitate the access to a project contractor or subcontractor.
- *Incidents response policy*: It defines the procedures to be implemented when a certain security incident is detected. An incident can be a virus detected in a corporate computer, an intrusion in the DMZ or in the internal network, an unsecure transmission of an e-mail with classified information to outside of the corporate, and so on. Two different procedures can be adopted: contention or reaction. Contention aims to isolate an affected area, avoiding its propagation, for minimizing the damages. Reaction aims to combat the source of the affected system or equipment, for recovering from the incident. The incidents response policy defines who does what and when, in case an incident occurs. Moreover, it also defines the types of reports that must be filled out in case of an incident.

- *Disaster recovery policy*: This is normally associated with a document (disaster recovery plan), and aims to define the actions that must be undertaken by different entities to recover from a disaster when it occurs. An example of a disaster can be a flooding or an earthquake. A disaster recovery plan comprises different items such as an isolated equipment, systems, data centers, and full sites. Depending on the affected item and the type and depth of the disaster, two different approaches can be implemented: repair of the affected item (recovery plan) or full replacement (contingency plan). The disaster recovery plan may include, for example, a backup data center that becomes operational in case a main data center is placed out of order. Moreover, the disaster recovery plan defines the functionalities that must be kept operational, and those that can be left out of order.

A security plan can be created using a template, using a security plan of another organization as an example, or written from the scratch. Nevertheless, one must take into account the organizational culture, the specificities and type of organization, as well as the organization structure and its processes. The International Organization for Standardization (ISO) created the ISO 27000 series of standards specifically reserved for information security matters. The ISO 27001 standard specifies an information security management system. Organizations that meet this standard may be accredited by an independent authorized institution. Moreover, ISO 27002 consists of a code of practice for information security, outlining hundreds of potential controls and control mechanisms, which may be implemented, in theory, subject to the guidance provided within ISO 27001. The ISO 27002 standard includes the following topics:

- Organization of information security
- Human resources security
- Access control mechanisms
- Cryptography
- Physical and environmental security
- Communications security
- Information systems acquisition
- Development and maintenance
- Information security incidents management
- Information security aspects of business continuity

Moreover, ISO/IEC 15408 defines a common criterion, establishing a group of procedures and mechanisms that assure a certain level of confidence in computers security (COMPUSEC).

It is worth noting that the creation of a security plan is not enough. It also needs to be implemented. Moreover, its implementation needs to be monitored, logged, and tested. Then, the security plan needs to be modified to take into account the results of periodical tests, or analysis that results from an incident. Therefore, it can be said that a security plan is a live document. Audits can be implemented as a result of an incident or in irregular periods. An audit may include intrusion tests, which is an important tool to evaluate the effectiveness of the security policy procedures and mechanisms.

It is worth noting that training needs to be given to personnel, taking into account their roles. Therefore, the training given to network administrators must be different from the one provided to users, and different from the training given to programmers. Training also draws the attention of people to the importance to adhere to security rules, acting as indoctrination.

16.8   PROTECTIVE MEASURES

Depending on the information attribute to protect, or security service, there are different protective mechanisms that can be implemented. Typically, confidentiality is protected with cryptography, whereas the digital signature provides protection from attacks against integrity, as well as against authenticity. Finally, redundancy is the keyword for allowing available protection. There are three main types of cryptography:

- Symmetric
- Asymmetric
- Combined

The symmetric cryptography uses the same key for ciphering and deciphering the messages, using one or more predefined number of elementary operations based on substitution and transposition. It requires the previous distribution of keys, which presents a certain level of vulnerability. Another cryptographic scheme, used very often by terrorist organizations, is called *stenography*. It consists of a technique that hinds messages within images [Klein 2003].

On the other hand, the asymmetric cryptography[*] makes use of two different keys, namely a private key and a public key. The private key is kept secret by its owner, whereas the public key is openly distributed. The advantage is that since these keys are different, the distribution of the public key does not involve a high risk. Asymmetric cryptography solves the vulnerability problem of key distribution that occurred in symmetric cryptography.

Using an asymmetric cryptography, a message is ciphered with the sender's private key, being deciphered with the sender's public key. Alternatively, a message can be ciphered with the receiver's public key, while deciphered with the receiver's private key. Nevertheless, as a rule of thumb, for a given message size, the processing time required to cipher and decipher a message using asymmetric cryptography is about a thousand times higher than that using symmetric cryptography, which is a disadvantage of asymmetric cryptography. The combined cryptography presents the advantages of the two previous schemes, while eliminating the corresponding disadvantages.

A combined cryptographic scheme requires the processing time of a symmetric cryptography, while solving the corresponding key's distribution vulnerability. The price to pay is a higher level of complexity, as it makes simultaneous use of symmetric and asymmetric schemes.

Sections 16.8.1, 16.8.2 and 16.8.6 describe each of these cryptographies.

16.8.1 SYMMETRIC CRYPTOGRAPHY

The development of an effective cryptography is a difficult task. Once such system is implemented, its successive operation relies on the use of different keys. Consequently, even though if an attacker has knowledge about the algorithm being used, as long as the key is unknown (secret), a successful attack against the confidentiality of a message (in transit, processing, stored, etc.) is not possible.

Symmetric cryptography has been used for centuries. One such example is the Caesar's cipher, which relies on a simple substitution of letters. Such technique is known as *mono-alphabetic substitution cipher*, where the key relies on the substitution of each alphabet letter by another randomly selected one. The effectiveness of such technique is limited, as long as the text length available to be analyzed by a cryptanalyst[†] is long enough. In fact, the most frequent letter for each language is usually known. When analyzing a long cryptogram,[‡] it is easy to find the most frequent letter, and its correspondence to the plaintext is therefore known. Then, such process can be repeated for the second-most frequent letter, and so on.

In order to solve the main vulnerability of the mono-alphabetic cipher, the polyalphabetic substitution cipher, generically referred to as *Vigenère cipher*, was introduced in the sixteenth century.

---

[*] Asymmetric cryptography is also referred to as *public key cryptography*.

[†] A cryptanalyst is someone who performs cryptanalysis. Cryptanalysis is the action of analyzing cryptographic algorithms (using intercepted cryptograms), for the purpose of identifying vulnerabilities and extracting the key.

[‡] A cryptogram is the sequence of text that results from the encryption of an original message (clear text). In other words, it consists of a message's ciphertext.

Such cipher requires a key that is used to convert the plaintext into the ciphertext and reciprocally. As a result of using a key, such cipher does not rigidly replace a certain letter by another fixed one. This improves the confusion capabilities of the cipher. Nevertheless, such cipher is also possible to be broken using computational capabilities. This cipher can also be improved by using long length key. In the limit, having a key length corresponding to the length of the message leads us to the *one-time pad*, whose resulting cryptogram presents the maximum confusion. Nevertheless, this requires a complex generation and distribution of keys. Since the key is as long as the message, the decoding process is a difficult task.

Afterward, symmetric algorithms started using a combination of two elementary operations:

- Substitution
- Transposition*

Since the Enigma cryptography machine used during World War II by Germany and the Purple machine used by Japan, until the currently used cryptography machines, they all work based on a combination of these two elementary operations. Substitution consists of replacing a character by another one (currently, this is digitally performed by applying a mathematic function to the bits). Transposition consists of changing the sequence of letters or bits within a message (as can be seen from Figure 16.12, write characters in columns and read them in rows). These operations are performed based on a key, which changes periodically.

Symmetric cryptography presents the advantage that the required processing for ciphering and deciphering messages is much lower than that of asymmetric cryptography. However, since the key used for ciphering and deciphering is the same, there is the need to distribute the keys along the potential users (senders and receivers of messages). This makes symmetric systems vulnerable.

Figure 16.13 depicts the generic block diagram of a symmetric cryptography. In this figure, SIMK stands for symmetric key. Moreover, Message??? represents the ciphered ciphertext version of the message.

G	O	O
D	–	M
O	R	N
I	N	G

FIGURE 16.12   Transposition. Cleartext: GOOD MORNING; Ciphertext: GDOIO_RNOMNG.

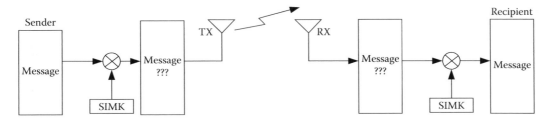

FIGURE 16.13   Generic block diagram of symmetric cryptography.

---

* Transposition is also known as *permutation*.

16.8.1.1 Symmetric Cryptographic Systems

Currently, most of the existing cryptographic systems are binary based. This means that, instead of implementing substitution of one character by another using a certain key, the ciphered data is the result of a logic function* applied to an input data, that is, applied to a block of the clear text message. Moreover, the binary transposition operation consists of changing the sequence order of bits (this operation is similar to the interleaving operation), whereas the opposite operation is performed at the receiver side. Most of the cryptographic algorithms perform complex iterative sequences of these two binary elementary operations.

As can be seen from Figure 16.14, the basic parameters of cryptographic algorithms are the block size and the key length. In this figure, plaintext data stands for clear data (before encryption), whereas ciphertext data refers to ciphered data.†

Moreover, the definition of the number and sequence of elementary operations (binary version of substitution and transposition) that compose the encryption operation is also an important issue of cryptographic algorithms.

Among the currently used symmetric cryptographic algorithms, there is the digital encryption standard (DES), the triple data encryption algorithm (TDEA or Triple DES), the advanced encryption standard (AES), the international data encryption algorithm (IDEA), the Blowfish, and the RC5 algorithm [Stallings and Brown 2008].

The DES algorithm presents a 64-bit block size and a length of 56-bit key. Its cryptographic operations rely on the application of successive transpositions to the *plaintext bits*. Due to its reduced key length, it is considered vulnerable to brute force attacks with existing computation power. Consequently, the TDEA algorithm was developed (commonly referred to as *3DES algorithm*). This comprises the same block size, whereas the key length presents an effective length of

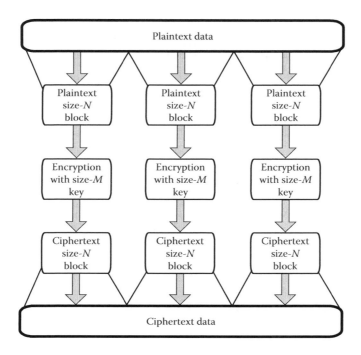

FIGURE 16.14 Division of plaintext data into blocks, before encryption operation is applied.

---

* For example, binary addition (OR), multiplication (AND), bitwise, and eXclusive OR (XOR).
† The whole ciphered message is commonly referred to as a *cryptogram*.

168 bits, corresponding to three times that of the DES. In fact, TDEA consists of the application of the DES cryptographic algorithm three times to each block of data. At first, encryption is performed with the first key, decryption with the second key and, finally, re-encryption is performed with the third key, before the message is sent. This results in a much more resistant cryptographic algorithm from brute force attacks, without having to develop a complete new one.

The IDEA algorithm was initially described in 1991, and it corresponds to an improvement of the proposed encryption standard. The IDEA algorithm presents a 64-bit block size and a length of 128-bit key, and is freely available for noncommercial use. The IDEA is used as a session (symmetric) key by the pretty good privacy (PGP) v.2.0, which is a combined encryption protocol.

Another symmetric algorithm is the AES. The AES presents a 128-bit block size, whereas the length of its key varies depending on the AES version. Three AES options exist based on the length of the key: 128-, 192-, and 256-bit, which correspond, respectively, to the AES-128, AES-192, and AES-256. This results in an algorithm that is resistant to brute force attacks. The AES algorithm was previously published as the Rijndael* algorithm, having been selected by the National Institute of Standards and Technology (NIST), in 2002, by the United States, for governmental use. It is worth noting that the AES was approved by the U.S. national security agency for encryption of data up to the classification of Top Secret.

The Blowfish algorithm is another symmetric cryptographic algorithm. It presents a 64-bit block size, with a variable key of length up to 448 bits, being easy to implement by software. The Blowfish algorithm was developed as an alternative to DES. Nevertheless, its encryption capabilities are much below those of the AES. Finally, the RC5 is a very attractive cryptographic algorithm because of its simplicity. Initially developed by Ronald Rivest, the RC5 corresponds to an update of the previous RC4 version. Note that the acronym RC (in RC5) stands for River Cipher. The RC5 has a variable block size of 32, 64, or 128 bits, whereas the key length is fixed to 255 bits. The RC5 is currently used as session (symmetric) key by combined protocols such as SSL or TLS.

16.8.1.1.1 Diffie–Hellman Protocol

The aim of this protocol relies on allowing the exchange of symmetric keys to be used in a transaction as session key, without their explicit exchange through the network. Therefore, this protocol mitigates the most important vulnerability associated to the symmetric cryptography: the distribution of keys.

The Diffie–Hellman protocol was invented in 1976, just after the presentation of the asymmetric cryptography concept. Note that in addition to its ability to implicitly generate a session key (symmetric key), the Diffie–Hellman protocol can be used as asymmetric cryptography, as an alternative to the RSA protocol. It is worth noting that this protocol is employed as an option by the SSL and TLS protocols (to define the session key), as well as by the Internet key exchange, which is the protocol used to establish the session keys to be used by the IPsec.

The processing associated with the Diffie–Hellman protocol is described as follows (see Figure 16.15):

- Choose a long prime number p.
- Choose an integer g, such that $0 < g < p$.
  - Note that p and g can be pre-known by different entities or can be exchanged in clear through the network. It is assumed that an attacker has knowledge about p and g.
- An entity that intends to initiate a connection (let us call her Alice) chooses a random number a, such that $1 < a < p-1$.
  - Alice computes and sends $A = g^a (\bmod\ p)$ to Bob (the second party of the connection).
- Bob chooses another random number b, such that $1 < b < p-1$.
  - Bob computes and sends $B = g^b (\bmod\ p)$ to Alice.

---

* Rijndael stands for the combination of the names of the two Belgian inventors: Joan Daemen and Vincent Rijmen.

FIGURE 16.15 Processing associated with the Diffie–Hellman algorithm.

- Alice computes the secret key (symmetric key) as $S = B^a (\text{mod } p) = (g^b)^a (\text{mod } p) = g^{ba} (\text{mod } p)$.
- Bob computes the same secret key (symmetric key) as $S = A^b (\text{mod } p) = (g^a)^b (\text{mod } p) = g^{ab} (\text{mod } p)$.

Note that the secret key generated by the two entities is the same, and the two parties generated it without their explicit exchange through the network.

Once we have determined the secret key S, it is time to describe both encryption and decryption, whose procedures are defined as follows:

- Take the whole characters alphabet and map each character to a different number m, such that $0 \leq m \leq p$.
- At the transmitter side, taken m, one can calculate the ciphertext c as $c = m \times S$, where S stands for the secret key.
- At the receiver side, from the ciphertext c, one can calculate m as $m = c \times S^{-1}$. In a second stage, one should consult the table that maps the variable m into the character M.

Note that the Diffie–Hellman protocol can also be used as asymmetric cryptography. In this sense, the variable A and B can be viewed as public keys, whereas the variables a and b comprise the private keys.

Before the description of the Diffie–Hellman protocol is finalized, it is worth analyzing the level of protection and the vulnerabilities associated with this protocol. An interceptor sees $A = g^a (\text{mod } p)$ and $B = g^b (\text{mod } p)$, but from A and B, he is not able to compute $S = g^{ab} (\text{mod } p)$, because he does not have knowledge about a and b. Therefore, it can be said that the Diffie–Hellman protocol gives protection from attacks against the interception. Nevertheless, as can be seen from Figure 16.16, a potential man-in-the-middle attacker (let us call him Eve) is able to receive A from Alice, remove it from the network, generate a', compute $A' = g^{a'} (\text{mod } p)$, and is able to forward it to Bob. Similarly, Eve can remove B received from Bob, generate b', compute $B' = g^{b'} (\text{mod } p)$, and can forward it to Alice. Now, Alice computes the secret $S' = B'^a (\text{mod } p) = g^{b'a} (\text{mod } p)$ and Bob computes the secret $S'' = A'^b (\text{mod } p) = g^{a'b} (\text{mod } p)$. These two entities are not able to communicate directly. Nevertheless, Eve has knowledge about both S' and S''. Therefore, Eve is able to receive the message from one party, decipher it with the corresponding secret key, modify the message, re-cipher it with the other secret key, and send it to the other party.

The solution to mitigate this vulnerability relies on the encrypting the handshaking procedure with an asymmetric cryptography.

FIGURE 16.16 Man-in-the-middle attack associated with the Diffie–Hellman algorithm.

16.8.2 Asymmetric Cryptography

The concept of asymmetric cryptography was initially proposed by Diffie and Hellman, in 1976. The main purpose of asymmetric cryptography relies on solving two main issues:

- Mitigate the limitations that results from the distribution of symmetric keys.
- Allow the implementation of digital signature, which provides mechanisms to protect from attacks against the integrity and authenticity.

Asymmetric cryptography is also referred to as *public key cryptography*. Since the public key cryptography makes use of a public key, which is widely distributed, and a private key, which is kept secret, there is no need to implement a complex and vulnerable key's distribution system, as required for symmetric cryptography. Moreover, since the encryption operation can be implemented by any of the keys[*] (private of public), this system is well fitted for both data encryption and digital signature.

The generation of public and private keys is mathematically based on the factorization of numbers into two prime numbers. The generation of asymmetric keys' pairs, as well as the ciphering and deciphering processes, is very computationally demanding, which translates in high time-consuming processes.

As a rule of thumb, for a given message size, the processing time required to cipher and decipher a message using an asymmetric cryptographic system is about a thousand times higher than using symmetric cryptography. As described in Section 16.8.6, the solution to this limitation is overcome by combined schemes, at the cost of a higher system complexity.

Figure 16.17 depicts the generic block diagram of an asymmetric cryptography. In this figure, PUK stands for public key and PRK stands for private key. Furthermore, PUK2 stands for the public key of the message recipient, whereas PRK2 stands for the private key of the recipient. Note that a message is ciphered by the sender, with the public key of the recipient, being deciphered with the private key of the recipient.

Although better than symmetric cryptography, the generation and distribution of public keys still presents some vulnerabilities, as there is the risk that someone distributes a public key from someone else, claiming he is the owner. As described in Section 16.8.5, the solution for this complex problem is provided by the public key infrastructure (PKI) and digital certificates.

16.8.2.1 Asymmetric Cryptographic Systems

There are different types of algorithms used to generate asymmetric keys' pairs. The RSA[†] is the most common one. RSA was the first publicly described asymmetric algorithm that allowed the implementation of asymmetric cryptography, as well as digital signature. The RSA1024 is a frequent

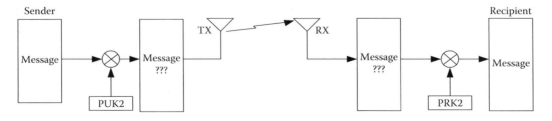

FIGURE 16.17   Generic block diagram of an asymmetric cryptography.

---

[*] As long as the deciphering operation is performed with the other key.
[†] RSA stands for Rivest, Shamir, and Adleman, the inventors' names of such algorithm.

version of the RSA algorithm, where 1024 stands for the key length. Currently, the RSA2048 is already available and recommended for many applications.

Another public key algorithm based on elliptic curves is available. Elliptic curves algorithm provides similar security protection results as those of the RSA, but with shorter key lengths. Consequently, this algorithm tends to be faster than the RSA.

16.8.2.1.1 RSA Protocol

The RSA protocol comprises a public key and a private key. The generation of the public and the private keys are as follows:

- Choose two different prime numbers p and q (should be of similar bit-length).
- Calculate $n = p \times q$. Note that the length of the RSA protocol (e.g., 1024 or 2048) corresponds to the length of n, expressed in number of bits. The value n is shown in the public key field of digital certificates, being presented in hexadecimal format, after the word *modulus*.
- Calculate $\phi(n) = (p-1) \times (q-1)$.
- Choose an integer e, entitled *public key exponent*, such that $1 < e < \phi(n)$, and such that e and $\phi(n)$ are co-prime.
- Calculate the private key exponent d, such that[*] $d \times e \equiv 1 \mod[\phi(n)] \Leftrightarrow (d \times e) \mod[\phi(n)] \equiv 1$.
- The set of the two variables e and n corresponds to the public key, whereas the set d and n corresponds to the private key.[†]

It is worth noting that p, q, and $\phi(n)$ must be kept secret, as these figures can be used to calculate d (the private key).

Once we have determined the public and the private keys, it is time to describe the encryption and decryption, whose procedures are defined as follows:

- Take the whole characters alphabet and map each character to a different number m, such that $0 \leq m \leq n$.
- At the transmitter side, taken m, one can calculate the ciphertext c as $c = m^e \mod(n)$, using the destination public key. Note that one can share its public key composed of the variables e and n.
- At the receiver side, from the ciphertext c, one can calculate m as $m = c^d \mod(n)$, using its private key. In the second stage, the receive party should consult the table that maps the variable m into the character M.

16.8.3 DIGITAL SIGNATURE

After receiving a ciphered text with a symmetric cryptography, the subject cannot be sure that the text was not modified. In case the message is a human readable text, changing the ciphertext while in transit implies that the result of the deciphering process does not lead to the original plaintext. In fact, this text modification may originate that part or the whole text is not readable. However, in case the message is not text, the detection of the integrity violation becomes more difficult. Such detection is even more difficult in the case of machine-to-machine communication. Therefore, another solution for preserving integrity needs to be implemented.

---

[*] $X \mod(y)$ corresponds to the rest of the division of X by y.
[†] The letter e stands for encrypt and d for decrypt, as normally employed in asymmetric cryptography.

The Hash function contributes to protect the message integrity from attacks, while notifying the receiver in case such attack is executed. It consists of an advanced checksum.[*] A Hash function needs to comply with the following basic properties:

- The function is a *one-way encryption*, that is, it only works in the *encryption* direction, as it is not possible to obtain the clear message from the Hash sum. Note that the Hash function is not exactly an encryption function; it is a well-known function that does not make use of a key. However, due to its similarities with cryptography, we may refer to this as an *encryption function*.
- Taking a message M and its Hash sum H(M), it is not computationally possible to find another message that presents the same Hash sum.[†]
- The Hash sums of two slightly different messages should be completely different.

Consequently, if one captures a Hash sum and intends to find its corresponding *clear text*, he needs to compute a brute force attack (try all different possible combination of characters, with different lengths, in a password request). Since the original text is normally very long, such computation is almost impossible.

The most known type of Hash functions are the MD5 (message digest) and the SHA-1 (secure hashing algorithm). The MD5 is widely used in the Internet, and it has a length of 128 bits. On the other hand, the SHA-1 has been standardized by the NIST, and it presents different options depending on the Hash sum length: 160, 256, 384, and 512 bits. Combined encryption protocols such as SSL and TLS comprise the digital signature using a customized Hash function and length (negotiated during the initial handshaking).

The Hash function is used as a protection mechanism from attacks against integrity as follows: the message's sender applies a Hash function to the original message and sends the message together with its Hash sum. The message receiver also applies the Hash function to the received message and compares the resulting Hash sum against the received Hash sum. If they are the same, one may, in principle, conclude that the message was not modified.

Since the Hash function is a well-known function, an attacker may capture the message, modify it, compute the Hash function of the modified message, and send the new message together with the new Hash sum. In this scenario, the receiver is not able to detect such message modification. The solution to this problem relies on encrypting one of the two elements: either the message or its Hash sum. For the sake of processing saving, it is preferable to encrypt the Hash sum, as its length (e.g., 128 bits) is typically lower than the message. Encrypting the Hash sum with the asymmetric sender's private key, allows it being deciphered, at the receiver side, with the sender's public key. Once the message and its ciphered Hash sum are received, the following operations are performed:

- Decipher the Hash sum with the sender's public key.
- Apply the Hash function to the received message.
- Compare the received deciphered Hash sum against the computed Hash sum.

If the two Hash sums are equal, then the receiver may take two conclusions:

- The message was created by its legitimate author.[‡] This corresponds to authenticity.
- The message was not modified. This corresponds to integrity.

---

[*] A Hash function may be implemented, for example, using shift registers.

[†] A Hash sum is a string obtained from applying the Hash function to a message. If a Hash sum is 128-bit long, the number of possible combinations for the Hash sum is $2^{128} \approx 3,40^{138}$, that is, the possibility of having two messages with the same Hash sum is very low ($1/2^{128}$).

[‡] As only the legitimate author knows, the private key is used to cipher the Hash sum.

Note that the digital signature mechanism, while not avoiding an attack against authenticity or integrity, allows its detection, in case the integrity or authenticity has been violated.

The above-described sequence of operations comprises the concept of digital signature, which is depicted in Figure 16.18, where PUK1 stands for the public key of the message sender, and PRK1 stands for the private key of the sender. Note that the sequence of data designated in Figure 16.18 as Hash ??? is sometimes referred to as *keyed hashed message authentication code*.

It is worth noting that the digital signature ensures authenticity (which is an attribute), while not comprising authentication (which is a process). Note that even though if a third party sends a previously recorded message (replay attack), the receiver is not able to detect it, as authenticity was not violated (i.e., the claimed author of the message is its legitimate). In order to avoid such nonlegitimate operation, an authentication process such as the previously described CHAP needs to be implemented. Note that combined cryptographic protocols, such as the SSL or the TLS, incorporate their own authentication processes, in addition to the authenticity provided by the digital signature.

Figure 16.19 depicts the simplified block diagram of a system composed of a combination of the digital signature combined with an asymmetric cryptography. In this figure, PUK1 and PUK2 stand for the public keys of the message sender and of the message recipient, respectively, whereas PRK1 and PRK2 stand for the private keys of the message sender and of the message recipient, respectively.

Since both schemes make use of public key cryptography (public and private keys), it makes sense to implement these protection mechanisms together. This resulting system incorporates protection mechanisms from attacks against confidentiality, integrity, and authenticity.

16.8.4 DIGITAL CERTIFICATES

Asymmetric cryptography considers the open and wide distribution of public keys, without any level of confidentiality. Nevertheless, it is of high importance to keep the mapping between a public key and its owner. This mapping is achieved with the use of digital certificates (see Figure 16.20).

An attacker that generates a pair of asymmetric keys and advertises the public key, stating it belongs to a third party, can start sending messages signed by such third party (source spoofing). The destination of the messages believes such messages were sent by the legitimate third party, while it was sent by an attacker. Moreover, the intercepted messages sent by the third party to any other parties can easily be deciphered by such attacker. To avoid this type of attacks, the digital certificate is employed.

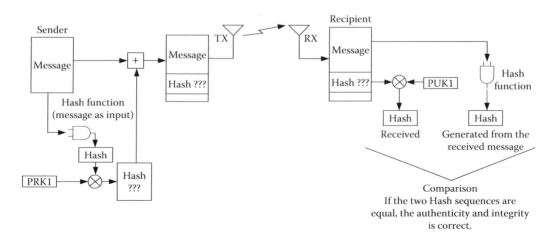

FIGURE 16.18 Generic block diagram of a digital signature.

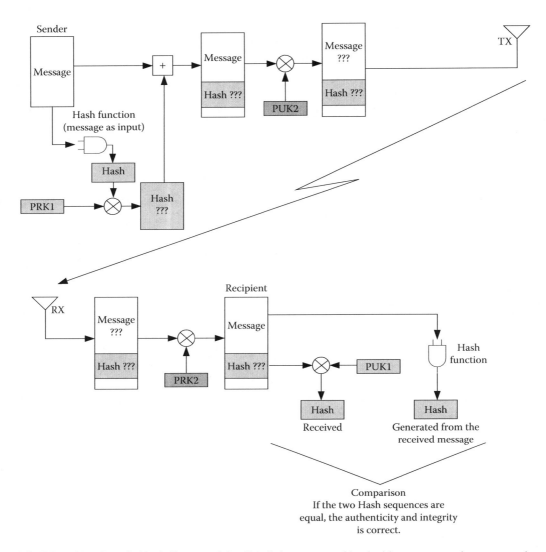

FIGURE 16.19 Generic block diagram of the digital signature combined with an asymmetric cryptography.

A digital certificate consists of a document emitted by a certification authority (CA) that maps a public key to its owner.* A digital certificate includes a set of information such as website address or e-mail address, the validation date of the certificate, the identification of the CA and its digital signature, the owner entity, and its public key (used to initiate the ciphered transaction using asymmetric cryptography).

In order to ensure that the certificate was not modified, the digital certificate is digitally signed by a CA. The application of a client (web browser, e-mail application, etc.) has a list of accepted certification authorities, together with their public keys. When a user wants to initiate a ciphered session with a specific server, he receives the server's digital certificate, and the validation of the certificate is automatically performed by the client's application. This validation is done by applying the stored CA's public key in the client's application to the CA's digital signature located in the digital certificate. Note that the certificate has been previously signed, that is, encrypted, with the

---

* A digital certificate works like an ID card, establishing the relationship between an identification and an entity.

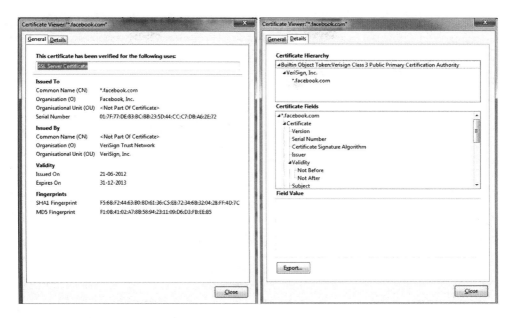

FIGURE 16.20 An example of a digital certificate.

private key of the CA. Therefore, this can only be deciphered with the CA's public key, stored in the user's application. If the digital signature of the CA is validated, and the CA belongs to the list of accepted authorities, the encryption handshaking proceeds. Alternatively, if the digital certificate of the CA does not belong to the client's accepted list of CAs, the client may accept the new authority as an accepted CA. Otherwise, the encryption handshaking ends.

It is worth noting that the CA keeps a list of all signed and revoked certificates, possible to be accessed by any entity (client or server).

Finally, let us analyze the situation where an attacker sends a known server's digital certificate, claiming he is the owner (source spoofing). Before exchanging ciphered data, the handshaking takes place. Let us consider the exchange of data in the client–server direction. Since the asymmetric cryptography (and combined cryptography) includes the data (or session key) ciphered with the destination's (server) public key, and since such attacker does not have the corresponding private key, he is not able to decode the received data (or the session's key). Therefore, even though if an attacker sends a third party a digital certificate claiming he is such entity, he does not succeed to implement an attack against confidentiality.

16.8.5 Public Key Infrastructure

The PKI is meant to allow a secure public keys' distribution system. It consists of an infrastructure using CIS, as well as databases, providing a service to allow the generation of digital certificates, implementing the corresponding keys management and distribution system.

As previously described, the trust chain is achieved with the use of digital certificates, together with certification authorities. Moreover, the PKI also includes registration authorities (RA), validation authorities, certificate revocation lists, and valid certificate lists.

Each CA stores its signed digital certificates in a public repository (database), which is generally accessible to the public. Similarly, when the certificate ends the validation or if an error is detected in any distributed certificate, the CA revokes the digital certificate, and advertises it using a certificate revocation list. These databases are accessible by entities using the online certificate status protocol.

The PKI follows a hierarchical structure, with several levels of CAs. Upper layer CAs are normally used to validate the work developed by lower layer CA, as well as to validate the certificates generated or signed by lower level CA. Consequently, these entities are also referred to as *validation authorities*. On the other hand, the lower level CA, responsible for directly interacting and identifying end entities, are referred to as RA. In this scenario, only the lower level RA is responsible for sending digital certificates directly to the entities, on a daily basis, while such certificates have to be sent from an RA to an upper layer CA, for validation. As can be seen from the example of digital certificate plotted in Figure 16.20, the certified entity is the www.Facebook.com, whereas the Verisign, Inc. is the RA. Moreover, the Verisign, Inc. (RA) is certified by a top-level CA, that is, the Verisign Class 3 Public Primary Certification Authority.

Before an entity starts using a digital certificate (e.g., the owner of a web server), he first needs to have the digital certificate properly certified and signed by a CA. This process can be carried out using one out of two procedures:

- *Procedure 1*: The CA produces and distributes private and public keys. In this case, the entity sends a certificate request, together with its personal data (see Figure 16.21). The CA sends the signed certificate, the private key, and the public key to the entity. Moreover, the CA stores the public key in a CA public repository for public consultation. The drawback of this procedure relies on how to be sure that the CA does not keep a copy of the generated private key, using it illegitimately.
- *Procedure 2*: The entity produces a private key and a public key. In this scenario, the entity signs his public key and his identification information with the corresponding Hash function and his private key (see Figure 16.22). Then, the entity sends his identification in clear, his public key in clear, as well as the corresponding digital signature to the CA, which verifies the identification information with the received clear public key. Once it has been verified, the CA creates a certificate and signs it with the CA private key, and returns the new certificate to the entity, posting it in a public repository. The drawback of this procedure relies on how to be sure that the entity is exactly who he is claiming to be. This procedure can easily be improved by using one or more authentication mechanisms such as using personal presentation of ID card, or other identification documents. Note that procedure 2 has overcome the drawback of procedure 1, as the CA does not have knowledge about the private keys of the entities.

The most used digital certificate standard, also adopted by the IETF, is the ITU-T X.509 standard. This standard defines a format for the content of the several certificate's fields, including version, serial number, owner identification, CA digital signature, validation data, as well as the entity's public key.

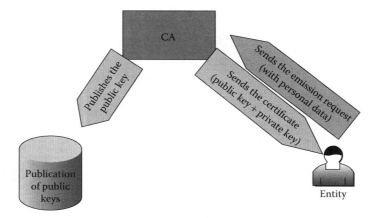

FIGURE 16.21   Procedure 1 for generation of digital certificates.

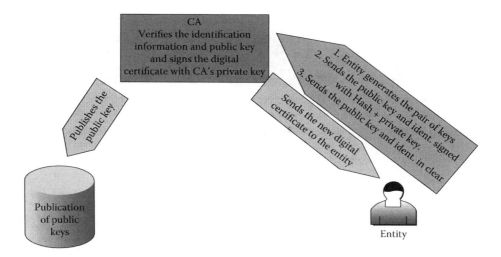

FIGURE 16.22 Procedure 2 for generation of digital certificates.

A PKI can be implemented within an organization (for internal use) or as an open infrastructure for generation, distribution, and storage of digital certificates over the Internet.

16.8.6 COMBINED CRYPTOGRAPHY

Due to its highly demanding mathematical computation, the processing time required to cipher and decipher a message using asymmetric cryptography is approximately a thousand times higher than that of symmetric cryptography. On the other hand, symmetric cryptography presents a vulnerability associated to key's distribution. Combined schemes aim to, simultaneously, exploit the advantages of symmetric and asymmetric cryptography, while overcoming their disadvantages. This is achieved by using both symmetric and asymmetric keys in an efficient manner. As can be seen from Figure 16.23, the combined cryptography works as follows:

- When a message* needs to be sent, the sender generates a symmetric key, and ciphers the message with such key (session key).
- Then, the symmetric key is ciphered with the receiver's public key, and the result is added to the previously ciphered message. These two components are sent together to the receiver.

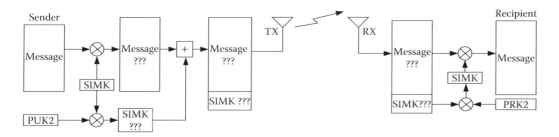

FIGURE 16.23 Generic block diagram of a combined cryptography.

---

* In fact, sending a message should be understood as the establishment of a session.

- The receiver takes the ciphered symmetric (session) key and deciphers it with its private key.
- Then the *plaintext* symmetric key is finally used to decipher the ciphertext message.

Similar to asymmetric cryptography, the combined cryptography requires the previous distribution of public keys.

Note that all described operations are automatically performed by the browser or by the e-mail application.

Similar to the diagrams of the symmetric and asymmetric cryptography, in Figure 16.23, SIMK stands for symmetric key, PUK stands for public key, and PRK stands for private key. Furthermore, PUK1 and PUK2 stand for the public keys of message sender and of the message recipient, whereas PRK1 and PRK2 stand for the private keys of the sender and recipient, respectively.

With the combined cryptography, since the message is ciphered with a symmetric key, the required processing time is approximately a thousand times lower than that of asymmetric cryptography. Moreover, encrypting the symmetric key with the receiver's public key is not time demanding, as the key length is much lower than the message length. Although combined schemes use symmetric keys for encrypting messages, the previous distribution of such keys has been avoided. This translates in a measure that mitigates the vulnerabilities associated to the distribution of symmetric keys. Consequently, the combined cryptography takes advantage of the reduced ciphering and deciphering processing time provided by the symmetric cryptography, while taking advantage of the simple and secure way public keys are distributed. Moreover, since symmetric key is used in a single session, this type of cryptography is very reliable.

Figure 16.24 depicts the generic block diagram of a system with a combined cryptography associated with a digital signature.

Examples of combined cryptographic protocols include the SSL, TLS, and the IPsec,[*] being defined in Sections 16.8.6.1 and 16.8.6.2.

16.8.6.1   SSL and TLS

SSL and TLS [RFC 5246; 6066] are two examples of important combined cryptographic protocols widely used in the Internet. These protocols make use of different symmetric and asymmetric algorithms, as well as Hash functions, in a customized manner. The combination of the three selected algorithms is called cipher suit, and an authentication process is also included in these protocols.

The initial handshaking between the client and the server includes the negotiation of the protocols possible to be used by both parties. In case one of the parties does not accept an initially proposed protocol, they have to re-negotiate the protocols to use or the secure connection cannot be established. The server generally chooses the strongest cipher suite possible to be used by both parties (client and server). As an example, an SSL or a TLS session may include the AES128 symmetric algorithm, used as a session key, together with the RSA1024 asymmetric algorithm, along with the 128-bit SHA-1 as a Hash function. Note that in online secure connections, the keys are negotiated during setup of each secure session.

As can be seen from Figure 16.25, the SSL/TLS protocols are used to implement an end-to-end encryption, interacting directly with the application layer. Some authors refer to the SSL/TLS protocols as transport protocols, while others refer to these protocols as application layer protocols. This latter approach is justified by the fact that the SSL/TLS protocols implement HTTPS, POP3S, and SMTPS secure protocols[†] in the application layer (which includes authentication process, cryptography, and digital signature). The implementation of these protocols by the application layer facilitates the implementation of high-level security functions such as user authentication.

---

[*] The PGP is another combined protocol.
[†] These protocols aim to provide secure services like web browsing and e-mail.

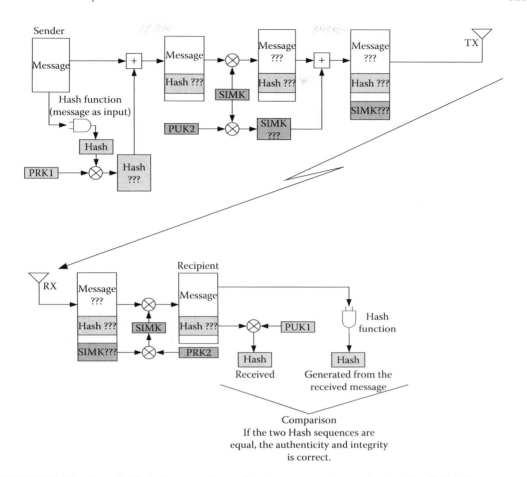

FIGURE 16.24   Generic block diagram of a combined cryptography associated with a digital signature.

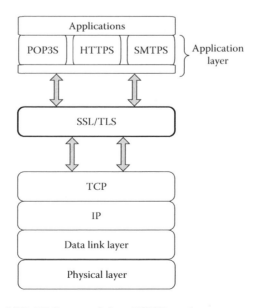

FIGURE 16.25   Location of SSL/TLS protocols into TCP/IP stack.

As can be seen from Figure 16.25, the application layer protocols implemented in association with the SSL/TLS protocols are similar to those of the clear mode, but with a suffix S added to its designation (e.g., POP3 becomes POP3S, HTTP becomes HTTPS, and SMTP becomes SMTPS). Moreover, the well-known ports assigned to the secure layer 4 protocols (in the server side) are different from those assigned to non-secure protocols (e.g., HTTP uses, by default, the port 80, whereas the HTTPS uses the port 443).

The implementation of the HTTPS secure protocol is achieved following a client–server architecture, using a browser in the client side, and a web server in the server side. Similar process is implemented for secure e-mail. Current browsers and e-mail applications support, by default, the ability to establish secure sessions using SSL/TLS.

The SSL protocol was created in 1994 by Netscape Communications Corporation, in order to provide some security to financial transactions through the Internet. In order to improve security functions, the Netscape Communications Corporation deployed, in 1996, the SSL v3.0. During that year, the IETF used the SSL v3.0 as the basis for the development of the TLS v1.0,[*] which is the most used cryptographic protocol in the Internet. Although both TLS v1.0 and SSL v3.0 are supported by most web browsers, the TLS v1.0 is normally the default secure protocol. Note that the protocol used in a secure session is negotiated by both parties.

The following paragraphs provide a generic description of the SSL v3.0 protocol (generically referred to as the *SSL protocol*). Since the TLS v1.0 is very similar, for the sake of simplicity this latter protocol is not here described.

Figure 16.26 depicts the SSL handshaking that is implemented before the exchange of ciphered application data takes place. The description of each step of the handshaking is described in the following (the reader should refer to the numbers in Figure 16.26):

1. The client application[†] that aims to establish a secure session with a server application initializes the handshaking by sending a *Hello* message followed by a 28-byte random number R_{client},[‡] as well as a list of cryptographic algorithms supported (asymmetric, symmetric, and Hash functions).
2. The server[§] also sends a *hello* message followed by another 28-byte random number R_{server}, its list of cryptographic algorithms supported, as well as a session identification (session ID). The cipher suite selected for the ciphered session is the one that provides the highest level of security possible to be implemented by both parties. In addition, the server also sends its public key along with its ITU-T X.509 digital certificate. Note that the validation of the digital certificate needs to be confirmed by the client.
3. If the server requires a digital certificate[¶] from the client, for allowing the establishment of the secure session, a *digital certificate request* message is sent to the client, along with the list of acceptable CAs.
4. If the server requires a digital certificate, the client sends it, which includes the client's public key. If such digital certificate is required but not available at the client's side, a *no digital certificate* message is sent, which may result in ending the session.

---

[*] TLS v1.0 is also known as SSL v3.1.

[†] For example, the e-mail or web browsing application.

[‡] This random number, together with the server's random number and with the premaster secret key S is used to generate the master secret key. The master secret key is then used to generate the session's key (symmetric). Note that these random sequence numbers do not consist of any authentication process.

[§] For example, e-mail or Web server.

[¶] Currently, most bank transactions do not require a digital certificate from the client. Normally, only companies have digital certificates. Nevertheless, some countries are including a digital certificate embedded in the ID card for interaction with the public sector institutions.

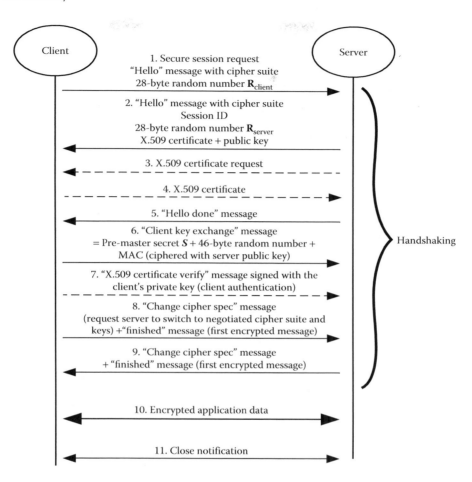

FIGURE 16.26 SSL handshaking, secure data exchange, and session closeout.

5. After all of the previously described operations, the server sends a *hello done* message.

6. The client verifies the validation of the server's digital certificate and verifies all parameters proposed by the server. Afterward, the client sends to the server the *client key exchange* message that contains the following items: the 48-byte pre-master secret key* S, together with a 46-byte random number,[†] and the message authentication code (MAC)[‡] key. All these three items are ciphered with the server's public key. Note that the master secret key K is generated by both the client and the server independently, from random numbers and pre-master secret key S (which is generated by the client, and sent to the server encrypted). The master secret key K is used by both parties to generate their own session keys (symmetric keys). Note that the master secret key K is never explicitly exchanged. The master secret is computed in SSL 3.0 as

---

* This key is computed by the client and is sent ciphered to the server, using the server's public key.

† The 46-byte random number is used in the computation of the pre-master secret key S.

‡ The MAC keys are used to ensure the authenticity and integrity of the exchanged messages. It consists of a digital signature, using Hash functions.

$$master\_secret = MD5(pre\_master\_secret + SHA('A' + pre\_master\_secret$$
$$+ ClientHello.random + ServerHello.random))$$
$$+ MD5(pre\_master\_secret + SHA('BB' + pre\_master\_secret$$
$$+ ClientHello.random + ServerHello.random))$$
$$+ MD5(pre\_master\_secret + SHA('CCC' + pre\_master\_secret$$
$$+ ClientHello.random + ServerHello.random))$$

7. In case the server requires a digital certificate from the client, the client sends a *digital certificate verify* message signed with the client's private key. This corresponds to the authentication process of the client, assuring the authenticity and legitamacy of the client's digital certificate. In case such digital certificate is not required by the server, the authentication of the client is only performed after the establishment of the secure session (i.e., once the session is already ciphered), using a protocol such as the CHAP or simply using username and password. Note that the server does not need to send the digital certificate to the client properly signed with the server's private key. This server's authentication process is not required because in case the server does not have a private key corresponding to the previously sent pre-master secret key S, it cannot decipher it[*] and the server cannot create the master secret key.[†] In such cases, the handshake fails.

8. Once the master secret has been generated, the client sends the *change cipher spec* message in order to switch to the accepted cipher suite. Then, the client sends the *finished* message that is the first ciphered message, and that finalizes the SSL handshaking in this direction.

9. The server also sends the *change cipher spec* message along with the ciphered *finished* message. This ends the SSL handshaking in both directions.

10. The ciphered session is finally established and both parties can exchange data in cipher mode.

11. When one of the parties wants to finalize the ciphered session, it sends a close notification message to the other party.

OpenSSL is a widely used open source toolkit responsible for implementing the SSL/TLS algorithms in a server. OpenSSL is a collaborative project that implements a commercial grade SSL/TLS. The *Heartbleed* corresponds to a vulnerability detected in OpenSSL versions 1.0.1 to 1.0.1f, being of *buffer overflow* type. The *Heartbleed* vulnerability results from an improper use of the memory that takes place during the *keep alive* handshaking implemented between both parties when they want to keep the session open even when they do not have any data to exchange, as defined in RFC 6520. The damages of the exploitation of this vulnerability may include message contents (eavesdropping), client credentials, session keys, or even copies of the server's private keys. The immunization to this vulnerability relies on upgrading the OpenSSL to a version equal or newer than OpenSSL 1.0.1g.

16.8.6.2 Security Architecture for an Internet Protocol

In Section 16.8.6.1, it was described that the SSL and TLS consist of cipher protocols implemented between the transport layer and the application layer of a TCP/IP connection.[‡] Consequently, these two protocols are utilized to provide an end-to-end security. Conversely, if a point-to-point

---

[*] This is ciphered with the server's public key.

[†] This rule is applicable to all transactions where a digital certificate is used. Even though if an attacker sends a known server's digital certificate, claiming to be the known entity, since the asymmetric and combined cryptography includes the data or session's key ciphered with the destination's (server) public key, and since such attacker does not have the corresponding private key, he will not be able to decode the transmitted data or the session's key.

[‡] This facilitates the user authentication process.

encryption in a certain path of a connection[*] is necessary to cipher all data that crosses it, the IPsec is the appropriate protocol (in tunnel mode). Although the IPsec may also be used to provide end-to-end (host-to-host) encryption (in transport mode), since the original IP headers are exchanged in clear (in this mode), the level of protection is reduced.

The IPsec was developed to provide security to the IPv6, while kept as an option for IPv4 RFC 4301 and RFC 4309. As can be seen from Figure 16.27, the IPsec protocol is implemented between layer 3 and 4, being transparent to the application layer.[†] Consequently, user interaction is not included, which may also present some advantages. In fact, the encryption can even be implemented without the knowledge of the end users. The IPsec is directly implemented in the operating system. This way, the security service is provided to all applications that interact with the transport layer (TCP or user datagram protocol [UDP]). Therefore, IPsec is well fitted for uses such as remote access of an intranet from the Internet or for the interconnection of office delegations with the headquarters through a VPN. This IPsec utilization can be seen from Figure 16.41, as well as from Figure 16.42.

The reader should refer to Section 16.13 for a detailed description of VPN, and to figures therein.

The security services provided by the IPsec can be of two types:

- *Authentication header (AH)*: It provides authenticity and integrity to the exchanged packets, protecting the point-to-point connections from attacks such as replay attack, spoofing, and connection hijacking. This is achieved using the integrity check value (ICV), together with the identification of packets with a sequence number (see Figure 16.28).
- *Encapsulation security payload (ESP)*: It provides authenticity, integrity, and confidentiality to the exchanged packets, protecting the point-to-point connections from attacks such as replay attack, packets partitioning, and packets sniffing. This is achieved using authentication and integrity protection mechanisms (provided by the ICV field, as well as with the identification of packets with a sequence number), together with combined cryptography[‡] (see Figure 16.29).

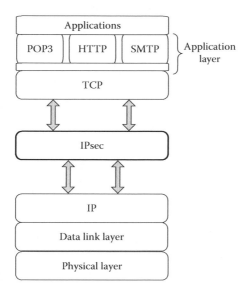

FIGURE 16.27 Location of the IPsec protocol into TCP/IP stack.

---

[*] For example, between two different LANs or between a host and a LAN.

[†] Note that, in the case of the IPsec protocol, the application layer protocols are not added with the S suffix (contrary to SSL/TLS protocols).

[‡] Asymmetric protocol is combined with DES or 3DES symmetric (session) protocol.

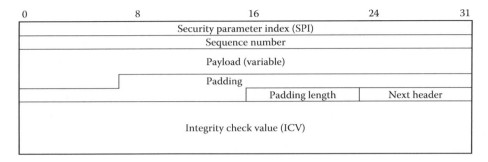

FIGURE 16.28   Header of the IPsec packet using the AH mode.

Security parameter index (SPI)		
Sequence number		
Payload (variable)		
Padding		
	Padding length	Next header
Integrity check value (ICV)		

0   8   16   24   31

FIGURE 16.29   Header of the IPsec packet using the ESP mode.

Optionally, these two protocols, AH and ESP, can be combined. In this case, the ESP is encapsulated into the AH protocol (externally), allowing the destination station to verify the authenticity of the packet, before it is deciphered.

The IPsec can be implemented to protect one or more connections between a pair of hosts, between two security gateways,[*] or between a host and a gateway. The protection is achieved through the creation of a security association (SA) that is identified by three different parameters:

- *Security parameter index (SPI)*: A unique number defined during the previous handshaking of a connection that identifies the SA. The parties involved in the ciphered connection need to use the SPI during the data exchange.
- *Destination IP address*: Although the type of connection can be unicast, multicast, or broadcast, the IPsec utilizes a unicast address, assuming the other cases as an extension of the unicast.
- *Protocol identification (AH or ESP)*: Includes the identifier 50 for the ESP and 51 for the AH protocol.

In case both the AH and ESP protocols are used together, two SA must be defined.

In addition to the AH and ESP protocols, the Internet key exchange protocol [RFC 2409] is utilized to create, manage, and exchange different keys between hosts or security gateways.

The IPsec can be implemented either in transport or in tunnel mode:

- *Transport mode*: Only the packet payload, that is, the segment,[†] is protected. This mode establishes an end-to-end SA,[‡] inserting the IPsec header between the IP header and the higher level protocol header (TCP or UDP). Finally, the IP payload is encapsulated into the IPsec payload (see Figure 16.30).

---

[*] A firewall or a router running the IPsec protocol.

[†] The segment is the message format of the transport layer. Since only the segment is protected, this is mode is called the *transport mode*.

[‡] In fact, two SAs can be established, in case the AH and ESP are used together.

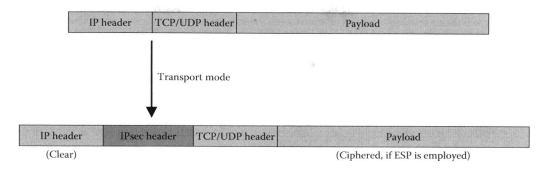

FIGURE 16.30   Packet format of IPsec in transport mode.

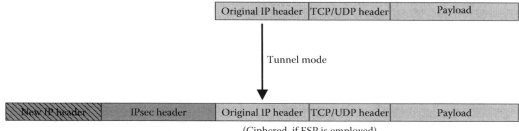

FIGURE 16.31   Packet format of IPsec in tunnel mode.

- *Tunnel mode*: Both the payload and the header of the packet are protected. In this mode, an SA is established between two security gateways,* or between a host and a security gateway located at the entrance of a network. In the tunnel mode, the whole IP packet (including its header) is encapsulated into the IPsec payload (including the insertion of a new IPsec header). Then, the whole IPsec packet (including its header) is encapsulated into another IP packet. The source and destination IP addresses of this external IP packet correspond to the source and destination IP addresses of the security gateways, whereas the source and destination IP addresses included in the internal IP header specifies the real source and destination of the IP packet (see Figure 16.31).

16.9   SECURITY IN IEEE 802.11 WIRELESS NETWORKS

A wireless access point (AP) of an IEEE 802.11 network acts similar to an Ethernet hub for wireless devices that join a basic service set (BSS). This means that the NIC of a wireless device receives the signals sent by an AP, regardless of their destinations. In case the AP transmissions are in clear mode (open) or use a cipher common to all wireless devices (WEP), the frames sent to all BSS wireless devices are monitored by all other stations. Then, in normal mode, the NIC discards the frames whose destination address is not this wireless device. When the NIC is placed in promiscuous mode, and with the aid of a packet sniffing software, the wireless device can capture all the traffic, and eavesdropping can be executed. Therefore, a major concern in IEEE 802.11 networks relies on executing authentication of wireless devices that want to join the BSS. Moreover, data encryption

---

* A security gateway can be either a firewall or a router, which runs the IPsec protocol. It is typically located at the entrance of a LAN, such that the interconnection between two LANs (through unsecure WAN or MAN), or between a remote host and a LAN.

is also an important requirement. Otherwise, the data exchanged with the AP will be available not only to those wireless devices that are part of the BSS but also to other terminals that can listen to the AP transmissions.

16.9.1 WIRED EQUIVALENT PRIVACY

In the initial IEEE 802.11 standard, the Wi-Fi Alliance comprised two possibilities for wireless devices to join a BSS:

- *Open*: This corresponds to the absence of authentication, that is, the AP always permits wireless devices to join the BSS.
- *WEP*: This comprises authentication and encryption, but the level of flexibility is low, which is also highly susceptible to be cracked. This comprised, initially, a 32-bit pre-shared key (PSK), shared between the client (wireless device) and the AP, with manual introduction in the client and in the AP, before the authentication process starts.

As can be seen from Figure 16.32, the WEP authentication handshaking comprises four steps. It starts with the message *authentication request*, being sent by the client to the AP. Then, the AP responds with a random text, entitled *challenge*, which is to be ciphered by the client. At a third stage, the client sends the challenge encrypted using the RC4 algorithm and using the PSK previously introduced manually in the client. Then, the AP takes the clear text challenge previously sent to the client and performs the same encryption with its own shared key. If the shared keys are the same, the result of this encryption equals the ciphered text received from the client at the third stage. In this case, the AP finalizes the authentication process by sending the *authentication verified* message. Otherwise, the authentication process is not finalized, the wireless device cannot join the BSS, and the exchange of data does not take place. Once the authentication has been positively verified, the exchange of data takes place, being ciphered using the RC4 algorithm, and with the same PSK key that was utilized for the authentication handshaking. This is the reason why the WEP is considered vulnerable: a hacker that captures the clear text challenge and its encrypted version during the authentication procedure can easily extract the PSK key. Then, he decrypts the data exchange using such computed key. Placing the NIC in promiscuous mode, and with the aid of a packet sniffer, the eavesdropping can be executed.

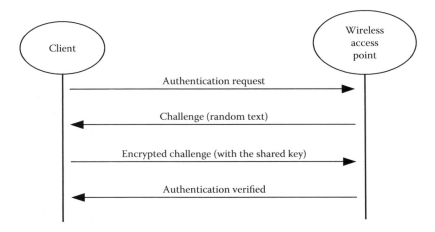

FIGURE 16.32   WEP authentication handshaking.

Another vulnerability of the WEP relies on the fact that the key utilized for ciphering the frames are common to all wireless devices. Therefore, the clear mode frames sent to a certain wireless device are captured by all other stations that belong to the same BSS. In normal mode, the NIC discards the frames whose destination address is not this wireless device. When the NIC is placed in promiscuous mode, and with the aid of a packet sniffing software, the wireless device can capture all the traffic, and eavesdropping can be executed.

16.9.2   WI-FI PROTECTED ACCESS

WPA security method comprises over-the-air authentication and encryption specifically designed to protect data transmitted over IEEE 802.11 networks. WPA has been released in 2003, with the objective of improving the security level of IEEE 802.11 networks, relating to the previous widely used WEP. While the WEP was included in IEEE 802.11 standard, WPA is utilized in IEEE 802.11 networks, but it consists of an independent security standard.

Before the development of the WPA took place, many manufacturers tried multiple techniques to improve the authentication and encryption characteristics, relating to the WEP. The temporal key integrity protocol (TKIP) was one of these techniques, having been developed in 2002 and was the baseline for the WPA. The TKIP consists of a cryptographic algorithm that modifies its key in every 10,000 packets exchanged by the NIC. This way, an attacker that succeeds to obtain the key, can only decipher the message for a short period. While the 128-bit key of TKIP is longer than that utilized in WEP, and is automatically modified periodically, the basic RC4 algorithm is still used. Nevertheless, the used key consists of a combination between a PSK between the client and the AP, and the NIC MAC address of the wireless device. This makes the key unique, and dependent of each wireless device. Consequently, using the WPA, a certain wireless device cannot decipher the frames sent to other wireless devices. This key is entitled *temporal key*, being recomputed and modified after intervals. In addition to providing protection from attacks against authenticity and confidentiality (authentication and cryptography), the TKIP also provides protection from attacks against integrity by implementing the message integrity checksum (MIC) algorithm.

In 2003, the Wi-Fi Alliance took the TKIP and adopted it as the WPA standard.

16.9.3   WI-FI PROTECTED ACCESS 2 AND THE IEEE 802.1I

Although WPA provided a level of security much higher than that of WEP, the Wi-Fi Alliance wanted to further improve the security features. Moreover, in parallel with the work being developed by the Wi-Fi Alliance, the IEEE was also progressing in the search for new wireless security protocols. In this scope, the IEEE developed, in 2004, the IEEE 802.1i standard, comprising a follow up to the previous WPA, but where the RC4 algorithm previously adopted was replaced by a new algorithm based on a combination between the AES and the RC4, with a key length of 256 bits. This new standard was also adopted by the Wi-Fi Alliance with the designation Wi-Fi protected access 2 (WPA2). This new combined cryptographic algorithm based on AES provides much better security protection, but the level of cryptographic processing also increases. Therefore, depending on the processing capabilities of the used terminal and of the security requirements, one may choose between WPA and WPA2.

The WPA2 can be utilized in the personal mode using a PSK, or in the enterprise mode, where an authentication infrastructure is required to manage the authentication process. The PSK corresponds to a symmetric key that is previously shared between the two parties: the client and the AP. Since this is a vulnerability, the length of this key is normally an important issue to increase the security level.

It is worth noting that both the IEEE 802.1i and the WPA2 can be associated to optional standards and algorithms, in addition to the baseline standards. This includes the extensible authentication protocol (EAP) and the IEEE 802.1x, as defined in the following paragraphs.

16.9.4   EAP AND IEEE 802.1x

The level of protection provided by WPA2 and IEEE 802.1i from attacks against confidentiality and integrity was already good. Nevertheless, the use of a PSK was still a vulnerability associated with the authentication process due to the need to previously share the key between multiple users. In small size networks, since the users' universe is limited, the level of vulnerability is low. Nevertheless, in high dimension wireless networks, such as corporate applications, or when the level of security requirements is a key issue, the PSK alone is not an acceptable solution. A solution to this problem relies on using an additional authentication process between the client and the AP (authenticator), but properly certified by an authentication, authorization, and accounting (AAA) server (commonly referred to as the *authentication server*). The most used AAA server is the remote authentication dial-in user service database (RADIUS), and the authentication procedure relies on the EAP framework. It is worth noting that the use of IEEE 802.1x or EAP is not limited to the wireless environment. In fact, 802.1x/EAP can also be utilized in wired networks, such as to allow or to deny a new terminal to connect to a wired switch.

As can be seen from Figure 16.33, the authentication process starts after a wireless device has been associated with the AP. At this stage, the AP creates a virtual port for communicating with the wireless device, only accepting 802.1x traffic, and blocking all other traffic types, that is, data traffic cannot go beyond the AP. Once the wireless association has been performed, the AP, acting as authenticator, sends to the wireless device (client) an *identity request* message. The client responds with the *identity response* message that includes the authenticating credentials. These credentials may include a username and password, a smart card, and a digital certificate if PKI is established. Note that this corresponds to a second level of authentication, in addition to the PSK. The authenticator forwards such credentials to the authentication server (typically a RADIUS database) for verification. If the authentication procedure succeeds, the server sends an EAP success message to the authenticator, and this message is forwarded to the client. In this case, the client is then authorized to send data traffic through the virtual port of the AP, after the establishment of a data link encryption between the client and this specific virtual port of the AP. Note that for the sake of security, each authenticated wireless device exchanges data with the AP using a different virtual port. Finally, when the wireless device intends to logoff, it sends an *EAP logoff* message to the authenticator, and the virtual port is modified to disabled mode, blocking again non-802.1x traffic.

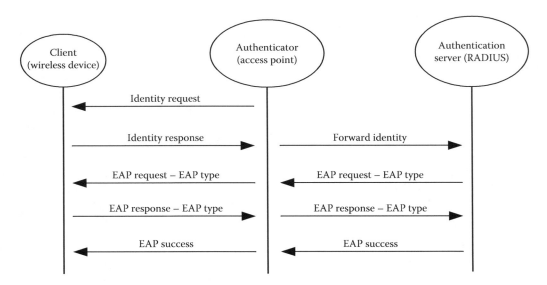

FIGURE 16.33   IEEE 802.1x/EAP authentication handshaking.

From the above description, it can be stated that while TKIP and AES protocol deal with confidentiality, IEEE 802.1x/EAP is responsible for implementing the authentication process. The wireless authentication process widely implemented in corporate environments is the *enterprise mode*. The *enterprise mode* corresponds to the use of a PSK, together with IEEE 802.1x/EAP protocol (i.e., an additional authentication process). Alternatively, in domestic applications, the most common authentication process simply relies on the use of the PSK.

16.10   CISCO SWITCH SECURITY

In order to avoid an unauthorized access to a Cisco switch, or to a switch port, few rules and procedures should be implemented:

- Configure a ciphered access password to the privileged mode and another to access the lines.
- Configure the switch to allow remote management and configuration only through SSH (not through telnet).
- Configure port security in the switch interfaces.

16.10.1   PASSWORD CONFIGURATION

The configuration of a ciphered password to access the privileged mode in a Cisco switch is performed using the following Cisco IOS syntax:

- Switch(config)#enable secret *pass_word* (*pass_word* stands for a ciphered password)

Moreover, the configuration of a password to access the telnet ports is performed with the following:

- Switch(config)#line vty 0 15
- Switch(config-line)#password *secret_word*
- Switch(config-line)#login (makes the system requesting a line login password)

The command line "line vty 0 15" activates and configures 16 telnet access ports: 0 to 15. For security reason, a single telnet port may be activated, while de-activating the remaining 15. In this case, the telnet configuration becomes

- Switch(config)#line vty 0
- Switch(config-line)#password *secret_word*
- Switch(config-line)#login
- Switch(config-line)#line vty 1 15
- Switch(config-line)#no password
- Switch(config-line)#login (the command "no password," followed by "login," deactivates the telnet port)

16.10.2   SSH CONFIGURATION

A Cisco switch can be remotely configured through telnet or through SSH. An advantage of using SSH over telnet relies on the use of authentication and encryption. The configuration of a switch through SSH is performed using a sequence of instructions (similar procedure is used to remotely control a router through SSH):

- Switch(config)#hostname *hostname* (the hostname configuration is required)
  - *Example*:
    - Switch(config)#hostname S1

- Switch(config)#ip domain-name *domain*
  - *Example*:
    - Switch(config)#ip domain-name domain_example.net
- Switch(config)#crypto key generate rsa (to generate a certificate using the RSA algorithm; the user should select 1024 bits to use a more advanced SSH version)

Then, a username and password should be configured in the switch for SSH access, as follows:

- Switch(config)#username *username* priv 15 secret *pass_word* (*pass_word* stands for a ciphered password)
- Switch(config)#line vty 0 15
- Switch(config-line)#transport input ssh (this command disables telnet, allowing only SSH access in these lines)

Still, in the scope of the SSH configuration, an IP address must be assigned to the management VLAN to allow the switch to be remotely configured through SSH (or telnet). It is worth noting that the ports assigned to the management VLAN should be secured.* Otherwise, any device could get remote access to the switch, and could remotely configure it, which might represent a security violation. The creation of a management VLAN is performed using the following steps:

- Switch(config)#vlan *management_vlan_number* (*management_vlan_number* is the number of the new VLAN to be used as management VLAN)
- Switch(config-vlan)#name management
- Switch(config-vlan)#exit
- Switch(config)#interface vlan *management_vlan_number*
- Switch(config-if)#IP address *ip_address subnet_mask* (the IP address and subnet mask of the management VLAN that is then utilized to remotely access the switch)
- Switch(config-if)#no shutdown

16.10.3 PORT SECURITY CONFIGURATION

The configuration of port security in Cisco switch interfaces can be implemented using a number of ways. The switch performs layer 2 switching based on the destination MAC address contained in a frame, and taking into account the information present in a MAC address table. A MAC address table maps MAC addresses into output ports. Initially, this table is empty. Consequently, the equipment forwards a frame received in one of the ports through all ports, except the port from where the frame was received. When a switch receives frames from the different ports, the corresponding source MAC addresses are registered and associated to certain ports (switch interfaces), resulting in the construction of the MAC address table. Most of the Cisco switches (e.g., Catalyst 2960 series switches) that make use of Cisco IOS allow checking MAC address tables, from the privileged mode, using the following syntax:

- Switch#show mac-address-table

Medium-to-high dimension networks have typically a topology with several layers of switches, resulting in redundant paths (mesh configuration), and originating an improved resistance from failures. In this case, a switch has to decide about which output port to use, among several possibilities. These decisions are taken by switches using the spanning tree protocol, which results in an efficient

---

* For example, using the command "switchport port-security" (as described in Section 16.10.3) and creating a secret password to access the telnet line.

manner of building of MAC address tables. In this case, the output ports of a switch that belong to a selected route are placed in forwarding state, whereas the others can be placed in blocking state. The IEEE 802.1D is a widely known the STPr standard.

Port security restricts the devices that are allowed to access a certain switch port. This can be performed using a number of different ways, as described in the following paragraphs.

The configuration of port security in a switch is initially performed using the following syntax:

- Switch(config)#interface [*interfaceinterface_range*]*
- Switch(config-if-range)#switchport port-security
  - *Example*:
    - Switch(config)#int fa0/1-10
    - Switch(config-if)#switchport port-security

After having entered into the port security mode, we can statically configure the hosts that are allowed to connect to a certain switch port. This configuration is added to the running configuration, and can be saved to the startup configuration (see Chapter 10). This is performed by entering the hosts' MAC addresses, using the following syntax:

- Switch(config-if-range)#switchport port-security mac-address *static_mac_address* (where *static_mac_address* is the mac-address† of the host(s) that is allowed to access the switch port)
  - *Example*:
    - Switch(config-if)#switchport port-security mac-address 0010.5A44.12B5

Then, we can configure a certain the switch port to shutdown, in case the host connected is not the one previously configured statically with the MAC address. This is performed using the following syntax:

- Switch(config-if-range)#switchport port-security violation shutdown

Instead of statically configuring the MAC address that is allowed to access a certain switch port, one can configure a switch port (or a range of switch ports) to dynamically learn the MAC addresses that are connected to a certain port (or range of ports), and to add them to the list of allowed MAC addresses. This is performed by placing a switch port in sticky mode, using the syntax:

- Switch(config-if-range)#switchport port-security mac-address sticky

One can also configure a switch port to limit the maximum number of hosts that can connect to it. This is performed using the following syntax:

- Switch(config-if-range)#switchport port-security maximum *max_number* (where *max_number* stands for a number)

For the sake of security, it is a good procedure to disable the switch ports that are not used. This is performed using the command "shutdown" in the interface:

---

* *Interface_range* stands for a range of interfaces (e.g., int fa0/1-4) to be configured with port security. Alternatively, this can be performed independently for a single interface (e.g., int fa0/1).

† The MAC address is inserted using the notation 0123.4567.89AB (three groups of four characters, and separated by dot), instead of the regular MAC address notation 01:23:45:67:89:AB (six groups of two characters, and separated by colon).

- Switch(config)#interface [*interface* | *interface_range*]
- Switch(config-if)#shutdown

Finally, one can verify the port security configuration, from the privileged mode, using the following Cisco IOS syntax:

- Switch#show interfaces *interface* switchport

16.11 NETWORK ARCHITECTURES

Depending on the services and security levels, there are different network architectures that can be implemented within an organization such as in a company. Two different network types may exist within an organization:

- *Intranet*: This is only used by the employees of an organization.
- *Extranet*: This is used by both customers and business partners.

16.11.1 SINGLE NETWORK ARCHITECTURE

Figure 16.34 depicts a generic network architecture normally implemented in small companies. In this case, a single network is used as both the intranet and the extranet, and the firewall protection is normally placed at the output of the network. In fact, most common architectures such as small business networks include a common device that acts as both router and firewall. This architecture is very vulnerable, as the access to the servers is achievable from the Internet (to be used by the customers and partners). Since the servers used by both customers (extranet) and employees (intranet) are the same, the level of vulnerability of this network architecture is very high.

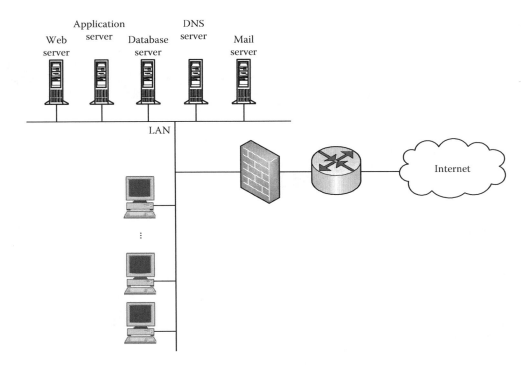

FIGURE 16.34 Generic corporate architecture composed of a single network.

When the web service is used to access an application (as in the case of e-banking), a chain of hierarchical servers are utilized, as depicted in Figure 16.35. A web server is responsible for establishing HTTP connections with remote client hosts. In the above-described scenario, the service required by a web client is indirectly provided by an application server. In fact, the client only connects to the web server, but this server makes use of the application server to provide the desired service to the client (e.g., a bank transfer). In addition, the application server processes data that is stored in a database server, and a service provided to a client may result in a change in a database. Such example is a bank transfer, where an HTTP connection is established between a client and a web server. The transaction itself is carried out using the application server, and the application server requests the database server to check whether or not the client has enough balance in his current account. In case the response is positive, the transaction is carried out by the application server, and the database server reduces the current account balance by an amount corresponding to the amount transferred. During this transaction, the database server provided a service to the application server, and the application server provided a service to the web server. The client only establishes a connection with the web server, and all other functions are performed in a transparent manner to the client using the hierarchical structure of the web service. These three servers are included in the several network architectures presented in this chapter. Nevertheless, in many cases, the different server functionalities are implemented by a single server or two servers.[*]

It is worth noting that the simple presentation of data (e.g., for marketing purposes) does not require this hierarchical structure of servers, as an application is not required. In this case, a simple web server is required.

16.11.2 DOUBLE NETWORK ARCHITECTURE

In order to improve the security level, a common procedure relies of separating the intranet from the extranet using a firewall. This can be seen from Figure 16.36. This architecture includes the duplication of servers, some of them dedicated to the intranet, and others dedicated to the extranet. In the depicted architecture, the database server only exists in the intranet. In this case, the external application server connects to it every time it is needed. Such permission needs to be properly included in the list of firewall rules that allow connections. Moreover, the two e-mail servers are normally synchronized, which needs also to be allowed by the firewall. The architecture shown in Figure 16.36 already includes a proxy (placed in the extranet) that avoids the direct connection from the intranet to the Internet. A host in the intranet that aims to establish a connection with a web server placed in the Internet initiates a connection with the proxy and, in order to provide the desired web service to the host, the proxy establishes a second connection with the destination server. Therefore, the proxy establishes two different HTTP connections. Then, the proxy *downloads* the desired HTML file from the web server and, in the second step, it uses the HTTP connection established with the

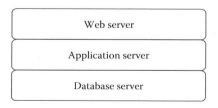

FIGURE 16.35 Hierarchical structure of servers for an application carried out through a web interface.

---

[*] For example, using the Internet information system (IIS) server or the Apache server, which include both the web and the application server functionalities.

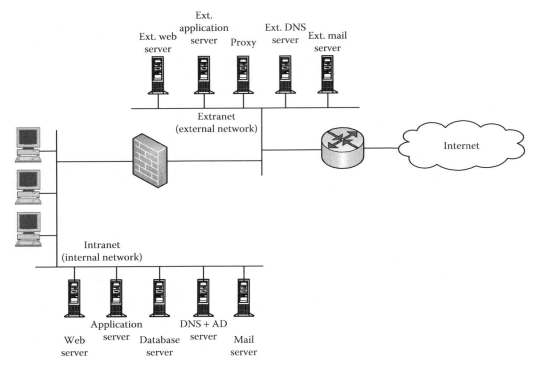

FIGURE 16.36 Generic corporate architecture composed of an intranet and an extranet.

host to transfer such file. Note that the proxy stores HTML files during a certain period, which can be accessed by internal hosts, without having to get access to the destination server.

From Figure 16.36, the existence of the active directory server in the intranet is also seen. This is responsible for allowing or denying requests (permissions) from hosts connected in the intranet, in accordance with the established organizational security policy.

16.11.3 Network Architecture with DMZ

In order to further improve the security level, the external access to the client servers needs to be protected by a firewall. This is achieved by creating a DMZ, as depicted in Figure 16.37. A DMZ consists of a network that accommodates the external servers, and whose access needs to be properly authorized by the firewall. Contrarily, in the previous architecture, Internet users could get access to all external servers. Therefore, using the DMZ architecture, the extranet is left empty. It is worth mentioning that using this architecture, direct connection from/to the intranet into/from the Internet is commonly not allowed. Note that the firewall acts as a router, as it enables connectivity among three different networks (the intranet, the DMZ, and the extranet).

This architecture could even be improved by using two firewalls, instead of a single one. As depicted in Figure 16.38, one firewall is placed between the intranet and the DMZ, whereas the second is placed between the DMZ and the empty external network (which then gives access to the Internet). Moreover, inside the internal network (intranet), some implementations include an additional firewall between the intranet clients and the intranet servers. The aim is to restrict the access from internal users to internal servers. This improves the security level because it is known that most of the attacks come from the interior of the organizations.

Another possible architecture implemented by organizations that want to provide higher level of security makes use of a Honey Net, which acts as a decoy for attackers (see Figure 16.39).

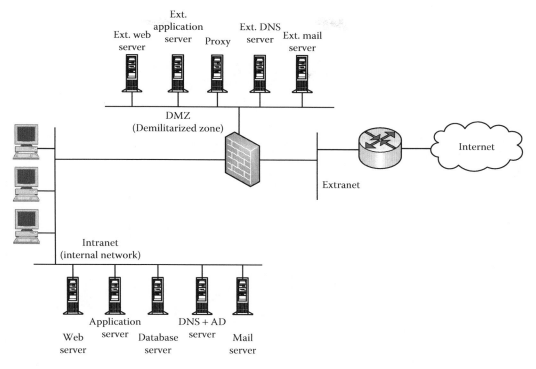

FIGURE 16.37 Generic corporate architecture with DMZ.

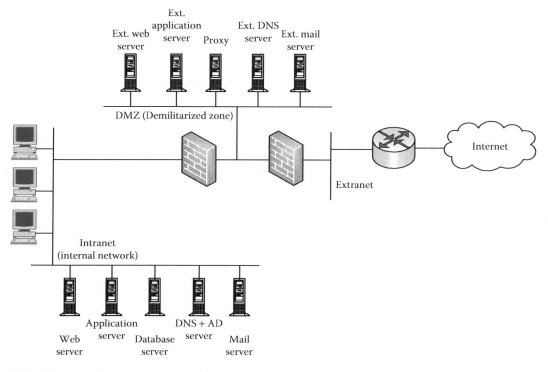

FIGURE 16.38 Generic corporate architecture with advanced DMZ.

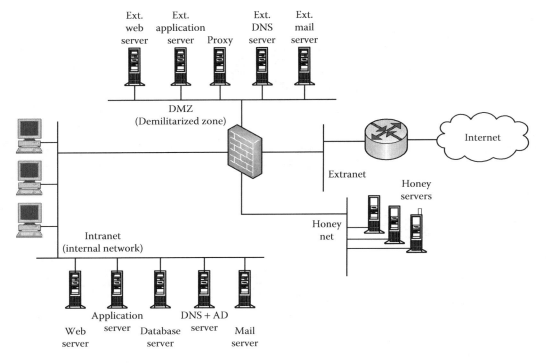

FIGURE 16.39 Generic corporate architecture with DMZ and Honey Net.

The firewall permissions to the Honey Servers are made easier, such that an attacker thinks he is hacking the organizational servers, while only getting access to decoy servers. In addition, the network may use other mechanisms such as an IDS to detect such intrusion, identify the author, register his actions, and track him.

16.12 INTRUSION DETECTION SYSTEM

An IDS aims to detect malicious activities or procedure violations in a host or a network. In case of a host, the IDS is referred to as *host-based IDS* (HIDS), whereas in the case of the network it is known as *network IDS* (NIDS).

The detection of malicious activities or procedure violations can be achieved using different techniques:

- *Statistical anomaly-based IDS*: It comprises the procedure used by an IDS that relies on the analysis of the instantaneous bandwidth consumption, the accessed devices, the used protocols, and the port addresses used to establish connections.
- *Signature-based IDS*: This corresponds to an IDS that keeps track of the accessed files and of the packets exchanged in the network, comparing them against a predefined attack profile, known as *signature*.

It is worth noting that in addition to monitoring and reporting the procedure violations and malicious procedures, a few types of IDS can also implement counter-measures, that is, presenting an active behavior, aiming at disabling such malicious activities. In this case, the device is referred to as an *intrusion preventive system* (IPS). Note that an IPS is commonly combined with IDS. In this case, the device is normally known as *intrusion detection and preventive system* (IDPS).

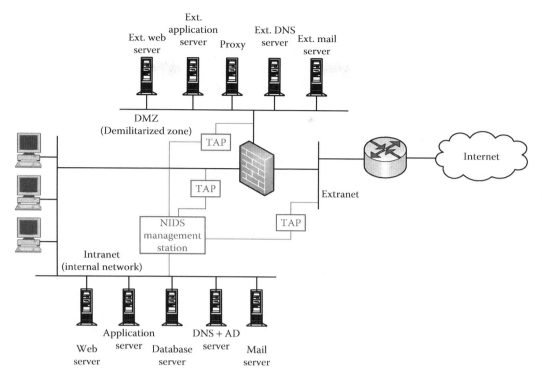

FIGURE 16.40 Generic corporate architecture with an NIDS.

A network IDS, that is, an NIDS monitors the network activities by accessing to different network zones using sensors, known as *taps*. A tap captures the network traffic and analyzes it using a technique such as the statistical anomaly- or signature-based IDS. Moreover, the NIDS also keeps track of the network hosts activities. Different NIDS taps are connected to different network borders of a firewall, such as the intranet, extranet, DMZ, and the external area (connected to the Internet). Moreover, an NIDS may also be connected to a switch, a hub, a router, and multiple servers (e.g., proxy server). The NIDS interconnection is implemented in a transparent manner to the network, that is, in *stealthy mode*, without any IP address and without responding to ping commands. The data collected by all NIDS sensors (taps) is centralized in an NIDS management station, located in the intranet, which processes it and generates the required reports and alerts/alarms. This can be seen from Figure 16.40 for a network architecture with a DMZ. In case of a network IDPS (NIDPS), the detection of a possible incident can be followed by the modification of firewall or proxy rules.

Finally, a host IDS, that is, an HIDS consists of an application installed in a host that monitors the modification of files, the files integrity, the logs, the ACL, passwords, and accessed services. Moreover, an HIDS may monitor the content of packets.

16.13 VIRTUAL PRIVATE NETWORKS

A VPN can be used by both employees and partners to allow access to a public communications network (typically, from the Internet) to the intranet or to a partners' DMZ. A partner's DMZ is commonly referred to as *partners' network*. As can be seen from Figure 16.41, this consists of an additional network configured in the firewall, in parallel with the regular DMZ. The partners' DMZ is the location where the servers used by partners are installed. The information stored in these servers is part of the information stored in internal servers (intranet). This is achieved by synchronization of the desired servers through the firewall.

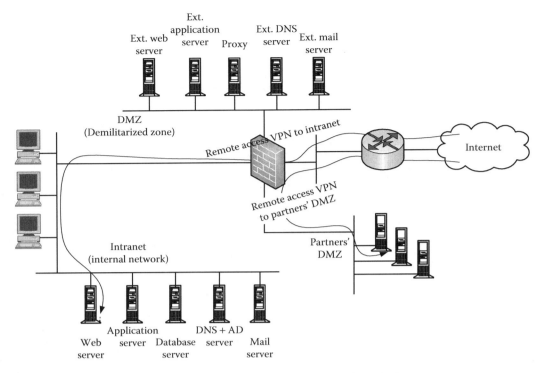

FIGURE 16.41   Remote access VPN to intranet or to partners' DMZ from the Internet.

The remote access VPN can be secure or unsecure. The ability for the employees to get access to intranet and for the partners to get access to partners' DMZ from the Internet, while preventing Internet attackers from getting access to the exchanged data, is achieved by using a cryptographic tunnel configured in firewalls or routers that implement the IPsec cryptographic protocol. These mechanisms ensure confidentiality, authenticity, and integrity of exchanged data through the Internet.

Currently, organizations, such as banks, have headquarters where all important servers are located, and a high number of remote delegations spread out over the territory (close to the customers). The remote delegations are typically connected to the headquarters through the Internet, using a secure VPN. The IPsec in tunneling mode (using the ESP) is the most used cryptographic protocol adopted to allow such connectivity. In this case, a secure IPsec tunnel is configured at both the headquarters' and remote delegations' firewall. The firewalls act as security gateways, encapsulating the original IP packet (including the original IP header) into an IPsec packet. The IPsec packet is then encapsulated into the payload of another IP packet, whereas its source and destination IP address refers to the IP address of the security gateways (firewalls) of the remote delegation and headquarters. This is depicted in Figure 16.42.

A VPN can also be established between an individual user from a public communications network (Internet)* and the headquarters. In this case, the same IPsec VPN in tunneling mode can be implemented, but the tunnel is established directly between the remote host and the headquarters' security gateway.

The exchange of data (connectivity) can be assured by any transmission means depending on the service requirements, such as the required data rates and accepted delays. A remote user from home may access the intranet using the regular home network (e.g., a wireless LAN connected to

---

* Physically, a remote user can be anywhere (e.g., at home or in a hotel).

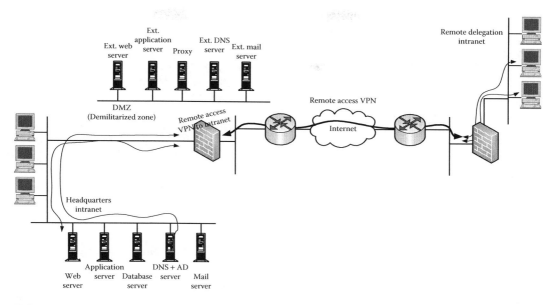

FIGURE 16.42 Remote access VPN configured between a remote delegation's and headquarters' firewall.

the Internet through xDSL modem). A very small remote delegation may opt for a low cost xDSL or cable modem connection, whereas larger remote delegations may use an MPLS connectivity to get access to the headquarters' router with high-speed connectivity.

16.14 FIREWALLS

A firewall is an access control mechanism used to separate two or more networks, while implementing a group of rules to verify the conformity of the traffic between different networks with the established security rules, procedures, and policies.

As can be seen from Figure 16.37, a firewall can be used to separate the intranet from the DMZ and from the extranet. In this case, the firewall also acts as a router, while implementing a number of rules to permit or deny the traffic between two adjacent networks.

While a firewall acts as a router, many of the existing routers also implement firewall functionalities. In the case of Cisco routers, such firewall functionalities are typically implemented using ACLs[*] or using context-based access control (CBAC).[†] Therefore, even in a network without a conventional firewall, some level of protection can be implemented using a router.

A firewall can be either implemented above a traditional operating system in a regular computer or implemented as a dedicated and stand-alone system. In order to reduce the vulnerabilities, firewall implementation using a dedicated and stand-alone system tends to be more resilient. This results from the fact that increasing the amount and software complexity results in more vulnerabilities.[‡] Note that a typical network attack consists of exploring the vulnerabilities of operating systems. In order to solve a detected problem, a network administrator normally

---

[*] ACLs are used by Cisco routers for the implementation of packet filtering in stateless or statefull mode (depending on the Cisco IOS version).

[†] Some Cisco routers allow CBAC, implementing a number of functionalities similar to a firewall application layer gateway, using statefull inspection.

[‡] On an average, a software program presents a certain number of bugs per million of instructions. Therefore, reducing the software complexity results in a more resilient system.

switches off applications, which may leave room for an attacker to penetrate in the target network and to implement the desired actions.

In order to better understand the way traffic is permitted or denied by a firewall, let us focus on the architecture depicted in Figure 16.37, where three different network areas exist: the internal network (intranet), the DMZ, and the extranet. Since the extranet is empty, it is only used to get access to the Internet. Therefore, in the following analysis this third area is simply referred to as the *Internet*.

Table 16.2 lists a number of possible rules that can be configured in the firewall of Figure 16.37 (between the internal network, the DMZ, and the Internet). First, it is worth noting that, by default, a firewall denies all the traffic. For the sake of clarity, such implicit rules are also included in Table 16.2 (see rows with action deny). Any traffic that is intended to be allowed needs to be explicitly authorized through a firewall rule. Moreover, the example of security policy implemented by this firewall does not allow direct connectivity between the internal network and the Internet. Therefore, there is no rule allowing such connectivity, falling in the default rule that denies all traffic not explicitly authorized. An internal host that intends to access the Internet does it through the proxy server of the DMZ (for HTTP and FTP), or using the mail server of the DMZ (for the mail service). This can be seen from the two initial rules of Table 16.2. The 7th and 13th rules authorize any host from the Internet (typically a corporate employee) to have access to the mail server of the DMZ using the SMTPS/POP3S (for uploading/downloading e-mails), whose protocols comprise authentication and encryption using SSL/TLS. From the penultimate row, it can be seen that a rule authorizes the application server located in the DMZ to access the database server located in the internal network. Note that the DMZ could have its own database server to be utilized by the application server but, in such scenario, the database servers of the DMZ and internal networks would have to be synchronized.

TABLE 16.2
Possible List of Rules to Configure in the Firewall of Figure 16.37

Source Interface	Destination Interface	Source Host	Destination Host	Protocol	Action
Internal network	DMZ	Any	DMZ proxy	HTTP/FTP	Permit
Internal network	DMZ	Internal mail server	DMZ mail server	SMTP	Permit
Internal network	DMZ	Any	DMZ web server	HTTP/FTP	Permit
Internal network	Any	Any	Any	Any	Deny
Internet	DMZ	Any	DMZ web server	HTTP/FTP	Permit
Internet	DMZ	Internet mail ISP	DMZ mail server	SMTP	Permit
Internet	DMZ	Any	DMZ mail server	SMTPS	Permit
Internet	DMZ	Any	DMZ DNS server	DNS	Permit
Internet	Any	Any	Any	Any	Deny
DMZ	Internet/internal network	DMZ proxy	Any	HTTP/FTP	Permit
DMZ	Internet/internal network	DMZ web server	Any	HTTP/FTP	Permit
DMZ	Internet	DMZ mail server	Internet mail ISP	SMTP	Permit
DMZ	Internet	DMZ mail server	Any	POP3S	Permit
DMZ	Internet	DMZ DNS server	Any	DNS	Permit
DMZ	Internal network	DMZ mail server	Internal mail server	SMTP	Permit
DMZ	Internal network	DMZ application server	Internal database server	Client/server	Permit
DMZ	Any	Any	Any	Any	Deny

    Different firewall rules must be grouped depending on the source or the destination network and whether they are applied to a certain interface in the inbound* or in the outbound† direction. Cisco routers group these rules into ACLs. Moreover, each interface of a Cisco router may have simultaneously two ACLs: one in the inbound and another in the outbound direction (see Figure 16.43).

    Focusing on the example of rules listed in Table 16.2, and considering Figure 16.43, the initial four rules (internal network to outside) should be included in one ACL, applied to the interface FastEthernet 0/0 (Fa0/0), in the inbound direction. The following five rules (from the Internet) should be included in another ACL, being applied to the interface FastEthernet 0/1 (Fa0/1), in the inbound direction. Finally, the last eight rules (from the DMZ) should be grouped into another ACL, and applied to the interface FastEthernet 1/0 (Fa1/0) in the inbound direction. Note that ACLs can also be applied in the outbound direction. In this case, the interface to where the ACL is applicable must be different. For example, the ACL corresponding to the initial three rules listed in Table 16.2 could also be applied to the interface FastEthernet 1/0 (Fa1/0), but in the outbound direction (instead of Fa0/0 in the inbound direction). This latter rational relies on selecting the interface based on the network where the destination host is located (instead of based on the source host). Nevertheless, it is always preferable to filter the traffic before it enters the firewall/router, as undesirable traffic may cause damages in the device. Therefore, most of the time, the best option relies on applying packet filtering rules in the inbound direction.

    The way traffic is filtered by a firewall/router is as follows (see Figure 16.44):

1. The router or firewall starts by comparing the packet against the first rule. In case the inspected packet fits with the rule, the rule action is applied (permit or deny) and the inspection is terminated.
2. In case the previous inspected packet did not apply the action rule, compare the packet against the following rules by descending order until it finds a rule that fits with the inspected packet. In this case, the packet is subject to the rule action and the inspection is terminated.
3. In case the inspected packet does not fit with any of the rules, the packet is submitted to the default rule action: the packet is denied (blocked). Note that, by default, the traffic to be authorized must be explicitly configured in a firewall/router.

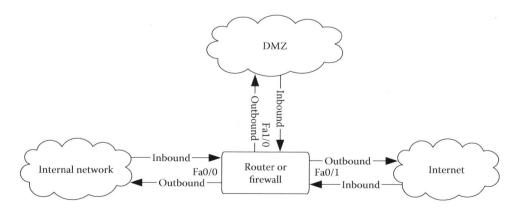

FIGURE 16.43   Application of ACLs to interfaces and inbound/outbound directions.

---

* Entering the router.
† Exiting the router.

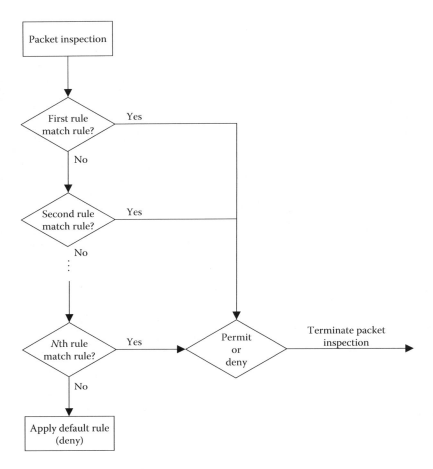

FIGURE 16.44   Packet filtering of ACLs.

For the sake of reducing router's processing, the most frequent rules should be placed at the top of the ACL list. Moreover, at least one permit rule should exist in the ACL; otherwise, all traffic is blocked, as the implicit rule is to block traffic. In case one intends to block a specific traffic and permit all other traffic, these two rules must be explicitly configured, as the default implicit rule is to block traffic.

The above description applies direction to the inbound ACLs. Nevertheless, in case of outbound ACLs, the router has first to decide whether or not the packet is routable. If, based on packet destination IP address and on the routing table, the packet is not routable, then the packet is dropped. In case the packet is routable, it forwards to the output interface. Based on such selected output interface, the router decides whether or not an ACL exists. In case there is an ACL in the output interface, then the generic packet filtering is performed as shown in Figure 16.44.

The order of the rules is an important factor. This results from the fact that a packet is authorized or denied according to the lower order inspected rule.

Depending on the depth and methodology used to verify the conformity of the traffic with the established policy, firewalls can be grouped into three categories:

- *Packet filtering—stateless*: It inspects layer 3 and 4 headers (source and destination IP and port addresses, sequence numbers, and traffic type), allowing or denying the establishment of connections based on these parameters. Note that it does not keep the state of the connections (stateless), that is, it does not take into account the established connections to allow or deny new connections. Moreover, it does not interrupt connections (contrary to the proxy firewall).

- *Packet filtering—statefull*: Similar to the stateless firewall type, it inspects layer 3 and 4 headers. However, unlike stateless inspection, the statefull inspection keeps the state of the established connections, taking into account this information in the decision to allow or deny new connections. An example of a firewall implementing packet filtering in statefull inspection is a firewall that implicitly authorizes traffic coming from an Internet HTTP server into a client located in the intranet only if such TCP connection has been previously established, and requested by the client to the HTTP server (port 80).
- *Application layer gateway—statefull inspection*: This firewall type works at the application layer, performing a deep inspection of the packets contents (instead of only layer 3 and 4 headers), while taking into account the information about the state of the established connections to allow or deny new connections. Moreover, this firewall type interrupts the connections, that is, it establishes two independent connections,* to allow the establishment of a connection request. For this reason, this firewall type is also referred to as *proxy firewall* (similar to the proxy server).

Depending on the objectives, security level, available budget, and available technologies, the access control mechanisms can be implemented in different ways. The three most important technologies are as follows:

- Packet filtering in Cisco routers
- Firewalls Cisco ASA (adaptive security appliances)
- Netfilter architecture (Linux operating system)

16.14.1 PACKET FILTERING CONFIGURATION

The border between a firewall and a router is nowadays not clear. Currently, the firewalls implement routers functionalities, whereas routers also implement firewall functionalities. Cisco routers allow certain types of packet filtering and network address translation (NAT) translation. The extension of packet filtering capabilities depends on the Cisco IOS version, as follows:

- Cisco routers with Cisco IOS version lower than 11.3 support ACLs that implement packet filtering in stateless mode.
- Cisco routers with Cisco IOS version equal to or higher than 11.3 support reflexive ACLs. These ACLs implement packet filtering using statefull inspection.
- Cisco routers whose Cisco IOS includes the firewall feature set support CBAC. Cisco routers with CBAC implement a number of functionalities similar to the firewall application layer gateway—statefull inspection.

16.14.1.1 ACLs—Stateless Inspection

Cisco routers group a number of packet filtering rules into ACLs. Moreover, each ACL is applied to a certain router's interface, in the inbound or the outbound direction.

An ACL may have two different formats: normal or extended. A standard ACL simply allows packet filtering based on the source IP address.

As described in Chapter 10, the configuration of Cisco system devices is performed using the command line interface or using the setup configuration, through the console port or telnet.† Using

---

* In the outgoing direction, the *proxy firewall* establishes, for example, the first connection from a client located in the intranet up to the firewall, and the second connection from the firewall up to a server located in the Internet.
† As described in Chapter 10, the auxiliary port only permits access to the user mode.

the command line interface, the configuration of an ACL is performed in the configuration mode. Therefore, once in the user mode, the command "enable" must be issued in the prompt in order to enter the privileged mode. Then, the configuration is accessed by typing "configure terminal." Once in the configuration mode, a standard ACL can be created using one of the following syntaxes (see Figure 16.45):

- Access-list *list_number* [permit|deny] *source_address* [*wild_card_mask*]
  - *Example*: access-list 10 permit 11.20.0.0 0.0.255.255
  - *Example*: access-list 20 deny 11.20.1.2

An extended ACL allows packet filtering based on a number of parameters:

- Protocol type (IP, TCP, UDP, etc.)
- Source IP address
- Destination IP address
- Source port
- Destination port
- Date and time

Therefore, an ACL includes a number rules to filter the traffic based on layer 3 and 4 addresses, protocol type, and date/time. Since extended ACLs are mostly implemented in Cisco routers, the current description focuses on this type of ACLs. Since a standard ACL does not specify the destination address, this should be placed as close as possible to the destination device. On the contrary, in order to reduce the network usage, since an extended ACL specify the destination address, this should be included as close as possible to the source device.

An extended ACL can be created using one of the following syntaxes (see Figure 16.45):

- Router(config)#access-list extended *list_number*
  - *Example*:
    - Router(config)#access-list extended 120
- Router(config)#**ip** access-list extended *list_name*
  - *Example*:
    - Router(config)#ip access-list extended intranet_to_internet

The first option is utilized when an extended ACL is to be identified by a number, whereas the second option is used to identify an ACL by a name. Note that the "**ip** access-list" command is employed to create an ACL identified by a name. Such a name is alphanumeric and it should not begin with a number. Moreover, *list_number* or *list_name* refers to the number or name given to the ACL. In case of standard ACLs, *list_number* must be within the range 1 and 99 or within 1300 and 1999, whereas for extended ACLs, this number must be within the range 100 and 199 or within 2000 and 2699. Moreover, in case of standard ACLs, the command "extended" is omitted. With the example above, the user is creating an extended ACL with the name intranet_to_internet.

FIGURE 16.45   Example of a router with ACLs used to filter traffic between an intranet and the Internet.

An ACL can be removed by using the word "no," before the sequence of commands used to create the ACL (e.g., **no** ip access-list extended intranet_to_internet).

Once an ACL has been created, the user may start configuring ACL rules, using the following syntax (see Figure 16.45):

- [Permit | deny] *protocol source_address* [*source_port*] *destination_address* [*dest_port*] [log]
  - *Example*: Router(config)#permit tcp 194.13.14.0 0.0.0.255 host 193.12.13.14 eq 80[*]
  - *Example*: Router(config)#deny icmp host 194.13.14.1 193.12.13.0 0.0.0.255 log
  - *Example*: Router(config)#permit icmp 193.12.13.0 0.0.0.255 host 194.13.14.1 echo-reply

The command log is optional, and enables logging when the ACL condition is verified. In the above example, a log is created every time a packet comes from the host 194.13.14.1 to the range of hosts 193.12.13.0–193.12.13.255, using the ICMP. Such packet is denied and a log is generated. Moreover, the command "echo-reply" associated with "permit icmp" means that ping replies are allowed.

The syntax used to define layer 3 and 4 addresses is as follows:

- Source or destination IP address
  - *Any*: It represents any host with any IP address.
  - *host x.y.z.w*: It represents a specific host with the IP address x.y.z.w.
  - *x.y.z.w mask*: It represents a range of hosts with IP addresses starting from x.y.z.w up to x.y.z.w + mask (note that the mask must use the wildcard mask notation[†]).
- Source or destination port address
  - *eq x*: It represents a port address equal to x.
  - *neq x*: It represents a port address not equal to x.
  - *gt x*: It represents a port address greater than x.
  - *lt x*: It represents a port address lower than x.
  - *range x y*: It represents a port address within the range x and y.

From the exposed syntax, one is already able to understand the meaning of the rule exposed above as an example. This allows packets carrying the TCP, from the range of hosts with IP addresses within 194.13.14.0 and 194.13.14.255, from any port (because the source port is not defined) to the host with IP address 193.12.13.14, and with the destination port equal to 80 (HTTP server). Note that the source and destination ports may be omitted. In the above example, the source port was omitted, meaning that it can be any source port. On the contrary, the source and destination IP addresses have always to be explicitly defined. In case the source or destination IP address is any, the word "any" can be used (e.g., deny ip any host 193.12.13.14; this denies IP packets from any IP address, and any port, to the host with the IP address 193.12.13.14, with any port).

Note that rules are recorded by the same order they are inserted. In case an ACL is identified by a name, the different rules are identified by a line number (by default, the rules numbering is 10, 20, 30…). In case one intends to insert a rule in a specific position, the required rule must be preceded by a number (e.g., **20** permit tcp 194.13.14.0 0.0.0.255 host 193.12.13.14 eq 80). Moreover, a rule can be removed by using the command "no," preceding the rule (e.g., **no** 20 permit tcp 194.13.14.0 0.0.0.255 host 193.12.13.14 eq 80).

---

[*] In case one does not remember the port number of a service, the following sequence can be typed: "Router(config)#permit tcp 194.13.14.0 0.0.0.255 host 193.12.13.14 eq ?". The prompt lists all possible options.

[†] The reverse of the conventional mask, being computed as wildcard = 255.255.255.255-subnet mask. For example, a subnet mask 255.255.252.0 corresponds to the wildcard 0.0.3.255.

Once an ACL has been created, and the corresponding rules have been added, one can verify the ACL and the corresponding rules using the following syntax (from the privileged mode):

- Router#show access-list [*list_name|list_number*]
  - *Examples*:
    - Router#show access-list (lists all ACLs, and corresponding rules configured in the router).
    - Router#show access-list intranet_to_internet (shows the rules associated to the ACL with the name intranet_to_internet).

Finally, after having configured an ACL (standard or extended), such ACL needs to be applied to a certain interface, in a certain direction (see Figure 16.43). The syntax used for this purpose comprises two command lines (see Figure 16.45):

- Router(config)#interface *interface*
- Router(config)#ip access-group [*list_name|list_number*] [in|out]*
  - *Example*:
    - Router(config)#interface FastEthernet 0/0
    - Router(config-if)#ip access-group intranet_to_internet in

Finally, if one intends to eliminate an ACL from an interface, the word "no" must be typed before the command line used to apply an ACL to an interface, as follows:

- Router(config)#interface *interface*
- Router(config)#**no** ip access-group [*list_name|list_number*] [in|out]
  - *Example*:
    - Router(config)#interface FastEthernet 0/0
    - Router(config-if)#**no** ip access-group intranet_to_internet in

Figure 16.46 shows an example of creation of an ACL entitled intranet_to_internet, the configuration of four ACL rules, and its application to the interface FastEthernet 0/0, in the inbound direction. It is assumed that the lines and interfaces have already been configured and connected to the devices. The list of commands in Figure 16.46 corresponds to the configuration of an ACL used in the diagram of Figure 16.45, to allow HTTP, HTTPS, SMTP, and POP3 traffic between the intranet and the Internet.

Let us now consider that using the stateless mode, all traffic from one area to another are aimed to be blocked, except the responses to the hosts from the opposite direction. As an example, let us consider that one intends to authorize traffic coming from an HTTP server into a client located in the intranet only if such TCP connection has been previously established, and requested by the client to the HTTP server (port 80). In this case, the syntax used to permit traffic is also applicable, but followed by the word "established":

- Router(config-ext-nacl)#[permit|deny] *protocol source_address* [*source_port*] *destination_address* [*dest_port*] **established**
  - *Example*: Router(config-ext-nacl)#permit tcp host 193.12.13.14 eq 80 194.13.14.0 0.0.0.255 **established**

Note that, by using the command "established," all traffic is blocked, except the one defined in the rule, as long as it has previously been established in the opposite direction.

---

* *in* refers to applying the ACL in the inbound direction (entrance into the router), whereas *out* refers to the outbound direction (router's exit).

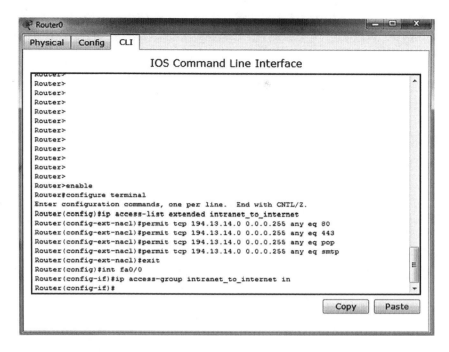

FIGURE 16.46 Example of creation of an ACL and its rules (corresponds to Figure 16.45, between the intranet and the Internet).

Focusing on the example depicted in Figure 16.45, another ACL needs to be created to filter traffic coming from the Internet into the intranet. It is assumed that it is aimed to block all traffic from the Internet to the intranet, except the responses to the intranet hosts, using the stateless mode. Such ACL, entitled internet_to_intranet, should preferably be applied to the interface FastEthernet 0/1, being listed in Figure 16.47.

When configuring a router with address translation, ACL and CBAC*, it is important to keep in mind which function is performed first: address translation or packet filtering? In case address translation is performed first, the packet filtering rules have to specify the translated address, instead of the original address.

Table 16.3 shows the sequence order of the different operations performed by a router.

From Table 16.3 it is viewed that inbound ACL is performed before NAT, allowing ACL rules including original (private) addresses. In all other situations, the translated addresses must be considered in the rules.

Appendix IV lists the configuration of ACLs using the stateless inspection in a router, to filter traffic between three different areas: the intranet, the DMZ, and the Internet.

16.14.1.2   Reflexive ACLs

Cisco routers with Cisco IOS version equal to or higher than 11.3 include reflexive ACLs. Reflexive ACLs allow packet filtering in statefull mode. Reflexive ACLs keep the state of the established connections in the opposite direction, taking into account this information in the decision to allow or deny new connections. An example is an ACL that implicitly authorizes traffic coming from an Internet HTTP server into a client located in an intranet only if such TCP connection has been previously established and requested by the client to the HTTP server (port 80). Note that, unlike to stateless ACLs, reflexive ACLs can only be defined by names (not by numbers).

---

* See Chapter 16 for a detailed description of ACL and CBAC.

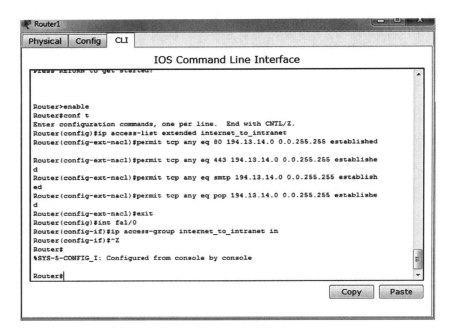

FIGURE 16.47 Example of creation of an ACL and its rules (corresponds to Figure 16.45, between the Internet and the intranet, in stateless mode).

TABLE 16.3

Sequence Order of Operations Performed by a Router

Sequence Order	Outgoing Traffic
1	Inbound ACL
2	NAT
3	Outbound ACL
4	CBAC
5	TCP intercept

Most of the description performed for ACLs in stateless inspection is also valid for reflexive ACLs. The current description focuses on their differences.

Associated with each connection initiated in a certain direction, a router with reflexive ACLs automatically generates a temporary ACL rule in the opposite direction (reflexive), allowing the entrance of the traffic for the duration of the previously established logical link.

The generation of an ACL rule in the opposite direction is performed adding the instruction "reflect" at the end of the rule line, followed by a name given to this rule. The syntax becomes as follows:

- Router(config-ext-nacl)#[permit | deny] *protocol source_address* [*source_port*] *destination_ address* [*dest_port*] **reflect** *name*
  - *Example*: Router(config-ext-nacl)#permit tcp 194.13.14.0 0.0.0.255 host 193.12.13.14 eq 80 **reflect** reverse_traffic

In the case of TCP traffic, the initiation of a TCP connection is performed using the *SYN* message and its finalization is done using the *FIN* message. Therefore, the router keeps track of the established TCP connections, allowing traffic in the opposite direction. In the case of UDP traffic, such traceability is not so straightforward, but the statefull inspection can still be applied by Cisco routers with reflexive ACLs. In this case, the rule authorizing UDP traffic should be followed by "timeout x," where x stands for a number, expressed in seconds, within which it is permitted UDP traffic after the previous UDP datagram in the opposite direction. In case of UDP traffic, the syntax comes:

- Router(config-ext-nacl)#[permit | deny] *protocol source_address* [*source_port*] *destination_ address* [*dest_port*] reflect *name* **timeout *time_in_seconds***
  - *Example*: Router(config-ext-nacl)#permit udp 194.13.14.0 0.0.0.25 host 193.12.13.14 eq domain reflect reverse_traffic **timeout 30**

Note that the command "timeout x" can also be applied to other types of traffic (e.g., TCP). The main advantage of using reflexive ACLs relies on the fact that one does not need to configure all rules in the opposite direction followed by word "established," as in the stateless mode. Using reflexive ACLs the traffic in the opposite direction is simply authorized by using the following syntax in the ACL of the opposite direction:

- Router(config-ext-nacl)#evaluate *name*
  - *Example*: Router(config-ext-nacl)#evaluate reverse_traffic

Focusing again on the example depicted in Figure 16.45, assuming that, in addition to the previous rules considered for the stateless example, it is intended to add a rule authorizing DNS traffic for the duration of 30 seconds after the outgoing traffic, the complete reflexive ACLs configuration becomes

```
Router>enable
Router#conf t
Router (config)#ip access-list extended intranet_to_internet
Router (config-ext-nacl)#permit tcp 194.13.14.0 0.0.0.255 any eq 80 reflect reverse_tcp_traffic
Router (config-ext-nacl)#permit tcp 194.13.14.0 0.0.0.255 any eq 443 reflect reverse_ tcp_traffic
Router (config-ext-nacl)#permit tcp 194.13.14.0 0.0.0.255 any eq pop reflect reverse_ tcp_traffic
Router (config-ext-nacl)#permit tcp 194.13.14.0 0.0.0.255 any eq smtp reflect reverse_
    tcp_traffic
Router (config-ext-nacl)#permit udp 194.13.14.0 0.0.0.255 any eq domain reflect reverse_udp_
    traffic timeout 30
Router (config-ext-nacl)#exit
Router (config)#int fa0/0
Router (config-if)#ip access-group intranet_to_internet in
Router (config-if)#exit
Router (config)#ip access-list extended internet_to_intranet
Router (config-ext-nacl)#evaluate reverse_tcp_traffic
Router (config-ext-nacl)#evaluate reverse_udp_traffic
Router (config-ext-nacl)#exit
Router (config)#int fa0/1
Router (config-if)#ip access-group internet_to_intranet in
Router (config-if)#Ctrl-Z
```

Appendix IV lists the configuration of ACLs using reflexive ACLs (statefull inspection) in a router, to filter traffic between three different areas: intranet, DMZ, and Internet.

16.14.1.3 Context-Based Access Control

Cisco routers with CBAC implement a number of functionalities similar to the firewall application layer gateway—statefull inspection. Cisco routers whose Cisco IOS includes the firewall feature set support CBAC. While the previously described ACLs simply inspect layer 3 and 4 headers, the application layer inspection includes the analysis of the packet and segment contents (payload data). Moreover, since this inspection is performed in statefull mode, this means that the new connections are permitted or denied taking into account the previously established connections.

An important advantage of the CBAC relies on the ability to perform a forecast of the application data traffic that needs to be authorized. This is done based on the service under consideration. This feature is especially important in case of specific services, such as IP telephony or file transfer. In the case of IP telephony, it is known that the session initiation protocol requires two channels: one TCP channel used for control, and another one based on the UDP for data exchange (using the real time protocol [RTP]). Therefore, using the CBAC, once the system has knowledge that the service is the IP telephony, using the SIP, it automatically gives permission to the associated protocols. Moreover, similar to reflexive ACLs, CBACs also keep track of the connections established in a certain direction, allowing the corresponding traffic in the opposite direction.

In addition to the previously described capabilities, since the CBAC technology inspects the application layer data, it can detect, register, and block a number of threats, such as Java applets or DoS attacks.

The syntax associated to a CBAC rule that inspects and authorizes the traffic associated to a certain protocol is as follows:

- Router(config)#ip inspect name *name protocol*
  - *Example*: Router(config)#ip inspect name *control_traffic* sip

Afterward, the inspect rule needs to be activated in an interface, and in a certain direction, using the following syntax:

- Router(config)#interface *interface*
- Router(config-if)#ip inspect *name* [in | out]
  - *Example*:
    - Router(config)#interface FastEthernet 0/0
    - Router(config-if)#ip inspect *control_traffic* in

This inspection is activated in one direction, while the return traffic is automatically authorized in the opposite direction.

Focusing again on the example depicted in Figure 16.45, with the same policy rules considered for the reflexive ACLs, the new router's configuration using CBACs comes:

```
Router>enable
Router#conf t
Router (config)#ip access-list extended intranet_to_internet
Router (config-ext-nacl)#permit tcp 194.13.14.0 0.0.0.255 any eq 80
Router (config-ext-nacl)#permit tcp 194.13.14.0 0.0.0.255 any eq 443
Router (config-ext-nacl)#permit tcp 194.13.14.0 0.0.0.255 any eq pop
Router (config-ext-nacl)#permit tcp 194.13.14.0 0.0.0.255 any eq smtp
Router (config-ext-nacl)#permit udp 194.13.14.0 0.0.0.255 any eq domain
Router (config-ext-nacl)#exit
Router (config)#ip access-list extended internet_to_intranet
Router (config-ext-nacl)#deny ip any any
```

```
Router (config-ext-nacl)#exit
Router (config)#ip inspect name control_traffic http
Router (config)#ip inspect name control_traffic https
Router (config)#ip inspect name control_traffic pop
Router (config)#ip inspect name control_traffic smtp
Router (config)#ip inspect name control_traffic domain timeout 30
Router (config)#int fa0/1
Router (config-if)#ip access-group intranet_to_internet out
Router (config-if)#ip inspect control_traffic out
Router (config-if)#ip access-group internet_to_intranet in
Router (config-if)#Ctrl-Z
```

16.14.2 FIREWALL CISCO ASA

The firewall Cisco ASAs replaced the previously commercialized firewall Cisco private Internet exchange (PIX) appliances, which were limited to packet filtering and address translation (NAT). The firewall Cisco ASAs groups the functionalities of the following individual equipment:

- Cisco PIX appliances, which includes packet filtering and address translation
- Cisco VPN concentrators, which include the implementation of VPNs, using encryption algorithms such as SSL/TLS, IPsec, and AES/3DES
- Cisco IPS series, including intrusion detection and preventive systems
- Authentication, authorization, and accounting configuration, to implement the accountability functionalities (other functionalities than access control, which is implemented by the firewall), as well as the implementation of authentication processes and logging
- Additional optional modules, such as the content security and control, which performs inspection of the traffic, detecting, and giving protection against threats such as virus, spyware, phishing, and spam

The firewall ASA can be accessed through the console port, through telnet, or using a browser. In case of telnet access, the firewall ASA includes the ability to implement the SSH algorithm, which brings an added value in terms of security. The most used method employed to configure a firewall ASA comprises the adaptive security device manager. This consists of a user interface with a graphic application that can be opened in a browser. The different rules can be configured in an intuitive way. A user that is familiar with the configuration of packet filtering and NAT, in Cisco routers' ACLs, can easily configure a firewall ASA through the adaptive security device manager interface.

CHAPTER SUMMARY

This chapter provided a view about the network security. A description about the security services was given, as well as the corresponding attack types.

Confidentiality was the first described security service. It was viewed that confidentiality can be protected using cryptography. Moreover, it was described that eavesdropping, snooping, interception, and trust exploitation are four types of attacks against confidentiality. Eavesdropping, also referred to as *sniffing*, or as *packet sniffing*, is an attack to data in transit without interfering with it. On the other hand, it was viewed that the snooping comprises the visualization of e-mails, passwords, and phone calls, and can also include more advanced mechanisms, such as the remote monitoring of terminals. It was described that the interception consists of an attack to the data in transit, assuming an active behavior in the middle of the communication, acting as the source (source spoofing), or as the recipient of a communication (destination spoofing).

It was described that the integrity is another security service, whose most important attack against integrity, is the man-in-the-middle. It was described that the man-in-the-middle comprises the modification of the content of a communication, while in transit.

We have viewed that availability is also an important security service, and the DoS is an important attack against availability. The DoS aims to block the use of information or data, systems, or resources from their legitimate users. Three types of DoS attacks were studied: the SYN attack, the smurf attack, and the DDoS.

It was studied that authenticity is a service that aims to ensure that a certain information or data was effectively generated by the declared author or entity. It was described that while authenticity is an attribute, or a security service, authentication is a process. We have viewed that the most important attack against authenticity is the replay attack, where an attacker saves the authentication data, and uses it at a later stage to get access to illegitimate resources.

It was described that confidentiality is normally protected with cryptography, while both integrity and authenticity are commonly protected with the digital signature. Finally, availability is typically protected using redundancy.

We have viewed that accountability is a security service that complements the four basic security services. Its main purpose is to ensure that the entity is the legitimate, and that the event logs show the truth of the facts. It was studied that accountability includes several services, which can be split into two groups:

- The I3A, which comprises identification, authentication, authorization, and access control
- The MRA, which comprises monitoring, registration, and auditing functions

The different types of malware were described, including virus, worm, and the Trojan horse.

Then, the physical and environmental security was introduced, representing a threat to the network, in parallel with the other attack types.

We have viewed that risk management comprises a process for monitoring the risk, that is, for monitoring the potential loss of assets and values, which are subject to threats that exploit vulnerabilities in a system, organization, or persons. It was described that the risk can be quantified by the multiplication of the value of goods by the vulnerabilities, and by the threats. Moreover, it was described that the risk can be mitigated, by minimizing the vulnerabilities and the threats.

The security plan was detailed, as well as the security policy, security rules, and procedures. An introduction to the ISO 27002 standard was also performed under the topic of security plan.

We have viewed that the protective measures include a myriad of mechanisms that protect the information attributes. Depending on the information attribute to protect, there are different protective measures that can be implemented. Typically, confidentiality is protected with cryptography.

Different cryptographies were described, including the symmetric and asymmetric cryptography. It was viewed that symmetric cryptography uses the same key to cipher and decipher messages. This requires the previous distribution of keys, which represents a vulnerability. It was described that asymmetric cryptography aims to mitigate this vulnerability, by using two different keys: a public (widely distributed) and a private key (kept secret). Nevertheless, an important disadvantage relies on the fact that ciphering and deciphering messages with asymmetric cryptography takes about 1000 times more time than with symmetric cryptography. It was viewed that using asymmetric cryptography, a message is commonly ciphered with the recipient's public key and deciphered by the recipient with its private key.

It was viewed that the digital signature is a protective measure that provides protection from attacks against integrity, as well as against authenticity. The digital signature is typically implemented using a Hash function applied to the message, which is ciphered by the sender's private key, sent together with the message, and deciphered at the destination's side, with the sender's public key.

It was described that the digital certificate is an important measure that keeps the mapping between a public key and its owner. PKI is meant to allow a secure, public keys' distribution system. PKI includes a hierarchical structure of certification authorities where the higher authorities are entitled validation authorities and the lowers are typically referred to as *registration authorities*.

Combined cryptography aims to exploit simultaneously the advantages of symmetric and asymmetric cryptography, while overcoming their disadvantages. This is achieved by using both symmetric and asymmetric keys in an efficient manner, nevertheless, at the cost of a higher complexity. Three combined cryptographic systems were described: IPsec, SSL, and TLS. It was viewed that SSL and TLS work between the transport layer and the application layer, facilitating the authentication of entities, being used for end-to-end encryption. On the other hand, it was described that IPsec works at the network layer and presents different modes. The tunnel model comprises the encryption of the whole IP packet, being well suited for creating a VPN between a headquarters and a remote delegation, or for creating a VPN between a headquarters and a host.

Then, the topic security in IEEE 802.11 wireless networks was studied. This includes the description of WEP, as well as WPA and WPA2 protocols. Moreover, the protocols EAP and IEEE 802.1x were also studied.

Depending on the services and security level, this chapter described the different network architectures that can be implemented within an organization, such as in a company.

It was described that a DMZ is an external area of a network, which is protected by a firewall. A description about VPN was provided in this chapter. It was described that a VPN is used by both employees and partners to allow access from a public communications network to the intranet, or to a partners' DMZ. Finally, firewall was introduced, as well as the configuration of ACLs in Cisco routers.

The different type of firewalls were described, including the packet filtering stateless, packet filtering statefull, the application layer gateway, and statefull inspection.

REVIEW QUESTIONS

1. Enumerate the studied types of attacks against confidentiality.
2. How is the authentication process implemented in the SSL protocol?
3. Which security subareas are included in the INFOSEC concept?
4. What is the difference between authentication and authenticity?
5. Describe the CHAP.
6. Describe the way accountability protects attacks against confidentiality, integrity, and authenticity.
7. What is a distributed denial of service (DDoS) attack?
8. What are the mechanisms that can be used to protect the information from attacks against confidentiality?
9. Define combined cryptography.
10. What are the mechanisms that can be used to protect the information from attacks against integrity?

11. What is the difference between a digital signature and a digital certificate?
12. What are the command lines that should be used to configure an ACL rule to block TCP traffic from the host with IP addresses 194.13.14.16, from any source port, to access an HTTP and HTTPS servers with any IP address?
13. Describe the man-in-the-middle attack.
14. Which procedures can be used to generate digital certificates? What are their drawbacks?
15. Which types of authentication mechanisms do you know?
16. What is the command line that should be used to configure of an ACL rule to allow TCP traffic from the range of hosts with IP addresses 194.13.14.0 up to 194.13.14.255, from any source port, to access an HTTP server with any IP address?
17. What is the difference between authenticity and nonrepudiation?
18. How is the risk managed, quantified, and what is the residual risk?
19. What are the mechanisms that can be used to protect the information from attacks against authenticity?
20. What does PKI stand for? What is its purpose?
21. What are the elementary operations used in symmetric cryptography?
22. Why is it important to employ long passwords?
23. Which combined protocols do you know?
24. What is a brute force attack?
25. What is a digital signature? What does it protect from?
26. What is the meaning of public key cryptography?
27. In the scope of the risk analysis, what is the difference between vulnerability and threat? Give examples of both.
28. What is the disadvantage of asymmetric cryptography, relating to symmetric?
29. What does VPN stand for?
30. What are the advantages and disadvantages of asymmetric cryptography, relating to symmetric?
31. What is a DMZ?
32. How can a VPN be established between a remote delegation and a headquarters?
33. What is the command line that should be used to configure an ACL rule to permit TCP traffic from the range of hosts with IP address 11.1.0.0 up to 11.1.255.255, from any source port, to access a POP3 server with the IP address 15.1.2.3?
34. What are the command lines that should be used to configure an ACL rule to block TCP traffic from the host with IP address 10.1.1.1, from any source port, to access an HTTPS server with any IP address?
35. What are the differences between a standard ACL and an extended ACL?
36. What is the advantage of using the SSL/TLS relating to the IPsec?
37. Which type of firewalls do you know?
38. What is the difference between an intranet and an extranet?

LAB EXERCISES

1. Consider the network depicted in the figure below. Configure the plotted network using the Cisco Packet Tracer simulator, assuming the following stateless ACL rules (include configuration of NAT translations):
   a. Hosts located in the internal network (192.168.0.0/24) can only access the DMZ servers, namely HTTP, HTTPS, POP3, SMTP, and FTP.
   b. DMZ servers can access the Internet servers, namely HTTP, HTTPS, POP3, and SMTP (194.1.2.253/24).
   c. DMZ servers cannot be accessed from the Internet (except responses to DMZ requests).

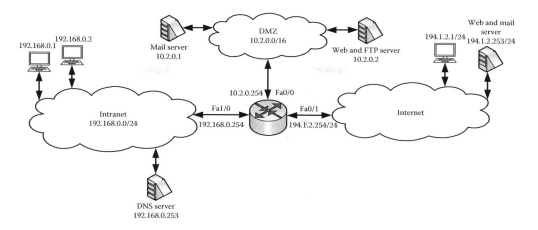

2. With regard to the previous exercise, list the instructions that are different in case a reflexive ACL is adopted (instead of a stateless ACL).

3. Consider the ACL rules listed in Table 16.2 and the network depicted in the figure below. Configure them using the Cisco Packet Tracer simulator, assuming a stateless ACL (do not configure the rules relating to the DMZ application server, nor to the internal database server). Before starting, *clean* the startup configuration, and then re-load the router with the *cleaned* startup configuration. Moreover, configure a ciphered access password to the privileged mode and another to the console port, a hostname, and a motd.

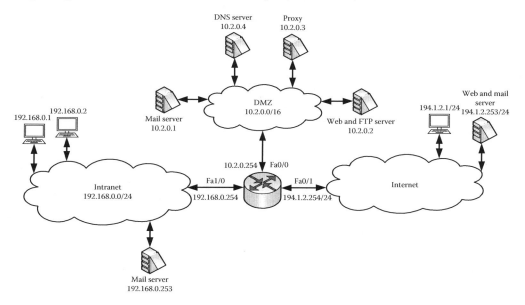

4. Consider the network depicted in the figure below. Configure the plotted network using the Cisco Packet Tracer simulator. Configure the PC connected through the wireless access point WRT300N to receive an IP address from the DHCP server (router D). Router A performs NAT overload for the hosts using private IP addresses. The EIGRP routing protocol is utilized between routers. Configure the following stateless ACL rules:

a. Hosts located in the internal network (192.168.0.0/24) and all other hosts in the other networks can only access the DMZ servers, namely HTTP, HTTPS, POP3, SMTP, and FTP, as well as all other hosts in the other networks, except those in the Internet.

b. DMZ servers can access the Internet servers, namely HTTP, HTTPS, POP3, and SMTP (194.1.2.253/24).

c. DMZ servers cannot be accessed from the Internet (except responses to DMZ requests).

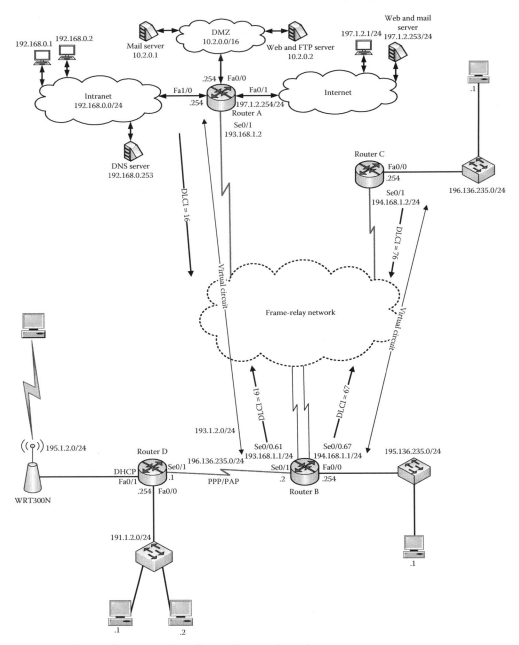

5. Repeat the previous exercise using real network equipment.

6. Consider the network depicted in the figure below. Configure the plotted network using the Cisco Packet Tracer simulator. Configure the PC connected through the wireless access point WRT300N to receive an IP address from the DHCP server (router D). Router A performs NAT overload for the hosts using private IP addresses. The OSPF routing protocol is utilized between routers. Configure the following stateless ACL rules:

a. Hosts located in the internal network (192.168.0.0/24) and all other hosts in the other networks can only access the DMZ servers, namely HTTP, HTTPS, POP3, SMTP, and FTP, as well as all other hosts in the other network, except those in the Internet.

b. DMZ servers can access the Internet servers, namely HTTP, HTTPS, POP3, and SMTP (194.1.2.253/24).

c. DMZ servers cannot be accessed from the Internet (except responses to DMZ requests).

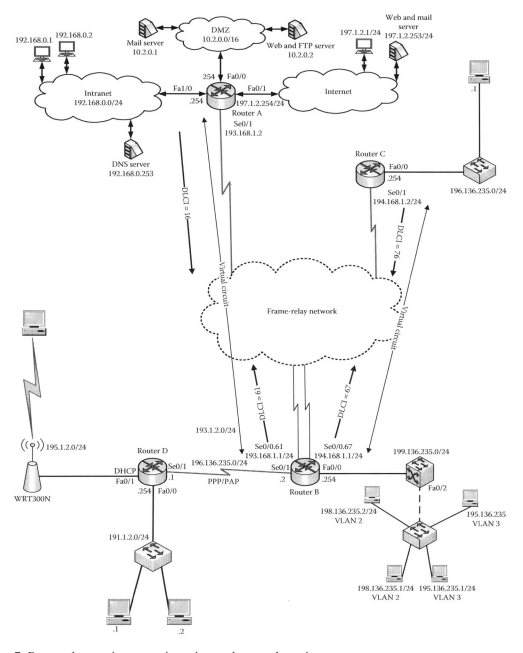

7. Repeat the previous exercise using real network equipment

Appendix I
Fourier Transforms

The Fourier transform of the time domain variable $x(t)$ is mathematically defined by

$$X(f) = \mathcal{F}\left[x(t)\right]$$

$$= \int_{-\infty}^{+\infty} x(t) \cdot e^{-j2\pi ft} dt \tag{I.1}$$

Similarly, the inverse Fourier transform of the frequency domain variable $X(f)$ is mathematically defined by

$$x(t) = \mathcal{F}^{-1}\left[X(f)\right]$$

$$= \int_{-\infty}^{+\infty} X(f) \cdot e^{j2\pi ft} df \tag{I.2}$$

Note that the lower and upper case signal variables correspond to time and frequency domain variables, respectively. The mapping between one and the other is achieved through Fourier transform $\left(\mathcal{F}[x]\right.$ denotes *Fourier transform* of $x)$ and inverse Fourier transform operations $\left(\mathcal{F}^{-1}[x]\right.$ denotes *inverse Fourier transform* of $x)$, that is, $X = \mathcal{F}[x]$ and $x = \mathcal{F}^{-1}[X]$.

A list of Fourier transforms of common functions is listed in Table I.1. As an example, the Fourier transform of the sin c function corresponds to the rectangular pulse, that is, $\mathcal{F}\left[\sin c(2Wt)\right] = (1/2W)\Pi(f/2W)$ (see Table I.1). This means that, having the sin c function in the time domain, its frequency spectrum is the rectangular pulse. The Fourier transform of a signal gives us information about which frequency components are present in the signal, as well as their relative strength.

As defined by Equation I.1, the channel's frequency response $H(f)$ is related to the channel's impulse response $h(t)$ (plotted in Figure I.1) through the Fourier transform defined by $H(f) = \int_{-\infty}^{+\infty} h(t)e^{-j2\pi ft} dt$.

The channel's impulse response $h(t)$ is obtained at the output of the channel when a Dirac function is injected at the channel's input. The resulting frequency domain signal at the channel's output becomes

$$V_R(f) = V_E(f) \cdot H(f) \tag{I.3}$$

Using Equation I.2, the time domain signal at the channel's output can be obtained performing the inverse Fourier transform to the frequency domain signal as follows:

$$v_R(t) = \int_{-\infty}^{+\infty} V_R(f)e^{j2\pi ft} df \tag{I.4}$$

TABLE I.1
Common Fourier Transforms

Function	$v(t)$	$V(f)$
Rectangular	$\Pi\left(\dfrac{t}{\tau}\right)$	$\tau \operatorname{sinc}(f\tau)$
Triangular	$\Lambda\left(\dfrac{t}{\tau}\right)$	$\tau \operatorname{sinc}^2(f\tau)$
Gaussian	$e^{-\pi(bt)^2}$	$\left(\dfrac{1}{b}\right)e^{-\pi(f/b)^2}$
Causal exponential	$e^{-bt}u(t)$	$\dfrac{1}{b+j2\pi f}$
Symmetric exponential	$e^{-b|t|}$	$\dfrac{2b}{b^2+(2\pi f)^2}$
Sinc	$\operatorname{sinc}(2Wt)$	$\dfrac{1}{2W}\Pi\left(\dfrac{f}{2W}\right)$
Sinc squared	$\operatorname{sinc}^2(2Wt)$	$\dfrac{1}{2W}\Lambda\left(\dfrac{f}{2W}\right)$
Constant	1	$\delta(f)$
Phasor	$e^{j(w_c t+\phi)}$	$e^{j\phi}\delta(f-f_c)$
Sinusoid	$\cos(w_c t+\phi)$	$\dfrac{1}{2}\left[e^{j\phi}\delta(f-f_c)+e^{-j\phi}\delta(f+f_c)\right]$
Impulse	$\delta(t-t_d)$	e^{-jwt_d}
Sampling	$\displaystyle\sum_{k=-\infty}^{+\infty}\delta(t-kT_s)$	$\displaystyle f_s\sum_{n=-\infty}^{+\infty}\delta(f-nf_s)$
Signum	$\operatorname{sgn}(t)$	$\dfrac{1}{j\pi f}$
Step	$u(t)$	$\dfrac{1}{j2\pi f}+\dfrac{1}{2}\delta(f)$

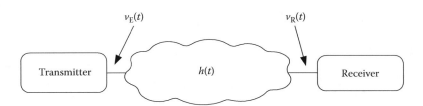

FIGURE I.1   Generic communication system with the signals depicted in the time domain.

Instead of using Equation I.3 followed by Equation I.4 in the computation of the signal at the channel's output, this can also be obtained mathematically in the time domain by performing the convolution operation of the input signal $v_E(t)$ with the channel's impulse response $h(t)$, defined by

$$v_R(t) = \int_{-\infty}^{+\infty} v_E(\tau)h(t-\tau)d\tau \tag{I.5}$$

TABLE I.2
Theorems of Fourier Transforms

Operation	Function	Transform
Superposition	$a_1 v_1(t) + a_2 v_2(t)$	$a_1 V_1(f) + a_2 V_2(f)$
Time delay	$v(t - t_d)$	$V(f)e^{-jwt_d}$
Scale change	$v(\alpha t)$	$\dfrac{1}{|\alpha|}V\left(\dfrac{f}{\alpha}\right)$
Conjugation	$v^*(t)$	$V^*(-f)$
Duality	$V(t)$	$v(-f)$
Frequency translation	$v(t)e^{jw_c t}$	$V(f - f_c)$
Modulation	$v(t)\cos(w_c t + \phi)$	$\dfrac{1}{2}\left[V(f - f_c)e^{j\phi} + V(f + f_c)e^{-j\phi}\right]$
Differentiation	$\dfrac{d^n v(t)}{dt^n}$	$(j2\pi f)^n V(f)$
Integration	$\displaystyle\int_{-\infty}^{t} v(\lambda)\,d\lambda$	$\dfrac{1}{j2\pi f}V(f) + \dfrac{1}{2}V(0)\delta(f)$
Convolution	$v * w(t)$	$V(f)W(f)$
Multiplication	$v(t)w(t)$	$V * W(f)$
Multiplication by t^n	$t^n v(t)$	$(-j2\pi)^{-n}\dfrac{d^n V(f)}{df^n}$

The computation of $v_R(t)$ may be simpler to perform in the frequency domain using Equation I.3, as it consists of a multiplication, while the convolution operation defined by Equation I.5 may be very demanding for complex channel impulse responses.

Table I.1 lists the most common Fourier transforms, while Table I.2 lists the important Fourier transforms theorems.

Appendix II
Common Functions Used in Telecommunications

TABLE II.1

Common Functions Used in Telecommunications

Gaussian probability

$$Q(k) = \frac{1}{\sqrt{2\pi}} \int_k^\infty e^{-\lambda^2/2} d\lambda$$

Exponential

$$\exp(t) = e^t$$

Sin c

$$\sin c(t) = \frac{\sin(\pi t)}{\pi t}$$

Sign

$$\mathrm{sgn}(t) = \begin{cases} 1 & t \geq 0 \\ -1 & t < 0 \end{cases}$$

Step

$$u(t) = \begin{cases} 1 & t \geq 0 \\ 0 & t < 0 \end{cases}$$

Rectangle

$$\Pi\left(\frac{t}{\tau}\right) = \begin{cases} 1 & |t| < \tau/2 \\ 0 & |t| > \tau/2 \end{cases}$$

Triangle

$$\Lambda\left(\frac{t}{\tau}\right) = \begin{cases} 1 - \dfrac{|t|}{\tau} & |t| < \tau \\ 0 & |t| > \tau \end{cases}$$

Appendix III
Example of Configuration of Cisco Routers 1800 Series and Cisco Switches Catalyst 2960 Using Cisco IOS

This appendix lists the configuration commands in different Cisco routers and switches for the internetwork depicted in Figure III.1. As can be seen, the internetwork comprises three Cisco routers and five Cisco switches. Figure III.1 considers that hosts belonging to the same VLAN are connected to switches 0 and 1. In this case, the interfaces used to interconnect these two switches need to be configured in trunk mode (and crossover cable is used). The connectivity between the Switch 1 and Router 1 needs also to be configured in trunk mode, and with as many subinterfaces in the router side as the number of VLANs. Moreover, a multilayer switch (L3Switch) is connected to Router 3, which is intended to perform layer 3 switching.

It is assumed that the lines are already configured and connected to the devices.

In Router1:

Router>enable
Router#conf t
Router(config)#hostname router1
Router1(config)#int se0/0
Router1(config-if)#ip address 196.136.235.1 255.255.255.0
Router1(config-if)#clock rate 64000 (clock rate configured in the DCE terminal of serial cable)
Router1(config-if)#no shutdown
Router1(config-if)#int fa0/0
Router1(config-if)#no ip address (the interface has no IP address)
Router1(config-if)#no shutdown
Router1(config-if)#exit
Router1(config)#int fa0/0.2 (creation of a sub-interface)
Router1(config-subif)#encapsulation dot1q 2 (sub-interface is connected to VLAN 2, using IEEE802.1q)
Router1(config-subif)#ip address 193.136.235.254 255.255.255.0
Router1(config-subif)#int fa0/0.3
Router1(config-subif)#encapsulation dot1q 3
Router1(config-subif)#ip address 195.136.235.254 255.255.255.0
Router1(config-subif)#int fa0/0.99
Router1(config-subif)#encapsulation dot1q 99 native
Router1(config-subif)#ip address 194.136.235.254 255.255.255.0
Router1(config-subif)#exit
Router1(config)#router eigrp 100
Router1(config-router)#network 193.136.235.0 0.0.0.255
Router1(config-router)#network 195.136.235.0 0.0.0.255
Router1(config-router)#passive-interface default (alternatively, one would have to configure fa0/0.2, fa0/0.3 and fa0/0.99 in passive mode)
Router1(config-router)#no passive-interface se0/0

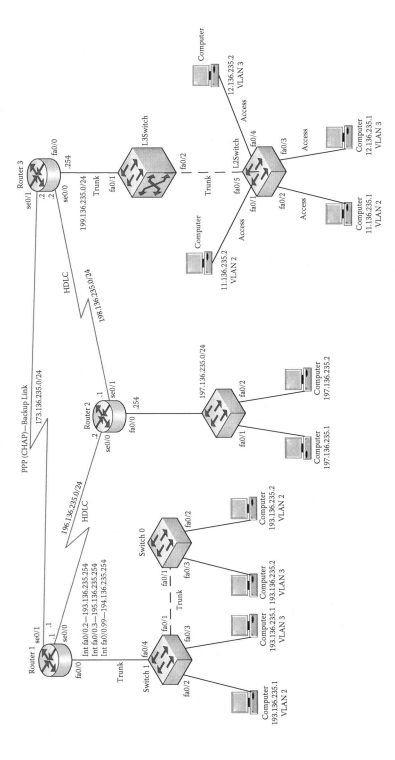

FIGURE III.1 Example of an internetwork configuration.

Router1(config-router)#no auto-summary
Router1(config-router)#exit
Router1(config)#username Router3 password xyz (PPP configuration with CHAP)
Router1(config)#int se0/1
Router1(config)#ip address 173.136.235.1 255.255.255.0
Router1(config)#no shutdown
Router1(config)#encapsulation ppp
Router1(config)#ppp authentication chap
Router1(config)#ip route 0.0.0.0 0.0.0.0 se0/1 254 (floating/backup static route to Router3)
Router1(config)#CTRL-Z
Router1#

In Switch0:

Switch>enable
Switch#conf t
Switch(config)#hostname switch0
switch0(config)#vtp version 2
switch0(config)#vtp mode server
switch0(config)#vtp pruning
switch0(config)#vtp domain Laboratory
switch0(config)#vtp password Xyz@_123
switch0(config)#vlan 2
switch0(config-vlan)#name finances
switch0(config-vlan)#vlan 3
switch0(config-vlan)#name human_resources
switch0(config-vlan)#vlan 98
switch0(config-vlan)#name management
switch0(config-vlan)#vlan 99
switch0(config-vlan)#name native
switch0(config-vlan)#exit
switch0(config)#int fa0/2
switch0(config-if)#switchport mode access
switch0(config-if)#switchport access vlan 2
switch0(config-if)#int fa0/3
switch0(config-if)#switchport mode access
switch0(config-if)#switchport access vlan 3
switch0(config-if)#int fa0/1
switch0(config-if)#switchport mode trunk
switch0(config-if)#switchport trunk native vlan 99
switch0(config-if)#int vlan 98 (configure an IP address to allow the remote management of the switch)
switch0(config-if)#ip address 191.136.235.1 255.255.255.0
switch0(config-if)#CTRL-Z
switch0#

In Switch1:

Switch>enable
Switch#conf t
Switch(config)#hostname switch1
switch1(config)#vtp version 2

switch1(config)#vtp mode client
switch1(config)#vtp pruning
switch1(config)#vtp domain Laboratory
switch1(config)#vtp password Xyz@_123
switch1(config-vlan)#exit
switch1(config)#int fa0/2
switch1(config-if)#switchport mode access
switch1(config-if)#switchport access vlan 2
switch1(config-if)#int fa0/3
switch1(config-if)#switchport mode access
switch1(config-if)#switchport access vlan 3
switch1(config-if)#int fa0/1
switch1(config-if)#switchport mode trunk
switch1(config-if)#switchport trunk native vlan 99
switch1(config-if)#int fa0/4
switch1(config-if)#switchport mode trunk
switch1(config-if)#switchport trunk native vlan 99
switch1(config-if)#int vlan 98 (configure an IP address to allow the remote management of the switch)
switch1(config-if)#ip address 191.136.235.2 255.255.255.0
switch1(config-if)#CTRL-Z
switch1#

In Router2:

Router>enable
Router#config t
Router(config)#hostname Router2
Router2(config)#int fa0/0
Router2(config-if)#ip address 197.136.235.254 255.255.255.0
Router2(config-if)#no shutdown
Router2(config-if)#exit
Router2(config)#int se0/0
Router2(config-if)#ip address 196.136.235.2 255.255.255.0
Router2(config-if)#no shutdown
Router2(config-if)#int se0/1
Router2(config-if)#ip address 198.136.235.1 255.255.255.0
Router2(config-if)#clock rate 64000
Router2(config-if)#no shutdown
Router2(config-if)#exit
Router2(config)#router eigrp 100
Router2(config-router)#network 196.136.235.0 0.0.0.255
Router2(config-router)#network 197.136.235.0 0.0.0.255
Router2(config-router)#network 198.136.235.0 0.0.0.255
Router2(config-router)#passive-interface fa0/0
Router2(config-router)#no auto-summary
Router2(config-router)#CTRL-Z
Router2#

Note that the configuration of the switch attached to Router 2 is not listed, but interfaces remain in the default access mode and without VLANs, and therefore, only the hostname can be configured. This configuration is straightforward.

In Router3:

Router>enable
Router#conf t
Router(config)#hostname Router3
Router3(config)#int fa0/0
Router3(config-if)#ip address 199.136.235.254 255.255.255.0
Router3(config-if)#no shutdown
Router3(config-if)#int se0/0
Router3(config-if)#ip address 198.136.235.2 255.255.255.0
Router3(config-if)#no shutdown
Router3(config-if)#exit
Router3(config)#ip route 0.0.0.0 0.0.0.0 199.136.235.1 (default route to Router3)
Router3(config)#router eigrp 100
Router3(config-router)#network 198.136.235.0 0.0.0.255
Router3(config-router)#redistribute static (to advertise the static route through EIGRP)
Router3(config-router)#passive-interface default
Router3(config-router)#no passive-interface se0/0
Router3(config-router)#no auto-summary
Router3(config-router)#exit
Router3(config)#username Router1 password xyz (PPP configuration with CHAP)
Router3(config)#int se0/1
Router3(config)#ip address 173.136.235.2 255.255.255.0
Router3(config)#no shutdown
Router3(config)#encapsulation ppp
Router3(config)#ppp authentication chap
Router3(config)#ip route 193.136.235.0 255.255.255.0 se0/1 254 (floating static route to Router1)
Router3(config)#ip route 195.136.235.0 255.255.255.0 se0/1 254
Router3(config)#ip route 196.136.235.0 255.255.255.0 se0/1 254
Router3(config)#ip route 197.136.235.0 255.255.255.0 se0/1 254
Router3(config)#CTRL-Z
Router3#

In L3Switch:

Switch>enable
Switch#conf t
Switch(config)#hostname L3Switch
L3Switch(config)#vtp version 2
L3Switch(config)#vtp mode server
L3Switch(config)#vtp pruning
L3Switch(config)#vtp domain Solaris
L3Switch(config)#vtp password Vwx@_456
L3Switch(config)#vlan 2
L3Switch(config-vlan)#name operations
L3Switch(config-vlan)#vlan 3
L3Switch(config-vlan)#name strategy
L3Switch(config-vlan)#vlan 99
L3Switch(config-vlan)#name native
L3Switch(config-vlan)#exit
L3Switch(config)#interface fa0/1

L3Switch(config-if)#switchport mode trunk
L3Switch(config-if)#no switchport (interface in route-port mode)
L3Switch(config-if)#ip address 199.136.235.1 255.255.255.0 (assign an IP address to interface)
L3Switch(config-if)#no shutdown
L3Switch(config)#interface fa0/2
L3Switch(config-if)#switchport mode trunk
L3Switch(config-if)#switchport trunk native vlan 99
L3Switch(config-if)#exit
L3Switch(config)#ip routing (activation of layer 3 switching)
L3Switch(config)#ip route 0.0.0.0 0.0.0.0 199.136.235.254 (default static route to Router3)
L3Switch(config)#int vlan 2
L3Switch(config-if)#ip address 11.136.235.254 255.255.255.0 (assign an IP address to VLAN)
L3Switch(config-if)#no shutdown
L3Switch(config-if)#int vlan 3
L3Switch(config-if)#ip address 12.136.235.254 255.255.255.0
L3Switch(config-if)#no shutdown
L3Switch(config-if)#int vlan 99
L3Switch(config-if)#ip address 13.136.235.254 255.255.255.0
L3Switch(config-if)#no shutdown
L3Switch(config-if)#exit
L3Switch(config-if)#int range fa0/3-24
L3Switch(config-if)#shutdown (for security reasons, shutdown interfaces not utilized)
L3Switch(config-if)#CTRL-Z
L3Switch#

In L2Switch:

Switch>enable
Switch#conf t
Switch(config)#hostname L2Switch
L2Switch(config)#vtp version 2
L2Switch(config)#vtp mode client
L2Switch(config)#vtp pruning
L2Switch(config)#vtp domain Solaris
L2Switch(config)#vtp password Vwx@_456
L2Switch(config)#int fa0/1
L2Switch(config-if)#switchport mode access
L2Switch(config-if)#switchport access vlan 2
L2Switch(config)#int fa0/2
L2Switch(config-if)#switchport mode access
L2Switch(config-if)#switchport access vlan 2
L2Switch(config)#int fa0/3
L2Switch(config-if)#switchport mode access
L2Switch(config-if)#switchport access vlan 3
L2Switch(config)#int fa0/4
L2Switch(config-if)#switchport mode access
L2Switch(config-if)#switchport access vlan 3
L2Switch(config)#int fa0/5
L2Switch(config-if)#switchport mode trunk

L2Switch(config-if)#switchport trunk native vlan 99
L2Switch(config-if)#int range fa0/6-24
L2Switch(config-if)#shutdown (for security reasons, interfaces not utilized should be shutdown)
L2Switch(config-if)#CTRL-Z
L2Switch#

Appendix IV
Examples of ACL Configuration Using Cisco IOS

This appendix presents two examples of access control list (ACL) configuration in a Cisco router 1800 series. The first corresponds to the stateless inspection, while the second performs the same tasks but using reflexive ACLs (statefull inspection). Both configurations correspond to the Cisco router of the architecture depicted in Figure IV.1, necessary for the creation of the ACLs to implement the following policy rules:

- The hosts in network 10.2.0.0/16 (intranet) should have access to the following protocols (services) of the demilitarized zone (DMZ): hyper text transfer protocol (HTTP), hyper text transfer protocol secure (HTTPS), file transfer protocol (FTP), post office protocol (POP3), and simple mail transfer protocol (SMTP). The hosts in network 10.2.0.0/16 should not have direct connectivity with the Internet.
- The DMZ web and mail servers should have access to the Internet for the corresponding services.
- The DMZ servers should not be accessed from the Internet, except to responses to DMZ server's accesses.

It is assumed that the lines and interfaces are already configured and connected to the devices.

Stateless configuration:

Router#configure terminal
Router(config)#ip access-list extended intranet_to_dmz (creates an ACL)
Router(config-ext-nacl)#permit tcp 10.2.0.0 0.0.255.255 host 192.168.0.2 eq 80
Router(config-ext-nacl)#permit tcp 10.2.0.0 0.0.255.255 host 192.168.0.2 eq 443
Router(config-ext-nacl)#permit tcp 10.2.0.0 0.0.255.255 host 192.168.0.2 eq ftp
Router(config-ext-nacl)#permit tcp 10.2.0.0 0.0.255.255 host 192.168.0.1 eq pop
Router(config-ext-nacl)#permit tcp 10.2.0.0 0.0.255.255 host 192.168.0.1 eq smtp
Router(config-ext-nacl)#exit
Router(config)#interface FastEthernet 0/0
Router(config-if)#ip access-group intranet_to_dmz in (activates ACL in Fa0/0, inbound)
Router(config-if)#exit
Router(config)#ip access-list extended dmz_to_internet_and_intranet (creates an ACL)
Router(config-ext-nacl)#permit tcp host 192.168.0.2 eq 80 10.2.0.0 0.0.255.255 **established** (from DMZ to Intranet)
Router(config-ext-nacl)#permit tcp host 192.168.0.2 any eq 80 (from DMZ to Internet)
Router(config-ext-nacl)#permit tcp host 192.168.0.2 eq **ftp** 10.2.0.0 0.0.255.255 **established**
Router(config-ext-nacl)#permit tcp host 192.168.0.2 eq 443 10.2.0.0 0.0.255.255 **established**
Router(config-ext-nacl)#permit tcp host 192.168.0.2 any eq 443
Router(config-ext-nacl)#permit tcp host 192.168.0.1 eq pop 10.2.0.0 0.0.255.255 **established**

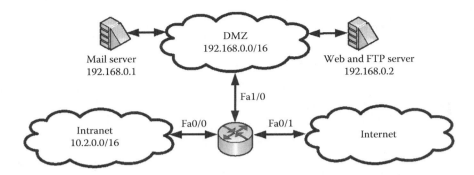

FIGURE IV.1   Example of a Cisco router in a network to be configured with ACLs.

Router(config-ext-nacl)#permit tcp host 192.168.0.1 any eq pop
Router(config-ext-nacl)#permit tcp host 192.168.0.1 eq smtp 10.2.0.0 0.0.255.255 **established**
Router(config-ext-nacl)#permit tcp host 192.168.0.1 any eq smtp
Router(config-ext-nacl)#exit
Router(config)#interface FastEthernet 1/0
Router(config-if)#ip access-group dmz_to_internet_and_intranet in (activates ACL in Fa1/0, inbound)
Router(config-if)#exit
Router(config)#ip access-list extended internet_to_intranet_and_dmz (creates an ACL inbound in fa0/1)
Router(config-ext-nacl)#permit tcp any eq 80 host 192.168.0.2 **established**
Router(config-ext-nacl)#permit tcp any eq 443 host 192.168.0.2 **established**
Router(config-ext-nacl)#permit tcp any eq pop host 192.168.0.1 **established**
Router(config-ext-nacl)#permit tcp any eq smtp host 192.168.0.1 **established**
Router(config-ext-nacl)#exit
Router(config)#interface FastEthernet 0/1
Router(config-if)#ip access-group internet_to_intranet_and_dmz in (activates ACL in Fa0/1, inbound)
Router(config-if)#Ctrl-Z

Reflexive ACLs (statefull configuration):

Router#configure terminal
Router(config)#ip access-list extended intranet_to_dmz (creates an ACL)
Router(config-ext-nacl)#permit tcp 10.2.0.0 0.0.255.255 host 192.168.0.2 eq 80 reflect reverse_traffic_intranet
Router(config-ext-nacl)#permit tcp 10.2.0.0 0.0.255.255 host 192.168.0.2 eq 443 reflect reverse_traffic_intranet
Router(config-ext-nacl)#permit tcp 10.2.0.0 0.0.255.255 host 192.168.0.2 eq ftp reflect reverse_traffic_intranet
Router(config-ext-nacl)#permit tcp 10.2.0.0 0.0.255.255 host 192.168.0.1 eq pop reflect reverse_traffic_intranet
Router(config-ext-nacl)#permit tcp 10.2.0.0 0.0.255.255 host 192.168.0.1 eq smtp reflect reverse_traffic_intranet
Router(config-ext-nacl)#exit
Router(config)#interface FastEthernet 0/0
Router(config-if)#ip access-group intranet_to_dmz in (activates ACL in Fa0/0, inbound)
Router(config-if)#exit
Router(config)#ip access-list extended dmz_to_internet_and_intranet (creates an ACL)
Router(config-ext-nacl)#evaluate reverse_traffic_intranet (from DMZ to Intranet)

Router(config-ext-nacl)#permit tcp host 192.168.0.2 any eq 80 reflect reverse_traffic_dmz (from DMZ to Internet)

Router(config-ext-nacl)#permit tcp host 192.168.0.2 any eq 443 reflect reverse_traffic_dmz

Router(config-ext-nacl)#permit tcp host 192.168.0.1 any eq pop reflect reverse_traffic_dmz

Router(config-ext-nacl)#permit tcp host 192.168.0.1 any eq smtp reflect reverse_traffic_dmz

Router(config-ext-nacl)#exit

Router(config)#ip access-list extended internet_to_intranet_and_dmz (creates an ACL)

Router(config-ext-nacl)#evaluate reverse_traffic_dmz (from Internet to DMZ)

Router(config-ext-nacl)#exit

Router(config)#interface FastEthernet 1/0

Router(config-if)#ip access-group dmz_to_internet_and_intranet in (activates ACL in Fa1/0, inbound)

Router(config-if)#exit

Router(config-ext-nacl)#exit

Router(config)#interface FastEthernet 0/1

Router(config-if)#ip access-group internet_to_intranet_and_dmz in (activates ACL in Fa0/1, inbound)

Router(config-if)#Ctrl-Z

Appendix V
Responses to Review Questions

CHAPTER 1

1. Digital signals present several advantages, relating to analog, which are listed as follows:
   a. *Error control is possible in digital signals*: Corrupted bits can be detected and/or corrected.
   b. Since they only present two discrete values, the consequences of channel impairments can be more easily detected and avoided (as compared to analog signals).
   c. Digital signals can be regenerated, almost eliminating the effects of channel impairments. Contrarily, the amplification process of analog signals results in the amplification of signals, noise, and interferences, keeping the signal-to-noise ratio (SNR) relationship unchanged.
   d. The digital components are normally less expensive than analog.
   e. Digital signals facilitate cryptography and multiplexing.
   f. Digital signals can be used to transport different sources of information (voice, data, multimedia, etc.), in a transparent manner.

   However, digital signals present an important disadvantage:
   a. For the same information source, the bandwidth required to accommodate a digital signal is typically higher than in the analog counterpart. This results in a higher level of attenuation and distortion.

2. A modem is used to transmit bits (digital signals) in an analog transmission medium. Moreover, a modem can also be used when the aim is to transmit a carrier-modulated signal (i.e., a signal around a certain carrier frequency), instead of a baseband signal (i.e., a signal around the null carrier frequency).

3. A simplex communication consists of a communication between two or more entities where the signals only flow in a single direction. In this case, one entity only acts as a transmitter and the other(s) as a receiver. Note that the transmitter may be transmitting signals to more than one receiver. When the signals flow in a single direction, but with alternation in time, it is stated that the communication is half-duplex. Therefore, although both entities act simultaneously as transmitter and as receiver (at different time instants), instantaneously, each host acts either as a transmitter or as a receiver. Finally, when the communication flows simultaneously in both directions, it is in full-duplex mode. In this case, two or more entities act simultaneously as both transmitter and receiver. Full-duplex communications require, normally, two parallel transmission mediums (e.g., two pairs of wires): one for transmission and another for reception.

4. There are two different types of topologies: physical and logical. The physical topology refers to the real cabling distribution along the network, while the logical topology stands for the way the data is exchanged in the network. A physical star topology with a repeater (a hub) as a central node presents a common medium and is shared by all network hosts. In such case, the logical topology is the bus (common and shared medium).

5. Unicast stands for a communication whose destination is a single station. In case the destination of data is all the network stations, the communication is referred to as a broadcast. Very often broadcast communications are established in a single direction (i.e., there is no

feedback from the receiver into the transmitter). Finally, when the destination of the data is more than a single station, but less then all network stations, the communication is referred to as *multicast.*

6. Analog signals present a continuous amplitude variation over time. Example of an analog signal is the voice. Contrarily, digital signals present amplitude discontinuities (e.g., voltages or light pulses). An example of digital data is the bits generated in a workstation. The text is another example of digital data.

7. A LAN consists of a network that covers a reduced area like a home, an office, or a small group of buildings (e.g., an airport), using high-speed data rates. A MAN consists of a backbone (transport network) used to interconnect different LANs within a coverage area of a city, a campus, or similar. Such a backbone is typically implemented using high-speed data rates. Finally, a WAN consists of a transport network (backbone) used to interconnect different LAN and MAN, whose area of coverage typically goes beyond 100 km. While the transmission medium used in a LAN is normally the twisted pair, optical fiber, or wireless, the optical fiber is among the most used transmission medium in a MAN and a WAN.

8. Depending on the end-to-end service provided, the connection modes through networks can be of two types: connectionless and connection-oriented. These modes are used in any of the layers of a network architecture, such as in the open system interconnection reference model (OSI-RM), or in the transmission control protocol/Internet protocol (TCP/IP) stack. In order to provide a connection-oriented service, there is the need to previously establish a connection before data is exchanged, and to terminate it after data exchange. The connection is established between entities, incorporating the negotiation of the QoS and cost parameters of the service being provided. The communication is bidirectional, and the data is delivered with reliability. Moreover, in order to prevent a faster transmitter to overload a slower receiver, flow control is employed (to prevent overflow situations). Contrarily, the connectionless mode does not require the previous connection setup, before data is exchanged. Consequently, the data is delivered based on the best effort, without error control and without flow control. While the TCP is connection oriented, the UDP and the IP are connectionless protocols.

9. A point-to-point communication establishes a direct connection (link) between two adjacent stations, between two adjacent network nodes (e.g., routers), or between an end station and an adjacent node. A network can be viewed as a concatenation of point-to-point communications, composed of several nodes and end stations, where each node is responsible for switching the data, such that an end-to-end connection is established between two end stations. An end-to-end network connection consists of a concatenation of several point-to-point links, where each of these links can be implemented using a different transmission medium (e.g., satellite and optical fiber).

10. Circuit switching establishes a permanent physical path between the origin and the destination. This is the switching mode used in classic telephone networks. Only after startup, it is allowed a synchronous exchange of data. This end-to-end path (circuit) is permanently dedicated until the connection ends. The time to establish the connection is high, but it is assured a delay only due to the propagation speed of signals. This kind of switching is ideal for delay-sensitive communications, such as voice. With the introduction of data services, the notion of packet switching has arrived, which considers the segmentation of a message into parts, where each part is referred to as a packet (with fixed or variable length). Packets are forward and switched independently through the nodes of a network, between the origin and the destination. Each packet transports enough information to allow its routing (end destination address included in a header). While the nodes of a circuit-switching network establish a permanent shunt between one input and one output, since packet switching considers a number of bits grouped into a packet, the nodes of a packet switching network only switch data for the duration of a packet transmission. The following packet that uses the same input

or output of a node may belong to a different end-to-end connection. Consequently, packet switching networks make much better usage of the network resources (nodes) than circuit switching. Note that a node of a packet switching network is typically a router. There are different packet switching protocols, such as ATM, IP, frame relay, and X.25.

11. Different services need different transmission rates, different margin of latencies and jitter, different performances, or even fixed or variable transmission rates. Different services present different quality of service requirements:

    a. Voice communications are delay sensitive, but present low sensitivity to data loss, and require low data rate but approximately constant.

    b. Iterative multimedia communications (e.g., web browsing) are sensitive to data loss, requiring considerable data rate, with a variable transmission rate, and are moderately delay sensitive.

    c. Pure data communications (e.g., database access, file transfer) are highly sensitive to data loss, requiring relatively variable data rate, without sensitivity to delay.

    Jitter is defined as the delay variation through the network. Depending on the application, jitter can be a problem, or jitter issues can be disregarded. For instance, data applications that only deliver their information to the user if the data is completely received (reassembling of data) pay no attention to the jitter issues (e.g., file transfer). This is totally different if voice and video applications are considered; those applications degrade immediately if jitter occurs.

12. The convergence of telecommunications can be viewed in different ways. It can be viewed as the convergence of services, that is, the creation of a network able to support different types of service, as voice, data (e-mail, web browsing, database access, files transfer, etc.), and multimedia, in an almost transparent way to the user. The convergence can also be viewed as the complement between telecommunications, information systems, and multimedia in a way to achieve a unique objective: make the information available to the user with reliability, speed, efficiency, and at a low price. According to Gilder's law, the speed of telecommunications will increase three times every year, in the next 20 years, and, according to Moore's law, the speed of microprocessors will duplicate every 18 months. The convergence can also be the integration of different networks into a single one, in a transparent way to the user. The convergence can also be viewed as the convergence between fixed and mobile concepts, as the mobile is covering indoor environments, allowing data and television/multimedia services, traditionally provided by fixed services, whereas fixed telecommunications are giving mobility with the cordless systems, whose example is the DECT standard.

13. Refer to the concept discussed in point 4. On the other hand, a physical star topology with a switch as a central node corresponds, as well, to a logical star topology. Moreover, a logical ring topology corresponds to a physical star topology with a central node that rigidly switches the data to the adjacent host (left or right).

14. Media is classified into the following three groups:

    a. Text: plaintext, hypertext, ciphered text, and so on

    b. Visuals: images, cartography, videos, VTC, graphs, and so on

    c. Sounds: music, speech, other sounds, and so on

    While the text is inherently digital data (mostly represented using a string of 7-bit ASCII characters), the visuals and sounds are typically analog signals, that need to be digitized first, in order to allow its transmission through a digital network, such as an IP-based network (e.g., the Internet or an intranet). The multimedia is simply the mixture of different types of media, such as speech, music, images, text, graphs, and videos.

15. When media sources are being exchanged through a network, it is generically referred to as traffic. The traffic can be considered as real-time (RT) or nonreal-time (NRT). While RT traffic is delay sensitive, NRT media is not. An example of RT traffic is the telephony

or the VTC, whereas a file transfer or the web browsing can be viewed as NRT traffic. RT traffic can also be classified as continuous or discrete. Continuous RT traffic consists of a stream of elementary messages with interdependency. An example of continuous RT traffic is the telephony, whereas the chat is an example of discrete RT traffic. Finally, RT continuous traffic can still be classified as delay tolerant or delay intolerant. RT continuous delay-tolerant traffic can accommodate a certain level of delay in signals, without sudden performance degradation. Such tolerance to delays results from the use of a buffer that stores in memory the difference between the received data and the played data. In case the transfer of data is suddenly delayed, the buffer accommodates such delay, and the media presented to the user does not translate such delay introduced by the network. The video streaming is an example of a delay-tolerant media. Contrarily, the performance of delay-intolerant traffic degrades heavily when the data transfer is subject to delays (or variation of delays). An example of RT continuous and delay-intolerant media is the telephony or the VTC. The Internet telephony (IP telephony) or the VTC allows a typical maximum delay of 200 ms, in order to achieve an acceptable performance.

16. While the convergence approach was based on the ability to allow information sharing using a common network infrastructure, the new approach consists of the use of the network as an enabler to allow sharing of knowledge. It consists of the ability to provide the right information to the right person at the right time. For this to be possible, a high level of interactivity made available to each Internet user is required. In parallel, business intelligence are important platforms that allow decision makers to receive the filtered information required for the decision to be made in a correct moment. We observe, nowadays, an explosion of ad-hoc applications that allows any Internet user to inject nonstructured information (e.g., Wikipedia) into the Internet world, in parallel with an increase of peer-to-peer applications such as Torrent, eMule, and IP telephony. Social networks are currently being used by millions of people that allow the exchange of unmanaged multimedia by groups of people just to share information or by groups interested in the same subject. Note that this multimedia exchange can be text, audio, video, multiplayer games, and so on. This can only be possible with the ability of the IP to support all types of services in parallel with the provision of quality of service by the network, that is, with the convergence as a support platform. This is the new paradigm of the modern society: the collaborative age. The collaborative age of the Internet can also be viewed as the transformation of man-to-man communication into man-to-machine and machine-to-machine communication, using several media, and where the source or destination party can be a group instead of a single entity (person or equipment).

CHAPTER 2

1. Both the TCP and the UDP work at the transport layer of the TCP/IP model. While the TCP is a connection-oriented protocol (performs error control and flow control), the UDP is a connectionless protocol, and therefore it does require the previous connection setup before data is exchanged.

2. The network layer of the OSI reference model is the layer responsible for forwarding the packets along the several nodes of the network. This is performed using the address or the virtual channel identifier (VCI) contained in the NSDU (network service data unit) present in each packet and the routing tables present in each of the network nodes (routers).

3. Since a LAN works at the data link layer, it makes use of frames, instead of packets, and the forwarding of frames is performed using the destination medium access control (MAC) address, instead of the destination IP address. Therefore, before a host or a router sends frames to a certain destination, such station needs to find the MAC address that corresponds to the destination IP address included in the packet. This is required because a

packet is encapsulated into a frame, for local transmission, and one of the frame's overhead is the destination MAC address. Such mapping is listed in the ARP table. If the IP address is not listed in the ARP table then the ARP procedure needs to be implemented as follows: when a host has a packet to send or to relay, it tries to find the destination IP address in the ARP table, in order to extract the corresponding MAC address. In case there is no entry table corresponding to such address, it broadcasts (in the LAN) an ARP packet that contains information about a desired IP address. The station with such IP address answers with a hello packet, and the station extracts its MAC address, inserting a new line into the ARP table with the mapping. This entry to this table is kept for a certain period of time. After a certain period without traffic to be passed to (or received from) this station, the entry to this table is removed, and the procedure is re-started, when required.

4. The TCP is a connection-oriented protocol. Therefore, it requires the previous connection setup, before data is exchanged. The data is exchanged using error control and flow control. Moreover, the TCP assures that the packets are delivered in the correct sequence, packet duplication is avoided, and the lost packets are detected and corrected.

5. While the switch performs layer 2 switching based on the destination MAC address, the hub simply repeats in all other outputs the bits received in one input. Therefore, the medium becomes common, and when a station sends data, all other stations in the same LAN receive such data. When a hub is utilized, the CSMA-CD is employed to define when a host is allowed to transmit within a LAN. Contrarily, with a switch working in half-duplex, we can have half of the stations transmitting to the other half of the stations, without collisions. This results in a much better network efficiency.

6. Routers make use of routing tables. In the datagram mode, a routing table stores information about the output interface through which packets should be forwarded in order to reach a certain destination address. In the virtual circuit mode, a routing table stores information about the output interface that corresponds to a certain virtual circuit. Note that the virtual circuit mode allows data to be forwarded faster, but the construction of the routing table is more complex (and requires higher level of overhead) than in the case of datagram. The IP is based on datagram mode.

7. A router works at layer 3 of the OSI reference model, and switches packets along the internetwork, that is, between different networks (e.g., between a LAN and WAN). This is performed using the destination IP address and using the routing table that is stored in each router. On the other hand, the switch is the central node of a network, and acts at layer 2 of the OSI reference model. The switch forwards frames within a LAN based on the destination MAC address. A router has typically two or more interfaces, and the interfaces are normally of different type. On the other hand, a switch has typically a wide number of interfaces (12 or 24), and these interfaces are typically of the same type.

8. The ARP is used to map destination IP addresses into the destination MAC addresses. When a router or a host that is connected to a LAN receives a packet, it consults an ARP table to find out the MAC address that corresponds to the destination IP address of the packet. This is performed because the LAN works at the data link layer. In case the ARP table does not have an entry corresponding to such IP address, it implements the ARP protocol to discover the corresponding MAC address.

9. Both the TCP and the UDP work at the transport layer of the TCP/IP model. While the TCP is a connection-oriented protocol (performs error control and flow control), the UDP is a connectionless protocol, and therefore it does require the previous connection setup before data is exchanged. Moreover, contrary to UDP, the TCP assures that the packets are delivered in the correct sequence, packets duplication is avoided, and lost packets are detected and corrected. Nevertheless, the TCP introduces latency in signals (not used VoIP traffic) and cannot be used when the link is unidirectional (e.g., broadcast satellite link). The file transfer is normally implemented using the FTP, which is based on the TCP.

On the other hand, the network management is normally implemented using the SNMP, which is based on the UDP.

10. The network layer is responsible for the end-to-end forwarding of data. This is performed using routers between different networks, whose switching is performed using the destination address or the virtual circuit identifier, and consulting the routing table present in the router.

11. The problem of interconnecting terminals in a network is a complex task. The approach of trying to solve all the problems without segmentation of functions in groups becomes an equation with a very difficult solution. Therefore, the traditional solution is to group functionalities into different layers and allocate each group to a different layer. This is called network protocol architecture, also commonly known as *network architecture*. This approach only defines *what* is to be done by each layer, but not *how* such functionalities are to be implemented by the layer, whose responsibility belongs to the protocol of the layer individually. This approach leaves room for a layer to improve (due to, e.g., technological evolutions), without implications in the remaining layers, as long as the interface between a certain layer and its adjacent layers are kept as specified by the network protocol architecture. In this sense, the network architecture defines the number of layers, what is to be done by each layer, and the interface between different layers. Note that a network architecture not based on layers would not allow changing the *how to do* without changing the architecture itself and without changing the remaining functions of the network architecture. The TCP/IP architecture is the most implemented network architecture model based on layers.

CHAPTER 3

1. The main sources of interferences can be grouped in four main categories—intersymbol interference (ISI), multiple access interference, co-channel interference, and adjacent channel interference. Besides interference, there are other channel impairments, such as the attenuation, noise, and distortion.

2. In both cases, the channel impairments originate at a degradation of the received signal-to-noise plus interference ratio (SNIR). The SNIR needs to be higher than a certain threshold to be possible to extract the data from the received signal, which is valid for either analog or digital signals. In the case of digital signals, the degradation of the SNIR originates from a degradation of the bit error rate.

3. The most important types of noise can be grouped as follows: external noise (which can be split into atmospheric and man-made noise), extraterrestrial noise, and internal noise (which can be split into thermal and electronic noise).

4. ISI occurs in digital transmissions of symbols when the channel is characterized by the existence of several paths, where the delay of relevant signal replicas arrives at the receiver's antenna with a delay higher than the symbol period. ISI exists when the signal is propagated through a channel whose RMS delay spread of the channel is higher than the symbol period. In this case, this effect can be viewed in the frequency domain as having two sinusoids with frequency separation greater than the channel coherence bandwidth, being affected differently by the channel (in terms of attenuation and delay/phase shift). This corresponds to distortion, but applied to digital signals.

5. Adjacent channel interference consists of an inadequate bandwidth overlapping of adjacent signals. This is due to inadequate frequency control, transmission with spurious, broadband noise, intermodulation distortion, transmission with a bandwidth greater than the one to which the operator is authorized, and so on. Guard bands are normally used to minimize the inadequate frequency control. On the other hand, co-channel interference occurs when two different communications using the same channel interfere with each other. In a cellular environment, this occurs when a communication is interfered by another

communication being transmitted in the same carrier frequency, but typically coming from an adjacent cell. In cellular networks using time division multiple access (TDMA)/ frequency division multiple access (FDMA), this type of interference can be mitigated by avoiding the use of the same frequency bands in adjacent cells, introducing the concept of frequency reuse factor higher than 1. Therefore, the difference between adjacent channel interference and co-channel interference relies on the fact that while the former considers an interference generated from signals that overlap the spectrum (normally partially); the latter also refers to a spectrum overlap but coming from different locations (normally different cells).

6. Multiple access interference occurs in networks that make use of multiple access techniques. This type of interference is experienced when there is no perfect orthogonality between signals from different users, viewed at the receiver's antenna of a certain user. In TDMA networks, this orthogonality is normally assured through guard periods, which avoids the overlapping of signals transmitted in different time slots (from different users). In FDMA networks, this orthogonality is assured by using guard bands and filters that reject undesired in-band interferences.

7. For roll off factor $\alpha = 0$ and bandpass SSB transmission, we obtain the minimum bandwidth capable of transmitting signals with zero ISI defined by

$$B_{\min} = (1/2T_S) = (R_S/2) = (R_b/2)\log_2 M$$

where:

R_S stands for the symbol rate

R_b stands for the bit rate

$\log_2 M$ stands for the number of bits transported in each symbol, where M stands for the symbols constellation order

Specifically, the minimum bandwidth of a baseband transmission of absolute values $B_{\min}$ is defined by $B_{\min} = R_S/2$. Naturally, the transfer function that leads to the minimum bandwidth with $\alpha = 0$ is not physically realizable. Consequently, the transmission bandwidth is always higher than the minimum bandwidth $B_{\min}$.

8. The bandwidth of a baseband transmission of absolute values B_T is defined by $B_T' = \left[(1+\alpha)/(2T_S)\right]$. However, in the case of bandpass transmissions (carrier modulated), the previously negative baseband part of the spectrum becomes positive, being also transmitted. Therefore, the transmitted bandwidth is $f \in \left[-W + W\right]$ (i.e., it is doubled), being defined by $B_T = 2 \cdot B_T'$ becoming $B_T = 2 \cdot \left[(1+\alpha)/2T_S\right]$. Consequently, the minimum bandwidth of a bandpass transmission (double side band [DSB]) is obtained making $\alpha = 0$, originating $B_{\min} = 2R_S/2 = R_S$.

9. Transmitted signals are not composed of a single frequency. In contrast, signals are composed of a myriad of frequency components, presenting a certain bandwidth. As an example, an audible spectrum spans from around 20 Hz up to around 20 kHz. Transmission mediums tend to introduce different attenuations at different frequencies. As a rule of thumb, the attenuation level tends to increase with the increase of the frequency. Moreover, channels tend to introduce different delays and nonlinear phase shifts at different frequency components. Therefore, the signal after the propagation through a medium is subject to different attenuations and nonlinear phase shifts at different frequency components, and hence, the received signal is different from the transmitted one. This effect is known as *distortion*.

10. There are two different types of distortion: attenuation and phase shift. Attenuation distortion stands for a signal whose bandwidth is subject to different attenuations in different frequency components. Moreover, phase shift distortion stands for a signal whose bandwidth is subject to nonlinear phase shifts in different frequency components.

11. Spurious is a particular type of direct interference, which is normally generated in transmitters. A spurious consists of an undesired transmission in a frequency band different from the one reserved to send the signal. A receiver located at short distance from such transmission with spurious may result in a high power interfering signal that may block one or more channels. The measure that can be used to mitigate this direct interferences consists of keeping transmit and receive antennas sufficiently spaced apart, to assure the required isolation. Moreover, spurious transmission filtering is normally mandatory from frequency management regulators. Pre- and post-selector filters may also mitigate the negative effects of spurious.

12. Thermal noise is experienced inside electrical conductors (wires, electrolytes, resistors, etc.), being caused by thermal agitation of charges at the amplifier's input resistance. In case of radio communications, thermal noise presents a wide variation of amplitude, depending on the temperature viewed by the receive antenna. The frequency profile of the thermal noise presents a power spectral density approximately constant along the frequency spectrum, that is, the thermal noise is approximately white. The noise power P_n captured by its amplifier's input resistance is given by

$$P_n = k_B T_n B$$

where:

k_B is the Boltzmann's constant, with $k_B = 1.3806503 \times 10^{-23}$ J/K (expressed in Joules per degree Kelvin)

T_n is the resistor's absolute temperature (expressed in kelvin)

B is the receiver's bandwidth (expressed in Hertz)

13. Free space path loss is quantified by

$$\text{FSPL} = \left(4\pi df/c \right)^2$$

where:

d stands for the distance from the transmitter

f represents the frequency

c the speed of light

Therefore, in the special case of electromagnetic waves, and assuming free space propagation, the attenuation increases with the square of the distance.

14. The capacity of any telecommunications system is taken to be the resulting throughput obtained through the full usage of the allowed spectrum. For an additive white Gaussian noise channel, Claude Shannon derived, in 1948, the Shannon capacity formula given by $C = W \log_2 \left(1 + (S/N) \right)$ (bit/s). This equation provides information about maximum theoretical rate at which the transfer of information bits can be achieved, with an acceptable quality, in a certain transmission medium which has a channel bandwidth W (in Hz), power of noise N, and a transmit signal power S (both in watts).

15. Using the Shannon capacity formula $C = W \log_2 \left(1 + (S/N) \right)$ (bits/n), assuming $W = 1$ MHz and $S/N = 5$ db, we obtain the maximum speed of information bits $C = 16,609,655$ bps $\simeq$ 16.6 Mbps. Note that $S/N = 5$ db first needs to be translated into linear units, before it is inserted into the Shannon capacity formula.

16. Two or more signals present at a nonlinear element (such as in a transmit amplifier, receiver, low noise amplifier, multicoupler) are processed, and additional signals are generated, at sum and difference of multiple frequencies, being known as intermodulation

products (IMPs). IMPs can be of third order, fifth order, seventh order, and so on. Assuming that two isolated carriers f_1 and f_2 are present at a nonlinear element, the generated third-order IMPs becomes $(2f_1 - f_2)$ and $(2f_2 - f_1)$. IMP is a type of indirect interference that degrades the SNIR and, in case of digital signals, degrades the bit error rate. This can be mitigated with the use of post-selector filters (at the transmitter), as well as pre-selector filters (at the receiver).

17. Atmospheric noise consists of an electromagnetic disturbance, being caused by natural atmospherics' phenomenon, such as lightning discharges in thunderstorms. It consists of cloud-to-ground and cloud-to-cloud flashes. On the other hand, man-made noise is electromagnetic, being caused by human activity, namely by the use of electrical equipment, such as car ignitions, domestic equipment, or vehicles. The intensity of this kind of noise varies substantially with the region. Urban man-made noise tends to be more intense than rural noise. This noise is characterized by the emission of low duration and high power pulses, when the corresponding source is activated (e.g., when the car ignition is activated).

18. The noise factor is defined as the ratio between the received noise power P_n and the noise power delivered by a charge with the reference noise temperature of 300 K (T_o) and bandwidth B (k_B is the Boltzmann's constant, with $k_B = 1.3806503 \times 10^{-23}$ J/K). The noise factor is defined by $f_a = (P_n/kT_oB)$. Expressing this value in logarithmic units leads us to the noise figure, defined by $F_a = 10\log_{10} f_a$.

19. In case of a cascade of N electronic devices (e.g., amplifiers and filters), the resulting noise factor f_{OUT} becomes $f_{OUT} = f_1 + ((f_2 - 1)/g_1) + ((f_3 - 1)/g_1g_2) + ((f_4 - 1)/g_1g_2g_3) + \cdots + ((f_N - 1)/\prod_{i=1}^{N-1} g_i)$, where g_i stands for the device gain and $f_i : i = 1\ldots N$ for the noise factor of the ith electronic device.

20. The ideal frequency response of a channel, in terms of attenuation, is one that has a flat response, that is, presents the same attenuation for all frequency components. Moreover, the ideal frequency response of a channel, in terms of phase shift, is one that has a linear response, that is, presents a linear phase shift along the frequency spectrum.

21. Such nonideal frequency response that translates in attenuation and phase shift distortion can be mitigated using an equalizer at the receiver side. The equalizer aims to transform the nonlinear phase shift introduced by the channel into a linear phase shift introduced by the combined system composed of the propagation channel and the equalizer. Similarly, the equalizer frequency response should be such that the attenuation of the combined system that results from cascading the channel and the equalizer is a straight line, that is, a continuous attenuation over the bandwidth of interest (absence of distortion).

22. According to the Nyquist sampling theorem, the minimum sampling rate that can be employed to digitize a signal with a spectrum in the range 8–60 kHz is $f_{min} = 2 \times B_{max}$, where B_{max} is the maximum frequency component present in the signal, that is, 60 kHz. This makes the minimum sampling rate $f_{min} = 2 \times 60$ kHz $= 120$ kHz.

CHAPTER 4

1. The advantages of optical fibers, relating to twisted pairs, are as follows:
   a. Extremely high bandwidth, which translates in high throughputs (typically 150 THz).
   b. Low attenuation coefficient, as compared to twisted pairs.
   c. Moreover, as opposed to other cable transmission mediums, the attenuation coefficient of an optical fiber varies very smoothly with the frequency, which translates in low level of distortion.
   d. Longer repeating distances, which results from low level of attenuation.
   e. Immunity to electromagnetic interferences, that results from the fact that optical fibers are typically made of silica glass (S_iO_2) fiber. Since this substance is not a metallic conductor, optical fibers do not suffer from crosstalk.

    f. *Small dimensions and low weight*: An optical fiber presents typically a diameter 10 times smaller than that of a coaxial cable and its weight is typically 30 times lower than that of a coaxial cable.

    g. *Greater capacity*: This results from the fact that a duct previously used for coaxial cables can accommodate 10 times more optical fibers, and because the capacity of an optical fiber is much higher than that of a coaxial cable (typically 300,000 times higher).

    h. *Reduced cost*: Silica is one of the most abundant materials in the earth, which makes it much less expensive than copper.

2. The most important cable transmission mediums studied are twisted pairs, coaxial cables, and optical fibers.

3. Twisted pairs are normally grouped as unshielded twisted pairs, foiled twisted pairs, shielded twisted pairs, or as screened shielded twisted pair—S/STP. As the name refers, the UTP cabling is not surrounded by any shielding, whereas STP presents a shielding with a metallic braid or sheathing, applied to each individual pair of wires that protects wires from noise and interferences. When the shielding is applied to multiple pairs, instead of a single pair of wires, it is referred to as screening. This is the case of FTP, being also referred to as the *screened unshielded twisted pair*—S/UTP. It consists of a UTP cabling whose shielding surrounds the cable (screened), not presenting shielding in each pair of copper wires. Consequently, while this cabling presents good resistance to interferences originated from outside of the cable, the crosstalk properties (interference between different pairs of the cabling) is typically poorer than STP.

   Finally screened foiled twisted pair cabling—S/FTP, also referred to as S/STP, presents a shielding surrounding both individual pairs and the entire group of copper pairs, and therefore, it is both externally and internally protected from adjacent pairs (crosstalk).

4. Twisted pair cables can be grouped into different categories, from 1 to 7. Increased cabling category results in a higher bandwidth and data rate. The better performance is achieved at the cost of better and thicker copper wires, isolation, improved shielding, or improved twisting. Consequently, higher bandwidths tend to correspond to higher costs. With the exception of the voice-graded twisted pair, cables of the other categories comprise four pairs of conductors.

5. NEXT stands for near end crosstalk. Let us consider a host connected to a switch in full-duplex mode. Such host needs to use two conductor pairs: one for transmission and another for reception. However, when the host starts a transmission, depending on the isolation/coupling properties, a certain amount of the transmitted signal is inducted in the adjacent pair, which is employed to receive another signal in the opposite direction. Naturally, this interfering signal (transmitted signal) is superimposed on the received one, degrading its SNR. This effect is referred to as interference, being quantified by the NEXT factor, which refers to the attenuation, expressed in decibel, between the transmitting pair and the receiving one. Note that the receiver of reference is located close to the transmitter of reference. Metallic conductors of better quality correspond to a higher NEXT value.

6. STP presents a shielding with a metallic braid or sheathing applied to each individual pair of wires that protects wires from noise and interferences. When the shielding is applied to multiple pairs, instead of a single pair of wires, it is referred to as *screening*. This is the case of FTP, being also referred to as the *screened unshielded twisted pair*—S/UTP. It consists of a UTP cabling whose shielding surrounds the cable (screened), not presenting shielding in each pair of copper wires. Consequently, while this cabling presents good resistance to interferences originated from outside of the cable, the crosstalk properties (interference between different pairs of the cabling) is typically poorer than STP.

7. The power sum ELFEXT, PSELFEXT, corresponds to the ELFEXT, but where the FEXT is replaced by the PSFEXT, being given by $PSELFEXT = PSFEXT_{dB} - Att_{dB}$. This parameter

should be taken into account in a configuration that comprises the parallel transmission of signals using multiple metallic cable transmission mediums (e.g., multiple twisted pairs). Since we are considering the power sum, PS, the number of parallel transmission lines should be more than one.

8. STP presents a shielding with a metallic braid or sheathing, applied to each individual pair of wires that protects wires from noise and interferences. S/STP presents a shielding surrounding both individual pairs and the entire group of copper pairs, and therefore, it is protected both externally and internally from adjacent pairs (crosstalk).

9. The bandwidth and attenuation frequency response of a twisted pair cable depends on a number of parameters, such as the type of protection (e.g., quality of shielding or screening), twisting length, isolation, and copper quality. Improving the type of protection (e.g., using STP, instead of UTP), reduces the crosstalk effect and increases the bandwidth of the cable. The choice of the proper twisted length is also a mechanism that improves the bandwidth of a cable.

10. The power sum ACR, PSACR, corresponds to the ACR, but where the NEXT is replaced by the PSNEXT, being quantified by $PSACR_{dB} = PSNEXT_{dB} - Att_{dB}$. PSACR is used in scenarios where the NEXT generated from multiple parallel pairs must be taken into account. An example, is the scenario with a cable of four pairs: one is used for reception and three are used for transmission. In this scenario, the three pairs used for transmission generate crosstalk interference in the received one, whose value is quantified by $PSNEXT_{dB}$, instead of by $NEXT_{dB}$. Consequently, one must consider $PSACR_{dB}$, instead of ACR_{dB}.

11. The link distance and the bandwidth of a cable are two variables that are used when dimensioning a cable transmission medium. A certain cable has a specific attenuation coefficient profile, as a function of the frequency. The attenuation corresponds to the attenuation coefficient multiplied by the distance. Therefore, increasing the link distance results in an increased attenuation, which may become critical for certain frequency components. Consequently, increasing the link distance typically reduces the maximum bandwidth that can be used.

12. When the most important requirement is mechanical protection instead of capacity, then the preferable optical cable is the tight buffer. The level of mechanical protection provided by the tight buffer is higher than that of the loose tube. This is achieved by sacrificing the number of optical fibers available in a cable. The individual buffer and the individual mechanical and anti-humidity protection give extra resistance, as compared to the loose tube.

13. Let us focus on the scenario where two parallel lines are employed to split a high rate transmission into two smaller rate transmissions. Along the transmitting chain, the transmitted signal (desired signal) reaches the reference receiver with attenuation. Moreover, it is superimposed with the interfering signal (crosstalk), as quantified by the FEXT. Therefore, the performance index similar to ACR but measured far from the transmitter of reference is referred to as *equal level far end crosstalk*. The ELFEXT refers to the difference, in decibel, between the FEXT and the attenuation, that is, $ELFEXT_{dB} = FEXT_{dB} - Att_{dB}$.

14. Multimode optical fibers are typically characterized by the modal bandwidth. The modal bandwidth characterizes the effect of modal dispersion. This factor relates the available bandwidth with the optical fiber length, that is, with the link distance. Considering a modal bandwidth of 300 MHz·km we may write $BW \times L = 300$ MHz·km, where BW stands for the bandwidth and L for the length. Considering a link length of 3 km, the maximum bandwidth that this fiber can support becomes $BW = 300 / L$ MHz, that is, 100 MHz.

15. There are two main types of optical fibers: single mode and multimode. The former type of fibers supports a single propagation mode, whereas the latter supports several modes. In the single mode, a single path exists over the fiber length, between the emitter of the light pulse and the corresponding receiver. In this case, there are no rays propagated in

any direction different from the normal to the fiber section. Conversely, in the multimode fiber, several paths exist between the light emitter and the receiver. There are two different types of multimode fibers: step index and graded index. The core of multimode step index fiber presents a single refraction index, whereas the graded index fiber presents an index that decreases from the center to the extremities of the core. In the graded index fiber, different paths still exist, but they converge to the same point. Therefore, as in the case of a step index fiber, the receiver of a graded index fiber detects a signal that is composed of the superimposition of several replicas (i.e., the same signal from different paths), but these replicas tend to be aligned in time, that is, they tend to be synchronized. The level of alignment is never perfect, resulting in a certain level of pulse dispersion. However, the level of pulse dispersion is smaller than in the case of step index fiber, resulting in an increased bandwidth.

16. The single-mode optical fiber supports a single propagation mode, whereas the multimode supports several modes. In the single mode, a single path exists over the fiber length, between the emitter of the light pulse and the corresponding receiver. In this case, there are no rays propagated in any direction different from the normal to the fiber section. Conversely, in the multimode fiber, several paths exist between the light emitter and the receiver (reflected in the border between the core and the cladding), and the rays arrive at the receiver with dispersion originating, in the case of digital signals, ISI. This effect is almost neglected in the case of single-mode optical fibers. Consequently, the bandwidth made available by a single-mode optical fiber is much higher than that of a multimode.

17. There are two different types of multimode fibers: step index and graded index. The core of multimode step index fiber presents a single refraction index, whereas the graded index fiber presents an index that decreases from the center to the extremities of the core. In the graded index fiber, different paths still exist, but they converge to the same point. Therefore, as in the case of a step index fiber, the receiver of a graded index fiber detects a signal that is composed of the superimposition of several replicas (i.e., the same signal from different paths), but these replicas tend to be aligned in time, that is, they tend to be synchronized. The level of alignment is never perfect, resulting in a certain level of pulse dispersion. However, the level of pulse dispersion is smaller than in the case of step index fiber, resulting in an increased bandwidth.

18. In the case of the multimode optical fiber, the typical diameter of the cladding is 125 μm, whereas the core diameter may span from 50 up to 65 μm.

19. The FEXT must be utilized when we have two parallel lines (twisted pairs) transmitting simultaneously. In this scenario, the receiver of reference is far from the transmitter. This is why this parameter is referred to far end crosstalk.

20. In the case of the single-mode optical fiber, the typical diameter of the cladding is 125 μm, whereas the core diameter may span from 3 up to 10 μm.

CHAPTER 5

1. Fading is characterized by random variations of the received signal level. This is caused by several factors, such as atmospheric turbulences, movement of the receiver or the transmitter, movement of the environment that surrounds the receive antenna, variations of the atmospheric refraction index, and so on.

   Depending on the depth of the received signal fluctuations, and of their reasons, there are two types of fading:
   a. Slow fading (shadowing)
   b. Fast fading (multipath fading)
   Slow fading (shadowing) is characterized by slow variations of the received signal level, being mainly caused by the terrain contour between the transmitter and receiver, being

directly related to the presence of obstacles in the signal path that avoid line-of-sight propagation. This effect can be compensated for with power control schemes. Fast fading (multipath fading) is characterized by fast variations of the received signal level, being caused by the reflection of the signal in various objects (buildings, trees, vehicles, etc.) that originate multiple replicas of the signal reaching the receive antenna through different paths. These replicas arrive with different delays and attenuations, superimposed in such a way that they will interfere with each other, either constructively or destructively. Due to the mobility of the transmitter or receiver, and of the surrounding objects, the replicas are subject to variations on their paths, and hence in their delays and attenuations, leading to great oscillations on the envelope of the received signal.

2. Depending on the orbit altitude and attitude, there are different types of orbits: geostationary earth orbit (GEO), medium earth orbit (MEO), low earth orbit (LEO), and highly elliptical orbit (HEO). The LEO altitude varies between 300 and 2000 km, whereas the MEO orbit corresponds to an altitude between 5,000 and 15,000 km. The GEO altitude is typically 35,782 km. Finally, the HEO presents an elliptical orbit with perigee at very low altitudes (typically 1,000 km) and with the apogee at high altitudes (between 39,000 and 53,600 km).

3. The LEO is the one that has an orbit at lower altitude, which varies between 300 and 2000 km.

4. The LEO communications can operate with lower power levels and with lower antenna gains than GEO. Typically, LEO communications use an omnidirectional antenna, whereas GEO communications make use of directional antennas (parabolic type). Moreover, communications provided by LEO constellations present lower delays than those of GEO. Nevertheless, the coverage of the GEO (approximately one-third of the earth's surface) is much higher than that of the MEO, and the MEO requires frequent handovers (higher control complexity).

5. The carrier-to-noise ratio (C/N) corresponds to the relationship between the carrier and the noise powers: $(C/N) = (P_R/P_N)$. In this expression, the power of the carrier $C = P_R = \text{EIRP} \cdot g_R \cdot A_{tt}$ and $N = P_N = k_B T_n B$ stands for the power of noise at the receiver. The C/N differs from the SNR because the former refers to the power of a modulated carrier, whereas the latter refers to the signal power after carrier demodulation (baseband), that is SNR refers to the relationship between the baseband signal and the noise powers. The conversion between C/N and SNR depends on the modulation scheme under consideration.

6. The SNR is normally the performance indicator adopted in analog communications, whereas in digital communications the performance indicator considered is E_b/N_0. The bit error rate can also be utilized in digital communications as a performance index.

7. A received electromagnetic wave can be viewed as the result of several propagation effects, namely line-of-sight, reflection, diffraction, and scattering. When multiple components are present, the received signal is composed of the sum of all these components. Reflection consists of a change in the wave's propagation direction as a result of a collision into a surface. Electromagnetic waves are typically reflected in buildings, vehicles, or streets. Diffraction occurs when a wave faces an obstacle that does not allow it reaching the receive antenna in line-of-sight. In this case, even in the absence of line-of-sight, a bending effect of waves is experienced, allowing the waves reaching the receive antenna, but properly attenuated. Scattering occurs when a wave is reflected by an obstacle that is not flat. Since the incident wave covers a certain area (a group of points in the surface), and since each point of such area has a different normal to the obstacle, the scattering effect corresponds to an amount of successive reflections, each one in each point of the surface's obstacle covered by the incident wave.

8. Long-range radio communications can be achieved by different means: the modern type of long-range communications is normally achieved with satellite communications. Nevertheless, this can also be achieved using the denominated short wave or high frequency

(HF) communications, whose waves propagate at long range using the ionosphere. In fact, the skywave propagation can be the mode to support a long-range communication link using frequencies from few hundreds of kHz up to few dozens of MHz. Skywave propagation consists of successive refraction in the ionosphere layers and successive reflection in the earth's surface.

9. Groundwave propagation is the sum of several elementary waves: (a) the surface component (wave), whose electromagnetic waves are guided over the earth's surface; (b) the direct wave; and (c) the reflected wave in the ground. On the other hand, surface component can be viewed as the result of diffraction of low frequency electromagnetic waves by the earth's surface. As known from the knife-edge model, diffraction effect is experienced with higher intensity at lower frequencies (as lower frequencies are less subject to attenuation by objects). The surface wave propagates mainly using the vertical polarization, as horizontal polarization experiences high attenuation levels.

10. Skywave propagation can be the mode to support for a long-range communication link using frequencies from few hundreds of kHz up to few dozens of MHz. Skywave propagation consists of successive refraction in the ionosphere layers and successive reflection in the earth's surface. On the other hand, surface component can be viewed as the result of diffraction of low frequency electromagnetic waves by the earth's surface. As known from the knife-edge model, diffraction effect is experienced with higher intensity at lower frequencies (as lower frequencies are less subject to attenuation by objects). The surface wave propagates mainly using the vertical polarization, as horizontal polarization experiences high attenuation levels.

11. During the day, the typical layers are D layer (between 50 and 90 km), E layer (between 90 and 140 km), F1 layer (between 140 and 200 km), and F2 layer (above 230 km). Note that F2 layer is the most important, as it is present even in the absence of sun (during the night). Furthermore, since it is placed at higher altitudes, it also allows establishing communication links at longer distances. Moreover, it refracts higher frequencies. During the night, F1 and F2 layers merge, creating a single F layer. Moreover, while the E layer tends to remain present during the night, the D layer is normally absent. In addition to the above-described layers present during the day and night, E sporadic may also be present under certain conditions of the ionosphere (during the day or night). This layer may refract the same frequencies as the F layer.

12. Ionospheric layers of higher altitudes allow establishing communication links at longer distances. Moreover, higher altitude layers refract higher frequencies. Therefore, if one intends to achieve higher ranges, a higher frequency should be utilized, as these frequencies are not refracted by lower layers, crossing it.

13. Ionospheric layers of higher altitudes allow establishing communication links at longer distances. Moreover, higher altitude layers refract higher frequencies. Therefore, if one intends to achieve higher ranges, a higher frequency should be utilized, as these frequencies are not refracted by lower layers, crossing it.

14. The relationship between E_b/N_0 and C/N is given by $(E_b/N_0) = (C/N) \cdot (B/R_b)$, where B/R_b stands for the inverse of the minimum spectral efficiency.

15. Taking into account one reflected ray, the received power strength presents a decay rate of 4 with the distance whereas, in the free space model, the received power presents a decay rate of 2. Consequently, one can conclude that the presence of reflected waves represents a negative effect in the received signal strength. Establishing a link over a ground that presents bad reflection properties makes the negative effect of reflected waves less visible, reducing the decay rate of the received power strength with the distance.

16. The diffraction effect is normally quantified using the knife-edge model (without line-of-sight between the transmitter and the receiver). The knife-edge model considers a semi-infinite plan, located between a transmit and a receive antenna, in the object's position.

The knife-edge presents a number of formulas that allows quantifying the received signal strength, as a function of the carrier frequency, of the distance from the transmitter and the receiver to the semi-infinite plan, and as a function of the depth of the receiver, relating to the surface of the semi-infinite plan. This model is very useful in many different scenarios. One common application of this model is when one wants to quantify the attenuation introduced by an obstacle or by the earth curvature in a microwave line-of-sight link.

17. The received signal level increases with the decrease of the carrier frequency, the increase of the horizontal distance between the receive antenna and the semi-infinite plan (which represents the obstacle) and with the decrease of the depth of the receive antenna.

18. The level of interference generated by the reflected signal depends on several factors such as the antenna directivity. The use of an antenna with low antenna gain in the direction of the reflected wave reduces the level of interferences and, consequently, the signal fluctuations caused by fading, as well as the decay rate of the received signal strength with the distance. Alternatively, the selection of a path that blocks the reflected wave also leads to a reduction in the level of interference, reducing the decay rate, as well. Moreover, for longer distances between a transmit and a receive antenna, the signal fluctuations tend to decrease, but the level of attenuation tends to be higher than that in free space propagation (i.e., only direct path). Transmitting electromagnetic waves over a soil that presents low refraction index also leads to low level of interference between the direct and reflected path, and lower decay rate. Finally, using diversity such as multiple-input multiple-output (MIMO) systems avoids the fading effects, improving very much the performance of communications.

19. For both scenarios, the received power strength decreases with the increase of the distance. Taking into account one reflected ray, the received power strength presents a decay rate of 4 with the distance, whereas in the free space model, the received power presents a decay rate of 2. Consequently, one can conclude that the presence of reflected waves represents a negative effect in the received signal strength. Establishing a link over a ground that presents bad reflection properties makes the negative effect of reflected waves less visible, reducing the decay rate of the received power strength with the distance.

20. The FSPL refers to the losses between the radiated power $P_E g_E$ and the EIRP at a certain distance d, being defined by $FSPL = \left(4\pi d/\lambda\right)^2$.

21. The bit energy E_b is defined by $E_b = P_R \cdot T_B = (P_R/R_B)$, where R_B is the transmitted bit rate, T_B is the transmitted bit period, and P_R stands for the received power.

22. Depending on the depth of the fast fading, there are two statistical models characterizing these effects:

    a. *Rayleigh model—fast and deep variations*: It is typically employed when there is no line-of-sight between the transmitter and the receiver, that is, there is only interference between the several reflected, diffracted, and scattered multipaths. This is normally experienced in urban environments. Considering that, for each delay τ_i, a large number of scattered waves arrive from random directions then, in accordance with the central limit theorem, $a_i(t)$ can be modeled as a complex Gaussian process with zero mean. This means that the phase $\theta_i(t)$ will follow a uniform distribution in the interval $\left[0 \ 2\pi\right]$ and the fading amplitude $|h(\tau,t)|$ will follow a Rayleigh distribution.

    b. *Rice model—fast but weak variations*: It is typically employed when in the presence of a line-of-sight between the transmitter and the receiver, to which several multipaths are added at the receiver. In this case, there is interference between the line-of-sight and the several reflected paths. This effect is defined by a Rice distribution. It consists of a sum of a constant component (line-of-sight) with several reflected paths (defined by a Rayleigh distribution). This effect is typical of rural or indoor environments. Assuming the presence of a line-of-sight component with amplitude A arriving at the receiver, then $a_i(t)$ will be a complex Gaussian process with nonzero mean and thus the fading amplitude $|h(\tau,t)|$ will follow a Rice distribution.

23. Slow fading, or shadowing, is experienced when there is no direct line-of-sight between the transmitter and the receiver, and therefore the propagation is characterized by diffraction. The average value of the received signal level follows a log-normal distribution (the logarithm of the amplitude of the field follows a normal distribution). Higher attenuations have been experienced in urban zones with higher building densities. The standard deviation s increases with an increase in the considered area, an increase in the buildings proportions, and an increase in the carrier frequency. Typical values for s, in cellular environments, are between 6 and 18 dB.

24. Silence zone corresponds to the area between the maximum range covered by the ground wave propagation and the skip distance. The skip distance corresponds to the start of the skywave coverage.

25. According to the Friis formula, the received power becomes $P_R = \text{EIRP} \cdot g_R = g_E S(\lambda^2/4\pi)g_R = P_E g_E (\lambda/4\pi d)^2 g_R$. Alternatively, one could express the received power as a function of the frequency as $P_R = P_E g_E (c/4\pi f d)^2 g_R$. Assuming a 1 kW transmit power, and a 10 km distance, the received power becomes $P_R = 1,000 \times (c/4\pi f \times 10,000)^2$. Knowing the carrier frequency, the received power strength can be calculated in a straightforward manner.

26. The transmitting power of 10 W is 20 dB below the 1 kW reference transmitter. The voice communication requires typically a SNR of 9 dB, which means that the signal needs to be 9 dB above the noise level, that is, $32 + 9 = 41$ dBuV/m. Assuming a 3 dB of fading margin, this value becomes 44 dBuV/m. Since the transmitting power is 20 dB below the reference considered by the curves, and entering with the correction factor, the level becomes $44 + 20 = 64$ dBuV/m. Finally, entering with such level into the 2 MHz curve as seen in Figure 5.17, we obtain an approximate range of 140 km.

CHAPTER 6

1. A modem is used to transmit bits (digital signals) in an analog transmission medium. Moreover, a modem can also be used when the aim is to transmit a carrier-modulated signal (i.e., a signal around a certain carrier frequency), instead of a baseband signal (i.e., a signal around the null carrier frequency). On the other hand, a line coder comprises the transmission of signals in baseband, simply performing an adaptation of the signal levels to the transmission medium. In the case of the line coder, the transmitted signals are discrete, and therefore correspond to digital signals.

2. We have studied multiple classes of line encoding techniques: unipolar, polar, bipolar, and biphase. Moreover, the following line encoding techniques were studied: return to zero, nonreturn to zero, nonreturn to zero inverted, bipolar AMI, pseudoternary, Manchester, differential Manchester, and 2B1Q.

3. Multiplexing consists of a mechanism that allows the sharing of communication resources among different channels. It includes a multiplexer (MUX) at the transmitter and a demultiplexer (DEMUX) at the receiver side. The multiplexing is the operation of encapsulating different channels into a single structured signal for transmission into a common transmission medium. Depending on whether different channels are transported in different frequency subcarriers or different time slots, the multiplexing technique is referred to as frequency division multiplexing (FDM) or as time division multiplexing. Independent channels, transmitted in different frequency bands or different time slots, are uncorrelated. The uncorrelation is assured by making use of guard bands or guard times between adjacent subcarriers or time slots.

4. Depending on whether different channels are transported in different frequency subcarriers or different time slots, the multiplexing technique is referred to as FDM or as time division multiplexing. Independent channels, transmitted in different frequency bands or different

time slots, are uncorrelated. The uncorrelation is assured making use of guard bands or guard times between adjacent subcarriers or time slots. In case the resources are directly shared by different users, instead of channels, the multiplexing terminology is equivalent, but the designation is followed by the word access, that is, FDMA or TDMA. When different users transmit simultaneously, using the same bandwidth, but using the spread spectrum transmission technique with different spreading codes, the multiple access technique is referred to as *code division multiple access*. Another multiple access technique widely employed in LAN and MAN is the carrier sense multiple access with collision detection or the carrier sense multiple access with collision avoidance. These multiple access techniques are employed to perform a statistical multiplexing of the resources, instead of rigidly allocating resources to users (i.e., at a certain moment, may not need to use).

5. Scrambling consists of an operation that aims to improve the signal quality by changing a sequence and the logic state of bits. This is achieved by splitting a long sequence of bits into groups, and applying the scrambling operation to each group of bits individually. A scrambler can be used for the following purposes:

  a. *Synchronism*: Since it breaks long sequence of bits with the same logic state, it increases the number of logic state transitions. This results in an improved capability of the receiver to extract the clock signal from the received signal.

  b. *Error control*: After scrambling of signals, some sequence of bits become impossible. Detecting an impossible sequence of bits gives the receiver the ability to detect an error.

  c. *Security*: A third party who intercepts a message cannot decode the data without having knowledge about the generator polynomial. Therefore, the scrambling operation can be viewed as a type of encryption.

6. The coding efficiency of a symbol employed in a line coding or a modem consists of the quotient between the source bit rate R_B and the modulation frequency f_M used to encode the source bit rate. The modulation frequency corresponds to the number of different discrete levels per second (amplitude, frequency, phase, or a combination of these elements) that is used to encode the source bit rate.

7. From the equation $B_{min} = R_S/2$, we know that a signal with a symbol rate of 2 Msymbol/s requires a minimum baseband bandwidth of $B_{min} = 2 \times 10^6/2 = 1$ MHz.

8. In case of bandpass (carrier modulated) signal (signal at the output of a modem), the negative baseband part of the spectrum is also transmitted. Therefore, the minimum bandwidth of a bandpass signal is given by the following expression: $B_{min} = R_S$. From this, we conclude that the minimum bandwidth of bandpass required to accommodate a carrier-modulated signal that comprises a symbol rate of 2 Msymb/s is 2 MHz.

9. An FDM transmitter consists of modulating different channels with different subcarriers $(f_1, f_2, ..., f_N)$, followed by an adder module (i.e., summer at signal level). Afterward, the resulting signal is carrier modulated around the certain carrier frequency f_C, and bandpass filtered in order to remove the (negative) frequencies below the carrier frequency f_C.

10. The first operation consists of bandpass filtering the received signal with a bandwidth corresponding to approximately the bandwidth occupied by all transported channels. This filter is also referred to as the receiving filter, and its main objective consists of removing all the noise present outside the band of the signal of interest. After the filtering operation, the carrier demodulation operation is performed. This consists of performing a translation of the bandpass signal from the carrier frequency f_C into the baseband (i.e., frequency zero). This is followed by an operation of filtering centered in the subcarrier frequency. This isolates each of the signals, after which independent subcarrier demodulation is performed, in order to recover the replicas of the transmitted signals (corresponding to different channels).

11. SSB signals only consider the transmission of the lower side band or the upper side band, which results in a power and spectrum saving. A regular bandpass signal comprises the

two sidebands, entitled DSB. The generation of a SSB signal from the DSB comprises the use of a sideband filter to reject one of the sidebands.

12. A modem can be utilized in an analog transmission medium, performing digital to analog conversion, that is, a modem is used when the data is digital but with an analog transmission medium. On the other hand, a line encoder is employed when both the data and the medium are digital. Moreover, a modem sends carrier-modulated (bandpass) signals, whereas a line coder sends baseband signals. An advantage of using a modem relies on the ability to select the frequency band (carrier frequency) to send the signals, such that it presents low level of distortion.

13. The bandpass filter is used in a FDM receiver, with a bandwidth corresponding to approximately the bandwidth occupied by all transported channels. This filter is also referred to as the *receiving filter*, and its main objective consists of removing all the noise present outside the band of the signal of interest.

14. The Euclidian distance is the distance between two constellation points of a modulation scheme. For a lower minimum Euclidian distance, a lower level of noise is enough to originate a corrupted bit.

15. For a lower minimum Euclidian distance, a lower level of noise is enough to originate a corrupted bit. Maximizing the minimum Euclidian distance, the SNR is maximized, as the amount of noise or interferences necessary to originate a corrupted bit is increased.

16. The Manchester line coding belongs to the group of biphase line coding technique. It is characterized by representing the logic state one with a positive transition at half of the bit duration (from 0 to +V) and the logic state zero with a negative transition at half of the bit duration (from +V to 0). Alternatively, the transitions can be reversed for the two logic states, or the considered voltage can be negative, instead of positive. An important advantage of the Manchester technique relies on its inherent synchronism capability that results from the fact that at least one transition exists per bit duration. Moreover, an important disadvantage of the Manchester scheme relies on its excessive required bandwidth that results from the excessive number of transitions.

17. The bipolar alternate mark inversion (Bipolar AMI) line coding belongs to the group of bipolar line coding technique. This is characterized by representing the logic state zero with an absence of voltage and the logic state one by an alternating positive and negative voltage (+V or −V). Two important characteristics of this line coding techniques rely on its inherent ability to detect errors (two successive +V or −V are impossible conditions) and on its zero mean value.

18. The nonreturn-to-zero inverted (NRZ-I) technique belongs to the group of unipolar line coding techniques. It is characterized by representing the logic state one with a transition (from 0 to V or from V to 0) and the logic state zero by an absence of transition. The transition, or its absence, occurs at the start of the bit duration. Since the encoded signal is a function of the difference between the previous and following logic state, this technique is also referred to as differential. Alternatively, the representation of the two logic states can be reversed. Note that the voltage V can be a positive or a negative value.

19. These three techniques are used in modems. The amplitude shift keying, frequency shift keying, and phase shift keying are elementary modulation schemes that use amplitude, frequency, or phase of a carrier, respectively, as the parameter to encode bits in a carrier. Advanced modulation techniques use a combination of these elementary modulation schemes. The 16 quadrature amplitude modulation is an example of a modulation technique that combines amplitude and phase, using a total of 16 constellation points.

20. Advanced modulation techniques use a combination of these elementary modulation schemes. The 16 quadrature amplitude modulation is an example of a modulation technique that combines amplitude and phase, using a total of 16 constellation points.

21. Differential codec is an alternative solution to encode (digitization process) a source analog signal based on encoding the difference between the actual signal sample and the previous signal sample (or, alternatively, the predicted signal sample). This is especially useful for signals such as voice or video, as those signals tend to suffer a small variation from sample to sample. Since the variation range of the difference signal is lower than that of the samples, a lower number of bits are necessary to encode the signal, in order to achieve the same performance. A widely known differential voice coding technique is the delta modulation. This transmits a single bit to encode a sample. Its logic state is 1 if the signal increases its level (comparing to the previous sample) or 0 otherwise. In order to achieve an acceptable performance, the sampling frequency needs to be higher than that of the Nyquist sampling theorem.

22. It is known that lower amplitudes present a higher probability of occurrence than higher amplitudes. In order to minimize the quantization noise without increasing the number of quantization levels in PCM, a solution is to consider narrower quantization intervals at lower amplitudes and wider quantization intervals at higher amplitudes. This adapts the quantization levels to the statistic of the voice signal, and therefore, it allows reducing the quantization noise without having to increase the number of quantization levels. Alternatively, this allows a reduction in the source data rate for the same level of quantization noise. This technique is known as nonuniform PCM. Such nonuniform quantization is obtained using a logarithmic compression characteristic, as applied in the real PCM systems normalized by the ITU-T Recommendation G.711. Two main techniques exist: the μ-law, used in the United States and Japan, and the A-law, adopted by Europe (among other countries). The difference between these two techniques relies on the position of the quantization intervals. The μ-law leads to a slightly less amount of the quantization noise than the A-law, for lower signal levels, whereas the A-law tends to achieve better performance at higher amplitudes. Comparing logarithmic A-law or μ-law against the linear algorithm, a reduction in the dynamic range from 12 into 8 bits is achieved. This translates in a throughput reduction from 96 into 64 kbps.

23. It is known that lower amplitudes present a higher probability of occurrence than higher amplitudes. In order to minimize the quantization noise without increasing the number of quantization levels, a solution is to consider narrower quantization intervals at lower amplitudes and wider quantization intervals at higher amplitudes. This adapts the quantization levels to the statistic of the voice signal, and therefore, it allows reducing the quantization noise without having to increase the number of quantization levels. Alternatively, this allows a reduction in the source data rate for the same level of quantization noise. This technique is known as nonuniform PCM. Such nonuniform quantization is obtained using a logarithmic compression characteristic, as applied in the real PCM systems normalized by the ITU-T Recommendation G.711, resulting in a minimization of the quantization noise, relating to uniform codec.

24. The A law of ITU-T G.711 standard uses 13 segments, with seven positives and seven negatives. Those two segments that cross the origin are linear, and therefore, they are accounted for as a single segment. Each 8-bit PCM word is encoded as follows:
    a.   The first bit represents the polarity of the input analog sample.
    b.   The following three bits represent the segment.
    c.   The following (and last) four bits encode the linear quantization within each segment. Comparing logarithmic A-law or μ-law against the linear algorithm, a reduction in the dynamic range from 12 into 8 bits is achieved. This translates in a throughput reduction from 96 into 64 kbps.

25. Similar to the ITU-T G.711 codec used for voice, there are several codecs used for video. MPEG is the mostly known video algorithm, presenting different versions. The MPEG comprises a suite of standards that define the way video is digitized and compressed.

These standards encode the video using a differential procedure. Different MPEG standards present different bandwidth requirements, compression procedures, and resolutions. On the other hand, the DVB defines the physical layer and data link layer employed in the distribution of digital video. In this context, it is worth referring that the MPEG-2 is a video compression algorithm using data rates from 3 up to 100 Mbps, comprising different resolutions from 352×240 up to 1920×1080. This standard is used in DVB and HDTV.

26. The European video system is entitled phase alternation line (PAL). The PAL considers a total of 625 transmitted lines, from which only 575 lines are visible. Moreover, the PAL system considers 25 images per second. The number of transmitted image elements per second are $M = ABC$, where A corresponds to the number of lines (625), B corresponds to the number of pixels per line (572), and C stands for the number of images per second (25). This leads to $M = 625 \times 572 \times 25 = 8{,}937{,}500$ image elements per second. Since each image element can be viewed as a signal sample, according to the Nyquist sampling theorem we can conclude that the PAL system requires a minimum bandwidth of $B = M/2 = 4.47$ MHz. Taking into account the return of the beam, the required bandwidth becomes 6 MHz for luminance. Adding to this value the bandwidth required for audio and color, the required bandwidth of a video system using the PAL standard becomes approximately 8 MHz.

27. Knowing that a bandpass filtered voice signal has a spectrum in the range of 300 Hz to 3.4 kHz, according to the Nyquist theorem the minimum sampling rate becomes $f_a \geq 2 \times 3.4 = 6.8$ ksamples/s. This results in a minimum sampling period of $T_a \leq (1/(6.8 \times 10^3)) \approx 147.06\,\mu s$. In voice signals, the sampling period of $T_a = 125\,\mu s$ is widely adopted, translating in a sampling period of $f_a = 8$ ksamples/s, that is, $f_a = 8$ kHz. The sampling rate of 8 kHz is employed in PCM.

28. Considering the QPSK modulation ($M = 4$), and a bit rate $R_B = 20$ Mbps, the resulting symbol rate becomes $R_S = (R_B/(\log_2 M)) = ((20 \times 10^6)/\log_2 4) = 10^7 = 10$ Msymb/s. Since the QPSK modulation is used by a modem, the resulting minimum bandwidth becomes $B_{\min} = R_S = 10$ MHz.

29. Considering the 64-QAM modulation ($M = 64$), and a bit rate $R_B = 10$ Mbps, the resulting symbol rate becomes $R_S = (R_B/(\log_2 M)) = ((10 \times 10^6)/\log_2 64) = (10^7/6) = 1666$ Msymb/s. Since the 64-QAM modulation is used by a MODEM, the resulting minimum bandwidth comes $B_{\min} = R_S = 1666$ MHz.

30. Considering the unipolar nonreturn-to-zero line coding technique ($M = 2$), and a bit rate $R_B = 20$ Mbps, the resulting symbol rate comes $R_S = (R_B/\log_2 M) = ((20 \times 10^6)/\log_2 4) = 10^7 = 10$ Msymb/s. Since the unipolar nonreturn-to-zero line coding technique is used by a line encoder, the resulting minimum bandwidth becomes $B_{\min} = R_S/2 = 10^7/2 = 5$ MHz.

CHAPTER 7

1. Multiplexing consists of a mechanism that allows the sharing of communication resources among different channels. It includes a MUX at the transmitter, and a DEMUX at the receiver side. The multiplexing is the operation of encapsulating different channels into a single structured signal for transmission into a common transmission medium. Depending on whether different channels are transported in different frequency subcarriers or different time slots, the multiplexing technique is referred to as FDM or as time division multiplexing (TDM). Independent channels, transmitted in different frequency bands or different time slots, are uncorrelated. The uncorrelation is assured making use of guard bands or guard times between adjacent subcarriers or time slots. In case the resources are directly shared by different users, instead of channels, the multiplexing terminology is equivalent, but the designation is followed by the word access, that is, FDMA or TDMA. When different users transmit simultaneously, using the same bandwidth, but using the

spread spectrum transmission technique with different spreading codes, the multiple access technique is referred to as code division multiple access (CDMA).

2. When channels corresponding to different users are transported in different frequency subcarriers, the multiple access technique is referred to as FDMA. On the other hand, the OFDM technique splits the symbols of a certain user in several lower rate streams, which are then transmitted in orthogonal parallel subcarriers. Consequently, the symbol period is increased, making the signal less sensitive to ISI. To avoid interference between subcarriers, the several streams should be transmitted in orthogonal subcarriers. It is known that sinusoids with frequencies spaced by $1/T$ form an orthogonal basis set in a T-duration interval, and a periodic signal with period T can be represented as a linear combination of the orthogonal sinusoids. This means that the orthogonality between subcarriers is assured by using the discrete Fourier transform (DFT) and inverse DFT (IDFT). In practice, OFDM is normally implemented through an efficient technique called fast Fourier transform (FFT) and inverse FFT (IFFT). Therefore, OFDM signals are commonly generated by computing the N-point IFFT, where the input of the IFFT is the frequency domain representation of OFDM signal. The output of the IFFT is the time domain representation of the OFDM signal, the N-point IFFT out is defined as a useful OFDM symbol. Although with OFDM signals the symbol stream is split into several parallel substreams with lower rate (each one associated to a different subcarrier), ISI can still occur within each substream. To mitigate the effects of ISI caused by channel delay spread, each block of N IDFT coefficients is typically preceded by a cyclic prefix (CP) or a guard interval consisting of Ng samples (Ng stands for the number of samples at the CP), such that the length of the CP is, at least, equal to the time span of the channel (channel length). The CP is simply a repetition of the last Ng time-domain symbols.

3. The typical WCDMA receiver is a RAKE receiver, which has several fingers to detect different multipaths of the channel. Each finger comprises each of its own decorrelator, associated to an equalizer. In multipath environment, since each finger of the RAKE receiver discriminates a different multipath, the combination of the signals from different fingers with a maximum ratio combiner tends to achieve a performance improvement compared to a single decorrelator. For this reason, it is normally stated that a WCDMA system jointly with RAKE receiver is able to exploit multipath diversity. As the spreading factor increases, the resolution of the RAKE receiver also increases, allowing better discrimination of the several propagation paths, increasing the diversity order and, potentially, improving the performance.

4. RAKE receiver is associated with WCDMA signals.

5. The basic concept behind MIMO techniques relies on exploiting the multiple propagation paths of signals between multiple transmit and multiple receive antennas. The use of multiple antennas at both the transmitter and receiver aims to improve the performance or symbol rate of systems, without an increase of the spectrum bandwidth, but it usually requires higher implementation complexity. In the case of frequency selective fading channel, different symbols suffer interference from each other, whose effect is usually known as the intersymbol interference. This effect tends to increase with the increase of the symbol rate. By exploiting diversity, multiantenna systems can be employed to mitigate the effects of ISI. The antenna spacing must be larger than the coherence distance to ensure independent fading across different antennas. Alternatively, different antennas should use orthogonal polarizations to ensure independent fading across different antennas. The various configurations are referred to as single input single output (SISO), multiple input single output (MISO), single input multiple output (SIMO), or MIMO. SIMO and MISO architectures are a form of receive and transmit diversity (TD) schemes, respectively, comprising a single antenna at the transmitter and multiple antennas at the receiver, whereas MISO consists of the opposite. Moreover, MIMO architectures can be used for

combined transmit and receive diversity, as well as for the parallel transmission of data or spatial multiplexing (SM). When used for SM, MIMO technology promises high bit rates in a narrow bandwidth, therefore, it is of high significance to spectrum users. MIMO systems transmit different signals from each transmit element so that receive antenna array receives a superposition of all the transmitted signals. MIMO schemes are implemented based on multiple-antenna techniques. These multiple-antenna techniques can be of different forms:

a. Space-time block coding (STBC)
b. Multilayer transmission
c. Space division multiple access
d. Beamforming

STBC is essentially a MISO system. Nevertheless, the use of receive diversity makes it a MIMO, which corresponds to the most common configuration for this type of diversity. STBC-based schemes focus on achieving a performance improvement through the exploitation of additional diversity, while keeping the symbol rate unchanged. Open loop TD schemes achieve diversity without previous knowledge of the channel state at the transmitter side. In contrast, closed loop TD requires knowledge about the CSI at the transmitter side. STBC, also known as *Alamouti scheme*, is the most known open loop technique. The selective TD (STD) is a closed loop TD. This is not exactly an STBC scheme. Nevertheless, since this is also a TD scheme, this subject is dealt with alongside STBC. On the other hand, multilayer transmission and SDMA belong to another group, entitled SM, whose principles are similar but whose purposes are quite different. The goal of the MIMO based on multilayer transmission scheme relies on achieving higher symbol rates in a given bandwidth, whose increase rate corresponds to the number of transmit antennas. Using multiple transmit and receive antennas, together with additional post-processing, multilayer transmission allows exploiting multiple and different flows of data, increasing the throughput and the spectral efficiency. In this case, the number of receive antennas must be equal or higher than the number of transmit antennas. The increase of symbol rate is achieved by *steering* the receive antennas to each one (separately) of the transmit antennas, in order to receive the corresponding symbol stream. This can be achieved by using the nulling algorithm. Finally, beamforming is implemented by antenna array with certain array elements at the transmitter or receiver being closely located to form a beam array (typically separated half wavelength). Since it steers the transmit (or receive) beam toward the receive (or transmit) antenna, this scheme is an effective solution to maximize the SNR. As a result, an improved performance or coverage can be achieved with beamforming.

6. Multilayer transmission and SDMA belong to the same MIMO group, entitled SM, whose principles are similar but whose purposes are quite different. The goal of the MIMO based on multilayer transmission scheme relies on achieving higher symbol rates in a given bandwidth, whose increase rate corresponds to the number of transmit antennas. Using multiple transmit and receive antennas, together with additional post-processing, multilayer transmission allows exploiting multiple and different flows of data, increasing the throughput and the spectral efficiency. In this case, the number of receive antennas must be equal or higher than the number of transmit antennas. The increase of symbol rate is achieved by *steering* the receive antennas to each one (separately) of the transmit antennas, in order to receive the corresponding symbol stream. This can be achieved by using the nulling algorithm. On the other hand, the beamforming is implemented by antenna array with certain array elements at the transmitter or receiver being closely located to form a beam array (typically separated half wavelength). Since it steers the transmit (or receive) beam toward the receive (or transmit) antenna, this scheme is an effective solution to maximize the SNR. As a result, an improved performance or coverage can be achieved with beamforming.

7. Multiresolution techniques are interesting for applications where the data being transmitted is scalable, that is, it can be split into classes of different importance. For example, in case of video transmission, the data from the video source encoders may not be equally important. The same happens in the transmission of coded voice. The introduction of multiresolution in a broadcast cellular system deals with source coding and transmission of output data streams. In a broadcast cellular system, there is a heterogeneous network with different terminal capabilities and connection speeds. For the particular case of video, scalable video consists of a common strategy presented in the literature that adapts its content within a heterogeneous communications environment. Scalable video provides a base layer for minimum requirements, and one or more enhancement layers to offer improved quality at increasing bit/frame rates and resolutions. Therefore, this method significantly decreases the storage costs of the content provider. Scalable video may consist of three QoS regions, where the first region receives all the information, whereas the second and third regions receive the most important data. The QoS regions are associated to the geometry factor that reflects the distance of the UE from the base station (BS) antenna. Scalable video transmission can also be implemented using different techniques. One possibility relies on the use of hierarchical constellations. Strong bit blocks are QPSK demodulated, medium bit blocks are 16-QAM demodulated, and weak bit blocks are 64-QAM demodulated. Furthermore, hierarchical constellations may also be combined with different channel coding rates. This corresponds to the concept of adaptive modulation and coding.

8. The introduction of multiresolution in a broadcast cellular system deals with source coding and transmission of output data streams. In a broadcast cellular system, there is a heterogeneous network with different terminal capabilities and connection speeds. For the particular case of video, scalable video consists of a common strategy presented in the literature that adapts its content within a heterogeneous communications environment. Scalable video provides a base layer for minimum requirements, and one or more enhancement layers to offer improved quality at increasing bit/frame rates and resolutions. Therefore, this method significantly decreases the storage costs of the content provider. Scalable video consists of three QoS regions, where the first region receives all the information, whereas the second and third regions receive the most important data. The QoS regions are associated to the geometry factor that reflects the distance of the UE from the BS antenna. Scalable video transmission can also be implemented using different techniques. One possibility relies on the use of hierarchical constellations. Strong bit blocks are QPSK demodulated, medium bit blocks are 16-QAM demodulated, and weak bit blocks are 64-QAM demodulated. Furthermore, hierarchical constellations may also be combined with different channel coding rates. This corresponds to the concept of adaptive modulation and coding. Another possibility to implement scalable video transmission (multiresolution) relies on the use of SM MIMO technique, where each transmit antenna sends a different data stream. The first data stream (most powerful) may include the base layer, whereas the enhancement layer may be sent by a second antenna (less powerful data stream). Depending on the power and channel conditions, a certain UE may successfully receive either the two streams or only the base layer.

9. Digital communications employing spread spectrum signals are characterized by using a bandwidth, B_{wd}, much greater than the information bit rate. This means that for a spread spectrum signal we have $SF = B_{wd}/R_b \gg 1$, where R_b is the information bit rate and SF is the bandwidth expansion factor (called spreading factor or processing gain). This bandwidth expansion can be accomplished through different types of spread spectrum techniques: direct sequence (DS), frequency hopping (FH), time hopping, and hybrid techniques. In the case of DS, spread spectrum is implemented by multiplying each symbol by a spreading sequence, composed of a number of chips. One common application of DS spread spectrum signals is CDMA communications where several users share the same

channel bandwidth for transmitting information simultaneously. If the spreading of each user is the joint effect of the channel encoder and the spreading code then the transmission is designated as DS-CDMA. If the bandwidth expansion is performed by the channel encoder alone then the transmission is referred to as CS-CDMA. All the users can transmit in the same frequency band, and at the same time, and may be distinguished from each other by using a different pseudorandom sequence.

10. MIMO schemes are implemented based on multiple-antenna techniques. These multiple-antenna techniques can be of different forms:
    a. STBC
    b. Multilayer transmission
    c. Space division multiple access
    d. Beamforming
    STBC is essentially a MISO system. Nevertheless, the use of receive diversity makes it a MIMO, which corresponds to the most common configuration for this type of diversity. STBC-based schemes focus on achieving a performance improvement through the exploitation of additional diversity, while keeping the symbol rate unchanged. Open loop TD schemes achieve diversity without previous knowledge of the channel state at the transmitter side. In contrast, closed loop TD requires knowledge about the CSI at the transmitter side. STBC, also known as Alamouti scheme, is the most known open loop technique. The STD is a closed loop TD.

11. OFDM technique splits the symbols of a certain user in several lower rate streams, which are then transmitted in orthogonal parallel subcarriers. Therefore, the symbol period is increased, making the signal less sensitive to ISI. To avoid interference between subcarriers, the several streams should be transmitted in orthogonal subcarriers. It is known that sinusoids with frequencies spaced by $1/T$ form an orthogonal basis set in a T-duration interval, and a periodic signal with period T can be represented as a linear combination of the orthogonal sinusoids. This means that the orthogonality between subcarriers is assured by using the discrete Fourier transform (DFT) and inverse DFT (IDFT). In practice, OFDM is normally implemented through an efficient technique called *fast Fourier transform* (FFT) and *inverse FFT* (IFFT). Therefore, OFDM signals are commonly generated by computing the N-point IFFT, where the input of the IFFT is the frequency domain representation of OFDM signal. The output of the IFFT is the time domain representation of the OFDM signal, the N-point IFFT out is defined as a useful OFDM symbol. Although with OFDM signals the symbol stream is split into several parallel substreams with lower rate (each one associated to a different subcarrier), ISI can still occur within each substream. To mitigate the effects of ISI caused by channel delay spread, each block of N-IDFT coefficients is typically preceded by a CP or a guard interval consisting of Ng samples (Ng stands for the number of samples at the CP), such that the length of the CP is, at least, equal to the time span of the channel (channel length). The CP is simply a repetition of the last Ng time-domain symbols. Although OFDM are the most popular block transmission techniques, the same concept can be used with single-carrier modulations. In fact, single-carrier modulation using FDE is an alternative approach based on this principle. As with OFDM, with single carrier—frequency domain equalization (SC-FDE) the data blocks are preceded by a cyclic prefix, long enough to cope with the overall channel length. Owing to the lower envelope fluctuations of the transmitted signals (and, implicitly a lower PMEPR), SC-FDE schemes are especially interesting when a low-complexity and efficient power amplification is required. While the OFDM transmitter includes the computation of the IDFT, the SC-FDE transmitter is a regular time-domain one with the exception that the CP is added. Nevertheless, after reception (and after removing the cyclic prefix), the SC-FDE receiver computes the DFT, before performing the frequency domain equalization, followed by the IDFT computation. As the signal is transmitted in blocks (to which the CP is added and to

which the equalization is performed), SC-FDE transmission is also considered as a block transmission technique.

12. The diversity system must optimally combine the receiver-diversified waveforms so as to maximize the resulting signal quality. There are several ways of combining the signals received from the different types of diversity, namely selection combining, maximal ratio combining, equal ratio combining (EGC), and the mean square error (MSE)-based combining. In addition, this combining may be performed either at the received signal level or at the symbol level. Selection combiner selects the strongest signal from the total signals at its input. This criteria is used by the STD scheme, where the receiver selects one of the several transmit antennas (instead of one of the several receive antennas). Maximal ratio combiner performs a weighted sum of all signals at its input. This criterion is optimum only in the presence of noise, since it maximizes the SNR, while it minimizes the noise. The resulting SNR level of the signal at the combiner's output corresponds to the sum of the elementary SNRs (combiner's input SNRs). If the signal from one branch has a deep fade, this effect is felt in the resulting signal since the combining weights are equal in the several branches. This scheme presents a great advantage as knowledge about the channel state information (CSI) is not required, and therefore, estimation circuits are not needed, leading to a very simple combiner. MSE-based combining criterion tends to lead to better performances in the presence of interference, namely, multipath interference or MAI. The combining weights are calculated based on information provided by some way. A traditional way consists of using a pilot or a training sequence that allows the receiver to periodically evaluate a coefficient that corresponds to some difference (MSE) between the transmitted signal and the received one. Let us consider that we have available N receive branches that are intended to be combined to provide diversity (e.g., output from the N fingers from L multipaths or output from the N receive antennas).

13. STBC, also known as Alamouti scheme, is the most known open loop MIMO technique whose purpose relies on the exploitation of diversity to improve the bit error rate performance. The Alamouti scheme consists of a TD, whose symbols are sent through different antennas using STBCs. Adding receive diversity transforms a TD scheme into a MIMO scheme. On the other hand, multilayer transmission belongs to another group, entitled SM, whose principles are similar but whose purposes are quite different. The goal of the MIMO based on multilayer transmission scheme relies on achieving higher symbol rates in a given bandwidth, whose increase rate corresponds to the number of transmit antennas. Using multiple transmit and receive antennas, together with additional post-processing, multilayer transmission allows exploiting multiple and different flows of data, increasing the throughput and the spectral efficiency. In this case, the number of receive antennas must be equal or higher than the number of transmit antennas. The increase of symbol rate is achieved by *steering* the receive antennas to each one (separately) of the transmit antennas, in order to receive the corresponding symbol stream. This can be achieved by using the nulling algorithm.

CHAPTER 8

1. The conversion between alphanumeric names and IP addresses carried out by the DNS protocol can be performed in two different methods:
   a. *Recursive*: A host sends a mapping request to a DNS server. If the interrogated server does not have the response, it searches the response in other servers and returns it to the caller.
   b. *Iterative*: A host sends a mapping request to a DNS server. If the interrogated server does not have the response, it returns to the host a pointer toward the following DNS server to which the request should be sent.

2. The DNS operation comprises the existence of a number of key elements, namely
    a. *Domain name space*: This corresponds to a list of alphanumeric names (which corresponds to a set of IP addresses).
    b. *DNS database*: This stores the mapping between alphanumeric names and IP addresses.
    c. *Name resolver*: A program that runs in a server, or in a local host, in order to respond to DNS requests.
    d. *Name server*: A server that stores the DNS database and the name resolver, and responds to DNS requests to perform the mapping.
3. The DNS protocol is used to perform the conversion between alphanumeric names and IP addresses, which can be performed in two different directions:
    a. *Binding*: It corresponds to the conversion of alphanumeric names (typed by users) into IP addresses (inserted into packet headers, before transmission). This is the most common DNS operation.
    b. *Reverse mapping*: It corresponds to the translation of IP addresses into alphanumeric names. In some circumstances, this operation may also be required.
4. The three most important SNMP messages exchanged between a network management station and the agents are as follows:
    a. *Get*: It is used for network monitoring purposes, namely for the management station to request the value of an object from an agent.
    b. *Set*: It is used by the management station to send a value of an object to an agent. Consequently, it is employed for network control purposes.
    c. *Notify*: Sent by an agent to the management station to report a failure event (monitoring purposes).
5. In order to allow the remote management of equipment, four different elements need to be implemented:
    a. *Agent*: It is the device that is intended to be remotely controlled and/or monitored (managed) through the network. For the device to be subject to remote management, it needs to be equipped with a SNMP agent, which comprises a local database with actual and historic variables (object states).
    b. *Management station*: It corresponds to a computer or workstation that runs specific software that allows sending and receiving network management messages to/from an agent.
    c. *Network management protocol*: The communication between an agent and a management station is carried out using the SNMP network management protocol.
    d. *Management information base* (*MIB*): It consists of a hierarchical database, stored in the management station, which compiles the data stored in all different agents (the set of all local databases).
6. The SNMP is an IP-based network management protocol, supported on the UDP. The reason for using the connectionless mode is not because data reliability is not a requirement but because such data reliability can be assured through the repetition of the transmitted data. Since the SNMP comprises the periodic retransmission of messages. In case a message is received corrupted, the following messages assure that the information is correctly interpreted by the destination.
7. The SMTP is the TCP/IP application layer protocol that allows a client uploading an e-mail to an e-mail server (user mailbox) or exchanging an e-mail between different e-mail servers, through an IP-based network, such as the Internet. Moreover, the SMTP is also employed by an e-mail server to send e-mail messages to clients. Note that the connection between the source's server and the recipient's server can be either established directly or may go through one or more intermediate servers.
8. The SMTP is used by a server to exchange e-mail messages between successive e-mail servers or to deliver a message to a client, or by a client, to upload an e-mail message.

In addition, there is the post office protocol (POP), which allows a client receiving an e-mail message from an e-mail server (download), using authentication. The well-known port of POP version 3 (POPv3) is 110.

9. The SMTP is used to upload e-mails from a client into a server, or to support the exchange of e-mails between different servers. Since the SMTP does not support authentication, the download of e-mails is normally performed using the POP or using the IMAP, as follows:

   a. *Post office protocol*: It allows a client to receive an e-mail message from an e-mail server (download), using authentication. The well-known port of POPv3 is 110.

   b. *Internet message access protocol*: It presents some additional features than those available with the POP, namely the ability to read e-mails located in a server (through downloading or webmail), while leaving a copy in the mail server. It also allows the share of folders among different users, as well as the search of e-mails in a server. IMAP servers are listed on well-known TCP port 143.

10. The SMTP message format comprises text type data, consisting of a string of 7-bit ACSII characters. In case a message contains other types of media (pictures, videos, executable files, or others than 7-bit ASCII characters code), the MIME format is utilized. Note that when a message contains annexes, the message format is automatically switched into the MIME format. The mapping between MIME and SMTP messages is performed using e-mail clients or servers.

11. From the ISO reference model perspective, the application layer deals with the task, which are specific of the application program, whereas the session and presentation layers take care of the tasks, which are common to different application programs. From the TCP/IP model perspective, the application layer takes care of all tasks defined for the application layer, presentation layer, and session layer of the OSI reference model.

12. The session layer comprises a group of functionalities that involves the management of a session. Such functionalities can be split into the following four categories:

    a. *Setup of sessions*: A session consists of a logical link that is established between end entities. It has a meaning different from a physical link, which is established and managed by the transport layer. In case a physical link that supports a certain logical link fails, a new physical link is established and the session (logical link) continues with its exchange of data (SPDU).

    b. *Synchronization*: It considers the periodic transmission of synchronization points that are properly acknowledged by the receiving entity. In case the transport link fails, the session data does not need to be retransmitted from the beginning, but from the last synchronization point correctly acknowledged. Note that, from the session layer perspective, an acknowledgment means that the received data has been properly processed.

    c. *Dialog management*: Although the physical link may support full-duplex operation, some user processes only support, from the logical link perspective, half-duplex operation. The management of which entity is allowed to send data is managed at this layer, using a token that is transferred from one entity into another.

    d. *Activities management*: The application data is typically split into a set of elementary activities. Each activity is properly managed and signalized using marks in order to register that it has initiated or finalized. These marks are important in order to allow the receiving entity to identify the borders of an activity. An example of a sequence of two elementary activities can be a copy of a file followed by its deletion. The data cannot be deleted (second activity) without being sure that the copy (first activity) has been successfully performed. The ISO 8327/CCITT X.225 is an example of a standard, which defines the session layer.

13. The MIB consists of a hierarchical database, stored in the management station, which compiles the data stored in all different agents (the set of all local databases).

14. There are two kinds of indirect HTTP connections:
    a. *Proxy*: It consists of a server located at the edge of an internal network. All connections, established between an internal host (host located in the intranet) and an external host (e.g., a server located in the Internet), are broken down into two successive connections for the purpose of avoiding direct connectivity, as well as for the purpose of traffic filtering (i.e., network security). From the client perspective, the proxy acts as a server. Contrarily, from the server perspective, the proxy acts as a client. This represents a protective measure from certain types of network attacks.
    b. *Gateway*: It is typically a network device that changes a protocol into another. As an example, a host located in a Microsoft network, which needs to have access to a server located in an IP network needs a gateway to perform the translation between the two network protocols. A gateway may also be employed in the implementation of the IPsec cryptographic protocol. In this case, the hosts from outside of an organization connect to an IPsec gateway, located in the interior of an organization, through the Internet using the IPsec cryptographic protocol, while the gateway connects to a server in plaintext (in clear). Contrarily, the connections between internal hosts (intranet) are performed in plaintext.
15. A data service comprises a functionality that is provided to the end user, making use of a specific program (application program). Examples of data services include the IP telephony, the web browsing, the e-mail, the file transfer, and so on. On the other hand, an application program is a program, external to the network architecture, which allows getting access to a certain service. For example, the Microsoft Outlook, the Microsoft Outlook Express, and the Groupwise are three examples of specific application programs that can be employed to have access to the e-mail service. The web browsing is also a service that can be accessed making use of the Internet Explorer, Mozilla Firefox, Google Chrome, and so on. This corresponds to a user's program, external to the network architecture. Finally, the application layer is the layer of the OSI or TCP/IP stack, which provides some functionalities to the application program, making use of specific application layer protocols, such as the SMTP, FTP, and telnet. Note that the application program above described has a meaning different from the application layer of the network architecture.
16. The application layer is the layer of a network architecture that provides some network functionalities to the application program, making use of specific application layer protocols, such as the SMTP, FTP, and telnet.
17. The presentation layer refers to the way the data is represented and exchanged in the interaction among different ending entities. Moreover, it provides means for the establishment and use of abstract syntax, which allows the exchange of data. An important syntax commonly employed among different entities is the abstract syntax notation 1 (ASN.1). The task of this layer can be viewed as the definition of a language (standard) to use in the communication among different hosts. Instead of allowing the communication being performed with the internal language of a host (e.g., computer), a standardized language is established, and all hosts must use such language in the interaction with the others. This avoids the need to have successive translators (gateways) among all different possible host languages. This way, it is only needed to perform the translation between the host's internal language and a common standardized language. Note that the presentation layer also deals with encryption and compression of data, as these two functionalities change the language format used in communications. Such functionalities need to be properly negotiated and accepted, before the exchange of data takes place. The ISO 8823/CCITT X.226 standard is an example of a presentation layer standard. From the ISO reference model perspective, the application layer deals with the task that are specific of the application program, whereas the session and presentation layers take care of the tasks that are common to different application programs. From the TCP/IP model perspective, the application layer

takes care of all tasks defined for the application layer, presentation layer, and session layer of the OSI reference model.

18. In order to implement the services, the ITU-T H.323 comprises four different network component:

    a. *Terminals*: It consist of the equipment (hardware and software) used for human interfacing. This can be a personal computer, a smart phone, or a personal digital assistant (PDA) running a softphone or a packet switching-based telephone.

    b. *Gatekeeper*: It is responsible for performing address resolution, namely for mapping a telephone number or username/e-mail address into a transport address. It performs a task similar to the DNS mapping protocol. In case the terminals have knowledge about the transport addresses of destination endpoints, a gatekeeper may not be necessary. Such mode of operation is referred to as *direct mode*. Nevertheless, a gatekeeper also performs additional functionalities such as admission control and bandwidth management. Therefore, the use of a gatekeeper is essential to avoid congestion, especially in case of medium to large networks.

    c. *Gateway*: In case interoperability is required between different network types, a gateway is employed. Such network device assures voice and video codec translation, data format translation, as well as signaling translation. This network component is required only when translation is necessary.

    d. *Multipoint control unit (MCU)*: This component allows H.323 point-to-multipoint communications, that is, conferencing. Without the MCU, only point-to-point communication is possible.

19. The SIP signaling protocol comprises different network component types:

    a. *User agents*: It consist of the equipment (hardware and software) used in the client/server interaction, for the purpose of connection admission and control. The client endpoint is called user agent client (UAC), and can be a personal computer or a PDA running a softphone or a packet-based telephone. The server endpoint is referred to as user agent server (UAS), being employed to receive session initiation requests and to return the corresponding responses.

    b. *Registration server (registrar)*: It consists of a server that keeps the location of different SIP users. A certain user with a certain telephone number or username/e-mail address can physically move along an IP-based network. This information is stored in the registrar, as well as the corresponding transport layer address. Similar to the gatekeeper of the ITU-T H.323 protocol, the registrar is responsible for translating telephone numbers or username/e-mail addresses into transport layer addresses.

    c. *Proxy server*: It is responsible for forwarding SIP requests and responses to their destination. A LAN implementing a multimedia service using the SIP has its own proxy server.

    d. *Redirect server*: It is responsible for returning the location of another SIP agent, to which the call forwarding must be transferred.

    e. *Gateway*: Similar to the ITU-T H.323 protocol, it assures interoperability among different network types.

    f. *MCU*: Similar to the ITU-T H.323 protocol, it allows point-to-multipoint multimedia communications.

20. Both the ITU-T H.323 and the SIPs are used in IP telephony and VTC. After an initial setup, the exchange of IP telephony traffic is performed in peer-to-peer mode. While the H.323 protocol uses the UDP to support the transfer of multimedia traffic, the SIP may support the transfer of multimedia traffic on either TCP or UDP transport protocol. The gatekeeper is the network component of the H.323 protocol, responsible for performing address resolution, namely for mapping a telephone number, or username, or e-mail address into a transport address. In the SIP, this task is performed by the registrar.

21. The real-time protocol (RTP) is responsible for controlling a session, and for allowing the exchange of voice/video traffic. Since H.323 voice and video traffic is transferred using the UDP transport protocol, the RTP is an intermediate protocol, which adds a header, with a number of fields. This complements the UDP service by adding some level of reliability to the exchanged data. The RTP fields are as follows:

    a. *Sequencing of packets*: It is implemented through the addition of a sequence number field into the RTP header. This allows the delivery of packets in the correct sequence, the detection of lost packets, and the avoidance of packets duplication.

    b. *Payload identification*: This field allows the identification of the audio and/or video codec, which can be employed for different purposes, such as bandwidth management.

    c. *Frame identification*: It is used to signalize the beginning and end of frames.

    d. *Source identification*: It is used to identify the originator of packets. This is relevant in case of point-to-multipoint communications.

    e. *Intramedia synchronization*: It is used to implement buffering and avoid a certain level of jitter.

22. The real-time control protocol (RTCP) consists of a control protocol that allows the RTP performing its tasks. It indicates the identity of a caller, and provides a variety of network information, which can be useful for keeping the network performing well (e.g., average delay, jitter, and the rate of lost packets).

23. While the H.323 protocol uses the UDP to support the transfer of multimedia traffic, the SIP may support the transfer of multimedia traffic on either TCP or UDP transport protocol.

24. The different protocols used by the ITU-T H.323 are as follows:

    a. *RTP*: It is responsible for controlling a session, and for allowing the exchange of voice/video traffic. Since H.323 voice and video traffic is transferred using the UDP transport protocol, the RTP is an intermediate protocol, which adds a header, with a number of fields. This complements the UDP service by adding some level of reliability to the exchanged data. The RTP fields are as follows:

        i. *Sequencing of packets*: It is implemented through the addition of a sequence number field into the RTP header. This allows the delivery of packets in the correct sequence, the detection of lost packets, and the avoidance of packets duplication.

        ii. *Payload identification*: This field allows the identification of the audio and/or video codec, which can be employed for different purposes, such as bandwidth management.

        iii. *Frame identification*: It is used to signalize the beginning and end of frames.

        iv. *Source identification*: It is used to identify the originator of packets. This is relevant in case of point-to-multipoint communications.

        v. *Intramedia synchronization*: It is used to implement buffering and avoid a certain level of jitter.

    b. *RTCP*: It consists of a control protocol that allows the RTP performing its tasks. It indicates the identity of a caller, and provides a variety of network information, which can be useful for keeping the network performing well (e.g., average delay, jitter, and the rate of lost packets).

    c. *H.225 (RAS)*: It is used by terminals to register with the gatekeeper and to request it a call.

    d. *Q.931 signaling protocol*: It is used for call setup and termination between two or more terminals.

    e. *H.245*: It is used in a communication for the negotiation of audio and/or video codec.

    f. *T.120*: It consists of a protocol that manages the real-time data conferencing such as instant messaging, file transfer, application sharing, and so on. Note that these types of data communications are performed using the TCP, instead of the UDP transport protocol (employed in voice and video communications).

25. T.120 consists of a protocol used in ITU-T H.323 that manages the real-time data conferencing such as instant messaging, file transfer, and application sharing. Note that these types of data communications are performed using the TCP, instead of the UDP transport protocol (employed in voice and video communications).

26. ITU-T H.323 and the SIPs belong to the application layer of the TCP/IP architecture.

27. In ITU-T H.323, the exchange of voice/video traffic is carried out using the UDP, whereas the data traffic is implemented using the TCP. The protocol T.120 is used in ITU-T H.323 to manage the real-time data conferencing, such as instant messaging, file transfer, and application sharing.

28. Registration server (registrar) consists of a server used in SIP to keep the location of different SIP users. A certain user with a certain telephone number or username/e-mail address can physically move along an IP-based network. This information is stored in the registrar, as well as the corresponding transport layer address. Similar to the gatekeeper of the ITU-T H.323 protocol, the registrar is responsible for translating telephone numbers or username/e-mail addresses into transport layer addresses.

29. The SIP presents a much lower level of complexity than ITU-T H.323. This results from the fact that the handshaking used in SIP, necessary to establish and to maintain a call, is much simpler than that of the ITU-T H.323.

CHAPTER 9

1. The transport layer of the TCP/IP architecture is an end-to-end layer that aims to provide, as much as possible, the required QoS to the application layer, at the lowest cost. In order to assure this functionality, the transport layer aims to achieve a trade-off between the QoS parameters requested by the application layer, and the QoS available at the network layer (which is a function of the instantaneous traffic). The quality of service parameters dealt with and negotiated by the transport layer includes requested bit rate in each direction, bit error rate, maximum end-to-end delay, maximum end-to-end jitter, priority rules, probability of failure in a connection establishment, maximum time to establish a transport connection, maximum time to terminate a transport connection, cost, and security protection mechanisms.

2. Both IntServ and DiffServ are techniques that can be employed to provide QoS in IP networks. The same applies to the MPLS protocol. The need to achieve such QoS requirements results from the integration of different services into a common network infrastructure. IntServ aims to provide QoS by implementing a group of mechanisms on the flow-by-flow basis, that is, by applying individual reservations to a group of routers in a network. Moreover, IntServ comprises the provision of QoS by applying individual reservations defined by flow specifications, combined with the resource reservation protocol—RSVP, to a group of routers in a network. On the other hand, DiffServ provides QoS in high dimension networks, while reducing the required resources, such as bandwidth, overhead, and processing. While the IntServ operates over individual flows of data, DiffServ is applicable to high volumes of data, reducing the required signaling, reducing the granularity and, consequently, the complexity. Finally, DiffServ is implemented based on SLAs previously negotiated by each organization and using the datagram field DSCP.

3. The TCP is a connection-oriented protocol, providing error control and flow control, and delivering data to the destination in the correct sequence, avoiding duplication and lost packets, and detecting packet losses. As a connection-oriented protocol the TCP requires the connection setup (three-way handshaking), before data is exchanged. The exchange of data is performed using the sliding window protocol. On the other hand, the UDP is a connectionless protocol, and therefore, data reliability is not provided. The UDP only provides a minimum service to its upper layer, acting as a simple interface between the

application layer and the network layer. The UDP does not require the previous connection establishment before data is exchanged. Although presenting several advantages in terms of data reliability, the TCP introduces delays into signals, which can be problematic in delay-sensitive services. In this case, the UDP can be more useful.

4. Example of application layer protocols that may be supported on the UDP due to its data redundancy is the SNMP, as well as the DNS protocol. On the other hand, application layer protocols such as HTTP, POP3, SMTP, and FTP are supported on the TCP transport protocol.

5. The TCP is connection-oriented, providing end-to-end error control and flow control, and delivering the packets to the destination in the correct sequence, avoiding duplication and lost packets and detecting packet losses. As a connection-oriented protocol the TCP requires the connection setup (three-way handshaking), before data is exchanged. The exchange of data is performed using the sliding window protocol.

6. The IntServ aims to provide QoS in networks taking into account the different service requirements. The IntServ is a model that implements a group of mechanisms on the flow-by-flow basis. The provision of QoS is achieved by applying individual reservations defined by flow specifications, combined with the RSVP to a group of routers in a network. The packets belonging to an end-to-end connection that pass through a group of routers, benefit from a resource reservation that may avoid latency, or packets discard. Consequently, packets belonging to a service such as audio or video streaming may benefit from higher priority in intermediate routers. To allow the provision of resources, IntServ classify the different traffic into a wide range of levels of granularity. The exchange of traffic is preceded by a connection establishment phase that requests a certain level of QoS from the network (between two end entities), along a certain route. The intermediate routers receive such request and, based on the previously allocated resources, on the available resources, and on the type of data, decide whether or not it is possible to assign the required resources (buffers, bandwidth, etc.) to the new connection. Such resource reservation is implemented using the RSVP.

7. The UDP only provides a minimum service to its upper layer, acting as a simple interface between the application layer and the network layer. The UDP does not require the previous connection establishment before data is exchanged. Although presenting several advantages in terms of data reliability, the TCP introduces delays into signals, which can be problematic in delay-sensitive services, such as in IP telephony or VTC. In this case, the UDP can be more useful.

8. The establishment of a TCP connection is a three-way handshaking: the initial step comprises the establishment request, by the client to the server, using the SYN message. The second step corresponds to the response from the server, acknowledging the establishment request, that is, accepting the request to establish the connection. This second step is signalized using the SYN and ACK messages, together. Finally, the third and last step corresponds to the acknowledgment, sent by the client, using the ACK message.

9. The termination of a TCP connection is as follows: the client sends a segment to the server with the FIN flag activated, and this entity responds initially with an ACK message, stating that the message was correctly received. Then, in order to accept the request to terminate the connection, a segment is sent by the server with the ACK and FIN flags activated. This corresponds to the confirmation from the server. Finally, the client acknowledges the FIN message received from the server.

10. Error control is provided by the TCP, using the positive acknowledgment with retransmission (PAR) procedure as follows:

   a. The transmitting entity sends a number of segments corresponding to the length allowed by the sliding window protocol. The transmitting entity activates a timer, and waits for the acknowledgment (ACK) by the recipient entity (correct reception of data) within a certain time period.

b. If the segments are correctly received by the recipient, its ACK is sent back to the transmitter (service confirmation). The error detection technique is implemented by the TCP using CRC codes. Note that although the data link layer also makes use of error control mechanisms, some residual errors exist. Moreover, packets can be discarded by intermediate routers and packets can be duplicated or can arrive out of order. Consequently, the TCP does not assume a reliable service provided by the lower layers. Consequently, the TCP has mechanisms to deal with such impairments.

c. In case the transmitting entity does not receive the ACK within the expected time period, it assumes that the amount of data did not reach the recipient correctly, and repeats its transmission.

d. Then the transmitter proceeds with the transmission of the following data.

Note that in addition to error control, this procedure allows flow control. When the receiver informs the transmitter that a certain amount of data was correctly received (error control), it is also informing that it is ready to receive more data (flow control).

In order to make an efficient use of the available transmission medium, while providing flow control, the TCP makes use of the sliding window protocol. Therefore, instead of sending a segment at a time (as in the case of the stop-and-wait protocol), a group of segments are sent together. Moreover, signaling data (e.g., acknowledgments) are often exchanged using piggybacking, where a segment is used simultaneously to send data and to acknowledge data previously received from the opposite direction. The sliding window protocol is implemented using number of octets as a reference. Before the data is exchanged, the receiver informs the transmitter about the number of octets that can be sent without acknowledgments. This maximum number of octets is signalized in a TCP header field entitled window (the window size is expressed in number of octets). Such number of octets (window size) is encapsulated into several segments. Moreover, the maximum number of octets that can be transported in a single segment is known as maximum segment size (MSS).

11. The sequence number is used to assure that packets are delivered to the application layer in the correct sequence. Note that packets may reach the receiver out of order, due to the use of the datagram method. Re-ordering of packets is possible using the sequence number, being one of the fields of the TCP segment header. Moreover, the sequence number is also used in error control, to allow identifying the transmitted packets, as well as to identify the packets acknowledgments.

12. DiffServ is a protocol that can be implemented to tentatively allow the provision of QoS in high dimension networks, while reducing the required resources (overhead, bandwidth, processing, etc.). While the IntServ operates over individual flows of data, DiffServ is applicable to high volumes of data, reducing the required signaling, reducing the granularity, and, consequently, the complexity. DiffServ handles different connections of the same type using the same principles. This implies the prior negotiation of resources reservation for the traffic generated by a certain customer. Based on the customer's payment, an ISP offers a certain SLA that defines the maximum amount of data for each class of traffic (IP telephony, video streaming, file transfer, web browsing, etc.), as well as the type of warranties that are offered to each traffic class. Different traffic classes are identified by the DSCP field of the datagrams' headers. When a packet enters the ISP cloud, intermediate routers read the DSCP field, and the packet is treated according to the negotiated SLA. If the amount of traffic exceeds that negotiated in the SLA, packets can be dropped or delayed (or an additional fee may be charged to the customer). Moreover, in case of congestion, for the same traffic class (identified by the DSCP field), packets belonging to a customer with higher SLA have precedence. Remind, for a certain SLA, packets with lower DSCP field

have priority. DiffServ is the mechanism currently widely implemented in the Internet for the provision of QoS, when MPLS is not the implemented solution.

13. In contrast to the TCP, the UDP is a connectionless protocol, and therefore, data reliability is not provided. The UDP only provides a minimum service to its upper layer, acting as a simple interface between the application layer and the network layer. The UDP does not require the previous connection establishment before data is exchanged. Although presenting several advantages in terms of data reliability, the TCP introduces delays into signals, which can be problematic in delay-sensitive services. In this case, the UDP can be more useful. Moreover, the UDP is a much simpler transport layer protocol.

14. The Internet Assigned Numbers Authority assigns port numbers between 0 and 1023, called well-known ports. These port numbers are to be used by servers and not to be used by hosts. Port number between 1,024 and 49,151 are also registered ports but are used for lesser known services. Finally, port numbers between 49,152 and 65,535 are port numbers to be used by the hosts.

15. Jitter corresponds to delay variation. While certain services can still handle a certain level of delay (e.g., VoIP), they degrade heavily when subject to jitter, as a number of bits are lost when the delay variation occurs.

16. Basic QoS requirements are as follows:
    a. *Delay*: Services such as audio or video streaming can accommodate a low level of delay (are delay sensitive).
    b. *Jitter*: This corresponds to a variation in delay. While a small amount of delay can still be supported by audio or video streaming, its variation rapidly degrades the quality of these services.
    c. *Throughput*: The bit rate that is required to support a specific service. This can be fixed or variable. While the video streaming requires a high and fixed bit rate, the web browsing is not very demanding in terms of throughput, but is a service that requires a variable bit rate.
    d. *Packet loss*: The increase of the number of lost packets corresponds to a degradation of the bit error rate. There are other factors that may degrade the bit error rate, such as the errors generated in the physical and data link layers. Nevertheless, errors generated in physical and data link layer do not rely on the scope of the provision of QoS by transport layer. While the voice telephony can perform well with a bit error rate up to 10^{-3}, a file transfer typically requires a bit error rate better (lower) than 10^{-6}.

17. Although the data link layer may use error control, its effectiveness is not completely accurate, and it is only implemented in a certain link of a connection. On the other hand, the TCP implements error control at end-to-end level, which comprises a number of intermediate data link layer links, some of them with reliability and others without it. Moreover, due to layer 3 protocol problems, packets can be lost (because they achieve time-to-live 0) or packets can be duplicated. The TCP aims to mitigate these impairments.

18. DiffServ handles different connections of the same type using the same principles. This implies the prior negotiation of resources reservation for the traffic generated by a certain customer. Based on the customer's payment, an ISP offers a certain SLA that defines the maximum amount of data for each class of traffic (IP telephony, video streaming, file transfer, web browsing, etc.), as well as the type of warranties that are offered to each traffic class. Different traffic classes are identified by the DSCP field of the datagrams' headers. When a packet enters the ISP cloud, intermediate routers read the DSCP field, and the packet is treated according to the negotiated SLA. If the amount of traffic exceeds that negotiated in the SLA, packets can be dropped or delayed (or an additional fee may be charged to the customer). Moreover, in case of congestion, for the same traffic class (identified by the DSCP field), packets belonging to a customer with higher SLA have precedence. For a certain SLA, packets with lower DSCP field have priority. DiffServ is the mechanism

currently widely implemented in the Internet for the provision of QoS, when MPLS is not the implemented solution.

19. A HTTP connection with a server is supported on the TCP at the transport layer (connection-oriented), which is the protocol that requires the previous connection establishment, before data is exchanged. The establishment of a TCP connection is a three-way handshaking: the initial step comprises the establishment request, by the client to the server, using the SYN message. The second step corresponds to the response from the server, acknowledging the establishment request, that is, accepting the request to establish the connection. This step is signalized using the SYN and ACK messages, together. Finally, the third and last step corresponds to the acknowledgment, sent by the client, using the ACK.

CHAPTER 10

1. The subnet mask /12 corresponds to 255.240.0.0. Moreover, since this is a class A address, it belongs to the class 10.0.0.0. The base number becomes $256 - 240 = 16$. Consequently, the host 10.96.2.33 belongs to the subnetwork with IP address 10.96.0.0, and the following subnetwork has the IP address 10.112.0.0. Consequently, the broadcast address of the subnetwork where the host with the IP address 10.96.2.33/12 is located becomes 10.111.255.255.

2. The IP version 6 (IPv6) includes the following advantages, relating to IP version 4 (IPv4):

   a. *Improved QoS mechanisms*: With the integration of the different services into a common telecommunications infrastructure, there was the need to make routing faster and smarter, with a more efficient and effective way of implementing packets differentiation. This is a characteristic of the IPv6, allowing a better handling of packets belonging to different services. Some mechanisms were improved in order to reduce the resources consumed by intermediate routers. An example is the lower number of fields of an IPv6 header, as compared to the number of IPv4 fields (despite the longer mandatory IPv6 header, namely due to longer IPv6 addresses). This translates in lower resources consumption by intermediate nodes, as well as faster routing. Another example is the IPv6 source routing (optional mechanism), which presents several improvements relating to the IPv4 source routing. This is a mechanism that allows specifying some of the routers that will be visited by packets belonging to a certain flow of data. In addition, the flow label field (mandatory header field) allows prioritized handling by intermediate routers. These mechanisms allow improving the QoS for different types of services.

   b. *Improved security functions*: While the IPsec is an IPv4 option, the exchange of packets in IPv6 networks is always carried out using the IPsec framework. This mechanism prevents attacks against the confidentiality, integrity, and authenticity.

   c. *Support for higher user mobility*: The host (interface) identification of an IPv6 address is unique, and can be automatically generated from the MAC address. Consequently, the movement of a user across different networks gives him the possibility to keep connectivity (using the same interface part of the IPv6 address) without administrative intervention. With such automatic MAC/host address conversion, the ARP is no longer needed. Moreover, the Internet control message protocol version 6 (ICMPv6) used by IPv6 improves the autoconfiguration of hosts, allowing an easier and quicker discovery of neighboring nodes.

   d. *Faster and easier routing*: The IPv6 address is hierarchical, consisting of a global routing prefix, subnet identification, and interface identification. The global routing prefix is normally assigned to sites by geographical regions. Since the IPv4 addresses were not organized by regions, interdomain routers had to consult enormous routing tables to allow the selection of the output interface to forward a datagram. Conversely, since most of the IPv6 routing is performed inside a certain hierarchical level, this results in shorter routing tables that allows a more efficient use of the routers resources (CPU, memory, bandwidth, etc.), enabling a faster routing.

3. Link-local addresses are known as automatic private Internet protocol addressing (APIPA) in Windows operating systems. Similar to the private addresses, APIPA addresses are only to be used inside private networks. Moreover, a host using an APIPA address that needs to exchange data with the Internet world must use NAT/PAT address translation. The link-local addresses (APIPA) can be used as an alternative to the dynamic host configuration protocol (DHCP), consisting of a mechanism that can be employed by a network to allow hosts getting IP addresses. In case a host contacts the DHCP server for obtaining an IP address without any response, and if the APIPA is configured in the NIC, the operating system automatically assigns an IP address within the APIPA range. The APIPA protocol assures that this address is unique within the LAN. This procedure allows obtaining an IP address without making use of the DHCP server or without having to manually configuring it. This procedure is very useful in small-size networks.

4. The network 194.123.3.4, used in classful mode, belongs to class C. Therefore, its network address is 194.123.3.0, as the three leftmost octets identify the network.

5. The IPv6 address is hierarchical, consisting of a global routing prefix, subnet identification, and interface identification. The global routing prefix is normally assigned to sites by geographical regions. Since the IPv4 addresses were not organized by regions, inter-domain routers had to consult enormous routing tables to allow the selection of the output interface to forward a datagram. Conversely, since most of the IPv6 routing is performed inside a certain hierarchical level, this results in shorter routing tables that allows a more efficient use of the routers resources (CPU, memory, bandwidth, etc.), enabling a faster routing.

6. A workstation, located in a LAN with Internet access, may have either a public IP address or a private IP address, as long as the router performs the address translation from a private into a public (for outgoing datagrams). This address translation can be implemented using one of the following two methodologies:

    a. *Network address translation*: The router that interconnects the LAN with a MAN or WAN for Internet access, is running the NAT application, being responsible for mapping private into public addresses. Depending on the number of simultaneous workstations that need to have access to the Internet, a single or multiple public IP addresses are used by the NAT. The NAT changes the source IP address in every outgoing datagrams, whereas the destination IP address are changed in every incoming datagrams.

    b. *Port address translation*: The router that interconnects the LAN with a MAN or WAN for Internet access, is running the PAT application, being responsible for mapping private IP addresses into a single public IP address. The PAT changes the source IP address of a datagram from a private address into a unique public IP address. Moreover, the PAT changes the outgoing segment's source port into one out of a range of ports used by the PAT router in the Internet world. Consequently, although the outgoing datagrams from different workstations present the same source IP address, they are differentiated by different source ports. The opposite operation is performed by the PAT for incoming datagrams. Consequently, for incoming traffic, the PAT converts different destination's ports in segments (and a single public IP address) into the different private IP addresses used by different LAN workstations. The PAT is also referred to as NAT overload.

7. The assignment of an IP address to a host can be either static or dynamic. In the former case, the network administrator uses the network settings of a host to assign an IP address.

Nevertheless, assigning IP addresses statically does not allow using such addresses when those hosts are switched off, or disconnected from the network. A more efficient mechanism relies on the use of DHCP. The DHCP is another example of a protocol that contributes to achieve a better usage of the depleted IPv4 address space. The DHCP assigns dynamically IP addresses to hosts. Most of the home routers run the DHCP server, but a computer server can also be employed for this task. The DHCP allows a DHCP server (router or computer) to dynamically assign IP addresses to different LAN workstations. The assigned IP addresses can be either private or public, depending on the specific environment and use. The DHCP is also normally used by ISPs, such that every time a client powers on a router, it receives a new IP address. Note that, similar to the NAT and PAT, the dynamic modification of IP addresses performed by the DHCP also represents an added value from the security point of view.

8. An IPv6 address has a total of 128 bits (16 octets). For the sake of simplicity, an IPv6 address is displayed in eight groups of four hexadecimal digits, separated by colons. Each group of four hexadecimal digits is within the range 0000-FFFF. Moreover, within an IPv6 address, zeros can be omitted. The position of the omitted zeros is marked with double colons. Therefore, the complete IPv6 address becomes 3D0F:00F3:0000:0000:0000:0000:0000:0000:0000: 0000:0000:0000:0000:0000:5A3C, as it has a total of 16 octets.

9. The IP address space in classful mode is very inefficient, since the rigid allocation of bits to host or network part of the IP address leaves many IP addresses unused. Therefore, due to this reason, and due to the rapid growth of the number of Internet hosts, the available IPv4 addressing space is depleted. The measure that can mitigate this limitation is making use of the IPv4 address is classless mode. In the classless mode, although bits are left pre-assigned to classes, we add another level of flexibility through the use of a subnet mask. In classless mode, the border between the network and the host identification is determined by the subnet mask. Mask of a network is an address with one logic state bits in the position that identifies the network part of an IP address, and zero logic state bits in the position that belongs to the host part of the IP address. Using the classless mode, the border between network and host part of the IP address is no longer defined by the complete octets, that is, by 8-bit segments. Classless mode can use any number of bits to address network and host parts of addresses, and the address range is not divided into classes. The IP range previously defined for any of the classes (e.g., class A) can now be utilized without any restriction between network and host assignments.

10. This syntax is used to configure a ciphered access password to the privileged mode. The access password is the word that follows *enable secret.*

11. The classless mode of IPv4 address scheme is one of the mechanisms that can be used to maximize the limited IPv4 address space. Other mechanisms include the use of private addresses inside LANs, the NAT, the PAT, as well as the DHCP.

12. The interface ID of a unicast address is 64-bit long and can be generated using the EUI-64 notation. It is directly generated from the NIC MAC address. The resulting EUI-64 address consists of the MAC address subject to the hexadecimal sequence *FFFE* added in the middle of the MAC address. Taking the previous MAC address as an example, the resulting EUI-64 address comes 0213:45FF:FE67:98BA.

13. When a serial cable is used to interconnect two routers, the configuration must consider one router acting as data terminal equipment (DTE) and the other acting as data communication equipment (DCE). The DCE is the device responsible for sending a synchronism clock to the DTE, required for the communication between these devices. For the communication to be possible, the router acting as DCE needs to send to synchronism clock to the router acting as DTE, which is performing using the Cisco IOS command *clock rate*, followed by the clock rate, expressed in Hertz.

14. The base number comes $256 - 192 = 64$. Consequently, the subnetwork addresses are the multiples of the base number (in the octet incomplete), that is, 10.0.0.0 (b/c address is 10.0.63.255); 10.0.64.0 (b/c address is 10.0.63.255); 10.0.128.0 (b/c address is 10.0.191.255); 10.0.192.0 (b/c address is 10.0.255.255); 10.1.0.0 (b/c address is 10.1.63.255); 10.1.64.0 (b/c address is 10.1.127.255); ...; 10.255.192.0 (b/c address is 10.255.255.255).

15. When configuring an interface in a router, after inserting its IP address, the command *no shutdown* is used to activate the interface.

16. The most important reason was the depletion of the IPv4 address space.

17. The base number is $256 - 192 = 64$. Therefore, the first subnetwork address is 126.12.0.0, and the second subnetwork address is 126.12.0.64.

18. The subnet mask /27 corresponds to 255.255.255.224. The base number becomes $256 - 224 = 32$. Consequently, the subnetwork addresses are 193.123.10.0 (b/c address is 193.123.10.31), 193.123.10.32 (b/c address is 193.123.10.63), 193.123.10.64 (b/c address is 193.123.10.95), 193.123.10.96 (b/c address is 193.123.10.127), ..., 193.123.10.224 (b/c address is 193.123.10.255).

19. The IP address with 193 value in the leftmost octet belongs to class C.

20. The base number is $256 - 192 = 64$. Consequently, the subnetwork addresses becomes 10.0.0.0 (b/c address 10.0.0.63), 10.0.0.64 (b/c address 10.0.0.127), 10.0.0.128 (b/c address 10.0.0.191), and 10.0.0.192 (b/c address 10.0.0.255).

21. This IPv6 address is using the prefix /64 to separate the network and subnetwork part (64 bits) of an address from the interface part of an address. Consequently, the address 31CF:0123:4567:89AB:CDEF:0123:4567:89AB/64 has a total of 64 bits to identify the network and subnetwork part of the address (with the address 31CF:0123:4567:8 9AB:0000:0000:0000:0000), and that the remaining 64 bits are used to identify the interface within the network (CDEF:0123:4567:89AB).

22. The class A has a total of eight bits assigned to the network, but the leftmost bit is fixed to zero. Therefore, seven bits are used to identify networks ($2^7 = 128$), but the all zero value bits cannot be used (reserved for the identification of the own IP address) and the decimal value 127 is reserved for loopback, leaving a total of $2^7 - 2 = 126$ existing class A networks.

23. The base number is $256 - 192 = 64$. Consequently, the subnetwork addresses becomes 10.0.0.0 (b/c address 10.63.255.255), 10.64.0.0 (b/c address 10.127.255.255), 10.128.0.0 (b/c address 10.191.255.255), and 10.192.0.0 (b/c address 10.255.255.255).

24. The class B, used in classful mode, use two octets to address hosts (32 bits), making $2^{16} = 65,536$. Nevertheless, the all zero value bits cannot be used (reserved for the identification of the own IP address), and the all one value bits cannot be used (reserved for the identification of the broadcast address). Consequently, each class B network address, used in classful mode, can accommodate $2^{16} - 2 = 65,536 - 1 = 65,534$.

25. The subnet masks possible to be used with a class B network are /16, /17, /18, /19, /20, /21, /22, /23, /24, /25, /26, /27, /28, /29, and /30.

26. The correct statement is – Eight sub-networks can be created.

27. The MAC address is obtained from EUI-64 address removing the characters FF and FE, making 0123:4567:89AB.

28. The IP is the most important network protocol of the TCP/IP stack. It provides a basic service to the delivery of datagrams, enabling a wide list of application and services. The basic functionalities of the IP are as follows:
    a. Definition of the address space in the Internet
    b. Routing of datagrams across the network nodes
    c. Fragmentation and reassembling of datagrams

29. Completing the logic AND operation between the binary version of an IP address and the binary version of the subnet mask lead us to the network and subnetwork parts of an IP address. Let us examine a host with an IP address 10.88.1.2 and with subnet mask /11 that corresponds to 255.224.0.0. The subnet mask in binary representation is 11111111.11100000.00000000.00000000. Performing the AND operation between the IP address and the subnet mask, we conclude that the host is in the network address 10.64.0.0.

30. The interface ID of a unicast address is 64-bit long and can be generated using the EUI-64 notation. It is directly generated from the NIC MAC address. The resulting EUI-64 address consists of the MAC address subject to the hexadecimal sequence *FFFE* added in the middle of the MAC address. Taking the previous MAC address as an example, the resulting EUI-64 address comes 0123:45FF:FE67:89AB.

31. The host 10.76.1.2/11 is in a subnetwork that uses 3 bits used to address subnetworks, within the class A address. Moreover, it comprises a total of $32 - 11 = 21$ bits to address hosts, within each subnetwork. This allows addressing a total of $2^{21} - 2 = 2,097,150$ hosts.

32. An IPv6 address comprises a total of 128 bits. For the sake of simplicity, an IPv6 address is displayed in eight groups of four hexadecimal digits, separated by colons. Each group of four hexadecimal digits is within the range 0000-FFFF.

33. The IPv6 address is hierarchical, consisting of a global routing prefix, subnet identification, and interface identification. The global routing prefix is normally assigned to sites by geographical regions. Since the IPv4 addresses were not organized by regions, inter-domain routers had to consult enormous routing tables to allow the selection of the output interface to forward a datagram. Conversely, since most of the IPv6 routing is performed inside a certain hierarchical level, this results in shorter routing tables that allows a more efficient use of the routers resources (CPU, memory, bandwidth, etc.), enabling a faster routing.

34. An IPv6 address is composed of 128 bits (16-octets). For the sake of simplicity, an IPv6 address is displayed in eight groups of four hexadecimal digits, separated by colons. Each group of four hexadecimal digits is within the range 0000-FFFF.

35. The DHCP assigns dynamically IP addresses to hosts. Most of the home routers run the DHCP server, but a computer server can also be employed for this task. The DHCP allows a DHCP server (router or computer) to dynamically assign IP addresses to different LAN workstations. The assigned IP addresses can be either private or public, depending on the specific environment and use. The DHCP is also normally used by ISPs, such that every time a client powers on a router, it receives a new IP address. Note that, similar to the NAT and PAT, the dynamic modification of IP addresses performed by the DHCP also represents an added value from the security point of view. When a host connects to a network, by default it assigns to itself the IP address 0.0.0.0. Then, it contacts the DHCP server, requesting the assignment of an IP address.

36. The organization requires three subnetworks: one subnetwork with 16 hosts, another with 2 hosts, and a third one with 12 hosts. Based on the procedure described for VLSM, we have:
    a. Ordering the required number of hosts by descending order, we obtain 16, 12, and 2 hosts. Taking into account the subnetwork and broadcast addresses, we obtain the following address requirements for different subnets, by descending order:
        i.   16 hosts → 18 addresses
        ii. 12 hosts → 14 addresses
        iii. 2 hosts → 4 addresses

b. Deducting the required number of logic state 0's in the subnet mask that supports such requirements, this comes:

    i. 18 addresses → the exponent of two upper layer corresponds to 32, which requires five 0's in the subnet mask (twenty-seven 1's in the subnet mask) → 11111111 11111111 11111111 11100000.

    ii. 14 addresses → the exponent of two upper layer corresponds to 16, which requires four 0's in the subnet mask (twenty-eight 1's in the subnet mask) → 11111111 11111111 11111111 11110000.

    iii. 4 addresses → two 0's in the subnet mask which will make a total of 4 available addresses (thirty 1's in the subnet mask) → 11111111 11111111 11111111 11111100.

c. Defining the subnet mask for each subnetwork, we obtain:

    i. 18 addresses → 255.255.255.224

    ii. 14 addresses → 255.255.255.240

    iii. 4 addresses → 255.255.255.252

d. Defining the IPv4 address space for each subnetwork, we obtain:

    i. 18 addresses → address space → 192.168.1.0 up to 192.168.1.31 (base number is $256 - 224 = 32$, making the address range bounded between 0 and $32 - 1 = 31$).

    ii. 14 addresses → address space → 192.168.1.32 up to 192.168.1.47 (base number is $256 - 240 = 16$, making the address range bounded between 32 and $32 + 16 - 1 = 47$).

    iii. 4 addresses → address space → 192.168.1.48 up to 192.168.1.51 (base number is $256 - 252 = 4$, making the address range bounded between 48 and $48 + 4 - 1 = 51$).

37. The organization requires three subnetworks: one subnetwork with 12 hosts, another with 2 hosts, and a third one with 250 hosts. Based on the procedure described for VLSM, we have:

a. Ordering the required number of hosts by descending order, we obtain 250, 12, and 2 hosts. Taking into account the subnetwork and broadcast addresses, we obtain the following address requirements for different subnets, by descending order:

    i. 250 hosts → 252 addresses

    ii. 12 hosts → 14 addresses

    iii. 2 hosts → 4 addresses

b. Deducting the required number of logic state 0's in the subnet mask that supports such requirements, this becomes

    i. 252 addresses → the exponent of two upper layer corresponds to 256, which requires eight 0's in the subnet mask (twenty-four 1's in the subnet mask) → 11111111 11111111 11111111 00000000.

    ii. 14 addresses → the exponent of two upper layer corresponds to 16, which requires four 0's in the subnet mask (twenty-eight 1's in the subnet mask) → 11111111 11111111 11111111 11110000.

    iii. 4 addresses → two 0's in the subnet mask which will make a total of 4 available addresses (thirty 1's in the subnet mask) → 11111111 11111111 11111111 11111100.

c. Defining the subnet mask for each subnetwork, we obtain

    i. 252 addresses → 255.255.255.0

    ii. 14 addresses → 255.255.255.240

    iii. 4 addresses → 255.255.255.252

d. Defining the IPv4 address space for each subnetwork, we obtain

    i. 252 addresses → address space → 10.20.0.0 up to 10.20.0.255 (base number is $256 - 0 = 256$, making the address range bounded between 0 and $256 - 1 = 255$).

    ii. 14 addresses → address space → 10.20.1.0 up to 10.20.1.15 (base number is $256 - 240 = 16$, making the address range bounded between 0 and $0 + 16 - 1 = 15$).

    iii. 4 addresses → address space → 10.20.1.16 up to 10.20.1.19 (base number is $256 - 252 = 4$, making the address range bounded between 16 and $16 + 4 - 1 = 19$).

CHAPTER 11

1. The administrative distance considered by RIP is 120.
2. a.

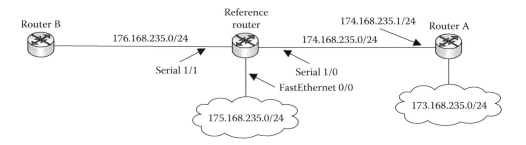

   b. According to the first row of the routing table, the next router to which the datagrams should be forwarded in order to reach a host with the destination IP address 173.168.235.3 (that belongs to the network 173.168.235.0) has the address 174.168.235.1 (reached through the interface Serial 1/0).
3. This Cisco IOS command is used to configure a router to forward all datagrams, regardless of their destination IP addresses and subnet masks, to the next router with the IP address 194.136.235.1.
4. The MTU is the largest number of bytes that can be carried in a frame payload. The payload length of the data link layer frame is normally variable, depending on the amount of data to transport between two adjacent nodes. If a router is switching a packet from a data link layer with a longer MTU into a data link layer with a shorter MTU, and if the datagram length is longer than the shorter MTU, fragmentation is implemented. In this case, each packet is decomposed into a group of smaller pieces. Conversely, when a router is switching a packet from a data link layer with a shorter MTU into a longer MTU, reassembling occurs, that is, the small pieces are re-grouped so as to re-create the packet with the original size.
5. Routing algorithms can be grouped into two categories: adaptive and nonadaptive routing algorithms. Moreover, in the case of nonadaptive algorithms, these can be of two types: static or flooding algorithm. With nonadaptive algorithms, the entries into the routing tables are not calculated using network topologies, traffic, or bandwidth metrics. The selection of the path to utilize is previously computed, and pre-loaded into the router. A static entry into the routing table consists of an action manually performed by the network administrator. One disadvantage results from the fact that, in case such path is suddenly interrupted (or subject to bad conditions), the traffic is not automatically switched over to another path. Using static route, such re-routing of the traffic can only be performed through human intervention, that is, changing the required entry into the routing table. In this case, the traffic is interrupted during a certain time period. Flooding is another nonadaptive algorithm. Using the flooding algorithm, a router forwards an input datagram into all different outputs. Consequently, the datagrams will arrive at the destination with redundancy (the same datagram will be repeated), and through all possible paths. Using the flooding algorithm, it is assured that one of the paths is the optimum one. An important disadvantage results from the fact that the network becomes overloaded with many replicas of each datagram, which may degrade its performance. Moreover, routing protocols can be distance vector or link state based. Depending on whether the information in a router is simply the distance or the full network topology, then the adaptive routing protocol is classified respectively as distance vector or link state. Another classification for routing

protocols relies on whether they are classful or classless. Adaptive routing protocols allow the automatic construction of routing tables by exchanging routing information with neighboring routers.

6. Depending on whether the information in a router is simply the distance or the full network topology, the adaptive routing protocol is classified respectively as distance vector or link state.

7. Using the RIP, a maximum of 15 hops is allowed for a datagram to be delivered to a destination network. If the metric distance is higher than this value, the packet is considered as unreachable, being discarded. Moreover, similar to other routing protocols, the RIP supports equal cost load balancing, up to four different routes. This means that when a router has up to four different routes to the same destination network with the same cost (metric distance), each packet of each group of four consecutive packets is sent through a different route.

8. The MTU is the largest number of bytes that can be carried in a frame payload. The payload length of the data link layer frame is normally variable, depending on the amount of data to transport between two adjacent nodes. If a router is switching a packet from a data link layer with a longer MTU into a data link layer with a shorter MTU, and if the datagram length is longer than the shorter MTU, fragmentation is implemented. In this case, each packet is decomposed into a group of smaller pieces. Conversely, when a router is switching a packet from a data link layer with a shorter MTU into a longer MTU, reassembling occurs, that is, the small pieces are re-grouped so as to re-create the packet with the original size.

9. This Cisco IOS command is used to configure a router to forward the datagrams whose destination IP address is 193.123.213.0 with the subnet mask 255.255.255.0 to the next router with the IP address 194.136.235.1.

10. An IPv4-mapped-IPv6 is generated from an IPv4 address as 0:0:0:0:FFFF:<IPv4 address>. Consequently, the IPv4-mapped-IPv6 corresponding to the IPv4 address 194.136.235.1 comes 0:0:0:0:FFFF:C288:EB01.

11. The MTU is the largest number of bytes that can be carried in a frame payload. The payload length of the data link layer frame is normally variable, depending on the amount of data to transport between two adjacent nodes. If a router is switching a packet from a data link layer with a longer MTU into a data link layer with a shorter MTU, and if the datagram length is longer than the shorter MTU, fragmentation is implemented. In this case, each packet is decomposed into a group of smaller pieces. Conversely, when a router is switching a packet from a data link layer with a shorter MTU into a longer MTU, reassembling occurs, that is, the small pieces are re-grouped so as to re-create the packet with the original size.

12. Administrative distance is a value allocated to each routing protocol or mechanism that translates its reliability. As lower an administrative distance is, the more reliable it is. For example, the RIP has an administrative distance of 120, whereas the OSPF has an administrative distance of 110. This means that the OSPF is more reliable than the RIP. In case there is more than one path to reach a certain destination, the router selects the route with the lower administrative and/or metric distance, as defined in the following.

13. The metric distance (also referred to as *cost*) consists of a value that ranks each path to a certain destination. The preferable path has the lowest metric distance, whereas the worst path has the highest metric distance. The path with the lowest metric distance is considered as the preferred route, and included into the routing table. Finally, in case there is more than one path with the same lowest metric distance, equal cost load balancing is implemented, that is, the traffic is split into these routes with equal metric distances. The parameters taken into account in the calculation of the metric distance depends on the routing protocol.

14. a.

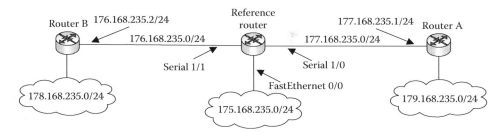

Router B 176.168.235.2/24   Reference   177.168.235.1/24 Router A
                                             router
                    176.168.235.0/24   177.168.235.0/24

                              Serial 1/1   Serial 1/0

                                          FastEthernet 0/0
       178.168.235.0/24   179.168.235.0/24
                              175.168.235.0/24

   b. According to the fourth row of the routing table, the next router to which the datagrams should be forwarded in order to reach a host with the destination IP address 178.168.235.3 (that belongs to the network 178.168.235.0) has the address 176.168.235.2 (reached through the interface Serial 1/1).

15. The metric utilized by the OSPF protocol is metric $= \sum_{n=1}^{\text{no. of hops}} (10^8/\text{bandwidth [bps]})$, where bandwidth (bps) stands for the bandwidth configured in the router's interface, expressed in bits per second. The value for bandwidth used in the metric calculation does not correspond to the link bandwidth, but consists of a static bandwidth value configured in the router's interfaces, in the outgoing direction. Both sides of a serial link should be configured with the same bandwidth. Moreover, while in EIGRP the metric is calculated taking into account the lowest bandwidth of the multiple outgoing routers' interfaces in a route, using the OSPF routing protocol, one must calculate the metric for each hop and the metric route is the cumulative sum of the independent metrics for different hops.

16. Link state routing protocols compute a topologic database in each router, with the overall map of network connectivity, whereas distance vectors protocols simply compute the distances to each remote node. The map has the form of a graph, showing all network nodes and their interconnection. Link state protocols can use configured bandwidth in the interfaces or delay (among other parameters), as metric parameters.

CHAPTER 12

1. Virtual local area networks (VLANs) can be created in switches to allow different logical networks with a single physical network. VLANs are used for different purposes, namely
   a. For providing access to different networks with a single switch.
   b. For creating different broadcast domains with a single switch.
   c. For achieving logical isolation among different hosts (security).
2. Similar to the upper layers that make use of the services made available by the lower layers, the data link layer makes use of the service provided by the physical layer. Since the physical layer is responsible for a nonreliable exchange of bits, the DLL adds the required reliability. Consequently, the LLC sublayer is responsible for allowing a reliable exchange of data between two adjacent nodes of a network. This is achieved implementing the following functionalities:
   a. Error control
   b. Flow control
   c. Grouping isolated bits into frames
   The most common LLC sublayer protocol is the IEEE 802.2.
3. A MAC determines when a station is allowed to transmit within a LAN. Note that this sublayer only exists in case of a LAN (and some types of MAN), that is, in multiuser networks. When stations share the transmission medium in a LAN, it is said that the access method is with collisions (e.g., Ethernet). In this case, the MAC sublayer is responsible for defining

when a station is allowed to transmit in such a way that collisions among transmissions from different stations is avoided (which originate errors). On the other side, when stations of a LAN do not share the transmission medium, it is said that the method is without collisions (e.g., Token Ring). This sublayer is implemented using protocols such as the IEEE 802.3, IEEE 802.4, IEEE 802.5, or IEEE 802.11.

4. The IEEE 802.3 is much simpler than the IEEE 802.5, without the need to have a management station responsible for the control of the token. Moreover, the IEEE 802.3 network does not present a long minimum waiting time as that of the IEEE 802.5 (corresponding to time required for the token to go around the ring). In any case, the IEEE 802.3 tends to be the preferable standard due to its simplicity, and due to its good performance under low traffic load conditions.

5. The contention mechanism employed by the IEEE 802.3 standard is the carrier sense multiple access—collision detection (CSMA-CD), defined as follows:

   a. A host that intends to initiate a frame transmission within the network starts by listening to the channel (carrier sense).

   b. If the shared medium is idle (i.e., if there isn't any host transmitting), the host may initiate the transmission. Otherwise, if another host is transmitting (medium is busy), it waits a backoff period of time. After this period, the host re-checks the channel availability, in order to start the transmission.

   c. Albeit the channel could be idle, and a host could have started the transmission, another host could have started the transmission in a very close moment. This originates a collision. In order to assure that such collisions are detected, a host needs to keep listening to the medium (receiving), while transmitting. If a host detects a collision, the host must proceed with the transmission such that the minimum frame size (512 bits) is transmitted and such that the transmission is finalized with the jamming signal. Then, the host must interrupt its transmission. Otherwise, the host transmits immediately the jamming signal, and then interrupts the transmission. This procedure ensures that all receivers detect a collision by checking the received corrupted CRC, including those located in the other physical extreme of the network. After the jamming signal, both hosts must interrupt their transmissions and try to re-start after a listening strategy defined as follows:

      i. *Nonpersistent*: If the medium is idle, transmit immediately. If the medium is busy, wait for a backoff period and listen again to the medium. Note that it may happen that the medium is already idle, but the host is still waiting a random period.

      ii. *1-persistent*: If the medium is idle, transmit immediately. If the medium is busy, keep waiting and listen to the medium. Then, when it becomes idle, transmit immediately. If more than a single host is waiting for the medium to be idle, the transmissions of these waiting hosts will certainly collide. This represents a disadvantage of the 1-persistent scheme.

      iii. *P-persistent*: If the medium is idle, the transmit probability is p, while the probability to delay one time unit is $(1 - p)$. If the medium is busy, wait for being idle, and then use the previous procedure. If the transmission is delayed, after the waiting time use the previous procedure. The value of p varies, but typical values are between 0.5 and 0.75.

6. Error control is normally used by the data link layer (more precisely in the logical link control sublayer), in point-to-point mode, to provide reliability to the bits exchanged by the physical layer. Moreover, the transport layer can also implement error control mechanisms (e.g., using the TCP of the TCP/IP architecture). Nevertheless, in this case, the error control is performed end-to-end (instead of point-to-point, as adopted by the data link layer).

7. Determine the sequence of bits to transmit.

   The information bits consist of the sequence 11011 and the generator polynomial is $P(x) = x^2 + 1$. Based on the above description, the processing of the transmitting entity becomes

   a. Since the degree of the generator polynomial is two, the number of zeros added to the information sequence is also two. The resulting sequence becomes 1101100.

   b. The generator polynomial $P(x) = x^2 + 1$ is mathematically equivalent to $P(x) = 1 \times x^2 + 0 \times x + 1 \times x^0$. Then, the sequence of bits that results from the generator polynomial is 101. Performing the operation 1101100:101 results in the sequence 11100 and the rest of this division comes 00 (the number of bits taken from the rest of the division corresponds to the degree of the generator polynomial). Note that the binary difference operation performed to calculate the rest consists of the module 2 adder without carry.

   c. The transmitting sequence that consists of the concatenation of the information bits with the rest becomes 1101100.

8. The processing of the receiving entity becomes:

   a. The received block of bits (1101101) is divided by the bits that results from the generator polynomial (101). Performing the operation 1101101:101 results in 11100 and the rest of this division comes 01 (the number of bits taken from the rest of the division corresponds to the degree of the generator polynomial). Consequently, since the rest is not zero, we may conclude that the received block is corrupted.

9. Depending on the adopted ARQ procedure there are two basic versions of flow control protocol that can be implemented, namely

   a. *Stop and wait*: This type of flow control is implicitly and automatically performed when the stop-and-wait ARQ protocol is implemented. It is worth remembering that the ARQ uses positive confirmation. Similar to the corresponding ARQ protocol, this procedure is performed independently for every frame. The transmitter sends a single frame and, when the receiver is ready to receive more data, it sends the receive ready message, as described for the stop-and-wait ARQ. The acknowledgment is achieved using the receive ready message.

   b. *Sliding window*: The sliding window flow control is associated to the go back N ARQ or to the selective reject ARQ protocols. When any of these ARQ protocols are implemented, the sliding window flow control is implicitly implemented. The objective of the transmitting window is to allow that the transmitter keeps sending data (successive frames), without having to stop and wait for the reception of confirmations. Ideally, if everything occurs as expected, the transmitting entity receives confirmations while still transmitting data. This allows the maximization of the transmission medium usage. An important aspect of this protocol relies on the fact that the frames need to be numbered. The transmitting window N is a function of the relationship between the propagation time and the frame duration. The window size N corresponds to the maximum number of frames that can be transmitted without confirmation. If the confirmations are delayed, the transmitter may also have to delay the transmission of the following frames. Nevertheless, if such delay is within the window size, the transmitter has permission to proceed with the transmission of more data. The lower window edge (LWE) corresponds to the position of the last confirmed frame, and the upper window edge (UWE) is an upper bound for the last frame that can be transmitted. These two parameters are related by $UWE = LWE + N$.

10. Stop-and-wait ARQ, go back N ARQ, and selective reject ARQ.

11. Both are error control protocols associated to the sliding window flow control. Using the go back N ARQ, in case an error is detected, the transmitting entity retransmits the corrupted frame, while retransmitting the following frames that have already been corrected

received. On the other hand, using the selective reject ARQ, in case an error is detected, the transmitting entity retransmits the corrupted frame, but then returns the previous transmission sequence, not retransmitting the following frames that have already been corrected received.

12. This can be performed using the puncturing operation. The puncturing is an operation that can be applied to the output of the error correction encoder in order to, artificially, increase its code rate. This operation is achieved by periodically removing some bits from a codeword at the transmitting side, whereas the receiver inserts a zero value bits in the corresponding predefined positions. The puncturing operation should be performed in such a way that the error correction code should be able to correct the inserted zero value bits to the correct logic states.

13. In case the transmission medium is highly subject to channel impairments (e.g., wireless medium), the choice goes normally to the use of error correction, instead of error detection. In this latter case, the level of overhead per frame is higher, but it avoids successive repetitions, which also translates in a decrease of overhead. Moreover, error correction can also be used when the transmission is unidirectional or when latency must be avoided (the handshaking associated to error detection and retransmission introduces delays).

14. Both consists of physical layer technologies used with IEEE 802.3 protocol, and implemented with twisted pairs. Nevertheless, while 10BASE-T consists of a 10 Mbps network, 1000BASE-T consists of a 1 Gbps network. Moreover, while 10BASE-T uses four pairs of UTP Cat. 3, 4, or 5, 1000BASE-T uses four pairs (in this case, negotiated for transmission or reception) of UTP Cat. 5. Finally, while 10BASE-T uses Manchester line coding technique, 1000BASE-T employs 4D-PAM5 with scrambling and forward error control.

15. Puncturing is an operation that can be applied to the output of the error correction encoder in order to, artificially, increase its code rate. This operation is achieved by periodically removing some bits from a codeword at the transmitting side, whereas the receiver inserts a zero value bits in the corresponding predefined positions. The puncturing operation should be performed in such a way that the error correction code should be able to correct the inserted zero value bits to the correct logic states.

16. It is known that the bits are normally corrupted in bursts. This is the result of channel impairments such as a deep fading, an impulsive noise or even an instantaneous strong interference. The error correction codes are able to correct up to a certain number of bits. Beyond this number, error correction codes are not able to correct those bits. In order to improve the error correction capability, the error correction is normally associated to interleaving. The interleaving is used to remove the sequential properties of errors and to allow the error correction codes to perform better. The aim of an interleaver simply relies on modifying the sequential position of the bits, splitting a sequence of corrupted bits into several frames, at the transmitting side. The de-interleaver, located at the receiver side, performs the opposite operation, repositioning the bits into the original sequence. This way, the number of corrupted bits that appears in each frame becomes possible to be corrected by an error correction code.

17. The Hamming distance is the minimum number of bits between any two codewords (blocks of encoded bits). The error correction of Hamming codes are deducted from the Hamming distance, as follows: the maximum number of corrupted bits r that can be corrected is obtained from the Hamming distance d as $r = (d-1)/2$. Therefore, increasing the Hamming distance, we are also increasing the maximum number of bits that can be corrected by an error correction code based on block codes, and consequently, we are improving its error correction capability.

18. The most used encapsulation protocol employed in VLANs is the IEEE 802.1Q.

19. The start and end of frame can be signalized using three different procedures, namely
    a. *Delimiting character string*: In this case, the receiver detects the start of the frame through the reception of a sequence of two predefined characters. These two characters are the data link escape (DLE) and the start of text (STX). Moreover, the receiver detects the end of a frame by receiving another group of two predefined characters, namely the DLE and the end of text (ETX). A limitation of this procedure comes from the fact that the bits corresponding to the DLE character may be part of the payload data. In this case, to avoid that the receiver becomes confused, instead of sending a single DLE in the payload data, the transmitter sends twice the DLE character, and the receiver removes one of these characters.
    b. *Flag*: This is the most common frame synchronization method. In this case, the receiver detects the start and end of frame through the reception of a predefined sequence of bits. It is composed of a sequence of bits with low probability to occur within the payload data part of a frame. The PPP and the HDLC protocols use the sequence 01111110 as the flag. A common limitation of this procedure comes from the fact that a part of the payload data may have a sequence of bits equal to the flag. In such cases, the receiver could interpret it as start or end of a frame. To avoid this, a procedure known as bit stuffing is implemented, defined as follows for the HDLC protocol: every time the transmitter detects a sequence of five *1* logic state bits in the data, it inserts a *0* logic state bit as the sixth bit. The receiver performs the reverse operation. With the bit stuffing procedure, it is assured that a sequence of six *1* logic state bits are never transmitted within the data field.
    c. *Violation of the line coding mechanism*: In this case, the receiver detects the start and end of a frame through the reception of a predefined sequence of signal levels that is not allowed to occur within the normal transmission of data. As an example, the biphase Manchester line coding technique considers always a transition (from 0 to 1 or from 1 to 0) at the middle of the bit period. The absence of such transition may be used as a signaling for the start and/or end of a frame.
20. Both are physical layer technologies used to implement IEEE 802.3 network at 1 Gbps. Nevertheless, 1000BASE-CX is implemented with two pairs of STP, with an MSS of 25 m (DB9 or HSSDC connectors), whereas 1000BASE-SX uses two pairs of single-mode optical fibers, and an MSS of 5 km (SC connectors).
21. The Hamming distance is the minimum number of bits between any two codewords (blocks of encoded bits). The error detection of Hamming codes are deducted from the Hamming distance, as follows: The maximum number of corrupted bits s that can be detected is obtained from the Hamming distance d as $s = d - 1$. Therefore, increasing the hamming distance, we are also increasing the maximum number of bits that can be detected by an error detection code based on block codes, and consequently, we are improving its error detection capability.
22. The 1000BASE-T implements an IEEE 802.3 network at 1 Gbps by employing four pairs of UTP category 5 (negotiated for transmission or reception), the start physical topology, the 4D-PAM5 encoding technique associated to scrambling and forward error correction (FEC). The MSS of the 1000BASE-T is 100 m (200 m with repeater).
23. There may be multiple paths between two network points. Multiple paths exist in the case of a network with a mesh topology. In this case, the bridge has to decide about which one to use. Such discovery and decision is performed making use of the spanning tree protocol, whose most implemented protocol corresponds to IEEE 802.1D. The obtained information is utilized in order to remove loops, as well as to keep information about redundant paths. In case a path is interrupted, the bridge switches to the backup path. Moreover, there are situations where a bridge uses two or more paths simultaneously to perform load

balancing. Moreover, the protocol IEEE 802.1D is also used to fill in the MAC address table of a switch.

24. The maximum collision domain diameter (MCDD) is defined as the maximum distance between the two furthest network nodes. From this parameter we may calculate the total time it takes for the shortest frame (minimum frame size) to travel round trip between the two furthest network nodes in that domain. Note that this time takes into consideration the round trip time (double) in order to allow the collision (interfering) signal returning to the transmitting source.

25. At low speeds, the maximum network size is normally constrained by the MSS, instead of the MCDD. This is the case of a 10BASET network, where the MSS is limited to 100 m, while the MCDD is 280 m. Nevertheless, for speeds of 1 Gbps or above, the MCDD becomes the constraint. In this case, a mechanism called carrier extension is employed, in order to overcome such limitation. The carrier extension consists of a number of bits transmitted after and together with the regular frame, when its length is shorter than 4096 bit. The objective is to assure that the resulting frame length (original frame plus carrier extension) is a minimum of 4096 bit long (used in the calculation of the 1 Gbps IEEE 802.3 slot time). Naturally, in case the frame length is longer than 4096 bit, the carrier extension is not added.

26. Using the CSMA-CD, a host that intends to initiate a frame transmission within the network starts by listening to the channel (carrier sense). If the shared medium is idle (i.e., if there is not any host transmitting), the host may initiate the transmission. Otherwise, if another host is transmitting (medium is busy), it waits for a backoff period of time. After this period, the host re-checks the channel availability, in order to start the transmission. The backoff period used in IEEE 802.3 is referred to as the *binary exponential backoff*. It consists of a random waiting period, whose mean value is doubled during the initial 10 retransmission attempts. The mean value of the random waiting period is kept unchanged for six more additional retransmission attempts. In case 16 unsuccessful collisions occur, the host gives up attempting and sends an error to the upper layer. Then, the cycle restarts after a larger backoff (this value increases with the increase of the number of groups of 16 attempts).

27. The MCDD consists of the maximum length (diameter) of a domain, where a collision can be handled by the MAC sublayer. It is dimensioned for half of the slot time period (round trip), resulting in 256 bits duration. Nevertheless, to take into account additional delay in network devices (e.g., repeaters) the one-way calculations are performed for a total of 232 bits. Considering a 10 Mbps IEEE 802.3 network, the time for transmitting 232 bits becomes $T = (\text{number of bits/bit rate}) = (232/10^7) = 23.2\,\mu\text{s}$. From the physics, the MCDD becomes $MCDD = v \cdot T$, where v is the propagation speed of signals in the transmission medium (coaxial cable, twisted pair). Taking into account the approximate propagation speed in the coaxial cable of $v = 1.22 \cdot 10^8$ m/s (note that the light speed in the vacuum is $v = 3 \cdot 10^8$ m/s), the MCDD comes 2800 m.

28. There are three listening strategies employed in CSMA-CD:
    a. *Nonpersistent*: If the medium is idle, transmit immediately. If the medium is busy, wait a backoff period and listen again to the medium. Note that it may happen that the medium is already idle, but the host is still waiting a random period.
    b. *1-persistent*: If the medium is idle, transmit immediately. If the medium is busy, keep waiting and listening to the medium. Then, when it becomes idle, transmit immediately. If more than a single host is waiting for the medium to be idle, the transmissions of these waiting hosts will certainly collide. This represents a disadvantage of the 1-persistent scheme.
    c. *P-persistent*: If the medium is idle, the transmit probability is p, while the probability to delay one time unit is $(1 - p)$. If the medium is busy, wait for being idle, and

then use the previous procedure. If the transmission is delayed, after the waiting time use the previous procedure. The value of p varies, but typical values are between 0.5 and 0.75.

29. Convolutional codes encode an arbitrary group of input bits, using a certain logic function. When there are no output bits fed back to the input, this belongs to the group of non-recursive codes. Contrarily, in case there are output bits fed back to the input, the convolutional code is referred to as recursive.

30. The logical topology comprised by the IEEE 802.5 standard is the ring, where a token circulates in a certain direction. When a host receives the token, it is allowed to send data to the common transmission medium for a certain maximum period of time. After such period, the host must interrupt the transmission and the token is forwarded to the adjacent host. In the next round, the host may continue its transmission. After having received the token, in case the host has no data to transmit, it immediately forwards the token to the following adjacent host, without consuming a rigid time period. This represents an advantage relating to the TDMA, where fixed time slots are allocated to different users, regardless of whether they have or do not have data to transmit.

31. Similar to the bridge, the switch performs the forwarding of frames based on the destination's MAC address. Therefore, this device works at layer two of the OSI model. The forwarding of frames is performed based on a MAC address table, which maps the destination MAC addresses into output ports. Similar to the bridge, the MAC address table is filled in as the switch receives frames from the corresponding interfaces (ports). Moreover, there may be more than one path between two network nodes. The Spanning Tree protocol (IEEE 802.1D) normally resolves that by using a metric based on the lower number of hops, while the other paths are kept as redundancy. Load balancing is also possible in high-dimension LANs. Contrary to the bridge, which performs the switching of frames by software, the switch performs its functionality through hardware. Consequently, the latency introduced by a switch is typically much lower than that of a bridge. A bridge has normally two interfaces, whereas a switch has a high number of interfaces.

32. A hub is simply a repeater of bits that retransmits in all outputs the bits received in one input, and therefore, it works at the physical layer. On the other hand, the switch forwards frames from one input into a certain output based on the destination MAC address. Therefore, it works at the data link layer. Nowadays, the switch tends to be the central node of the network, instead of the previously used hub. While the corresponding logical topology of a network employing a hub is the bus, the star is the logical topology that results from the use of the switch as the central node. This results in a high improvement of the network performance. A network using a hub as a central node only allows a host transmitting at a time. Contrarily, a switch allows up to half of the network hosts transmitting to the other half of the network (half-duplex). Assuming that a LAN works at 10 Mbps in half-duplex mode, and that it has a total of 10 hosts, then we may have up to 5 hosts transmitting at 10 Mbps to the other 5 hosts. This results in a cumulative network throughput of 50 Mbps. In case the network works in full-duplex mode, then the maximum cumulative network throughput becomes 100 Mbps.

33. The switch is the central node of a network (LAN), whereas the router is used to interconnect different networks. The switch forwards frames from one input into a certain output based on the destination MAC address. Therefore, it works at the data link layer. The router forwards packets from one input into a certain output based on the destination IP address. Therefore, the router works at the network layer. Moreover, while the multiple interfaces of a switch are typically of the side type, the different interfaces of a router can be of different types (to interconnect networks of different types, such as a LAN with a WAN).

34. The output sequence becomes

$$k_1 = n_1 \oplus n_0 \oplus n_{-1} = 1 \oplus 0 \oplus 1 = 0$$
$$k_2 = n_1 \oplus n_0 = 1 \oplus 0 = 1$$
$$k_3 = n_1 \oplus n_{-1} = 1 \oplus 1 = 0$$

35. VLANs can be created in switches to allow different logical networks with a single physical network. VLANs are used for different purposes, namely
    a. For providing access to different networks with a single switch.
    b. For creating different broadcast domains with a single switch.
    c. For achieving logical isolation among different hosts (security).
36. Frame bursting can be adopted by 1 Gbps or faster IEEE 802.3 networks, instead of spending bandwidth at transmitting a carrier extension. If the transmitting host has several short frames to transmit, these frames can be sent together, linked with the interframe gap (IFG). The IFG consists of a predefined bit pattern. This improves the network performance, relating to the carrier extension use. The initial frame is always sent using the carrier extension, whereas others can be transmitted using the frame bursting procedure. The maximum length of the linked short frames is limited to 56,536 bits. The use of the frame bursting results in a higher minimum frame size, which translates in a higher MCDD.
37. The sliding window protocol is used to implement flow control, being associated to the Go Back N ARQ or to the Selective Reject ARQ protocols. When any of these ARQ protocols are implemented, the sliding window flow control is implicitly implemented. The objective of the transmitting window is to allow that the transmitter keeps sending data (successive frames), without having to stop and wait for the reception of confirmations. Ideally, if everything occurs as expected, the transmitting entity receives confirmations while still transmitting data. This allows the maximization of the transmission medium usage. An important aspect of this protocol relies on the fact that the frames need to be numbered. The transmitting window N is a function of the relationship between the propagation time and the frame duration. The window size N corresponds to the maximum number of frames that can be transmitted without confirmation. If the confirmations are delayed, the transmitter may also have to delay the transmission of the following frames. Nevertheless, if such delay is within the window size, the transmitter has permission to proceed with the transmission of more data. The LWE corresponds to the position of the last confirmed frame, and the UWE is an upper bound for the last frame that can be transmitted. These two parameters are related by $UWE = LWE + N$.
38. The error correction codes are generically referred to as FEC. There are two main types of FEC:
    a. Convolutional codes
    b. Block codes
    In some applications, these two types of error correction codes are combined, originating the so-called concatenated codes. Convolutional codes encode an arbitrary group of input bits, using a certain logic function. Block codes encode a fixed group of input bits to generate another fixed group of output bits. They use n input bits to generate a codeword of length k from a certain alphabet. Reed–Solomon codes are among the most used block codes. The Hamming code is another type of block code.
39. When the transmitted block of an error control technique includes the original information bits, to which some redundant bits is added, the code is referred to as systematic. An example of a systematic error control technique is the CRC. On the other hand, error control codes whose generated codewords does not directly include the original information bits are called nonsystematic.

40. In case of positive confirmation, the receiving entity sends a feedback message in case the frame is free of errors. This procedure is commonly referred to as PAR and the feedback message is known as ACK. The positive confirmation procedure can be resumed as follows:

    a. *With errors*: If the ACK is not received by the transmitting entity within a certain time period (controlled by the transmitter's timer), it assumes an error and retransmits the frame.

    b. *Without errors*: If the ACK is received by the transmitting entity within the expected time period it assumes correct reception of the frame and proceeds with the transmission of the following frame.

    In case of negative confirmation, the receiving entity only sends a feedback message in case of errors. This procedure is commonly referred to as negative acknowledgment (NAK). The negative confirmation procedure can be resumed as follows:

    a. *With errors*: If a feedback message is received by the transmitting entity within a certain time period informing that the frame was in the presence of errors, this entity repeats the frame.

    b. *Without errors*: If a feedback message is not received within a certain time period informing the transmitting entity about the presence of errors, this entity assumes correct reception of the frame and proceeds with the transmission of the following frame.

    The advantage of negative confirmation relies on the lower amount of data exchanged, as opposed to the positive confirmation. Note that, in both cases, the frames are numbered for use by the positive or negative confirmation handshaking.

41. The code rate R_C of error detection and error correction codes is defined as $R_C = (n/k)$, where n stands for the number of information bits and k for the number of transmitted bits. An error control algorithm considers a block of n information bit to which m redundant bits are added, to allow the error detection or error correction capability. These redundant bits are calculated from the original n information bits. The output of the error control algorithm comprises a total of $k = n + m$ bits (codeword), which are transmitted.

42. The IEEE 802.1Q protocol is used for coordinating trunks on IEEE 802.3 interfaces. An IEEE 802.3 frame header does not provide any information about the VLAN. This simply transports a packet, encapsulated into the frame. When in trunk mode, IEEE 802.3 frames need to be tagged with information about which VLAN does the transported frame refers to. Subsequently, when Ethernet frames are placed on a trunk they need additional information about the VLANs they belong to. This is accomplished by using the 802.1Q encapsulation header. This header adds a tag to the original Ethernet frame specifying the VLAN to which the frame belongs to.

43. Trunks enable the extension of VLANs over the whole network. Alternatively, one could configure the interconnection between two switches or between a router and a switch in access mode but using as many cables as the number of VLANs. An important characteristic of VLAN relies on the inability to exchange traffic between switch's interfaces connected to different VLANs. This advantage can be used for security purposes (e.g., to separate voice and data traffic). Let us suppose that each corporate department corresponds to a different VLAN. In this case, inter-VLAN connectivity is normally required. Then, a layer 3 device needs to be connected to the switch, while the switch interface connected to the layer 3 device needs to be configured in trunk mode. A trunk is an interconnection between two switches, or between a router and a switch, used for the transportation of data traffic of more than one VLAN and control traffic. Trunks enable the extension of VLANs over the whole network. The IEEE 802.1Q protocol is used for coordinating trunks on IEEE 802.3 interfaces. An IEEE 802.3 frame header does not provide any information about the VLAN. This simply transports a packet, encapsulated into the frame. When in trunk mode, IEEE 802.3 frames need to be tagged with information about which VLAN

does the transported frame refers to. Subsequently, when Ethernet frames are placed on a trunk they need additional information about the VLANs they belong to. This is accomplished by using the 802.1Q encapsulation header. This header adds a tag to the original Ethernet frame specifying the VLAN to which the frame belongs to.

CHAPTER 13

1. The basic idea behind the structured cabling concept relies on using a single cabling for different services (such as telephony, computer networks, security systems, VTC, and telemetry) and to allow an easy adaptation to new service requirements and technologies. This is materialized by using a modular cabling system specified by standards and independent of the manufacturers, and for use in a wide variety of application services. Moreover, the structured cabling concept is oriented by performances, taking redundancy as an important requirement, and keeping flexibility to support future additional network requirements. The structured cabling consists of a way networks are organized, with the aim of facilitating installation, maintenance, and administration. Therefore, the structured cabling must be generic to support a wide range of communication technologies and contents (data, voice, VTC, etc.), and sufficiently flexible to accommodate the normal technologies' evolution of communication systems and to support an eventual growth of the organization, without the need to radically modify the existing cabling infrastructure.

2. We can calculate the aggregate bandwidth in a building (vertical) backbone cabling by summing the individual bandwidths consumed by terminal stations, where each of these parcels should be properly weighted by a simultaneousness coefficient. This can be expressed mathematically by

$$\text{Aggregate\_BW} = \sum_{i=1}^{N} \left[\text{Indiv\_BW}_i * \text{simultan\_coef}_i \right]$$

where:

Aggregate_BW stands for the aggregate bandwidth, also referred to as total throughput
N is the number of individual trunks under consideration in the aggregate bandwidth
Indiv_BW$_i$ refers to the bandwidth of the ith trunk
simultan_coef$_i$ stands for the simultaneousness coefficient of the ith trunk

3. When the structured cabling is only composed of optical fibers (i.e., twisted pairs are not employed), the horizontal subsystem can be omitted, by implementing a centralized optical architecture. In this case, optical fibers connect the BD directly with TOs. One or more cables, composed of multiple optical fibers, leave the BD toward multiple TOs. Since optical fibers commonly support the distance between the BD and TOs, the regeneration performed by the FD is not required. Moreover, although the cable that leaves the BD carries multiple optical fibers, since the width of optical fibers is much thinner than twisted pairs, the width of the cable composed of a high number of optical fibers is still acceptable. Due to the lower number of distributors, an installation and maintenance cost reduction is achieved with this architecture. The optical fiber length employed in successive connections depends on the type of optical fiber. In case of multimode optical fibers, attention should be paid to the modal bandwidth factor, which relates the maximum bandwidth with the maximum length of an optical link.

4. The horizontal subsystem is used to interconnect the floor distributor with the telecommunication outlets, comprising these two types of functional elements, as well as the horizontal cabling.

5. The workstation subsystem is used to interconnect the telecommunication outlets with the terminal equipment, including devices such as patch cords, adapters, and connectors.

It also includes the workstation cabling. Moreover, the workstation cabling is used to allow the interconnection between a TO and a terminal equipment (workstation, printer, server, IP telephone, etc.).

6. We studied about the North American standards ANSI/TIA/EIA-568, ANSI/TIA/EIA-568-A, ANSI/TIA/EIA-568-B, and ANSI/TIA/EIA-568-C. The structured cabling concept was initially standardized by ANSI/TIA/EIA, in 1991, as ANSI/TIA/EIA-568. The aim was to define a generic and flexible cabling concept, independent of the manufacturers' technologies, able to support different traffic types and data rates, and easy to be modified. The definition of subsystems and functional elements of a structured cabling, the type of transmission medium to be used in each module, the corresponding maximum length and bandwidth, as well as the type of connectors to use was established in this standard. The other North American standards are updates to ANSI/TIA/EIA-568.

7. As a follow up to the standardization that took place in North America, the International Organization for Standardization, jointly with the International Electrotechnical Commission, released, in 1995, the standard ISO/IEC 11801. This initial structured cabling standard by ISO/IEC corresponds approximately to the standard ANSI/TIA/EIA-568-A, but with some additional requirements, namely the cabling that may optionally present shielding or screening, while North American standard comprises unshielded twisted pairs without any shielding. Moreover, this standard also states that the cable sheathing may present low smoke zero halogen (LSZH) property, as an option. In case of fire, a cable with LSZH characteristics burns with low emissions of smoke, and with zero emissions of halogen.

8. The maximum length allowed by ISO/IEC 11801amend2 for the building backbone cabling is 500 m.

9. The maximum length allowed by ISO/IEC 11801amend2 for the horizontal cabling is 90 m.

10. The simultaneousness coefficient that can be considered in a LAN is normally of the order of 0.8. On the other hand, when it comes to WAN interconnection calculation, the simultaneousness coefficient becomes of the order of 0.3.

11. While ANSI/EIA/TIA-568-A does not require the use of LSZH, the standard EN 50173 imposes LSZH as a requirement.

12. The international standard ISO/IEC 11801 considers twisted pairs optionally with shielding or screening (i.e., UTP, STP, FTP), whereas the European standard EN 50173 comprises twisted pairs with screening (mandatory) or shielding (optional) (i.e., FTP or S/FTP).

13. The maximum length of an optical fiber is a function of the used bandwidth, being calculated from the modal bandwidth factor for multimode optical fibers. In case the available modal bandwidth factor (of multimode fibers) does not allow using multimode optical fibers for a certain bandwidth/distance, then single-mode optical fibers should be employed.

14. The optical patch panels should be placed at the top of a rack, whereas the active pieces of equipment are placed in the middle part. The patch panel is responsible for allowing the flexible connectivity between the terminal equipment and the required positions of the switch, hub, or PABX.

15. The height of a rack is normally performed after a market survey that determines the available equipment that fits the requirements. This market survey allows making a cost estimate of the distributor and getting knowledge about the size of each piece of equipment. With this figures, the engineer can calculate the required space in a rack, with especially attention to the required height. Then, a rack choice must be performed, from those available in the market. For the sake of future expansion, it is important to leave some empty space in a rack.

16. The distributor rack should have the capability to work, even in case of electrical power failure. This can be achieved using either an emergency electrical source for the whole building or an individual UPS only for the distributor. In the former case, the emergency

energy can either be provided using only an UPS system (e.g., the whole building), or the UPS system can simply be employed to assure the provision of electrical power only for an interim time period, until the moment an emergency generator is switched on.

17. A patch panel aims to provide flexibility between the positions of the switch, hub, or PABX panel and the conductors of the horizontal cabling that come from the TOs. The conductors of the horizontal cabling that come from the TOs, connect rigidly to the posterior part of the patch panel (in an FD). Then, the patch cords, which connect in the front part of the patch panel, allow the corresponding flexible interconnectivity with the required position of the switch, hub, or PABX panel.

18. PoE power can be directly provided by a switch (when it presents this capability) or by using a PoE injector that is connected in the middle between a switch and the terminal equipment.

19. The North American standard ANSI/EIA/TIA-568-A does not specify any type of electromagnetic isolation to twisted pairs (i.e., it considers UTP), whereas the international standard ISO/IEC 11801 considers, optionally, the use of twisted pairs with shielding or screening (i.e., it can employ UTP, STP, or FTP).

CHAPTER 14

1. ATM adaptation layer (AAL) aims to adapt the requirements of the different classes of services provided by the higher layers to the ATM layer. Consequently, the AAL is service dependent. The rate of generated blocks of data and their corresponding sizes is service dependent, being performed by the AAL. Then, these blocks of different size are encapsulated into cells by the ATM layer. In order to support different services, with different requirements, some service-dependent and service-independent functions have to be provided. Consequently, the AAL is split into two sublayers: convergence sublayer (CS), which provides services to the upper layer, and segmentation and reassembling sublayer (SAR), which segments the data received by the CS, in order to allow its encapsulation into ATM cells. To allow these functionalities, the AAL provide the following four different types of service:

    a. *Type 1 AAL*: Supports constant bit rate services (e.g., circuit switching emulation) from the higher layer (class A), keeping the synchronism information between the source and the destination. In addition, it manages errors (lost cells, corrupted cells, cells in wrong sequence, duplication of cells, etc.).

    b. *Type 2 AAL*: Multiplexes low data rate channels, such as mobile communications.

    c. *Type 3/4 AAL*: Supports classes C and D variable bit rate services with error detection.

    d. *Type 5 AAL*: While the type 3/4 AAL considers the multiplexing of small blocks of data, the type 5 AAL groups the data into large blocks, allowing a more efficient multiplexing (reducing the overhead). Consequently, the type 5 AAL is more efficient for data communications.

2. ATM is associated to a concept entitled broadband-ISDN (B-ISDN). Nevertheless, while ATM is a protocol, B-ISDN is a concept, which is materialized with a reference model.

3. In order to provide different services, the B-ISDN reference model is split into different layers with the following functionalities:

    a. *Physical layer*: It includes two different groups of functionalities:

        i. *Physical medium*: It includes functions that are dependent of the physical medium.

        ii. *Transmission convergence*: It takes care of the generation and recovery of frames, header error control, cell rate decoupling, etc.

    b. *ATM layer*: It is responsible for the multiplexing and demultiplexing of cells, flow control, insertion and removal of headers, and so on. It encapsulates blocks of data with different sizes, generated by the ATM adaptation layer, into cells (fixed size).

c. *AAL*: It adapts the requirements of the different classes of services provided by the higher layers to the ATM layer. Consequently, the AAL is service dependent. The rate of generated blocks of data and their corresponding sizes is service dependent, being performed by the AAL. Then, these blocks of different size are encapsulated into cells by the ATM layer. In order to support different services, with different requirements, some service-dependent and service-independent functions have to be provided. Consequently, the AAL is split into two sublayers: convergence sublayer (CS), which provides services to the upper layer, and segmentation and reassembling sublayer (SAR), which segments the data received by the CS, in order to allow its encapsulation into ATM cells. To allow these functionalities, the AAL provides the following four different types of service:

    i. *Type 1 AAL*: It supports constant bit rate services (e.g., circuit switching emulation) from the higher layer (class A), keeping the synchronism information between the source and the destination. In addition, it manages errors (lost cells, corrupted cells, cells in wrong sequence, duplication of cells, etc.).

    ii. *Type 2 AAL*: It multiplexes low data rate channels, such as mobile communications.

    iii. *Type 3/4 AAL*: It supports classes C and D variable bit rate services with error detection.

    iv. *Type 5 AAL*: While the type 3/4 AAL considers the multiplexing of small blocks of data, the type 5 AAL groups the data into large blocks, allowing a more efficient multiplexing (reducing the overhead). Consequently, the type 5 AAL is more efficient for data communications.

d. *Higher layers*: Interface with different classes of service and provide the required services to these different classes.

4. ATM is commonly referred to as cell switching because this corresponds to a variation of the packet switching. The reason for calling it cell switching, instead of packet switching, relies on the use of fixed size packets (cells), instead of variable size packets (MPLS, IP, etc.). ATM cells are 53-octet long, being composed of a 48-octets long payload data field and of a 5-octets long header field.

5. A virtual channel comprises the layer 3 switching of a group of different cells that are identified by a common VCI. On the other hand, a virtual path comprises the layer 3 switching of a group of different virtual channels that are identified by a common virtual path identifier (VPI), and that are equally switched in a certain path of the route.

6. Similar to the SDH layers, depending on device types, the physical layer of the ATM protocol is split into different sublayers, each one with each own overhead:

a. *Transmission path sublayer*: Corresponds to the SDH path layer, and comprises the ATM end-to-end exchange of data, from the local where the data is encapsulated into cells up to the local where those bits are removed from cells.

b. *Digital section sublayer*: Corresponds to the exchange of data between, for example, two adjacent SDH multiplexers.

c. *Regeneration section sublayer*: Corresponds to the exchange of data between two adjacent regenerators.

7. Depending on the type of interface where ATM cells are employed, the headers present two different formats: UNI stands for user–network interface and NNI for network–network interface. The contents of different header's fields are as follows:

a. *Generic control field (GCF)*: With the default value 0000.

b. *VPI*: Composed of 8 bits (UNI) or 12 bits (NNI).

c. *VCI*: Composed of 16 bits.

d. *Payload type (PT)*: To signalize the type of data transported in the cell.

e. *Reserved (RES)*: For future use.

f. *Cell loss priority* (*CLP*): In case of congestion, cells with this bit active are discarded first.

g. *Header error control* (*HEC*): Composed of 8 bits used to allow checking the presence of errors in the cell header.

8. A VP switch performs the switching of different VPs, translating the incoming VPIs into other outgoing VPIs. A VP switch performs the switching of multiple VCns, whose inputs and outputs are common. Similarly, a VCn switch performs the switching of different VCns, translating the incoming VCIs into other outgoing VCIs.

9. The B-ISDN reference model includes different classes of services, grouped as follows:

a. *Class A*: Circuit switching emulation, audio and video of constant bit rate.

b. *Class B*: Compressed audio and video of variable bit rate.

c. *Class C*: Connection-oriented data (such as file transfer of web browsing).

d. *Class D*: Connectionless data (such as network management protocols or DNS).

10. A certain label switching path (LSP) can be used to transport different types of traffic (voice, video streaming, file transfer, etc.). For the sake of QoS provisioning, different traffic types must be properly identified with different forward equivalent classes (FECs). An MPLS label identifies a pair of LSP/FEC. When a new FEC is created, it is assigned to a certain LSP. Note that different FECs of a communication between the same two end points may belong to the same LSP or to different LSPs.

11. The MPLS is a connection-oriented protocol, whose forwarding of packets is performed using the virtual circuits method (similar to the ATM protocol). The MPLS protocol was developed taking the weaknesses of the ATM protocol, while maximizing the QoS capabilities in transport networks. An important limitation of the ATM protocol relies on the fact that the overhead is high (5-octets), as compared to the cells size (48-octets), which also results in a low network efficiency. The current existing physical layer infrastructures allow the exchange of data at a much higher speeds than before. This allows exchanging longer frames without introducing delays to the supported services. Consequently, contrary to the ATM, the MPLS protocol allows the exchange of connection-oriented packets with different sizes.

12. Contrary to the ATM protocol that comprises a certain packet (cell) format, the MPLS limits to add a certain header to the transported layer 3 packets (ATM, IPv4, IPv6, IPX, etc.).The MPLS header (shim header) is typically added between the layer 2 and the layer 3 headers. The shim header comprises the following fields:

a. *Label*: It is employed to identify the LSP (virtual path), as well as the FEC (type of traffic being transported in the packet).

b. *Class of service* (*CoS*): It is used for QoS provisioning and for explicit congestion notification purposes.

c. *Stack* (*S*): In case multiple labels are inserted into a frame, it indicates a hierarchy. The value one means that the label is the last of the stack. Optionally, MPLS may implement a hierarchy of virtual paths, requiring multiple shim headers in a MPLS packet.

d. *Time to live* (*TTL*): Corresponds to the TTL field of IPv4 packets, being assigned to the last label of a stack.

13. When the transported protocol has fields that identify virtual circuits, the MPLS shim header is directly inserted into the corresponding fields. This is the case of ATM and frame-relay protocols:

a. *ATM*: The shim header is inserted into the VPI/VCI fields.

b. *Frame relay*: The shim header is inserted into the data link channel identifier (DLCI) field.

Alternatively, the shim header is placed between the layer 2 and the layer 3 headers. This latter procedure is employed in protocols such as Ethernet, PPP, and token ring.

14. The MPLS virtual circuit is designated, in the MPLS world, as LSP. A certain LSP can be used to transport different types of traffic (voice, video streaming, file transfer, etc.).

For the sake of QoS provisioning, different traffic types must be properly identified with different FECs. A label identifies a pair LSP/FEC. When a new FEC is created, it is assigned to a certain LSP. Note that different FECs of a communication between the same two end points may belong to the same LSP or to different LSPs.

15. The MPLS protocol comprises a number of mechanisms that facilitate the provisioning of QoS:

   a. Makes use of virtual circuits method (instead of datagram method), which allows a faster switching in intermediate routers.

   b. Accommodates variable length frames, and with a reduced overhead.

   c. The use of FEC, and the header field CoS, both contribute for differentiating traffic of different types, and the corresponding handling.

16. The MPLS shim header comprises the following fields:

   a. *Label*: It is employed to identify the LSP (virtual path), as well as the FEC (type of traffic being transported in the packet).

   b. *CoS*: It is used for QoS provisioning and for explicit congestion notification purposes.

   c. *Stack (S)*: In case multiple labels are inserted into a frame, it indicates a hierarchy. The value one means that the label is the last of the stack. Optionally, MPLS may implement a hierarchy of virtual paths, requiring multiple shim headers in a MPLS packet.

   d. *TTL*: Corresponds to the TTL of IPv4 packets, being assigned to the last label of a stack.

17. The calculation of the output interface to be used by an MPLS router in order to forward a packet is performed taking into account the incoming label (i.e., the virtual circuit), as well as the information contained in a routing table. Routing tables are built in the same manner as the IP routing table, using static or dynamic protocols. In the MPLS world, the routing tables, that is, the lookup tables, are referred to as label forwarding information base (LFIB) or as forwarding information base (FIB). While an LSR makes use of the LFIB table to route packets (MPLS to MPLS), an ingress LER makes use of the FIB table (IP to MPLS), and an egress LER makes use of the LFIB table as well (MPLS to IP).

18. The different MPLS routers are called label switching router (LSR). Nevertheless, the LSRs that are located at the edge of the MPLS cloud (network), interfacing with the customer's equipment router (CE Router), are referred to as label edge router (LER), or simply as Edge LSR. The LER is a provider equipment router (PE Router), as it is owned by the transport service provider.

19. CE router stands for customer's equipment router, whereas PE router refers to provider's equipment router. A PE router is a router that is supplied by the service provider (company). On the other hand, a CE router is a router whose propriety belongs to the customer (he owns it). The regular IP router is a CE router, whereas the LER is a PE router.

20. An IP packet is received by the ingress LER as a regular IP packet. This router translates the IP packet into the MPLS format by simply adding a shim header. The shim header consists of a 32-bit header, composed of four fields. The most important field is the label (20 bits), being employed to identify the virtual circuit and the CoS of the IP packet. It is important noting that, similar to the VPI/VCI identifiers employed in ATM networks, the label has a local meaning. This means that each LSR performs the removal of the label corresponding to the previous point-to-point MPLS connection, adding a new label corresponding to the following point-to-point connection. This function is referred to as label swap (performed in LSRs), whereas the addition or removal of a label performed by LERs are referred to as push (in ingress LER) or pull (in egress LER), respectively. The calculation of the output interface to use by an MPLS router in order to forward a packet is performed taking into account the incoming label (i.e., the virtual circuit), as well as the information contained in a routing table. Routing tables are built in the same manner as the IP routing table, using static or dynamic protocols. In the MPLS world, the routing tables, that is, the lookup tables, are

referred to as LFIB or as FIB. While an LSR makes use of the LFIB table to route packets (MPLS to MPLS), an ingress LER makes use of the FIB table (IP to MPLS), and an egress LER makes use of the LFIB table as well (MPLS to IP).

21. The MPLS was designed to support any layer 3 protocol (IPv4, IPv6, IPX, ATM, etc.). Moreover, it can be implemented over any type of layer 2 protocol, such as SDH, SONET, Ethernet, PPP, and so on. Consequently, it is normally stated that the MPLS protocol belongs to the layer 2, 5 of the OSI reference model.

22. The SDH network comprises different layers. Each layer refers to the communication between certain types of devices that includes a number of functionalities. Each layer has its own overhead type, being processed by the corresponding device. The path is considered end-to-end. Consequently, the path overhead is inserted by the initial multiplexer (LTM or ADM), being removed by the final multiplexer (LTM or ADM). The multiplexing section consists of the parts of the path between adjacent multiplexers, including those intermediates (ADM or SDXC). The intermediate multiplexer (ADM or SDXC) removes the initial multiplexing section overhead at its input, and inserts another multiplexing section overhead at its output, corresponding to the second multiplexing section. Identical rational applies to the regeneration section overhead. Note that a regenerator only processes the regeneration section overhead (removed at the input of the device, and inserted at its output), whereas an intermediate MUX processes both the regeneration section overhead and the multiplexing section overhead.

23. Frame justification consists of the addition or removal of some bits, in order to allow the correct operation of multiplexers and demultiplexers when the different tributaries rate is subject to fluctuations, and therefore it differs from the nominal rate. When the rate is higher than the nominal, the justification comprises the addition of a bit without information (from time to time), in order to adjust the rates. Some bits are preallocated in frames, for justification purposes. In addition, when bit justification is employed, this needs to be properly signalized using a justification indication bit, which is also a preallocated bit in the frame composition. Due to successive insertions and removals of headers in different devices (regenerators, multiplexors, etc.), the payload (SPE/VC) may travel faster than the corresponding frame. Another problem that originates similar effects is when the incoming clock rate is lower than the outgoing clock. In order to overcome such fluctuation, an extra octet is inserted into the last octet of the pointer. This is known as negative justification. In this case, the value of the pointer needs to be decreased by one. Contrarily, when the incoming clock rate is higher than the outgoing clock rate, the frame flows faster than the payload. In order to overcome such fluctuation, a stuff octet is transmitted after the pointer. This is known as *positive justification.*

24. This mechanisms are positive or negative justification. Due to successive insertions and removals of headers in different devices (regenerators, multiplexors, etc.), the payload (SPE/VC) may travel faster than the corresponding frame. Another problem that originates similar effects is when the incoming clock rate is lower than the outgoing clock. In order to overcome such fluctuation, an extra octet is inserted into the last octet of the pointer. This is known as *negative justification.* In this case, the value of the pointer needs to be decreased by one. Contrarily, when the incoming clock rate is higher than the outgoing clock rate, the frame flows faster than the payload. In order to overcome such fluctuation, a stuff octet is transmitted after the pointer. This is known as positive justification.

25. The European plesiochronous digital hierarchy (PDH) system is entitled the European conference of postal and telecommunications (CEPT), having been normalized by the International Telecommunications Union (ITU) as ITU-T G.732. The PDH system used in the North America is entitled digital signal (DS), having also been normalized by the ITU as ITU-T G.733. All of the hierarchies are composed of a number of ITU-T G.711

voice channels (TDM-PCM), properly multiplexed. A higher hierarchy is generated using a multiplexer. The generation of a nth order hierarchy requires a stack of n multiplexer. Similarly, the extraction of an nth order tributary requires a stack of n (de)multiplexers. The primary European multiplexing hierarchy is called CEPT-1, whereas the North America and Japanese versions are referred to as DS-1. Note that the interface provided by a CEPT-1 is referred to as E1, whereas the interface provided by a DS-1 is named as T1. From the transportation of data point of view, the nomenclature CEPT-x or DS-x is employed, whereas the nomenclature employed in interfaces is E-x, T-x, or J-x. Both CEPT-1 and DS-1 use word interposition, with 8 bits per word, and a rate of 8000 frames per second. The voice codec comprised by the CEPT-1 is the A law and its frame has 32 time slots. Similarly, the DS-1 considers the μ law and its frame has a total of 24 time slots. From the 32 time slots of the CEPT-1, 30 are used to transport 64 kbps telephone channels. In addition, the time slot 0 of odd frames is employed for framing, and the time slot 16 is used for signaling. The CEPT-1 comprises a cumulative throughput of $32 \times 8 \times 8000 = 2.048$ Mbps. The DS-1 frame comprises a total of $24 \times 8 + 1 = 193$ bits, corresponding to twenty-four 64 kbps telephone channels plus an additional bit (per frame) for framing purposes. The cumulative throughput of the DS-1 frame is $193 \times 8000 = 1.544$ Mbps. The signaling is transported in the data time slots. Specifically, the signaling bits are transmitted in the 6th–12th frame of each multiframe, using the eighth bit of each time slot. This results in a PCM word of seven bits, leading to a small degradation of the PCM voice channels. Higher order tributaries are obtained from the multiplexing of a number of immediately lower order tributaries. With the exception of the primary multiplexing tributary order, all upper multiplexing tributary orders are multiplexed using bit interposition.

26. Different SDH and SONET hierarchies are formed by octet interposition of the lower hierarchy tributaries.

27. All of the PDH hierarchies are composed of a number of ITU-T G.711 voice channels (TDM-PCM), properly multiplexed. Higher order tributaries are obtained from the multiplexing of a number of immediately lower order tributaries. With the exception of the primary multiplexing tributary order that uses word interposition, all upper multiplexing tributary orders are multiplexed using bit interposition.

28. The inaccuracy of independent clocks employed in PDH did not allow throughputs higher than 140 Mbps. These throughputs were not enough to face the new information exchange requirements. This issue was the main motivation for the development of a synchronous hierarchy, optimized for optical fiber transmission mediums. Moreover, the level operation, administration and maintenance (OA&M), as well as the level standardization in PDH systems were very low. These capabilities were highly improved in synchronous systems. Synchronous hierarchies consider atomic clocks that allow the exchange of throughputs at a rate much higher than those possible with PDH systems. Moreover, the higher level of standardization comprised by SDH/SONET between equipment of different manufacturers lead to improved interoperability. Another important innovation of synchronous transport network systems relies on the ability to insert or extract a tributary from any other tributary of any other order, without the need to have a stack of multiplexers or demultiplexers, as required for the PDH.

29. The elementary physical topology employed in SDH/SONET networks is the ring. Since optical fibers are unidirectional, a cable composed of, at least, two optical fibers (one pair) is employed to allow simultaneous bidirectional communications. In case of two optical fibers, both are employed for operation and backup using different time slots. The external fiber allows the exchange of data in the clockwise direction, whereas the internal fiber performs the same in the anticlockwise direction. Each node (ADM) receives the same data twice, coming respectively by the external and by the internal fiber. The destination node (ADM) combines the data coming from the two directions in order to provide

diversity. The combining technique is the selective combining. This is performed using the path overhead. The resultant signal presents better performance than the two independent signals. In case the optical fiber cable is cut, such impairment is detected by the two closest ADM and bridges are introduced in those ADMs. This is performed using the multiplexing overhead. Consequently, a switch to the backup time slots is performed by the network. Note that this self-recovery capability represents a great advantage of the synchronous systems, as compared to plesiochronous systems.

30. In case the SDH/SONET ring is implemented with four optical fibers, all time slots of two optical fibers are allocated for operation, while the other two optical fibers (one pair) are reserved for backup. In this case, redundancy is not assured with TDM, but with an extra optical fiber pair. The closest ADM insert bridges, but those bridges are responsible for forwarding the signals to the redundant fibers in the opposite direction.

31. Each layer has its own overhead type, being processed by the corresponding device. The path is considered end-to-end. Consequently, the path overhead is inserted by the initial multiplexer (LTM or ADM), being removed by the final multiplexer (LTM or ADM). The multiplexing section consists of the parts of the path between adjacent multiplexers, including those intermediates (ADM or SDXC). Consequently, the path is composed of a number of multiplexing sections. In this case, the intermediate multiplexer (ADM or SDXC) removes the initial multiplexing section overhead at its input, and inserts another multiplexing section overhead at its output, corresponding to the second multiplexing section. Identical rational applies to the regeneration section overhead. Note that a regenerator only processes the regeneration section overhead (removed at the input of the device and inserted at its output), whereas an intermediate MUX processes both the regeneration section overhead and the multiplexing section overhead.

32. The data over cable service interface specification (DOCSIS) can be viewed as an alternative to xDSL, but using coaxial cables as transmission medium (instead of twisted pairs). Alternatively, hybrid fiber-coaxial, or just optical fiber, may also be utilized as transmission medium. DOCSIS is currently widely employed by ISPs to provide Internet access to domestic users. This service is normally provided together with the cable television. When telephony service is also provided, the three services are commonly known as triple play. In this case, the telephony service is provided using voice over IP over DOCSIS modems (employed between the subscriber home and the ISP). The DOCSIS modem shares the spectrum with regular analog or digital video distribution, using FDMA (downlink). The uplink channels are implemented by using a TDMA (DOCSIS 1.0) or a CDMA (DOCSIS 2.0 and 3.0). In the uplink, the higher modulation order and bandwidth of DOCSIS 2.0 and 3.0 allows a higher data rate per channel, as compared to DOCSIS 1.0. In the downlink, all three DOCSIS versions allow the same data rate per channel. Nevertheless, an important difference of the DOCSIS 3.0 as compared to DOCSIS 2.0 relies on its ability to aggregate multiple channels (in both uplink and downlink), which results in a data rate that is N times higher. Typically, four channels are aggregated in DOCSIS 3.0, resulting in a data rate of 122.88 Mbps in the uplink and 171.52 Mbps in the downlink.

33. The DOCSIS modem shares the spectrum with regular analog or digital video distribution, using FDMA (downlink). The uplink channels are implemented by using a TDMA (DOCSIS 1.0) or a CDMA (DOCSIS 2.0 and 3.0). In the uplink, the higher modulation order and bandwidth of DOCSIS 2.0 and 3.0 allows a higher data rate per channel, as compared to DOCSIS 1.0. In the downlink, all three DOCSIS versions allow the same data rate per channel. Nevertheless, an important difference of the DOCSIS 3.0 as compared to DOCSIS 2.0 relies on its ability to aggregate multiple channels (in both uplink and downlink), which results in a data rate that is N times higher. Typically, four channels are aggregated in DOCSIS 3.0, resulting in a data rate of 122.88 Mbps in the uplink and 171.52 Mbps in the downlink.

34. The DOCSIS modem shares the spectrum with regular analog or digital video distribution, using FDMA (downlink). The uplink channels are implemented using TDMA (DOCSIS 1.0) or CDMA (DOCSIS 2.0 and 3.0). In the uplink, the higher modulation order and bandwidth of DOCSIS 2.0 and 3.0 allows a higher data rate per channel, as compared to DOCSIS 1.0. In the downlink, all three DOCSIS versions allow the same data rate per channel. Nevertheless, an important difference of the DOCSIS 3.0 as compared to DOCSIS 2.0 relies on its ability to aggregate multiple channels (in both uplink and downlink), which results in a data rate that is N times higher. Typically, four channels are aggregated in DOCSIS 3.0, resulting in a data rate of 122.88 Mbps in the uplink and 171.52 Mbps in the downlink.

35. The PPP includes error control, authentication, compression, encryption, link control protocol (LCP), and network control protocol (NCP) capabilities. A NCP runs directly over the data link layer protocol, being used to negotiate options and specific parameters of the network layer (which also runs over the data link layer). Examples of NCP are the PPP IP control protocol (IPCP), the IPv6 control protocol over PPP (IPv6CP), the PPP Apple Talk control protocol, and so on. For connection-oriented protocols, the NCP is responsible for performing the connection establishment, as well as the connection termination. In addition, it is responsible for configuring and supporting the operation of the network layer. The NCP is only initiated after the LCP of the data link layer had been successfully established.

36. The high bit rate digital subscriber line (HDSL), defined by ITU-T G.911.1 performs the transmission in baseband, using line coding 2B1Q. It transmits at a maximum uplink rate of 1.544 or 2.048 Mbps, and at a maximum downlink rate of 2.048 Mbps.

37. DOCSIS consists of a set of standards that allows high-speed data communications over coaxial cables, typically installed for the provision of television service. The DOCSIS can be viewed as an alternative to xDSL, but using coaxial cables as transmission medium (instead of twisted pairs). Alternatively, hybrid fiber-coaxial, or just optical fiber, may also be utilized as transmission medium.

38. In order to mitigate the effects of the channel impairments, most xDSL modems have embedded an equalization module, which is an effective measure to minimize the effects of ISI. This kind of interference is also mitigated by using high-order modulations (e.g., M-QAM) or by employing the OFDM transmission technique. Some type of xDSL modems use a variation of the OFDM transmission technique in the downlink (more demanding direction), entitled discrete multitone (DMT). With DMT transmission technique, the downlink spectrum is split into multiple subcarriers. Similar to OFDM, the aim is the mitigation of the negative effects of ISI. The flow of data is split into a lower rate transmission, and each lower rate transmission is independently transmitted in a different subcarrier. The DMT is very similar to the OFDM technique with the difference that some subcarriers are removed and the modulation order of different subcarriers can be different (as a function of the channel impairments experienced by each subcarrier). Similar to the OFDM technique, each DMT subcarrier tends to suffer from flat fading (i.e., nonfrequency selective fading), instead of frequency selective fading. In order to mitigate the remaining fading effects, each subcarrier is subject to an equalization process at the receiver side. This results in a better signal quality, which translates in a more efficient use of the spectrum (expressed in bit/s/Hz).

39. An alternative to the preallocation of different uplink and downlink bandwidths in xDSL modems (frequency division duplexing) consists of using a bandwidth simultaneously for the uplink and downlink. In such case, the downlink bandwidth includes the uplink bandwidth, and an echo cancellation (EC) mechanism is employed to avoid interferences between the uplink and the downlink spectrum. The echo cancellation is implemented using an adaptive equalizer to reject the non-desired signal.

40. Typical uplink frequency band is between 25 and 140 kHz, whereas the typical downlink band is between 150 kHz and 1 MHz. Latest xDSL standards utilize higher frequencies.

An example is the very high data rate DSL (VDSL) whose frequency band utilized for data goes from 25 kHz up to 12 MHz.

41. A permanent circuit is normally leased from a telecommunications operator. Alternatively, a circuit may be implemented and established by an operator or institution as a request to a specific need (using, for example, terrestrial microwave systems or optical fibers). Logical LCPs that can be ran over this type of connectivity include the synchronous data link control (SDLC), high level data link control (HDLC), and the point-to-point protocol (PPP).

42. The control field of the HDLC protocol is used to control the connection, and can be 8 or 16 bits long. Depending on the type of frame (I, S, or U), and of the control field length, its content varies. The use of 8 or 16 bits in the control field needs to be negotiated during setup phase (setup extended mode corresponds to employing 16 bits in the control field). The 16-bits control field format is used when the sequence of the frame numbering is high (e.g., long window size in the sliding window mechanism). Note that only I and S frames can be 16-bits long (U frames are always 8-bits long). Some bits are variable, whose meanings are defined as follows:

    a. $N(S)$: Sending sequence number. It is used to identify the sequence number of the current transmitted frame.

    b. $N(R)$: Receiving sequence number. It is used for control purposes to identify the sequence number following a previously correctly received frame. This number corresponds to the sequence number of the frame that the receiving entity is ready to receive.

    c. S_n: Supervisory bits, used for flow control and error control, with the following meanings for S_1 and S_2 bits:

        i. *00*: Receive ready (RR). This corresponds to the acknowledgment (RR followed by the number of the frame $N[R]$ that the entity is ready to receive).

        ii. *01*: Reject (REJ). This is used to reject a frame with errors.

        iii. *10*: Receive not ready (RNR). This is used by the receiving entity when it is not ready to receive more data.

        iv. *11*: Selective reject (SR). This is used by the receiving entity to reject a specific frame using the selective reject ARQ protocol.

    d. M_n: Unnumbered bits, used for multipurpose functions, such as setting up and finalizing a connection.

    e. *P/F* (pool/final bit): When the *P* bit is active (1), the transmitting entity is querying the receiving entity about which frame is it ready to receive. The *F* bit active means that it is a response to a previous query (*P*).

43. The HDLC protocol considers the following three different types of frames:

    a. *Information* (I): It is used to transport upper layer (Internet layer) information data, with error control data piggybacked (in the control field).

    b. *Supervisory* (S): It is employed for error control purposes when piggybacking is not employed, that is, when the receiving entity has no information data to transmit.

    c. *Unnumbered* (U): It consists of frames not numbered, being used for several functions such as connection establishment or connection termination, as well as to define the mode of the link to be established, as defined in the following.

44. The protocol field of a PPP header identifies the type of protocol being handled by the payload data of the frame. Possible protocols being handled by the frame includes IPv4 data, IPv6 data, IPX data, LCP, NCP, authentication, encryption control protocol (ECP), compression control protocol (CCP), and so on.

45. A PPP LCP is responsible for establishing and configuring a data link layer, as well as for operating and terminating a link. Different data link layers use different LCPs, with different message formats. Note that the PPP authentication is performed after the establishment and configuration of the LCP, being invoked by this protocol. When

authentication is required, the NCP link establishment does not start before authentication process succeeds.

46. Supervisory (S) frames are used for error control purposes, when piggybacking is not employed, that is, when the receiving entity has no information data to transmit.

47. The PPP is an LLC connection-oriented protocol employed in either permanent circuits or circuit switching (e.g., dial-up or xDSL connection). It can establish links in either half- or full-duplex modes, using synchronous or asynchronous mode, being however limited to point-to-point connections. The PPP is the successor of the serial line Internet protocol (SLIP), consisting of a very simple protocol whose functionalities were almost limited to a basic framing, without any advanced functionalities. Conversely, the PPP includes error control, authentication, compression, encryption, LCP, and NCP capabilities.

CHAPTER 15

1. The difference refers to the dimension of the cells. A macrocell is bigger than a microcell, and both are bigger than a picocell. A macrocell may include, inside its area, multiple micro- and picocells. Cellular coverage is related to capacity. The cellular capacity is viewed as the number of simultaneous calls within a cell. Regardless of the cell dimension, a cell accommodates a certain number of simultaneous calls. Increasing the cell dimension (e.g., macrocell, instead of a microcell) allows accommodating a lower number of calls per square kilometer while corresponding to a higher cellular coverage. The decision on whether or not to implement a lower hierarchy cell relies on the rate of calls expected.

   Allocating M carriers per cell allows accommodating $M \times N$ calls, with N the number of calls per carrier. Note that the number M that relates to the number of carriers allocated per cell and the resulting $M \times N$ maximum simultaneous calls are independent of whether it is a macro-, micro-, or a picocells. The above equivalence assumes time division duplexing. In case of frequency division duplexing, the number of calls per cell is halved.

2. The advantage of using a hierarchical cellular structure relates to the network capacity. Cellular coverage is related to capacity. The cellular capacity is viewed as the number of simultaneous calls within a cell. Regardless of the cell dimension, a cell accommodates a certain number of simultaneous calls. Increasing the cell dimension (e.g., macrocell, instead of a microcell) allows accommodating a lower number of calls per square kilometer while corresponding to a higher cellular coverage. The decision on whether or not to implement a lower hierarchy cell relies on the rate of calls expected. Allocating M carriers per cell allows accommodating $M \times N$ calls, with N the number of calls per carrier. Note that the number M that relates to the number of carriers allocated per cell and the resulting $M \times N$ maximum simultaneous calls are independent on whether it is a macro-, micro-, or a picocells. The above equivalence assumes time division duplexing. In case of frequency division duplexing, the number of calls per cell is halved.

3. The dimension of the cell to be covered depends on the population density and on the propagation environment. In case of rural environment, the coverage area of a cell tends to be higher than in urban scenarios, as the expected rate of calls is lower in rural areas. Note that the fading typically experienced in rural areas is modeled by a Rice distribution. It consists of a Rayleigh fading model (Rayleigh distribution) to which a line-of-sight component is summed. Due to the existence of line-of-sight, the shadowing effect is normally not experienced, resulting in a lower rate of path loss in rural areas, as compared to urban environments. Consequently, a rural environment is much easier to cover and the cell size is typically higher than in urban areas.

4. The dimension of the cell to be covered depends on the population density and on the propagation environment. An urban environment is typically characterized by the absence of line-of-sight component, where the shadowing effect is normally experienced, and whose

path loss rate is higher than in the rural environments. It is modeled by a Rayleigh fading model (Rayleigh distribution). Note that due to the expected low rate of calls in rural areas, other lower size cells (e.g., micro- and picocells) are normally not implemented.

5. Picocells are designed for indoor or outdoor highly populated environments, such as hotels, offices, and shopping centers. In addition, the area to be covered varies from few meters up to around 200 m. Since the line-of-sight component is normally present in pico-cells, the type of fading experienced is typically the multipath Rice fading. In picocells, the SNR is normally high and the available throughput is maximized. Consequently, the number of calls per square area is maximized. The implementation of a picocell requires a high investment in infrastructures, being only justified in highly populated areas. Since walls originate high attenuation levels, the implementation of an indoor picocell solves the problem of covering indoor environments from outdoor BS. In order to avoid co-channel interference, the spectrum made available in lower dimension cells must be different from that in higher dimension cells. Finally, due to the low distances involved, the battery use in picocells is much lower than in higher order cells, translating into an energy saving cell.

6. A femtocell is the lower hierarchical structure of a cellular environment that typically covers indoor environments, being implemented making use of the existing cabled infra-structures to provide indoor and small-size environment cellular access. On the other hand, a picocell is also a low dimension cell, but it is implemented using a regular BS (can be installed indoor or can cover indoor areas from outdoors). A femtocell makes use of a small-size BS that typically interconnects with the cellular network through xDSL or cable modem. Femtocells solve coverage problems and enable a reduction in the electrical consumption. It is viewed as economically more effective than picocells. While a low cost solution, it enables a high SNR in the interior of domestic houses, offices, or other indoor buildings, resulting in high throughput available to the end user, providing a wide range of services. Similar to a picocell, since the penetration rate of electromagnetic waves over walls is limited, a femtocell solves the problem of covering indoor environments from outdoor BS. Note that, due to the low distances involved, the battery use in femtocells is minimized, representing another important advantage of this cell configuration.

7. Cellular coverage is related to capacity. The cellular capacity is viewed as the number of simultaneous calls within a cell. Regardless of the cell dimension, a cell accommodates a certain number of simultaneous calls. Increasing the cell dimension (e.g., macrocell, instead of a microcell), allows accommodating a lower number of calls per square kilome-ter while corresponding to a higher cellular coverage. The decision on whether or not to implement a lower hierarchy cell relies on the rate of calls expected. Allocating M carriers per cell allows accommodating $M \times N$ calls, with N the number of calls per carrier. Note that the number M that relates to the number of carriers allocated per cell and the resulting $M \times N$ maximum simultaneous calls are independent on whether it is a macro-, micro-, or a picocells. The above equivalence assumes time division duplexing. In case of frequency division duplexing, the number of calls per cell is halved.

8. The cell geometry consists of a hexagon. Instead of considering a cell consisting of a circle around the center (BS), the hexagon is adopted. With such configuration, the anten-nas from different BSs are equidistant. Each cell uses a different set of frequency bands. In order to avoid co-channel interference, the same bands are not assigned to adjacent cells. A common reuse factor employed in the cellular environment is the reuse factor seven, where each group of seven cells uses a group of seven sets of frequency bands, and the repetition of frequency bands only occur in groups of seven cells. In fact, CDMA networks (such as the UMTS) make normally use of all bands in all cells, leading to reuse factor one. In this case, the resulting residual interference is mitigated by employing mul-tiuser detectors.

9. Allocating M carriers per cell allows accommodating $M \times N$ calls, with N the number of calls per carrier. Note that the number M that relates to the number of carriers allocated per cell and the resulting $M \times N$ maximum simultaneous calls are independent of whether it is a macro-, micro Token Ring, or a picocells. The above equivalence assumes time division duplexing. In case of frequency division duplexing, the number of calls per cell is halved.

10. The second generation of cellular networks (2G), like the global system for mobile communications (GSM), was widely used between 1992 and 2003. This introduced the digital technology in the cellular environment, with a much better performance, better reliability, higher capacity, and even with the roaming capability between operators, due to its high level of standardization and technological advancements. The multiple access technique used by GSM was TDMA, where signals generated by different users were transmitted in different (orthogonal) time slots. Narrowband CDMA system was adopted in the 1990s by IS-95 standard, in the United States. IS-95 was also a 2G system. Afterward, the UMTS, standardized in 1999 by 3GPP release 99, proceeded with its utilization, in this particular case using the wideband CDMA. The UMTS consists of a third generation cellular system (3G). The CDMA concept relies on different spread spectrum transmissions, each one associated to a different user's transmission, using a different (ideally orthogonal) spreading sequence.

11. The UMTS, standardized in 1999 by 3GPP release 99 was developed based on WCDMA transmission technique. The UMTS consists of a 3G system. The CDMA concept relies on different spread spectrum transmissions, each one associated to a different user's transmission, using a different (ideally orthogonal) spreading sequence. The third generation of cellular system is composed of different evolutions. The latest UMTS version, entitled HSPA+, achieves 28 Mbps in the downlink (3GPP release 7). In order to respond to the increased speed demands of the emergent services, higher speeds became possible with the already deployed LTE, supporting 160 Mbps in the downlink (as defined by 3GPP release 8), and even higher speeds with some additional improvements to LTE baseline introduced in 3GPP release 9 (e.g., advanced MIMO systems). The LTE air interface was the result of a study item launched by 3GPP named Evolved UTRAN (E-UTRAN). The goal was to face the latest demands for voice, data, and multimedia services, improving spectral efficiency by a factor 2–4, as compared to HSPA release 7. The LTE can be viewed as a cellular standard for 3.9G (3.9 generation). The LTE air interface relies on a completely new concept that introduced a number of technological evolutions as a means to support the performance requirements of this new standard. This includes block transmission technique using multicarriers, multiantenna systems (MIMO), BS cooperation, multihop relaying, as well as the all-over IP concept. The air interface of LTE considers the OFDMA transmission technique in the downlink and SC-FDMA in the uplink. Depending on the purpose, different types of MIMO systems are considered in 3GPP release 8. The modulation employed in LTE comprises quadrature phase shift keying (QPSK), 16-QAM or 64-QAM (quadrature amplitude modulation), using adaptive modulation and coding (AMC). When in the presence of noisy channels, the modulation order is reduced and the code rate is increased. The opposite occurs, when the channel presents better conditions. The LTE comprises high spectrum flexibility, with different spectrum allocations of 1.4, 3, 5, 10, 15, and 20 MHz. This allows a more efficient spectrum usage and a dynamic spectrum allocation based on the bandwidths/data rates required by the users.

12. The LTE can be viewed as a cellular standard for 3.9G (3.9 generation). The LTE air interface relies on a completely new concept that introduced a number of technological evolutions as a mean to support the performance requirements of this new standard. This includes block transmission technique using multicarriers, multiantenna systems (MIMO), BS cooperation, multihop relaying, as well as the all-over IP concept. The air interface of LTE considers the OFDMA transmission technique in the downlink and SC-FDMA in the uplink.

Depending on the purpose, different types of MIMO systems are considered in 3GPP release 8. On the other hand, 4G aims to support the emergent multimedia and collaborative services, with the concept of *anywhere* and *anytime* facing the latest bandwidth demands. The LTE-Advanced (standardized by 3GPP) consists of a 4G system. Based on LTE, the LTE-Advanced presents an architecture using the all-over IP concept. The support for 100 Mbps in vehicular and 1 Gbps for nomadic access is achieved with the following mechanisms:

a. Carrier aggregation composed of multiple bandwidth components (up to 20 MHz) in order to support transmission bandwidths of up to 100 MHz.

b. Advanced antenna systems increasing the number of downlink transmission layers to eight and uplink transmission layers to four. Moreover, LTE-Advanced introduced the concept of multiuser MIMO, in addition to the single-user MIMO previously considered by the LTE.

c. Multihop relay (adaptive relay, fixed relay stations, configurable cell sizes, hierarchical cell structures, etc.) in order to achieve a coverage improvement and/or an increased data rate.

d. Advanced intercell interference cancellation (ICIC) schemes.

e. Advanced BS cooperation, including macrodiversity.

f. Multiresolution techniques (hierarchical constellations, MIMO systems, OFDMA multiple access technique, etc.).

Standardization of LTE-Advanced is part of 3GPP release 10 (completed in June 2011), and enhanced in its release 11 (December 2012) and release 12 (March 2013).

The IMT-Advanced refers to the international 4G system, as defined by the ITU-R. Moreover, the LTE-Advanced was ratified by the ITU as an IMT-Advanced technology in October 2010.

13. The OFDM is a block transmission technique that is utilized by both LTE and LTE-Advanced. This allows mitigating the negative effects of ISI by splitting the spectrum utilized by a certain transmission into small pieces of spectrum (that are orthogonal, and that correspond to a serial to parallel conversion of data), and where each piece is independently equalized. OFDM results in an improved spectral efficiency, as compared to TDMA (used in 2G), or even to CDMA (used in 3G).

14. While LTE is a cellular system, WiMAX (IEEE 802.16) consists of a technology that implements a wireless metropolitan area network (WMAN). The basic idea of WiMAX relies on providing wireless Internet access to the last mile, with a range of up to 50 km. Therefore, it can be viewed as a complement or competitor of the existing asynchronous digital subscriber line or cable modem, providing the service with the minimum effort in terms of required infrastructures. On the other hand, fixed WiMAX can also be viewed as a backhaul for Wi-Fi (IEEE 802.11), cellular BS, or mobile WiMAX. In terms of throughputs and coverage, using WiMAX these two parameters are subject to a trade-off: typically, mobile WiMAX provides up to 10 Mbps per channel (symmetric), over a range of 10 km in rural areas (LOS environment) or over a range of 2 km in urban areas (non-LOS environment). With the fixed WiMAX, this range can normally be extended. Mobile version considers an omnidirectional antenna, whereas the fixed WiMAX uses a high gain antenna (directional). Note that, contrary to WiMAX, the LTE does not comprise the fixed WiMAX, which can be viewed as a microwave line-of-sight system. The WiMAX version currently available (IEEE 802.16-2005) incorporates most of the techniques also adopted by LTE. Such examples of techniques are OFDMA, MIMO, and advanced turbo coding. In addition, the inclusion of multihop relay capabilities (IEEE 802.16j) aims to improve the speed of service delivery and coverage by a factor of 3–5. Moreover, IEEE 802.16m integrates and incorporates several advancements in transmission techniques that meet the IMT-Advanced requirements, including 100 Mbps mobile and 1 Gbps nomadic access, as defined by ITU-R.

15. Both LTE and LTE-Advanced comprise IP-based networks. Since WiMAX only defines the physical layer and MAC sublayer, it can be used associated to either IPv4 or IPv6. Therefore, WiMAX is not an IP-based network.

16. The HSPA+ (3GPP release 7), LTE (3GPP release 8 and 9), and LTE-Advanced (3GPP release 10, 11, and 12) comprise MIMO systems. Moreover, the WiMAX version IEEE 802.16-2005 also incorporates MIMO systems.

17. The support for 100 Mbps in vehicular and 1 Gbps for nomadic access is achieved in 4G with the following mechanisms:

    a. Carrier aggregation composed of multiple bandwidth components (up to 20 MHz) in order to support transmission bandwidths of up to 100 MHz.

    b. Advanced antenna systems increasing the number of downlink transmission layers to eight and uplink transmission layers to four. Moreover, LTE-Advanced introduced the concept of multiuser MIMO, in addition to the single-user MIMO previously considered by the LTE.

    c. Multihop relay (adaptive relay, fixed relay stations, configurable cell sizes, hierarchical cell structures, etc.) in order to achieve a coverage improvement and/or an increased data rate.

    d. Advanced intercell interference cancellation schemes.

    e. Advanced BS cooperation, including macrodiversity.

    f. Multiresolution techniques (hierarchical constellations, MIMO systems, OFDMA multiple access technique, etc.).

18. Besides allowing connectivity in infrastructure network mode (using a WAP), IEEE 802.11 also allows ad-hoc networks (peer-to-peer interconnection). This means that wireless devices can interconnect directly, without using an IEEE 802.11 WAP. In addition, by using two wireless bridges, an IEEE 802.11 link can be established to interconnect two cable LANs, as long as the two bridges are within Wi-Fi wireless range. The IEEE 802.11 networks are formed by cells known as basic service set (BSS). The BSS are created using a WAP (in infrastructure network mode) or without a WAP (in ad-hoc mode).

19. The initial version of Wi-Fi, invented in 1991 by NCR Corporation/AT&T, in the Netherlands, was standardized as IEEE 802.11 supporting 1 or 2 Mbps in the 2.4 GHz band, using either frequency hopping spread spectrum (FHSS) or direct sequence spread spectrum (DSSS). This version was upgraded by the following newer versions:

    a. IEEE 802.11a, as a standard version that consists of an extension to IEEE 802.11 that allows up to 54 Mbps in the 5 GHz band using OFDM transmission technique, with an approximated range of the order of 35 m.

    b. IEEE 802.11b, allowing a data rate of 11 Mbps in the 2.4 GHz band, using DSSS due to its relative immunity to interference (instead of OFDM), with an approximated range of the order of 35 m.

    c. IEEE 802.11g, as an extension to IEEE 802.11b that allows up to 54 Mbps in the 2.4 GHz band, using either OFDM or DSSS transmission techniques, with an approximated range of the order of 35 m.

    d. IEEE 802.11n, as an upgrade in order to allow over 100 Mbps in the 5 GHz band, by using both OFDM transmission technique and the multistreaming MIMO scheme, with an approximated range of the order of 70 m.

    e. IEEE 802.11ac, as an upgrade to IEEE 802.11n, in order to support up to 500 Mbps in the 5 GHz band, using a wider bandwidth of up to 160 MHz, higher number of parallel MIMO streams and higher modulation orders, with an approximated range of the order of 70 m.

20. The additional messages of the CSMA-CA allow it reaching a performance improvement, as compared to the CSMA-CD, by alerting the other stations that a frame transmission is going to take place. In addition, this solves the hidden terminal problem that can be

experienced in ad hoc mode (making use of an access point) and in infrastructure network mode (with an access point). Just as in an Ethernet LAN, having more users results in a reduction of throughput (within the coverage area). Therefore, its efficiency is limited to a reduced number of users and/or reduced traffic.

21. CSMA-CA is normally utilized in wireless networks, such as in IEEE 802.11. This solves the hidden terminal problem that can be experienced in ad hoc mode (making use of an access point) and in infrastructure network mode (with an access point).

CHAPTER 16

1. The studied attacks against the confidentiality comprise eavesdropping, snooping, interception, and trust exploitation.

2. The initial part of the SSL handshaking process that precedes an encrypted transfer of data corresponds to the authentication process are as follows:

   a. The client application that aims to establish a secure session with a server application initializes the handshaking by sending a *Hello* message followed by a 28-byte random number R_{client}, as well as a list of cryptographic algorithms supported (asymmetric, symmetric, and Hash functions).

   b. The server also sends a *hello* message, followed by another 28-byte random number R_{server}, its list of cryptographic algorithms supported, as well as a session identification (session ID). In addition, the server also sends its public key along with its ITU-T X.509 digital certificate. Note that the validation of the digital certificate needs to be confirmed by the client.

   c. If the server requires a digital certificate from the client, for allowing the establishment of the secure session, a *digital certificate request* message is sent to the client, along with the list of acceptable certification authorities. In this case, the client sends the digital certificate, which includes the client's public key. If such digital certificate is required but not available at the client's side, a *no digital certificate* message is sent, which may result in ending the session.

   d. After all of the previously described operations, the server sends a *hello done* message.

   e. The client verifies the validation of the server's digital certificate and verifies all parameters proposed by the server. Moreover, a number of other parameters relating to cipher keys is also exchanged.

   f. In case the server requires a digital certificate from the client, the client sends a *digital certificate verify* message signed with the client's private key. This corresponds to the authentication process of the client, assuring the authenticity and legitimacy of the client's digital certificate. In case such digital certificate is not required by the server, the authentication of the client is only performed after the establishment of the secure session (and therefore, once the session is already ciphered) using a protocol such as the CHAP, or simply using a username and a password. Note that the server does not need to send the digital certificate to the client properly signed with the server's private key. This server's authentication process is not required because in case the server does not have a private key corresponding to the previously sent pre-master secret key S, it cannot decipher it, and the server cannot create the master secret key from where the encryption of data is carried out. In such case, the handshake fails.

3. The INFOSEC includes the following subareas:

   a. Computers security (COMPUSEC)

   b. Communications security (COMSEC)

   c. Network security (NETSEC)

   d. Emanations security (EMSEC)

   e. Physical security

Each of these subareas needs to be independently protected. It is worth noting that INFOSEC is a process, not a product or technology. It may use various human means, procedures, and/or technologies to enable its implementation, including antivirus, access control, firewalls, authentication mechanisms (passwords, smart cards, tokens, biometrics, etc.), intrusion detection systems, policy management, vulnerability scanning, encryption, and physical security.

Computers security relies on protecting the processing and storage devices. The standard defines a common criteria, whose purpose relies on establishing a group of procedures and mechanisms that assure a certain level of confidence. These mechanisms include the whole information system life cycle, from the project phase, operation, up to destruction. A group of computer mechanisms, such as access control or antivirus system, aims to ensure that security policies are followed, and therefore, can be viewed as COMPUSEC mechanisms.

Communications security refers to the protection applied to data in transit between a transmitter and a receiver. The most common type of COMSEC protection relies on using cryptography.

Network security refers to the security of the network infrastructure, except the channel protection (covered by COMSEC). NETSEC refers to the protection of routing, DHCP, DNS, and so on. A common type of NETSEC protection relies on ciphering the exchange of the routing data.

Emanations security aims to protect the electromagnetic and acoustic radiations that may reveal the original message or information that is being processed or transmitted. A common type of EMSEC protection relies on using tempest or separating the cabling that carries classified data, from those that carries unclassified. Moreover, since an optical fiber is not a metallic medium and therefore it does not radiate electromagnetic signals, it can be used as an EMSEC protection.

Finally, physical security refers to the physical protection of resources. This aims to keep the physical survivability of resources, covering areas such as the electrical power supply, air conditioning, or the physical access control to areas where the systems and equipment are located.

4. The authentication is a process, whereas the authenticity is an attribute. The authentication comprises a number of physical or technological mechanisms to assure that the entity is legitimate. On the other hand, authenticity corresponds to a security attribute that assures that the data was effectively generated by the claimed author.

5. The challenge handshaking authentication protocol (CHAP) is an authentication protocol that is used by the PPP, being executed during the connection establishment and periodically (from a server request). Once the authenticator server has received a request to connect from the client, the CHAP is executed as follows (three-way handshaking):

a. *Challenge*: The server sends a variable length random stream.

b. *Response*: The client sends to the server the Hash function applied to the digest. The digest corresponds to the concatenation of the challenge (previously received from the server) with the password. This Hash sum is sent to the server, together with the username.

c. *Success or failure*: The authenticator server verifies the response from the client against its own calculation. If the response fits with its own calculation, the server sends a success message to the client, otherwise a failure message is sent.

Note that the CHAP requires that both the client and the server know the plaintext password, although it is never explicitly exchanged. The procedure of encrypting the password with a Hash function is widely used. An example of its use is the authentication process of computers.

While not perfect, the CHAP is much better than its predecessor password authentication protocol (PAP), where usernames and passwords are exchanged clearly. Note the CHAP considers periodic authentication, whereas the PAP only requires authentication during setup phase.

6. Accountability complements the basic security services, such as confidentiality, integrity, and authenticity. The accountability services do not bring any direct added value, but its inexistence represents a great reduction in the effectiveness of confidentiality, integrity, and authenticity services. The main purpose of this service is to ensure that the entity (user or process) is the legitimate, and that the events log revel the truth of the facts. These complementary services require using additional resources such as processing, memory, bandwidth, and so on. Using these complementary services as protective mechanisms, an attack becomes more difficult to be implemented, and possible to be registered and traced *a posteriori*. Therefore, a successful attack against the confidentiality or integrity needs also to include an attack against the authentication process, as well as against registration (events log). A lack in one of these complementary functions represents a lack of the basic security services.

The accountability can be brokendown into the following two groups:

a. I3A—identification, authentication, authorization, and access control

b. MRA—monitoring, registering, and auditing

I3A intends to find out who is an entity, to prove its authenticity, check its permissions, and allow or deny the access to resources. The MRA aims to register events, errors, and accesses to identify who performed those actions, and to register the moment when those actions were executed. To be effective, it requires an exact log of events, as well as a precise timing reference used in registration.

7. A distributed denial of service (DDoS) is an attack where the perpetrator does not directly act against the target. It makes use of intermediate machines for this purpose. The most common DDoS is the Botnet. In the Botnet attack, an attacker typically spreads out a worm file along different hosts. The worm is typically propagated in chain along the network, and the target is typically a server. These intermediate hosts are known as *zombies*, and can comprise as much as hundreds or even thousands of hosts. The attack is normally triggered (preprogrammed by the worm) to take place at a certain time, being executed against the target server. The attack can be of SYN type, Smurf attack, or any other (e.g., buffer overflow). In the Botnet attack, the attacker is difficult to be identified. Note that, most of the time, these third-party hosts do not even have knowledge that they are materializing such attacks. There are other types of DDoS attacks. The peer-to-peer attack can be viewed as a DDoS that uses a peer-to-peer service, which presents vulnerabilities. Such vulnerabilities facilitate the use of malicious tools, which allows implementing the DDoS in peer-to-peer mode. A popular peer-to-peer service used in DDoS relies on the file sharing. The attacker takes advantage of conception vulnerabilities to control the clients, forcing them to disconnect from the legitimate masters, and trying to connect to the target machine. In addition to the redundancy previously referred, a preventive protection also consists of keeping antivirus and other software up to date.

8. Cryptography is the most important protective measure from attacks against confidentiality. But there are other mechanisms. Accountability is a mechanism that complements cryptography to provide protection from attacks against confidentiality. Protective measures such as firewalls, intrusion detection systems, and auditing tools give protection from snooping (keyloggers).

9. Due to its highly demanding mathematical computation, the processing time required to cipher and decipher a message using asymmetric cryptography is approximately a 1000 times higher than that of symmetric cryptography. On the other hand, symmetric

cryptography presents a vulnerability associated to key's distribution. Combined schemes aim to, simultaneously, exploit the advantages of symmetric and asymmetric cryptography, while overcoming their disadvantages. This is achieved by using both symmetric and asymmetric keys in an efficient manner. Combined cryptography works as follows:

a. When a message needs to be sent, the sender generates a symmetric key, and ciphers the message with such key (session key).

b. Then, the symmetric key is ciphered with the receiver's public key, and the result is added to the previously ciphered message. These two components are sent together to the receiver.

c. The receiver takes the ciphered symmetric (session) key and deciphers it with its private key.

d. Then the *plaintext* symmetric key is finally used to decipher the ciphertext message.

Similar to asymmetric cryptography, the combined cryptography requires the previous distribution of public keys.

10. The effectiveness of the protective measures from attacks against the integrity depends on the combination of a Hash (digital signature) function with the accountability service (access control, authentication, identification, etc.), namely to prevent the creation, modification, or deletion of information and data (e.g., permissions to write data). The digital signature is the most important protective measure from attacks against the integrity.

11. The digital signature gives protection from attacks against authenticity and integrity. The digital signature comprises a Hash function applied to a message, properly signed by the private key of the sender. The receiver deciphers the Hash function with the sender's public key and compares it against the result of a Hash function applied to the received message. If the two Hashes are the same, than it means that the message is integer and that it was effectively generated by the claimed author. Note that the digital signature mechanism, while not avoiding an attack against the authenticity or integrity, it allows its detection, in case the integrity or authenticity has been violated. On the other hand, the digital certificate is an important measure that keeps the mapping between a public key and its owner. The digital certificate comprises a document signed, by a certification authority (CA), which verifies that a certain public key effectively belongs to a certain entity.

12. a.   deny tcp host 194.13.14.16 any eq 80
    b.   deny tcp host 194.13.14.16 any eq 443

13. The man-in-the-middle is the most common attack against integrity. It comprises the modification of the content, such as a financial transaction. In this case, the attacker captures the bits in transit and modifies it. Such modification can be the destination bank account of a financial transaction. This attack is of difficult execution, as the attacker has to be in line between the source and the recipient. Moreover, the attacker must have software that allows him taking charge of the session, such that the legitimate parties do not detect. An attack against integrity may also consist of adding or removing words from a message. Sometimes, adding or removing part of a message may change dramatically the sense of a message.

14. Before an entity starts using a digital certificate (e.g., the owner of a web server), he first needs to have the digital certificate properly certified and signed by a CA. This process can be carried out using one out of two procedures:

a. *Procedure 1*: The CA produces and distributes private and public keys. In this case, the entity sends a certificate request, together with its personal data. The CA sends the signed certificate, the privat,e and the public keys to the entity. Moreover, the CA stores the public key in a CA public repository for public consultation. The drawback of this procedure relies on how to be sure that the CA does not keep a copy of the generated private key, using it illegitimately.

b. *Procedure 2*: The entity produces a private and a public key. In this scenario, the entity signs his public key and his identification information with the corresponding Hash function and his private key. Then, the entity sends his identification in clear, his public key in clear, as well as the corresponding digital signature to the CA, which verifies the identification information with the received clear public key. Once it has been verified, the CA creates a certificate and signs it with the CA private key, and returns the new certificate to the entity, posting it in a public repository. The drawback of this procedure relies on how to be sure that the entity is exactly who he is claiming to be. This procedure can easily be improved by using one or more authentication mechanisms such as using personal presentation of ID card, or other identification documents. Note that procedure 2 has overcome the drawback of procedure 1, as the CA does not have knowledge about the entities' private keys.

15. The authentication function intends to prove that the entity (user or process) who wants to perform a certain action is the legitimate one, and effectively corresponds to the claimed identification. Note that authentication differs from authenticity, as the former is a process, whereas the latter is a security service. There are different mechanisms used to implement identification and authentication (I&A) such as:
    a. Something that can be known (password, PIN).
    b. Something that can be possessed (smart card, token).
    c. Something that can be seen, that is, biometrics (speech, eye, footprint).
    A more effective I&A process includes a combination of two or more of these mechanisms. Authentication may use a ciphered password with a Hash function. This procedure prevents the authenticator machine from having to store passwords in clear. This way, an intrusion attack against such machine, does not enable him to obtain such password(s). It is also worth noting that it is widely known that most of the attacks come from the interior of the organizations. Therefore, preventing employees from having access to passwords is a basic principle. Note that computers normally store passwords ciphered with Hash codes.

16. permit tcp 194.13.14.0 0.0.0.255 any eq 80

17. Authenticity aims to ensure that the author of information or data is the declared author. On the other hand, nonrepudiation aims to ensure that the author does not come in due course to deny the authorship of an action. The mechanisms that better ensure the authenticity and nonrepudiation consists of using the digital signature.

18. A risk is generally defined as a potential loss of assets and values, which are subject to threats that exploit vulnerabilities in a system, organization, or persons. The risk management comprises a process, for the risk monitoring. A risk is often mathematically quantified as Risk = Threat × Vulnerability × Value of Goods. With regard to the value of a good, this does not refer to its monetary value, but consists of the impact to the organization, in case such good fails as a result of an attack. The risk management should also include the identification of the mitigation measures that can be applied to minimize the risk, as well as the establishment of foundations for a security plan. Each risk is associated to a certain probability and impact, which needs to be noted at this stage. The residual risk corresponds to the initial risk, to which the countermeasures are subtracted (patches, firewall, intrusion detection systems, redundancy of systems, cryptography, procedures, policies, etc.). There are measures that aim to minimize the vulnerabilities of information systems (patching, hardening, training, etc.) and measures to protect from threats (firewall, cryptography, backups, etc.). The residual risk should be accepted by the management of the organization.

19. The effectiveness of the protective measures from attacks against the authenticity depends on the combination of a Hash function (digital signature) with the accountability service (access control, authentication, identification, etc.). The digital signature is the most important protective measure from attacks against the authenticity. The authentication

mechanisms also give protection from attacks against the authenticity. There are different mechanisms used to implement I&A such as the following:

a. Something that can be known (password and PIN)

b. Something that can be possessed (smart card and token)

c. Something that can be seen, that is, biometrics (speech, eye, and footprint)

A more effective I&A process includes a combination of two or more of these mechanisms.

20. Public key infrastructure consists of an infrastructure that aims to generate digital certificates properly certified and signed by a CA. This process can be carried out using one out of two procedures:

a. *Procedure 1*: The CA produces and distributes private and public keys. In this case, the entity sends a certificate request, together with its personal data. The CA sends the signed certificate, the private, and the public keys to the entity. Moreover, the CA stores the public key in a CA public repository for public consultation. The drawback of this procedure relies on how to be sure that the CA does not keep a copy of the generated private key, using it illegitimately.

b. *Procedure 2*: The entity produces a private and a public key. In this scenario, the entity signs his public key and his identification information with the corresponding Hash function and his private key. Then, the entity sends his identification in clear, his public key in clear, as well as the corresponding digital signature to the CA, which verifies the identification information with the received clear public key. Once it has been verified, the CA creates a certificate and signs it with the CA private key, and returns the new certificate to the entity, posting it in a public repository. The drawback of this procedure relies on how to be sure that the entity is exactly who he is claiming to be. This procedure can easily be improved by using one or more authentication mechanisms such as using personal presentation of ID card, or other identification documents. Note that procedure 2 has overcome the drawback of procedure 1, as the CA does not have knowledge about the entities' private keys.

21. The symmetric cryptography uses the same key for ciphering and deciphering the messages, using one or more predefined number of elementary operations based on substitution and transposition.

22. A brute force attack consists of trying all different possible characters, with different lengths, in a password request. The success of such attack is warranted, as long as there is enough time to perform such attack. Depending on the processing speed, a brute force attack against an authentication system using a password with eight characters may take as much as several months. Naturally, reducing the password's length leads to a successful brute force attack in much shorter time frame. Therefore, it is very important keeping long passwords.

23. The IPsec, TLS, and SSL are examples of combined protocols. The PGP is another combined protocol widely used.

24. A brute force attack consists of trying all different possible characters, with different lengths, in a password request. The success of such attack is warranted, as long as there is enough time to perform such attack. Depending on the processing speed, a brute force attack against an authentication system using a password with eight characters may take as much as several months. Naturally, reducing the password's length leads to a successful brute force attack in much shorter time frame. Therefore, it is very important keeping long passwords.

25. The digital signature gives protection from attacks against the authenticity and integrity. The digital signature comprises a Hash function applied to a message, properly signed by the private key of the sender. The receiver deciphers the Hash function with the sender's public key and compares it against the result of a Hash function applied to the received message. If the two Hashes are the same, then it means that the message is an integer and

that it was effectively generated by the claimed author. Note that the digital signature mechanism, while not avoiding an attack against the authenticity or integrity, it allows its detection, in case the integrity or authenticity has been violated.

26. Public key cryptography is another designation for asymmetric cryptography. Public key cryptography makes use of a public key widely distributed, and a private key, that is kept secret. Consequently, using the public key cryptography, there is no need to implement a complex and vulnerable key's distribution system, as required for symmetric cryptography. Moreover, since the encryption operation can be implemented by any of the keys (private of public), this system is well fitted for both data encryption and digital signature.

27. A vulnerability consists of a potential route of exploitation to the threats. A malicious software can be viewed as a threat, whereas an operating system without the appropriate patches can be viewed as a vulnerability. A risk is generally defined as a potential loss of assets and values, which are subject to threats that exploit vulnerabilities in a system, organization, or persons. The risk management comprises a process, for the risk monitoring. A risk is often mathematically quantified as Risk = Threat × Vulnerability × Value of Goods. With regard to the value of a good, this does not refer to its monetary value, but consists of the impact to the organization, in case such good fails as a result of an attack. The risk management should also include the identification of the mitigation measures that can be applied to minimize the risk, as well as the establishment of foundations for a security plan. Each risk is associated to a certain probability and impact, which needs to be noted at this stage. The residual risk corresponds to the initial risk, to which the countermeasures are subtracted (patches, firewall, intrusion detection systems, redundancy of systems, cryptography, procedures, policies, etc.). There are measures that aim to minimize the vulnerabilities of information systems (patching, hardening, training, etc.) and measures to protect from threats (firewall, cryptography, backups, etc.).

28. An important disadvantage of asymmetric cryptography, relating to symmetric, relies on the fact that, ciphering and deciphering messages with asymmetric cryptography takes about 1000 times more time than with symmetric cryptography.

29. A virtual private network (VPN) can be used by both employees and partners to allow access from a public communications network (typically, from the Internet) to the intranet or to a partners' DMZ. Using a VPN, a user can remotely access a network, using the Internet for this purpose. Since the Internet presents multiple vulnerabilities, a VPN is normally implemented using cryptography. The IPsec is widely used to implement secure VPNs.

30. The symmetric cryptography uses the same key for ciphering and deciphering the messages. The symmetric cryptography requires the previous distribution of keys, which presents a certain level of vulnerability. Asymmetric cryptography mitigates this vulnerability, as it makes use of two keys, a public (widely distributed) and a private (kept secret). An important disadvantage of asymmetric cryptography, relating to symmetric, relies on the fact that, ciphering and deciphering messages with asymmetric cryptography takes about 1000 times more time than with symmetric cryptography.

31. A DMZ consists of a network area that accommodates the external servers, and whose access needs to be properly authorized by the firewall. Contrary, using an extranet, Internet users could get access to all external servers. Commonly, using the DMZ architecture, the extranet is left empty.

32. The remote delegations are typically connected to the headquarters through the Internet, using a secure VPN. The IPsec in tunneling mode (using the ESP) is the most used cryptographic protocol adopted to allow such connectivity. In this case, a secure IPsec tunnel is configured at both the headquarters' and remote delegations' firewall. The firewalls act as security gateways, encapsulating the original IP packet (including the original IP header) into an IPsec packet. The IPsec packet is then encapsulated into the payload of

another IP packet, whereas its source and destination IP address refers to the IP address of the security gateways (firewalls) of the remote delegation and headquarters. A VPN can also be established between an individual user from a public communications network (Internet) and the headquarters. In this case, the same IPsec VPN in tunneling mode can be implemented, but the tunnel is established directly between the remote host and the headquarters' security gateway. The exchange of data (connectivity) can be assured by any transmission means depending on the service requirements, such as the required data rates and accepted delays. A remote user from home may access the intranet using the regular home network (e.g., a wireless LAN connected to the Internet through xDSL modem). A very small remote delegation may opt for a low cost xDSL or cable modem connection, whereas larger remote delegations may use MPLS connectivity to get access to the headquarters' router with high-speed connectivity.

33. permit tcp 11.1.0.0 0.0.255.255 host 15.1.2.3 eq 110

34. deny tcp host 10.1.1.1 any eq 443

35. A standard ACL simply allows packet filtering based on the source IP address, whereas an extended ACL allows packet filtering based on a number of parameters, namely
    a. Protocol type (IP, TCP, UDP, etc.)
    b. Source IP address
    c. Destination IP address
    d. Source port
    e. Destination port
    f. Date and time

    Therefore, an ACL includes a number rules to filter the traffic based on layer 3 and 4 addresses, protocol type, and date/time. Since a standard ACL does not specify the destination address, this should be placed as close as possible to the destination device. In contrast, in order to reduce the network usage, since an extended ACL specify the destination address, this should be included as close as possible to the source device.

36. SSL and TLS work between the transport layer and the application layer, facilitating the authentication of entities, being used for end-to-end encryption. On the other hand, since IPsec works at the network layer it encrypts all packets that cross a certain path of the route, but authentication mechanisms are not taken into account.

37. Depending on the depth and methodology used to verify the conformity of the traffic with the established policy, firewalls can be grouped into three categories:
    a. *Packet filtering—stateless*: It inspects layer 3 and 4 headers (source and destination IP and port addresses, sequence numbers, and traffic type), allowing or denying the establishment of connections based on these parameters. Note that it does not keep the state of the connections (stateless), that is, it does not take into account the established connections to allow or deny new connections. Moreover, it does not interrupt connections (contrary to the proxy firewall).
    b. *Packet filtering—statefull*: It is similar to the stateless firewall type, it inspects layer 3 and 4 headers. However, unlike stateless inspection, the statefull inspection keeps the state of the established connections, taking into account this information in the decision to allow or deny new connections. An example of a firewall implementing packet filtering in statefull inspection is a firewall that implicitly authorizes traffic coming from an Internet HTTP server into a client located in the intranet only if such TCP connection has been previously established, and requested by the client to the HTTP server (port 80).
    c. *Application layer gateway—statefull inspection*: This firewall type works at the application layer, performing a deep inspection of the packets contents (instead of only layer 3 and 4 headers), while taking into account the information about the state of the established connections to allow or deny new connections. Moreover, this firewall

type interrupts the connections, that is, it establishes two independent connections, to allow the establishment of a connection request. For this reason, this firewall type is also referred to as *proxy firewall* (similar to the proxy server).

38. Depending on the services and security levels, there are different network architectures that can be implemented within an organization such as in a company. Two different network types may exist within an organization: an intranet is only used by the organizational employees, whereas an extranet is used by both customers and business partners.

Appendix VI
List of Experiments Using Emona TIMS-301C and Net*TIMS

Experiment Reference	Experiment Designation	Chapter 1	Chapter 3	Chapter 4	Chapter 5	Chapter 6	Chapter 7	Chapter 12	Chapter 14	Chapter 15
	Emona TIMS Student Text Lab Manual—Volume A1—Fundamental Analog Experiments									
A1-01	Introduction to modeling with TIMS	X								
A1-02	Modeling an equation	X								
A1-03	DSBSC generation					X				
A1-04	Amplitude modulation					X				
A1-05	Envelopes					X				
A1-06	Envelope recovery					X				
A1-07	SSB generation—the phasing method					X				
A1-08	Product demodulation—synch and asynch					X				
A1-09	SSB demodulation—the phasing method					X				
A1-10	The sampling theorem		X							
A1-11	PAM and time division multiplex					X				
A1-12	Power measurements		X							
	Emona TIMS Student Text Lab Manual—Volume A2—Further and Advanced Analog Experiments									
A2-01	Amplitude modulation—method 2					X				
A2-02	Weaver's SSB generator					X				
A2-03	Weaver's demodulator					X				
A2-04	Carrier acquisition and the PLL					X				
A2-05	Spectrum analysis—the WAVE ANALYSER		X							
A2-06	Amplifier overload		X							
A2-07	Frequency division multiplex					X				
	Emona TIMS Student Text Lab Manual—Volume D1—Fundamental Digital Experiments									
D1-01	PRBS generation						X			
D1-02	Eye patterns		X							
D1-03	The noisy channel model		X							
D1-04	Detection with the DECISION MAKER		X							

(Continued)

Experiment Reference	Experiment Designation	Chapter 1	Chapter 3	Chapter 4	Chapter 5	Chapter 6	Chapter 7	Chapter 12	Chapter 14	Chapter 15
D1-05	Line coding					X				
D1-06	ASK—amplitude shift keying					X				
D1-07	FSK—frequency shift keying					X				
D1-08	BPSK—binary phase shift keying					X				
D1-09	Signal constellations					X				
D1-10	Sampling with SAMPLE and HOLD		X							
D1-11	PCM encoding					X				
D1-12	PCM decoding					X				
D1-13	Delta modulation					X				
D1-14	Delta demodulation					X				
D1-15	Adaptive delta modulation					X				
	Emona TIMS Student Text Lab Manual—Volume D2—Further and Advanced Digital Experiments									
D2-01	BER measurement in the noisy channel		X							
D2-03	Bit clock regeneration		X							
D2-06	PCM TDM					X				
D2-07	Block coding and decoding							X		
D2-08	Block coding and coding gain							X		
D2-09	Convolutional coding							X		
D2-12	QAM and 4-PSK					X				
D2-13	Multilevel QAM & PSK					X				
D2-14	Spread spectrum—DSSS and CDMA						X			
	Emona TIMS Student Text Lab Manual—Volume D3—Advanced Digital Experiments									
D3-01	ISI: PAM and ASK over band-limited channels					X				
D3-02	Equalization for ISI				X					
D3-03	Pulse shaping for band-limited channels			X						
D3-04	Base-line wander and line coding					X				
D3-05	Timing jitter in band-limited channels				X					
D3-06	Introduction to OFDM principles						X			

(Continued)

Experiment Reference	Experiment Designation	Chapter 1	Chapter 3	Chapter 4	Chapter 5	Chapter 6	Chapter 7	Chapter 12	Chapter 14	Chapter 15
D3-07	Additive noise in digital baseband channel						X			
D3-08	Additive noise in block coded channel							X		
D3-09	Introduction to FHSS using FSK						X			
D3-10	FHSS: fast and slow hopping						X			
D3-11	FHSS and bit error rate performance						X			
D3-13	SONET PCM data frame								X	
D3-14	SONET STS-1 demultiplexing								X	
D3-15	SONET STS-1 transmission via an optical link with bit recovery								X	
Emona TIMS Student Text Lab Manual—Volume D4—Further Advanced Digital Experiments										
D4-01	BER measurement of unipolar NRZ signals in a baseband distortionless channel					X				
D4-02	BER measurements of bipolar NRZ signals in a baseband distortionless channel					X				
D4-03	BER measurement of coherent BPSK signaling in an ideal distortionless channel					X				
Emona TIMS Student Text Lab Manual—Volume D5—Basic Spread Spectrum Experiments										
D5-01	Spread spectrum—analysis of direct sequence						X			
D5-02	Spread spectrum—analysis of frequency hop						X			
D5-03	Spread spectrum—analysis of time hop						X			
D5-04	Spread spectrum—analysis of hybrid FH-DS						X			
Emona TIMS Student Text Lab Manual—Volume D6—Advanced Spread Spectrum Experiments										
D6-01	DS-CDMA/BPSK—3-channel basic system						X			X
D6-02	DS-SS baseband system—processing gain measurement						X			
D6-03	FH-CDMA/BFSK						X			
Emona TIMS Student Text Lab Manual—Volume D8—Multipath and OFDM Experiments										
D8-01	Time-invariant fading channel characteristics				X					

(Continued)

Experiment Reference	Experiment Designation	Chapter 1	Chapter 3	Chapter 4	Chapter 5	Chapter 6	Chapter 7	Chapter 12	Chapter 14	Chapter 15
D8-02	ISI rejection in DS SS						X			
D8-03	OFDM in band-limited, multipath, time-invariant channel with BER measurement						X			
	Emona TIMS Student Text Lab Manual—TIMS Signals and Systems-V1 Experiments									
S1-01	Special signals—characteristics and applications		X							
S1-02	Modeling linear and nonlinear systems with TIMS		X							
S1-04	Comparing responses in the time and frequency domains		X							
S1-06	Spectrum analysis of various signal types		X							
S1-08	Sampling and aliasing		X							
S1-09	Getting started with analog–digital conversion					X				
	Emona TIMS LabSheets—Concise, 2-Page, Quick-Start Experiment Guides									
L-01	Introduction to TIMS		X							
L-02	Modeling equations		X							
L-03	DSB—generation					X				
L-04	Product demodulation					X				
L-05	Amplitude modulation-1					X				
L-06	Amplitude modulation-2					X				
L-07	Envelope detection					X				
L-08	SSB generation					X				
L-09	SSB demodulation					X				
L-10	ISB—independent sideband					X				
L-16	PAM and TDM					X				
L-17	FDM—frequency division multiplex					X				
L-22	Spectra using a WAVE ANALYZER				X					
L-24	PCM encoding					X				
L-25	PCM decoding					X				
L-26	ASK generation					X				
L-27	ASK demodulation					X				

(Continued)

Experiment Reference	Experiment Designation	Chapter 1	Chapter 3	Chapter 4	Chapter 5	Chapter 6	Chapter 7	Chapter 12	Chapter 14	Chapter 15
L-28	BPSK modulation					X				
L-29	BPSK demodulation					X				
L-30	QPSK generation					X				
L-31	QPSK demodulation					X				
L-32	FSK generation					X				
L-33	FSK envelope demodulation					X				
L-34	Constellations					X				
L-35	DSSS spread spectrum						X			
L-36	Eye patterns				X					
L-38	Detection with the DECISION MAKER		X							
L-39	Noisy channel		X							
L-40	BER instrumentation		X							
L-41	BER measurement introduction		X							
L-42	Line coding and decoding					X				
L-43	Delta modulation					X				
L-45	Adaptive delta modulation					X				
L-47	Bit clock regeneration					X				
L-48	QAM generation					X				
L-49	QAM demodulation					X				
L-50	BPSK					X				
L-51	Broadcasting	X								
L-52	Fiber optic transmission			X						
L-53	Multichannel FDM digital fiber link			X						
L-54	PCM-TDM-T1 implementation								X	
L-60	Matched filter detection		X							
L-62	CDMA introduction									X
L-63	CDMA processing gain									X
L-64	CDMA—2 channel									X
L-65	CDMA—multichannel									X

(Continued)

Experiment Reference	Experiment Designation	Chapter 1	Chapter 3	Chapter 4	Chapter 5	Chapter 6	Chapter 7	Chapter 12	Chapter 14	Chapter 15
L-68	Nonlinearity and distortion		X							
L-70	Speech in telecommunications					X				
L-71	Binary data via voiceband									X
L-72	Multilevel data via voiceband									X
L-73	Data rates and voiceband modems—transmission									X
L-74	Data rates and voiceband modems—demodulation									X
L-78	Block code encoding (method 1)							X		
L-79	Block code encoding (method 2)							X		
L-80	Block code decoding							X		
L-81	Error correcting with block coding							X		
L-86	Convolutional coding I							X		
L-87	Convolutional coding II							X		
L-88	Optical signal splitting and combining			X						
L-89	Fiber optic bidirectional communication			X						
L-90	WDM—wave division multiplex			X						
L-92	The SONET PCM data frame								X	
L-93	SONET STS-1 demultiplexing								X	
L-94	SONET transmission via an optical link								X	
L-95	SONET STS-3 multiplexing								X	
L-97	SONET STS-3 demultiplexing								X	
L-98	Intro to FHSS using FSK						X			
L-99	FHSS slow and fast hopping						X			
L-100	FHSS and BER performance						X			
L-101	Multichannel FHSS with BER						X			
L-102	Hybrid DSSS FHSS system						X			
L-103	Introduction to OFDM generation						X			
L-104	Introductory PAM-TDM					X				
L-106	Introduction to pulse shaping		X							
L-107	Noise generation using binary sequences		X							
L-108	Principles of spread spectrum						X			

(*Continued*)

Experiment Reference	Experiment Designation	Chapter 1	Chapter 3	Chapter 4	Chapter 5	Chapter 6	Chapter 7	Chapter 12	Chapter 14	Chapter 15
L-112	SSB linear amplifier measurements					X				
L-113	SNR—SSB compared with DSBSC					X				
L-114	AM demodulation and SNR					X				
L-115	4/8/16-QPSK and 4/8/16-QAM with BER in a noisy channel					X				
L-135	Spread spectrum—analysis of THSS						X			
L-136	Spread spectrum—analysis of FHSS-SFH						X			
L-137	Spread spectrum—analysis of hybrid FH-DSSS						X			
L-138	Spread spectrum—analysis of hybrid FH-DSSS						X			
L-139	Spread spectrum—analysis of FHSS						X			
L-140	Spread spectrum—analysis of DSSS						X			
L-141	DSSS baseband system-processing gain measurement—broadband jamming									X
L-142	DSSS baseband system-processing gain measurement—pulse jamming									X
L-150	ISI rejection in DS SS systems									X
L-151	IDFT, complex exponent, and complex quad signals						X			
L-152	OFDM, cyclic prefix, and PAPR						X			
L-153	OFDM and channel equalization with BER measurement						X			
L-154	OFDM in band-limited, multipath, time-invariant channel with BER measurement						X			
L-161	BER of coherent QPSK in distortionless channel									X

References

CHAPTER 1

[Khanvilkar et al. 2005] Khanvilkar, S. et al., *Multimedia Networks and Communication*, Academic Press, Burlington, MA, 2005.

[Marques da Silva et al. 2010] Marques da Silva, M., Correia, A., Dinis, R., Souto, N., Silva, J. C., *Transmission Techniques for Emergent Multicast and Broadcast Systems*, 1st edition, CRC Press, Boca Raton, FL, 2010.

[Monica 1998] Monica, P., *Comunicação de Dados e Redes de Computadores*, ISCTEC, Lisbon, Portugal, 1998.

[Raj et al. 2010] Raj, M., Narayan, A., Datta, S., Das, S., Fixed mobile convergence: Challenges and solutions, *IEEE Communications Magazine*, 48(12), 26–34, December 2010.

[Stallings 2010] Stallings, W., *Data and Computer Communications*, 8th edition, Prentice Hall, Upper Saddle River, NJ, 2010.

CHAPTER 2

[Forouzan 2007] Forouzan, A. F., *Data Communications and Networking*, 4th edition, McGraw-Hill, New York, 2007.

[Marques da Silva 2012] Marques da Silva, M., *Multimedia Communications and Networking*, 1st edition, CRC Press, Boca Raton, FL, March 2012.

[Marques da Silva et al. 2010] Marques da Silva, M., Correia, A., Dinis, R., Souto, N., Silva, J. C., *Transmission Techniques for Emergent Multicast and Broadcast Systems*, 1st edition, CRC Press, Boca Raton, FL, May 2010.

[RFC 826] Plummer, D. C., *RFC 826—An Ethernet Address Resolution Protocol*, Network Working Group, November 1982.

[RFC 2460] Deering, S., Hinden, R., *RFC 2460—Internet Protocol Version 6 (IPv6) Specification*, Network Working Group, December 1998.

[Shannon 1948] Shannon, C. E., A mathematical theory of communication, *Bell System Technical Journal*, 27, 379–423, 623–656, October 1948.

[Stallings 2010] Stallings, W., *Data and Computer Communications*, 8th edition, Prentice Hall, Upper Saddle River, NJ, 2010.

CHAPTER 3

[Benedetto et al. 1987] Benedetto, S., Biglieri, E., Castellani, V., *Digital Transmission Theory*, Prentice Hall, Englewood Cliffs, NJ, 1987.

[Burrows 1949] Burrows, C. R., *Radio Wave Propagation*, Academic Press, New York, 1949.

[Carlson 1986] Carlson, A. B., *Communication Systems*, 3rd edition, McGraw-Hill, New York, 1986.

[CCIR 322] Lawrence, D. C., *CCIR Report 322 Noise Variation Parameters*, Technical Document 2813, June 1995.

[Foschini and Gans 1998] Foschini, G. J., Gans, M. J. On limits of wireless communications in fading environments when using multiple antennas, *Wireless Personal Communications*, 6, 315–335, March 1998.

[Marques da Silva et al. 2010] Marques da Silva, M., Correia, A., Dinis, R., Souto, N., Silva, J. C., *Transmission Techniques for Emergent Multicast and Broadcast Systems*, 1st edition, CRC Press, Boca Raton, FL, 2010.

[Marques da Silva et al. 2012] Marques da Silva, M., Correia, A., Dinis, R., Souto, N., Silva, J. C., *Transmission Techniques for 4G Systems*, 1st edition, CRC Press, Boca Raton, FL, 2012.

[Proakis 1995] Proakis, J. G., *Digital Communications*, 3rd edition, McGraw-Hill, New York, 1995.

[Shannon 1948] Shannon, C. E., A mathematical theory of communication, *Bell System Technical Journal*, 27, 379–423, 623–656, October 1948.

[Theodore 1996] Theodore, S., *Rappaport—Wireless Communications*, Prentice Hall, Englewood Cliffs, NJ, 1996.

CHAPTER 4

[ANSI/TIA/EIA-568] ANSI/TIA/EIA-568, *Commercial Building Telecommunications Standard*, 1991.
[ANSI/TIA/EIA-568-A] ANSI/TIA/EIA-568-A, *Commercial Building Telecommunications Standard*, 1995.
[ANSI/TIA/EIA-568-B] ANSI/TIA/EIA-568-B, *Commercial Building Telecommunications Standard*, 2001.
[ISO/IEC 11801] ISO/IEC 11801, *International Standard—Information Technology—Generic Cabling for Customer Premises*, 2nd edition, 2002.
[ITU-T G.621] ITU-T Recommendation G.621—*Transmission Media Characteristics of 0.7/2.9 mm Coaxial Cable Pairs*, 1993.
[ITU-T G.622] ITU-T Recommendation G.622—*Transmission Media Characteristics of 1.2/4.4 mm Coaxial Cable Pairs*, 1993.
[ITU-T G.623] ITU-T Recommendation G.623—*Transmission Media Characteristics of 2.6/9.5 mm Coaxial Cable Pairs*, 1993.
[Keiser 1991] Keiser, G., *Optical Fiber Communications*, 2nd edition, McGraw-Hill, New York, 1991.
[Stallings 2010] Stallings, W., *Data and Computer Communications*, 8th edition, Prentice Hall, Upper Saddle River, NJ, 2010.
[Winzer 2010] Winzer, P., Bell Labs, Alcatel-Lucent, "beyond 100G Ethernet," *IEEE Communications Magazine*, 48(7), 26–30, July 2010.

CHAPTER 5

[Burrows 1949] Burrows, C. R., *Radio Wave Propagation*, Academic Press, New York, 1949.
[Carlson 1986] Carlson, A. B., *Communication Systems*, 3rd edition, McGraw-Hill, New York, 1986.
[Fernandes 1996] Fernandes, C. A., *Aspectos de Propagação na Atmosfera—Anexo*, Secção de Propagação e Radiação IST, Lisbon, Portugal, 1996.
[Ha 1990] Ha, T. T., *Digital Satellite Communications*, 2nd edition, McGraw-Hill Communications Series, McGraw-Hill, New York, 1990.
[ITU-R P.368-7] ITU-R Recommendation P.368-7—*Ground-Wave Propagation Curves for Frequencies Between 10 kHz and 30 MHz*, 1992.
[Kadish and East 2000] Kadish, J. E., East, T. W. R., *Satellite Communications Fundamentals*, Artech House, Norwood, MA, 2000.
[Marques da Silva et al. 2010] Marques da Silva, M., Correia, A., Dinis, R., Souto, N., Silva, J. C., *Transmission Techniques for Emergent Multicast and Broadcast Systems*, 1st edition, CRC Press, Boca Raton, FL, May 2010.
[NBS 1967] Rice, P. L. et al., *Transmission Loss Predictions for Tropospheric Communication Circuits*, NBS Technical Note 101, Vol. II, U.S. Department of Commerce, National Bureau Standards, Washington, DC, 1967.
[Parsons 2000] Parsons, J. D., *The Mobile Radio Propagation Channel*, 2nd edition, John Wiley & Sons, Chichester, England, 2000.
[Proakis 1995] Proakis, J. G., *Digital Communications*, 3rd edition, McGraw-Hill, New York, 1995.

CHAPTER 6

[Carlson 1986] Carlson, A. B., *Communication Systems*, 3rd edition, McGraw-Hill, New York, 1986.
[Chamberlain 2001] Chamberlain, M. W., A 600 bps MELP vocoder for use on HF channels, *Military Communications Conference*, 1, 447–453, 2001.
[Goldsmith and Chua 1997] Goldsmith, A., Chua, S. G., Variable-rate variable power M-QAM for fading channels, *IEEE Transactions on Communications*, 45(10), 1218–1230, October 1997.
[ITU-T G.711] ITU-T Recommendation G.711—*Pulse Code Modulation of Voice Frequencies*, 1972.
[ITU-T G.726] ITU-T Recommendation G.726—*40, 32, 24, 16 kbit/s Adaptive Differential Pulse Code Modulation*, 1990.
[Marques da Silva et al. 2009] Marques da Silva, M., Correia, A., Dinis, R., On transmission techniques for multi-antenna W-CDMA systems, *European Transactions on Telecommunications*, 20(1), 107–121, January 2009.
[MIL-STD-3005] Department of Defense—Telecommunications Systems Standard, *Analog-to-Digital Conversion of Voice by 2,400 Bit/Second Mixed Excitation Linear Prediction*, December 20, 1999.
[Proakis 1995] Proakis, J. G., *Digital Communications*, 3rd edition, McGraw-Hill, New York, 1995.
[Wang et al. 2000] Wang, T., Koishida, K., Cuperman, V., Gersho, A., Collura, J. S., A 1200 bps speech coder based on MELP, *Acoustics, Speech, and Signal Processing*, 3, 1376, 2000.

[Watkinson 2001] Watkinson, J., *The MPEG Handbook: MPEG-I, MPEG-II, MPEG-IV*, Focal Press, Waltham, MA, 2001.

[Webb and Hanzo 1994] Webb, W. T., Hanzo, L., *Modern Quadrature Amplitude Modulation: Principles and Applications for Fixed and Wireless Channels*, IEEE, New York, 1994.

[Webb and Steele 1995] Webb, W. T., Steele, R., Variable rate QAM for mobile radio, *IEEE Transactions on Communications*, 43(7), 2223–2230, July 1995.

CHAPTER 7

[3GPP 2003a] 3GPP, TR 25.214-v5.5.0, *Physical Layer Procedures (FDD)*, European Telecommunications Standards Institute, France, 2003a.

[3GPP 2003b] 3GPP, TR 25.211-v5.2.0, *Physical Channels and Mapping of Transport Channels onto Physical Channels (FDD)*, European Telecommunications Standards Institute, France, 2003b.

[3GPP 2004] 3GPP, TS 25.213-v6.1.0, *Spreading and Modulation (FDD)*, European Telecommunications Standards Institute, France, December 2004.

[3GPP 2010a] 3GPP, TR 36.912-v9.2.0, *Feasibility Study for Further Advancements for E-UTRA (LTE-Advanced)*, European Telecommunications Standards Institute, France, 2010a.

[3GPP 2010b] 3GPP, TR 36.806-v9.0.0, *Evolved Universal Terrestrial Radio Access (E-UTRA); Relay Architectures for E-UTRA (LTE-Advanced)*, European Telecommunications Standards Institute, France, May 2010b.

[Alamouti 1998] Alamouti, S. M., A simple transmitter diversity scheme for wireless communications, *IEEE JSAC*, 16(8), 1451–1458, October 1998.

[Benvenuto and Tomasin 2002] Benvenuto, N., Tomasin, S., Block iterative DFE for single carrier modulation, *IEE Electronic Letters*, 39(19), 1144–1145, September 2002.

[Cimini 1985] Cimini, L., Analysis and simulation of a digital mobile channel using orthogonal frequency division multiplexing, *IEEE Transactions on Communications*, 33, 665–675, July 1985.

[Correia and Marques da Silva 2014] Correia, A., Marques da Silva, M., Link and system level simulations for MIMO, *MIMO Processing for 4G and Beyond: Fundamentals and Evolution*, (eds.) M. Marques da Silva, F. A. Monteiro, CRC Press, Boca Raton, FL, pp. 405–447, June 2014.

[Correia et al. 2010] Correia, L., Zeller, D., Blume, O., Ferling, D., Jading, Y., István, G., Auer, G., Perre, L., Challenges and enabling techniques for energy aware mobile radio networks, *IEEE Communications Magazine*, 48, 66–72, November 2010.

[Cover 1972] Cover, T., Broadcast channels, *IEEE Transactions on Information Theory*, IT-18, 2–14, January 1972.

[Diehm et al. 2010] Diehm, F., Marsch, P., Fetweeis, G., The FUTON prototype: Proof of concept of coordinated multi-point in conjunction with a novel integrated wireless/optical architecture, *Proceedings of the IEEE WCNCW*, Sydney, New South Wales, Australia, pp. 1–4, April 2010.

[Dinis et al. 2003] Dinis, R., Gusmão, A., Esteves, N., On broadband block transmission over strongly frequency-selective fading channels, *Wireless Conference 2003*, Calgary, Canada, pp. 261–269, July 2003.

[Dogan et al. 2004] Dogan, S. et al., Video content adaptation using transcoding for enabling UMA over UMTS, *Proceedings of the WIAMIS*, Lisbon, Portugal, April 2004

[Ericsson 2007] Ericsson, Sustainable energy use in mobile communications, *White Paper*, August 2007. http://www.ericsson.com, accessed in December 23, 2011.

[Falconer et al. 2002] Falconer, D., Ariyavisitakul, S., Benyamin-Seeyar, A., Eidson, B., Frequency domain equalization for single-carrier broadband wireless systems, *IEEE Communications Magazine*, 4(4), 58–66, April 2002.

[Ferling et al. 2010] Ferling, D., Bohn, T., Zeller, D., Frenger, P., Gódor, I., Jading, Y., Tomaselli, W., Energy efficiency approaches for radio nodes, *Proceedings of the Future Networks Mobile Summit*, Florence, Italy, June 2010.

[Foschini 1996] Foschini, G. J., Layered space-time architecture for wireless communication in a fading environment when using multiple antennas, *Bell Laboratories Technical Journal*, 1(2), 41–59, Autumn 1996.

[Foschini and Gans 1998] Foschini, G. J., Gans, M. J., On limits of wireless communications in fading environments when using multiple antennas, *Wireless Personal Communications*, 6, 315–335, March 1998.

[Glisic and Vucetic 1997] Glisic, S., Vucetic, B., *Spread Spectrum CDMA Systems for Wireless Communications*, Artech House, Norwood, MA, 1997.

[Holma and Toskala 2000] Holma, H., Toskala, A., *WCDMA for UMTS*, John Wiley & Sons, New York, 2000.

[Holma and Toskala 2007] Holma, H., Toskala, A., *WCDMA for UMTS: HSPA Evolution and LTE*, 4th edition, John Wiley & Sons, New York, 2007.

[Hottinen et al. 2003] Hottinen, A., Tirkkonen, O., Wichman, R., *Multi-Antenna Transceiver Techniques for 3G and Beyond*, John Wiley & Sons, Chichester, West Sussex, 2003.

[Johansson and Svensson 1999] Johansson, A. L., Svensson, A., On multi-rate DS/CDMA schemes with interference cancellation, *Journal on Wireless Personal Communications*, 9(1), 1–29, January 1999.

[Kudoh and Adachi 2003] Kudoh, E., Adachi, F., Transmit power efficiency of a multi-hop virtual cellular system, *Proceedings of the IEEE Vehicular Technology Conference*, Orlando, FL, Vol. 5, pp. 2910–2914, October 6–9, 2003.

[Lee et al. 2012] Lee, D., Seo, H., Clerckx, B., Hardouin, E., Mazzarese, D., Nagata, S., Sayana, K., Coordinated multipoint transmission and reception in LTE-Advanced deployment: Scenarios and operational challenges, *IEEE Wireless Communications Magazine*, 50(2), 148–155, February 2012.

[Li 2001] Li, W., Overview of fine granularity scalability in MPEG-4 video standard, *IEEE Transactions on CSVT*, 11(3), 301–317, March 2001.

[Liu and Li 2005] Liu, H., Li, G., *OFDM-Based Broadband Wireless Networks*, John Wiley & Sons, Hoboken, NJ, 2005.

[Liu et al. 2003] Liu, J., Li, B., Zhang, Y.-Q., Adaptive video multicast over the internet, *IEEE Multimedia*, 10(1), 22–33, January–March 2003.

[Marques da Silva and Correia 2000] Marques da Silva, M., Correia, A., Parallel interference cancellation with commutation signaling, *IEEE International Conference on Communications—ICC2000 Spring*, New Orleans, LA, June 18–22, 2000.

[Marques da Silva and Correia 2001] Marques da Silva, M., Correia, A., Space time diversity for the downlink of WCDMA, *IEEE Wireless Personal and Mobile Communications*, Aalborg, Denmark, September 9–12, 2001.

[Marques da Silva and Correia 2002a] Marques da Silva, M., Correia, A., Space time block coding for 4 antennas with coding rate 1, *IEEE International Symposium on Spread Spectrum Techniques and Application*, Prague, Czech Republic, September 2–5, 2002a.

[Marques da Silva and Correia 2002b] Marques da Silva, M., Correia, A., Space time coding schemes for 4 or more antennas, *IEEE International Symposium on Personal Indoor and Mobile Radio Communications*, Lisbon, Portugal, September 16–18, 2002b.

[Marques da Silva and Correia 2003a] Marques da Silva, M., Correia, A., Joint multi-user detection and intersymbol interference cancellation for WCDMA satellite UMTS, *International Journal of Satellite Communications and Networking* (Special Issue on Interference Cancellation, Wiley), 21(1), 93–117, January 2003a.

[Marques da Silva and Correia 2003b] Marques da Silva, M., Correia, A., Combined transmit diversity and beamforming for WCDMA, *IEEE EPMCC*, Glasgow, Scotland, April 2003b.

[Marques da Silva and Correia 2014] Marques da Silva, M., Correia, A., MIMO techniques and applications, *MIMO Processing for 4G and Beyond: Fundamentals and Evolution*, (eds.) M. Marques da Silva, F. A. Monteiro, CRC Press, Boca Raton, FL, pp. 1–45, June 2014.

[Marques da Silva et al. 2005a] Marques da Silva, M., Correia, A., Dinis, R., A decorrelating MUD approach for the downlink of UMTS considering a RAKE in the receiver, *Proceedings of the 16th IEEE Personal Indoor and Mobile Radio Communications*, Berlin, Germany, September 11–14, 2005a.

[Marques da Silva et al. 2009] Marques da Silva, M., Correia, A., Dinis, R., On transmission techniques for multi-antenna W-CDMA systems, *European Transactions on Telecommunications*, 20(1), 107–121, January 2009.

[Marques da Silva et al. 2014] Marques da Silva, M., Correia, A., Dinis, R., Montezuma, P., On coordinated multi-point transmission for cellular environments, *Progress in Electromagnetics Research Symposium*, Guangzhou, China, August 25–28, 2014.

[Marques da Silva et al. 2010] Marques da Silva, M., Correia, A., Dinis, R., Souto, N., Silva, J., *Transmission Techniques for Emergent Multicast and Broadcast Systems*, 1st edition, CRC Press, Boca Raton, FL, June 2010.

[Marques da Silva et al. 2012a] Marques da Silva, M., Correia, A., Dinis, R., Souto, N., Silva, J., *Transmission Techniques for 4G Systems*, 1st edition, CRC Press, Boca Raton, FL, November 2012a.

[Marques da Silva et al. 2004] Marques da Silva, M., Correia, A., Silva, J. C., Souto, N., Joint MIMO and parallel interference cancellation for the HSDPA, *Proceedings of the IEEE International Symposium on Spread Spectrum Techniques and Applications*, Sydney, Australia, September 2004.

[Marques da Silva et al. 2012b] Marques da Silva, M., Correia, A., Souto, N., Seguro, J., Gomes, P., Dinis, R., On the Multi-resolution techniques for LTE-advanced, *Wireless Personal Communications*, 66(4), 833–853, DOI: 10.1007/s11277-011-0366-8, September 2012b.

[Marques da Silva and Dinis 2011] Marques da Silva, M., Dinis, R., Iterative frequency-domain detection and channel estimation for space-time block codes, *European Transactions on Telecommunications*, 22(7), 339–351, DOI: 10.1002/ett.1484, November, 2011.

[Marques da Silva et al. 2005b] Marques da Silva, M., Dinis, R., Correia, A., A V-BLAST detector approach for W-CDMA signals with frequency-selective fading, *Proceedings of the 16th IEEE Personal Indoor and Mobile Radio Communications*, Berlin, Germany, September 11–14, 2005b.

[Marques da Silva et al. 2013] Marques da Silva, M., Dinis, R., Montezuma, P., Iterative frequency-domain packet combining techniques for UWB systems with strong interference levels, *Wireless Personal Communications*, 70(1), 501–517, DOI: 10.1007/s11277-012-0704-5, May 2013.

[Marques da Silva and Monteiro 2014] Marques da Silva, M., Monteiro, F. A., *MIMO Processing for 4G and Beyond: Fundamentals and Evolution*, CRC Press, Boca Raton, FL, June 2014.

[Montezuma et al. 2014] Montezuma, P., Marques da Silva, M., Dinis, R., Frequency-domain packet combining techniques for UWB, *MIMO Processing for 4G and Beyond: Fundamentals and Evolution*, (eds.) M. Marques da Silva, F. A. Monteiro, CRC Press, Boca Raton, FL, pp. 377–404, June 2014.

[Nam and Lee 2002] Nam, S. H., Lee, K. B., Transmit power allocation for an extended V-BLAST system, *PIMRC'2002*, Lisbon, Portugal, September 2002.

[Ojanpera and Prasad 1998] Ojanperä, T., Prasad, R., *Wideband CDMA for Third Generation Mobile Communications*, Artech House, Norwood, MA, 1998.

[Papadogiannis et al. 2009] Papadogiannis, A., Hardouin, E., Gesbert, D., Decentralising multicell cooperative processing: A novel robust framework, *EURASIP Journal on Wireless Communications and Networking*, 2009, 1–10, August 2009.

[Patel and Holtzman 1994] Patel, P., Holtzman, J., Analysis of simple successive interference cancellation scheme in DS/CDMA, *IEEE Journal on Selected Areas in Communications*, 12(5), 796–807, June 1994.

[Price and Green 1958] Price, R., Green, P. E., A communication technique for multipath channels, *Proceedings of the IRE*, 46(3), 555–570, 1958.

[Proakis 2001] Proakis, J., *Digital Communications*, 4th edition, McGraw-Hill, New York, 2001.

[Reis et al. 2014] Reis, C., Correia, A., Souto, N., Marques da Silva, M., Coordinated multi-point MIMO processing for 4G, *Progress in Electromagnetics Research Symposium*, Guangzhou, China, August 25–28, 2014.

[Rooyen et al. 2000] Rooyen, P. V., Lötter, M., Wyk, D., *Space-Time Processing for CDMA Mobile Communications*, Kluwer Academic, Boston, MA, 2000.

[Sari et al. 1994] Sari, H. et al., An analysis of orthogonal frequency-division multiplexing for mobile radio applications, *IEEE VTC*, Stockholm, Sweden, June 1994.

[Schacht et al. 2002] Schacht, M., Dekorsy, A., Jung, P., Downlink beamforming concepts in UTRA FDD, *Kleinheubacher Tagung 2002*, Kleinheubach, Germany, September 2002.

[Schoene et al. 2008] Schoene, R., Halfmann, R., Walke, B. H., An FDD multihop cellular network for 3GPP-LTE, *VTC 2008*, Singapore, pp. 1990–1994, May 2008.

[Silva et al. 2003] Silva, J. C., Souto, N., Rodrigues, A., Cercas, F., Correia, A., Conversion of reference tapped delay line channel models to discrete time channel models, *Proceedings of the IEEE Vehicular Technology Conference*, Orlando, FL, October 2003.

[Souto et al. 2007] Souto, N., Dinis, R., Silva, J. C., Iterative decoding and channel estimation of MIMO-OFDM transmissions with hierarchical constellations and implicit pilots, *International Conference on Signal Processing and Communications*, Dubai, United Arab Emirates, November 24–27, 2007.

[Stencel et al. 2010] Stencel, V., Muller, A., Frank, P., LTE advanced: A further evolutionary step for next generation mobile networks, *20th International Conference on Radioelektronika*, Brno, Czech Republic, pp. 19–21, April 2010.

[Sydir and Taori 2009] Sydir, J., Taori, R., An evolved cellular system architecture incorporating relay stations, *IEEE Communications Magazine*, 47(6), 150–155, June 2009.

[Tarokh et al. 1999] Tarokh, V. et al., Space-time block codes from orthogonal designs, *IEEE Transactions on Information Theory*, 45(5), 1456–1467, July 1999.

[Telatar 1995] Telatar, I. E., Capacity of multiantenna Gaussian channels, *AT&T Bell Labs Technical Memorandum*, June 1995.

[Telatar 1999] Telatar, I. E., Capacity of multiantenna Gaussian channels, *European Transactions on Telecommunications*, 10(6), 585–595, 1999.

[Tuchler et al. 2002] Tuchler, M., Koetter, R., Singer, A., Turbo equalization: Principles and new results, *IEEE Transactions on Communications*, 50, 754–767, May 2002.

[Varanasi and Aazhang 1990] Varanasi, M. K., Aazhang, B., Multistage detection in asynchronous CDMA communications, *IEEE Transactions on Communications*, 38(4), 509–519, April 1990.

[Vetro et al. 2003] Vetro, A., Christopoulos, C., Sun, H., Video transcoding architectures and techniques: An overview, *IEEE Signal Processing Magazine*, 20(2), 18–29, March 2003.

[Viterbi 1990] Viterbi, J., Very low rate convolutional codes for maximum theoretical performance of spread-spectrum multiple-access channels, *IEEE Journal on Selected Areas in Communications*, 8(4), 641–649, April 1990.

[Vitthaladevuni and Alouini 2001] Vitthaladevuni, P. K., Alouini, M.-S., BER computation of 4/M-QAM hierarchical constellations, *IEEE Transactions on Broadcasting*, 47(3), 228–239, September 2001.

[Vitthaladevuni and Alouini 2004] Vitthaladevuni, P. K., Alouini, M.-S., A closed-form equation for the exact BER of generalized PAM and QAM constellations, *IEEE Transactions on Communications*, 52(5), 698–700, May 2004.

CHAPTER 8

[ITU-T H.310] ITU-T Recommendation H.310, *Broadband Audiovisual Communication Systems and Terminals*, 1995.

[ITU-T H.320] ITU-T Recommendation H.320, *Narrow-Band Visual Telephone Systems and Terminal Equipment*, 2004.

[ITU-T H.321] ITU-T Recommendation H.321, *Adaptation of H.320 Visual Telephone Terminals to B-ISDN Environments*, 1998.

[ITU-T H.323] ITU-T Recommendation H.323, *Packet-Based Multimedia Communications Systems*, Version 7, 2009.

[ITU-T H.324] ITU-T Recommendation H.324, *Terminal for Low Bit-Rate Multimedia Communication*, 2009.

[Liu and Mouchtaris 2000] Liu, H., Mouchtaris, P., Voice over IP signaling: H.323 and beyond, *IEEE Communications Magazine*, 38(10), 142–148, October 2000.

[Minoli and Minoli 2002] Minoli, D., Minoli, E., *Delivering Voice Over IP Networks*, 2nd edition, John Wiley & Sons, New York, 2002.

[RFC 114] Bhushan, A., *RFC 114—A File Transfer Protocol*, Network Working Group, April 1971.

[RFC 326] Westheimer, E., *RFC 326—Network Host Status*, Network Working Group, April 1972.

[RFC 959] Postel, J., Reynolds, J., *RFC 959—File Transfer Protocol (FTP)*, Network Working Group, October 1985.

[RFC 1034] Mockapetris, P., *RFC 1034—Domain Names—Concepts and Facilities*, Network Working Group, November 1987a.

[RFC 1035] Mockapetris, P., *RFC 1035—Domain Names—Implementation and Specification*, Network Working Group, November 1987b.

[RFC 1225] Rose, M., *RFC 1225—Posts Office Protocol—Version 3*, Network Working Group, May 1991.

[RFC 1448] Case, J. et al., *RFC 1448—Protocol Operations for Version 2 of the Simple Network Management Protocol (SNMPv2)*, Network Working Group, April 1993.

[RFC 2045] Freed, N., Borenstein, N., *RFC 2045—Multipurpose Internet Mail Extensions (MIME), Part One: Format of Internet Message Bodies*, Network Working Group, November 1996.

[RFC 2543] Handley, M. et al., *RFC 2543—SIP: Session Initiation Protocol*, Network Working Group, March 1999.

[RFC 2616] Fielding, R. et al., *RFC 2616—Hypertex Transfer Protocol—HTTP/1.1*, Network Working Group, June 1999.

[RFC 2821] Klensin, J., *RFC 2821—Simple Mail Transfer Protocol*, Network Working Group, April 2001.

[RFC 3261] Rosenberg, J. et al., *RFC 3261—SIP: Session Initiation Protocol*, Network Working Group, June 2002.

[Schulzrinne and Rosenberg 2000] Schulzrinne, H., Rosenberg, J., The session initiation protocol: Internet-centric signaling, *IEEE Communications Magazine*, 38(10), 134–141, October 2000.

[Thom 1996] Thom, G.A., H.323: The multimedia communications standard for local area networks, *IEEE Communications Magazine*, 34(12), 52–56, December 1996.

CHAPTER 9

[RFC 1633] Braden, R., Clark, D., Shenker, S., *RFC 1633—Integrated Services in the Internet Architecture: An Overview*, Network Working Group, June 1994.

[RFC 2205] Braden, R., Berson, S., Herzog, S., *RFC 2205—Resource ReSerVation Protocol (RSVP)*, Network Working Group, September 1997.

[RFC 2211] Wroclawski, J., *RFC 2211—Specification of the Controlled-Load Network*, Network Working Group, September 1997.

[RFC 2212] Shenker, S., Partridge, C., Guerin, R., *RFC 2212—Specification of Guaranteed Quality of Service*, Network Working Group, September 1997.

[RFC 2213] Baker, F., Krawczyk, J., Sastry, A., *RFC 2213—Integrated Services Management Information Base Using SMIv2*, Network Working Group, September 1997.

[RFC 2215] Krawczyk, J., *RFC 2215—General Characterization Parameters for Integrated Service Network Elements*, Network Working Group, September 1997.

[RFC 2475] Blake, S. et al., *RFC 2475—An Architecture for Differentiated Services*, Network Working Group, December 1998.

CHAPTER 10

[RFC 790] Postel, J., *RFC 790—Assigned Numbers*, Network Working Group, September 1981.

[RFC 1519] Fuller, V., Li, T., Yu, J., Varadhan, K., *RFC 1519—Classless Inter-Domain Routing (CIDR): an Address Assignment and Aggregation Strategy*, Network Working Group, September 1993.

[RFC 1752] Bradner, S., Mankin, A., *RFC 1752—The Recommendation for the IP Next Generation Protocol*, Network Working Group, January 1995.

[RFC 2131] Droms, R., *RFC 2131—Dynamic Host Configuration Protocol*, Network Working Group, March 1997.

[RFC 2373] Hinden, R., Deering, S., *RFC 2373—IP Version 6 Addressing Architecture*, Network Working Group, July 1998.

[RFC 2374] Hinden, R., O'Dell, O., Deering, S., *RFC 2374—An IPv6 Aggregatable Global Unicast Address Format*, Network Working Group, July 1998.

[RFC 2460] Deering, S., Hinden, R., *RFC 2460—Internet Protocol, Version 6 (IPv6) Specification*, Network Working Group, December 1998.

[RFC 2474] Nichols, K., Blake, S., Baker, F., Black, D., *RFC 2474—Definition of the Differentiated Services Field (DS Field) in the IPv4 and IPv6 Headers*, Network Working Group, December 1998.

[RFC 3168] Ramakrishnan, K., Floyd, S., Black, D., *RFC 3168—The addition of Explicit Congestion Notification (ECN) to IP*, Network Working Group, September 2001.

[RFC 3927] Cheshire, S., Aboba, B., Guttman, E., *RFC 3927—Dynamic Configuration of IPv4 Link-Local Addresses*, Network Working Group, May 2005.

[RFC 4260] McCann, P., *RFC 4260—Mobile IPv6 Fast Handovers for 802.11 Networks*, Network Working Group, November 2005.

CHAPTER 11

[Blanchet 2006] Blanchet, M., *Migrating to IPv6: A Practical Guide to Implementing IPv6 in Mobile and Fixed Networks*, John Wiley & Sons, Chichester, West Sussex, 2006.

[Davies 2008] Davies, J., *Understanding IPv6*, 2nd edition, Microsoft Press, Washington, DC, 2008.

[Dijkstra 1959] Dijkstra, E. W., A note on two problems in connection with graphs, *Numerische Mathematik*, 1, 269–271, 1959.

[RFC 792] Postel, J., *RFC 792—Internet Control Message Protocol*, Network Working Group, September 1981.

[RFC 1058] Hedrick, C., *RFC 1058—Routing Information Protocol*, Internet Engineering Task Force, June 1988.

[RFC 1247] Moy, J., *RFC 1247—OSPF Version 2*, Internet Engineering Task Force, July 1991.

[RFC 1723] Malkin, G., *RFC 1723—RIP Version 2 Carrying Additional Information*, Internet Engineering Task Force, November 1994.

[RFC 2328] Moy, J., *RFC 2328—OSPF Version 2*, Internet Engineering Task Force, April 1998.

[RFC 2740] Coltun, R., Ferguson, D., Moy, J., *RFC 2740—OSPF for IPv6*, Network Working Group, December 1999.

CHAPTER 12

[ANSI X3.139] Information Systems—Fiber Distributed Data Interface (FDDI)—Token Ring Media Access Control (MAC) Replaces ANSI X3.139, 1987 R(1997).

[ANSI X3.148] Information Systems—Fiber Distributed Data Interface (FDDI)—Token Ring Physical Layer Protocol (PHY) Replaces ANSI X3.148, 1988 R(1999).

[Benedetto et al. 1997] Benedetto, S., Biglieri, E., Castellani, V., *Digital Transmission Theory*, Prentice Hall, Upper Saddle River, NJ, 1997.

[IEEE 802.1Q] WG802.1—Higher Layer LAN Protocols Working Group, 802.1Q-2011—IEEE Standard for Local and Metropolitan Area Networks—Media Access Control (MAC) Bridges and Virtual Bridged Local Area Networks, IEEE, 2011.

[IEEE 802.2] IEEE Standard for Information Technology: Telecommunications and Information Exchange between Systems; Local and Metropolitan Area Networks; Specific Requirements for Logical Link Control, Revision 2003, May 7, 1998.

[IEEE 802.3] IEEE Standard for Information Technology—Specific Requirements: Carrier Sense Multiple Access with Collision Detection (CSMA/CD) Access Method and Physical Layer Specifications, Revision 2005, December 26, 2008.

[IEEE 802.5] IEEE Standard for Information Technology—Specific Requirements: Token Ring Access Method and Physical Layer Specifications, 1998.

[ISO 9314-1] International Standard, Information Processing Systems—*Fibre Distributed Data Interface (FDDI)—Part 1: Token Ring Physical Layer Protocol (PHY)*, 1989.

[ISO 9314-2] International Standard, Information Processing Systems—*Fibre Distributed Data Interface (FDDI)—Part 2: Token Ring Media Access Control (MAC)*, 1989.

[ISO/IEC 8802-2:1998] ISO/IEC Standard for Information Technology: Telecommunications and Information Exchange between Systems; Local and Metropolitan Area Networks; *Specific Requirements for Logical Link Control*, 1998.

[ISO/IEC 8802-5:1998] ISO/IEC Standard for Information technology—*Specific Requirements: Token Ring Access Method and Physical Layer Specifications*, 1998.

[Watkinson 2001] Watkinson, J., *The MPEG Handbook: MPEG-I, MPEG-II, MPEG-IV*, Focal Press, London, 2001.

CHAPTER 13

[ANSI/TIA/EIA-568] ANSI/EIA/TIA-568, *Commercial Building Telecommunications Standard*, 1991.

[ANSI/TIA/EIA-568-A] ANSI/TIE/EIA-568-A, *Commercial Building Telecommunications Standard*, 1995.

[ANSI/TIA/EIA-568-B] ANSI/TIE/EIA-568-B, *Commercial Building Telecommunications Standard*, 2001.

[ANSI/TIA/EIA-568-C] ANSI/TIE/EIA-568-C, *Commercial Building Telecommunications Standard*, 2009.

[EN 50173] CENELEC, *Information Technology—Generic Cabling Systems*, 1995.

[EN 50173-1] CENELEC, *Information Technology—Generic Cabling Systems*, 1997.

[EN 50173-2] CENELEC, *Information Technology—Generic Cabling Systems*, 2007.

[IEC 60297-3-100] IEC, *Mechanical Structures for Electronic Equipment—Dimensions of Mechanical Structures of the 482,6 mm (19 in) Series*, 2009.

[IEEE 802.3af] IEEE Standard for Information Technology—Telecommunications and Information Exchange Between Systems—Local and Metropolitan Area Networks—Specific Requirements—Part 3: Carrier Sense Multiple Access with Collision Detection (CSMA/CD) Access Method and Physical Layer Specifications—Data Terminal Equipment (DTE) Power Via Media Dependent Interface (MDI), amendment to IEEE 802.3-2002, 2003.

[ISO/IEC 11801] ISO/IEC 11801, *Information Standard—Information Technology—Generic Cabling for Customer Premises*, 1st edition, 1995.

[ISO/IEC 11801amend2] ISO/IEC, *International Standard—Information Technology—Generic Cabling for Customer Premises*, Amendment 2 to the 2nd edition, 2010.

[ISO/IEC 11801ed2] ISO/IEC 11801, *Information Standard—Information Technology—Generic Cabling for Customer Premises*, 2nd edition, 2002.

[Monteiro 2011] Monteiro, E., Boavida, F., *Engenharia de Redes Informáticas*, 10th edition, Editora FCA, Lisbon, Portugal, 2011.

[Stallings 2010] Stallings, W., *Data and Computer Communications*, 8th edition, Prentice Hall, Upper Saddle River, NJ, 2010.

CHAPTER 14

[ANSI T1.105] ANSI Standard T1.105.07—Synchronous Optical Network (SONET)—Sub-STS-1 Interface Rates and Formats Specifications 1996, Revised 2005.

[Costa 1997] Costa, C., *Apontamentos de Redes de Telecomunicações—ATM*, IST—Departamento de Engenharia Electrotécnica e de Computadores, Lisbon, Portugal, 1997.

[Handel and Hubber 1991] Handel, R., Hubber, M., *Integrated Broadband Networks. An Introduction to ATM-Based Networks*, Addison-Wesley, Munich, Germany, June 1991.

[ISO 3309] ISO 3309, ISO International Standard (IS) 3309, *Data Communications—High-Level Data Link Control Procedures—Frame Structure*, 1979.

[ISO 4335] ISO International Standard (IS) 4335, *Data Communications—High-Level Data Link Control Procedures. Specification for Elements of Procedures*, 1979.

[ITU-T G.707] ITU-T Recommendation G.707, *Network Node Interface for the Synchronous Digital Hierarchy (SDH)*, 2007.

[ITU-T G.732] ITU-T Recommendation G.732, *Characteristics of Primary PCM Multiplex Equipment Operating at 2048 kbit/s*, 1988.

[ITU-T G.733] ITU-T Recommendation G.733, *Characteristics of Primary PCM Multiplex Equipment Operating at 1544 kbit/s*, 1988.

[ITU-T I.113] ITU-T Recommendation T.113, *Vocabulary of Terms for Broadband Aspects of ISDN*, 1997.

[ITU-T I.121] ITU-T Recommendation T.121, *Integrated Services Digital Network (ISDN) General Structure and Service Capabilities*, 1991.

[Prycker 1991] Prycker, M., *Asynchronous Transfer Mode*, Ellis Horwood, New York, 1991.

[RFC 1663] Rand, D., *RFC 1663—PPP Reliable Transmission*, Network Working Group, July 1994.

[RFC 2702] Awduche, D. et al., *RFC 2702—Requirements for Traffic Engineering Over MPLS*, Network Working Group, September 1999.

[RFC 3031] Rosen, E., Viswanathan, A., Callon, R., *RFC 3031—Multi-Protocol Label Switching Architecture*, Network Working Group, January 2001.

[RFC 3270] Le Faucheur, F. et al., *RFC 3270—MPLS Support of Differentiated Services*, Network Working Group, May 2002.

[Sexton and Reid 1992] Sexton, M., Reid, A., *Transmission Networking: SONET and the Synchronous Digital Hierarchy*, Artech House, Norwood, MA, 1992.

CHAPTER 15

[Andrews et al. 2007] Andrews, J. G., Gosh, A., Muhamed, R., *Fundamentals of WiMAX: Understanding Broadband Wireless Networking*, Prentice-Hall, Upper Saddle River, NJ, 2007.

[Astely et al. 2009] Astely, D., Dahlman, E., Furuskar, A., Jading, Y., Lindstrom, M., Parkvall, S., LTE: The evolution of mobile broadband, *IEEE Communications Magazine*, 47(4), 44–51, April 2009.

[Bhat et al. 2012] Bhat, P., Nagata, S., Campoy, L., Berberana, I., Derham, T., Liu, G., Shen, X., Zong, P., Yang, J., LTE-advanced: An operator perspective, *IEEE Communications Magazine*, 50(2), 104–114, February 2012.

[Eklund et al. 2002] Eklund, C. et al., IEEE standard 802.16: A technical overview of the wireless MAN air interface for broadband wireless access, *IEEE Communications Magazine*, 40(6), 98–107, June 2002.

[Ferro and Potorti 2005] Ferro, E., Potorti, F. Bluetooth and Wi-Fi wireless protocols: A survey and a comparison, *IEEE Wireless Communications*, 12(1), 12–26, March 2005.

[Geier 2002] Geier, J., *Wireless LANs*, SAMs Publishing, Indianapolis, IN, 2002.

[IEEE 802.11] IEEE 802.11, Wireless LAN Medium Access Control (MAC) and Physical Layer (PHY) Specifications, ANSI/IEEE Std 802.11; 1999 (E) Part 11, ISO/IEC 8802-11, 1999.

[IEEE 802.16] IEEE 802.16, Relay Task Group, 2008; http://www.ieee802.org/16/relay.

[IEEE 802.16-2004] IEEE Standard 802.16-2004, Air Interface for Fixed Broadband Wireless Access Systems, 2004.

[IEEE 802.16e] IEEE Standard 802.16e-2005, Amendment to Air Interface for Fixed and Mobile Broadband Wireless Access Systems—Physical and Medium Access Control Layers for Combined Fixed and Mobile Operations in Licensed Bands, 2005.

[IEEE 802.16m] IEEE Standard 802.16m-07/002r8, IEEE 802.16m System Requirements, January 2009. http://ieee802.org/16/tgm/index.html, accessed January 5, 2012.

[ITU 2010] ITU, *ITU Paves Way for Next-Generation 4G Mobile Technologies*, ITU press release, October 21, 2010.

[ITU-R 2008] ITU-R Recommendation M.2133, *Requirements, evaluation criteria and submission templates for the development of IMT-Advanced*, 2008.

[Marques da Silva et al. 2010] Marques da Silva, M., Correia, A., Dinis, R., Souto, N., Silva, J., *Transmission Techniques for Emergent Multicast and Broadcast Systems*, 1st edition, CRC Press, Boca Raton, FL, June 2010.

[Marques da Silva et al. 2012] Marques da Silva, M., Correia, A., Dinis, R., Souto, N., Silva, J., *Transmission Techniques for 4G Systems*, 1st edition, CRC Press, Boca Raton, FL, November 2012.

[Ohrtman 2008] Ohrtman, F., *WiMAX Handbook*, McGraw-Hill, New York, 2008.

[Ohrtman and Roeder 2003] Ohrtman, F., Roeder, K. *Wi-Fi Handbook: Building 802.11b Networks*, McGraw-Hill, New York, 2003.

[Oyman et al. 2007] Oyman, O., Laneman, J.N., Sandhu, S., Multihop relaying for broadband wireless mesh networks: From theory to practice, *IEEE Communications Magazine*, 45(11), 116–122, November 2007.

[Peters and Heath 2009] Peters, S. W., Heath, R. W., The future of WiMAX: Multihop relaying with IEEE 802.16j, *IEEE Communications Magazine*, 47(1), 104–111, January 2009.

[Yarali and Rahman 2008] Yarali, A., Rahman, S., *WiMAX Broadband Wireless Access Technology: Services, Architecture and Deployment Models*, IEEE CCECE2008, Niagara Falls, ON, Canada, pp. 77–82, May 4–7, 2008.

[Zhang and de la Roche 2010] Zhang, J., de la Roche, G., *Femtocells—Technologies and Deployment*, 1st edition, Wiley, Chichester, 2010.

CHAPTER 16

[Arquilla 1997] Arquilla, J., *Athena's Camp: Preparing for Conflict in the Information Age*, RAND publisher, Santa Monica, CA, 1997.

[Cross et al. 2003] Cross, M. et al., *Security +*, Syngress Publishing, Rockland, MA, 2003.

[IEEE 802.11] IEEE 802.11, Wireless LAN Medium Access Control (MAC) and Physical Layer (PHY) Specifications, ANSI/IEEE Std 802.11; 1999 (E) Part 11, ISO/IEC 8802-11, 1999.

[ISO 27001] *ISO—Information Technology—Security Techniques—Information Security Management Systems—Requirements*, 2013.

[ISO 27002] *ISO—Information Technology—Security Techniques—Code or Practice for Information Security Controls*, 2013.

[ISO/IEC 15408] ISO/IEC 15408, *Information Technology—Security Techniques—Evaluation Criteria for IT Security*, 2009.

[Klein 2003] Klein, J., CyberTerrorism, movie from www. HistoryChannel.com, 2003, accessed in May 23, 2010.

[Maiwald 2003] Maiwald, E., *Network Security—A Beginner's Guide*, 2nd edition, McGraw-Hill, New York, 2003.

[McClure and Scambray 2005] McClure, S., Scambray, J., *Hacking Exposed: Network Security, Secrets & Solutions*, McGraw-Hill, New York, 2005.

[RFC 2196] Fraser, B., *RFC 2196—Site Security Handbook*, Network Working Group, September, 1997.

[RFC 2409] Harkins, D., Karrel, D., *RFC 2409—The Internet Key Exchange (IKE)*, Network Working Group, November 1998.

[RFC 4301] Kent, S., Seo, K., *RFC 4301—Security Architecture for the Internet Protocol*, Network Working Group, December 2005.

[RFC 4309] Wousley, R., *RFC 4309—Using Advanced Encryption Standard (AES) CCM Mode with IPsec Encapsulating Security Payload (ESP)*, Network Working Group, December 2005.

[RFC 5246] Dierks, T., Rescorla, E., *RFC 5246—The Transport Layer Security (TLS) Protocol Version 1.2*, Network Working Group, August 2008.

[RFC 6066] Eastlake, D., *RFC 6066—The Transport Layer Security (TLS) Extensions: Extension Definitions*, Network Working Group, January 2011.

[RFC 6520] Seggelmann, R., Tuexen, M., Williams, M., *RFC 6520—Transport Layer Security (TLS) and Datagram Transport Layer Security (DTLS) Heartbeat Extension*, Network Working Group, February 2012.

[Stallings and Brown 2008] Stallings, W., Brown, L., *Computer Security—Principles and Practice*, Pearson Education International, New York, 2008.

Index

Note: Locator followed by 'f' and 't' denotes figure and table in the text

1G cellular networks. *See* First generation (1G) cellular networks
2G cellular networks. *See* Second generation (2G) cellular networks
6to4 tunnel, 337f, 338
8P8C modulator connectors (*RJ45*)
    T568A termination of, 76t
    T568B termination of, 77t
64-QAM hierarchical constellation, 213, 214f
64-QAM modulation, 152, 153f

A

Access control lists (ACLs), 551, 551f, 583
    creation of, 557f, 558f
    extended, 554
    to implement policy rules, 583
    to interfaces and inbound/outbound directions, 551f
    packet filtering of, 552f
    reflexive, 557–559, 584–585
    router with, 554f
Access layer, 412–413
Acknowledgment (ACK) message, 9, 240, 357
Adaptive differential PCM, 142
Adaptive modulation and coding (AMC), 485
Adaptive routing protocols
    and algorithms, 307–335
    classification of, 309
    distance vector protocols and their configuration, 309–325
        EIGRP, 316–325
        RIP, 310–316
    link-state protocols and their configuration, 325–335
Add and drop multiplexer (ADM), 437
Additive white Gaussian noise channel, 46
Address resolution protocol (ARP), 41
Adjacent cells, 482–483
Adjacent channel interference, 69–71, 69f, 70f
Advanced encryption standard (AES), 517–518
Alamouti scheme, 192
A-law, 139, 141f
Albeit galvanic coupling, 80
Alternate current (AC), 427
American National Standards Institute (ANSI), 416
Amplitude shift keying (ASK), 149, 149f
Analog signals, 2–3, 2f
Analog telephony, 137
Antenna gain, 100–101
Application layer protocol, 219
Application PDU (APDU), 24
ARP cache, 274
Asymmetric coding mechanism, 143
Asymmetric cryptography, 515, 520–521, 520f
Asymmetric DSL (ADSL), 447
Asynchronous balanced mode, 458
Asynchronous communication systems, 4

Asynchronous digital subscriber line (ADSL), 3, 486
Asynchronous response mode, 458
Asynchronous transfer mode (ATM), 437, 449–453, 449f, 453f
    B-ISDN reference model, 449f, 450–451, 451f
        layers of, 451, 451f
    cell format, 453, 453f
    network, 452–453, 452f
ATM adaptation layer (AAL), 451
ATM cell format, 453, 453f
ATM network, 452–453, 452f
Atmospheric noise, 50–52, 51t, 53f
Attenuation, 48–49, 48f
Attenuation to crosstalk ratio (ACR), 77, 91f
Authentication, authorization, and accounting (AAA) server, 538
Authentication header (AH), 533–534
Automatic private IP addressing (APIPA), 252
Automatic repeat request (ARQ), 18, 367–371
    Go-Back-N ARQ, 369–370, 370f
    selective reject, 370–371, 371f
    stop-and-wait, 368–369
Autonomous system (AS), 297
Avalanche photodiode (APD), 85

B

Backup DR (BDR), 333–334
Backup policy, 513
Backward explicit congestion notification bit, 468
Bandpass signals, 136, 136f, 156
Bandwidth requirement, 14, 15f
Baseband signal, 136, 136f
Base station (BS), 477
Base station cooperation, 203–207
    architectures, 205
    combined with downlink MIMO, 204f
    CoMP transmission, 204–205
    downlink, 203f
    macrodiversity, 205–207
Basic service set (BSS), 490
Beamforming, 200–201, 201f
Bellman–Ford algorithm, 309, 317
Binary phase shift keying (BPSK), 45, 150, 150f
    spread-spectrum signal, 168f
Binary transposition operation, 517
Bipolar alternate mark inversion (bipolar AMI), 146–147, 147f
B-ISDN reference model, 449f, 450–451, 451f
Bit error probability, 101
Bit error rate (BER), 1–2, 45
Bit rate, 237
Bit stuffing, 459
Block codes, 365
Blowfish algorithm, 518
Border gateway protocol (BGP), 297

Botnet attack, 505, 506*f*
Bridge, 347–348, 347*f*
Bridge identification (BID), 352
Bridge protocol data units (BPDUs), 352
Bristol, 22–23
Broadband-ISDN (B-ISDN), 449
Broadcast communication, 6
Broadcast storms, 351
Brute force attack, 509
Brute force detector, 199
Building backbone cabling, 414
Building backbone subsystem, 415
Building distributor (BD), 414
Bus topology, 10–11, 10*f*, 347

C

Cable modem, 3
Cable service interface specification, 448, 448*f*, 448*t*
Cable transmission mediums, bandwidths, 74*t*
Cache memory, 222
Campus backbone cabling, 414
Campus backbone subsystem, 415
Campus distributor (CD), 414
Canonical format identifier (CFI), 397
Carbon footprint reduction, 214
Carrier extension, 380*f*
Carrierless amplitude phase modulation (CAP), 447
Carrier sense multiple access-collision avoidance
    (CSMA-CA), 489
Carrier-to-noise ratio *(C/N)*, 100–101
Cell loss priority (CLP), 453
Cell switching (ATM), 432
Cellular architecture, 114
Cellular communications, 477
Cellular network concept, 477–483, 478*f*, 479*f*, 480*f*
    femtocell, 482
    macrocell, 481
    microcell, 480*f*, 481
    picocell, 481–482
    power control, 482–483
Centralized optical architecture, 415–416, 416*f*
Certification authority (CA), 524–526
Challenge handshaking authentication protocol (CHAP),
    506–507
Change cipher spec message, 532
Channel
    distortion, 57–58, 58*f*
    equalization, 58–60, 59*f*, 60*f*
    frequency response, 55, 58
    impairments, 45
    transfer function, 55
    transmission influence, 55–60, 56*f*
        delay and phase shift, 56–57, 56*f*, 57*f*
        distortion, 57–58, 58*f*
        equalization, 58–60, 59*f*, 60*f*
Channel impulse response (CIR), 169, 205, 206*f*
Circuit switching, 6–7, 7*f*, 431
Circuit switching transport networks, 432–448
    cable service interface specification, 448, 448*f*, 448*t*
    digital subscriber line, 443–448, 445*f*, 446*f*, 447*f*, 447*t*
    frequency division multiplexing hierarchy, 433, 433*t*
    plesiochronous digital hierarchy, 433–435, 434*f*, 435*t*

synchronous digital hierarchies, 436–443, 436*f*, 437*f*
    SDH/SONET frame format, 441–443, 442*f*, 443*f*,
        444*f*, 445*f*
    SDH/SONET network, 437–441, 438*f*, 439*f*, 440*f*,
        441*f*, 441*t*
Cisco discovery protocol (CDP), 339–340, 500
Cisco equipment, 79–80
Cisco internetwork operating system (IOS), 273–292
    client host of, 286
    configuration
        of Router1 with EIGRP, 319*f*
        of Router1 with OSPF, 331*f*
    configuration of routers and switches, 279–283
        configuration mode, 279–280
        interface configuration submode, 280–283
        interface description, 281*f*
        interface with IP address/clock rate, 281*f*
        line configuration submode, 280
        setting time and date, 279*f*
    configuring IP address, 277*f*
    connectivity parameters, 275*f*
    dynamic host configuration protocol, 283–286
    help command, 278*f*
    introduction to, 274–279
    memories, 273–274
    NAT and PAT, 286–292
    syntax, 276, 307, 313
Cisco private Internet exchange (PIX) appliances, 561
Cisco proprietary protocol, 500
Cisco routers, 575–581
    in network to configure with ACLs, 584*f*
Cisco switches, 575–581
Cisco switch security, 539–542
    password configuration, 539
    port security configuration, 540–542
    SSH configuration, 539–540
Classful routing protocol, 309
Classless interdomain routing (CIDR), 255
Classless routing protocol, 309
Class of service, 457
Client/server protocol, 221
Coaxial cables, 80–81
    attenuation coefficient of, 81*f*
    bandwidth, 81
    composed of isolated and concentric conductors, 80*f*
    concentric conductors, 80
    dimensions of normalized, 81*t*
    structure, 80
Co-channel interference, 68–69, 69*f*
Codec device, 138
Code division multiple access (CDMA), 63, 156,
    167–177, 478
    general model, 169–171
    in multiuser environment, 172*f*
    narrowband, 171–174
    wideband, 174–177
Code spread (CS), 166
Coding efficiency of symbol, 154
Cognitive radio, 214–215
Coherence bandwidth, 114–115
Collaborative age
    of network applications, 16–17, 17*f*
    transition towards, the, 17–19, 18*f*

Combined cryptography, 515, 527–535
  associated with digital signature, 529*f*
  security architecture for Internet protocol,
    532–535
  SSL and TLS, 528–532
Combined encryption protocols, 522
Comité Européen de Normalisation Electrotechnique
  (CENELEC), 416
Command line interface (CLI), 274–275
Committed information rate, 465
Communication chain including modem, 149*f*
Communications, 6, 6*f*
Communications and information systems
  (CIS), 495
Communications security, 496
Communication system, 1, 2*f*
  conventional client/server, 96*f*
  generic chain, 45, 46*f*
  with line-of-sight propagation, 98*f*
  point-to-multipoint satellite, 127*f*
  point-to-point satellite, 126*f*
Compression control protocol (CCP), 463
Computer usage policy, 513
Confirmed service, 9–10
Connectionless mode, 9–10
Connection modes, 8–10
  connectionless, 9–10
  connection-oriented service, 8–9
Connection-oriented service, 8–9
Consolidation point (CP), 415
Contention mechanism, 375
Context-based access control (CBAC), 549, 560–561
Control, 459–460
Control station, 125
Conventional cellular network, 209*f*
Conventional client/server communication system, 96*f*
Convergence, 15–16
Convergence sublayer (CS), 451
Convolutional codes, 364–365
Coordinated multipoint (CoMP), 202
  transmission, 204–205
Core layer, 412
Coverage hole, 487
Crossover cable, 79
Crosstalk
  ACR, 91, 91*f*
  common performance index of, 89
  ELFEXT, 91
  FEXT, 90–91, 90*f*
  NEXT, 90, 90*f*
  other performance indexes of, 92
Crosstalk interference, 89
Cryptography, 498
  asymmetric, 515, 520–521, 520*f*
  symmetric. *See* Symmetric cryptography
CSMA-CD contention mechanism, 378*f*
CSMA-CD mechanism, 381
Cyclic prefix insertion, 178
Cyclic redundancy check (CRC) codes, 26, 240, 345,
  361, 460
  decoding with errors, 363*f*
  decoding without errors, 363*f*
  encoding, 361*f*

D

Data communication equipment (DCE), 282, 460, 466
Data communication systems, 1, 2*f*
Datagram mode, 27–28
Data link escape (DLE), 355
Data link layer (DLL), 26–27, 26*f*, 345
  sublayers, 345
Data terminal equipment (DTE), 282, 464
Decision feedback equalizers (DFEs), 183
Decoders, types of, 199
De-encapsulation, 249
Default gateway, 303
Default route, 303
  configuration, 306–307, 306*f*
Delay shift, 56–57, 56*f*, 57*f*
Delimiting character string, 355, 355*f*
Delta modulation, 141
Demand assignment multiple access, 375
Demilitarized zone (DMZ), 501
  corporate architecture, 545*f*
    with DMZ and Honey Net, 546*f*
  network architecture with, 544–546
Demodulation, 135
Demodulator, 3, 3*f*
Demultiplexer (DEMUX), 155
Denial of service (DoS) attack, 503–505, 504*f*
Designated port, 354
Designated router (DR), 328
Destination address (DA), 387
Destination IP address, 534
Destination unreachable message, 335
Differential Manchester technique, 147–148, 148*f*
Differential source coding techniques, 141–142
Differential transmission, 143–144, 144*f*
  2B1Q, 148, 148*f*, 148*t*
  bipolar AMI, 146–147, 147*f*
  differential Manchester, 147–148
  Manchester, 147
  nonreturn to zero, 146
  nonreturn to zero inverted, 146, 146*f*
  pseudoternary, 147
  return to zero, 146
Differential voice coding technique, 141
Differentiated services (DiffServ), 245
Differentiated services code point (DSCP), 264
Diffie–Hellman protocol, 518–519
  man-in-the-middle attack associated with, 519*f*
  processing associated with, 519*f*
Diffraction effect, 108*f*, 112
Digital baseband modulation, 144
Digital certificates, 523–525, 525*f*, 526*f*, 527*f*
  request message, 530
  verify message, 532
Digital coding, 135, 144
Digital communication chain, 136*f*
Digital encoding technique, 145
Digital encryption standard (DES), 517
Digital European Cordless Telephone standard, 15
Digital modem, 3
Digital section sublayer, 452
Digital signal (DS), 2–3, 2*f*, 434
Digital signature, 521–523, 523*f*, 524*f*

Digital subscriber line, 443–448, 445f, 446f, 447f, 447t
Digital video broadcast (DVB), 143, 388
Dijkstra algorithm, 326, 327f
Direct current (DC), 428
Direct sequence spread spectrum (DSSS), 489
Direct sequence (DS) technique, 166, 167f
Direct wave propagation, 95–102
    bit error probability calculation, 101–102
    carrier-to-noise ratio calculation, 100–101
    FSPL, 96–97
    line-of-sight/reflected component, 96f
    link budget calculation, 97–100
Disaster recovery policy, 514
Discard eligibility bit, 468
Discontiguous subnetworks, 311, 311f
Discrete Fourier transform (DFT), 177
Discrete multitone (DMT), 446
Distance vector protocols, 310
Distortion channel, 57–58, 58f
Distortion effect, 115
Distributed DoS (DDoS), 504–505
Distribution layer, 412
Diversity combining algorithms, 185–188
    EGC, 187
    MRC, 186–187, 186f
    MSE-based combining, 187–188
    selection combining, 186
Domain, 233
Domain name system (DNS), 223, 232–234
    mapping protocol, 232
    operation, 232f, 233, 233f
Double network architecture, 543–544
Downlink, 128–130
Downlink base station cooperation, 202
DS-CDMA communication link, 169f
DUAL algorithm, 317, 322
Dual stacking, 337
Duplex communications, 5, 5f
Duplicate unicast frames, 351
Dynamic address translation, configuration of, 288–289
Dynamic host configuration protocol (DHCP), 283–286
    communication between client and, 284, 284f
    configuration, 285–286
    forwarding, 285f
    starvation, 500

E

Earth station, satellite's, 129f
Eavesdropping, 497–500, 498f
Echo request/reply message, 335
Edge LSR, 455
Eight groups of four hexadecimal numbers, 232
Electrical supply for hardware, 426f, 427–428
Electromagnetic beam, 121
Electromagnetic waves
    propagation using ionospheric layers, 119f
    refraction in ionospheric layers, 118f
Electronic Industries Alliance (EIA), 416
Electronic mail (e-mail), 223
    policy, 513
Electronic noise, 54–55
Elementary modulation schemes, 149–150

Emanations security (EMSEC), 496
Encapsulation security payload (ESP), 533–534
Encryption control protocol (ECP), 463
Ending delimiter (ED), 388
End of frame sequence (EFS), 388
End-to-end network, 6
End-to-end transmission, 249
Enhanced interior gateway routing protocol (EIGRP), 316–325
    PDM of, 316
    routing table after configuration of
        neighbor table of Router1, 321f
        Router1, 318f
        Router2, 319f
        topology table of Router1, 321f
    topology designations, 321t
Enhanced SNMP, 230
Enigma cryptography machine, 516
Equal gain combining (EGC) algorithm, 187
Equalization channel, 58–60, 59f, 60f
Equal level far end crosstalk (ELFEXT), 91
Equivalent isotropic radiated power (EIRP), 96
Error control techniques, 356–367
    employed by, 357
    encoder, 356f
    error correction codes, 363–367
        adaptive modulation and coding, 366
        block codes, 365
        convolutional codes, 364–365
        interleaving, 366, 367f
        puncturing, 366–367, 367f, 368f
    error detection codes, 357–363
        CRC, 361
        hamming codes, 358–360
        parity bits, 360–361, 360f, 361f
Error detection technique, 240
Ethernet cabling, 79
European Conference of Postal and Telecommunications (CEPT), 434
European standards, 424
European Telecommunications Standards Institute (ETSI), 487
Evolution from 3G systems into long-term evolution, 484t, 485–486
Evolution of cellular systems and new paradigm of 4G, 483–488, 484t
    evolution from 3G systems into long-term evolution, 484t, 485–486
    IEEE 802.16 protocol (wiMaX), 486–488
    LTE-A and IMT-advanced, 488
Evolved-MBMS (E-MBMS), 211
    transmission gaps in, 212f
Evolved MBMS (eMBMS), 486
Evolved Node-B (eNode-B), 203, 209
Evolved packet core (EPC), 209
Evolved UTRAN (E-UTRAN), 485
Extended address (EA) bit, 468
Extended information rate, 465
Extended service set (ESS), 490
Extensible authentication protocol (EAP), 537
Exterior gateway protocols (EGPs), 297, 298f
Extranet, 542, 544f
Extraterrestrial noise, 52–53, 53f

F

Fading, 110–115
    multipath fading, 113–115
    shadowing fading, 112
Far end crosstalk (FEXT), 90–91, 90*f*
Fast fading, 111
Fast Fourier transform (FFT), 177
Feasibility distance (FD), 320
Feasible successor (FS), 322
Femtocell, 482
Fiber distribution data interface (FDDI) protocol,
    387–388, 387*f*
File transfer, 224
File transfer access and management (FTAM), 219
File transfer protocol (FTP), 224, 224*f*
Firewalls, 549–561
    firewall Cisco ASA, 561
    packet filtering configuration, 553–561
    rules to configure in, 550*t*
    way traffic filtered by, 551
First generation (1G) cellular networks, 483
Flag, 355, 356*f*, 459
Floating route configuration, 305–306
Floor distributor (FD), 414–415
Flow control techniques, 371–372
Foiled twisted pairs (FTPs), 75
Forward equivalent class (FEC), 456
Forward error correction (FEC) codes, 364
Forward explicit congestion notification bit, 468
Forwarding information base, 455
Four groups of decimal numbers, 232
Fourier transform, 569
    common, 570*t*
    theorems of, 571*t*
    time domain, 569, 570*t*
Fragmentation and reassembling, 335–337
Frame, 24, 26, 160
    decomposition, 26, 26*f*
Frame bursting, 381*f*
Frame check sum (FCS), 460, 463, 491
Frame format, 355*f*
Frame relay (FR), 449, 465–471, 466*f*, 467*f*, 469*f*
    address fields, 468
    configuration using Cisco IOS, 468–471,
        469*f*, 470*f*
    map, 467
Frame status (FS), 388
Framing, 433
Free space path loss (FSPL), 48, 96–97
Frequency division duplexing (FDD), 480
Frequency division multiple access (FDMA),
    67, 167, 483
Frequency division multiplexing (FDM), 156–160, 156*f*,
    433, 433*t*
    receiver with signals, 159*f*
    transmitter with signals, 157*f*
Frequency-domain equalization (FDE) techniques, 165
Frequency hopping spread spectrum (FHSS), 489
Frequency hopping (FH) technique, 166
Frequency shift keying (FSK), 149–150, 150*f*
Fresnel coefficients, 104
Full-cosine rolloff characteristic, 66

Full-duplex communications, 5, 5*f*
Full-meshed topology, 467*f*, 468
Fundamentals of communications, 1–14
    analog and digital signals, 2–3, 2*f*
    classification of media and traffic, 13–14
    communications and networks, 6, 6*f*
    connection modes, 8–10
        connectionless, 9–10
        connection-oriented service, 8–9
    modulator and demodulator, 3, 3*f*
    network coverage areas, 10
    network topologies, 10–13
        bus, 10–11, 10*f*
        mesh, 11, 13*f*
        ring, 11, 11*f*
        star, 11, 12*f*
        tree, 11, 12*f*
    simplex and duplex communications, 5, 5*f*
    switching modes, 6–8
        circuit switching, 6–7, 7*f*
        packet switching, 7–8, 7*f*, 8*f*
    synchronous and asynchronous communication
        systems, 4
    transmission mediums, 4

G

Galactic noise, 52
Gateway, 229
Generic control field, 453
Geostationary earth orbit (GEO), 121, 123–124
    footprints, 125*f*
Gigabits per second (Gbps), 4
Gilder law, 15
Global system for mobile communications (GSM),
    142, 483
Global unicast address, 268
Go-Back-N ARQ, 369–370, 370*f*
Graded index multimode fibers, 83
Gray mapping, 151
Green radio communications, 213–216
Groundwave propagation, 115–116, 117*f*
Guard band, 156

H

Half-duplex communications, 5, 5*f*
Half-duplex transmission in LAN, 349*f*
Hamming codes, 358–360
Hardware and accessories, 424–428
    electrical supply, 426*f*, 427–428
    labeling distributors and cables, 428
    patch panel, 425–427, 426*f*, 427*f*
    rack, 424–425, 425*t*, 426*f*
Hash function, 502, 505, 522
HDLC protocol, 458–461, 459*f*, 460*t*, 461*f*
    configuration using Cisco IOS, 460–461
    forms of, 458
    frame format, 458–460, 459*f*
    frame types, 458
    modes of, 458
    station types, 458
Header error control (HEC), 453

Hierarchical network design, 412–413, 412*f*
High bit rate DSL (HDSL), 447
High-definition television (HDTV), 484
High-level data link control (HDLC), 431
Highly elliptical orbit (HEO), 121, 126
High speed packet access (HSPA), 483–484
HiperMAN, 487
Horizontal cabling, 415
Horizontal polarization, 103*f*
    Fresnel coefficient for, 104
Horizontal subsystem, 415
Host-based IDS (HIDS), 546
Hub, 346–347, 347*f*
Hub-and-spoke topology, 467*f*, 468
Hybrid techniques, 166
Hypertext transfer protocol (HTTP), 221–222, 222*f*

I

I3A functions, accountability, 508, 508*f*
IEEE (Institute of Electrical and Electronics
        Engineers), 486
IEEE 802.1Q protocol, 396
IEEE 802.2 LLC sublayer, 387
IEEE 802.2 protocol, 373–374, 373*f*
    PDU format, 374*f*
    services to Internet layer, 374
IEEE 802.3 protocol, 376–383
    advantages/disadvantages of IEEE 802.5 relating
        to, 384
    backoff time used in, 379
    frame format, 377*f*
        with IEEE 802.1Q encapsulation, 397*f*
    identification of, 382*f*
    MCDD, 379–381
    minimum frame size of, 380
    physical characteristics used in, 382*t*, 383*t*
    physical layer employed in, 381–383
    standard, versions and technologies, 376*f*
IEEE 802.5 protocol, 383–387
    advantages/disadvantages relating to IEEE 802.3, 384
    frame format, 386*f*
    operation of MAC sublayer, 385–386
    physical/logical topologies of, 385*f*
    token format, 386*f*
IEEE 802.11 MAC data fields, 490–491, 490*f*
IEEE 802.11 protocol (Wi-Fi), 488–491, 490*f*
IEEE 802.16e, 487
IEEE 802.16 protocol (wiMaX), 486–488
Incidents response policy, 513
Information gathering, 496–497
Information policy, 512–513
Information security
    attributes, 497–498
    risk and its mitigation, 512*f*
    services, 497
Information systems projects implementation policy, 513
Information systems security (INFOSEC), 495–496, 511
Injection laser diode (ILD), 85
Integrated services (IntServ), 244–245
Integrated services digital network (ISDN), 431
Interference parameters in metallic conductors, 89–92
    ACR, 91, 91*f*
    ELFEXT, 91

FEXT, 90–91, 90*f*
NEXT, 90, 90*f*
other performance indexes of crosstalk, 92
Interference sources, 61–71
    adjacent channel, 69–71, 69*f*, 70*f*
    co-channel, 68–69, 69*f*
    intersymbol, 61–63, 61*f*, 62*f*
        nyquist ISI criterion, 63–67, 63*f*, 65*f*, 66*f*
    multiple access, 67–68, 68*f*
Interframe gap (IFG), 381
Interior gateway protocols (IGP), 297, 298*f*
Intermodal dispersion, 85
Intermodulation distortion (IMD), 69
Intermodulation product (IMP), 70–71, 70*f*
International data encryption algorithm (IDEA),
        517–518
International Electrotechnical Commission (IEC), 416
International mobile telecommunications-Advanced
        (IMT-Advanced), 484
International Organization for Standardization (ISO), 21,
        416, 514
International standards, 417*f*, 418–424, 419*f*, 420*t*, 421*t*,
        422*f*, 423*f*
International Telecommunications Union (ITU), 434
International Telecommunications Union-Radio
        communications (ITU-R), 484
Internet, 495
    policy, 513
    queries, 497
Internet Assigned Numbers Authority (IANA), 38–39,
        250, 266
Internet control message protocol (ICMP), 335
Internet Engineering Task Force (IETF), 432
Internet message access protocol (IMAP), 223
Internet protocol (IP)
    address
        host part of, 254*f*
        net/subnet part of, 254*f*
        reserved, 251*t*
    datagram, 263
    functionalities of, 249
Internet protocol of next generation (IPng), 37
Internet service provider (ISP), 39
Internetwork, 304*f*
    example of, 313*f*
    with floating route, 306*f*
Internetwork configuration, 576*f*
Intersymbol interference, 61–63, 61*f*, 190
    nyquist ISI criterion, 63–67, 63*f*, 65*f*, 66*f*
Intersymbol interference (ISI), 19
Inter-VLAN connectivity, 398
    configuration of, 399
    with multilayer switch, 401–403, 402*f*
    with router, 399–401
    routing with multiple cables, 399*f*
    routing with router, 400*f*
Inter-VLAN routing, 396–403
    configuration of native VLAN, 398
    connectivity with multilayer switch, 401–403
    connectivity with router, 399–400
Intramodal dispersion, 85
Intranet, 542, 544*f*
Intrusion detection and preventive system (IDPS), 546
Intrusion detection system (IDS), 546–547

Intrusion preventive system (IPS), 546
Inverse address resolution protocol (inverse ARP), 467
Inverse DFT (IDFT), 177, 180
Inverse FFT (IFFT), 177–178
Ionospheric layers
    propagation of electromagnetic waves using, 119f
    refraction of electromagnetic waves in, 118f
    representation of, 120f
IP multimedia protocols, 225f
IPsec cryptographic protocol, 222
IPsec protocol into TCP/IP stack, 533f
IP telephony, 73
IP telephony and IP videoteleconference, 224–230
    ITU-T H.323, 225–228, 226f
    session initiation protocol, 228–230
IPv4-mapped-IPv6 address, 338
IPv6 protocol, 232
IP version 4 (IPv4), 8, 250–265
    classful addressing, 250–253, 250f
    classless addressing, 253–263
    in classless mode, 257f
    datagram, 263–265
    header fields, 263–265, 264f
    routing table, 298f
IP version 6 (IPv6), 8, 265–273
    addressing, 266–270
    in binary and hexadecimal notation, 266f
    fragmentation, 336
    global unicast address, 267, 267t, 269t
    header
        generic decomposition of, 271f
        mandatory part of, 272–273, 272f
    hierarchical layers, 268
    interface ID of, 270
    manually configured IPv6 tunnel, 338f
    multicast address, 270t
    packet, 270–273
    prefix, 267
    reserved address ranges, 268t
    space and routing, 269f
    transition and configuration, 337–339
        RIPng configuration using Cisco IOS, 338–339
        transition from IPv4 into IPv6, 337–338
    types of address, 267
ISO 27002 standard, 514
ISO/IEC 15408 standard, 496
Iterative block-DFE (IB-DFE) receivers, 183, 183f
ITU-T, 225
ITU-T H.323, 225–228
    network component types, 225–226, 226f
    protocol stack for IP-based networks, 227f
    protocols used by, 226–227
    voice/video call with corresponding protocols, 228f

J

Jamming signal, 379
Jitter, 18

K

Key and video monitor (KVM), 424
Keyed hashed message authentication code, 523
Knife-edge model, 108, 109f

L

Label, 457
    swap, 455
    switching protocol, 454
Label distribution protocol (LDP), 456
Label edge router (LER), 455
Label forwarding information base (LFIB), 455
Labeling distributors and cables, 428
Label stacking, 457
Label switching path (LSP), 455
Label switching router (LSR), 454–455
LAN standard IEEE 802.3, 77
Legacy cabling systems, 411
Legacy communication systems, 137
Light emitting diode (LED), 85
Linear predictive coding (LPC), 142
Linear suboptimal detectors, 176
Line coding, 135
    schemes, 144–148
    symbols used by, 145
Line coding mechanism, violation of, 356
Line encoder, 19
Line encoder/decoder, 145f
Line-of-sight (LOS), 481
Line terminal multiplexer (LTM), 437
Link access procedures balanced (LAPB), 432
Link access procedures over D channel
        (LAPD), 431
Link control protocol (LCP), 461–463, 462f
Link-local addresses, 252, 253t
Link-local unicast address, 268
Link-state ACKnowledgment (LSAck), 328
Link-state advertisement packets (LSPs), 326
Link-state request (LSR), 328
Link-state routing protocols, 325–335
Link-state update (LSU), 328
Local area network (LAN), 73, 431, 488
    devices, 346–354
        bridge, 347–348, 347f
        hub, 346–347, 347f
        STPr, 350–354
        switch, 348–350
    half-duplex transmission with switch, 349f
Local management interface (LMI), 469–470
Local oscillator offset, 180
Logical link control (LLC), 40, 345, 431
    protocols, 373–374
    sublayer, 354–372
        ARQ. See Automatic repeat request
            (ARQ)
        error control techniques, 356–367
        flow control techniques, 371–372
Logical topology, 12–13
Log likelihood ratios (LLRs), 184
Log-normal distribution, 112
Long-range communications, 116
Long-term evolution (LTE), 15, 482
Loose-tube optical fiber cable, 88, 88f
Low-bandwidth twisted pairs, 73
Low earth orbit (LEO), 121, 123–125
Lower window edge (LWE), 372
Low smoke zero halogen (LSZH), 419
LSP *versus* FEC, 456, 456f

LTE-A and IMT-advanced, 488
LTE-Advanced (LTE-A), 484
Lucent connector (LC), 87

M

Macrocell, 114, 481
Malware, 510
Manchester line coding scheme, 147, 147*f*
Man-in-the-middle attack, 502–503, 503*f*
  associated with Diffie–Hellman algorithm, 519*f*
Man-made noise, 52, 52*f*
Martyn's law, 118
Maximal ratio combining (MRC) algorithm, 186–187, 186*f*
Maximum collision domain diameter (MCDD), 379–381
Maximum likelihood sequence estimator (MLSE)
  detector, 199
Maximum segment size (MSS), 240, 346
Maximum transmission unit (MTU), 238, 335
  for data link layer, 336*t*
MBSFN area, 486
MD5, Hash functions, 522
Media classification, 13–14, 13*f*
Medium access control (MAC), 30, 345, 486
  address, 348–349
  address flooding, 499
  protocols, 375–388
    DVB, 388
    FDDI, 387–388
    IEEE 802.3 protocol, 376–383
    IEEE 802.5 protocol, 383–387
  sublayer, 374–375
    defined by IEEE 802.5 standard, 385–386
Medium earth orbit (MEO), 121, 123–125
  footprints, 125*f*
Mesh topology, 11, 13*f*
Message integrity checksum (MIC) algorithm, 537
Message transfer agents, 223
Metallic conductors, interference parameters in, 89–92
  ACR, 91, 91*f*
  ELFEXT, 91
  FEXT, 90–91, 90*f*
  NEXT, 90, 90*f*
  other performance indexes of crosstalk, 92
Metropolitan area network (MAN), 431
Microcell, 114, 480*f*, 481
Microwave systems, 131
Millington method, 116
M-law, 139, 141
Mobile ad-hoc networking (MANET), 214
Mobile stations (MS), 477
Modal bandwidth, 83
Modal dispersion effect, 83
MODEM, 3, 3*f*
Modulation schemes, 148–153
  ASK, 149, 149*f*
  FSK, 149–150, 150*f*
  M-QAM, 151–153
  PSK, 150–151
Modulator, 3, 3*f*
Mono-alphabetic substitution cipher, 515
Moore law, 15
Most significant bit (MSB), 39, 251
Moving picture experts group (MPEG), 142–143

Multiaccess network, 334
Multiantenna systems (MIMO), 485
Multicast communication, 6
Multicell transmission, 486
Multifrequency networks (MFNs), 205
Multihop relay technique, 207–211, 208*f*
  in 3GPP, 209–211
  adaptive relaying, 207–208
  architecture for LTE-Advanced, 210*f*
  configurable virtual cell sizes, 208
  in E-UTRAN, 210*f*
Multilevel quadrature amplitude modulation (M-QAM)
    constellations, 151–153, 151*f*
Multimedia broadcast and multicast service (MBMS),
    165, 486
Multimode optical fiber, 82–83
  diameter, 85
  types of, 83
Multipath fading, 111*f*, 113–115, 185
  discrete impulsive response of, 112
  propagation environment with, 114*f*
Multiple access interference, 67–68, 68*f*, 169
Multiple antennas configurations, 191*f*
Multiple input multiple output (MIMO), 190–202,
    191*f*, 483
  2 × 2 multilayer, 198*f*
  advanced applications, 202–216
    base station cooperation, 203–207
    green radio communications, 213–216
    multihop relay, 207–211
    multiresolution transmission schemes, 211–213
  beamforming, 200–201
  with fixed relay station, 215*f*
  M × N multilayer, 197*f*
  multilayer transmission, 197–199
  multiple-antenna techniques, 192
  MU-MIMO, 202
  SDMA, 199
  space-time coding, 193–196
    STBC for four antennas, 195–196
    STBC for two antennas, 194
  STD, 196–197, 196*f*
Multiplexer (MUX), 155
Multiplexing mechanism, 155–161
  FDM, 156–160
  link using, 155*f*
  TDM, 160–161
Multiplexing section overhead, 442
Multipoint control unit, 229
Multiprotocol label switching (MPLS), 244, 431–432,
    453–457, 454*f*
  network, 454–456, 455*f*, 456*f*
  packet format, 456–457, 457*f*
Multipurpose Internet mail extension (MIME) format, 222
Multiresolution techniques, 212
Multiuser detectors (MUD), 68, 176
Multiuser MIMO (MU-MIMO), 202
Mutual inductance effect, 89*f*

N

Names' resolution, 232–234
Narrowband CDMA, 168, 171–174
Native VLAN, configuration of, 398

Near end crosstalk (NEXT), 90, 90*f*
Negative acknowledgment (NAK), 9, 358, 368
Negative justification, 443
Network, 6, 6*f*
    access layer, 33–34
    architecture, 21, 22*f*, 542–546
        with DMZ, 544–546
        double, 543–544
        single, 542–543, 542*f*
    coverage areas, 10
    layer, 25*f*, 27–28
    with loops, 350*f*
    physical topology, 43
    protocol architectures of
        concept, 21–23
        defined, 21
    root bridge of, 352
    segmentation with bridge, 347*f*
    topologies, 10–13
        bus, 10–11, 10*f*
        mesh, 11, 13*f*
        ring, 11, 11*f*
        star, 11, 12*f*
        tree, 11, 12*f*
Network address translation (NAT), 286–292, 287*f*
    dynamic configuration, 287–291
    overload, 287, 290
    static configuration, 291–292, 291*f*
Network control protocol (NCP), 461–463, 462*f*
Network IDPS (NIDPS), 547
Network IDS (NIDS), 546
    corporate architecture with, 547*f*
Network interface card (NIC), 10–11, 39, 252, 499
    router with, 253*f*
Network management, 230–232
Network–network interface (NNI), 453
Network PDU (NPDU), 24
Network SAP (NSAP), 30
Network security (NETSEC), 495–496
Newton's law of universal gravitation, 122
Next generation Internet protocol (IPng), 37
Noise
    factor, 54
    sources, 48*f*, 49–55, 49*f*, 50*f*
        atmospheric noise, 50–52, 51*t*, 53*f*
        electronic noise, 54–55
        extraterrestrial noise, 52–53, 53*f*
        man-made noise, 52, 52*f*
        thermal noise, 53–54
Nonbroadcast multiple access (NBMA), 469
Nonconfirmed service, 9–10
Nondesignated port, 354
Nondifferential transmission, 143–144, 144*f*
    2B1Q, 148, 148*f*, 148*t*
    bipolar AMI, 146–147, 147*f*
    differential Manchester, 147–148
    Manchester, 147
    nonreturn to zero, 146
    nonreturn to zero inverted, 146, 146*f*
    pseudoternary, 147
    return to zero, 146
Non-real time (NRT), 14
Nonrecursive convolutional encoder, 365*f*
Nonreturn-to-zero line coding, 146, 146*f*

Nonuniform PCM, 139
Nonvolatile random access memory (NVRAM),
    403–404
Normal response mode, 458
North American standards, 416–418, 417*t*, 418*f*
North Atlantic Treaty Organization (NATO), 503
*n*th layer protocol control information (*n*-PCI), 24
*n*th layer service data unit (*n*-SDU), 24
Nyquist ISI criterion, 63–67, 63*f*, 65*f*, 66*f*
Nyquist pulse shaping criterion, 170–171, 173
Nyquist sampling theorem, 47, 143

O

One-way encryption, 522
Open loop transmitter scheme, 192*f*
Open shortest path first (OSPF) protocol, 298, 327–335
    configuration
        neighbor table of Router1 after, 331*f*
        routing table of Router1 after, 329*f*
        routing table of Router2 after, 329*f*
    metric distances and routing priorities using, 301*f*
    metric utilized by, 332
    syntax, 330
OpenSSL, 532
Open system interconnection (OSI), 219, 220*f*
Operation, administration, and maintenance (OA&M), 436
Optical fibers, 81–88
    attenuation coefficient of, 84*f*, 86*f*
    categories, 82–87, 83*f*
    characteristics, 82
    composed of core and cladding, 83*f*
    connectors and cables, 87–88
    dielectric guide, 82
    emergency of, 82
Orbits
    advantages/disadvantages, 126*t*
    altitudes and radius, 123*t*
    characteristics of, 123–126
        GEO, 121, 123–124
        HEO, 126
        medium and low earth orbit, 124–125
    physical analysis of satellite, 121–123, 121*f*
    types of, 121
Organizationally unique identifier, 377
Orthogonal FDMA (OFDMA), 181, 483
Orthogonal frequency division multiplexing (OFDM), 19,
    165, 177–181
    implicit baseband processing of, 179*f*
    single-carrier signal *vs.*, 178*f*
    transmission technique, 215–216
    transmitted, 180*f*
OSI reference model, 21, 22*f*, 23*f*, 24–33, 24*f*, 219,
    230, 308
    defined, 25
    service access point, 29–33, 30*f*, 31*f*, 32*f*, 33*f*
    seven-layers of, 22*f*, 24*f*, 25–29, 25*f*
        application layer, 29
        data link layer, 26–27, 26*f*
        network layer, 25*f*, 27–28
        physical layer, 25–26, 25*f*
        presentation layer, 29
        session layer, 29
        transport layer, 28–29

P

Packet filtering configuration, 553–561
    ACL, 553–557
    CBAC, 561–561
    reflexive ACLs, 557–559
Packet sniffing, 499
Packet switching, 7–8, 7f, 8f, 431–432
Packet switching transport protocol, 449–471
    asynchronous transfer mode (ATM), 449–453, 449f, 453f
        B-ISDN reference model, 449f, 450–451, 451f
        cell format, 453, 453f
        network, 452–453, 452f
    defined, 454
    frame relay, 465–471, 466f, 467f, 469f
        address fields, 468
        configuration using Cisco IOS, 468–471, 469f, 470f
        map, 467
    HDLC protocol, 458–461, 459f, 460t, 461f
        configuration using Cisco IOS, 460–461
        forms of, 458
        frame format, 458–460, 459f
        frame types, 458
        modes of, 458
        station types, 458
    multiprotocol label switching (MPLS), 453–457, 454f
        network, 454–456, 455f, 456f
        packet format, 456–457, 457f
    point-to-point protocol (PPP), 461–465, 462f, 463f, 464f
        configuration using Cisco IOS, 464–465
        troubleshooting, 465
Parallel interference cancellation (PIC) detector, 176
Parity bits, 360–361, 360f, 361f
Partners' network, 547
Password authentication protocol (PAP), 507
Patch panel, 425–427, 426f, 427f
Path overhead, 441
Payload data, 387, 460
Payload type (PT), 453
Peak to average power ratio (PAPR), 216
Periodic synchronization, 4
Permanent circuit, 431
Phase alternation line (PAL), 142–143
Phase shift, 56–57, 56f, 57f
Phase shift keying (PSK), 150–151
Photodiode, 85
Physical layer, 25–26, 25f
Physical layer convergence procedure (PLCP), 490
Physical security, 496
Physical topology of network, 12–13, 43
Picocell, 114, 481–482
Ping sweeps, 497
P intrinsic N (PIN) photodiode, 85
Pixels, 142
Plaintext bits, 517, 517f
Plesiochronous digital hierarchy, 433–435, 434f, 435t
Point-to-multipoint satellite communication system, 127f
Point-to-point communication, 6, 6f
Point-to-point protocol (PPP), 431, 461–465, 462f, 463f, 464f
    configuration using Cisco IOS, 464–465
    troubleshooting, 465

Point-to-point satellite communication system, 126f
Port address translation (PAT), 286–292, 288f
    dynamic configuration, 287–291
    static configuration, 291–292
Port forwarding, 291
Port numbers, 238
Port scans, 497
Port security configuration, 540–542
Positive acknowledgment with retransmission (PAR), 9, 35, 357, 357f
Positive justification, 443
Possible authentication protocol (PAP), 464
Post office protocol (POP), 223
Power control, 482–483
Power delay profile (PDP), 111
Power over Ethernet (PoE), 413
Power spatial density of electromagnetic wave, 104
Power spectral density of noise, 102
Power sum ACR (PSACR), 92
Power sum ELFEXT (PSELFEXT), 92
Power sum FEXT (PSFEXT), 92
Power sum NEXT (PSNEXT), 92
Pre-coding, 202
Present and future of telecommunications, 14–19
    collaborative age of network applications, 16–17, 17f
    convergence, 15–16
    transition towards collaborative age, 17–19, 18f
Presentation layer protocol, 29, 221
Presentation PDU (PPDU), 24
Presentation SAP (PSAP), 30
Printer policy, 513
Private automatic branch exchange (PABX), 6, 431
Private (site local) IP address ranges, 253t
Private key (PRK), 520, 523
Probability density function (PDF), 113
Protocol data unit (PDU), 24, 490
Protocol-dependent modules (PDMs), 316
Proxy server, 222, 229
Pseudoternary line coding technique, 147, 147f
Public key (PUK), 520, 523
Public key cryptography, 520
Public key infrastructure (PKI), 520, 525–527
Public switching transport network (PSTN), 431
Pulse amplitude modulation (PAM), 137, 138f, 433
Pulse code modulation (PCM), 138–141, 140f, 160
Pulse density modulation (PDM), 433
Pulse position modulation (PPM), 433

Q

Quadrature amplitude modulation (QAM), 485
Quadrature phase shift keying (QPSK), 45, 150, 150f, 485
Quality of service (QoS), 8, 237, 449, 484
    improved mechanisms, 265
Quantization process, 138

R

Rack, 424–425, 425t, 426f
Radio transmitter, 95
RAKE receiver, 168–169, 175f, 188–190, 189f

Rayleigh fading model, 481
Rayleigh model, 113, 129
Real-time control protocol (RTCP), 227
Real-time protocol (RTP), 226–227
Receiver's merit factor, 100
Receiving filter, 158
Reception of bits, 43
Recursive convolutional encoder, 366*f*
Redirect message, 335
Reflection effect, 103*f*
Reflexive ACLs, 557–559, 584–585
Refraction effect, 84*f*
Regeneration header, 442
Regeneration section
    overhead, 442
    sublayer, 452
Regenerator (REG), 438
Registration authorities (RA), 525–526
Registration server (registrar), 229
Reliable transport protocol (RTP), 317
Remote access policy, 513
Remote authentication dial-in user service database
        (RADIUS), 538
Replay attack, 507–508, 507*f*
Reserved (RES), 453
Reserved IP addresses, 251*t*
Return-to-zero line coding, 146
Rice model, 114
Rijndael algorithm, 518
Ring topology, 11, 11*f*
Risk, security, 511
Rivest, Shamir, and Adleman (RSA) algorithm,
        520–521
RMS delay spread, 115
Rolloff factor, 64, 171
Root bridge election process, 352, 352*f*
Root mean square (RMS), 54
Root port, 354
Round robin mechanism, 375
Router, 249, 253, 274, 297, 302
    with ACLs, 554*f*
    configuring protocol RIPv2 in, 314*f*
    console and auxiliary lines in, 274*f*
    ID assigned to router, 333
    interconnecting IEEE802.11 LAN with ISDN
            segment, 336*f*
    inter-VLAN connectivity with, 399–401
    with NIC, 253*f*
    sequence order of operations performed, 558*t*
    with subinterfaces for VLAN
            interconnectivity, 401*f*
Route summarization, 301–302, 302*f*
Routing algorithms, 299
Routing information protocol (RIP), 300, 310–316
    internetwork for configuration of, 312*f*
    mechanisms, 310
    messages, 310
    RIPv1, 311
    RIPv2, 311
    versions of, 311
Routing protocols, 298, 307, 309
    classification of, 309*t*
RSA protocol, 521

S

Satellite communication systems, 121–131
    characteristics of orbits, 123–126
        GEO, 121, 123–124
        HEO, 126
        MEO and LEO, 124–125
    C/N ratio analysis, 126–131
    physical analysis of satellite orbits, 121–123, 121*f*
Scalable OFDMA (SOFDMA), 487
Scalable video transmission, 213*f*
Scattering effect, 110*f*
Scrambling decoder, 155*f*
Scrambling encoder, 154, 154*f*
Screened FTP (S/FTP), 75
Screened STP (S/STP), 75
Screened UTP (S/UTP), 75
SDH/SONET
    frame format, 441–443, 442*f*, 443*f*, 444*f*, 445*f*
    network, 437–441, 438*f*, 439*f*, 440*f*, 441*f*, 441*t*
Second generation (2G) cellular networks, 483
Secure shell (SSH), 499
    configuration, 539–540
Secure sockets layer (SSL), 508, 528–532, 529*f*
    handshaking, 530, 531*f*
    protocol, 530
Security
    evaluation phase, 512
    in IEEE 802.11 wireless networks, 535–539
        EAP and IEEE 802.1x, 538–539, 538*f*
        Wi-Fi protected access 2 and IEEE
            802.1i, 537
        wired equivalent privacy, 536–537
        WPA, 537
    physical and environmental, 510–511
    plan, 512–514
    policy, 513
    protective measures, 514–535
        asymmetric cryptography, 520–521
        combined cryptography, 527–535
        digital certificates, 523–525, 525*f*, 526*f*, 527*f*
        digital signature, 521–523, 523*f*, 524*f*
        PKI, 525–527
        symmetric cryptography, 515–519
    risk management, 511
    services and attack types. *See* Security services and
            attack types
Security association (SA), 534
Security parameter index (SPI), 534
Security services and attack types, 497–510, 497*t*
    accountability, 508–510
        access control, 509
        auditing, 510
        authentication, 508–509
        authorization, 509
        identification, 508
        monitoring, 509
        registration, 510
    authenticity, 505–508
    availability, 503–505
    confidentiality, 498–501
        eavesdropping, 498–500, 498*f*
        interception, 500–501

Security services and attack types (*Continued*)
  snooping, 500
    trust exploitation, 501, 501*f*
  integrity, 502–503
Segmentation and reassembling sublayer (SAR), 451
Selection combining algorithm, 186
Selective reject ARQ, 370–371, 371*f*
Selective transmit diversity (STD), 196–197, 196*f*
Semi-infinite plan, 108
Serial cable, 282
Service access point (SAP), 29–33, 30*f*, 31*f*, 32*f*, 33*f*
Service-level agreement (SLA), 245
Service primitives, 31, 32*f*
  confirmed connectionless services, 32, 33*f*
  connection-oriented services, 32, 32*f*
  non-confirmed connectionless services, 33, 33*f*
Service set identifier (SSID), 490
Session initiation protocol (SIP), 228–230
  components, 229*f*
  signaling protocol, 229
  voice/video call, 230*f*
Session layer protocol, 219–221
Session PDU (SPDU), 24
Session SAP (SSAP), 30
Seven-layers of OSI-RM, 22*f*, 24*f*, 25–29, 25*f*
  application layer, 29
  data link layer, 26–27, 26*f*
  network layer, 25*f*, 27–28
  physical layer, 25–26, 25*f*
  presentation layer, 29
  session layer, 29
  transport layer, 28–29
SHA-1 (secure hashing algorithm), 522
Shadowing fading, 111*f*, 112
Shannon, 26
Shannon capacity, 45–47, 46*f*
Shielded twisted pairs (STPs), 75
"Show ip protocols" command, 315*f*, 317*f*
Signal encoding, 43
Signal processing technique, 216
Signals, scrambling of, 154–155
Signal-to-interference plus noise ratio (SINR), 211
Signal-to-noise plus interference ratio (SNIR), 1, 45
Signal-to-noise ratio (SNR), 45, 100, 144, 481
Signature-based IDS, 546
Simple mail transfer protocol (SMTP), 35, 223, 223*f*
Simple network management protocol (SNMP), 36,
        231–232, 499
  exchanged messages, 232
  IP-based, 230
  network using, 231*f*
  into TCP/IP stack, 231*f*
Simplex communications, 5, 5*f*
Single-carrier-FDE (SC-FDE) schemes, 165, 182–185
Single-carrier FDMA (SC-FDMA), 483
Single-cell transmission, 486
Single frequency network (SFN), 486
Single-mode optical fiber, 82, 85
Single network architecture, 542–543, 542*f*
Single sideband (SSB), 64, 158, 158*f*
Single-user MIMO (SU-MIMO), 202
Site-local unicast address, 268
Skywave propagation, 116, 118–120
Sliding window flow control, 372, 372*f*

Sliding window protocol, 240
Slow fading, 111
Smurf attack, 504, 505*f*
Snooping, 500
Solar noise, 52
Source address (SA), 387
Source coding, 136–143
  video, 142–143
    analog, 142–143
    digital, 143
  voice, 137
    analog, 137
    digital, 137–142
Space division multiple access (SDMA), 199, 200*f*
Space-frequency block coding, 193
Space-time block coding (STBC), 192
  for four antennas, 195–196
  for two antennas, 194
Space-time coding, 193–196
Space wave propagation. *See* Direct wave propagation
Spanning tree algorithm (STA), 351
Spanning tree protocol (STPr), 350–354
  aim of, 351
  to overcome loops, 353*f*
Spectral efficiency, minimum, 102
Spherical wave, 104
Spoofing, 500, 501*f*
Spread-spectrum communications, 166–167
  transmitter, 168*f*
  vs. narrowband signal, 166*f*
Start of frame sequence (SFS), 387
Star topology, 11, 12*f*
Static routing and flooding, 302–307
  configuring, 304*f*
  default route configuration, 306–307
  in Ethernet network, 305
  floating route configuration, 305–306
  static route configuration, 303–305
  syntax, 303–305
Statistical anomaly-based IDS, 546
Stenography, 515
Step index multimode fibers, 83
Stop-and-wait ARQ, 368–369
Stop-and-wait flow control, 372
Straight-through cable, 79
Straight tip (ST) connector, 87
Structured cabling elements, 413–416, 414*f*
  centralized optical architecture, 415–416, 416*f*
  functional elements of, 414–415, 414*f*
  subsystems of, 415
Structured cabling standards and specifications, 416–424
  European standards, 424
  International standards, 417*f*, 418–424, 419*f*, 420*t*, 421*t*,
          422*f*, 423*f*
  North American standards, 416–418, 417*t*, 418*f*
Structured cabling system, 411
Subnet mask, 256, 256*t*, 257*t*
  VLSM, 262–263
Subnetwork, 258
  calculation, 261*t*
    breakdown of, 260*t*, 261*t*, 262*t*
    using decimal notation, 259*t*
Subscriber connector (SC), 87
Subsystems of structured cabling elements, 415

Successive interference cancellation (SIC) detector, 176
Successor's reported distance (SRD), 322
Switches, configuration of, 391–392, 398*f*
    within VTP domain, 403–404
Switching modes, 6–8
    circuit switching, 6–7, 7*f*
    packet switching, 7–8, 7*f*, 8*f*
Symbol, coding efficiency of, 154
Symmetric cryptography, 515–519
    block diagram of, 516*f*
    systems, 517–519
    transposition, 516, 516*f*
Synchronous communication systems, 4
Synchronous data link control (SDLC), 431
Synchronous digital cross connect (SDXC), 438
Synchronous digital hierarchy (SDH), 4, 49, 432, 436–443,
    436*f*, 437*f*
    SDH/SONET frame format, 441–443, 442*f*, 443*f*,
        444*f*, 445*f*
    SDH/SONET network, 437–441, 438*f*, 439*f*, 440*f*,
        441*f*, 441*t*
Synchronous optical network (SONET), 436
Synchronous payload envelope (SPE), 442–443
Synchronous transport module (STM), 436–437
Synchronous transport signal (STS), 436–437
SYN flooding attack, 504, 504*f*

T

Tag control information (TCI), 397
Taps, 547
TCP/IP. *See* Transmission control protocol/IP (TCP/IP)
TCP/IP architecture, overview, 33–43, 34*f*
    application layer, 35
    data link layer, 40–42, 40*f*, 41*f*, 42*f*
    internet layer, 34*f*, 37–39, 37*f*, 39*f*
    physical layer, 42–43
    transport layer, 34*f*, 35–36, 36*f*
Telecommunication outlet (TO), 415
Telecommunications, functions used in, 573*t*
Telecommunications Industry Association (TIA), 416
Telephone communication, 1
Telnet, 274
    attack, 499
Temporal dispersion, 111
Temporal key, 537
Temporal key integrity protocol (TKIP), 537
Terrestrial microwave systems, 131–133, 131*f*
    direct and reflected waves of, 132*f*
    link with dimensions, 132*f*
Thermal noise, 53–54
Third generation partnership project (3GPP), 482
Tight buffer optical fiber cabling, 88*f*
Time assignment speech interpolation, 137
Time division duplexing (TDD), 480
Time division multiple access (TDMA), 67, 167, 483
Time division multiplexing (TDM), 137, 156*f*, 160–161, 160*f*
Time domain OFDM signal, 178
Time hopping (TH) technique, 166
Time to live (TTL), 264
    exceeded message, 335
    field, 457
Token holding period (THP), 385
Token Ring protocol, 383–384

Topologies, 10–13
    bus, 10–11, 10*f*
    logical, 12–13
    mesh, 11, 13*f*
    physical, 12–13
    ring, 11, 11*f*
    star, 11, 12*f*
    tree, 11, 12*f*
Traffic, 13–14, 14*f*
    classification, 13–14, 14*f*
    flow security, 498
Transmission control protocol (TCP), 239–243
    connection establishment, 242*f*
    connections, 504
    connection termination, 243*f*
    data exchange, 242*f*
    error control, 240
    handshaking, 242–243
    port numbers, 238*t*
    properties, 239–240
    segment header format, 238*f*, 240–241
Transmission control protocol/Internet protocol (TCP/IP),
    8, 219, 220*f*
    address of, 238
    application layer protocol, 223
    connection, 238
    location of SNMP into, 231*f*
    protocols and technologies used by, 308*f*
    segment, 239
    stack, 225*f*
    transport layer of, 237
Transmission gaps, 211
    in E-MBMS, 212*f*
Transmission mediums, 4
Transmission of bits, 43
Transmission path sublayer, 452
Transmitter timer, 35
Transponder, 121
    satellite's, 129*f*
Transport layer, 28–29
Transport layer security (TLS), 508, 528–532, 529*f*
Transport mode, 534
Transport network, 4, 431, 432*f*
Transport PDU (TPDU), 24
Transport SAP (TSAP), 30
Transposition, 516, 516*f*
Tree topology, 11, 12*f*
Triple DES, 517
Triple play, 448
Trivial FTP (TFTP), 224, 278
Trojan horse, malware, 510
Tunnel mode, 535
Twisted pairs, 73–80
    attenuation coefficient of, 78*f*
    categories, 76–78, 76*t*
    characteristics, 74–75
    color code of four-pair, 79*t*
    composed of isolated copper wires, 74*f*
    connectors and cables, 78–80
    low-bandwidth, 73
    quality of, 74
    types of protection, 75, 75*t*, 89
    voice-grade, 74
Two binary one quaternary (2B1Q), 148, 148*f*, 148*t*

U

Uninterrupted power supply (UPS), 415
Unipolar nonreturn to zero technique, 146*f*
Unipolar return to zero technique, 146*f*
Universal mobile telecommunications system (UMTS), 17, 482
Unlicensed national information infrastructure (UNII), 489
Unshielded twisted pairs (UTPs), 75, 77
Uplink, 128–130
Upper window edge (UWE), 372
Urban macrocell, 114
User agent, 223
    client, 229
    server, 229
User datagram protocol (UDP), 10, 243
    datagram header format, 243, 244*f*
    port numbers, 239*t*
    properties, 243
User equipment (UE), 486
User-network interface (UNI), 453

V

Validation authorities, 526
Variable length subnet mask (VLSM), 262–263
Vertical backbone cabling. *See* Building backbone cabling
Vertical-Bell laboratories layered space-time (V-BLAST) detector, 199
Vertical polarization, 103*f*
    Fresnel coefficient for, 104
Very high data rate DSL (VDSL), 446
Vigenère cipher, 515–516
Virtual cellular network, 209*f*
Virtual channel identifier (VCI), 453
Virtual channel (VCn) sublayer, 453
Virtual circuit mode, 27–28, 466
Virtual container, 442
Virtual local area networks (VLANs), 388–406
    assigning interfaces to, 392*f*
    brief, 393*f*
    characteristic of, 391
    configuration of, 391–396
        default gateway, 396
        management VLAN, 395–396
    configured in switch, 389*f*
    creating, 392*f*
    creation of management, 395–396
    interfaces vlan *vlan_number*, 394*f*
    interface switchport, 394*f*
    inter-VLAN routing, 396–403
        configuration of native VLAN, 398
        connectivity with multilayer switch, 401–403
        connectivity with router, 399–400
    isolating networks with, 390*f*
    VTP. *See* VLAN trunking protocol (VTP)
Virtual path (VP) sublayer, 453
Virtual private networks (VPN), 453, 547–549, 548*f*, 549*f*

Virus, malware, 510
VLAN identifier (VLAN ID or VID), 397
VLAN trunking protocol (VTP), 403–406
    client, 403–404
    configuration of switches, 405
    password, 405–406
    protocol, 404*f*
    server, 403
    transparent, 404
VOCODER (VOice CODER), 1
Vocoder technique, 142
Voice-grade twisted pairs, 74
Vulnerability, security, 511

W

Wavelength division multiplexing (WDM), 87
Web browsing, 219, 221–222
Well-known ports, 238
    TCP port numbers, 238*t*
    UDP port numbers, 239*t*
Wide area network (WAN), 431
Wideband CDMA (WCDMA), 165, 168, 174–177, 483
Wi-Fi protected access (WPA), 537
WiMAX (Worldwide Interoperability for Microwave Access), 486
Window location offset, 180
Windows HyperTerminal, 274
Wired equivalent privacy (WEP), 499, 536–537
    authentication handshaking, 536*f*
    vulnerability of, 537
Wireless access point (WAP), 489
Wireless Fidelity (WiFi), 488
Wireless propagation, 95–120
    direct wave propagation, 95–102
        bit error probability calculation, 101–102
        carrier-to-noise ratio calculation, 100–101
        FSPL, 96–97
        link budget calculation, 97–100
    effects, 102–110, 103*f*
        diffraction, 107–110
        reflection, 102–107
        scattering, 110
    fading, 110–115
        multipath fading, 113–115
        shadowing fading, 112
    groundwave propagation, 115–116, 117*f*
    skywave propagation, 116, 118–120
Wireless systems, advances in, 165
Workstation cabling, 415
Workstation subsystem, 415
World Wide Web (www), 221
Worm, malware, 510

Z

Zero force (ZF) equalizer, 59–60
Zombies, 504